Lecture Notes in Computer Science   10416

Commenced Publication in 1973
Founding and Former Series Editors:
Gerhard Goos, Juris Hartmanis, and Jan van Leeuwen

More information about this series at http://www.springer.com/series/7408

Editor
J. Christopher Beck
University of Toronto
Toronto, ON
Canada

SN 0302-9743　　　　　　ISSN 1611-3349　(electronic)
cture Notes in Computer Science
BN 978-3-319-66157-5　　　ISBN 978-3-319-66158-2　(eBook)
)I 10.1007/978-3-319-66158-2

rary of Congress Control Number: 2017949522

CS Sublibrary: SL2 – Programming and Software Engineering

pringer International Publishing AG 2017

on acid-free paper

ringer imprint is published by Springer Nature
istered company is Springer International Publishing AG
istered company address is: Gewerbestrasse 11, 6330 Cham, Switzerland

J. Christopher Beck (Ed.)

Principles and Practice of Constraint Programming

23rd International Conference, CP 2017
Melbourne, VIC, Australia, August 28 – September 1,
Proceedings

 Springer

Preface

This volume contains the proceedings of the 23rd International Conference on the Principles and Practice of Constraint Programming (CP 2017) held from August 28 to September 1, 2017 in Melbourne, Australia and colocated with the 20th International Conference on Theory and Applications of Satisfiability Testing (SAT 2017) and the 33rd International Conference on Logic Programming. Detailed information about the CP 2017 conference with links to the colocated conferences can be found at http://cp2017.a4cp.org.

The CP conference is the annual international conference on all aspects of computing with constraints including theory, algorithms, environments, languages, models, systems, and applications such as decision making, resource allocation, scheduling, configuration, and planning. In addition to the main technical track and long-standing applications track, and as a continuation of the effort of the CP community to reach out to other research fields that intersect with constraint programming, CP 2017 featured thematic tracks in Machine Learning and CP, Operations Research and CP, Satisfiability and CP, and Test and Verification and CP. Each track had its own Track Chair(s) and Program Committee to ensure that the papers would be peer reviewed by expert reviewers with specific knowledge of the intersecting area.

The conference received 115 submissions across all tracks, including eight submissions to the Journal and Sister Conferences Track. Each paper was assigned to a Senior Program Committee member or the appropriate Track Chair and to three Program Committee members from either the technical track Program Committee or the relevant thematic track Program Committee. All papers received at least three reviews, following which the authors had an opportunity to respond. Detailed discussions were held on each paper by the PC members, led by the SPC member, Track Chair, and Program Chair. The Senior Program Committee, including Track Chairs, met in Padova, Italy, on June 5, 2017 with participation both in person and via video link. Each paper was discussed by the SPC with the decisions taken by consensus. The Journal and Sister Conferences Track papers followed a separate process, led by the Track Chair, to evaluate the relevance and significance of submitted papers that had been previously published in journals or other conferences. The Journal and Sister Conferences Track Program Committee met in Cupar, Scotland, UK on June 18, 2017 to make the final decisions. The final outcome of these meetings was the acceptance of 46 papers across all technical and thematic tracks, resulting in an acceptance rate of approximately 44%, and the acceptance of all eight papers submitted to the Journal and Sister Conferences Track.

The Senior Program Committee awarded four best paper awards, generously supported by Springer.

- Best Paper Award: Grigori German, Olivier Briant, Hadrien Cambazard, and Vincent Jost, "Arc Consistency via Linear Programming"

- Distinguished Paper Award: Fahiem Bacchus, Antti Hyttinen, Matti Järvisalo, and Paul Saikko, "Reduced Cost Fixing in MaxSAT"
- Best Student Paper Award: Adrian Goldwaser and Andreas Schutt, "Optimal Torpedo Scheduling"
- Distinguished Student Paper Award: Guillaume Derval, Jean-Charles Regin, and Pierre Schaus, "Improved Filtering for the Bin-Packing with Cardinality Constraint"

The Program Chair and the Journal-Publication-Fast-Track Chair, Louis-Martin Rousseau, invited four papers from across the technical and thematic tracks to participate in the *Constraints* journal fast-track process to publish an extended version in the journal at the same time as the conference, while also presenting the work at the conference. Due to the tight editorial deadlines, one paper accepted this invitation and so appears in this volume as an abstract with the full paper in *Constraints*.

The conference program included five invited talks in coordination with SAT 2017 and ICLP 2017 by Agostino Dovier, Holger Hoos, Nina Narodytska, Enrico Pontelli, and Mark Wallace. The conference also shared the workshop program with the two colocated conferences, resulting in seven workshops overseen by the Joint Workshop Chairs: Charlotte Truchet, Enrico Pontelli, and Stefan Rümmele. The tutorial program, also chaired by Charlotte Truchet with support from the SAT 2017 and ICLP 2017 Program Chairs, consisted of four tutorials on CP, SAT, Mixed Integer Nonlinear Programming, and Machine Learning and Data Science. The Doctoral Program, jointly organized by CP 2017 and ICLP 2017 and chaired by Chris Mears and Neda Saeedloei, hosted 24 students from around the world. The students had an opportunity to present their work, meet one-on-one with a senior researcher mentor, and attend invited talks targeted to the experiences of a PhD student.

The program for the conference is the result of a substantial amount of work by many people to whom I am grateful. I would like to thank the authors for their submission of high-quality scientific work and the substantial efforts of the Program Committees and external reviewers, who jointly prepared 341 high-quality reviews. The Senior Program Committee and Track Chairs played a crucial role in managing the reviews and discussions, in writing meta-reviews and recommendations for each submission, and in making the final decisions. I would like to specifically acknowledge the efforts of the Track Chairs to attract new contributors to the conference: Yael Ben-Haim and Yehuda Naveh (Satisfiability and CP Track Chairs), David Bergman and Andre Cire (Operations Research and CP Track Chairs), Ken Brown (Application Track Chair), Arnaud Gotlieb and Nadjib Lazaar (Test and Verification and CP Track Chairs), Tias Guns and Michele Lombardi (Machine Learning and CP Track Chairs), Karen Petrie (Journal and Sister Conferences Track Chair), Enrico Pontelli (Biology and CP Track Chair), and Louis-Martin Rousseau (Journal-Publication-Fast-Track Chair).

Beyond the peer review process, there is a substantial team that made the program and conference possible. I would like to particularly thank: Peter Stuckey and Guido Tack (CP 2017 Conference Chairs), Christopher Mears and Neda Saeedloei (Doctoral Program Chairs), Charlotte Truchet (Tutorial Chair), Charlotte Truchet, Enrico Pontelli, and Stefan Rümmele (Joint CP/SAT/ICLP Workshop Chairs), Tommaso Urli (Publicity Chair), Maria Garcia de la Banda (ICLP 2017 Conference Chair), Serge

Gaspers and Toby Walsh (SAT 2017 Conference and Program Chairs), and Ricardo Rocha and Tran Cao Son (ICLP 2017 Program Chairs).

I would also like to thank the sponsors of the conference for their generous support. At the time of writing, these sponsors include: the Artificial Intelligence Journal Division (AIJD) of IJCAI, the Association for Constraint Programming, the Association for Logic Programming, the City of Melbourne, CompSustNet, Cosling, Cosytec, CSIRO Data61, the European Association for Artificial Intelligence, IBM, Monash University, Satalia, Springer, and the University of Melbourne.

July 2017   Chris Beck

Tutorials and Workshops

Tutorials

Introduction to Constraint Programming - If You Already Know SAT or Logic Programming

Guido Tack   Monash University, Australia

An Introduction to Satisfiability

Armin Biere   Johannes Kepler University Linz, Austria

Introduction to Machine Learning and Data Science

Tias Guns   Vrije Universiteit Brussel, Belgium

Mixed Integer Nonlinear Programming: An Introduction

Pietro Belotti   FICO, UK

Workshops

Pragmatics of Constraint Reasoning

Daniel Le Berre   Université d'Artois, France
Pierre Schaus   UCLouvain, Belgium

Workshop on Answer Set Programming and Its Applications

Kewen Wang   Griffith University, Australia
Yan Zhang   Western Sydney University, Australia

Workshop on Constraint Solvers in Testing, Verification, and Analysis

Zakaria Chihani   Atomic Energy Commission (CEA), France

Workshop on Logic and Search

David Mitchell   Simon Fraser University, Canada

Progress Towards the Holy Grail

Eugene Freuder   University College Cork, Ireland

International Workshop on Constraint Modeling and Reformulation

Özgür Akgün   University of St Andrews, UK

Colloquium on Implementation of Constraint Logic Programming Systems

Jose F. Morales   IMDEA Software Institute, Spain
Nataliia Stulova   IMDEA Software Institute, Spain

———

Conference Organization

Program Chair

J. Christopher Beck   University of Toronto, Canada

Conference Chairs

Peter Stuckey   University of Melbourne, Australia
Guido Tack   Monash University, Australia

Application Track Chair

Ken Brown   University College Cork, Ireland

Biology and CP Track Chair

Enrico Pontelli   New Mexico State University, USA

Machine Learning and CP Track Chairs

Tias Guns   Vrije Universiteit Brussel, Belgium
Michele Lombardi   DISI, University of Bologna, Italy

Operations Research and CP Track Chairs

David Bergman   University of Connecticut, USA
Andre Augusto Cire   University of Toronto Scarborough, Canada

Satisfiability and CP Track Chairs

Yael Ben-Haim   IBM Research, Israel
Yehuda Naveh   IBM Research, Israel

Test and Verification and CP Track Chairs

Nadjib Lazaar   LIRMM, France
Arnaud Gotlieb   Simula Research Laboratory, Norway

Journal and Sister Conferences Track Chair

Karen Petrie   University of Dundee, UK

Journal-Publication-Fast-Track Chair

Louis-Martin Rousseau   École Polytechnique de Montréal, Canada

Doctoral Program Chairs

Christopher Mears   Redbubble, Australia
Neda Saeedloei   University of Minnesota Duluth, USA

Workshop and Tutorial Chair

Charlotte Truchet   University of Nantes, France

Publicity Chair

Tommaso Urli   CSIRO/Data61 and the Australian National University,
                              Australia

Senior Program Committee

Yael Ben-Haim   IBM Research, Israel
David Bergman   University of Connecticut, USA
Ken Brown   University College Cork, Ireland
Andre Augusto Cire   University of Toronto Scarborough, Canada
Sophie Demassey   CMA, MINES ParisTech, France
Bistra Dilkina   Georgia Institute of Technology, USA
Arnaud Gotlieb   Simula Research Laboratory, Norway
Tias Guns   Vrije Universiteit Brussel, Belgium
Nadjib Lazaar   LIRMM, France
Christophe Lecoutre   CRIL, Université d'Artois, France
Jimmy Lee   The Chinese University of Hong Kong, Hong Kong
Michele Lombardi   DISI, University of Bologna, Italy
Yehuda Naveh   IBM Research, Israel
Justyna Petke   University College London, UK
Karen Petrie   University of Dundee, UK
Enrico Pontelli   New Mexico State University, USA
Louis-Martin Rousseau   École Polytechnique de Montréal, Canada
Pierre Schaus   UCLouvain, Belgium
Andreas Schutt   Data61, CSIRO, and The University of Melbourne,
                              Australia

Technical Track Program Committee

Carlos Ansótegui   Universitat de Lleida, Spain
Nicolas Beldiceanu   IMT Atlantique (LS2N), France

David Cohen   Royal Holloway, University of London, UK
Pierre Flener   Uppsala University, Sweden
Emmanuel Hebrard   LAAS-CNRS, Université de Toulouse, France
John Hooker   Carnegie Mellon University, USA
Marie-José Huguet   LAAS-CNRS, Université de Toulouse, France
Said Jabbour   CRIL CNRS, Université d'Artois, France
Joris Kinable   Carnegie Mellon University, USA
Zeynep Kiziltan   University of Bologna, Italy
Philippe Laborie   IBM, France
Chavalit Likitvivatanavong   Thailand
Boon Ping Lim   NICTA, Australia
Andrea Lodi   École Polytechnique de Montréal, Canada
Samir Loudni   GREYC, CNRS UMR 6072, Université de Caen
                                   Basse-Normandie, France
Ines Lynce   INESC-ID/IST, Universidade de Lisboa, Portugal
Arnaud Malapert   Université Côte d'Azur, CNRS, I3S, France
Ciaran McCreesh   University of Glasgow, UK
Kuldeep S. Meel   Rice University, USA
Peter Nightingale   University of St Andrews, UK
Justin Pearson   Uppsala University, Sweden
Gilles Pesant   École Polytechnique de Montréal, Canada
Thierry Petit   Worcester Polytechnic Institute, USA
Patrick Prosser   University of Glasgow, UK
Claude-Guy Quimper   Université Laval, Canada
Jean-Charles Regin   University Nice-Sophia Antipolis/I3S/CNRS, France
Emma Rollon   Technical University of Catalonia, Spain
Francesca Rossi   IBM Research and University of Padova, Italy
Mohamed Siala   Insight Centre for Data Analytics, University College
                                   Cork, Ireland
Michael Trick   Carnegie Mellon University, USA
Tommaso Urli   CSIRO Data61 and the Australian National University,
                                   Australia
Peter van Beek   University of Waterloo, Canada
Pascal Van Hentenryck   University of Michigan, USA
Willem-Jan Van Hoeve   Carnegie Mellon University, USA
Petr Vilím   IBM, Czech Republic
Christel Vrain   LIFO, University of Orléans, France
Mohamed Wahbi   Insight, University College Cork, Ireland
Roland Yap   National University of Singapore, Singapore
William Yeoh   New Mexico State University, USA
Alessandro Zanarini   ABB Corporate Research, Switzerland
Roie Zivan   Ben Gurion University of the Negev, Israel
Stanislav Živný   University of Oxford, UK

Application Track Program Committee

Carmen Gervet	Université de Montpellier, France
Philip Kilby	Data61 and the Australian National University, Australia
Deepak Mehta	United Technologies Research Centre Ireland, Ireland
Laurent Michel	University of Connecticut, USA
Laurent Perron	Google, France
Christian Schulte	KTH Royal Institute of Technology, Sweden
Paul Shaw	IBM, France
Helmut Simonis	Insight Centre for Data Analytics, University College Cork, Ireland
Sylvie Thiébaux	The Australian National University, Australia

Biology and CP Track Program Committee

Nicos Angelopoulos	Wellcome Sanger Institute, UK
Alexander Bockmayr	Freie Universität Berlin, Germany
Simon De Givry	INRA - MIAT, France
Agostino Dovier	Università di Udine, Italy
Ferdinando Fioretto	University of Michigan, USA

Machine Learning and CP Track Program Committee

Christian Bessiere	CNRS, University of Montpellier, France
Bruno Cremilleux	Université de Caen, France
Thi-Bich-Hanh Dao	University of Orléans, France
Georgiana Ifrim	University College Dublin, Ireland
Michela Milano	DISI Università di Bologna, Italy
Siegfried Nijssen	Université Catholique de Louvain, Belgium

Operations Research and CP Track Program Committee

Serdar Kadioglu	Oracle Corporation, USA
Nick Sahinidis	Carnegie Mellon University, USA
Olivia Smith	IBM Research, Australia
Christine Solnon	LIRIS CNRS UMR 5205/INSA Lyon, France
Pascal Van Hentenryck	University of Michigan, USA
Tallys Yunes	University of Miami, USA

Satisfiability and CP Track Program Committee

Alan Frisch	University of York, UK
George Katsirelos	MIAT, INRA, France
Ian Miguel	University of St Andrews, UK
Nina Narodytska	Samsung Research America, USA

Steve Prestwich   Insight Centre for Data Analytics, Ireland
Ofer Strichman   Technion, Israel

Test and Verification and CP Track Program Committee

Sébastien Bardin   CEA LIST, France
Mats Carlsson   RISE SICS, Sweden
Roberto Castañeda Lozano   SICS, Sweden
Catherine Dubois   ENSIIE, Samovar, France
Vijay Ganesh   University of Waterloo, Canada
Anastasia Paparrizou   CRIL-CNRS, Université d'Artois, France
Marie Pelleau   LIP6, UPMC, France
Andreas Podelski   University of Freiburg, Germany
Michel Rueher   Université Côte d'Azur, CNRS, France
Pascal Van Hentenryck   University of Michigan, USA
Lebbah Yahia   University of Oran 1, Algeria

Journal and Sister Conferences Track Program Committee

Özgür Akgün   University of St Andrews, UK
Munsee Chang   University of St Andrews, UK
Ian Gent   University of St Andrews, UK
Christopher Jefferson   University of St Andrews, UK
Peter Nightingale   University of St Andrews, UK

Additional Reviewers

Rui Abreu   Ryo Kimura
Özgür Akgün   Philippe Laborie
Suguman Bansal   Javier Larrosa
Johannes Gerhardus Benade   Nadjib Lazaar
Clément Carbonnel   Olivier Lhomme
Mats Carlsson   Barnaby Martin
Supratik Chakraborty   Jacopo Mauro
Kenil Cheng   Ciaran McCreesh
Christel Christel   Jean-Noël Monette
Martin Cooper   Christian Muise
Thi-Bich-Hanh Dao   Cemalettin Ozturk
Daniel J. Fremont   Alexandre Papadopoulos
Luca Di Gaspero   Guillaume Perez
Chrysanthos Gounaris   Yash Puranik
Alban Grastien   Ashish Sabharwal
Peter Jeavons   Lakhdar Sais
Christopher Jefferson   Paul Scott

Thiago Serra
Gilles Simonin
Carsten Sinz
Friedrich Slivovsky
Atena Tabakhi
James Trimble

Gilles Trombettoni
Matt Valeriote
Christoph M. Wintersteiger
Ghiles Ziat
Roie Zivan
Ed Zulkoski

Journal Fast Track (Abstract)

Improved Filtering for the Bin-Packing with Cardinality Constraint

Guillaume Derval[1], Jean-Charles Régin[2], and Pierre Schaus[1]

[1] UCLouvain, Belgium
{guillaume.derval,pierre.schaus}@uclouvain.be
[2] University of Nice Sophia-Antipolis, France
jcregin@gmail.com

Previous research [2, 3] shows that a cardinality reasoning can improve the pruning of the bin-packing constraint, even when cardinalities are not involved in the original model. Our contribution is two-fold.

We first introduce a new algorithm, called BPCFlow, that filters both load and cardinality bounds on the bins, using a flow reasoning similar to the one used for the Global Cardinality Constraint.

Moreover, we detect impossible assignments of items by combining the load and cardinality of the bins using a new reasoning method called "too-big/too-small". This new method attempts to construct for each bin with load and cardinality bounds $[\underline{L}, \overline{L}]$ and $[\underline{C}, \overline{C}]$ a maximum-weighted set of $\overline{C} - 1$ items. Once this set is constructed, we detect that items with weight $w < \underline{L} - \sum_{i \in S} w_i$ cannot be assigned to the current bin. Similar arguments can be used to detect a maximum weight. The "too-big/too-small" reasoning is then adapted to the existing propagators, namely SimpleBPC [3], Pelsser's method [2] and BPCFlow.

We then experiment our four new algorithms on Balanced Academic Curriculum Problem and Tank Allocation Problem instances.

BPCFlow is shown to be indeed stronger than previously existing filtering, and more computationally intensive. We show that the new filtering is useful on a small number of hard instances, while being too expensive for general use.

Our results show the introduced "too-big/too-small" filtering can most of the time drastically reduce the size of the search tree and the computation time. This method is profitable in 88% of the tested instances.

This work is published in the Constraints journal [1].

References

1. Derval, G., Régin, J.C., Schaus, P.: Improved filtering for the bin-packing with cardinality constraint. In: Constraints. Springer (2017)
2. Pelsser, F., Schaus, P., Régin, J.C.: Revisiting the cardinality reasoning for binpacking Constraint. In: Schulte, C. (eds.) Principles and Practice of Constraint Programming, CP 2013. LNCS, vol 8124, pp. 578–586. Springer, Heidelberg (2013)

3. Schaus, P., Régin, J.C., Van Schaeren, R., Dullaert, W., Raa, B.: Cardinality reasoning for bin-packing constraint: application to a tank allocation problem. In: Milano, M. (eds) Principles and Practice of Constraint Programming. LNCS, vol 7514, pp. 815–822. Springer, Heidelberg (2012)

Journal and Sister Conference Tracks (Abstracts)

Ranking Constraints

Christian Bessiere[1], Emmanuel Hebrard[2], George Katsirelos[3],
Zeynep Kiziltan[4], and Toby Walsh[5]

[1] LIRMM, CNRS, Université de Montpellier
[2] LAAS-CNRS, Université de Toulouse
[3] MIAT, INRA
[4] University of Bologna
[5] University of New south Wales

Abstract. In many problems we want to reason about the ranking of items. For example, in information retrieval, when aggregating several search results, we may have ties and consequently rank orders. (e.g. [2, 3]). As a second example, we may wish to construct an overall ranking of tennis player based on pairwise comparisons between players. One principled method for constructing a ranking is the Kemeny distance [5] as this is the unique scheme that is neutral, consistent, and Condorcet. Unfortunately, determining this ranking is NP-hard, and remains so when we permit ties in the input or output [4]. As a third example, tasks in a scheduling problem may run in parallel, resulting in a ranking. In a ranking, unlike a permutation, we can have ties. Thus, 12225 is a ranking whilst 12345 is a permutation. To reason about permutations, we have efficient and effective global constraints. Regin [7] proposed an $O(n^4)$ GAC propagator for permutations. For BC, there is an even faster $O(n \log n)$ propagator [6]. Every constraint toolkit now provides propagators for permutation constraints. Surprisingly, ranking constraints are not yet supported. In [1], we tackle this weakness by proposing a global ranking constraint. We show that simple decompositions of this constraint hurt pruning. We then show that GAC can be achieved in polynomial time and we propose an $O(n^3 \log n)$ algorithm for achieving RC as well as an efficient quadratic algorithm offering a better tradeoff.

References

1. Bessiere, C., Hebrard, E., Katsirelos, G., Kiziltan, Z., Walsh. T.: Ranking constraints. In: Proceedings of the of IJCAI, pp. 705–711 (2016)
2. Brancotte, B., Yang, B., Blin, G., Denise S. Cohen-Boulakia, Hamel, S.: Rank aggregation with ties: experiments and analysis. In: Proceedings of the VLDB Endowment (PVLDB) (2015)
3. Fagin, R., Kumar, R., Mahdian, M., Sivakumar, D., Vee, E.: Comparing and aggregating rankings with ties. In: Proceedings of the PODS, pp. 47–58. ACM (2004)
4. Hemaspaandra, E., Spakowski, H., Vogel, J.: The complexity of kemeny elections. Theoret. Comput. Sci. **349**(3), 382–391 (2005)
5. Kemeny, J.G.: Mathematics without numbers. Daedalus **88**(4), 577–591 (1959)

6. Ortiz, A., Quimper, C.-G., Tromp, J., van Beek, P.: A fast and simple algorithm for bounds consistency of the all different constraint. In: Proceedings of the of IJCAI, pp. 245–250 (2003)
7. Régin. J.-C., A filtering algorithm for constraints of difference in CSPs. In: Proceedings of the AAAI, pp. 362–367 (1994)

Modeling with Metaconstraints and Semantic Typing of Variables

André Ciré[1], J.N. Hooker[2], and Tallys Yunes[3]

[1] University of Toronto
[2] Carnegie Mellon University
[3] University of Miami

Research in hybrid optimization shows that a combination of constraint programming and optimization technologies can significantly speed up computation. A key element of hybridization is the use of high-level metaconstraints in the problem formulation, which generalize the global constraints that are characteristic of constraint programming models. Metaconstraints aid solution by communicating problem structure to the solver.

Modeling with metaconstraints, however, raises a fundamental issue of variable management that must be addressed before its full potential can be realized. The solver frequently creates auxiliary variables as it relaxes and/or reformulates metaconstraints. Variables created for different constraints may actually have the same meaning, or they may relate in some more complicated way to each other and to variables in the original model. The solver must recognize these relationships among variables if it is to generate the necessary channeling constraints and formulate a tight overall continuous relaxation of the problem.

We address this problem systematically with a semantic typing scheme that reveals relationships among variables while allowing simpler, self-documenting models. We view a model as organized around user-defined, multiplace predicates that denote relations akin to those that occur in a relational database. A variable declaration is viewed as a database query that has the effect of assigning a semantic type to the variable. Relationships between variables are then deduced from their semantic types.

We develop this idea for a wide variety of constraint types, including systems of all-different constraints, employee scheduling constraints, general scheduling constraints with interval variables, sequencing problems with side constraints, disjunctions of linear systems, and constraints with piecewise linear functions. We develop three very general classes of channeling constraints that can be automatically inferred and are based on such relational database operations as projection. Finally, we discuss the advantages of semantic typing for error detection and model management.

This is an extended abstract of the full paper, which appears in *INFORMS Journal on Computing* **28** (2016) 1–13.

MaxSAT-Based Large Neighborhood Search for High School Timetabling

Emir Demirović and Nysret Musliu

Institute of Information Systems, Databases and Artificial Intelligence Group,
Vienna University of Technology, Vienna, Austria
{demirovic,musliu}@dbai.tuwien.ac.at

Extended Abstract

The problem of high school timetabling (HSTT) is to coordinate resources (e.g. rooms, teachers, students) with times in order to fulfill certain goals (e.g. scheduling lectures). It is a well known and widespread problem, as every high school requires some form of timetabling. Unfortunately, HSTT is hard to solve and just finding a feasible solution for simple variants of HSTT has been proven to be NP-complete. When solving hard combinatorial problems such as HSTT, there are two solving paradigms that are used often: local search algorithms, which usually find fast local optimal solutions, but cannot guarantee the optimality, and complete algorithms, which provide optimal results by exhaustively enumerating all solutions over longer periods of time.

In this paper [1], we aim to obtain the best of both worlds by combining the two strategies. More precisely, we develop a new anytime algorithm for HSTT which combines local search with a novel maxSAT-based large neighborhood search. A local search algorithm is used to drive an initial solution into a local optimum and then more powerful large neighborhood search (LNS) techniques based on maxSAT are used to further improve the solution. During the course of the algorithm, the solution is iteratively *destroyed*, by using one of the two neighborhood vectors, and *repaired* by maxSAT. The size of the neighborhood vectors is increased with time until the complete search space is explored, allowing the algorithm to prove optimality if given enough computational time.

The computational results demonstrate that we outperform the state-of-the-art solvers on numerous benchmarks and provide four new upper bounds. To the best of our knowledge, this is the first time maxSAT is used within a large neighborhood search scheme. In addition, we experiment with several variants to show the importance of each component of the algorithm. Furthermore, our algorithm is more efficient than a pure maxSAT-based approach for the given computational setting (20 min runtime).

Reference

1. Demirovic, E., Musliu, N.: Maxsat-based large neighborhood search for high school timetabling. Comput. OR **78**, 172–180 (2017)

Android Database Attacks Revisited

Behnaz Hassanshahi and Roland H.C. Yap

School of Computing, National University of Singapore, Singapore
b.hassanshahi@u.nus.edu
ryap@comp.nus.edu.sg

Many Android apps (applications) employ databases for managing sensitive data. In [1], we systematically study attacks targeting databases in benign Android apps and also study a new class of database vulnerabilities, which we call *private database* vulnerabilities.

We propose an analysis framework, extending the framework in [2], to find Android database vulnerabilities which are confirmed with a proof-of-concept (POC) exploit, i.e. zero-day. Our analysis combines static dataflow analysis, symbolic execution and constraint solving and finally dynamic testing to certify the exploit. In order, to generate a POC malware, our analysis uses an SMT solver to solve the path constraints in the program which together with the Android manifest is used to generate parameters for API calls which may exploit the app database vulnerabilities. Dynamic testing on the generated POC malware confirms whether or not the malware exploits the app database vulnerabilities, if not, alternative malware are generated.

In order to analyse how apps use databases, it is necessary to accurately handle URI objects and libraries which use them. We build accurate models for URI objects connecting them to appropriate constraints. Simple URI methods can be directly translated to SMT formulas while more complex URI methods are modelled using Symbolic Finite Transducers together with the SMT solver.

We evaluate our analysis on popular Android apps, successfully finding many database vulnerabilities. Surprisingly, our analyzer finds new ways to exploit previously reported and fixed vulnerabilities. We also propose a fine-grained protection mechanism which extends the Android manifest to protect against database attacks.

References

1. Hassanshahi, B., Yap, R.H.C.: Android database attacks revisited. In: ACM Asia Conference on Computer and Communications Security (ASIACCS), pp. 625–639, ACM (2017)
2. Hassanshahi, B., Jia, Y., Yap, R.H.C., Saxena, P., Liang, Z.: Web-to-application injection attacks on android: characterization and detection. In: 20th European Symposium on Research in Computer Security, LNCS, vol. 9327, pp. 577–598. Springer, Cham (2015)

This is a summary of paper [1].

Hybrid Optimization Methods
for Time-Dependent Sequencing Problems
(Abstract)

Joris Kinable[1,2], Andre A. Cire[3], and Willem-Jan van Hoeve[2]

[1] Robotics Institute, Carnegie Mellon University, 5000 Forbes Ave,
Pittsburgh, PA 15213, USA
jkinable@cs.cmu.edu
[2] Tepper School of Business, Carnegie Mellon University,
5000 Forbes Ave, Pittsburgh, PA 15213, USA
acire@utsc.utoronto.ca
[3] Department of Management, University of Toronto Scarborough,
1265 Military Trail, Toronto, ON M1C 1A4, Canada
vanhoeve@andrew.cmu.edu

Abstract. A large number of practical problems in manufacturing, transportation, and distribution require the sequencing of activities over time. Often activities in a sequencing problem are subject to operational constraints and optimization criteria involving *setup times*, i.e., the minimum time that must elapse between two consecutive activities in a sequence. A setup time typically models the time to change jobs in an assembly line or the travel time between two cities in traveling salesman problems. In classical sequencing problems, the setup time is only defined between pairs of activities. However, in many practical applications the setup time is also a function of the order of the activities in the sequence. Such *position-dependent* setup times are useful in modeling different states of a resource throughout a schedule, for example when the internal components of a machine degrade after performing a number of tasks.

In this paper, we introduce a novel optimization method for sequencing problems with position-dependent setup times. Our proposed method relies on a hybrid approach where a constraint programming model is enhanced with two distinct relaxations: A discrete relaxation based on multivalued decision diagrams, and a continuous relaxation based on linear programming, which are combined via the method of additive bounding. The relaxations are used to generate bounds and enhance constraint propagation. We conduct experiments on three variants of the time-dependent traveling salesman problem: the first considers no side constraints, the second considers time window constraints, and the third considers precedence constraints between pairs of activities. The experiments indicate that our techniques substantially outperform general-purpose methods based on mixed-integer linear programming and constraint programming models.

This paper appeared as "Joris Kinable, Andre A. Cire, and Willem-Jan van Hoeve. Hybrid Optimization Methods for Time-Dependent Sequencing Problems. *European Journal of Operational Research* 259(3):887–897, 2017".

Learning Rate Based Branching Heuristic
for SAT Solvers

Jia Hui Liang, Vijay Ganesh, Pascal Poupart,
and Krzysztof Czarnecki

University of Waterloo, Waterloo, Canada

Abstract. In this paper, we propose a framework for viewing solver branching heuristics as optimization algorithms where the objective is to maximize the *learning rate*, defined as *the propensity for variables to generate learnt clauses*. By viewing online variable selection in SAT solvers as an optimization problem, we can leverage a wide variety of optimization algorithms, especially from machine learning, to design effective branching heuristics. In particular, we model the variable selection optimization problem as an online multi-armed bandit, a special-case of *reinforcement learning*, to learn branching variables such that the learning rate of the solver is maximized. We develop a branching heuristic that we call *learning rate branching* or LRB, based on a well-known multi-armed bandit algorithm called *exponential recency weighted average* and implement it as part of MiniSat and CryptoMiniSat. We upgrade the LRB technique with two additional novel ideas to improve the learning rate by accounting for *reason side rate* and exploiting *locality*. The resulting LRB branching heuristic is shown to be faster than the VSIDS and conflict history-based (CHB) branching heuristics on 1975 application and hard combinatorial instances from 2009 to 2014 SAT Competitions. We also show that CryptoMiniSat with LRB solves more instances than the one with VSIDS. These experiments show that LRB improves on state-of-the-art. The original version of this paper appeared in the SAT 2016 proceedings [1].

Reference

1. Liang, J.H., Ganesh, V., Poupart, P., Czarnecki, K.: Learning rate based branching heuristic for SAT Solvers. In: Proceedings of the 19th International Conference on Theory and Applications of Satisfiability Testing, SAT 2016, Bordeaux, France, 5–8 July 2016, pp. 123–140 (2016)

Three Generalizations
of the FOCUS Constraint

Nina Narodytska[1], Thierry Petit[2,3], Mohamed Siala[4],
and Toby Walsh[5]

[1] Samsung Research America, Mountain View, USA
nina.n@samsung.com
[2] School of Business, Worcester Polytechnic Institute, Worcester, USA
tpetit@wpi.edu
[3] LINA-CNRS, Mines-Nantes, Inria, Nantes, France
thierry.petit@mines-nantes.fr
[4] Insight Centre for Data Analytics, Department of Computer Science,
University College Cork, Ireland
mohamed.siala@insight-centre.org
[5] UNSW, Data61 and TU Berlin, Sydney, NSW 2052, Australia
toby.walsh@data61.csiro.au

Abstract. The Focus constraint expresses the notion that solutions are concentrated. In practice, this constraint suffers from the rigidity of its semantics. To tackle this issue, we propose three generalizations of the Focus constraint. We provide for each one a complete filtering algorithm. Moreover, we propose ILP and CSP decompositions.

This work is published in [1, 2].

References

1. Narodytska, N., Petit, T., Siala, M., Walsh, T.: Three generalizations of the FOCUS constraint. In: Proceedings of the 23rd International Joint Conference on Artificial Intelligence, IJCAI 2013, Beijing, China, 3–9 August 2013, pp. 630–636 (2013)
2. Narodytska, N., Petit, T., Siala, M., Walsh, T.: Three generalizations of the FOCUS constraint. Constraints **21**(4), 495–532 (2016)

Conditions Beyond Treewidth for Tightness of Higher-Order LP Relaxations

Mark Rowland, Aldo Pacchiano, and Adrian Weller

UC Berkeley
pacchiano@berkeley.edu

We examine Boolean binary weighted constraint satisfaction problems without hard constraints, and explore conditions under which it is possible to solve the problem exactly in polynomial time [2]. We are interested in the problem of finding a configuration of variables $x = (x_1, \ldots, x_n) \in \{0, 1\}^n$ that maximizes a score function, defined by unary and pairwise rational terms $f(x) = \sum_{i=1}^{n} \psi_i(x_i) + \sum_{(i,j) \in E} \psi_{ij}(x_i, x_j)$. In the machine learning community, this is typically known as *MAP* (or *MPE*) *inference*.

In this work, we consider a popular approach which first expresses the MAP problem as an integer linear program (ILP) then relaxes this to a linear program (LP). If the LP optimum is achieved at an integral point we say the LP is tight. If the LP is performed over the marginal polytope, which enforces global consistency [1], then the LP will always be tight but exponentially many constraints are required. Sherali and Adams introduced a series of successively tighter relaxations of the marginal polytope: for any integer r, $\mathbb{L}_r$ enforces consistency over all clusters of variables of size $\leq r$. $\mathbb{L}_r$ is solvable in polynomial time and tight for graphs of treewidth $r - 1$ [1].

Most past work has focused on characterizing conditions for $\mathbb{L}_2$ and $\mathbb{L}_3$ tightness [3, 4]. Here we significantly improve on the result for $\mathbb{L}_3$ of [4], and provide important new results for when $LP + \mathbb{L}_4$ is tight, employing an interesting geometric perspective. The main result is to show that the relationship which holds between forbidden minors characterizing treewidth and $\mathbb{L}_r$ tightness for $r = 2$ and $r = 3$ breaks down for $r = 4$, hence demonstrating that treewidth is not precisely the right condition for analyzing tightness of higher-order LP relaxation.

References

1. Wainwright, M., Jordan, M.: Treewidth-based conditions for exactness of the Sherali-Adams and Lasserre relaxations. Technical report, University of California, Berkeley, 671:4 (2004)
2. Weller, A., Tang, K., Sontag, D., Jebara, T.: Understanding the Bethe approximation: when and how can it go wrong? In: Uncertainty in Artificial Intelligence (UAI) (2014)

This is a summary of the paper "M. Rowland, A. Pacchiano, A.Weller. Conditions Beyond Treewidth for Tightness of Higher-order LP Relaxations. *AISTATS 2017*"

3. Weller, A.: Characterizing tightness of LP relaxations by forbidding signed minors. In: Uncertainty in Artificial Intelligence (UAI) (2016)
4. Weller, A., Rowland, M., Sontag, D.: Tightness of LP relaxations for almost balanced models. In: Artificial Intelligence and Statistics (AISTATS) (2016)

Contents

Technical Track

A Novel Approach to String Constraint Solving 3
  Roberto Amadini, Graeme Gange, Peter J. Stuckey, and Guido Tack

Generating Linear Invariants for a Conjunction of Automata Constraints 21
  Ekaterina Arafailova, Nicolas Beldiceanu, and Helmut Simonis

AMONG Implied Constraints for Two Families of Time-Series Constraints . . . 38
  Ekaterina Arafailova, Nicolas Beldiceanu, and Helmut Simonis

Solving Constraint Satisfaction Problems Containing Vectors
of Unknown Size 55
  Erez Bilgory, Eyal Bin, and Avi Ziv

An Efficient SMT Approach to Solve MRCPSP/max Instances
with Tight Constraints on Resources 71
  Miquel Bofill, Jordi Coll, Josep Suy, and Mateu Villaret

Conjunctions of Among Constraints 80
  Víctor Dalmau

Clique Cuts in Weighted Constraint Satisfaction 97
  Simon de Givry and George Katsirelos

Arc Consistency via Linear Programming. 114
  Grigori German, Olivier Briant, Hadrien Cambazard, and Vincent Jost

Combining Nogoods in Restart-Based Search 129
  Gael Glorian, Frederic Boussemart, Jean-Marie Lagniez,
  Christophe Lecoutre, and Bertrand Mazure

All or Nothing: Toward a Promise Problem Dichotomy
for Constraint Problems 139
  Lucy Ham and Marcel Jackson

Kernelization of Constraint Satisfaction Problems: A Study
Through Universal Algebra 157
  Victor Lagerkvist and Magnus Wahlström

Defining and Evaluating Heuristics for the Compilation
of Constraint Networks 172
  Jean-Marie Lagniez, Pierre Marquis, and Anastasia Paparrizou

A Tolerant Algebraic Side-Channel Attack on AES Using CP   189
  Fanghui Liu, Waldemar Cruz, Chujiao Ma, Greg Johnson,
  and Laurent Michel

On Maximum Weight Clique Algorithms, and How They Are Evaluated. . . .   206
  Ciaran McCreesh, Patrick Prosser, Kyle Simpson, and James Trimble

MDDs: Sampling and Probability Constraints.   226
  Guillaume Perez and Jean-Charles Régin

An Incomplete Constraint-Based System for Scheduling
with Renewable Resources.   243
  Cédric Pralet

Rotation-Based Formulation for Stable Matching.   262
  Mohamed Siala and Barry O'Sullivan

Preference Elicitation for DCOPs   278
  Atena M. Tabakhi, Tiep Le, Ferdinando Fioretto, and William Yeoh

Extending Compact-Table to Basic Smart Tables   297
  Hélène Verhaeghe, Christophe Lecoutre, Yves Deville,
  and Pierre Schaus

Constraint Programming Applied to the Multi-Skill Project
Scheduling Problem.   308
  Kenneth D. Young, Thibaut Feydy, and Andreas Schutt

Application Track

An Optimization Model for 3D Pipe Routing with Flexibility Constraints . . .   321
  Gleb Belov, Tobias Czauderna, Amel Dzaferovic,
  Maria Garcia de la Banda, Michael Wybrow, and Mark Wallace

Optimal Torpedo Scheduling   338
  Adrian Goldwaser and Andreas Schutt

Constraint Handling in Flight Planning   354
  Anders Nicolai Knudsen, Marco Chiarandini, and Kim S. Larsen

NightSplitter: A Scheduling Tool to Optimize (Sub)group Activities   370
  Tong Liu, Roberto Di Cosmo, Maurizio Gabbrielli, and Jacopo Mauro

Time-Aware Test Case Execution Scheduling for Cyber-Physical Systems . . .   387
  Morten Mossige, Arnaud Gotlieb, Helge Spieker, Hein Meling,
  and Mats Carlsson

Integrating ILP and SMT for Shortwave Radio Broadcast Resource
Allocation and Frequency Assignment............................... 405
   Linjie Pan, Jiwei Jin, Xin Gao, Wei Sun, Feifei Ma, Minghao Yin,
   and Jian Zhang

Constraint-Based Fleet Design Optimisation for Multi-compartment
Split-Delivery Rich Vehicle Routing 414
   Tommaso Urli and Philip Kilby

Integer and Constraint Programming for Batch Annealing Process Planning ... 431
   Willem-Jan van Hoeve and Sridhar Tayur

Machine Learning and CP Track

Minimum-Width Confidence Bands via Constraint Optimization........... 443
   Jeremias Berg, Emilia Oikarinen, Matti Järvisalo, and Kai Puolamäki

Constraint Programming for Multi-criteria Conceptual Clustering 460
   Maxime Chabert and Christine Solnon

A Declarative Approach to Constrained Community Detection 477
   Mohadeseh Ganji, James Bailey, and Peter J. Stuckey

Combining Stochastic Constraint Optimization and Probabilistic
Programming: From Knowledge Compilation to Constraint Solving 495
   Anna L.D. Latour, Behrouz Babaki, Anton Dries, Angelika Kimmig,
   Guy Van den Broeck, and Siegfried Nijssen

Learning the Parameters of Global Constraints Using Branch-and-Bound.... 512
   Émilie Picard-Cantin, Mathieu Bouchard, Claude-Guy Quimper,
   and Jason Sweeney

CoverSize: A Global Constraint for Frequency-Based Itemset Mining 529
   Pierre Schaus, John O.R. Aoga, and Tias Guns

Operations Research and CP Track

A Column-Generation Algorithm for Evacuation Planning
with Elementary Paths... 549
   Mohd. Hafiz Hasan and Pascal Van Hentenryck

Job Sequencing Bounds from Decision Diagrams 565
   J.N. Hooker

Branch-and-Check with Explanations for the Vehicle Routing Problem
with Time Windows 579
   Edward Lam and Pascal Van Hentenryck

Solving Multiobjective Discrete Optimization Problems with Propositional
Minimal Model Generation.   596
  Takehide Soh, Mutsunori Banbara, Naoyuki Tamura,
  and Daniel Le Berre

Analyzing Lattice Point Feasibility in UTVPI Constraints   615
  K. Subramani and Piotr Wojciechowski

A Constraint Composite Graph-Based ILP Encoding of the Boolean
Weighted CSP   630
  Hong Xu, Sven Koenig, and T.K. Satish Kumar

Satisfiability and CP Track

Reduced Cost Fixing in MaxSAT   641
  Fahiem Bacchus, Antti Hyttinen, Matti Järvisalo, and Paul Saikko

Weight-Aware Core Extraction in SAT-Based MaxSAT Solving.   652
  Jeremias Berg and Matti Järvisalo

Optimizing SAT Encodings for Arithmetic Constraints   671
  Neng-Fa Zhou and Håkan Kjellerstrand

Test and Verification and CP Track

Constraint-Based Synthesis of Datalog Programs.   689
  Aws Albarghouthi, Paraschos Koutris, Mayur Naik, and Calvin Smith

Search Strategies for Floating Point Constraint Systems   707
  Heytem Zitoun, Claude Michel, Michel Rueher, and Laurent Michel

Author Index   723

Technical Track

A Novel Approach to String Constraint Solving

Roberto Amadini[1]([✉]), Graeme Gange[1], Peter J. Stuckey[1], and Guido Tack[2]

[1] University of Melbourne, Melbourne, Victoria, Australia
roberto.amadini@unimelb.edu.au
[2] Monash University, Melbourne, Australia

Abstract. String processing is ubiquitous across computer science, and arguably more so in web programming. In order to reason about programs manipulating strings we need to solve constraints over strings. In Constraint Programming, the only approaches we are aware for representing string variables—having bounded yet possibly unknown size—degrade when the maximum possible string length becomes too large. In this paper, we introduce a novel approach that decouples the size of the string representation from its maximum length. The domain of a string variable is dynamically represented by a simplified regular expression that we called a *dashed string*, and the constraint solving relies on propagation of information based on equations between dashed strings. We implemented this approach in G-STRINGS, a new string solver—built on top of GECODE solver—that already shows some promising results.

1 Introduction

Strings are fundamental datatypes in all the modern programming languages. String analysis [10, 23, 25] is needed in several real-life applications such as test-case generation [12], program analysis [8], model checking [17], web security [5], and bioinformatics [4]. Reasoning over strings requires the processing of constraints such as (in-)equality, concatenation, length, and so on.

A natural candidate to tackle string constraints is the *Constraint Programming* (CP) paradigm [19]. Unfortunately, practically no CP solver natively supports string constraints. To the best of our knowledge, the only exception is GECODE+S [29, 31], an extension of GECODE solver [18]. GECODE+S relies on *Bounded-Length Sequence* (BLS) string variables [31], implemented with dynamic lists of bitsets. Empirical results shows that GECODE+S is usually better than dedicated string solvers such as HAMPI [22], KALUZA [28], and SUSHI [14].

The *MiniZinc* [26] modelling language was recently extended to include string variables and constraints [1]. A MiniZinc library for converting MiniZinc models with strings into equivalent FlatZinc instances containing only integer variables has also been provided. In this way every solver supporting FlatZinc can now solve a MiniZinc model with strings, by converting each string of maximum length n into an array of n integer variables. This allowed the comparison of native string solvers like GECODE+S against state-of-the-art CP solvers using a

© Springer International Publishing AG 2017
J.C. Beck (Ed.): CP 2017, LNCS 10416, pp. 3–20, 2017.
DOI: 10.1007/978-3-319-66158-2_1

decomposition. Results indicate that native support for string variables usually pays off, but not always, in which case the technology of the best solver varies.

Having bounded-length strings is reasonable (note that satisfiability with unbounded-length strings is not decidable in general [16]) and enables finite-domain variables. The crucial issue here is to decide a maximum length ℓ for string variables. On the one hand, too small a value for ℓ may exclude solutions for important classes of string applications, e.g., where a variable represents a long XML string or part of a DNA string. On the other hand, too large a value for ℓ can significantly worsen performance even for relatively simple problems.

A common drawback, shared by both the GECODE+S solver and the approaches statically mapping string variables to arrays of integer variables, is that the solving process is coupled to the maximum string length ℓ. Indeed, the performance of these approaches degrade when ℓ becomes bigger and bigger even for relatively simple problems.

In this paper we address this problem by proposing a novel approach to string representation in CP solvers. The new representation is based on a restricted class of regular expressions, which we refer to as *dashed strings*. Given an alphabet Σ and a maximum string length ℓ, a dashed string consists of an ordered sequence $S_1^{l_1,u_1} S_2^{l_2,u_2} \cdots S_k^{l_k,u_k}$ of $0 < k \leq \ell$ *blocks*, where $S_i \subseteq \Sigma$ and $0 \leq l_i \leq u_i \leq \ell$ for $i = 1, \ldots, k$, and $\Sigma_{i=1}^k l_i \leq \ell$. Each block $S_i^{l_i,u_i}$ represents the set of all the strings of Σ having length in $[l_i, u_i]$ and characters in S_i. The idea of dashed strings takes inspiration from the *Bricks* abstract domain of [10]. In that paper, however, each block refers to a set of strings of Σ^* (while in our representation refers to a set of characters of Σ) and some workarounds are used in order to make the abstract domain a lattice.

We use dashed strings to model the domain of string variables. The propagators for string constraints rely on the notion of *equation* between dashed strings in order to possibly narrow each string domain to a concrete string (i.e., to a dashed string representing a single string of Σ^*). We also define a branching strategy that aims to select the strings with minimal length that satisfy all the constraints, using the lexicographic order for breaking ties.

Following the GECODE+S approach, we use GECODE [18] as a starting point for implementing our solver. The resulting solver, that we called G-STRINGS, already shows promising results. We compared its performance against: the aforementioned GECODE+S; the state-of-the-art CP solvers CHUFFED, GECODE, IZPLUS; the SMT solver Z3STR3 [34], a string theory plug-in built on top of Z3 solver. Results indicate that, despite still being in a preliminary stage, G-STRINGS often outperforms all such solvers. However, there are class of problems where it has worse performance. This leaves room for future enhancements.

The original contributions of this paper are: *(i)* new abstractions and algorithms for modelling and manipulating the domain of string variables; *(ii)* new propagators and branchers for string constraint solving; *(iii)* the implementation and the evaluation of a new string solver.

Paper Structure. Section 2 gives preliminary notions. Section 3 defines the dashed strings and the algorithms we used in Sect. 4 for implementing string variables and constraints. Section 5 provides an evaluation of our approach, before we conclude in Sect. 6.

2 Preliminaries

Given a finite alphabet $\Sigma = \{a_1, \ldots, a_n\}$, a string $x \in \Sigma^*$ is a finite sequence of $|x| \geq 0$ characters of Σ, where $|x|$ is the length of x. We omit the distinction between characters and strings of unary length. The interval $[a, b]$ will be denoted also with $\{a..b\}$.

The concatenation of $x, y \in \Sigma$ is denoted by $x \cdot y$ (or simply xy when not ambiguous) while x^n denotes the iterated concatenation of x for n times (where x^0 is the empty string ϵ). We generalise this definition to set of strings: given $X, Y \subseteq \Sigma$, we denote with $X \cdot Y = \{xy \mid x \in X, y \in Y\}$ (or simply with XY) their concatenation and with X^n the iterated concatenation of X for n times (where $X^0 = \{\epsilon\}$).

In this work we focus on bounded-length strings: fixed a maximum length ℓ, we consider only strings in the universe $\mathbb{S} = \bigcup_{i=0}^{\ell} \Sigma^i$. Clearly $\mathbb{S}$ is not closed under concatenation. We extend the canonical definition of *Constraint Satisfaction Problem* (CSP) by including string variables and constraints. Formally, a CSP is a triple $\langle \mathcal{V}, \mathcal{D}, \mathcal{C} \rangle$ consisting of a set of variables $\mathcal{V}$, each of which associated with a domain $\mathcal{D}(x) \in \mathcal{D}$ of values that $x \in \mathcal{V}$ could take, and a set of constraints $\mathcal{C}$ defining all the feasible assignments of values to variables. The goal is to find a solution, i.e., a variable assignment satisfying all the constraints of $\mathcal{C}$.

In addition to "standard" integer variables and constraints, in this paper we consider string variables x having domain $D(x) \subseteq \mathbb{S}$, and string constraints over string variables. We also consider constraints involving both string and integer variables, e.g., the length constraint $|x| = n$ or the power constraint $x^n = y$ where x, y are string variables and n is an integer variable.

3 Dashed Strings

A *dashed string* is a restricted regular expression denoting a finite set of concrete strings. The rationale behind this representation is to facilitate a compact and dynamic representation of set of strings of unknown length, without statically pre-allocating an arbitrarily large number of elements. Moreover, as we shall see later, dashed strings enable us to deal with concatenation—arguably the most common string operation—in a natural way.

Below we give the formal definition of dashed string, and then show how we propagate information over equations between dashed strings. Before that, we give an informal intuition of what a dashed string is. The name "dashed" comes from a graphical interpretation of $S = S_1^{l_1, u_1} S_2^{l_2, u_2} \cdots S_k^{l_k, u_k}$ where we imagine a block $S_i^{l_i, u_i}$ as a continuous segment of length l_i followed by a dashed segment of length $u_i - l_i$. The continuous segment indicates that exactly l_i characters of S_i *must* occur in each concrete string denoted by S; the dashed segment indicates that k characters of S_i, with $0 \leq k \leq u_i - l_i$, *may* occur. Consider Fig. 1, illustrating dashed string $S = \{\text{B},\text{b}\}^{1,1}\{\text{o}\}^{2,4}\{\text{m}\}^{1,1}\{\text{!}\}^{0,3}$. Each string represented by S starts with B or b, followed by 2 to 4 os, one m, then 0 to 3 !s.

Fig. 1. Graphical representation of $\{\text{B},\text{b}\}^{1,1}\{\text{o}\}^{2,4}\{\text{m}\}^{1,1}\{\text{!}\}^{0,3}$.

3.1 Definition

Let us fix the alphabet Σ, the maximum length ℓ, and the universe $\mathbb{S} = \bigcup_{i=0}^{n} \Sigma^\ell$. A *dashed string* of *length* k is defined by a concatenation of $0 < k \leq \ell$ blocks $S_1^{l_1,u_1} S_2^{l_2,u_2} \cdots S_k^{l_k,u_k}$, where $S_i \subseteq \Sigma$ and $0 \leq l_i \leq u_i \leq \ell$ for $i = 1,\ldots,k$, and $\Sigma_{i=1}^{k} l_i \leq \ell$. For block $S_i^{l_1,u_1}$, we call S_i the *base* and (l_i,u_i) the *cardinality*. $S[i]$ indicates the i-th block of dashed string S, and $|S|$ the number of blocks (i.e., the length of S). $\mathbb{DS}$ denotes the set of all dashed strings. We do not distinguish blocks from dashed strings of unary length.

Let $\gamma(S^{l,u}) = \{x \in S^* \mid l \leq |x| \leq u\}$ be the language denoted by block $S^{l,u}$. In particular the *null element* $\emptyset^{0,0}$ is such that $\gamma(\emptyset^{0,0}) = \{\epsilon\}$. We extend γ to dashed strings: $\gamma(S_1^{l_1,u_1} \cdots S_k^{l_k,u_k}) = (\gamma(S_1^{l_1,u_1}) \cdots \gamma(S_k^{l_k,u_k})) \cap \mathbb{S}$. A dashed string S is *known* if $|\gamma(S)| = 1$, i.e., it represents a single string.

We say $S^{l,u}$ is *coverable* by $T^{l',u'}$ if some string in $\gamma(S^{l,u})$ is a prefix of a string in $\gamma(T^{l',u'})$ (formally, if $l = 0 \vee (l \leq u' \wedge S \cap T \neq \emptyset)$). S and T are *incompatible* if neither S nor T is coverable by the other.

Given $S, T \in \mathbb{DS}$ we define the relation $S \sqsubseteq T \iff \gamma(S) \subseteq \gamma(T)$. Intuitively, operator $\sqsubseteq$ models the relation "is more precise than" between dashed strings.

Unfortunately, although $\sqsubseteq$ is a partial order over $\mathbb{DS}$, the structure $(\mathbb{DS}, \sqsubseteq)$ does not form in general a lattice. This means that it might not exist a greatest lower bound (or a least upper bound) for two given dashed strings $S, T \in \mathbb{DS}$. Proposition 1 proves this statement. Unlike other frameworks (e.g., Abstract Interpretation [11]), Constraint Programming does not require lattice structures to preserve the soundness of constraint solving. However, as we shall see, care must be taken in order to avoid leaks of feasible solutions or infinite propagations.

Proposition 1. *The structure* $(\mathbb{DS}, \sqsubseteq)$ *is not a lattice.*

Proof. Let $\Sigma = \{a,b\}$, $S = \{a\}^{1,1}\{b\}^{1,1}$, and $T = \{b\}^{1,1}\{a\}^{1,1}$. We prove that there is no least upper bound in $(\mathbb{DS}, \sqsubseteq)$ for S and T, nor a greatest lower bound for $S' = \{a\}^{0,1}\{b\}^{1,1}\{a\}^{0,1}$ and $T' = \{b\}^{0,1}\{a\}^{1,1}\{b\}^{0,1}$.

We first observe that S', T' and $\{a,b\}^{2,2}$ are the minimal elements greater than S, T according to $\sqsubseteq$. However, they are incomparable with $\sqsubseteq$ since $\gamma(S') = \{b, ab, aba\}$, $\gamma(T') = \{a, ba, bab\}$ and $\gamma(\{a,b\}^{2,2}) = \{aa, ab, ba, bb\}$. Thus, there not exist a least upper bound for S, T. The greatest lower bound of S', T' does not exists because the maximal elements smaller than S', T' are $\{a,b\}^{1,1}\{a,b\}^{0,2}$ and $\{a,b\}^{0,2}\{a,b\}^{1,1}$, which are incomparable according to $\sqsubseteq$. $\square$

The γ function is not injective. For example, for $S = \{a\}^{0,1}\{a\}^{0,1}$ and $T = \{a\}^{0,2}$ we have $\gamma(S) = \gamma(T) = \{\epsilon, a, aa\}$. To remove redundant configurations, and minimise the length of a dashed string, we introduce the notion of *normalisation*. A dashed string $S = S_1^{l_1,u_1} \cdots S_k^{l_k,u_k}$ is normalised if and only if:

(i) $S_i \neq S_{i+1}$,   for $i = 1, \ldots, k-1$.
(ii) $S_i = \emptyset \iff l_i = u_i = 0$,   for $i = 1, \ldots, k$;
(iii) $S = \emptyset^{0,0} \vee S_i \neq \emptyset$,   for $i = 1, \ldots, k$;

Condition (i) says that each adjacent base has to be distinct, since blocks $S^{l,u}$ and $S^{l',u'}$ are equivalent to $S^{l+l',u+u'}$. Condition (ii) avoid multiple configurations for the null element $\emptyset^{0,0}$. Condition (iii) forbids the redundant use of $\emptyset^{0,0}$, being in general $\gamma(B \cdot \emptyset^{0,0}) = \gamma(\emptyset^{0,0} \cdot B) = \gamma(B)$.

We omit the definition of the normalisation algorithm, that unsurprisingly has linear cost $O(|S|)$ for normalising a dashed string S. Note that if $S, S' \in \mathbb{DS}$ are normalised then $S = S' \iff \gamma(S) = \gamma(S')$.

Finally, we define the *size* $\|S^{l,u}\|$ of a block $S^{l,u}$ as:

$$\|S^{l,u}\| = \begin{cases} u - l + 1 & \text{if } |S| \leq 1 \\ \dfrac{|S|^{u+1} - |S|^l}{|S| - 1} & \text{otherwise.} \end{cases}$$

and we generalise this definition to dashed strings, i.e., $\|S\| = \Pi_{i=1}^{k}\|S_i^{l_i,u_i}\|$ for each dashed string $S = S_1^{l_1,u_1} \cdots S_k^{l_k,u_k}$.

The size of a dashed string gives a measure of the number of concrete strings it represents. Note that, while for a block $S^{l,u}$ we have that $\|S^{l,u}\| = |\gamma(S^{l,u})|$, for a generic dashed string $S \in \mathbb{DS}$ we have that $\|S\| \geq |\gamma(S)|$ but not $\|S\| = |\gamma(S)|$. For example, if $S = \{a\}^{0,1}\{a,b\}^{0,1}$, we have $|\gamma(S)| = |\{\epsilon, a, b, aa, ab\}| = 5$ while $\|S\| = \|\{a\}^{0,1}\| \cdot \|\{a,b\}^{0,1}\| = 2 \cdot 3 = 6$.

3.2   Equating Dashed Strings

We use dashed strings as a domain abstraction for string variables. Following the standard CP framework, each variable domain is iteratively narrowed until it becomes a single value, that will be assigned to the variable, or it becomes empty, meaning that the problem is unsatisfiable.

In this context, we have to iteratively "narrow" a dashed string S until it becomes known or the unsatisfiability is detected. Things are tricky here since $(\mathbb{DS}, \sqsubseteq)$ does not form a lattice. Consider for example two string variables x and y, having domain S' and T' as in the proof of Proposition 1. There is not an unique way to prune the domain of x and y when it comes to propagate the equality constraint $x = y$, since there is no greatest lower bound for S' and T'.

Regardless of the choice of how pruning, a propagator for a string constraint must be at least *sound* (it never prunes values that can appear in a solution) and *contracting* (is only allowed to remove values).

Algorithm 1. EQUATE algorithm

1: **function** EQUATE (S, T)
2:     **Input:** Dashed strings $S = S_1^{a_1,b_1} \cdots S_n^{a_n,b_n}$ and $T = T_1^{c_1,d_1} \cdots T_m^{c_m,d_m}$.
3:     **Output:** *true* if S and T are equatable; *false* otherwise.
4:     $Matches \leftarrow NoGoods \leftarrow \emptyset$
5:     CHECK$(S, 1, S_1^{a_1,b_1}, T, 1, T_1^{c_1,d_1}, Matches, NoGoods)$
6:     **if** $Matches = \emptyset$ **then**
7:         **return** *false*
8:     $SplitS, SplitT \leftarrow$ SPLIT$(S, T, Matches)$
9:     $\widetilde{S}, \widetilde{T} \leftarrow$ MERGE$(SplitS, SplitT)$
10:    UPDATE$(S, T, \widetilde{S}, \widetilde{T})$
11:    **return** *true*

The core algorithm that we adopted for string constraint propagation is based on the *equation* of two dashed strings. Informally, equating two dashed strings S and T means, firstly, to verify that there exists at least a concrete string shared by both $\gamma(S)$ and $\gamma(T)$ and, if so, to find a representation for S and T that includes all the strings of $\gamma(S) \cap \gamma(T)$ and removes the most values not belonging to $\gamma(S) \cap \gamma(T)$. More formally, this problem consists in finding, if feasible, two dashed strings S' and T' such that: *(i)* $S' \sqsubseteq S$, $T' \sqsubseteq T$; *(ii)* $\gamma(S') \cap \gamma(T') = \gamma(S) \cap \gamma(T)$. We could add a third condition stating that there not exist two dashed strings S'', T'' such that $S'' \sqsubset S'$ and $T'' \sqsubset T'$. However, this requirement makes the propagation too difficult.

We address this equation problem—that can be seen as a semantic unification problem—with a multiphase strategy, where dashed strings S are T in input are processed and possibly updated with two "refined" dashed strings S' and T'. These phases, namely *checking, splitting, merging,* and *updating,* are explained below. We use pseudo-code and we abstract as much as possible the technicalities, referring to a running example rather than going into the implementation details. The actual code we developed integrates and optimise these four stages that, for the sake of readability, here we present simplified and separately.

The main algorithm is summarised in Algorithm 1. Taking as input two dashed strings $S = S_1^{a_1,b_1} \cdots S_n^{a_n,b_n}$ and $T = T_1^{c_1,d_1} \cdots T_m^{c_m,d_m}$, that we assume already normalised, EQUATE initialises variables *Matches* and *NoGoods* to the empty set (we shall explain their meaning below) and then CHECK is called.

Checking. CHECK (Algorithm 2) both tests if S and T are equatable, and constructs a directed acyclic graph *Matches* encoding the set of solutions. SPLIT will then reconstruct *Matches* into dashed strings for S and T.

CHECK uses a top-down dynamic programming approach, recursively matching suffixes of S and T. In any matching, the first block of either S or T must finish first. If S, we compute what remains available of the T-block, and match the tail of S with the remnant of T (similarly for T) – this is done in lines 11–18. Lines 2–10 cover early termination, where S or T reached the end or have

---

Algorithm 2. CHECK algorithm

1: **function** CHECK $(S, i, S_i^{l_i, u_i}, T, j, T_j^{l_j, u_j}, Matches, NoGoods)$
2:     **if** $(i, l_i, u_i, j, l_j, u_j) \in NoGoods$ **then return** *false*
3:     **if** $i = |S| + 1$ **then**   ▷ Reached end of S
4:         **if** $l_j = c_{j+1} = \cdots = c_m = 0$ **then return** NEWMATCH(*Matches*)
5:         **else return** FAIL$(NoGoods, S_i^{l_i, u_i}, T_j^{l_j, u_j})$
6:     **else if** $j = |T| + 1$ **then**   ▷ Reached end of T
7:         **if** $l_i = a_{i+1} = \cdots = a_n = 0$ **then return** NEWMATCH(*Matches*)
8:         **else return** FAIL$(NoGoods, S_i^{l_i, u_i}, T_j^{l_j, u_j})$
9:     **else if** $l_i > 0 \wedge l_j > 0 \wedge S_i \cap T_j = \emptyset$ **then**   ▷ Incompatible blocks
10:        **return** FAIL$(NoGoods, S_i^{l_i, u_i}, T_j^{l_j, u_j})$
11:     **if** $l_i = 0 \vee (S_i \cap T_j \neq \emptyset \wedge l_i \leq u_j)$ **then**   ▷ $S_i^{l_i, u_i}$ coverable
12:        $Rem_T \leftarrow S_i \cap T_j \neq \emptyset$? $T_j^{\max(0, l_j - u_i), u_j - l_i}$: $T_j^{l_j, u_j}$
13:        $Check_S \leftarrow$ CHECK$(S, i + 1, S[i + 1], T, j, Rem_T, Matches, NoGoods)$
14:     **else** $Check_S \leftarrow$ *false*
15:     **if** $l_j = 0 \vee (S_i \cap T_j \neq \emptyset \wedge l_j \leq u_i)$ **then**   ▷ $T_j^{l_j, u_j}$ coverable
16:        $Rem_S \leftarrow S_i \cap T_j \neq \emptyset$? $S_i^{\max(0, l_i - u_j), u_i - l_j}$: $S_i^{l_i, u_i}$
17:        $Check_T \leftarrow$ CHECK$(S, i, Rem_S, T, j + 1, T[j + 1], Matches, NoGoods)$
18:     **else** $Check_T \leftarrow$ *false*
19:     **if** $\neg(Check_S \vee Check_T)$ **then**
20:        **return** FAIL$(NoGoods, S_i^{l_i, u_i}, T_j^{l_j, u_j})$
21:     **return** $Check_S \vee Check_T$

---

incompatible initial blocks. FAIL saves failed computations in *NoGoods* before returning *false*.

For a successful computation, the sequence of partial blocks consumed by calls to CHECK encode possible solutions to $S = T$. CHECK builds a directed acyclic graph representing the set of such sequences. Each sequence will be called a *match*. For simplicity, we elide details of how *Matches* is maintained – essentially, it amounts to recording the graph of successful CHECK calls.

CHECK defines a *match-tree*, i.e., a binary tree where: *(i)* each node is a pair of blocks (the root is $\langle S_1^{l_1, u_1}, T_1^{l_1, u_1} \rangle$); *(ii)* there is a branch from $\langle S_i^{l_i, u_i}, T_j^{l_j, u_j} \rangle$ to left child $\langle S_{i+1}^{l_{i+1}, u_{i+1}}, T_j^{l'_j, u'_j} \rangle$ if $S_i^{l_i, u_i}$ is coverable by $T_j^{l_j, u_j}$ and $T_j^{l'_j, u'_j}$ is the corresponding remnant (the dual definition applies to the right child); *(iii)* a leaf is either a success (a match is found) or a failure (due to incompatible blocks).

A match tree for $S = \{a..c\}^{0,30}\{d\}^{5,5}\{c..f\}^{0,2}$ and $T = \{b..d\}^{26,26}\{f\}^{1,1}$ is shown in Fig. 2 (ignoring for now dashed arrows). Failures are denoted with ×, while successes with ◇. A match identifies a path from root to ◇ representable with the coordinates $\langle i, j \rangle$ of each node $\langle S_i^{l_i, u_i}, T_j^{l_j, u_j} \rangle$. For each transition $\langle i, j \rangle \rightarrow \langle i', j' \rangle$ the invariant $(i' = i \wedge j' = j + 1) \vee (j' = j \wedge i' = i + 1)$ holds. We can thus see each transition as a move of length 1 in a $n \times m$ grid.

All the three matches of Fig. 2 are coloured in green. In particular the (partial) match $[\langle 1, 1 \rangle, \langle 2, 1 \rangle, \langle 2, 2 \rangle, \langle 3, 2 \rangle]$ is truncated. This is because the pair $\langle \{c..f\}_3^{0,2}, \{f\}_2^{1,1} \rangle$ has already been examined before and thus there is no need to rebuild the subtree again. Even if not explicitly detailed, our actual implementation defines a mechanism—similar to the recording of failures—that enables the caching of already visited nodes, and hence to prune redundant computations.

From Fig. 2 we can see for example that the rightmost subtree rooted in $\langle 1, 1 \rangle$ always fails. This is because if the block $\{b..d\}^{26,26}$ is entirely covered by $\{a..c\}^{0,30}$, then there is no other block in S that can cover $\{f\}^{1,1}$.

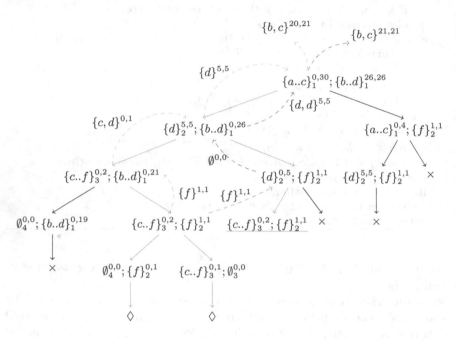

Fig. 2. Match tree for $S = \{a..c\}^{0,30}\{d\}^{5,5}\{c..f\}^{0,2}$ and $T = \{b..d\}^{26,26}\{f\}^{1,1}$. (Color figure online)

Lemma 1. *The worst case complexity of* EQUATE *is* $O(nm\max(n,m))$.

Proof. Each recursive call in CHECK$(S, i, S_i^{l_i,u_i}, T, j, T_j^{l_j,u_j}, Matches, NoGoods)$ removes one block completely from S or from T so one between $S_i^{l_i,u_i}$ and $T_j^{l_j,u_j}$ is an original block and the other one is a remnant block. If it is a remnant block it can only be changed $\max(n,m)$ times, since it runs out of blocks to cover. Hence the total number of different calls is $O(nm\max(n,m))$. □

Splitting. Suppose CHECK returned *true* (otherwise EQUATE terminates). Thus $\gamma(S) \cap \gamma(T) \neq \emptyset$. However, we would like to refine S and T in order to prune the most values not belonging to $\gamma(S) \cap \gamma(T)$. In this second phase we take advantage of the matches collected in *Matches* for possibly *splitting* the blocks of S and T. We aim to find (partial) maps $\sigma_S : [1, |S|] \to \mathbb{DS}$ such that $\sigma_S(i) \sqsubseteq S[i]$. In this way, by definition, $\sigma_S(1) \cdots \sigma_S(n) \sqsubseteq S$ (same applies for T).

Algorithm 3. SPLIT algorithm

1: **function** SPLIT $(S, T, Matches)$
2:     $k \leftarrow 1$; $splitS \leftarrow splitT \leftarrow [\ ]$; $Matches' \leftarrow \text{TRIM}(Matches)$
3:     **for** $M_k \in Matches'$ **do**
4:         $sl_i \leftarrow su_i \leftarrow sl_j \leftarrow su_j \leftarrow 0$; $split_S^k \leftarrow split_T^k \leftarrow \{\ \}$; $last_i \leftarrow -1$
5:         **for** $\langle S_i^{l_i, u_i}, T_j^{l_j, u_j} \rangle \in M_k$ **do**
6:             $R \leftarrow S_i \cap T_j$
7:             **if** $i = last_i$ **then**   ▷ direction $\searrow$
8:                 $l \leftarrow \max(l_i - su_j, l_j)$; $u \leftarrow \min(u_i - sl_j, u_j)$
9:                 **if** $l > u$ **then** $l \leftarrow u \leftarrow 0$
10:                 $split_S^k[i] \leftarrow \text{NORM}([R^{l,u}] + split_S^k[i])$; $split_T^k[j] \leftarrow [R^{l,u}]$
11:                 $sl_i \leftarrow l$; $su_i \leftarrow u$; $sl_j \leftarrow sl_j + l$; $su_j \leftarrow su_j + u$
12:             **else**   ▷ direction $\nearrow$
13:                 $l \leftarrow \max(l_j - su_i, l_i)$; $u \leftarrow \min(u_j - sl_i, u_i)$
14:                 **if** $l > u$ **then** $l \leftarrow u \leftarrow 0$
15:                 $split_T^k[j] \leftarrow \text{NORM}([R^{l,u}] + split_T^k[j])$; $split_S^k[i] \leftarrow [R^{l,u}]$
16:                 $sl_j \leftarrow l$; $su_j \leftarrow u$; $sl_i \leftarrow sl_i + l$; $su_i \leftarrow su_i + u$
17:             $last_i \leftarrow i$
18:         $splitS \leftarrow splitS + [split_S^k]$; $splitT \leftarrow splitT + [split_T^k]$; $k \leftarrow k + 1$
19:     **return** $splitS, splitT$

As mentioned, a match for $S = S_1^{a_1, b_1} \cdots S_n^{a_n, b_n}$ and $T = T_1^{c_1, d_1} \cdots T_m^{c_m, d_m}$ can be described by a path in the match tree. In particular, a sub-path of the form $[\langle i, j \rangle, \langle i, j+1 \rangle, \ldots, \langle i, j+k \rangle]$ enables us to split block $S_i^{l_i, u_i}$ of S into a concatenation of k blocks $(S_i \cap T_j)^{\alpha_k, \beta_k} (S_i \cap T_{j+1})^{\alpha_{k-1}, \beta_{k-1}} \cdots (S_i \cap T_{j+k})^{\alpha_1, \beta_1}$ where, for $h = 1, \ldots, k$, cardinalities α_h, β_h are computed iteratively by a bottom-up approach that we explain below.

Informally speaking, we "climb back up" the match-tree from the leaves to the root. Each move from a child node to its father has a *direction* that can be top-right (if it is a left child) or top-left (for a right child). If we "walk straight" in the same direction, for each node of the path there is always one block B that stays fixed, while the other blocks $B', B'', B''', \ldots$ vary along the way. So we can split B into sub-blocks thanks to the information given by $B', B'', B''', \ldots$, i.e., by all the blocks covered by B along the way. Care must be taken when computing the cardinality of the sub-blocks: we have to consider the cumulative cardinality of $B', B'', B''', \ldots$ and not only the block currently being examined. When the direction changes, we "turn" in the new direction. This process is repeated until the root is reached.

The SPLIT algorithm listed in Algorithm 3 performs the backward propagation from the leaves to the root. We consider each match $M_k \in Matches'$ where $Matches' = \text{TRIM}(Matches)$ and the TRIM function removes from $Matches$ all the pairs of the form $\langle \emptyset^{0,0}, B \rangle$ and $\langle B, \emptyset^{0,0} \rangle$ (useless in this context). For each M_k we have a map $split_S^k$ (resp., $split_T^k$) mapping each index $i \in [1, n]$ to a list of blocks $split_S^k[i]$ defining a splitting of $S[i]$ (resp., mapping each index $j \in [1, m]$ to $split_T^k[j]$). SPLIT returns two lists $splitS = [split_S^1, \ldots, split_S^p]$ and $splitT = [split_T^1, \ldots, split_T^p]$ where $p = |Matches'|$.

Each match M_k is already in reversed order, i.e., from leaf to root, since each match is registered following the stack of recursive calls to CHECK.

If we are going top-right (lines 7–11) then we are splitting on S_i. We then add at the head of the current split $split_S^k[i]$ the element $R^{l,u}$ with $R = S_i \cap T_j$, $l = \max(l_i - su_j, l_j)$ and $u = \min(u_i - sl_j, u_j)$. We store in variable sl_j (resp., su_j) the cumulative sum of the lower bounds (resp., upper bounds) encountered when walking in the same direction. The + operator is the concatenation between lists.

Note that splitting a block $S^{l,u}$ into $S' = (S \cap T_1)^{l_1, u_1} \ldots (S \cap T_k)^{l_k, u_k}$ always refines the base S, since $(S \cap T_1) \cup \cdots \cup (S \cap T_k) \subseteq S$, but in general does not ensure that S' is normalised and, most important, that $\gamma(S') \subseteq \gamma(S^{l,u})$. Consider matching $S = \{a, b, d\}^{2,3}$ with $T = \{a, c\}^{0,2}\{b, c\}^{0,2}$. After matching, we would obtain a split $S' = \{a\}^{0,2}\{b\}^{0,2}$ for $S[1]$. While S' refines the base of $S[1]$, the loss of cardinality information introduces new (spurious) strings (e.g., the string $aabb \in \gamma(S') \setminus \gamma(S)$). We must therefore consider the cardinality of the original block when splitting. This is performed by a function NORM that, when splitting $S^{l,u}$ into $S' = (S \cap T_1)^{l_1, u_1} \ldots (S \cap T_1)^{l_k, u_k}$, first checks if $\Sigma_{i=1}^k l_i \geq l$ and $\Sigma_{i=1}^k u_i \leq u$. If so, it returns the normalisation of S'. Otherwise, it returns the block $((S \cap T_1) \cup \cdots \cup (S \cap T_k))^{l,u}$.

The opposite direction (lines 12–16) is totally symmetric. To identify the direction it is enough to check the value of $last_i$, which is updated at each loop iteration at line 17. Line 18 updates the lists of the split for each new match; these lists are then returned in line 19.

To better understand how SPLIT works, consider again the match tree in Fig. 2. After CHECK algorithm, we have $Matches' = \{M_1, M_2\}$ where M_1 and M_2 correspond to paths $[\langle 3, 2\rangle, \langle 3, 1\rangle, \langle 2, 1\rangle, \langle 1, 1\rangle]$ and $[\langle 3, 2\rangle, \langle 2, 2\rangle, \langle 2, 1\rangle, \langle 1, 1\rangle]$ respectively. Let us consider M_1 (see the red dashed arrows). Its first node $\langle \{c..f\}^{0,2}, \{f\}^{1,1}\rangle$ propagates upward the block $(\{c..f\} \cap \{f\})^{\max(0,1), \min(2,1)} = \{f\}^{1,1}$. Then we change direction. Node $\langle \{c..f\}^{0,2}, \{b..d\}^{0,21}\rangle$ propagates upward $(\{c..f\} \cap \{b..d\})^{\max(0-1,0), \min(2-1,21)} = \{c, d\}^{0,1}$. Node $\langle \{d\}^{5,5}, \{b..d\}^{0,26}\rangle$ propagates $(\{d\} \cap \{b..d\})^{\max(5,0-0), \min(5,26-1)} = \{d\}^{5,5}$ and finally the root propagates $(\{a..c\} \cap \{b..d\})^{\max(0,26-5-1), \min(30,26-5-0)} = \{b, c\}^{20,21}$. The corresponding splits are then $split_S^1 = \{1 : \{b, c\}^{20,21}, 2 : \{d\}^{5,5}, 3 : \{c, d\}^{0,1}\{f\}^{1,1}\}$ and $split_T^1 = \{1 : \{b..d\}^{26,26}, 2 : \{f\}^{1,1}\}$. We observe that $split_T^1[1] = T[1]$ instead of $T' = \{b, c\}^{20,21}\{d\}^{5,5}\{c, d\}^{0,1}$ since, as explained above, $T' \not\subseteq T[1]$ (in particular, T' would compromise the soundness by allowing strings of length 25 and 27).

Similarly, we can construct $split_S^2 = \{1 : \{b, c\}^{21,21}, 2 : \{d\}^{5,5}\}$ and $split_T^2 = \{1 : \{b, c\}^{21,21}\{d\}^{5,5}, 2 : \{f\}^{1,1}\}$. Note that in the actual implementation the element $\{f\}^{1,1}$ coloured in violet in Fig. 2 does not need to be recomputed by SPLIT because it is already cached.

Merging. At this stage, we have two lists of splits $splitS = [split_S^1, \ldots, split_S^p]$ and $splitT = [split_T^1, \ldots, split_T^p]$ that can be used to refine S and T respectively. The question now is: how to actually refine each $S[i]$ and $T[j]$, having different splitting $split_S^k[i]$ and $split_T^k[j]$ for $k = 1, \ldots, p$? We have somehow to *merge* each split $split_S^1[i], \ldots, split_S^p[i]$ into a minimal dashed string $\widetilde{S}_i$ that "contains" each split, i.e., such that $\widetilde{S}_i \sqsupseteq split_S^1[i], \ldots, split_S^p[i]$ (analogously for each $\widetilde{T}_j$).

Unfortunately, we remark that $(\mathbb{DS}, \sqsubseteq)$ is not a lattice so there might not exist a least upper bound for $split_S^1[i], \ldots, split_S^p[i]$ (see Proposition 1). Even here we have thus to settle for a relaxed "join" operation $\sqcup$ returning a dashed string $\widetilde{S}_i = split_S^1[i] \sqcup \cdots \sqcup split_S^p[i]$ that over-approximates each split and it is a good compromise between precision and efficiency (same thing for $\widetilde{T}_j$). If some $split_S^k[i]$ is not defined, we simply ignore it.

In the general case, given $S = S_1^{a_1,b_1} \cdots S_n^{a_n,b_n}$ and $T = T_1^{c_1,d_1} \cdots T_m^{c_m,d_m}$ we define $S \sqcup T = R^{l,u}$ where $R = \bigcup_{i=1}^n \bigcup_{j=1}^m (S_i \cup T_j)$, $l = \min(\Sigma_{i=1}^n a_i, \Sigma_{j=1}^m c_j)$, and $u = \max(\Sigma_{i=1}^n b_i, \Sigma_{j=1}^m d_j)$. However, we also deal with particular cases to improve the precision (e.g., when $S = T$).

In the example of Fig. 2, having $split_S^1 = \{1 : \{b,c\}^{20,21}, 2 : \{d\}^{5,5}, 3 : \{c,d\}^{0,1}\{f\}^{1,1}\}$ and $split_S^2 = \{1 : \{b,c\}^{21,21}, 2 : \{d\}^{5,5}\}$, we get $\widetilde{S}_1 = \{b,c\}^{20,21}$, $\widetilde{S}_2 = \{d\}^{5,5}$, and $\widetilde{S}_3 = \{c,d\}^{0,1}\{f\}^{1,1}$. For T instead we simply get $\widetilde{T}_1 = T_1$ and $\widetilde{T}_2 = T_2$. Finally, we return $\widetilde{S} = \widetilde{S}_1 \ldots \widetilde{S}_n$ and $\widetilde{T} = \widetilde{T}_1 \ldots \widetilde{T}_m$.

Updating. In the last stage, we update the original dashed strings S and T trying to refine their blocks thanks to the information given by $\widetilde{S}$ and $\widetilde{T}$. To do so, we use a simple block-wise approach that compares each S_i with $\widetilde{S}_i$ and, in case $\|\widetilde{S}_i\| < \|S_i\|$, updates S_i with $\widetilde{S}_i$. For avoiding overflows, instead of $\|S\|$ we consider its logarithm $\log \|S\| = \Sigma_{i=1}^n \log \|S_i^{a_i,b_i}\|$. In particular, if $x = |S| > 1$, we compute $\log \|S^{l,u}\|$ as $\log \frac{x^{u+1} - x^l}{x-1} = \log \frac{x^l(x^{u-l+1} - 1)}{x-1} = \log(x^l(x^{u-l+1} - 1)) - \log(x-1) = l \cdot \log x + \log(x^{u-l+1} - 1) - \log(x-1)$. In the same way we possibly update each T_j with $\|\widetilde{T}_j\|$.

Considering again the example in Fig. 2, from the original dashed strings $S = \{a..c\}^{0,30}\{d\}^{5,5}\{c..f\}^{0,2}$ we get $S' = \{b,c\}^{20,21}\{d\}^{5,5}\{c,d\}^{0,1}$, while T remains unchanged. However, we observe that while $\|S\|$ is in the order of 10^{15}, the size of S' is 9437184. Note the size difference which results if we equate $S'' = \{a..c\}^{0,30M}\{d\}^{5M,5M}\{c..f\}^{0,2M}$ and $T'' = \{b..d\}^{26M,26M}\{f\}^{M,M}$, where M is an arbitrarily big parameter. A nice property of EQUATE algorithm is that in this case the complexity is totally independent from M: both EQUATE(S,T) and EQUATE(S'', T'') are solved instantaneously.

Finally, note that we could run EQUATE on S and T with the blocks reversed to determine different information. We do not consider this in our implementation since we will focus on extracting information about the earliest blocks which will be the most helpful when aligned with the search we perform.

4 Constraint Solving

In this Section we give an overview of how we applied the notions introduced in Sect. 3 in order to solve a CSP with string variables and constraints.

Given a CSP $\langle \mathcal{V}, \mathcal{D}, \mathcal{C} \rangle$, the domain of each string variable $x \in \mathcal{V}$ is a dashed string $\mathcal{D}(x) \in \mathbb{DS}$. Each constraint $C \in \mathcal{C}$ on string variables $x_1, \ldots, x_k$ has an associated *propagator* that aims to remove the inconsistent values from domains $\mathcal{D}(x_1), \ldots, \mathcal{D}(x_k)$. Since propagation is incomplete, we have to define *search* strategies that split the domain of strings to cause more propagation.

4.1   Constraints

The key property of dashed strings that makes them useful is that we can concatenate dashed strings in a natural way: given $S = S_1^{a_1,b_1} \cdots S_n^{a_n,b_n}$ and $T = T_1^{c_1,d_1} \cdots T_m^{c_m,d_m}$ we get $S{\cdot}T = S_1^{a_1,b_1} \cdots S_n^{a_n,b_n} T_1^{c_1,d_1} \cdots T_m^{c_m,d_m}$ without any effort. Analogously, we can easily define the iterated concatenation $S^k = S{\cdot}S^{k-1}$, where $S^0 = \emptyset^{0,0}$, and the reverse $S^{-1} = S_n^{a_n,b_n} \cdots S^{1,1}$. Hence we can define many propagators by simply relying on the dashed string concatenation and the EQUATE algorithm described in Sect. 3. To lighten the load of propagation, we defined CHECKEQUATE, a simplified version of EQUATE(S,T) that returns *true* if S and T have a match (and immediately returns), and *false* otherwise. CHECKEQUATE neither stores nor computes the matches.

We consider the following constraints, and the corresponding propagators:[1]

- *equality* $x = y$. Implemented by EQUATE$(\mathcal{D}(x), \mathcal{D}(y))$;
- *disequality* $x \neq y$. If CHECKEQUATE$(\mathcal{D}(x), \mathcal{D}(y)) = $ *false* the constraint is subsumed; otherwise we wait until both $\mathcal{D}(x)$ and $\mathcal{D}(y)$ are known;
- *half-reified* [13] equality $b \Rightarrow (x = y)$. If $b = $ *true*, the constraint is rewritten into $x = y$. If $b = $ *false*, the constraint is subsumed. Otherwise, if CHECKEQUATE$(\mathcal{D}(x), \mathcal{D}(y)) = $ *false* then b is set to *false*. We treat $b \Rightarrow (x \neq y)$ similarly. Full reification $b \Leftrightarrow (x = y)$ is encoded as the conjunction $(b \Rightarrow x = y) \wedge (\neg b \Rightarrow x \neq y)$.
- *length* $|x| = n$. If $\mathcal{D}(x) = S_1^{l_1,u_1} \cdots S_k^{l_k,u_k}$, it is implemented analogously to $x_1 + \cdots + x_k = n$ where x_i is an integer variable with domain $[l_i, u_i]$.
- *domain* $x :: S$, where $S \in \mathbb{DS}$. Implemented by a version of EQUATE$(\mathcal{D}(x), S)$ that only updates $\mathcal{D}(x)$. If $\mathcal{D}(x) \sqsubseteq S$, the constraint is subsumed.
- *concatenation* $z = x \cdot y$. Implemented by EQUATE$(\mathcal{D}(z), \mathcal{D}(x) \cdot \mathcal{D}(y))$, taking care of properly projecting the narrowing of $\mathcal{D}(x) \cdot \mathcal{D}(y)$ on $\mathcal{D}(x)$ and $\mathcal{D}(y)$.
- *iterated concatenation* $y = x^n$. If $\mathcal{D}(x) = S_1^{l_1,u_1} \cdots S_k^{l_k,u_k}$, it is propagated by EQUATE$(\mathcal{D}(y), \mathcal{D}(x)^{\underline{n}} \cdot (\bigcup_{i=1}^k S_i)^{0, \overline{n} - \underline{n}})$.
- *reverse* $y = x^{-1}$. Implemented by EQUATE$(\mathcal{D}(y), \mathcal{D}(x)^{-1})$, taking care of properly projecting the narrowing of $\mathcal{D}(x)^{-1}$ on $\mathcal{D}(x)$.
- *sub-string* $y = x[i..j]$. Rewritten in $l = |x| \wedge n = \max(1, i) \wedge m = \min(l, j) \wedge |y| = \max(0, m - n + 1) \wedge x = y' \cdot y \cdot y'' \wedge y' :: \Sigma^{\underline{n}-1, \overline{n}-1} \wedge y'' :: \Sigma^{\max(0, \underline{l} - \overline{m}), \overline{l} - \underline{m}}$.

This set of constraints is not fully exhaustive. In particular, the lack of *regular* constraint limits its expressiveness since we can not fully encode the Kleene star S^* when $S \subseteq \Sigma^*$ is a set of strings having length greater than one. However,

---

[1]   For conciseness, for integer variable x, we define $\underline{x} = \min(\mathcal{D}(x))$ and $\overline{x} = \max(\mathcal{D}(x))$.

thanks to reification and (iterated) concatenation we can often compensate this lack (and also define constraints that are not expressible with regular, i.e., see the SQL Injection problem introduced in [1] and evaluated in Sect. 5).

Each propagator is scheduled by *propagator events* that occur if and when the domain of a variable in the constraint changes. We consider the following events: *fail* (a domain became empty), *none* (domains unchanged), *value* (a domain became a singleton), *cardinality* (the cardinality of some block changed), *character* (the characters of some base changed), *domain* (cardinality or characters changed). For example, the propagator for $|x| = n$ can only narrow the length of $\mathcal{D}(x)$ and not its characters, hence it does not need to wake on character changes.

4.2 Search

Searching in string problems is very important since there are typically a very large number of solutions for each string variable x. The search strategy we implemented first chooses the string variable x with smallest domain (i.e., minimising $\log \|\mathcal{D}(x)\|$).

If the length of x is unknown it branches on the first unknown length block $S_i^{l_i,u_i}$ being equal to its minimal length or not (i.e., $S_i^{l_i,l_i}$ or $S_i^{l_i+1,u_i}$). This branching wakes up propagators dependent on the length of x.

Otherwise if the first non-zero length block $S_i^{l_i,l_i}$ is of length $l_i > 1$ it splits it into two fixed length blocks $S_i^{1,1} S_i^{l_i-1,l_i-1}$ (note this is not a branch point). If the first non-zero length block $S_i^{l_i,l_i}$ is of length 1 it branches on setting the block to its least value $a = \min(S_i)$ or not (i.e., $\{a\}$ or $S_i - \{a\}$). This branching wakes up propagators dependent of the contents of x.

Overall this search has the effect of enumerating the solutions of x in lexicographic order, as shown in Fig. 3 where we show the search tree when $\mathcal{D}(x) = \{0\}^{2,2}\{a,b,c\}^{0,1}\{1\}^{1,1}$. However, the branching can be generalised by defining proper heuristic to choose how to split an unknown-length block, and how to pick a value from the base of a known-length block.

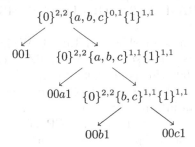

Fig. 3. Example of search tree.

5   Evaluation

We implemented our approach as an extension of GECODE [18], a mature CP solver written in C++. The resulting solver, that we called G-STRINGS, is publicly available at https://bitbucket.org/robama/g-strings.

G-STRINGS is a *copying solver*, i.e., during the search the domains are copied (and possibly restored) before a choice is committed. In this context the memory management becomes critical. We underline that G-STRINGS is still a prototype, and mainly relies on the GECODE built-in data structures. In a nutshell, a dashed string is currently implemented as a `DynamicArray` of blocks, where the base of each block is encoded by a `BndSet`, which represents finite set of integers as unions of disjoint ranges (see [18] for more details about this data structures). As a future work we plan to improve this implementation.

We compared G-STRINGS against the string CP solver GECODE+S [29,31], the string SMT solver Z3STR3 [34][2] and three state-of-the-art constraint solvers, namely: the aforementioned GECODE [18]; CHUFFED [9], a CP solver with lazy clause generation [27]; and IZPLUS [15], a CP solver that also exploits local search. For these three solvers we used the MiniZinc translation to integers [1] that statically maps string variables into arrays of integer variables. We did not compare against automata-based approaches like [21,24,32,33] since their limited effectiveness in our context (as an example, every single block $S^{l,u}$ has to be encoded by an automaton of exactly $u+1$ states).

As already noted in [1,20,29–31] there is unfortunately a lack of standardised and challenging string benchmarks. We decided to use the same string problems used in the evaluation of [1], namely: $a^n b^n$, ChunkSplit, HammingDistance, Levenshtein, StringReplace, SQLInjection. The only differences are: *(i)* the "HammingDistance" problem has been simplified since G-STRINGS does not yet support the regular constraint (for the other problems we have overcome this lack with (iterated) concatenation and reified equality); *(ii)* the "Palindrome" problem is omitted since neither G-STRINGS nor GECODE+S supports the new global cardinality constraint introduced in [1].

All these problems have no parameters, except for the maximum string length ℓ that we varied in $\{250, 1000, 10000\}$. We ran the experiments on Ubuntu 15.10 machines with 16 GB of RAM and 2.60 GHz Intel® i7 CPU by setting a solving timeout of $T = 1200\,\mathrm{s}$.

Comparative solving times are shown in Table 1. We ignore model construction time, which for the first three solvers using MiniZinc can be quite expensive (e.g., for SQL and $\ell = 10000$ this is almost 20 min). The results show that the G-STRINGS solver is (almost) independent of the maximum string length, and in particular it provides an instantaneous answer in all the NORN benchmarks. The performance of GECODE+S and the other CP solvers clearly degrade when increasing ℓ. Although being independent from ℓ, also Z3STR3 performs worse than G-STRINGS.

---

[2] We used the last stable release: https://sites.google.com/site/z3strsolver/.

Table 1. Results in seconds. 'n/a' indicates an abnormal termination while 't/o' means timeout. Unsatisfiable problems are marked with *, best performance are in bold font.

ℓ	CHUFFED			GECODE			IZPLUS			Z3STR3			GECODE+S			G-STRINGS		
	250	1000	10000	250	1000	10000	250	1000	10000	250	1000	10000	250	1000	10000	250	1000	10000
$a^n b^n$ *	0.1	1.3	483.83	1.81	129.32	t/o	0.74	16.45	t/o	t/o	t/o	t/o	0.31	29.46	t/o	**0.0**	**0.0**	**0.0**
ChunkSplit	1.62	t/o	26.33	0.39	12.93	61.91	1.81	11.56	116.61	0.6	0.6	0.6	1.28	182.24	t/o	**0.0**	**0.0**	**0.0**
Hamming *	0.32	1.69	61.27	0.16	0.77	19.58	0.24	1.89	49.95	1.22	1.2	1.2	0.0	0.12	129.8	**0.0**	**0.0**	**0.0**
Levenshtein	0.18	0.89	63.67	0.08	0.36	18.05	2.32	2.64	306.39	0.01	0.01	0.01	0.0	0.0	0.0	**0.0**	**0.0**	**0.0**
StringRep	1.46	43.75	n/a	0.56	19.28	n/a	0.7	8.54	673.18	2.58	2.59	2.62	0.06	2.25	t/o	**0.0**	**0.0**	**0.0**
SQLInj	10.78	t/o	n/a	0.81	375.17	t/o	130.68	613.16	n/a	t/o	t/o	t/o	0.01	0.2	**299.99**	0.09	72.58	t/o

Conversely, the SQL benchmark introduced in [1] illustrates a weakness of the current implementation. This problem involves a long fixed string of length ℓ, and our solver has worse performance than GECODE+S. In particular, G-STRINGS runs out of time when $\ell = 10000$. This points out that we need to specialise the EQUATE algorithm for parts of strings representable as fixed strings, and also switch to asymptotically faster propagation algorithms when the number of blocks becomes large.

6 Conclusions

In this work we introduced the dashed string representation to enable possibly very long strings to be represented succinctly, trying to decouple the complexity of constraint solving from the maximum length a string *may* have. Propagation of dashed strings is very efficient when the number of its blocks is small. Moreover, while dealing with large alphabets might be a problem for some approaches [31], this representation is weakly coupled to the size of the alphabet we are using.

Clearly dashed strings are not a universal panacea, since equating long dashed string representations can be too expensive. In other terms, this approach might fail when a string *must* be very long. Hence we need to develop weaker propagation algorithms to gracefully handle this case.

Using multiple representations for a string variable may be highly advantageous, where we choose the propagator for each string constraint which is most efficient to propagate. String abstract domains often combine different representations in this way (see, e.g., [3,10]). Although this may clearly reduce the propagation of information, it can avoid worst case behaviour. This hybrid approach can be implemented "internally", i.e., by building a *channeling* propagator between the representations, or "externally" via a *portfolio* approach [2] combining different solving strategies.

The introduction of dashed strings immediately opens several research branches. One of these concerns the definition of new propagators. We have already devised algorithms for propagating lexicographic comparisons, global cardinality, and regular constraints, although they are not implemented yet in G-STRINGS. Furthermore, the potentially huge search space suggests the exploration of different search approaches such as, e.g., Local Search [6,7]. Another interesting directions concerns the definition of *trailing* string solvers, i.e., solvers that store the domain changes instead of copying the entire domain during the search.

Acknowledgements. This work is supported by the Australian Research Council (ARC) through Linkage Project Grant LP140100437 and Discovery Early Career Researcher Award DE160100568.

References

1. Amadini, R., Flener, P., Pearson, J., Scott, J.D., Stuckey, P.J., Tack, G.: Minizinc with strings. In: Logic-Based Program Synthesis and Transformation - 25th International Symposium, LOPSTR 2016 (2016). https://arxiv.org/abs/1608.03650
2. Amadini, R., Gabbrielli, M., Mauro, J.: A multicore tool for constraint solving. In: Proceedings of the International Joint Conference on Artificial Intelligence, pp. 232–238. AAAI Press (2015)
3. Amadini, R., Jordan, A., Gange, G., Gauthier, F., Schachte, P., Søndergaard, H., Stuckey, P.J., Zhang, C.: Combining string abstract domains for javascript analysis: an evaluation. In: Legay, A., Margaria, T. (eds.) TACAS 2017. LNCS, vol. 10205, pp. 41–57. Springer, Heidelberg (2017). doi:10.1007/978-3-662-54577-5_3
4. Barahona, P., Krippahl, L.: Constraint programming in structural bioinformatics. Constraints 13(1–2), 3–20 (2008)
5. Bisht, P., Hinrichs, T.L., Skrupsky, N., Venkatakrishnan, V.N.: WAPTEC: white-box analysis of web applications for parameter tampering exploit construction. In: Proceedings of ACM Conference on Computer and Communications Security, pp. 575–586. ACM (2011)
6. Björdal, G.: String variables for constraint-based local search. Master's thesis, Department of Information Technology, Uppsala University, Sweden, August 2016. http://urn.kb.se/resolve?urn=urn:nbn:se:uu:diva-301501
7. Björdal, G., Monette, J.-N., Flener, P., Pearson, J.: A constraint-based local search backend for MiniZinc. Constraints 20(3), 325–345 (2015)
8. Bjørner, N., Tillmann, N., Voronkov, A.: Path feasibility analysis for string-manipulating programs. In: Kowalewski, S., Philippou, A. (eds.) TACAS 2009. LNCS, vol. 5505, pp. 307–321. Springer, Heidelberg (2009). doi:10.1007/978-3-642-00768-2_27
9. Chu, G.: Improving combinatorial optimization. Ph.D. thesis, Department of Computing and Information Systems, University of Melbourne, Australia (2011)
10. Costantini, G., Ferrara, P., Cortesi, A.: A suite of abstract domains for static analysis of string values. Softw.: Pract. Exp. 45(2), 245–287 (2015)
11. Cousot, P., Cousot, R.: Abstract interpretation: a unified lattice model for static analysis of programs by construction or approximation of fixpoints. In: Proceedings of the Fourth ACM Symposium on Principles of Programming Languages, pp. 238–252. ACM (1977)
12. Emmi, M., Majumdar, R., Sen, K.: Dynamic test input generation for database applications. In: Proceedings of the ACM SIGSOFT International Symposium on Software Testing and Analysis (ISSTA), pp. 151–162. ACM (2007)
13. Feydy, T., Somogyi, Z., Stuckey, P.J.: Half reification and flattening. In: Lee, J. (ed.) CP 2011. LNCS, vol. 6876, pp. 286–301. Springer, Heidelberg (2011). doi:10.1007/978-3-642-23786-7_23

14. Fu, X., Powell, M.C., Bantegui, M., Li, C.: Simple linear string constraints. Form. Asp. Comput. **25**(6), 847–891 (2013)
15. Fujiwara, T.: iZplus description (2016). http://www.minizinc.org/challenge2016/description_izplus.txt
16. Ganesh, V., Minnes, M., Solar-Lezama, A., Rinard, M.: Word equations with length constraints: what's decidable? In: Biere, A., Nahir, A., Vos, T. (eds.) HVC 2012. LNCS, vol. 7857, pp. 209–226. Springer, Heidelberg (2013). doi:10.1007/978-3-642-39611-3_21
17. Gange, G., Navas, J.A., Stuckey, P.J., Søndergaard, H., Schachte, P.: Unbounded model-checking with interpolation for regular language constraints. In: Piterman, N., Smolka, S.A. (eds.) TACAS 2013. LNCS, vol. 7795, pp. 277–291. Springer, Heidelberg (2013). doi:10.1007/978-3-642-36742-7_20
18. Gecode Team. Gecode: generic constraint development environment (2016). http://www.gecode.org
19. Golden, K., Pang, W.: Constraint reasoning over strings. In: Rossi, F. (ed.) CP 2003. LNCS, vol. 2833, pp. 377–391. Springer, Heidelberg (2003). doi:10.1007/978-3-540-45193-8_26
20. He, J., Flener, P., Pearson, J., Zhang, W.M.: Solving string constraints: the case for constraint programming. In: Schulte, C. (ed.) CP 2013. LNCS, vol. 8124, pp. 381–397. Springer, Heidelberg (2013). doi:10.1007/978-3-642-40627-0_31
21. Hooimeijer, P., Weimer, W.: StrSolve: solving string constraints lazily. Autom. Softw. Eng. **19**(4), 531–559 (2012)
22. Kiezun, A., Ganesh, V., Artzi, S., Guo, P.J., Hooimeijer, P., Ernst, M.D.: HAMPI: a solver for word equations over strings, regular expressions, and context-free grammars. ACM Trans. Softw. Eng. Methodol. **21**(4), Article 25 (2012)
23. Kim, S.-W., Chin, W., Park, J., Kim, J., Ryu, S.: Inferring grammatical summaries of string values. In: Garrigue, J. (ed.) APLAS 2014. LNCS, vol. 8858, pp. 372–391. Springer, Cham (2014). doi:10.1007/978-3-319-12736-1_20
24. Li, G., Ghosh, I.: PASS: string solving with parameterized array and interval automaton. In: Bertacco, V., Legay, A. (eds.) HVC 2013. LNCS, vol. 8244, pp. 15–31. Springer, Cham (2013). doi:10.1007/978-3-319-03077-7_2
25. Madsen, M., Andreasen, E.: String analysis for dynamic field access. In: Cohen, A. (ed.) CC 2014. LNCS, vol. 8409, pp. 197–217. Springer, Heidelberg (2014). doi:10.1007/978-3-642-54807-9_12
26. Nethercote, N., Stuckey, P.J., Becket, R., Brand, S., Duck, G.J., Tack, G.: MiniZinc: towards a standard CP modelling language. In: Bessière, C. (ed.) CP 2007. LNCS, vol. 4741, pp. 529–543. Springer, Heidelberg (2007). doi:10.1007/978-3-540-74970-7_38
27. Ohrimenko, O., Stuckey, P.J., Codish, M.: Propagation via lazy clause generation. Constraints **14**(3), 357–391 (2009)
28. Saxena, P., Akhawe, D., Hanna, S., Mao, F., McCamant, S., Song, D.: A symbolic execution framework for JavaScript. In: S&P, pp. 513–528. IEEE Computer Society (2010)
29. Scott, J.D.: Other things besides number: abstraction, constraint propagation, and string variable types. Ph.D. thesis, Department of Information Technology, Uppsala University, Sweden (2016). http://urn.kb.se/resolve?urn=urn:nbn:se:uu:diva-273311
30. Scott, J.D., Flener, P., Pearson, J.: Constraint solving on bounded string variables. In: Michel, L. (ed.) CPAIOR 2015. LNCS, vol. 9075, pp. 375–392. Springer, Cham (2015). doi:10.1007/978-3-319-18008-3_26

31. Scott, J.D., Flener, P., Pearson, J., Schulte, C.: Design and implementation of bounded-length sequence variables. In: Salvagnin, D., Lombardi, M. (eds.) CPAIOR 2017. LNCS, vol. 10335, pp. 51–67. Springer, Cham (2017). doi:10.1007/978-3-319-59776-8_5
32. Tateishi, T., Pistoia, M., Tripp, O.: Path- and index-sensitive string analysis based on monadic second-order logic. ACM Trans. Softw. Eng. Methodol. **22**(4), 33 (2013)
33. Yu, F., Alkhalaf, M., Bultan, T.: STRANGER: an automata-based string analysis tool for PHP. In: Esparza, J., Majumdar, R. (eds.) TACAS 2010. LNCS, vol. 6015, pp. 154–157. Springer, Heidelberg (2010). doi:10.1007/978-3-642-12002-2_13
34. Zheng, Y., Ganesh, V., Subramanian, S., Tripp, O., Dolby, J., Zhang, X.: Effective search-space pruning for solvers of string equations, regular expressions and length constraints. In: Kroening, D., Pǎsǎreanu, C.S. (eds.) CAV 2015. LNCS, vol. 9206, pp. 235–254. Springer, Cham (2015). doi:10.1007/978-3-319-21690-4_14

Generating Linear Invariants for a Conjunction of Automata Constraints

Ekaterina Arafailova[1]([✉]), Nicolas Beldiceanu[1], and Helmut Simonis[2]

[1] TASC (LS2N), IMT Atlantique, 44307 Nantes, France
{Ekaterina.Arafailova,Nicolas.Beldiceanu}@imt-atlantique.fr
[2] Insight Centre for Data Analytics, University College Cork, Cork, Ireland
Helmut.Simonis@insight-centre.org

Abstract. We propose a systematic approach for generating linear implied constraints that link the values returned by several automata with accumulators after consuming the same input sequence. The method handles automata whose accumulators are increased by (or reset to) some non-negative integer value on each transition. We evaluate the impact of the generated linear invariants on conjunctions of two families of time-series constraints.

1 Introduction

We present a compositional method for deriving linear invariants for a conjunction of global constraints that are each represented by an automaton with accumulators [10]. Since they do not encode explicitly all potential values of accumulators as states, automata with accumulators allow a constant size representation of many counting constraints imposed on a sequence of integer variables. Moreover their compositional nature permits representing a conjunction of constraints on a same sequence as the intersection of the corresponding automata [21,22], i.e. the intersection of the languages accepted by all automata, without representing explicitly the Cartesian product of all accumulator values. As a consequence, the size of such an intersection automaton is often quite compact, even if maintaining domain consistency for such constraints is in general NP-hard [8]; for instance the intersection of the 22 automata that restrict the number of occurrences of patterns of the Vol. II of the time-series catalogue [3] in a sequence has only 16 states. The contributions of this paper are twofold:

– First, Sects. 3 and 4 provide the basis of a simple, systematic and uniform preprocessing technique to compute necessary conditions for a conjunction of automata with accumulator constraints on the same sequence. Each necessary condition is a linear inequality involving the result variables of the different

E. Arafailova is supported by the EU H2020 programme under grant 640954 for the GRACeFUL project. N. Beldiceanu is partially supported by GRACeFUL and by the Gaspard-Monge programme. H. Simonis is supported by Science Foundation Ireland (SFI) under grant numbers SFI/12/RC/2289 and SFI/10/IN.1/I3032.

J.C. Beck (Ed.): CP 2017, LNCS 10416, pp. 21–37, 2017.
DOI: 10.1007/978-3-319-66158-2_2

automata, representing the fact that the result variables cannot vary independently. These inequalities are parametrised by the sequence size and are independent of the domains of the sequence variables. The method may be extended if the scopes of the automata constraints overlap, but not if the scopes contain the same variables ordered differently.
- Second, within the context of the time-series catalogue, Sect. 5 shows that the method allows to precompute in less than five minutes a data base of 7755 invariants that significantly speed up the search for time series satisfying multiple time-series constraints.

Adding implied constraints to a constraint model has been recognized from the very beginning of Constraint Programming as a major source of improvement [13]. Attempts to generate such implied constraints in a systematic way were limited (1) by the difficulty to manually prove a large number of conjectures [6,17], (2) by the limitations of automatic proof systems [12,15], or (3) to specific constraints like alldifferent, gcc, element or circuit [1,18,19]. Within the context of automata with accumulators, linear invariants relating consecutive accumulators values of a same constraint were obtained [14] using Farkas' lemma [11] in a resource intensive procedure.

2   Background

Consider a sequence of integer variables $X = \langle X_1, X_2, \ldots, X_n \rangle$, and a function $S \colon \mathbb{Z}^p \to \Sigma$, where Σ is a finite set denoting an alphabet. Then, the *signature* of X is a sequence $\langle S_1, S_2, \ldots, S_{n-p+1} \rangle$, where every S_i equals $S(X_i, X_{i+1}, \ldots, X_{i+p-1})$. Intuitively, the signature of a sequence is a mapping of p consecutive elements of this sequence to an alphabet Σ, where p is called the *arity* of the signature.

An *automaton* $\mathcal{M}$ with a memory of $m \geq 0$ integer accumulators [10] is a tuple $\langle Q, \Sigma, \delta, q_0, I, A, \alpha \rangle$, where Q is the set of *states*, Σ the *alphabet*, $\delta \colon (Q \times \mathbb{Z}^m) \times \Sigma \to Q \times \mathbb{Z}^m$ the *transition function*, $q_0 \in Q$ the *initial state*, I the m-tuple of *initial values* of the accumulators, $A \subseteq Q$ the set of *accepting states*, and $\alpha \colon \mathbb{Z}^m \to \mathbb{Z}$ the *acceptance function* – the identity in this paper –, transforming the memory of an accepting state into an integer. If the left-to-right consumption of the symbols of a word w in Σ^* transits from q_0 to some accepting state and the m-tuple C of final accumulator values, then the automaton *returns* the value $\alpha(C)$, otherwise it *fails*. An integer sequence $\langle X_1, X_2, \ldots, X_n \rangle$ is an *accepting sequence* wrt an automaton $\mathcal{M}$ if its signature is accepted by $\mathcal{M}$. The *intersection* [21] of k automata $\mathcal{M}_1, \mathcal{M}_2, \ldots, \mathcal{M}_k$ is denoted by $\mathcal{M}_1 \cap \mathcal{M}_2 \cap \cdots \cap \mathcal{M}_k$.

An automaton with accumulators can be seen as a checker for a constraint that has two arguments, namely (1) a sequence of integer variables X; (2) an integer variable R. Then, for a ground sequence X and an integer number R, the constraint holds iff after consuming the *signature* of X, the corresponding automaton returns R. In Example 1, we introduce two constraints with their automata that will be further used as a running example.

Example 1. Consider a sequence $X = \langle X_1, X_2, \ldots, X_n \rangle$ of integer variables. A *peak* (resp. *valley*) is a variable X_k of X (with $k \in [2, n-1]$) such that there exists an i where $X_{i-1} < X_i$ (resp. $X_{i-1} > X_i$) and $X_i = X_{i+1} = \cdots = X_k$ and $X_k > X_{k+1}$ (resp. $X_k < X_{k+1}$). For example, the sequence $\langle 1, 2, 6, 6, 7, 0, 4, 2 \rangle$ has two peaks, namely 7 and 4, and one valley, namely 0. Then, the $\mathtt{peak}(X, P)$ (resp. $\mathtt{valley}(X, V)$) constraint restricts P (resp. V) to be the number of peaks (resp. valleys) in the sequence X.

Both constraints can be represented by automata with one accumulator, which consume the signature of X, defined by the following conjunction of constraints: $S(X_i, X_{i+1}) = \text{`<'} \Leftrightarrow X_i < X_{i+1} \wedge S(X_i, X_{i+1}) = \text{`='} \Leftrightarrow X_i = X_{i+1} \wedge S(X_i, X_{i+1}) = \text{`>'} \Leftrightarrow X_i > X_{i+1}$. Figure 1 gives the automata for $\mathtt{peak}$, $\mathtt{valley}$, and their intersection. For a ground sequence X, and for an integer value P (resp. V), the constraint $\mathtt{peak}(X, P)$ (resp. $\mathtt{valley}(X, V)$) holds iff, after consuming the signature of X, the automaton in Part (A) (resp. Part(B)) of Fig. 1 returns P (resp. V). Given the sequence $X = \langle 1, 2, 6, 6, 7, 0, 4, 2 \rangle$, the constraints $\mathtt{peak}(X, 2)$ and $\mathtt{valley}(X, 1)$ hold since, after consuming $X = \langle 1, 2, 6, 6, 7, 0, 4, 2 \rangle$, the peak and valley automata of Fig. 1 return 2 and 1, respectively.   △

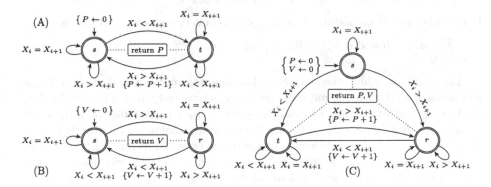

Fig. 1. Automata for (A) $\mathtt{peak}$, (B) $\mathtt{valley}$, and (C) their intersection; within each automaton accepting states are shown by double circles.

3   Generating Linear Invariants

Consider k automata $\mathcal{M}_1, \mathcal{M}_2, \ldots, \mathcal{M}_k$ over a same alphabet Σ. Let r_i denote the number of accumulators of $\mathcal{M}_i$, and let V_i designate its returned value. In this section we show how to systematically generate linear invariants of the form

$$e + e_0 \cdot n + \sum_{i=1}^{k} e_i \cdot V_i \geq 0 \text{ with } e, e_0, e_1, \ldots, e_k \in \mathbb{Z}, \tag{1}$$

which hold after the signature of a same input sequence $\langle X_1, X_2, \ldots, X_n \rangle$ is completely consumed by the k automata $\mathcal{M}_1, \mathcal{M}_2, \ldots, \mathcal{M}_k$. We call such invariant *general* since it holds regardless of any conditions on the result variables

$V_1, V_2, \ldots, V_k$. Stronger, but less general, invariants may be obtained when the result variables cannot be assigned the initial values of the accumulators.

Our method for generating invariants is applicable for a restricted class of automata with accumulators that we now introduce.

Property 1. An automaton $\mathcal{M}$ with r accumulators have the *incremental-automaton* property if the following conditions are all satisfied:

1. For every accumulator A_j of $\mathcal{M}$, its initial value α_j^0 is a natural number.
2. For every accumulator A_j of $\mathcal{M}$ and for every transition t of $\mathcal{M}$, the update
   of A_j upon triggering transition t is of the form $A_j \leftarrow \alpha_{j,0}^t + \sum\limits_{i=1}^{r} \alpha_{j,i}^t \cdot A_i$, with
   $\alpha_{j,0}^t \in \mathbb{N}$ and $\alpha_{j,1}^t, \alpha_{j,2}^t, \ldots, \alpha_{j,r}^t \in \{0, 1\}$.
3. The accumulator A_r is called the *main accumulator* and verifies the following three conditions:
   (a) the value returned by automaton $\mathcal{M}$ is the last value of its main accumulator A_r,
   (b) for every transition t of $\mathcal{M}$, $\alpha_{r,r}^t = 1$,
   (c) for a non-empty subset T of transitions of $\mathcal{M}$, $\sum\limits_{i=1}^{r-1} \alpha_{r,i}^t > 0, \ \forall t \in T$.
4. For all other accumulators A_j with $j < r$, on every transition t of $\mathcal{M}$, we have
   $\sum\limits_{i=1, i \neq j}^{r} \alpha_{j,i}^t = 0$ and if $\alpha_{r,j}^t > 0$, then $\alpha_{j,j}^t$ is 0.

The intuition behind the incremental-automaton property is that there is one accumulator that we name *main accumulator*, whose last value is the final value, returned by the automaton, (see 3a). At some transitions, the update of the main accumulator is a linear combination of the other accumulators, while on the other transitions its value either does not change or incremented by a non-negative constant, (see 3b and 3c). All other accumulators may only be incremented by a non-negative constant or assigned to some non-negative integer value, and they *may* contribute to the final value, (see 4). These accumulators are called *potential accumulators*. Both automata in Fig. 1 have the incremental-automaton property, and their single accumulators are main accumulators. Volumes I and II of the global constraint catalogue contain more than 50 such automata. In the rest of this paper we assume that all automata $\mathcal{M}_1, \mathcal{M}_2, \ldots, \mathcal{M}_k$ have the incremental-automaton property.

Our approach for systematically generating linear invariants of type $e + e_0 \cdot n + \sum\limits_{i=1}^{k} e_i \cdot V_i \geq 0$ considers each combination of signs of the coefficients e_i (with $i \in [0, k]$). It consists of three main steps:

1. Construct a non-negative function $v = e + e_0 \cdot n + \sum\limits_{i=1}^{k} e_i \cdot V_i$, which represents the left-hand side of the sought invariant (see Sect. 3.1).

2. Select the coefficients $e_0, e_1, \ldots, e_k$, called the *relative coefficients* of the linear invariant, so that there exists a constant C such that $e_0 \cdot n + \sum_{i=1}^{k} e_i \cdot V_i \geq C$ (see Sect. 3.2).
3. Compute C and set the coefficient e, called the *constant term* of the linear invariant, to $-C$ (see Sect. 3.3).

The three previous steps are performed as follows:

1. First, we assume a sign for each coefficient e_i (with $i \in [0, k]$), which tells whether we have to consider or not the contribution of the potential accumulators; note that each combination of signs of the coefficients e_i (with $i \in [1, k]$) will lead to a different linear invariant. Then, from the intersection automaton $\mathcal{I}$ of $\mathcal{M}_1, \mathcal{M}_2, \ldots, \mathcal{M}_k$, we construct a digraph called the *invariant digraph*, where each transition t of $\mathcal{I}$ is replaced by an arc whose weight represents the lower bound of the variation of the term $e_0 \cdot n + \sum_{i=1}^{k} e_i \cdot V_i$ while triggering t.
2. Second, we find the coefficients e_i (with $i \in [0, k]$) so that the invariant digraph does not contain any negative cycles.
3. Third, to obtain C we compute the shortest path in the invariant digraph from the node of the invariant digraph corresponding to the initial state of $\mathcal{I}$ to all nodes corresponding to accepting states of $\mathcal{I}$.

3.1 Constructing the Invariant Digraph for a Conjunction of Automaton Constraints Wrt a Linear Function

First, Definition 1 introduces the notion of *invariant digraph* $G_{\mathcal{I}}^{v}$ of the automaton $\mathcal{I} = \mathcal{M}_1 \cap \mathcal{M}_2 \cap \ldots \cap \mathcal{M}_k$ wrt a linear function v involving the values returned by these automata. Second, Definition 2 introduces the notion of *weight of an accepting sequence X wrt $\mathcal{I}$ in $G_{\mathcal{I}}^{v}$*, which makes the link between a path in $G_{\mathcal{I}}^{v}$ and the vector of values returned by $\mathcal{I}$ after consuming the signature of X. Finally, Theorem 1 shows that the weight of X in $G_{\mathcal{I}}^{v}$ is a lower bound on the linear function v.

Definition 1. *Consider an accepting sequence $X = \langle X_1, X_2, \ldots, X_n \rangle$ wrt the automaton $\mathcal{I} = \mathcal{M}_1 \cap \mathcal{M}_2 \cap \ldots \cap \mathcal{M}_k$, and a linear function $v = e + e_0 \cdot n + \sum_{i=1}^{k} e_i \cdot V_i$, where $(V_1, V_2, \ldots, V_k)$ is the vector of values returned by $\mathcal{I}$ after consuming the signature of X. The* invariant digraph *of $\mathcal{I}$ wrt v, denoted by $G_{\mathcal{I}}^{v}$ is a weighted digraph defined in the following way:*

- *The set of nodes of $G_{\mathcal{I}}^{v}$ is the set of states of $\mathcal{I}$.*
- *The set of arcs of $G_{\mathcal{I}}^{v}$ is the set of transitions of $\mathcal{I}$, where for every transition t the corresponding symbol of the alphabet is replaced by an integer weight,*

which is $e_0 + \sum_{i=1}^{k} e_i \cdot \beta_i^t$, where β_i^t is defined as follows, and where r_i denotes the number of accumulators of $\mathcal{M}_i$:

$$\beta_i^t = \begin{cases} \alpha_{r_i,0}^t \text{ of } \mathcal{M}_i, & \text{if } e_i \geq 0 \qquad (1) \\ \sum_{j=1}^{r_i} \alpha_{j,0}^t \text{ of } \mathcal{M}_i, & \text{if } e_i < 0 \qquad (2) \end{cases}$$

Definition 2. *Consider an accepting sequence $X = \langle X_1, X_2, \ldots, X_n \rangle$ wrt the automaton $\mathcal{I} = \mathcal{M}_1 \cap \mathcal{M}_2 \cap \ldots \cap \mathcal{M}_k$, and a linear function $v = e + e_0 \cdot n + \sum_{i=1}^{k} e_i \cdot V_i$, where $(V_1, V_2, \ldots, V_k)$ is the vector of values returned by $\mathcal{I}$ after consuming the signature of X. The walk of X in $G_{\mathcal{I}}^v$ is a path ω in $G_{\mathcal{I}}^v$ whose sequence of arcs is the sequence of the corresponding transitions of $\mathcal{I}$ triggered upon consuming the signature of X. The weight of X in $G_{\mathcal{I}}^v$ is the weight of its path in $G_{\mathcal{I}}^v$ plus a constant value, which is a lower bound on v corresponding to the initial values of the accumulators. It equals $e + e_0 \cdot (p-1) + \sum_{i=1}^{k} e_i \cdot \beta_i^0$, where p is the arity of the signature, and where β_i^0 is defined as follows, and where r_i denotes the number of accumulators of $\mathcal{M}_i$:*

$$\beta_i^0 = \begin{cases} \alpha_{r_i}^0 \text{ of } \mathcal{M}_i, & \text{if } e_i \geq 0 \qquad (1) \\ \sum_{j=1}^{r_i} \alpha_j^0 \text{ of } \mathcal{M}_i, & \text{if } e_i < 0 \qquad (2) \end{cases}$$

Example 2. Consider $\mathtt{peak}(\langle X_1, X_2, \ldots, X_n \rangle, P)$ and $\mathtt{valley}(\langle X_1, X_2, \ldots, X_n \rangle, V)$ introduced in Example 1. Figure 1 gives the automata for $\mathtt{peak}$, $\mathtt{valley}$, and their intersection $\mathcal{I}$. We aim to find inequalities of the form $e + e_0 \cdot n + e_1 \cdot P + e_2 \cdot V \geq 0$ that hold for every integer sequence X. After consuming the signature of $X = \langle X_1, X_2, \ldots, X_n \rangle$, $\mathcal{I}$ returns a pair of values (P, V), which are the number of peaks (resp. valleys) in X. The invariant digraph of $\mathcal{I}$ wrt $v = e + e_0 \cdot n + e_1 \cdot P + e_2 \cdot V$ is given in the figure on the right. Since both automata do not have any potential accumulators, the weights of the arcs of $G_{\mathcal{I}}^v$ do not depend on the signs of e_1 and e_2. Hence, for every integer sequence X, its weight in $G_{\mathcal{I}}^v$ equals $e + e_0 \cdot n + e_1 \cdot P + e_2 \cdot V$.   △

Theorem 1. *Consider an accepting sequence $X = \langle X_1, X_2, \ldots, X_n \rangle$ wrt the automaton $\mathcal{I} = \mathcal{M}_1 \cap \mathcal{M}_2 \cap \ldots \cap \mathcal{M}_k$, and a linear function $v = e + e_0 \cdot n + \sum_{i=1}^{k} e_i \cdot V_i$, where $(V_1, V_2, \ldots, V_k)$ is the vector of values return by $\mathcal{I}$. Then, the weight of X in $G_{\mathcal{I}}^v$ is less than or equal to $e + e_0 \cdot n + \sum_{i=1}^{k} e_i \cdot V_i$.*

Proof. Since, when doing the intersection of automata we do not merge accumulators, the accumulators of $\mathcal{I}$ that come from different automata $\mathcal{M}_i$ and $\mathcal{M}_j$ do not interact, hence the returned values of $\mathcal{M}_i$ and $\mathcal{M}_j$ are independent. By definition of the invariant digraph, the weight of any of its arc is $e_0 + \sum_{i=1}^{k} e_i \cdot \beta_i^t$, where β_i^t depends on the sign of e_i, and where t is the corresponding transition in $\mathcal{I}$. Then, the weight of X in $G_{\mathcal{I}}^v$ is the constant $e + e_0 \cdot (p-1) + \sum_{i=1}^{k} e_i \cdot \beta_i^0$ (see Definition 2) plus the weight of the walk of X, which is in total $e + e_0 \cdot (p-1) + \sum_{i=1}^{k} e_i \cdot \beta_i^0 + e_0 \cdot (n-p+1) + \sum_{j=1}^{n-p+1} \sum_{i=1}^{k} e_i \cdot \beta_i^{t_j} =$
$e + e_0 \cdot n + \sum_{i=1}^{k} e_i \cdot \left(\beta_i^0 + \sum_{j=1}^{n-p+1} \beta_i^{t_j} \right)$, where p is the arity of the considered signature, and $t_1, t_2, \ldots t_{n-p+1}$ is the sequence of transitions of $\mathcal{I}$ triggered upon consuming the signature of X. We now show that the value $e_i \cdot \left(\beta_i^0 + \sum_{j=1}^{n-p+1} \beta_i^{t_j} \right)$ is not greater than $e_i \cdot V_i$. This will imply that the weight of the walk of X in $G_{\mathcal{I}}^v$ is less than or equal to $v = e + e_0 \cdot n + \sum_{i=1}^{k} e_i \cdot V_i$.

Consider the $v_i = e_i \cdot V_i$ linear function. We show that the weight of X in $G_{\mathcal{I}}^{v_i}$, which equals $e_i \cdot \left(\beta_i^0 + \sum_{j=1}^{n-p+1} \beta_i^{t_j} \right)$, is less than or equal to $e_i \cdot V_i$. Depending on the sign of e_i we consider two cases.

Case 1: $e_i \geq 0$. In this case, the weight of every arc of $G_{\mathcal{I}}^{v_i}$ is e_i multiplied by $\alpha_{r_i,0}^t$, where t is the corresponding transition in $\mathcal{I}$, and r_i is the main accumulator of $\mathcal{M}_i$ (see Case 1 of Definition 1). If, on transition t, some potential accumulators of $\mathcal{M}_i$ are incremented by a positive constant, the real contribution of the accumulator updates on this transition to V_i is at least $\alpha_{r_i,0}^t$ since $e_i \geq 0$. The same reasoning applies to the contribution of the initial values of the potential accumulators to the final value V_i. Since this contribution is non-negative, it is ignored, and $\beta_i^0 = \alpha_{r_i}^0$ (see Case 1 of Definition 2). Hence,
$$e_i \cdot (\beta_i^0 + \sum_{j=1}^{n-p+1} \beta_i^{t_j}) = e_i \cdot (\alpha_{r_i}^0 + \sum_{j=1}^{n-p+1} \alpha_{r_i,0}^t) \leq e_i \cdot V_i.$$

Case 2: $e_i < 0$. In this case, the weight of every arc of $G_{\mathcal{I}}^{v_i}$ is e_i multiplied by the sum of the non-negative constants, which come from the updates of *every* accumulator of $\mathcal{M}_i$ (see Case 2 of Definition 1). The contribution of the potential accumulators is always taken into account, and since $e_i < 0$, it is always negative. The same reasoning applies to the contribution of the initial values of the potential accumulators to the returned value V_i. Since the initial values of the potential accumulators are non-negative, and $e_i < 0$, in order to obtain a lower bound on v we assume that the initial values of the potential accumulators always contribute to V_i (see Case 2 of Definition 2). Hence, $e_i \cdot (\beta_i^0 + \sum_{j=1}^{n-p+1} \beta_i^{t_j}) \leq e_i \cdot V_i.$ □

Note that if all the considered automata $\mathcal{M}_1, \mathcal{M}_2, \ldots, \mathcal{M}_k$ do not have potential accumulators, then for every accepting sequence $X = \langle X_1, X_2, \ldots, X_n \rangle$ wrt $\mathcal{I} = \mathcal{M}_1 \cap \mathcal{M}_2 \cap \ldots \cap \mathcal{M}_k$ and for any linear function $v = e + e_0 \cdot n + \sum\limits_{i=1}^{k} e_i \cdot V_i$, the weight of X in $G_{\mathcal{I}}^{v}$ is *equal* to v. If there is at least one potential accumulator for at least one automaton $\mathcal{M}_i$, then there may exist an integer sequence whose weight in $G_{\mathcal{I}}^{v}$ is strictly less than v.

3.2 Finding the Relative Coefficients of the Linear Invariant

We now focus on finding the relative coefficients $e_0, e_1, \ldots, e_k$ of the linear invariant $v = e + e_0 \cdot n + \sum\limits_{i=1}^{k} e_i \cdot V_i \geq 0$ such that, after consuming the signature of any accepting sequence by the automaton $\mathcal{I} = \mathcal{M}_1 \cap \mathcal{M}_2 \cap \ldots \cap \mathcal{M}_k$, the value of v is non-negative.

For any accepting sequence X wrt $\mathcal{I}$, by Theorem 1, we have that the weight w of X in $G_{\mathcal{I}}^{v}$ is less than or equal to v. Recall that w consists of a constant part, and of a part that depends on X, which involves the coefficients $e_0, e_1, \ldots, e_k$; thus, these coefficients must be chosen in a way that there exists a constant C such that $w \geq C$, and C does not depend on X. This is only possible when $G_{\mathcal{I}}^{v}$ does not contain *any* negative cycles. Let $\mathcal{C}$ denote the set of all simple circuits of $G_{\mathcal{I}}^{v}$, and let w_e denote the weight of an arc e of $G_{\mathcal{I}}^{v}$. In order to prevent negative cycles in $G_{\mathcal{I}}^{v}$, we solve the following minimisation problem, parameterised by $(s_0, s_1, \ldots s_k)$, the signs of $e_0, e_1, \ldots, e_k$:

$$\text{minimise} \ \sum_{c \in \mathcal{C}} W_c + \sum_{i=1}^{k} |e_i| \tag{3}$$

$$\text{subject to} \ W_c = \sum_{e \in c} w_e \qquad\qquad \forall c \in \mathcal{C} \tag{4}$$

$$W_c \geq 0 \qquad\qquad \forall c \in \mathcal{C} \tag{5}$$

$$s_i = \text{`$-$'} \Rightarrow e_i \leq 0, \quad s_i = \text{`$+$'} \Rightarrow e_i \geq 0 \qquad \forall i \in [0, k] \tag{6}$$

$$e_i \neq 0 \qquad\qquad \forall i \in [1, k] \tag{7}$$

In order to obtain the coefficients $e_0, e_1, \ldots, e_k$ so that $G_{\mathcal{I}}^{v}$ does not contain any negative cycles, it is enough to find a solution to the satisfaction problem (4)–(7). Minimisation is required to obtaining invariants that eliminate as many infeasible values of $(V_1, V_2, \ldots, V_k)$ as possible. Within the objective function (3), the term $\sum\limits_{c \in \mathcal{C}} W_c$ is for minimising the weight of every simple circuit, while the term $\sum\limits_{i=1}^{k} |e_i|$ is for obtaining the coefficients with the smallest absolute value. By changing the sign vector $(s_0, s_1, \ldots s_k)$ we obtain different invariants.

Example 3. Consider $\texttt{peak}(\langle X_1, X_2, \ldots, X_n \rangle, P)$ and $\texttt{valley}(\langle X_1, X_2, \ldots, X_n \rangle, V)$ from Example 2. The invariant digraph of the intersection of the automata for the Peak and Valley constraints wrt $v = e + e_0 \cdot n + e_1 \cdot P + e_2 \cdot V$ was given in Example 2. This digraph has four simple circuits, namely $s - s$, $t - t$, $r - r$, and $r - t - r$, which are labeled by 1, 2, 3 and 4, respectively. Then, the minimisation problem for finding the relative coefficients of the invariant $v \geq 0$, parameterised by (s_0, s_1, s_2), the signs of e_0, e_1 and e_2, is the following:

$$\text{minimise } \sum_{j=1}^{4} W_j + \sum_{i=0}^{2} |e_i|$$

$$\text{subject to } W_j = e_0, \qquad\qquad\qquad \forall j \in [1,3]$$
$$W_4 = e_0 + e_1 + e_2$$
$$W_j \geq 0 \qquad\qquad\qquad\qquad \forall j \in [1,4] \qquad (8)$$
$$s_i = {`-\text{'}} \Rightarrow e_i \leq 0, \quad s_i = {`+\text{'}} \Rightarrow e_i \geq 0 \;\; \forall i \in [0,2]$$
$$e_i \neq 0 \qquad\qquad\qquad\qquad \forall i \in [1,2]$$

Note that the value of e_0 must be non-negative otherwise (8) cannot be satisfied for $j \in \{1, 2, 3\}$. Hence, we consider only the combinations of signs of the form $({`+\text{'}}, s_1, s_2)$ with $s_1$ ans $s_2$ being either `$-$' or `$+$'. The following table gives the optimal solution of the minimisation problem for the considered combinations of signs:

(s_0, s_1, s_2)	$(+, -, -)$	$(+, -, +)$	$(+, +, -)$	$(+, +, +)$
(e_0, e_1, e_2)	$(1, -1, -1)$	$(0, -1, 1)$	$(0, 1, -1)$	$(0, 1, 1)$

$\triangle$

3.3   Finding the Constant Term of the Linear Invariant

Finally, we focus on finding the constant term e of the linear invariant $v = e + e_0 \cdot n + \sum_{i=1}^{k} e_i \cdot V_i \geq 0$, when the coefficients $e_0, e_1, \ldots, e_k$ are known, and when the digraph of the automaton $\mathcal{I} = \mathcal{M}_1 \cap \mathcal{M}_2 \cap \ldots \cap \mathcal{M}_k$ wrt v does not contain any negative cycles. By Theorem 1, the weight of any accepting sequence X wrt $\mathcal{I}$ in $G_{\mathcal{I}}^v$ is smaller or equal to v, then if the weight of X is non-negative, it implies that v is also non-negative. Since the digraph $G_{\mathcal{I}}^v$ does not contain any negative cycles, then the weight of X cannot be smaller than some constant C. Hence, it suffices to find this constant and set the constant term e to $-C$. The value of C is computed as the constant $e_0 \cdot (p-1) - \sum_{i=1}^{k} \beta_i^0$ (see Definition 2) plus the shortest path length from the node of $G_{\mathcal{I}}^v$ corresponding to the initial state of $\mathcal{I}$ to all the nodes of $G_{\mathcal{I}}^v$ corresponding to the accepting states of $\mathcal{I}$.

Example 4. Consider peak$(\langle X_1, X_2, \ldots, X_n \rangle, P)$ and valley$(\langle X_1, X_2, \ldots, X_n \rangle, V)$ from Example 2 with $n \geq 2$, i.e. the signature of $\langle X_1, X_2, \ldots, X_n \rangle$ is not empty. In Example 3, we found four vectors for the relative coefficients e_0, e_1,

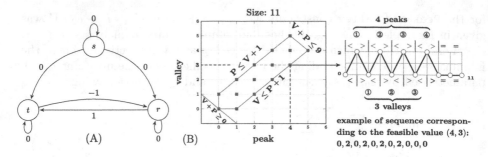

Fig. 2. (A) The invariant digraph of the automata for the `Peak` and the `Valley` constraints. (B) The set of feasible values of the result variables P and V of the `Peak` and the `Valley` constraints, respectively, for sequences of size 11.

e_2 of the invariant $e + e_0 \cdot n + e_1 \cdot P + e_2 \cdot V \geq 0$. For every found vector for the relative coefficients (e_0, e_1, e_2), we obtain a weighted digraph, whose weights now are integer numbers. For example, for the vector $(e_0, e_1, e_2) = (0, -1, 1)$, the obtained digraph is given in Part (A) of Fig. 2. We compute the length of a shortest path from the node s, which corresponds to the initial state of the automaton in Part (C) of Fig. 1 to every node corresponding to the accepting state of the automaton in Part (C) of Fig. 1. The length of the shortest path from s to s is 0, from s to t is 0, and from s to r is -1. The minimum of this values is -1, hence the constant term e equals $-(0 + (-1)) = 1$. The obtained invariant is $P \leq V + 1$.

In a similar way, we find the constant terms for the other found vectors of the relative coefficients (e_0, e_1, e_2), and obtain the following invariants: $V \leq P + 1$, $V + P \leq n - 2$, $V + P \geq 0$.

Part (B) of Fig. 2 gives the polytope of feasible points (P, V) when n is 11. Observe that three of the four found linear invariants are facets of the convex hull of this polytope, which implies that these invariants are sharp.   △

Example 5. We illustrate how the method presented in this section can also be used for generating linear invariants for non time-series constraints. Consider a sequence of integer variables $X = \langle X_1, X_2, \ldots, X_n \rangle$ with every X_i ranging over $[0, 3]$, four `among` [7] constraints that restrict the variables V_0, V_1, V_2, V_3 to be the number of occurrences of values $0, 1, 2, 3$, respectively, in X, as well as the four corresponding `stretch` [23] constraints restricting the stretch length in X to be respectively in $[1, 4]$, $[2, 5]$, $[3, 5]$, and $[1, 2]$. In addition assume that value 2 (resp. 1) cannot immediately follow a 3 (resp. 2). The intersection of the corresponding automata has 17 states and allows to generate 16 linear invariants, one of them being $2 + n + V_0 + V_1 - V_2 - 2 \cdot V_3 \geq 0$. Since the sum of all V_i is n, this invariant can be simplified to $2 + 2 \cdot n - 2 \cdot V_2 - 3 \cdot V_3 \geq 0$, which is equivalent to $2 \cdot (V_2 + V_3 - n) \leq 2 - V_3$. This inequality means that if X consists only of the values 2 and 3, i.e. $V_2 + V_3 - n = 0$, then $V_3 \leq 2$, which represents the conjunction of the conditions that the stretch length of $V_3 \in [1, 2]$ and $(X_i = 3) \Rightarrow (X_{i+1} \neq 2)$.   △

4 Conditional Linear Invariants

In Sect. 3, we presented a method for generating linear invariants linking the values returned by an automaton $\mathcal{I} = \mathcal{M}_1 \cap \mathcal{M}_2 \cap \ldots \cap \mathcal{M}_k$ after consuming the signature of a same accepting sequence $X = \langle X_1, X_2, \ldots, X_n \rangle$ wrt $\mathcal{I}$. In this section, we present several cases where the same method can be used for generating conditional linear invariants.

Quite often an automaton $\mathcal{M}_i$ (with i in $[1, k]$) returns its initial value only when the signature of X does not contain any occurrence of some regular expression σ_i. This may lead to a convex hull of points of coordinates $(V_1, V_2, \ldots, V_k)$ returned by $\mathcal{I}$ containing infeasible points, e.g. see Part (A) of Fig. 3. Some of these infeasible points can be eliminated by stronger invariants subject to the condition, called the *non-default value* condition, that no variable of the returned vector is assigned to the initial value of the corresponding accumulator. Section 4.1 shows to generate such invariants. Section 4.2 introduces the notion of *guard* of a transition t of $\mathcal{I}$, a linear inequality of the form $e + e_0 \cdot n + \sum_{i=1}^{k} e_i \cdot V_i \geq 0$, which is a *necessary condition* on the vector of values returned by $\mathcal{I}$ after consuming X for triggering the transition t upon consuming X.

4.1 Linear Invariants with the Non-default Value Condition

We first illustrate the motivation for such invariants.

Example 6. Consider the nb_decreasing_terrace($\langle X_1, X_2, \ldots, X_n \rangle, V_1$) and the sum_width_increasing_terrace($\langle X_1, X_2, \ldots, X_n \rangle, V_2$) constraints, where V_1 is restricted to be the number of maximal occurrences of Decreasing Terrace = ' $>=^+>$ ' in the signature of $X = \langle X_1, X_2, \ldots, X_n \rangle$, and V_2 is restricted to be the sum of the number of elements in subseries of X whose signatures correspond to words of the language of IncreasingTerrace = ' $<=^+<$ '. In Fig. 3, for $n = 12$, the squared points represent feasible pairs (V_1, V_2), while the circled points stand for infeasible pairs (V_1, V_2) inside the convex hull. The linear invariant $2 \cdot V_1 + V_2 \leq n - 2$ is a facet of the polytope, which does not eliminate the points $(1, 8), (2, 6), (3, 4), (4, 2)$. However, if we assume that both $V_1 > 0$ and $V_2 > 0$, then we can add a linear invariant eliminating these four infeasible points, namely $2 \cdot V_1 + V_2 \leq n - 3$, shown in Part (B) of Fig. 3. In addition, if we assume that $V_1 > 0$ and $V_2 > 0$, the infeasible points on the straight line $V_2 = 1$ will also be eliminated by the restriction $V_2 = 0 \lor V_2 \geq 2$ given in [3, p. 2598]. △

Consider that each automaton $\mathcal{M}_i$ (with i in $[1, k]$) returns its initial value after consuming the signature of an accepting sequence X wrt $\mathcal{M}_i$ iff the signature of X does not contain any occurrence of some regular expression σ_i over the alphabet Σ. Let $\mathcal{M}'_i$ denote the automaton which accepts the words of the language $\Sigma^* \sigma_i \Sigma^*$, where Σ^* denotes any word over Σ. Then, using the method of Sect. 3 we generate the invariants for $\mathcal{M}'_1 \cap \mathcal{M}'_2 \cap \ldots \cap \mathcal{M}'_k$. These invariants hold when the non-default value condition is satisfied.

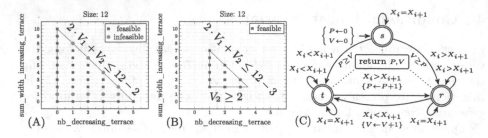

Fig. 3. Invariants on the result values V_1 and V_2 of nb_decreasing_terrace and sum_width_increasing_terrace for a sequence size of 12 (A) with the general invariants, and (B) with the non-default value condition. (C) Intersection automaton for Peak and Valley with the guards $P \geq V$ and $V \geq P$ on transitions $s \to t$ and $s \to r$ (as for the return statement, the P and V accumulators in the guards refer to the final values of the corresponding accumulators).

4.2 Generating Guards for Transitions of the Intersection of Several Automata

Consider k automata $\mathcal{M}_1, \mathcal{M}_2, \ldots, \mathcal{M}_k$ and let V_i (with $i \in [1, k]$) designate the value returned by $\mathcal{M}_i$. We focus on generating necessary conditions, called *guards*, introduced in Definition 3, for enabling transitions of the automaton $\mathcal{I} = \mathcal{M}_1 \cap \mathcal{M}_2 \cap \ldots \cap \mathcal{M}_k$. Further, we give a three-step procedure for generating guards for transitions of $\mathcal{I}$.

Definition 3. *Consider a transition t of the automaton $\mathcal{I} = \mathcal{M}_1 \cap \mathcal{M}_2 \cap \ldots \cap \mathcal{M}_k$. A guard of t is a linear inequality of the form $e + e_0 \cdot n + \sum_{i=1}^{k} e_i \cdot V_i \geq 0$ such that there does not exist any accepting sequence $X = \langle X_1, X_2, \ldots, X_n \rangle$ wrt $\mathcal{I}$ such that (1) after consuming the signature of X, the vector $(V_1, V_2, \ldots, V_k)$ returned by $\mathcal{I}$ satisfies the inequality $e + e_0 \cdot n + \sum_{i=1}^{k} e_i \cdot V_i < 0$, (2) and the transition t was triggered upon consuming the signature of X.*

The following example illustrates Definition 3.

Example 7. Given a sequence $X = \langle X_1, X_2, \ldots, X_n \rangle$, consider the peak$(X, P)$ and valley(X, V) constraints. The intersection $\mathcal{I}$ of the automata for peak and valley was given in Part (C) of Fig. 1. Observe that, if at the initial state s the automaton consumes '<' (resp. '>'), then the number of peaks (resp. valleys) in X is greater than or equal to the number of valleys (resp. peaks). Hence, we can impose the guard $P \geq V$ (resp. $V \geq P$) on the transition from s to t (resp. to r). Part (C) of Fig. 3 gives the automaton $\mathcal{I}$ with the obtained guards.   △

Guards for the transitions of an automaton $\mathcal{I} = \mathcal{M}_1 \cap \mathcal{M}_2 \cap \ldots \cap \mathcal{M}_k$ can be generated in three steps:

1. First, we identify the subset $\mathcal{T}$ of transitions of $\mathcal{I}$ such that, for any transition t in $\mathcal{T}$, upon consuming any sequence, t can be triggered at most once.

2. Second, for every transition t in $\mathcal{T}$, we obtain a new automaton $\mathcal{I}_t$ by removing from $\mathcal{I}$ all transitions of $\mathcal{T}$ different from t that start at the same state as t.
3. Third, using the technique of Sect. 3 on the invariant digraph $G^v_{\mathcal{I}_t}$, we obtain linear invariants that are guards of transition t.

5   Evaluation

To test the effectiveness of the generated invariants, we first try systematic tests on the conjunction of pairs of the 35 time-series constraints [5] of the *nb* and *sum_width* families for which the glue matrix constraints exist [2]. The *nb* constraints count the number of occurrences of some pattern in a time series, while the *sum_width* family constrains the sum of the width of pattern occurrences. Our intended use case is similar to [9], where constraints and parameter ranges of the problem are learned from real-world data, and are used to produce solutions that are similar to the previously observed data. It is important both to remove infeasible parameter combinations quickly, as well as helping to find solutions for feasible problems. Real world datasets often will only show a tiny subset of all possible parameter combinations, but as we don't know the data a priori, a systematic evaluation seems the most conservative approach.

For the experiments we use a database of generated invariants in a format compatible with the Global Constraint Catalogue [3]. Invariants are generated as Prolog facts, from which executable code, and other formats are then produced automatically. The time required to produce the invariants (5 min) is insignificant compared to the overall runtime of the experiments. For the 595 combinations of the 35 constraints we produce over 4100 unconditional invariants, over 3500 conditional invariants, and 86 guard invariants. In the test, we try each pair of constraints and try to find solutions for all possible pairs of parameter values. We compare four different versions of our methods: The *pure* baseline version uses the automata that were described in [4], the bounds on the parameter values for each prefix and suffix, and the glue matrix constraints as described in [2]. This version represents the state of the art before the current work. In the *invariant* version we add the generated invariants for the parameters of the complete time series. In the *incremental* version, we not only state the invariants for the complete time series, but also apply them for each suffix. The required variables are already available as part of the glue matrix setup, we only need to add the linear inequalities for each suffix length. In the *all* version, we add the product automaton of the conjunction of the two constraints, if it contains guard constraints, and also state some additional, manually derived invariants.

The test program uses a labeling routine that first assigns the signature variables, and only afterwards assigns values for the X_i decision variables. The variables in each case are assigned from left to right. For each pair of parameters values, defined by the product of the bounds from [2], we try to find a first solution with a timeout of 60 s.

We have tested the results for different time series length, Fig. 4 shows the result for length 18 and domain size 0..18, the largest problem size where we find solutions for each case within the timeout. All experiments were run on a laptop with Intel i7 CPU (2.9 GHz), 64 Gb main memory and Windows 10 64 bit OS using SICStus Prolog 4.3.5 utilizing a single core. For our four problem variants, we plot the percentage of undecided prob-

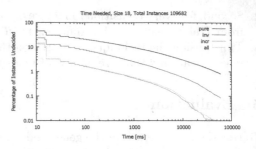

Fig. 4. Comparing constraint variants, undecided instances percentage for size 18 as a function of time, Timeout = 60 s

lem instances as a function of computation time. The plot uses log-log scales to more clearly show the values for short runtimes and for low number of undecided problems. The baseline *pure* variant solves around 55% of the instances immediately, and leaves just under one percent unsolved within the timeout. The *invariants* version improves on this by pruning more infeasible problems immediately. On the other hand, stating the invariants on the full series has no effect on feasible instances. When using the *incremental* version of the constraints, this has very little additional impact on infeasible problems, but improves the solution time for the feasible instances significantly. Adding (variant *all*) additional constraints further reduces the number of backtracks required, but these savings are largely balanced with the additional processing time, and therefore have no major impact on the overall results. After one second, around 9.5% of all instances are unsolved in the baseline, but only 0.5% in the *incremental* or *all* variant.

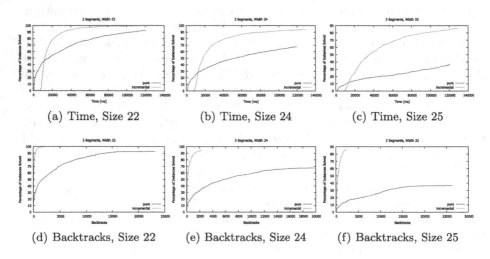

(a) Time, Size 22   (b) Time, Size 24   (c) Time, Size 25

(d) Backtracks, Size 22   (e) Backtracks, Size 24   (f) Backtracks, Size 25

Fig. 5. Percentage of problems solved for 3 overlapping segments of lengths 22, 24, and 25; execution time in top row, backtracks required in bottom row

To test the method in a more realistic setting, we consider the conjunction of all 35 considered time-series constraints on electricity demand data provided by an industrial partner. The time-series describes daily demand levels in half-hour intervals, giving 48 data points. To capture the shape of the time-series more accurately, we split the series into overlapping segments from 00–12, 06–18, and 12–24 h, each segment containing 24 data points, overlapping in 12 data points with the previous segment. We then setup the conjunction of the 35 time-series constraints for each segment, using the *pure* and *incremental* variants described above. This leads to $3 \times 35 \times 2 = 210$ automata constraints with shared signature and decision variables. The invariants are created for every pair of constraints, and every suffix, leading to a large number of inequalities. The search routine assigns all signature variables from left to right, and then assigns the decision variables, with a timeout of 120 s.

In order to understand the scalability of the method, we also consider time series of 44 resp. 50 data points (three segments of length 22 and 25), extracted from the daily data stream covering a four year period (1448 samples). In Fig. 5 we show the time and backtrack profiles for finding a first solution. The top row shows the percentage of instances solved within a given time budget, the bottom row shows the percentage of problems solved within a backtrack budget. For easy problems, the *pure* variant finds solutions more quickly, but the *incremental* version pays off for more complex problems, as it reduces the number of backtracks required sufficiently to account for the large overhead of stating and pruning all invariants. The problems for segment length 20 (not shown) can be solved without timeout for both variants, as the segment length increases, the number of time outs increases much more rapidly for the *pure* variant. Adding the invariants drastically reduces the search space in all cases, future work should consider if we can identify those invariants that actively contribute to the search by cutting off infeasible branches early on. Restricting the invariants to such an active subset should lead to a further improvement in execution time.

6   Conclusion

Future work may look how to extend the current approach to handle automata with accumulators that also allow the min and max aggregators for accumulator updates. It also should investigate the use of such invariants within the context of MIP. While MIP has been using linear cuts for a long time [16,20], no off-the-shelf data base of cuts in some computer readable format is currently available. Cuts are typically defined in papers and are then directly embedded within MIP solvers.

References

1. Appa, G., Magos, D., Mourtos, I.: LP relaxations of multiple all_different predicates. In: Régin, J.-C., Rueher, M. (eds.) CPAIOR 2004. LNCS, vol. 3011, pp. 364–369. Springer, Heidelberg (2004). doi:10.1007/978-3-540-24664-0_25

2. Arafailova, E., Beldiceanu, N., Carlsson, M., Flener, P., Francisco Rodríguez, M.A., Pearson, J., Simonis, H.: Systematic derivation of bounds and glue constraints for time-series constraints. In: Rueher, M. (ed.) CP 2016. LNCS, vol. 9892, pp. 13–29. Springer, Cham (2016). doi:10.1007/978-3-319-44953-1_2

3. Arafailova, E., Beldiceanu, N., Douence, R., Carlsson, M., Flener, P., Rodríguez, M.A.F., Pearson, J., Simonis, H.: Global constraint catalog, volume II, time-series constraints. CoRR abs/1609.08925 (2016). http://arxiv.org/abs/1609.08925

4. Arafailova, E., Beldiceanu, N., Douence, R., Flener, P., Francisco Rodríguez, M.A., Pearson, J., Simonis, H.: Time-series constraints: improvements and application in CP and MIP contexts. In: Quimper, C.-G. (ed.) CPAIOR 2016. LNCS, vol. 9676, pp. 18–34. Springer, Cham (2016). doi:10.1007/978-3-319-33954-2_2

5. Beldiceanu, N., Carlsson, M., Douence, R., Simonis, H.: Using finite transducers for describing and synthesising structural time-series constraints. Constraints 21(1), 22–40 (2016). Journal fast track of CP 2015: summary. LNCS, vol. 9255, p. 723. Springer, Heidelberg (2015)

6. Beldiceanu, N., Carlsson, M., Rampon, J.-X., Truchet, C.: Graph invariants as necessary conditions for global constraints. In: Beek, P. (ed.) CP 2005. LNCS, vol. 3709, pp. 92–106. Springer, Heidelberg (2005). doi:10.1007/11564751_10

7. Beldiceanu, N., Contejean, E.: Introducing global constraints in CHIP. Math. Comput. Model. 20(12), 97–123 (1994)

8. Beldiceanu, N., Flener, P., Pearson, J., Van Hentenryck, P.: Propagating regular counting constraints. In: Brodley, C.E., Stone, P. (eds.) AAAI 2014, pp. 2616–2622. AAAI Press, Palo Alto (2014)

9. Beldiceanu, N., Ifrim, G., Lenoir, A., Simonis, H.: Describing and generating solutions for the EDF unit commitment problem with the ModelSeeker. In: Schulte, C. (ed.) CP 2013. LNCS, vol. 8124, pp. 733–748. Springer, Heidelberg (2013). doi:10.1007/978-3-642-40627-0_54

10. Beldiceanu, N., Carlsson, M., Petit, T.: Deriving filtering algorithms from constraint checkers. In: Wallace, M. (ed.) CP 2004. LNCS, vol. 3258, pp. 107–122. Springer, Heidelberg (2004). doi:10.1007/978-3-540-30201-8_11

11. Boyd, S.P., Vandenberghe, L.: Convex Optimization. Cambridge University Press, Cambridge (2004)

12. Charnley, J.W., Colton, S., Miguel, I.: Automatic generation of implied constraints. In: ECAI 2006. Frontiers in AI and Applications, vol. 141, pp. 73–77. IOS Press (2006)

13. Dincbas, M., Simonis, H., Hentenryck, P.V.: Solving the car-sequencing problem in constraint logic programming. In: ECAI, pp. 290–295 (1988)

14. Rodríguez, M.A.F., Flener, P., Pearson, J.: Implied constraints for automaton constraints. In: Gottlob, G., Sutcliffe, G., Voronkov, A. (eds.) GCAI 2015. EasyChair Proceedings in Computing, vol. 36, pp. 113–126 (2015)

15. Frisch, A., Miguel, I., Walsh, T.: Extensions to proof planning for generating implied constraints. In: 9th Symposium on the Integration of Symbolic Computation and Mechanized Reasoning (2001)

16. Gomory, R.: Outline of an algorithm for integer solutions to linear programs. Bull. Am. Math. Soc. 64, 275–278 (1958)

17. Hansen, P., Caporossi, G.: Autographix: an automated system for finding conjectures in graph theory. Electron. Notes Discret. Math. 5, 158–161 (2000)

18. Hooker, J.N.: Integrated Methods for Optimization, 2nd edn. Springer Publishing Company, New York (2011). Incorporated

19. Lee, J.: All-different polytopes. J. Comb. Optim. 6(3), 335–352 (2002)

20. Marchand, H., Martin, A., Weismantel, R., Wolsey, L.A.: Cutting planes in integer and mixed integer programming. Discret. Appl. Math. **123**(1–3), 397–446 (2002)
21. Menana, J.: Automata and Constraint Programming for Personnel Scheduling Problems. Theses, Université de Nantes, October 2011. https://tel.archives-ouvertes.fr/tel-00785838
22. Menana, J., Demassey, S.: Sequencing and counting with the `multicost-regular` constraint. In: Hoeve, W.-J., Hooker, J.N. (eds.) CPAIOR 2009. LNCS, vol. 5547, pp. 178–192. Springer, Heidelberg (2009). doi:10.1007/978-3-642-01929-6_14
23. Pesant, G.: A filtering algorithm for the `stretch` constraint. In: Walsh, T. (ed.) CP 2001. LNCS, vol. 2239, pp. 183–195. Springer, Heidelberg (2001). doi:10.1007/3-540-45578-7_13

AMONG Implied Constraints
for Two Families of Time-Series Constraints

Ekaterina Arafailova[1(✉)], Nicolas Beldiceanu[1], and Helmut Simonis[2]

[1] TASC (LS2N), IMT Atlantique, 44307 Nantes, France
{Ekaterina.Arafailova,Nicolas.Beldiceanu}@imt-atlantique.fr
[2] Insight Centre for Data Analytics, University College Cork, Cork, Ireland
Helmut.Simonis@insight-centre.org

Abstract. We consider, for an integer time series, two families of constraints restricting the *max, and the sum, respectively, of the surfaces* of the elements of the sub-series corresponding to occurrences of some pattern. In recent work these families were identified as the most difficult to solve compared to all other time-series constraints. For all patterns of the time-series constraints catalogue, we provide a *unique per family parameterised* AMONG implied constraint that can be imposed on any prefix/suffix of a time-series. Experiments show that it reduces both the *number of backtracks/time spent* by up to 4/3 orders of magnitude.

1 Introduction

Going back to the work of Schützenberger [20], *regular cost functions* are quantitative extensions of regular languages that correspond to a function mapping a word to an integer value or infinity. Recently there has been renewed interest in this area, both from a theoretical perspective [14] with max-plus automata, and from a practical point of view with the synthesis of cost register automata [2] for data streams [3]. Within constraint programming, automata constraints were introduced in [18] and in [8,15], the latter also computing an integer value from a word.

This paper focusses on the $g\_\mathrm{SURF}\_\sigma(X, R)$ families of time-series constraints with g being either Max or Sum, and with σ being one of the 22 patterns of [5], as they were reported to be the most difficult in the recent work of [4]. Each constraint of one of the two families restricts R to be the result of applying the aggregator g to the sum of the elements corresponding to the occurrences of a pattern σ [3] in an integer sequence X, which is called a *time series* and corresponds to measurements taken over time. These constraints play an important role in modelling power systems [10]. If the measured values correspond to the power input/output, then the surface feature surf describes the

E. Arafailova is supported by the EU H2020 programme under grant 640954 for the GRACeFUL project. N. Beldiceanu is partially supported by GRACeFUL and by the Gaspard-Monge programme. H. Simonis is supported by Science Foundation Ireland (SFI) under grant numbers SFI/12/RC/2289 and SFI/10/IN.1/I3032.

J.C. Beck (Ed.): CP 2017, LNCS 10416, pp. 38–54, 2017.
DOI: 10.1007/978-3-319-66158-2_3

energy used/generated during the period of pattern occurence. The Sum aggregator imposes a bound on the total energy during all pattern occurences in the time series, the Max aggregator is used to limit the maximal energy during a single pattern occurence. Generating time series verifying a set of specific time-series constraints is also useful in different contexts like trace generation, i.e. generating typical energy consumption profiles of a data centre [16,17], or a staff scheduling application, i.e. generating manpower profiles over time subject to work regulations [1,6].

Many constraints of these families are not tractable, thus in order to improve the efficiency of the solving we need to address the combinatorial aspect of time-series constraints. We improve the reasoning for such time-series constraints by identifying implied AMONG constraints. Learning parameters of global constraints like AMONG [9] is a well known method for strengthening constraint models [11,12,19] with the drawback that it is instance specific, so this alternative was not explored here. Taking exact domains into account would lead to filtering algorithms rather than to implied constraints which assume the same minimum/maximum.

While coming up with implied constraints is usually problem specific, the theoretical contribution of this paper is a *unique per family* AMONG implied constraint, *that is valid for all regular expressions* of the time-series constraint catalogue [5] and that covers all the 22 time-series constraints of the corresponding family. Hence, it covers 44 time-series constraints in total. The main focus of this paper is on reusable necessary conditions that can be associated to a class of time-series constraints described with regular expressions. There have been several papers describing progress in propagation of a set of automata and time-series constraints. The techniques described in this paper are only one element required to make such models scale to industrial size.

Section 2 recalls the necessary background on time-series constraints used in this paper. After introducing several regular expression characteristics, Sect. 3 presents the main contribution, Theorems 1 and 2, while Tables 2 and 3 provide the corresponding derived concrete implied constraints for some subset of the MAX_SURF_σ and the SUM_SURF_σ time-series constraints, respectively, of the time-series constraint catalogue. Finally Sect. 4 systematically evaluates the impact of the derived implied constraints.

2   Time-Series Constraints Background

A time series constraint [7] imposed on a sequence of integer variables $X = \langle X_1, X_2, \ldots, X_n \rangle$ and an integer variable R is described by three main components $\langle g, f, \sigma \rangle$. Let $\mathcal{R}_\Sigma$ denote the set of regular expressions on $\Sigma = \{`<`, `=`, `>`\}$. Then, σ is a regular expression in $\mathcal{R}_\Sigma$, that is characterised by two integer constants a_σ and b_σ, whose role is to trim the left and right borders of the regular expression, and $\mathcal{L}_\sigma$ denotes the regular language of σ, while f is a function, called a *feature*. In this paper, we consider only the case when f is surf, which will be explained at the end of this paragraph. Finally g is also a function, called an *aggregator*, that is either Max or Sum. The *signature* $S = \langle S_1, S_2, \ldots, S_{n-1} \rangle$ of a time series X is defined by the following constraints: $(X_i < X_{i+1} \Leftrightarrow S_i = \text{'} < \text{'}) \wedge (X_i = X_{i+1} \Leftrightarrow S_i = \text{'} = \text{'}) \wedge (X_i > X_{i+1} \Leftrightarrow S_i = \text{'} > \text{'})$ for all $i \in [1, n-1]$. If a sub-signature $\langle S_i, S_{i+1}, \ldots, S_j \rangle$ is a maximal word matching σ in the signature of X, then the subseries $\langle X_{i+b_\sigma}, X_{i+b_\sigma+1}, \ldots, X_{j+1-a_\sigma} \rangle$ is called a σ-*pattern* and the subseries $\langle X_i, X_{i+1}, \ldots, X_{j+1} \rangle$ is called an *extended* σ-*pattern*. The *width* of a σ-pattern is its number of elements. The integer variable R is the aggregation, computed using g, of the list of values of feature f for all σ-patterns in X. The result of applying the surf feature to a σ-pattern is the sum of all elements of this σ-pattern. If there is no σ-pattern in X, then R is the *default value*, denoted by $\text{def}_{g,f}$, which is $-\infty$, or 0 when g is Max, or Sum, respectively. A time-series constraint specified by $\langle g, f, \sigma \rangle$ is named as g_f_σ. A time series is *maximal* for $g\_f\_\sigma(X, R)$ if it contains at least one σ-pattern and yields the maximum value of R among all time series of length n that have the same initial domains for the time-series variables.

Example 1. Consider the $\sigma = \text{DecreasingSequence} = \text{'}(>(>|=)^*)^*>\text{'}$ regular expression and the time series $X = \langle 4, 2, 2, 1, 5, 3, 2, 4 \rangle$ whose signature is $\text{'}>=><>><\text{'}$. A σ-pattern, called a *decreasing sequence*, within a time series is a subseries whose signature is a maximal occurrence of σ in the signature of X, and the surf feature value of a decreasing sequence is the sum of its elements. The time series X contains two decreasing sequences, namely $\langle 4, 2, 2, 1 \rangle$ and $\langle 5, 3, 2 \rangle$, shown in the figure on the right, of surfaces 9 and 10, respectively. Hence, the aggregation of their surfaces, obtained by using the aggregator Max, or Sum is 10, or 19 respectively. The corresponding time-series constraints are MAX_SURF_DECREASING_SEQUENCE, and SUM_SURF_DECREASING_SEQUENCE.   △

3   Deriving AMONG Implied Constraint

Consider a $g\_f\_\sigma(\langle X_1, X_2, \ldots, X_n \rangle, R)$ time-series constraint with g being either Sum or Max, with f being the surf feature, and with every X_i ranging over the same integer interval domain $[\ell, u]$ such that $u > 0$. For brevity, we do not consider here the case when $u \leq 0$, since it can be handled in a symmetric way. We derive an $\text{AMONG}(\mathcal{N}, \langle X_1, X_2, \ldots, X_n \rangle, \langle \underline{\mathcal{I}}^{\langle \ell, u \rangle}_{\langle g, f, \sigma \rangle}, \underline{\mathcal{I}}^{\langle \ell, u \rangle}_{\langle g, f, \sigma \rangle} + 1, \ldots, \overline{\mathcal{I}}^{\langle \ell, u \rangle}_{\langle g, f, \sigma \rangle} \rangle)$ implied constraint, where:

- For any value of R, $\mathcal{N}$ is an integer variable whose lower bound only depends on R, σ, f, ℓ, u, and n.
- The interval $\mathcal{I}^{\langle \ell, u \rangle}_{\langle g, f, \sigma \rangle} = [\underline{\mathcal{I}}^{\langle \ell, u \rangle}_{\langle g, f, \sigma \rangle}, \overline{\mathcal{I}}^{\langle \ell, u \rangle}_{\langle g, f, \sigma \rangle}]$ is a subinterval of $[\ell, u]$, which is called the *interval of interest* of $\langle g, f, \sigma \rangle$ *wrt* $\langle \ell, u \rangle$ and defined in Sect. 3.1.

Such an AMONG [13] constraint is satisfied if exactly $\mathcal{N}$ variables of $\langle X_1, X_2, \ldots, X_n \rangle$ are assigned a value in $\mathcal{I}^{\langle \ell, u \rangle}_{\langle g, f, \sigma \rangle}$. Before formally describing how to derive this implied constraint, we provide an illustrating example.

Example 2. Consider a MAX_SURF_$\sigma(\langle X_1, X_2, \ldots X_7 \rangle, R)$ time-series constraint with every X_i ranging over the same integer interval domain $[1, 4]$, and with σ being the DecreasingSequence regular expression of Example 1.

Let us observe what happens when R is fixed, for example, to 18. The table on the right gives the two distinct σ-patterns such that at least one of them appear in every ground time series $X = \langle X_1, X_2, \ldots, X_7 \rangle$ that yields 18 as the value of R. By inspection, we observe that for any ground time series X for which R equals 18, its single σ-pattern contains at least 4 time-series variables whose values are in $[3, 4]$. Hence, we can impose an AMONG$(\mathcal{N}, \langle X_1, X_2, \ldots, X_7 \rangle, \langle 3, 4 \rangle)$ implied constraint with $\mathcal{N} \geq 4$.   △

σ-pattern 1	σ-pattern 2
$\langle 4, 3, 3, 3, 3, 2 \rangle$	$\langle 4, 3, 3, 3, 2, 2, 1 \rangle$

We now formalise the ideas presented in Example 2 and systematise the way we obtain such an implied constraint even when R is *not initially fixed.*

- Section 3.1 introduces five characteristics of a regular expression σ, which will be used to obtain a parameterised implied constraint:
  - the *height of* σ (see Definition 1),
  - the *interval of interest of* $\langle g, f, \sigma \rangle$ *wrt* $\langle \ell, u \rangle$ (see Definition 2),
  - the *maximal value occurrence number of* $v \in \mathbb{Z}$ wrt $\langle \ell, u, n \rangle$ (see Definition 3),
  - the *big width of* σ *wrt* $\langle \ell, u, n \rangle$ (see Definition 4), and
  - the *overlap of* σ *wrt* $\langle \ell, u \rangle$ (see Definition 5).
- Based on these characteristics, Sect. 3.2 presents a systematic way of deriving AMONG implied constraints for the MAX_SURF_σ and the SUM_SURF_σ families of time-series constraints.

3.1   Characteristics of Regular Expressions

To get a unique per family AMONG implied constraint that is valid for any $g\_\text{SURF}\_\sigma(X, R)$ time-series constraint with g being either Sum or Max, we introduce five characteristics of regular expressions that will be used for parametrising our implied constraint. First, Definition 1 introduces the notion of height of a regular expression, that is needed in Definition 2, which defines the specific range of values on which the implied AMONG constraint focusses on.

Definition 1. *Given a regular expression σ, the height of σ, denoted by η_σ, is a function that maps an element of $\mathcal{R}_\Sigma$ to $\mathbb{N}$. It is the smallest difference between the domain upper limit u and the domain lower limit ℓ such that there exists a ground time series over $[\ell, u]$ whose signature has at least one occurrence of σ.*

Example 3. Consider the σ = DecreasingSequence regular expression of Example 1.

- When $u = \ell$, for any time-series length, there exists a single ground time series t whose signature is a word in the regular language of '$=^*$'. The signature of t contains no occurrences of the '$>$' symbol, and thus contains no words of $\mathcal{L}_\sigma$ either.

- But when $u - \ell = 1$, there exists, for example, a time series $t = \langle u - 1, u, u - 1, u - 1 \rangle$, depicted in Fig. 1a, whose signature '$<>=$' contains the word '$>$' of $\mathcal{L}_\sigma$. Hence, the height of σ equals 1.   △

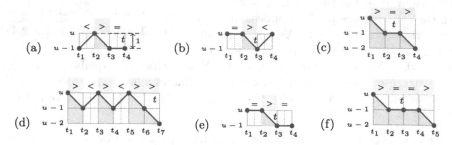

Fig. 1. For all the figures, σ is the `DecreasingSequence` regular expression. A time series t (a) with one σ-pattern such that the difference between its maximum and minimum is 1; (b) with one σ-pattern, which contains a single occurrence of value $u-1$; (c) with one σ-pattern, which contains 2 occurrences of value $u - 1$; (d) with the maximum number, 3, of σ-patterns, which all contain one occurrence of value $u - 1$, and only one contains an occurrence of value $u - 2$; (e) with one σ-pattern, which contains one occurrence of both u and $u - 1$; (f) with one σ-pattern, whose width is maximum among all other σ-patterns in ground time series of length 5 over the same domain $[u - 2, u]$.

Definition 2. *Consider a* $g\_f\_\sigma(X, R)$ *time-series constraint with* X *being a time series over an integer interval domain* $[\ell, u]$. *The interval of interest of* $\langle g, f, \sigma \rangle$ *wrt* $\langle \ell, u \rangle$, *denoted by* $\mathcal{I}_{\langle g,f,\sigma \rangle}^{\langle \ell,u \rangle}$, *is a function that maps an element of* $\mathcal{T} \times \mathbb{Z} \times \mathbb{Z}$ *to* $\mathbb{Z} \times \mathbb{Z}$, *where* $\mathcal{T}$ *denotes the set of all time-series constraints, and the result pair of integers is considered as an interval.*

- *The upper limit of* $\mathcal{I}_{\langle g,f,\sigma \rangle}^{\langle \ell,u \rangle}$, *denoted by* $\overline{\mathcal{I}}_{\langle g,f,\sigma \rangle}^{\langle \ell,u \rangle}$, *is the largest value in* $[\ell, u]$ *that can occur in a* σ-*pattern of a time series over* $[\ell, u]$. *If such value does not exist, then* $\mathcal{I}_{\langle g,f,\sigma \rangle}^{\langle \ell,u \rangle}$ *is undefined.*
- *The lower limit of* $\mathcal{I}_{\langle g,f,\sigma \rangle}^{\langle \ell,u \rangle}$, *denoted by* $\underline{\mathcal{I}}_{\langle g,f,\sigma \rangle}^{\langle \ell,u \rangle}$, *is the smallest value* v *in* $[\max(\ell, u - \eta_\sigma - 1), u]$ *such that for any* n *in* $\mathbb{N}$, *the number of occurrences of* v *in the union of the* σ-*patterns of any maximal time series for* g_f_σ *of length* n *over* $[\ell, u]$, *is a non-constant function of* n. *If such* v *does not exist, then* $\underline{\mathcal{I}}_{\langle g,f,\sigma \rangle}^{\langle \ell,u \rangle}$ *equals* $\overline{\mathcal{I}}_{\langle g,f,\sigma \rangle}^{\langle \ell,u \rangle} - \eta_\sigma$.

We focus on such intervals of interests because they consist of the largest values appearing in maximal time series for g_f_σ.

Example 4. Consider a $g\_f\_\sigma(X, R)$ time-series constraint with σ being the `DecreasingSequence` regular expression, with f being the `surf` feature, and with X being a time series of length $n \geq 2$ over an integer interval domain $[\ell, u]$

such that $u > 1$ and $u > \ell$. We consider different combinations of triples $\langle g, f, \sigma \rangle$ and their corresponding intervals of interest wrt $\langle \ell, u \rangle$. Note that the value of $\overline{\mathcal{I}}^{\langle \ell, u \rangle}_{\langle g, f, \sigma \rangle}$ depends only on σ, ℓ, and u and not on g and f. The largest value appearing in the σ-patterns of X is u, and thus $\overline{\mathcal{I}}^{\langle \ell, u \rangle}_{\langle g, f, \sigma \rangle} = u$. We compute the value of $\underline{\mathcal{I}}^{\langle \ell, u \rangle}_{\langle g, f, \sigma \rangle}$ wrt two time-series constraints:

- Let g be the Max aggregator.
  * If $u - \ell = 1$, then any σ-pattern of X has a signature '$>$', i.e. contains only two elements. Then, the maximum value of R is reached for a time series t that contains the $\langle u, u - 1 \rangle$ σ-pattern. The rest of the variables of t are assigned any value, e.g. all other variables have a value of u. Such a time series t for the length 4 is shown in Fig. 1b. Further, for any v in $[\ell, u]$, the number of occurrences of v in the union of the σ-patterns of t is at most 1, which is a constant, and does not depend on n. By definition
    $$\underline{\mathcal{I}}^{\langle \ell, u \rangle}_{\langle g, f, \sigma \rangle} = \overline{\mathcal{I}}^{\langle \ell, u \rangle}_{\langle g, f, \sigma \rangle} - \eta_\sigma = u - 1.$$
  * If $u - \ell > 1$, then any maximal time series t for g_f_σ contains a single σ-pattern whose signature is in the language of '$>=*>$'. If, for example, $n = 4$, then t has $n - 2 = 2$ time-series variables with the values $u - 1$, which is depicted Fig. 1c. In addition, the σ-pattern of t has a single occurrence of the value $u - 2$. Hence, $\underline{\mathcal{I}}^{\langle \ell, u \rangle}_{\langle g, f, \sigma \rangle} = u - 1$.
- Let g be the Sum aggregator.
  Any maximal time series t for g_f_σ contains $\lfloor \frac{n}{2} \rfloor$ σ-patterns, which contains u and $u - 1$, and at most one of them has the value $u - 2$. Such a time series t for the length $n = 7$ is depicted in Fig. 1d. Hence, $\underline{\mathcal{I}}^{\langle \ell, u \rangle}_{\langle g, f, \sigma \rangle} = u - 1$. △

The next characteristic, we introduce, is a function of ℓ, u and n related to the maximum number of value occurrences in a σ-pattern.

Definition 3. *Consider a regular expression σ, and a time series X of length n over an integer interval domain $[\ell, u]$. The maximum value occurrence number of v in $\mathbb{Z}$ wrt $\langle \ell, u, n \rangle$, denoted by $\mu^{\langle \ell, u, n \rangle}_\sigma(v)$, is a function that maps an element of $\mathcal{R}_\Sigma \times \mathbb{Z} \times \mathbb{Z} \times \mathbb{N}^+ \times \mathbb{Z}$ to $\mathbb{N}$. It equals the maximum number of occurrences of the value v in one σ-pattern of X.*

Example 5. Consider the $\sigma = $ DecreasingSequence regular expression and a time series X of length $n \geq 2$ over an integer interval domain $[\ell, u]$ such that $u > \ell$. We compute the maximum value occurrence number of v in $\mathbb{Z}$ wrt $\langle \ell, u, n \rangle$. If v is not in $[\ell, u]$, then $\mu^{\langle \ell, u, n \rangle}_\sigma(v) = 0$. Hence, we focus on the case when $v \in [\ell, u]$.

- If $u - \ell = 1$, then any σ-pattern of X has a signature '$>$', and thus it may have at most one occurrence of any value v in $[\ell, u]$. Hence, for any v in $[\ell, u]$, $\mu^{\langle \ell, u, n \rangle}_\sigma(v) = 1$.
- If $u - \ell > 1$, then we consider two subsets of $[\ell, u]$:

* For either v in the set $\{\ell, u\}$, the value of $\mu_\sigma^{\langle \ell, u, n \rangle}(v)$ is 1, since in any σ-pattern the lower and upper limits of the domain, namely ℓ and u, can appear at most once, as it illustrated in Fig. 1e for the length $n = 4$.
* For any v in $[\ell+1, u-1]$, the value of $\mu_\sigma^{\langle \ell, u, n \rangle}(v)$ is $\max(1, n-2)$, since v can occur at most $n - 2$ times in a σ-pattern of X. The time series in Fig. 1c has a single σ-pattern, namely $\langle t_1, t_2, t_3, t_4 \rangle$, which has $n - 2 = 4 - 2 = 2$ occurrences of the value $u - 1$. $\triangle$

The next characteristic, we introduce, is the largest width of a σ-pattern in a time series.

Definition 4. *Consider a regular expression σ, and a time series X of length n over an integer interval domain $[\ell, u]$. The* big width *of σ wrt $\langle \ell, u, n \rangle$, denoted by $\beta_\sigma^{\langle \ell, u, n \rangle}$, is a function that maps an element of $\mathcal{R}_\Sigma \times \mathbb{Z} \times \mathbb{Z} \times \mathbb{N}^+$ to $\mathbb{N}$. It equals the maximum width of a σ-pattern in X. If X cannot have any σ-patterns, then $\beta_\sigma^{\langle \ell, u, n \rangle}$ is 0.*

Example 6. Consider the $\sigma = \texttt{DecreasingSequence}$ regular expression and a time series X of length n over an integer interval domain $[\ell, u]$.

- If $n \leq 1$, then X cannot have any σ-patterns, since a minimum width σ-pattern contains at least two elements. Hence, $\beta_\sigma^{\langle \ell, u, n \rangle} = 0$.
- If $u - \ell = 0$, then, as it was shown in Example 3, no word of $\mathcal{L}_\sigma$ can appear in the signature of any ground time series over $[\ell, u]$, and thus X cannot have any σ-patterns. Hence, $\beta_\sigma^{\langle \ell, u, n \rangle} = 0$.
- If $u - \ell = 1$ and $n \geq 2$, then any σ-pattern of X has a signature '$>$'. The width of such a σ-pattern is 2. Hence, $\beta_\sigma^{\langle \ell, u, n \rangle} = 2$.
- If $u - \ell > 1$ and $n \geq 2$, then there exists a word in $\mathcal{L}_\sigma$ that is also in the language of '$>=^*>$' and whose length is $n - 1$. This word is the signature of some ground time series t of length n over $[\ell, u]$, which contains a single σ-pattern of width n. Such a time series t for the length $n = 5$ is illustrated in Fig. 1f. The width of a σ-pattern cannot be greater than n, thus $\beta_\sigma^{\langle \ell, u, n \rangle} = n$. $\triangle$

The last characteristic is the notion of maximum overlap of a regular expression wrt an integer interval domain. It will be used for deriving an implied AMONG constraint when the aggregator of a considered time-series constraint is Sum.

Definition 5. *Consider a regular expression σ and an integer interval domain $[\ell, u]$. The* overlap *of σ wrt $[\ell, u]$, denoted by $o_\sigma^{\langle \ell, u \rangle}$, is the maximum number of time-series variables that belong simultaneously to two extended σ-patterns of a time series among all time series over $[\ell, u]$. If such maximum number does not exist, then $o_\sigma^{\langle \ell, u \rangle}$ is undefined.*

Example 7. Consider the $\sigma = \texttt{DecreasingSequence}$ regular expression and an interval $[\ell, u]$ with $u > \ell$. For any time series over $[\ell, u]$, any of its two extended σ-patterns have no time-series variables in common, thus $o_\sigma^{\langle \ell, u \rangle} = 0$. $\triangle$

Table 1 gives the values of the four characteristics of regular expressions for some regular expressions of [5], while Tables 2 and 3 provide the intervals of interest for 12 time-series constraints.

Table 1. For every regular expression σ, $[\ell, u]$ is an integer interval domain, and n is a time series length, such that there is at least one ground time series of length n over $[\ell, u]$ whose signature contains at least one occurrence of σ. Then, η_σ is the height of σ, $\mu_\sigma^{\langle \ell, u, n\rangle}(v)$ is the maximum value occurrence number of $v \in [\ell, u]$ wrt $\langle \ell, u, n\rangle$, $\beta_\sigma^{\langle \ell, u, n\rangle}$ is the big width of σ wrt $\langle \ell, u, n\rangle$, and $o_\sigma^{\langle \ell, u\rangle}$ is the overlap of σ wrt $\langle \ell, u\rangle$.

σ	η_σ	$\mu_\sigma^{\langle \ell, u, n\rangle}(v)$	$\beta_\sigma^{\langle \ell, u, n\rangle}$	$o_\sigma^{\langle \ell, u\rangle}$
' >><>> '	2	$\begin{cases}1, & \text{if } v \in \{\ell, \ell+1, u-1, u\} \\ 2, & \text{if } v \in [\ell+2, u-2]\end{cases}$	3	3
'>'	1	$1, \forall v \in [\ell, u]$	2	1
'(>(>|=)*)*>'	1	$\begin{cases}1, & \text{if } v \in \{u, \ell\} \\ \max(1, n-2), & \text{if } v \in [\ell+1, u-1]\end{cases}$	$\begin{cases}2, & \text{if } u-\ell=1 \\ n, & \text{Otherwise}\end{cases}$	0
'(>(>|=)*)*><((<|=)*<)*'	1	$\begin{cases}0, & \text{if } v = u \\ n-3, & \text{if } v \in [\ell+1, u-1] \\ 1, & \text{if } v = \ell\end{cases}$	$\begin{cases}1, & \text{if } u-\ell=1 \\ n-2, & \text{Otherwise}\end{cases}$	1
'<(<|=)* (>|=)*>'	1	$\begin{cases}0, & \text{if } v = \ell \\ n-2, & \text{if } v \in [\ell+1, u]\end{cases}$	$n-2$	1
'(<>)⁺(< | <>)|(><)⁺(> | ><)'	1	$\left\lfloor \frac{n-1}{2}\right\rfloor, \forall v \in [\ell, u]$	$n-2$	$\begin{cases}0, & \text{if } u-\ell=1 \\ 1, & \text{Otherwise}\end{cases}$

3.2 Deriving an AMONG Implied Constraint for the MAX_SURF_σ and the SUM_SURF_σ Families

Consider a $g\_f\_\sigma(\langle X_1, X_2, \ldots, X_n\rangle, R)$ time-series constraint with every X_i ranging over the same integer interval domain $[\ell, u]$, with f being the surf feature, and with g being either Max or Sum. Our goal is to estimate a lower bound on $\mathcal{N}$, which is the number of time-series variables in the σ-patterns of $\langle X_1, X_2, \ldots, X_n\rangle$ that must be assigned a value in the interval of interest $\mathcal{I}_{\langle g, f, \sigma\rangle}^{\langle \ell, u\rangle}$ of $\langle g, f, \sigma\rangle$ wrt $\langle \ell, u\rangle$, in order to satisfy the $g\_f\_\sigma(\langle X_1, X_2, \ldots, X_n\rangle, R)$ constraint. Theorems 1 and 2 present such inequality for the cases when g is Max, and Sum, respectively, using the four characteristics introduced in Sect. 3.1. Example 8 first conveys the intuition behind Theorem 1.

Example 8. Consider a $g\_f\_\sigma(X, R)$ time-series constraint with g being Max, with f being surf, with σ being the DecreasingSequence regular expression, and with X being a time series of length $n = 9$ over the integer interval domain $[\ell, u] = [0, 4]$. Let us assign R to the value 24, and let us compute a lower bound on $\mathcal{N}$, the number of variables of X that must be assigned a value from $\mathcal{I}_{\langle g, f, \sigma\rangle}^{\langle \ell, u\rangle}$, which is $[3, 4]$ as it was shown in Example 4. Our aim is to show that for a σ-pattern in X, its number of time-series variables in $[3, 4]$ can be estimated as the difference between the value of the surface of this σ-pattern and some other value that is a function of σ, ℓ, u and n. In order to obtain this value, we construct a time series t of length $\beta_\sigma^{\langle \ell, u, n\rangle} = 9$ satisfying all the following conditions:

1. The number of time-series variables of t that are assigned to the value $\overline{\mathcal{I}}_{\langle g,f,\sigma \rangle}^{\langle \ell,u \rangle}$ equals $\mu_\sigma^{\langle \ell,u,n \rangle}(\overline{\mathcal{I}}_{\langle g,f,\sigma \rangle}^{\langle \ell,u \rangle}) = \mu_\sigma^{\langle 0,4,9 \rangle}(4) = 1$.

2. The number of time-series variables of t that are assigned to the value $\mathcal{I}_{\langle g,f,\sigma \rangle}^{\langle \ell,u \rangle}$, which is $\overline{\mathcal{I}}_{\langle g,f,\sigma \rangle}^{\langle \ell,u \rangle} - 1$, equals $\mu_\sigma^{\langle \ell,u,n \rangle}(\mathcal{I}_{\langle g,f,\sigma \rangle}^{\langle \ell,u \rangle}) = \mu_\sigma^{\langle 0,4,9 \rangle}(3) = n - 2 = 7$.

3. The rest of the time-series variables of t, namely $n - \mu_\sigma^{\langle \ell,u,n \rangle}(\overline{\mathcal{I}}_{\langle g,f,\sigma \rangle}^{\langle \ell,u \rangle}) - \mu_\sigma^{\langle \ell,u,n \rangle}(\underline{\mathcal{I}}_{\langle g,f,\sigma \rangle}^{\langle \ell,u \rangle}) = 1$ time-series variable, is assigned to the value $\mathcal{I}_{\langle g,f,\sigma \rangle}^{\langle \ell,u \rangle} - 1 = 2$.

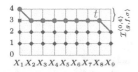

Figure on the left illustrates a ground time series t of length 9 over $[0,4]$ satisfying all the three conditions. By construction, the sum of elements of t is greater than or equal to the surface of any σ-pattern of X. Furthermore, for any σ-pattern of X, its number of time-series variables whose values are in $[3,4]$ is not greater than the number of such time-series variables of t.

Figure above on the left contains three type of points: circled, squared and diamond-shaped points; thus our goal is to evaluate the number of circles. The value of X_i is one plus the number of squared and diamond-shaped points under the point corresponding to X_i. Hence, the sum of all elements of t can be viewed as the total number of circled, squared and diamond-shaped points. Furthermore, the number of circles is the difference between the total number of points and the number of squared points, namely 27 minus 19, which is 8.

For any σ-pattern of X, its corresponding number of squared and diamond-shaped points is at most 19. Then, its number of time-series variables whose values are in $[3,4]$ can be estimated as the surface of the σ-pattern minus 19. Hence, when the surface of the σ-pattern is 24, a lower bound on $\mathcal{N}$ is 5. Figure on the right gives an example of a ground time series t' of length 9 over $[0,4]$ that contains a σ-pattern with a surface of 24. This σ-pattern has $6 \geq 5$ values in $[3,4]$, which agrees with our computed lower bound.   △

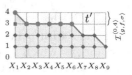

Theorem 1. *Consider a $g\_f\_\sigma(X,R)$ time-series constraint with $g =$ Max, $f =$ surf and X being a time series of length n over an integer interval domain $[\ell,u]$; then $\textsc{Among}(\mathcal{N},X,\mathcal{I})$ is an implied constraint, where $\mathcal{N}$ is restricted by*

$$\mathcal{N} \geq R - \max\left(0,\underline{\mathcal{I}} - 1\right) \cdot \beta - \sum_{v \in [\underline{\mathcal{I}}+1,\overline{\mathcal{I}}]} \mu_\sigma^{\langle \ell,u,n \rangle}(v) \cdot (v - \underline{\mathcal{I}}), \qquad (1)$$

where β (resp. $\mathcal{I}$) is shorthand for $\beta_\sigma^{\langle \ell,u,n \rangle}$ (resp. $\mathcal{I}_{\langle g,f,\sigma \rangle}^{\langle \ell,u \rangle}$), and $\underline{\mathcal{I}}$ (resp. $\overline{\mathcal{I}}$) denotes the lower (resp. upper) limit of interval $\mathcal{I}$.

Proof. We show that the right-hand side of the stated inequality is a lower bound on the number of time-series variables of a σ-pattern whose values are in $\mathcal{I}$, and the surface of the σ-pattern is R. In order to prove the lower bound on $\mathcal{N}$, we first compute a lower bound on the number $\mathcal{N}^{\underline{\mathcal{I}}}$ of time-series variables of the σ-pattern whose values are $\underline{\mathcal{I}}$, which is the smallest value of interval $\mathcal{I}$. We assume that for every $v > \underline{\mathcal{I}}$ in $\mathcal{I}$, the number of occurrences of v in the σ-pattern equals some $\mathcal{N}^v$. Note that the number of time-series variables in any σ-pattern is not greater than $\beta = \beta_\sigma^{\langle \ell, u, n \rangle}$. We state the following inequality:

$$R \leq \underbrace{\mathcal{N}^{\underline{\mathcal{I}}} \cdot \max(0, \underline{\mathcal{I}})}_{A} + \underbrace{\sum_{v \in [\underline{\mathcal{I}}+1, \overline{\mathcal{I}}]} \mathcal{N}^v \cdot \max(0, v)}_{B} \qquad (2)$$

$$\underbrace{+ \max(0, \underline{\mathcal{I}} - 1) \cdot (\beta - \mathcal{N}^{\underline{\mathcal{I}}} - \sum_{v \in [\underline{\mathcal{I}}+1, \overline{\mathcal{I}}]} \mathcal{N}^v),}_{C}$$

where A, B, and C correspond to the sums of elements of the σ-pattern that equal $\underline{\mathcal{I}}$, are in $\mathcal{I}$ and are greater than $\underline{\mathcal{I}}$, and are outside $\mathcal{I}_{\langle g, f, \sigma \rangle}^{\langle \ell, u \rangle}$ respectively. From Inequality (2) we obtain the following lower bound on $\mathcal{N}^{\underline{\mathcal{I}}}$:

$$\mathcal{N}^{\underline{\mathcal{I}}} \geq R - \sum_{v \in [\underline{\mathcal{I}}+1, \overline{\mathcal{I}}]} \mathcal{N}^v \cdot \max(0, v) - \max(0, \underline{\mathcal{I}} - 1) \cdot (\beta - \sum_{v \in [\underline{\mathcal{I}}+1, \overline{\mathcal{I}}]} \mathcal{N}^v). \quad (3)$$

In order to obtain a lower bound on $\mathcal{N}$ from the known lower bound on $\mathcal{N}^{\underline{\mathcal{I}}}$, we add $\sum_{v \in [\underline{\mathcal{I}}+1, \overline{\mathcal{I}}]} \mathcal{N}^v$ to both sides of Inequality (3). Further, we regroup some terms in Inequality (3), we eliminate $\sum_{v \in [\underline{\mathcal{I}}+1, \overline{\mathcal{I}}]} \mathcal{N}^v$ in the right-hand side of Inequality (3) by replacing it with $\sum_{v \in [\underline{\mathcal{I}}+1, \overline{\mathcal{I}}]} \mu_\sigma^{\langle \ell, u, n \rangle}(v)$, and obtain the inequality of the theorem. $\qquad \square$

Example 9. Consider the $g\_f\_\sigma(\langle X_1, X_2, \ldots, X_n \rangle, R)$ time-series constraint, with g being Sum, with f being surf, and with every X_i (with $i \in [1, n]$) ranging over the same domain $[\ell, u]$ with $u > 1$ and $u - \ell > 1$. We illustrate the derivation of AMONG implied constraints for two regular expressions.

- Consider the $\sigma = $ DecreasingSequence regular expression and $n \geq 2$. In Example 4, we computed the interval of interest of MAX_SURF_σ wrt $\langle \ell, u \rangle$, which is $[u - 1, u]$. In Example 5, we showed that $\mu_\sigma^{\langle \ell, u, n \rangle}(\ell) = \mu_\sigma^{\langle \ell, u, n \rangle}(u) = 1$, and for every value v in $[\ell + 1, u - 1]$, we have that $\mu_\sigma^{\langle \ell, u, n \rangle}(v)$ equals $\max(1, n - 2)$. Finally, in Example 6 we demonstrated that $\beta_\sigma^{\langle \ell, u, n \rangle} = n$. By Theorem 1, we can impose the AMONG$(\mathcal{N}, X, \langle u - 1, u \rangle)$ implied constraint with $\mathcal{N} \geq R - \mu_\sigma^{\langle \ell, u, n \rangle}(u) - \max(0, \underline{\mathcal{I}}_{\langle g, f, \sigma \rangle}^{\langle \ell, u \rangle} - 1) \cdot \beta_\sigma^{\langle \ell, u, n \rangle} = R - 1 - \max(0, u - 2) \cdot n$.

Turning back to Example 8 we observe that, in the obtained implied constraint, the term '1' corresponds to the number of squared points, and the term '$\max(0, u-2) \cdot n$' to the number of diamond-shaped points. The derived lower bound on $\mathcal{N}$ also appears in the third row of Table 2.

- Consider the $\sigma = \texttt{Peak} = \text{`} <(<|=)^* \; (>|=)^*> \text{'}$ regular expression whose values of a_σ and b_σ both equal 1, and $n \geq 3$. The maximum value in $[\ell, u]$ that appears in a σ-pattern is u. In addition, any maximal time series for $\langle g, f, \sigma \rangle$ contains a single σ-pattern whose values are all the same and equal u. Hence, the interval of interest of $\langle g, f, \sigma \rangle$ wrt $\langle \ell, u \rangle$ is $[u, u]$. Since both a_σ and b_σ equal 1, the smallest value in $[\ell, u]$ may not be in any σ-pattern and $\mu_\sigma^{\langle \ell, u, n \rangle}(\ell) = 0$. For any value $v \in [\ell + 1, u]$, we have $\mu_\sigma^{\langle \ell, u, n \rangle}(v) = n - 2$. By Theorem 2, we impose an AMONG$(\mathcal{N}, \langle X_1, X_2, \ldots, X_n \rangle, \langle u \rangle)$ implied constraint with $\mathcal{N} \geq R - \max(0, u - 1) \cdot (n - 2)$. The derived lower bound on $\mathcal{N}$ also appears in the fifth row of Table 2.   △

Table 2 gives for 6 regular expressions of [5] the corresponding intervals of interest of MAX_SURF_σ constraints wrt some integer interval domain $[\ell, u]$ such that $u > 1 \; \wedge \; u - \ell > 1$, as well as the lower bound LB on the parameter $\mathcal{N}$ of the derived AMONG constraint for time series that may have at least one σ-pattern.

Table 2. Regular expression σ, the corresponding interval of interest of MAX_SURF_σ (X, R) wrt an integer interval domain $[\ell, u]$ such that $u > 1$ and $u - \ell > 1$, and the lower bound LB on the parameter of the derived AMONG implied constraint. The value LB is obtained from a generic formula, which is parameterised by characteristics of regular expressions. The sequence X is supposed to be long enough to contain at least one σ-pattern.

σ	$\mathcal{I}_{\langle \text{MAX,SURF}, \sigma \rangle}^{\langle \ell, u \rangle}$	LB
'$>><>>$'	$[u-2, u]$	$R - \max(0, u-3) \cdot 3 - 3$
'$>$'	$[u-1, u]$	$R - \max(0, u-2) \cdot 2 - 1$
'$(>(>|=)^*)^*>$'	$[u-1, u]$	$R - \max(0, u-2) \cdot n - 1$
'$(>(>|=)^*)^*><((<|=)^*<)^*$'	$[u-1, u-1]$	$R - \max(0, u-2) \cdot (n-2)$
'$<(<|=)^* \; (>|=)^*>$'	$[u, u]$	$R - \max(0, u-1) \cdot (n-2)$
'$(<>)^+(<|<>)|(><)^+(>|><)$'	$[u-1, u]$	$R - \max(0, u-2) \cdot (n-2) - \lfloor \frac{n-1}{2} \rfloor$

Theorem 2. *Consider a* $g\_f\_\sigma(X, R)$ *time-series constraint with* $g = \text{Sum}$*,* $f = \texttt{surf}$ *and* X *being a time series of length* n *over an integer interval domain* $[\ell, u]$*; then* AMONG$(\mathcal{N}, X, \mathcal{I})$ *is an implied constraint, where* $\mathcal{N}$ *is restricted by*

$$
\begin{aligned}
\mathcal{N} \geq \; & R - \max\left(0, \overline{\mathcal{I}} - \underline{\mathcal{I}}\right) \cdot \left(n - a_\sigma - b_\sigma + (p_o - 1) \cdot \max(0, o_\sigma^{\langle \ell, u \rangle} - a_\sigma - b_\sigma)\right) \\
& - \sum_{v \in [\underline{\mathcal{I}} + 1, \overline{\mathcal{I}}]} \mu_\sigma^{\langle \ell, u, n \rangle}(v) \cdot p_o \cdot (v - \underline{\mathcal{I}}) \\
& - (p_o - 1) \cdot \max(0, o_\sigma^{\langle \ell, u \rangle} - a_\sigma - b_\sigma),
\end{aligned}
\tag{4}
$$

where $\mathcal{I}$ is shorthand for $\mathcal{I}_{\langle g,f,\sigma\rangle}^{\langle \ell,u\rangle}$, $\underline{\mathcal{I}}$ (resp. $\overline{\mathcal{I}}$) denotes the lower (resp. upper) limit of $\mathcal{I}$, and p_o is 1 if every maximal time series has a single σ-pattern, and is the maximal number of σ-patterns in a time series of length n, otherwise.

Proof. To prove Theorem 2 we consider a time series with $p \geq 1$ σ-patterns, where σ-pattern i (with $i \in [1,p]$) has a width of ω_i and a surface of R_i, and where $R = \sum_{i \in [1,p]} R_i$. The proof consists of two steps:

1. First, for each σ-pattern i (with $i \in [1,p]$), we compute the minimum number $\mathcal{N}_i$ of time-series variables that must be assigned to a value within the interval of interest $\mathcal{I}$, in order to reach a surface of R_i.
2. Second, we take the sum of $\mathcal{N}_i$, and minimise the obtained value, which, in the end, will be a minimum value for $\mathcal{N}$.

First Step. We use Inequality (1) of Theorem 1 for a subseries X' of X of length $\omega_i' = \omega_i + a_\sigma + b_\sigma$, knowing that X' has a single σ-pattern and $\beta_\sigma^{\langle \ell,u,n\rangle}$ is ω_i. Then, by Theorem 1, we obtain the following estimation of $\mathcal{N}_i$:

$$\mathcal{N}_i \geq R_i - \omega_i \cdot \max(0, \underline{\mathcal{I}} - 1) - \sum_{v \in [\underline{\mathcal{I}}+1, \overline{\mathcal{I}}]} (v - \underline{\mathcal{I}}) \cdot \mu_\sigma^{\langle \ell,u,\omega_i'\rangle}(v). \qquad (5)$$

Second Step. We obtain the minimum value of $\mathcal{N}$, by taking the sum of the derived minimum values for $\mathcal{N}_i$ over all the values of i:

$$\mathcal{N} = \sum_{i=1}^{p} \mathcal{N}_i \geq \sum_{i=1}^{p} (R_i - A_i - B_i) - C = R - \sum_{i=1}^{p} A_i - \sum_{i=1}^{p} B_i - C, \qquad (6)$$

where for any $i \in [1,p]$, $A_i = \omega_i \cdot \max(0, \underline{\mathcal{I}} - 1)$ and $B_i = \sum_{v \in [\underline{\mathcal{I}}+1, \overline{\mathcal{I}}]} \mu_\sigma^{\langle \ell,u,\omega_i'\rangle}(v) \cdot$

$(v - \underline{\mathcal{I}})$, and $C = (p-1) \cdot \max(0, o_\sigma^{\langle \ell,u\rangle} - a_\sigma - b_\sigma)$. The terms A_i and B_i come from Inequality (5) and the term C is used because some variables may belong to two σ-patterns: in order to not count them twice we subtract a correction term. Let A (resp. B) denote $\sum_{i=1}^{p} A_i$ (resp. $\sum_{i=1}^{p} B_i$). In order to satisfy Condition 6, we need to find the upper bounds on the sum $A + B + C$ by choosing the value of p, and the sum of σ-patterns lengths. We consider two cases, but any additional information may be used for a more accurate estimation of these parameters:

- [EVERY MAXIMAL TIME SERIES HAS A SINGLE σ-pattern] Then, the maximum value of $A + B + C$ is reached for p being 1, and $\sum_{i=1}^{p} \omega_i$ being $n - b_\sigma - a_\sigma$. It implies that for any $v \in [\underline{\mathcal{I}}_{\langle g,f,\sigma\rangle}^{\langle \ell,u\rangle} + 1, \overline{\mathcal{I}}_{\langle g,f,\sigma\rangle}^{\langle \ell,u\rangle}]$, the value of $\sum_{i \in [1,p]} \mu_\sigma^{\langle \ell,u,\omega_i'\rangle}(v)$ equals $\mu_\sigma^{\langle \ell,u,n\rangle}(v)$.
- [THERE IS AT LEAST ONE MAXIMAL TIME SERIES WITH MORE THAN ONE σ-pattern] We give an overestimation: we assign the value of p to its maximum

value, which depends on σ, the value of $\sum_{i=1}^{p} \omega_i$ is overestimated by $n - a_\sigma -$ $b_\sigma + (p_o - 1) \cdot \max(0, o_\sigma^{\langle \ell, u \rangle} - a_\sigma - b_\sigma)$, and the value of $\sum_{i \in [1,p]} \mu_\sigma^{\langle \ell, u, \omega'_i \rangle}(v)$ is overestimated by $\mu_\sigma^{\langle \ell, u, n \rangle}(v) \cdot p_o$.

Hence, we obtain a lower bound for $\mathcal{N}$, which is the right hand side of the inequality stated by Theorem 2. □

Table 3. Regular expression σ, the corresponding interval of interest of SUM_SURF_σ (X, R) wrt an integer interval domain $[\ell, u]$ such that $u > 1$ and $u - \ell > 1$, and the lower bound LB on the parameter of the derived AMONG implied constraint. The value LB is obtained from a generic formula, which is parameterised by characteristics of regular expressions. The sequence X is supposed to be long enough to contain at least one σ-pattern.

σ	$\mathcal{I}_{\langle \text{SUM,SURF}, \sigma \rangle}^{\langle \ell, u \rangle}$	LB
' >><>> '	$[u - 2, u]$	$R - \max(0, u - 3) \cdot$ $(n - 3) - 3 \cdot \lfloor \frac{n-3}{3} \rfloor$
'>'	$[u - 1, u]$	$R - \max(0, u - 2) \cdot$ $(2 \cdot n - 2) - (2 \cdot n - 3)$
'(>(>|=)*)*>'	$[u - 1, u]$	$R - \max(0, u - 2) \cdot n - \lfloor \frac{n}{2} \rfloor$
'(>(>|=)*)*><((<|=)*<)*'	$[u - 1, u - 1]$	$R - \max(0, u - 2) \cdot (n - 2)$
'<(<|=)* (>|=)*>'	$[u, u]$	$R - \max(0, u - 1) \cdot (n - 2)$
'(<>)$^+$(< | <>)|(><)$^+$(> | ><)'	$[u - 1, u]$	$R - \max(0, u - 2) \cdot$ $(n - 2) - \lfloor \frac{n-1}{2} \rfloor$

Example 10. Consider the $g\_f\_\sigma(\langle X_1, X_2, \ldots, X_n \rangle, R)$ time-series constraint, with g being Sum, with f being surf and with every X_i (with $i \in [1, n]$) ranging over the same domain $[\ell, u]$ with $u > 1$ and $u - \ell > 1$. We illustrate the derivation of AMONG implied constraints for two regular expressions.

- Consider the $\sigma = $ DecreasingSequence regular expression and $n \geq 2$. In Example 4, we found that the interval of interest of $\langle g, f, \sigma \rangle$ wrt $\langle \ell, u \rangle$ is $[u - 1, u]$, and in Example 5, we showed that $\mu_\sigma^{\langle \ell, u, n \rangle}(\ell) = \mu_\sigma^{\langle \ell, u, n \rangle}(u) = 1$, and for every value v in $[\ell+1, u-1]$, we have that $\mu_\sigma^{\langle \ell, u, n \rangle}(v)$ equals $\max(1, n - 2)$. Every maximal time series for SUM_SURF_σ contains the maximum number of σ-patterns. Hence, in this case, the value of p_o equals the maximum number of decreasing sequences in a time series of length n, which is $\lfloor \frac{n}{2} \rfloor$. By Theorem 2, we impose an AMONG$(\mathcal{N}, \langle X_1, X_2, \ldots, X_n \rangle, \langle u-1, u \rangle)$ implied constraint with $\mathcal{N} \geq R - \lfloor \frac{n}{2} \rfloor - \max(0, u - 2) \cdot n$. The derived lower bound on $\mathcal{N}$ also appears in the third row of Table 3.
- Consider the $\sigma = $ Peak = '<(<\|=)* (>\|=)*>' regular expression and $n \geq 3$. The maximum value in $[\ell, u]$ that occurs in a σ-pattern is u. In addition,

any maximal time series for $\langle g, f, \sigma \rangle$ contains a single σ-pattern whose values are all the same and equal u. Hence, the interval of interest of $\langle g, f, \sigma \rangle$ wrt $\langle \ell, u \rangle$ is $[u, u]$, and the value of p_o equals 1. We showed in Example 9 that $\mu_\sigma^{\langle \ell, u, n \rangle}(\ell) = 0$ and for any $v \in [\ell + 1, u]$, we have $\mu_\sigma^{\langle \ell, u, n \rangle}(v) = n - 2$. The value of $o_\sigma^{\langle \ell, u \rangle}$ equals 1. By Theorem 2, we impose an AMONG$(\mathcal{N}, \langle X_1, X_2, \ldots, X_n \rangle, \langle u \rangle)$ implied constraint with $\mathcal{N} \geq R - \max(0, u - 1) \cdot (n - 2)$. The derived lower bound on $\mathcal{N}$ also appears in the fifth row of Table 3. $\triangle$

Table 3 gives for 6 regular expressions of [5] the corresponding intervals of interest of SUM_SURF_σ constraints wrt some integer interval domain $[\ell, u]$ such that $u > 1 \wedge u - \ell > 1$, as well as the lower bound LB on the parameter $\mathcal{N}$ of the derived AMONG constraint for time series that may have at least one σ-pattern.

4   Evaluation

The intended use case is a problem where we learn parameters for a conjunction of many time-series constraints from data, and use this conjunction to create new time-series that are "similar" to the existing ones. An example would be electricity production data for a day [10], in half hour periods (48 values), or manpower levels per week over a year (52 values). To solve the conjunction, we need strong propagation for each individual constraint. We therefore evaluate the impact of the implied constraint on both execution time and the number of backtracks for the time-series constraints of the MAX_SURF_σ and the SUM_SURF_σ families for which a glue constraint [4] exists, which are 38 out of 44 time-series constraints of the two families. These families of constraints were the most difficult to solve in the experiments reported in [4].

In the experiments for both families, we consider a single g_SURF_$\sigma(X, R)$ time-series constraint with g being either Sum or Max, for which we first systematically try out all potential values of the parameter R, and then either find a solution by assigning the X_i or prove infeasibility. We compare the best (Combined) approach from the recent work [4] to the new method, adding the implied AMONG constraint on every suffix of $X = \langle X_1, X_2, \ldots, X_n \rangle$, and also a *preprocessing procedure*. The preprocessing procedure is a useful, if minor, contribution of the paper for 8 out of 38 of the constraints in the families studied. The purpose of this procedure is to find all feasible values of R, when σ is such that any σ-pattern has all values being the same. Such values of R must satisfy the following constraint:

$$ R = \mathbf{def}_{g,f} \vee \left(\exists V \in [\ell', u'] \; \beta_\sigma^{\langle \ell, u, n \rangle} \cdot V \geq R \wedge R \bmod V = 0 \right), $$

where ℓ' and u' are the smallest and the largest value, respectively, that can occur in a σ-pattern over $[\ell, u]$.

Since the implied constraints are precomputed offline, posting one implied constraint takes a *constant time*, and the time and space complexity of the preprocessing procedure does not exceed the size of the domain of R, which is $O(n \cdot (u - \ell))$.

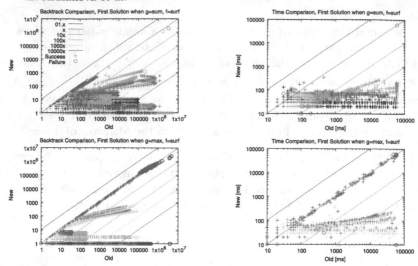

Fig. 2. Comparing backtrack count and runtime of the g_f_σ time-series constraints for previous best results (old) and new method for finding the first solution or proving infeasibility for time series of length 50 and domain $[0, 5]$. Colours of markers indicate the regular expression, the cross (resp. circle) marker type indicates success (resp. failure/timeout).

Figure 2 presents the results for the SUM_SURF_σ (upper plots) and the MAX_SURF_σ (lower plots) time-series constraints, where X is a time series of length 50 over the domain $[0, 5]$, when the goal is to find, for each value of R, the first solution or prove infeasibility. This corresponds to our main use case, where we want to construct time series with fixed R values. Our static search routine enumerates the time-series variables X_i from left to right, starting with the smallest value in the domain. Results for the backtrack count are on the left, results for the execution time on the right. We use log scales on both axes, replacing a zero value by one in order to allow plotting. A timeout of 60 s was imposed. We see that the implied constraints reduce backtracks by up to a factor exceeding 10,000 and runtime by up to a factor of 1,000, and they divide the total execution time of terminated instances by a factor of 5 and 45 times when g is Max and Sum, respectively. All experiments were run on a 2014 iMac 4 GHz $i7$ using SICStus Prolog.

The results for the case $g =$ Sum are better than for the case $g =$ Max because the aggregator Sum allows summing the surfaces of several σ-patterns, whereas for the Max aggregator, R is the surface of a single σ-pattern, the surfaces of other σ-patterns, if any, are absorbed.

5   Conclusion

In summary, based on 4 regular expression characteristics, we have defined a *single per family generic implied constraint* for all constraints of the MAX_SURF_σ

and SUM_SURF_σ families. The experimental results showed a good speed up in the number of backtracks and the time spent for the SUM_SURF_σ family.

References

1. Akşin, O.Z., Armony, M., Mehrotra, V.: The modern call center: a multidisciplinary perspective on operations management research. Prod. Oper. Manag. **16**(6), 665–688 (2007)
2. Alur, R., D'Antoni, L., Deshmukh, J.V., Raghothaman, M., Yuan, Y.: Regular functions and cost register automata. In: 28th Annual ACM/IEEE Symposium on Logic in Computer Science, LICS 2013, New Orleans, LA, USA, 25–28 June 2013, pp. 13–22. IEEE Computer Society (2013)
3. Alur, R., Fisman, D., Raghothaman, M.: Regular programming for quantitative properties of data streams. In: Thiemann, P. (ed.) ESOP 2016. LNCS, vol. 9632, pp. 15–40. Springer, Heidelberg (2016). doi:10.1007/978-3-662-49498-1_2
4. Arafailova, E., Beldiceanu, N., Carlsson, M., Flener, P., Francisco Rodríguez, M.A., Pearson, J., Simonis, H.: Systematic derivation of bounds and glue constraints for time-series constraints. In: Rueher, M. (ed.) CP 2016. LNCS, vol. 9892, pp. 13–29. Springer, Cham (2016). doi:10.1007/978-3-319-44953-1_2
5. Arafailova, E., Beldiceanu, N., Douence, R., Carlsson, M., Flener, P., Rodríguez, M.A.F., Pearson, J., Simonis, H.: Global constraint catalog, volume ii, time-series constraints. CoRR, abs/1609.08925 (2016)
6. Arafailova, E., Beldiceanu, N., Douence, R., Flener, P., Francisco Rodríguez, M.A., Pearson, J., Simonis, H.: Time-series constraints: improvements and application in CP and MIP contexts. In: Quimper, C.-G. (ed.) CPAIOR 2016. LNCS, vol. 9676, pp. 18–34. Springer, Cham (2016). doi:10.1007/978-3-319-33954-2_2
7. Beldiceanu, N., Carlsson, M., Douence, R., Simonis, H.: Using finite transducers for describing and synthesising structural time-series constraints. Constraints **21**(1), 22–40 (2016). Journal Fast Track of CP 2015. Summary. LNCS, vol. 9255, p. 723. Springer, Berlin (2015)
8. Beldiceanu, N., Carlsson, M., Petit, T.: Deriving filtering algorithms from constraint checkers. In: Wallace, M. (ed.) CP 2004. LNCS, vol. 3258, pp. 107–122. Springer, Heidelberg (2004). doi:10.1007/978-3-540-30201-8_11
9. Beldiceanu, N., Contejean, E.: Introducing global constraints in CHIP. Math. Comput. Model. **20**(12), 97–123 (1994)
10. Beldiceanu, N., Ifrim, G., Lenoir, A., Simonis, H.: Describing and generating solutions for the EDF unit commitment problem with the ModelSeeker. In: Schulte, C. (ed.) CP 2013. LNCS, vol. 8124, pp. 733–748. Springer, Heidelberg (2013). doi:10.1007/978-3-642-40627-0_54
11. Bessière, C., Coletta, R., Hébrard, E., Katsirelos, G., Lazaar, N., Narodytska, N., Quimper, C.-G., Walsh, T.: Constraint Acquisition via Partial Queries. In: IJCAI, Beijing, China, p. 7, June 2013
12. Bessière, C., Coletta, R., Petit, T.: Learning implied global constraints. In: IJCAI 2007, Hyderabad, India, pp. 50–55 (2007)
13. Bessière, C., Hébrard, E., Hnich, B., Kiziltan, Z., Walsh, T.: Among, common and disjoint constraints. In: Hnich, B., Carlsson, M., Fages, F., Rossi, F. (eds.) CSCLP 2005. LNCS (LNAI), vol. 3978, pp. 29–43. Springer, Heidelberg (2006). doi:10.1007/11754602_3

14. Colcombet, T., Daviaud, L.: Approximate comparison of functions computed by distance automata. Theory Comput. Syst. **58**(4), 579–613 (2016)
15. Demassey, S., Pesant, G., Rousseau, L.-M.: A `cost-regular` based hybrid column generation approach. Constraints **11**(4), 315–333 (2006)
16. Eeckhout, L., De Bosschere, K., Neefs, H.: Performance analysis through synthetic trace generation. In: 2000 ACM/IEEE International Symposium Performance Analysis Systems Software, pp. 1–6 (2000)
17. Kegel, L., Hahmann, M., Lehner, W.: Template-based time series generation with loom. In: EDBT/ICDT Workshops 2016, Bordeaux, France (2016)
18. Pesant, G.: A regular language membership constraint for finite sequences of variables. In: Wallace, M. (ed.) CP 2004. LNCS, vol. 3258, pp. 482–495. Springer, Heidelberg (2004). doi:10.1007/978-3-540-30201-8_36
19. Picard-Cantin, É., Bouchard, M., Quimper, C.-G., Sweeney, J.: Learning parameters for the sequence constraint from solutions. In: Rueher, M. (ed.) CP 2016. LNCS, vol. 9892, pp. 405–420. Springer, Cham (2016). doi:10.1007/978-3-319-44953-1_26
20. Marcel Paul Schützenberger: On the definition of a family of automata. Inf. Control **4**, 245–270 (1961)

Solving Constraint Satisfaction Problems Containing Vectors of Unknown Size

Erez Bilgory, Eyal Bin[✉], and Avi Ziv

IBM Research, Haifa, Israel
{erezbi,bin,aziv}@il.ibm.com

Abstract. Constraint satisfaction problems (CSPs) are used to solve real-life problems with inherent structures that contain vectors for repeating sets of variables and constraints. Often, the structure of the problem is a part of the problem, since the number of elements in the vector is not known in advance. We propose a method to solve such problems, even when there is no maximal length provided. Our method is based on constructing a vector size CSP from the problem description, and solving it to get the number of elements in the vector. We then use the vector size to construct and solve a CSP that has a specific number of elements. Experimental results show that this method enables fast solving of problems that cannot be solved or even constructed by existing methods.

Keywords: Constraint satisfaction problems · Unbounded vector size

1 Introduction

Constraint satisfaction problems (CSPs) [6] are used to model and solve many real-life problems. In many of these problems there is an inherent structure that includes repetitions (or vectors) of sub-problems. For example, in the configuration of hardware systems, the system, which is the target of the CSP, contains several, potentially different, racks and boxes [18]. In many cases, the size of the vectors is not known in advance, and therefore the role of the CSP solver is to find a correct vector size that enables the problem to be solved. In such problems, we can treat the vector size as a CSP variable on its own.

This problem cannot be handled by standard CSP techniques, which assume that the graph of the CSP (variables as nodes, constraints as hyperedges) is fully known before beginning to solve it. One possible solution to the problem is to apply external heuristics, or a simple random guess for the vector size, before constructing the CSP. This solution requires a different heuristic for each problem and may suffer from a low success rate, depending on the quality of the heuristic and the sparseness of the feasible vector size compared to its initial domain.

A second possible solution is to use conditional CSP techniques [9], where the existence of parts of the CSP is conditioned by other parts of it. The idea

© Springer International Publishing AG 2017
J.C. Beck (Ed.): CP 2017, LNCS 10416, pp. 55–70, 2017.
DOI: 10.1007/978-3-319-66158-2_4

of this solution is creating the CSP with maximal vector size, and marking elements in the vector beyond the vector size as inactive. This can be wasteful in terms of memory and compute time, especially if there is a substantial difference between the initial domain of the vector size and its feasible sizes. Moreover, an unbounded vector size requires a preprocessing step that estimates a bound.

This paper presents a method for solving CSPs that contain vectors of unknown sizes. Our method can handle problems with several vectors and hierarchies of vectors (i.e., vectors with unknown size that contain vectors of unknown size, etc.). The method is based on extracting information on the vector size from the problem description, and using this information to construct and solve a vector size CSP for each vector with unknown size. After we obtain the vector size with the vector size CSP, we can add the vector to the full CSP. Therefore, when the size of all the vectors is known, the construction of the full CSP is completed.

The heart of the method is the construction of the vector size CSP. This CSP has to include not only variables and constraints that explicitly (directly or indirectly) affect the vector size, but also constraints on the vector elements that can have an implicit effect on the size. For example, a constraint on the sum of the vector elements can affect the range of feasible sizes. To handle this implicit effect, we created projectors from the vector elements to its size, for commonly used constraints. Our projectors overapproximate the possible sizes, thereby maintaining the completeness of the solution procedure.

Our proposed method constructs the full CSP only when the sizes of all vectors are known. In addition, it does not require a predefined bound for the vector size. Therefore, it does not suffer from the waste of conditional CSPs. On the other hand, because it extracts all the available information from the problem description, it can significantly improve its success rate over the random selection of vector sizes. This is done without resorting to external heuristics, which are not always available. Another advantage of our method is that the modeler of the CSP does not have to add modeling to account for the solution method.

We implemented our method on top of our internal constraint solver [5,15]. The proposed method is currently in use in CSP-based systems in two separate domains: test program generation for processor verification [10], and data fabrication for the testing of database systems [1]. We present experimental results on real-life problems to illustrate the advantages of the proposed method over existing methods, in terms of success rate and resource consumption.

The rest of the paper is organized as follows. In Sect. 2, we provide a detailed description of the problem. Section 3 presents the proposed solution and provides details on specific aspects. Section 4 shows some experimental results. Finally, Sect. 5 concludes the paper.

2   Problem Description

A constraint satisfaction problem (CSP) is a mathematical formulation that is used to model and solve many real-life problems. Formally, a constraint satisfaction problem is defined as a triple (X, D, C), where

$X = \{X_1, \ldots, X_n\}$ is a finite set of variables,

$D = \{D_1, \ldots, D_n\}$ is a set of the respective domains of values, and

$C = \{C_1, \cdots, C_m\}$ is a finite set of constraints.

Each variable X_i can take on the values in the finite, nonempty domain D_i. Every constraint $C_j \in C$ is in turn a pair $\langle t_j, R_j \rangle$, where $t_j \subset X$ is a subset of k variables and R_j is a k-ary relation on the corresponding subset of domains D_j. A solution to the CSP is a consistent assignment of a single value to each variable X_i from its domain D_i, such that it does not violate any of the constraints. Note, this paper deals with finite CSPs.

Many problems that are modeled and solved as constraint satisfaction problems have an inherent structure. For example, in the configuration of hardware systems [18], the systems contain racks and each rack contains boxes of diverse types and capabilities. Another example is the generation of test programs for processors [10], where the generated test comprises instructions, which, in turn, comprise operands.

In such structured cases, it is natural and useful to use similar structures in the CSP that models the problem. Consequently, the full CSP is constructed using instances of structures or classes, each handling a sub-problem or sub-structure of the full system. A CSP sub-structure, or class, is a set of variables and constraints related to the variables, and in hierarchical cases, sub-classes. Constraints can be internal to an instance or can span classes and instances. For example, in the configuration of the hardware system, each rack in the system is a class with variables for type and dimensions, constraints on power and cooling capabilities, and sub-classes of boxes mounted in the rack.

In many cases, the structure of the problem is unknown ahead of time and has to be handled as part of the overall problem. If the exact types of instances in the overall model are not known in advance, object-oriented techniques [18] can be used to model the various sub-types. For example, in the configuration problem we can have sub-types or sub-classes for specific types of boxes in the system, such as compute boxes, memory boxes, and disk boxes.

This paper deals with a different type of unknown structure, where the system contains a vector of sub-structures, and the size of the vector is not known in advance. For example, the number of boxes in a rack depends on their dimensions and on their total power requirements compared to the power availability of the rack. In these cases, the size of the vector is related to or constrained by other variables in the problem. Therefore, the vector size can be modeled as a CSP variable.

The simple example in Fig. 1, which we use throughout the paper, illustrates the problem and the proposed solution. We used the constraint language of our internal CSP solver [5] for the problem description. The problem is taken from the domain of test program generation, where CSP is used to generate test programs that are valid, interesting, and fulfill specific user requests [4]. The example shows a CSP modeling of a test program template that verifies the memory sub-system of a processor. A test program generated from this template must include load and store instructions that access 4 K bytes of memory. Each instruction in the program is either a load instruction that reads from memory or a store instruction that writes to memory. The number of bytes accessed by each instruction is 64, 128, 256, or 512 bytes, with 512 bytes only possible in load instructions. A specific user request is that instruction 5 in the program reads 256 bytes.

```
1: class MemoryAccess
2:     variable int numOfBytes {64, 128, 256, 512};
3:     variable enum instructionType {load, store};
4:     constraint (numOfBytes = 512) → (instructionType = load);
5: end class;
6: variable int numMemInstructions;
7: vector MemoryAccess memAccessSeq[numMemInstructions];
8: constraint numMemInstructions < 100;
9: constraint SumOf(memAccessSeq[*].numOfBytes) = 4096;
10: constraint memAccessSeq[5].numOfBytes = 256;
11: constraint memAccessSeq[5].InstructionType = load;
```

Fig. 1. Simple program generation CSP

Lines 1–5 in the figure define the class of a memory accessing instruction. The class contains two variables, *numOfBytes* and *instructionType*, in lines 2 and 3, and a constraint that relates the number of bytes and the instruction type in line 4. The description of the problem contains a variable for the length of the program (line 6) and a vector of instructions for the generated program in line 7. The constraints in the description include a limit of 100 instructions on the program length in line 8, the specification of 4 K bytes access in line 9, and the specific user request in lines 10 and 11.

There are several known approaches to address such problems where the CSP solver has to find the number of elements in vectors of instances. The first possible solution is to use external heuristics to select the size of the vector before building the CSP. The simplest such heuristic is to randomly choose a vector length. It is also possible to use more sophisticated heuristics that are based on the analysis of the problem. For example, we deduce from the possible access length of the instructions in the program generation problem in Fig. 1 that the possible number of instructions is between 8 (if they all access 512 bytes) and 64 (if they all access 64 bytes). A more careful analysis can change this range from 9 to 61 instructions, because instruction 5 accesses 256 bytes. After applying the heuristics, we can randomly select the vector length from the set of possibilities it provides. The drawbacks of this approach are that it requires external analysis of the problem and it can lead to an infeasible vector length if the heuristics are not good enough.

Another possible solution is the use of conditional CSP [9]. In conditional CSP, the existence of some of the variables and constraints in the problem is dependent on the values of other variables. The concept of using conditional variables in CSPs was first introduced in [14]. This work was later enhanced and improved in many publications such as [9,11,20]. Sabin, Freuder, and Wallace suggest in [16] the use of a special null value in conditional variables as an alternative to the existence variables.

To model the program generation of Fig. 1 as a conditional CSP, we make each element in the instruction vector conditional and add two constraints that condition their existence, as shown in Fig. 2. The existence variable in line 2 is a special variable that indicates whether the element exists. If the element does not exist, the solver ignores all its variables and constraints. Lines 14 and 15 are existence constraints, which specify that consecutive instructions 0 to $numMemInstructions - 1$ exist, and that instruction 5 specifically exists. Note, in this paper we assume that constraints on specific elements in a vector (e.g., $memAccessSeq$ [5].$numOfBytes = 256$) imply their existence. This assumption is not required and it does not influence the proposed solution.

```
 1: class MemoryAccess
 2:     variable bool existence;
 3:     variable int numOfBytes {64, 128, 256, 512};
 4:     variable enum instructionType {load, store};
 5:     constraint (numOfBytes = 512) → (instructionType = load);
 6: end class;
 7: variable int numMemInstructions;
 8: vector MemoryAccess memAccessSeq[numMemInstructions];
 9: constraint numMemInstructions < 100;
10: constraint SumOf(memAccessSeq[*].numOfBytes) = 4096;
11: constraint memAccessSeq[5].numOfBytes = 256;
12: constraint memAccessSeq[5].InstructionType = load;
13: // Existence constraints:
14: constraint forEach (i, memAccessSeq[i].existence ⇔ (i < numMemInstructions));
15: constraint memAccessSeq[5].existence = true;
```

Fig. 2. Modeling the simple program generation as a conditional CSP

Using a conditional CSP is complete in the sense that it does not lead to the loss of solutions for the CSP. The drawback of this approach is that it can cause inefficiencies in constructing and solving the problem in terms of memory usage and compute time. This problem is significant, especially if the initial domain of the vector size is not tight and nested vectors exist. Moreover, this approach cannot handle cases when the initial domain of the vector size is unbounded. Handling unbounded sizes requires preprocessing that sets bounds.

Of course, the two approaches can be combined. If the heuristics provides a range of possible vector lengths, we create conditional elements of the vector only for the elements within the range. For example, if in our example the range provided by the heuristics is 6 to 10 elements, then we create 6 unconditional elements for elements 0–5, and 4 conditional elements for elements 6–9.

A third possible solution to the unknown vector size is the use of dynamic CSP techniques. Dynamic CSP allows us to add variables and constraints to the CSP after the solver starts the solution process. There are several dynamic CSP methods. For example, Mailharro [13] and Yokoo [21] present frameworks that use infinite domains to solve problems where the domains of the variables are not known in advance. Bessiere [3] enables adding and removing constraints during the solving of the CSP. The main disadvantage of dynamic CSP methods is their inability to propagate the constraints from vector elements that have not been added yet to the vector size. The proposed method overcomes this limitation by adding constraints that replace that global constraints of the missing elements to the skeleton CSP that determines that vector size.

A completely different solution involves converting the CSP with unknown vector size to an equivalent CSP over strings. Several propagation techniques that deal with strings of unknown length [19], or even unbounded length [8], have been proposed; some of them are used in CSP solvers. For example, support for strings is now part of MiniZinc [2]. Saxena et al. [17] use bounded length strings in the symbolic execution of JavaScripts.

The main idea in this method is to convert vectors of unknown size to strings. Each element in the vector is represented by a character in the string, and the alphabet of the string includes all the possible states of a single vector element. For example, to convert the program generation problem of Fig. 1 to a CSP over strings, the vector of instructions must be converted to a string with a varying size of up to 100 characters, using an alphabet containing 8 letters (7 if we include the constraint in line 4). This approach has two main disadvantages. First, it is not clear how to convert constraints in the original problem to efficient constraints on the string. For example, the *SumOf* constraint in line 9 needs to be converted to a long list of possible strings without an efficient way of pruning the string. The second disadvantage is that the size of the alphabet can be very large because of the number of possible solutions for a single vector element. Moreover, if vector elements contain vectors of unknown size, the size of the alphabet for the outer vector may be unknown or even unbounded.

3   Our Solution

The first step in solving a CSP is to read the input description or model of the problem and to create an equivalent data structure for the CSP solver. After the data structure is constructed, the CSP solver can be activated to solve it. The CSP construction step is not trivially possible when the description contains vectors of unknown or even unbounded size. Our method addresses this problem by constructing the full CSP in stages. Specifically, we construct the full CSP while solving sub-problems related to the unknown vector sizes.

Using a staged solution of CSPs is not new. It is used in many cases when the full CSP is too big or complex to be solved as a single problem [15]. The novelty of our proposed method lies in its application to CSP problems that cannot be constructed because of the unknown structure of the CSP and the unknown vector size.

Procedure 1 shows a high-level description of the proposed method. The first step (line 2 of the procedure) is to build a structural skeleton of the CSP. This skeleton is then expanded until it contains the full CSP with a known structure, or, in other words, all the unknown factors in the CSP structure are resolved and removed.

```
 1: procedure SOLVECSPWITHUNKNOWNVECTORSIZE(inputDescription)
 2:     root ← ConstructSkeleton(inputDescription);
 3:     ExpandKnownVectors(root);

 4:     lastNode ← root;
 5:     while ((lastNode ← DfsSearchUnknownVectors(lastNode) ) ≠ NULL) do
 6:         sizeProblem ← ConstructVectorSizeProblem (lastNode);
 7:         vectorSize ← SolveVectorSize(sizeProblem);
 8:         if (No solution found) then
 9:             backtrack;
10:         end if
11:         SetSize(lastNode, vectorSize);
12:         ExpandKnownVectors(lastNode);
13:     end while

14:     SolveFullCsp(root);
15:     if (no solution found) then
16:         backtrack;
17:     end if
18: end procedure
```

Procedure 1: Procedure for solving a CSP with vectors of unknown size

The structural skeleton is a tree of the structure of the CSP. It is only a skeleton because to begin with, it does not contain all the elements of the CSP. The root of the tree points to the entire problem, sub-trees starting at internal nodes are sub-structures in the problem, and the leaves are CSP variables. Constraints are hyperedges connecting leaves in the graph. Constraints are not part of the structure tree.

Vectors are special internal nodes in the graph. A vector node has two types of children: a leaf CSP variable for its size and nodes for special elements in the vector. These special nodes include a representative element of the vector, and one for each element that is explicitly referred to by name in the problem description. For example, $memAccessSeq[5]$ in Fig. 1. The representative element is a source of information and constraints that are used in the vector size CSPs. For example, the domain of the $numOfBytes$ variable in the representative element of $memAccessSeq$ is used when projecting the $SumOf$ constraint to the size of $memAccessSeq$. When building the skeleton graph, we do not need to know the vector size or even a bound on it, because only a fixed number of elements (that do not depend on the vector size) appear in the graph.

The construction of the skeleton graph is straightforward, therefore, we do not discuss it in the text. Figure 3 shows the skeleton graph for the simple program generation of Fig. 1. In the figure, ovals represent nodes in the structure tree and red rectangles represent constraints. The root of the problem contains two children, the *memAccessSeq* vector and *numMemInstructions* variable. The vector *memAccessSeq* has a *vecSize* child that is connected with the equality constraint to *numMemInstructions*, and two children for the vector elements of type *MemoryAccess*: one for the representative element and the second for *memAccessSeq*[5] that is specifically referred to in lines 10 and 11 of Fig. 1. Each *MemoryAccess* node has two leaf children for the *numOfBytes* and *instructionType* CSP variables in them.

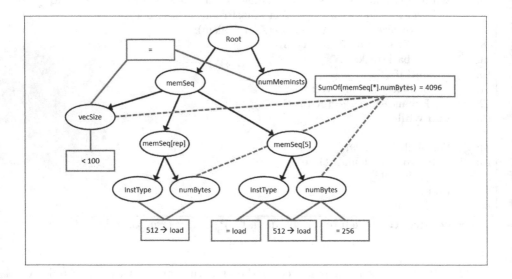

Fig. 3. CSP skeleton for the program generation problem

After construction of the skeleton graph, the next step is to expand all the vectors whose sizes were known during the construction of the graph to their actual size. This is done by the *ExpandKnownVectors* method in line 3. The method replaces the size node with a constant indicating the vector size, removes the representative element, and adds element nodes for each vector element that is not included in the original skeleton. This step may add new vectors with unknown size as descendants of vector elements.

Lines 4–13 are the main part of the algorithm. The "DfsSearchUnknown-Vectors" function in line 5 performs a pre-ordering DFS search for vectors with unknown sizes. The function searches for vector nodes in the graph whose size child is not yet determined. When such a node is found, the function returns it to the main algorithm. When the function does not find such a node, that is, when all the vectors have known sizes, the function returns NULL and the main loop terminates. After a vector with unknown size is found, we construct and solve a vector size CSP problem for the vector (lines 6–10). If we fail to find a solution, we backtrack to the previous vector and try again, or declare the problem unsatisfiable if this is the first vector. We describe the heart of the algorithm in detail in the next subsection.

After finding a possible size for the vector in the vector size problem, in lines 11–12, we set the size in the skeleton and expand the vector to include all elements, as described earlier. Note that setting the vector size may affect the size of other vectors whose sizes are unknown, for example via constraints relating the sizes of two vectors.

After the algorithm finishes the search for vectors with unknown size, the resulting skeleton graph becomes a graph containing the full CSP. For example, Fig. 4 shows the full CSP for the program generation problem with 10 instructions. Now, in lines 14–17 of the algorithm, we can solve the full CSP and obtain the requested solution.

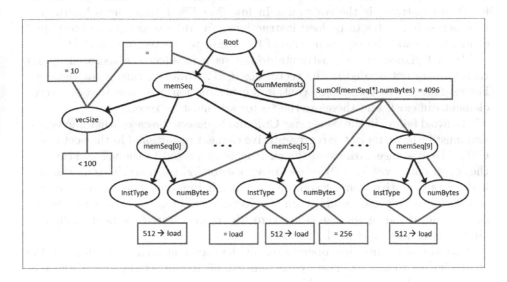

Fig. 4. Full CSP for the program generation problem

3.1 Constructing and Solving a Vector Size Sub-problem

The heart of our method is the extraction of a vector size CSP for a vector of unknown size. This CSP should include the factors that can affect the size. These factors can be divided into two main types: other variables in the problem outside the vector that affect the vector size, and constraints on elements in the vector. These constraints can be global constraints on all (or some) of the elements of the vector, or constraints on specific elements.

Other variables in the problem related to the vector size can obviously affect the vector size. For example, if a vector size N with initial domain 1–10 is related to another variable M with initial domain 1–5 with equality constraint, then the vector size cannot be greater than 5. It is easy to find the variables and constraints that directly or indirectly affect the vector size, by finding the transitive closure of CSP variables and constraints in the skeleton graph, for example using a BFS search from the vector size variable.

As stated in Sect. 2, we assume that a constraint on a specific element in a vector implies that this element in the vector exists, therefore, it affects the vector size. Hence, the constraint $memAccessSeq[5].numOfBytes = 256$ in the skeleton is reflected in the vector size CSP as the constraint $vecSize \; \dot{\iota} \; 5$ (we assume that the index of the first element is 0).

Global constraints on all the vector elements can be divided into two categories, constraints that operate on each element in the vector individually and constraints that operate on all the elements together. Constraints of the first category appear in the definition of the class of the vector elements. An example for such constraint is the constraint in line 5 of Fig. 1 that forces instruction that access 512 bytes to be load instructions. To add constraints on individual elements, we add the representative of the vector to the vector size CSP.

At first glance, these constraints do not seem to affect the vector size, and therefore are not needed in the vector size CSP. However, this is not the case. For example, if the constraints contain a contradiction, it means that no vector element can exist, and therefore the vector size must be zero.

To avoid failure in the vector size CSP and to detect the cases when the vector size must be 0, we make the representative element conditional in the vector size CSP. The existence variable of the element is connected to the vector size with the constraints $(vecSize > 0) \leftrightarrow vector[representative].existence$. This means that if the representative element contains a contradiction and its existence variable is set to false to avoid failure, then the vector size is forced to be zero. On the other hand, if the vector size is not zero, the representative element exists and all its constraints must hold.

Global constraints that operate on all the vector elements together are the most difficult to handle. These constraints implicitly affect the size of the vector in a 'chicken and egg problem' manner. Namely, the vector size depends on the vector elements, but we cannot construct the elements until we know the vector size. Our solution is to create a projection of the global constraints on the vector size and use these projections and information on the vector elements in the vector size CSP as constraints on the vector size. For example, the constraint $SumOf(memAccessSeq[*].numOfBytes) = 4096$ in line 10 of Fig. 1 and the minimum and maximum value of $numOfBytes$ imply that the vector size cannot be smaller than 8 or larger than 64. Therefore, the constraint $8 \leq memAccessSeq.vecSize \leq 64$ is added to the vector size problem.

The projections of these constraints to the vector size are not accurate. We use overapproximated projections, so that we do not lose solutions of possible vector sizes in the solution process.

The following examples show some commonly used global constraints on the vector elements that can affect the vector size and their projection on the vector size. Note that in the list below we use the simple expression consisting of a single variable in a vector element for clarity. More complex expressions can also be used.

Exists (predicate of vector.V). This operator gets a set of predicates on elements of the vector and ensures that there will be at least one vector element for which the predicate holds. The fact that the predicate for a vector element holds implies that this element exists, and thus, the vector is not empty. Therefore, the projection for this operator is the constraint $vecSize \geq 1$.

Similar projection is also used in other global operators that imply existence of at least one element in the vector, such as *MinOf* and *MaxOf* that calculate the minimum and maximum of variables in the vector.

AtLeast (predicate of vector.V, limit). This operator holds if the number of vector elements on which the predicate holds is at least *limit*. Therefore, its projection is $vecSize \geq limit$.

AllDiff (vector.V). This operator gets a set of variables as input and forces each of them to have a different value. This operator affects the size of the vector because it means that the size of the vector cannot be greater than the domain size of the variable. Therefore, when this operator is encountered, we replace it with the following constraint on the number of elements: $vecSize \leq |domain\ of\ variable|$. This projector assumes that the different input variables have the same domain.

SumOf (vector.V). This operator gets a set of variables as input and calculates their sum. It enables the modeler to write a constraint on the sum of their values (e.g., $SumOf(vector.V) = T$, where T is the target value). The relation between the minimal and maximal values of V and the minimal and maximal values of T can affect the range of feasible vector sizes. For example, if all the values in the domain of V and T are of the same sign, the following relation must hold: $\min(\frac{T}{V}) \leq vecSize \leq \max(\frac{T}{V})$, where $\min(\frac{T}{V})$ and $\max(\frac{T}{V})$ are the minimal and maximal values of all possible divisions of T and V. If V has both positive and negative values in its domain, or it has the value 0, then the *SumOf* constraint does not impose any upper bound on the vector size. The minimal vector size is the minimum of M_{pos} and M_{neg}, where M_{pos} (M_{neg}) is the minimal value of $\frac{T}{V}$ for only the positive (negative) values in V and T.

Note that the projections above are valid when the constraints have positive polarity. When the constraints are negated, their projection on the vector size is different. For example, the constraint *NotExists(predicate of vector.V)* (or *not(Exists(predicate of vector.V))*) does not imply anything on the vector size because this constraint holds if the vector is empty. Therefore, it is not included in the vector size CSP. Similarly, there are other global constraints that do not affect the vector size and therefore do not require a projection in the vector size CSP. Examples for these include:

AllSame (vector.V). This operator gets a set of variables and forces all of them to get the same value.

AtMost (predicate of vector.V, limit). This operator holds if the number of vector elements on which the predicate holds is at most limit.

Finally, there are global constraints with an overly complex projection to the vector size that prevent us from getting a good overapproximated bound. In these cases, we prefer not to project at all, which may increase the failure probability, over an under-approximated projection. For example, the $Monotonic(vector.V)$, which ensures that variables V in the vector elements is monotonically not decreasing, cannot be easily projected in an over-approximation projection. Therefore, we prefer to ignore this constraint instead of using an under-approximated projection.

Figure 5 shows the vector size CSP for the size of the $memAccessSeq$ vector derived from the skeleton in Fig. 3. Note the constraint $\frac{4096}{512} \leq vecSize \leq \frac{4096}{64}$ that replaced the $SumOf$ constraint in the CSP skeleton of Fig. 3, the new constraint $vecSize > 5$ caused by the specific reference to $memAccessSeq[5]$ in the problem description, and the existence variable added to the representative element.

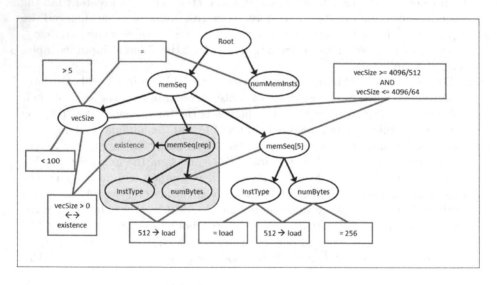

Fig. 5. Vector size CSP for the memAccessSeq vector

4   Experimental Results

We implemented the proposed method as part of our internal CSP engine [5,15]. It is currently being used in two applications: test-program generation and data fabrication for database systems. Our CSP engine uses a declarative language for describing CSP problems [5]. The language supports logical and arithmetic constraints over a rich set of data types, global operations on sets of variables, assembly of complex constraints, nullable variables, declaration of classes, including inheritance and other useful features. The examples in Figs. 1 and 2 use the syntax of this declarative language. The CSP solver [15] is based on the MAC algorithm [12], but contains many additions, adaptions, and optimizations needed for its main use, namely test and content generation. Our CSP solver provides random solutions to a CSP [4]. This feature is important because in many use cases the engine must provide a large number of different solutions to the same problem. Other key features include soft constraints, conditional CSP, and support of very large domains [4].

We could not test our method against standard CSP benchmarks, such as CSPLib [7], because these benchmarks do not include problems that contain vectors of unknown size. Instead, the experimental results presented in this section include three problems, one synthetic and two that are taken from the applications that use the proposed method. In each experiment, we compared the running time of the CSP engine and its memory consumption when the proposed method is used for the two alternatives of conditional CSP and random guess of vector sizes. In many of the problems we deal with, it is hard for the problem modeler to determine a limit to vector sizes because of the complexity of the constraints, or because they are highly dependent on user parameters. As a result, in almost all such problems, vector sizes are left unbounded. This makes the two alternative solutions unfeasible. To overcome this issue, we used a two-step solution method: first setting an upper bound to all vector sizes, and only then constructing and solving the CSP. The upper bounds that we used in the experiments are 10, 100, and 1000. We present the results for each bound. Note that our method does not require this initial step. This is one of the advantages of our method over other existing solutions.

The three problems we used in the experiments are:

Synthetic problem: This problem is like the one shown in Fig. 1, except it does not include line 8, which sets a bound to the vector size. This was done to match the other two problems. The CSP for this problem contains $2N + 1$ variables and $N + 3$ constraints, where N is the vector size.

Data fabrication problem: The problem involves fabricating data to test a database application for financial transactions. The data contains people and their financial transactions. Therefore, the problem contains a vector of persons with unknown size, and each entry in this vector contains a vector of unknown size for transactions. The fabricated data has to satisfy legality constraints and some statistic properties, all of which are described as constraints. Specifically, there are four constraints that sum up various variables in all the transactions in the problem, and four repeating constraints that sum variables for the transactions of each person.

Each transaction sub-problem contains 15 variables and 12 constraints. Each person contains, in addition to its transactions, 15 variables and 10 constraints. There are also 21 global constraints. Therefore, a problem with 10 people and an average of 13 transactions per person contains $21 + 10 \cdot (10 + 13 \cdot 12) = 1681$ constraints and $10 \cdot (15 + 13 \cdot 15) = 2100$ variables.

Test generation problem: The goal is to generate four instruction streams (note that this is a vector with fixed size). Each stream contains between 1 and 32 instructions. There are two types of instructions, which are reflected in two CSP sub-problems, one with one variable and one constraint and one with two variables and two constraints.

The CSP contains 1 global variable and 1 global constraint. Each stream contains 9 variables and 7 constraints in addition to the variables and constraints of the individual instructions. Therefore, a problem with an average stream size of 16 where the instructions are equally split between the 2 types contains $1 + 4 \cdot (9 + 16 \cdot 1.5) = 133$ variables and $1 + 4 \cdot (7 + 16 \cdot 1.5) = 125$ constraints.

The size of the problems is provided for the base problem. It does not include additions needed for the three solution methods, such as the vector size CSPs in the proposed method and the additional existence variable and constraints needed by the conditional CSP method (For example, see Fig. 2).

We ran these experiments on a virtual Linux machine with an x86_64 Intel processor that is part of a large server. While the server has 239 GB of memory, the CSP engine that was used in the experiments is limited to 4 GB. To avoid an infinite running time, we set several limits. First, we set a limited time and number of backtracks for each CSP the engine tried to solve. These limits were never reached in the experiments, meaning that each CSP either found a solution or reported that the problem is unsatisfiable. Second, we set a limit of 1000 to the number of attempts to select random vector sizes in the random size method.

Because our CSP engine provides random solutions to the CSPs, we ran each experiment six times. The reported results are the average of these runs. Tables 1 and 2 summarize the results. Table 1 shows the average execution time of each problem and Table 2 shows average memory consumption.

The results show that our method always outperformed the other two methods in both execution time and memory consumption when the bound on the vector sizes is high (100 or 1000). There are two reasons for this. The random size method can randomly select a vector size that is not feasible, leading to failure in the CSP solution and a retry. This issue becomes more severe when the bound on the vector size is not tight (as in the 1000 bound case where there are only a few legal solutions) or when the number of vectors with unknown sizes is large. The conditional CSP method has to construct and solve the CSP with the maximal number of vector elements. Therefore, it consumes more memory

Table 1. Experimental results - average execution time [seconds]

Method	Random size			Conditional CSP			Ours
Bound	10	100	1000	10	100	1000	
Synthetic	0.020	0.208	81.753	0.027	0.131	7.118	0.025
Data fabrication	6.448	4678.331	Timeout	24.321	N/A	N/A	1.674
Test generation	0.059	17.482	Timeout	0.077	0.315	2.998	0.112

Table 2. Experimental results - average memory consumption [MB]

Method	Random size			Conditional CSP			Ours
Bound	10	100	1000	10	100	1000	
Synthetic	6.2	6.6	10.4	6.4	7.9	22.9	6.3
Data fabrication	40.8	278.9	4096.0	84.0	Out-of-memory	Out-of-memory	38.0
Testgeneration	7.5	14.6	217.6	8.2	17.0	101.3	8.3

and takes longer to solve than our method. In fact, the memory requirements in the data fabrication experiment were so high, that it ran out of memory for bounds of 100 and 1000.

With a low bound of 10, the random size method outperformed our method in both the synthetic and test generation examples. This is because in these two problems, each size in the bound is feasible. The problem with this too-tight bound, which is extremely important in our domain that requires random solutions, is that many solutions with vector sizes greater than 10 cannot be reached. Note that because the data fabrication experiment contains only a single feasible size for the people vector (4), our method outperforms the random size with the tight bound.

5   Conclusions

We presented a method that enables solving CSPs that have vectors of unknown sizes, even when there is no obvious maximal length to the vector. The heart of our method lies in the construction of vector size CSPs that project the full problem to a smaller problem of finding the unknown size of a vector. The obtained vector sizes are then used to construct and solve a CSP with a fully known structure. We showed that in some real-life problems in which vector sizes are controlled by constraints, our method outperforms other existing solving methods.

Although we successfully deployed our proposed method, we continue to work primarily on improving the precision of projection of global constraints to the vector size in the vector size problem, adding new constraints that are currently not supported, and adding other features in our constraints language to this framework.

References

1. Adir, A., Levy, R., Salman, T.: Dynamic test data generation for data intensive applications. In: Eder, K., Lourenço, J., Shehory, O. (eds.) HVC 2011. LNCS, vol. 7261, pp. 219–233. Springer, Heidelberg (2012). doi:10.1007/978-3-642-34188-5_19
2. Amadini, R., Flener, P., Pearson, J., Scott, J.D., Stuckey, P.J., Tack, G.: Minizinc with strings. arXiv preprint arXiv:1608.03650 (2016)

3. Bessiere, C.: Arc-consistency in dynamic constraint satisfaction problems. In: Proceedings of the Ninth National Conference on Artificial Intelligence, pp. 221–226, July 1991
4. Bin, E., Emek, R., Shurek, G., Ziv, A.: Using a constraint satisfaction formulation and solution techniques for random test program generation. IBM Syst. J. **41**(3), 386–402 (2002)
5. Bin, E., Venezian, E.: Solving the address translation problem as a constraint satisfaction problem. In: CP Meets Verifiation Workshop of the 20th International Conference on Principles and Practice of Constraint Programming, September 2014
6. Dechter, R.: Constraint Processing. Morgan Kaufmann, Burlington (2003)
7. Gent, I., Walsh, T.: CSPLib: a problem library for constraints. http://www.csplib.org. Accessed 24 Apr 2017
8. Golden, K., Pang, W.: Constraint reasoning over strings. In: Rossi, F. (ed.) CP 2003. LNCS, vol. 2833, pp. 377–391. Springer, Heidelberg (2003). doi:10.1007/978-3-540-45193-8_26
9. Gottlob, G., Greco, G., Mancini, T.: Conditional constraint satisfaction: logical foundations and complexity. In: Proceedings of the Twentieth International Joint Conference on Artificial Intelligence, pp. 88–93, January 2007
10. Katz, Y., Rimon, M., Ziv, A.: Generating instruction streams using abstract CSP. In: Proceedings of the 2012 Design, Automation and Test in Europe Conference, pp. 15–20, March 2012
11. Keppens, J., Shen, Q.: Compositional model repositories via dynamic constraint satisfaction with order-of-magnitude preferences. J. Artif. Intell. Res. **21**, 499–550 (2004)
12. Mackworth, A.: Consistency in networks of relations. Artif. Intell. **8**(1), 99–118 (1977)
13. Mailharro, D.: A classification and constraint-based frame-work for configuration. Artif. Intell. Eng. Des. Anal. Manuf. J. **12**(4), 383–397 (1998)
14. Mittal, S., Falkenhainer, B.: Dynamic constraint satisfaction. In: Proceedings of the Eighth National Conference on Artificial Intelligence, pp. 25–32, July 1990
15. Naveh, Y., Rimon, M., Jaeger, I., Katz, Y., Vinov, M., Marcus, E., Shurek, G.: Constraint-based random stimuli generation for hardware verification. AI Mag. **28**(3), 13–30 (2007)
16. Sabin, M., Freuder, E.C., Wallace, R.J.: Greater efficiency for conditional constraint satisfaction. In: Rossi, F. (ed.) CP 2003. LNCS, vol. 2833, pp. 649–663. Springer, Heidelberg (2003). doi:10.1007/978-3-540-45193-8_44
17. Saxena, P., Akhawe, D., Hanna, S., Mao, F., McCamant, S., Song, D.: A symbolic execution framework for JavaScript. In: IEEE Symposium on Security and Privacy, pp. 513–528, May 2010
18. Schenner, G., Taupe, R.: Encoding object-oriented models in MiniZinc. In: Fifteenth International Workshop on Constraint Modelling and Reformulation, September 2016
19. Scott, J.D., Flener, P., Pearson, J.: Constraint solving on bounded string variables. In: Proceedings of the 12th International Conference on Integration of AI and OR Techniques in Constraint Programming, pp. 375–392, May 2015
20. Soininen, T., Gelle, E., Niemelä, I.: A fixpoint definition of dynamic constraint satisfaction. In: Jaffar, J. (ed.) CP 1999. LNCS, vol. 1713, pp. 419–433. Springer, Heidelberg (1999). doi:10.1007/978-3-540-48085-3_30
21. Yokoo, M.: Asynchronous weak-commitment search for solving distributed constraint satisfaction problems. In: Montanari, U., Rossi, F. (eds.) CP 1995. LNCS, vol. 976, pp. 88–102. Springer, Heidelberg (1995). doi:10.1007/3-540-60299-2_6

An Efficient SMT Approach to Solve MRCPSP/max Instances with Tight Constraints on Resources

Miquel Bofill, Jordi Coll$^{(\boxtimes)}$, Josep Suy, and Mateu Villaret

University of Girona, Girona, Spain
{miquel.bofill,jordi.coll,josep.suy,mateu.villaret}@imae.udg.edu

Abstract. The Multi-Mode Resource-Constrained Project Scheduling Problem with Minimum and Maximum Time Lags (MRCPSP/max) is a generalization of the well known Resource-Constrained Project Scheduling Problem. Recently, it has been shown that the benchmark datasets typically used in the literature can be easily solved by relaxing some resource constraints, which in many cases are dummy. In this work we propose new datasets with tighter resource limitations. We tackle them with an SMT encoding, where resource constraints are expressed as specialized pseudo-Boolean constraints and then translated into SAT. We provide empirical evidence that this approach is state-of-the-art for instances highly constrained by resources.

1 Introduction

The Resource-Constrained Project Scheduling Problem (RCPSP) is the problem of finding the start times of a set of non-preemptive activities while respecting some constraints, namely precedence relations between activities and correct usage of shared resources with limited capacity. There exist many extensions of this problem [5], e.g., involving multiple execution modes per activity (MRCPSP), or generalized precedence relations (RCPSP/max). The exact methods with the best known performance in such scheduling problems are based in Lazy Clause Generation [13,16], Failure-Directed Search (FDS) [17] and SAT Modulo Theories (SMT) [1,3,4].

In this work we address the Multi-mode Resource-Constrained Project Scheduling Problem with Minimal and Maximal Time Lags (MRCPSP/max),[1] which combines MRCPSP and RCPSP/max. The goal is to determine a start time and an execution mode for each activity, in order to obtain a schedule which satisfies all the resource and generalized precedence constraints, and which has minimum makespan (i.e., the total duration of the whole set of activities). Up to our knowledge, the best exact approaches for this problem are based in Constraint Integer Programming with cumulative constraint handlers [12] and FDS [17].

---

[1] This problem is denoted $MPS|temp|C_{max}$ in [5] and $m, 1|gpr|C_{max}$ in [8]. It is also known as the Multi-mode RCPSP with Generalized Precedence Relations.

© Springer International Publishing AG 2017
J.C. Beck (Ed.): CP 2017, LNCS 10416, pp. 71–79, 2017.
DOI: 10.1007/978-3-319-66158-2_5

Most of the recent works on MRCPSP/max have been evaluated using the benchmark datasets created in [14]. In [17] it was observed that, in those instances, the resource constraints are not the hardest component. In this work, we further analyse the reason why the resource constraints seem not to play an important role in those instances, concluding that such constraints are trivially satisfied in many cases. Therefore, we have crafted new instance sets, where resource constraints take a more important role. Moreover, we present an SMT-based system to solve the MRCPSP/max. We use Integer Difference Logic (IDL) to deal with generalized precedences, and recently introduced specialized pseudo-Boolean (PB) constraints [4] to deal with resource constraints. Such PB constraints allow compact encodings into SAT, and have been shown to improve performance in the MRCPSP. We provide experiments showing that our SMT approach performs better than other exact approaches on instances with tight constraints on resources.

2 The Multi-mode RCPSP with Minimum and Maximum Time Lags (MRCPSP/max)

The MRCPSP/max is defined by a tuple (V, M, p, E, g, R, B, b) where:

- $V = \{A_0, A_1, \ldots, A_n, A_{n+1}\}$ is a set of non-preemptive activities. A_0 and A_{n+1} are dummy activities representing the start and the end of the schedule, respectively. The set of non-dummy activities is defined as $A = \{A_1, \ldots, A_n\}$.
- $M \in \mathbb{N}^{n+2}$ is a vector of naturals, being M_i the number of modes in which activity i can be executed. $M_0 = M_{n+1} = 1$ and $M_i \geq 1, \forall A_i \in A$.
- p is a vector of vectors of naturals, being $p_{i,o}$ the duration of activity i when is executed in mode o, with $1 \leq o \leq M_i$, $p_{0,1} = p_{n+1,1} = 0$.
- E is a set of pairs of activities which have a time lag defined.
- g is a four dimensional vector of naturals, being $g_{i,j,o,o'}$ the time lag from activity A_i in mode o to activity A_j in mode o'. If $g_{i,j,o,o'} >= 0$, it is a minimum time lag, and means that if A_i is running in mode o and A_j is running in mode o', then A_j must start at least $g_{i,j,o,o'}$ units of time after the start of A_i. If $g_{i,j,o,o'} < 0$, it is a maximum time lag, and means that A_i must start at most $|g_{i,j,o,o'}|$ units of time after the start of A_j.
- $R = \{R_1, \ldots, R_{v-1}, R_v, R_{v+1}, \ldots, R_q\}$ is a set of resources. The first v resources are renewable, and the last $q - v$ resources are non-renewable.
- $B \in \mathbb{N}^q$ is a vector of naturals, being B_k the capacity of resource R_k.
- b is a matrix of naturals corresponding to the resource demands of activities per mode: $b_{i,k,o}$ represents the amount of resource R_k required by activity A_i in mode o, $b_{0,k,1} = 0$ and $b_{n+1,k,1} = 0, \forall k \in \{1, \ldots, q\}$. For renewable resources, this is the required amount per time step, whilst for non-renewable resources, it is the total amount required by the activity during its execution.

A schedule is a vector of naturals $S = (S_0, \ldots, S_{n+1})$ where S_i denotes the start time of activity A_i, $S_0 = 0$, and S_{n+1} is the makespan (completion time of the project). A schedule of modes is a vector of naturals $SM =$

$(SM_0, \ldots, SM_{n+1})$ where SM_i, satisfying $1 \leq SM_i \leq M_i$, denotes the mode of each activity A_i. A solution of the MRCPSP/max problem is a schedule of modes SM and a schedule S of minimal makespan S_{n+1} satisfying the generalized precedence (1), non-renewable (2) and renewable (3) resource constraints:

$$((SM_i = o) \wedge (SM_j = o')) \rightarrow (S_j - S_i \geq g_{i,j,o,o'})$$
$$\forall (A_i, A_j) \in E, \forall o \in [1, M_i], \forall o' \in [1, M_j] \quad (1)$$

$$\left(\sum_{A_i \in A} \sum_{o \in [1, M_i]} ite(SM_i = o; b_{i,k,o}; 0) \right) \leq B_k \quad \forall R_k \in \{R_{v+1}, \ldots, R_q\} \quad (2)$$

$$\left(\sum_{A_i \in A} \sum_{o \in [1, M_i]} ite((SM_i = o) \wedge (S_i \leq t) \wedge (t < S_i + p_{i,o}); b_{i,k,o}; 0) \right) \leq B_k$$
$$\forall R_k \in \{R_1, \ldots, R_v\}, \forall t \in H \quad (3)$$

where $ite(c; e_1; e_2)$ is an *if-then-else* expression denoting e_1 if c is true and e_2 otherwise, $H = \{0, \ldots, T\}$ is the scheduling horizon, and T (the length of the scheduling horizon) is an upper bound (UB) for the makespan.

3 Formulation

We propose a SAT modulo Integer Difference Logic formulation to solve the MRCPSP/max. This is a formulation for the decision version of the problem, i.e., it models the problem of finding a feasible schedule for an MRCPSP/max instance whose makespan is smaller or equal than a given UB. The optimization is achieved by successive calls to an SMT solver, as described in Sect. 4.

In our formulation, we use some precomputed values which serve to bound the execution times of the activities. Given an UB, these values can be recomputed according to different criteria [2]; here we consider time lag constraints between activities as done in [12]. By ES_i we denote the earliest time instant at which A_i can start, by LS_i the latest time instant at which A_i can start, and by LC_i the latest time instant at which A_i can end.

We use a set of integer variables $\{S_0, \ldots, S_{n+1}\}$ to denote the start time of each activity. We represent the schedule of modes with the set of Boolean variables $\{sm_{i,o} \mid 0 \leq i \leq n + 1, 1 \leq o \leq M_i\}$, being $sm_{i,o}$ true if and only if activity A_i is executed in mode o. The constraints are the following:

$$S_0 = 0 \qquad\qquad\qquad\qquad\qquad\qquad\qquad\qquad\qquad\qquad (4)$$

$$ES_i \leq S_i \leq LS_i \qquad\qquad\qquad \forall A_i \in \{A_1, \ldots, A_{n+1}\} \qquad (5)$$

$$(sm_{i,o} \wedge sm_{j,o'}) \rightarrow (S_j - S_i \geq g_{i,j,o,o'}) \quad \forall (A_i, A_j) \in E,$$
$$\forall o \in [1, M_i], \forall o' \in [1, M_j] \qquad (6)$$

$$\bigvee_{\forall o \in [1, M_i]} sm_{i,o} \qquad\qquad\qquad \forall A_i \in V \qquad\qquad\qquad (7)$$

$$\neg sm_{i,o} \vee \neg sm_{i,o'} \qquad\qquad \forall A_i \in A, \forall o, o' \in [1, M_i], o < o' \quad (8)$$

where (5) sets the earliest and latest start time of each activity, (6) encodes the generalized precedences and (7) and (8) ensure that each activity runs in exactly one mode—(8) is an at-most-one (AMO) constraint. The formulation of the constraints over resources is a time-indexed formulation [11], where we introduce the set of Boolean variables $x_{i,t,o}$, which are true if and only if activity A_i is being executed in mode o at time t in the schedule:

$$x_{i,t,o} \leftrightarrow (S_i \leq t < S_i + p_{i,o} \wedge sm_{i,o}) \quad \forall A_i \in A, \forall t \in [ES_i, LC_i), \forall o \in [1, M_i] \quad (9)$$

We can express resource constraints as pseudo-Boolean (PB) constraints. A PB constraint has the form $q_1 \cdot x_1 + \cdots + q_n \cdot x_n \# K$, where q_i and K are integer constants, x_i are 0/1 (false/true) variables, and $\# \in \{<, \leq, =, \geq, >\}$ [7]. A well-known approach to encode PB constraints is to represent them as Binary Decision Diagrams (BDDs), and then encode such BDDs into SAT [7]. This method was successfully applied to encode resource constraints for the MRCPSP in [3]. Recently, in [4] it was proposed a way to compactly encode PB constrains into SAT when their variables can be organized in groups that have AMO constraints already enforced. These PB constraints, called AMO-PB constraints, are built up from AMO-products.

Definition 1 (AMO-product). *We refer to an integer linear expression $q_1 \cdot x_1 + \cdots + q_m \cdot x_m$ over 0/1 variables $x_1, \ldots, x_m$, subject to the fact that at most one x_i is true, as an AMO-product. We conveniently express AMO-products as QX, where $Q = \langle q_1, \ldots, q_m \rangle$ and $X = \langle x_1, \cdots, x_m \rangle$.*

Definition 2 (AMO-PB). *We refer to an expression of the form $Q_1 X_1 + \cdots + Q_n X_n \leq K$, where $Q_i X_i$ are AMO-products and K is an integer constant, as a PB constraint with AMO relations (AMO-PB).*

Notice that an AMO-PB can be seen as a partial function, whose value is undefined if the AMO relation does not hold for some X_i.

The key idea in the use of AMO-PBs is that, if the AMO constraints over the variables of each AMO-product are already enforced, then the SAT encoding of the AMO-PB does not need to forbid the inconsistent assignments which do not satisfy the AMO constraints. In some formulations the needed AMO constraints are implicitly enforced, and hence there is no need to add additional clauses to the encoding. In our formulation of the MRCPCP/max, Constraints (8) explicitly enforce that at most one of all the variables $sm_{i,o}$ can be true for a particular activity A_i. Similarly, at most one of all the variables $x_{i,t,o}$ can be true for a particular activity A_i and time t, i.e., an activity can be running in at most one execution mode o at a time. Note, however, that the AMO relation of variables $x_{i,t,o}$ for an activity A_i and time t follows from the conjunction of Constraints (8) and (9). Interestingly, the generalized precedence relations (6) can introduce further implicit AMO relations between variables $x_{i,t,o}$. Let us consider a case in which $(A_i, A_j) \in E$, and it holds that $g_{i,j,o,o'} \geq p_{i,o}$, for all execution modes o and o'. In such cases we will say that A_i and A_j have an *end-start* precedence, meaning that A_j will always start after A_i has ended, so they will never be running at a same time. Therefore, in this case $x_{i,t,o}$ and $x_{j,t,o'}$ cannot be both true, because an AMO constraint is implicitly enforced by Constraints (6) and (9). We can take profit of all these explicit AMO constraints over $sm_{i,o}$, and the implicit AMO relations over $x_{i,t,o}$, to express the resource constraints as AMO-PBs.

Similarly to what is done in [4], we precompute a set $\mathcal{P}(t) = \{P_1, \ldots, P_m\}$ for each time t, where all P_j in $\mathcal{P}(t)$ are disjoint sets of activities, and $P_1 \cup \cdots \cup P_m$ contains all the activities A_i which can be running at time t according to their ES_i and LC_i. Moreover, all the activities in a set P_j are pairwise mutually exclusive due to end-start precedences. Hence, there is an AMO relation enforced over the variables $x_{i,t,o}$ of all activities A_i in each P_j. Then, the constraints over renewable resources can be formulated as AMO-PBs, where the j-th AMO-product contains the variables $x_{i,t,o}$ for all $A_i \in P_j$, $o \in [1, M_i]$:

$$\sum_{P_j \in \mathcal{P}(t)} Q(P_j, k) \cdot X(P_j, k, t) \leq B_k \qquad \forall R_k \in \{R_1, \ldots, R_v\}, \forall t \in [0, H] \quad (10)$$

where
$$Q(P_j, k) = \langle b_{i,k,o} \mid A_i \in P_j \wedge o \in [1, M_i] \wedge b_{i,k,o} > 0 \rangle$$

$$X(P_j, k, t) = \langle x_{i,t,o} \mid A_i \in P_j \wedge o \in [1, M_i] \wedge b_{i,k,o} > 0 \rangle$$

The non-renewable resource constraints 2 can also be represented using AMO-PB constraints. In this case, the i-th AMO-product will contain all the variables $sm_{i,o}$ for all $o \in [1, M_i]$:

$$\sum_{A_i \in A} Q(A_i, k) \cdot X(A_i, k) \leq B_k \qquad \forall R_k \in \{R_{v+1}, \ldots, R_q\} \quad (11)$$

where
$$Q(A_i, k) = \langle b_{i,k,o} \mid o \in [1, M_i] \rangle \qquad X(A_i, k) = \langle sm_{i,o} \mid o \in [1, M_i] \rangle$$

AMO-PB constraints (10) and (11) are encoded into SAT as described in [4].

4 Optimization Procedure

The formulation in Sect. 3 models whether it is possible to find a schedule whose makespan is not greater than a given UB. Some SMT solvers have built-in optimization mechanisms which let to specify an objective function [10,15]. We use the Yices SMT solver [6], because we have experienced that it performs very well in scheduling problems. However, this solver does not provide optimization. In order to get the optimal solution, we have implemented an optimization procedure that can use any off-the-shelf SMT solver as an oracle. Note that our formulation requires an UB in order to specify some constraints. An UB which is commonly used in scheduling problems is *trivialUB*, which is the makespan resulting from scheduling all the activities in a way such that only one runs at a time [2]. Since this UB tends to be significantly larger than the optimal makespan, we implement an optimization procedure similar to the one in [17]:

1. **Find a LB**. We optimize a relaxed version of the MRCPSP/max which does not include Constraints (10), by following a top-bottom search: starting from

trivialUB, we make successive satisfiability calls to the SMT solver, and after each call we set UB to be the makespan of the previous call minus 1. The procedure ends when the optimality is certified.

2. **Find an UB.** We optimize a single-mode version of the MRCPSP/max, by enforcing the execution modes to be the ones of the optimal solution found in step 1. We perform a bottom-top search, i.e., we try increasing upper bounds, starting from LB, until a model is found, or *trivialUB* is reached.

3. **Solve MRCPSP/max.** This last step is only required when UB > LB. In this case, we follow a top-bottom search starting at UB.

Steps 1 and 2 are intended to find tight bounds on the makespan. We perform these steps because reaching the optimal makespan of the MRCPSP/max decreasingly from *trivialUB* or increasingly from a LB, have shown in preliminary experiments to be far more time consuming than finding bounds by means of relaxed formulations. The main differences with respect to [17] are two: in [17], a MIP formulation in which the renewable resource constraints are relaxed using energetic reasoning (instead of completely ignoring them) is used in step 1; on the other hand, in [17] built-in optimization methods are used in all three steps.

5   New Benchmark Datasets for MRCPSP/max

Most of the recent approaches on solving the MRCPSP/max have been evaluated on the datasets generated in [14] and available in the PSPLib [9]. In particular, there are three datasets with 270 problem instances each one, namely mm30, mm50 and mm100 (instances have 30, 50 and 100 non-dummy activities). The number of execution modes ranges from 3 to 5, and there are 3 renewable and 3 non-renewable resources in each instance. The works of [12,17] have reported the best exact results, considering the mm30, mm50 and mm100 datasets. The former presented a handler for an extension of the cumulative constraint for the MRCPSP/max integrated into the SCIP optimization framework. The latter proposed a Failure-Directed Search Constraint Programming approach which shown to perform very well in different scheduling problems. Also, in [17] it was pointed out that, in the previously mentioned instances, resource constraints are not the hardest part of the problem. They used the relaxations on resource constraints mentioned in Sect. 4 to find a LB and an UB and it turned out that, in most of the cases, the LB and the UB were equal, and therefore an optimal solution was found without the need of encoding the whole original problem.

We have studied why the resources play such a minor role in those datasets. For 1432 out of the total 2430 non-renewable resources, it is trivially true that the demands do not exceed their capacity, i.e., any assignment of modes satisfies Constraint (2) for these resources. We have also checked if the relaxed optimal solutions obtained in step 1 of our optimization procedure satisfied the renewable resource constraints, which weren't enforced in this relaxation. We have observed that the relaxed optimal solution found for 593 out of the 810 instances satisfied the renewable resource constraints, although they were not encoded, and hence were also optimal solutions of the original problem. This number may be larger

because some instances timed out without having found the optimal solution of
the relaxation, and therefore have not been counted.

Table 1. Resource capacities of the new datasets. The capacity of each resource of each
instance have been generated uniformly and independently at random in the interval
indicated in the corresponding cell of the table.

Set name	mm30_1	mm30_2	mm50_1	mm50_2	mm100_1	mm100_2
Renewable capacity	[30,39]	[20,29]	[30,39]	[20,29]	[30,39]	[20,29]
Non-renewable capacity	[135,164]	[135,164]	[235,264]	[235,264]	[485,514]	[485,514]

These characteristics, in addition to the fact that all the mentioned instances
have been closed, suggest us that there is a need for new and more challenging
datasets, in particular regarding the hardness of resource constraints. The reason
why most renewable and non-renewable resources barely constrain the instances
is because the capacities are far large enough to supply the demands of the activities in a large amount of the possible combinations of mode assignments. For this
reason, we propose to use as a basis the same instances, but shrinking the capacities of the resources to amounts which make them non-dummy. For the case of
the renewable resources, we have conducted some preliminary experiments to
see approximately which capacity is needed to make the optimal makespan of
an instance increase. We have observed that, given the original demand values
and precedence network topologies, for capacities smaller than 30, rarely any
instance has an optimal makespan equal to the optimal of the MRCPSP/max
without resource constraints. Regarding the non-renewable resources, the original dataset was created with demands ranging from 1 to 10. Considering that all
the activities require the intermediate amount of 5 units for each non-renewable
resource, it would be needed a capacity of $5n$ to supply the demands, being n the
number of activities of the instance. Considering these facts, we have generated
two new versions of each one of the mm30, mm50 and mm100 datasets, namely
mm{30,50,100}_1 and mm{30,50,100}_2. They are the result of replacing the
resource capacities as stated in Table 1. The new mm{30,50,100}_2 datasets are
intended to be very constrained by resources. This is indeed the case, since we
have not been able to find any relaxed solution satisfying the renewable resource
constraints, and no non-renewable constraint is dummy. Sets mm{30,50,100}_1
are a bit softer regarding renewable resources.

6   Results and Conclusions

The experiments have been run on a 8 GB Intel® Xeon® E3-1220v2 machine
at 3.10 GHz. We have run our solver using Yices 2.4.2 [6] as the core SMT
solver, and compared the performance of our system with [17] (FDS) and [12]
(SCIP). We have run all three solvers in the same machine on both the old and

Table 2. Results for each solver (rows) and each dataset (columns), with a timeout of 600 s; *avg* denotes the average running time, in seconds, required to optimally solve the instances (computed on the instances which did not time out); *to* denotes the number of instances that timed out before reaching the optimum.

		30	50	100	30_1	50_1	100_1	30_2	50_2	100_2
FDS	avg	0.48	1.30	15.19	27.80	125.12	36.00	72.91	185.41	-
	to	0	0	3	1	113	267	24	216	270
SCIP	avg	14.49	26.74	134.03	88.97	163.79	-	178.33	281.23	-
	to	1	24	90	49	206	270	155	257	270
SMT	avg	0.87	17.14	67.97	16.39	120.38	-	54.36	224.47	-
	to	0	0	65	1	108	270	12	192	270

the new datasets. The results[2] are contained in Table 2. FDS is doubtlessly the best solver for the original datasets, with only 3 timeouts in the hardest dataset (mm100) and average runtimes one order of magnitude lower than the other approaches. This is because its MIP relaxation works very well with generalized precedence relations. It must be noted that SCIP does not start by solving a relaxed MRCPSP/max with respect to resources, which penalizes this approach in these datasets. On the other hand we can see that, in the new datasets, which are more constrained by resources, SMT is able to solve more instances than the other approaches in all cases except for the mm100_1 dataset. We remark that 2 out of the 3 instances that FDS solves in this dataset are optimally solved already in the relaxation solving steps. SMT is the best approach in sets mm{30,50}_2, which have the strongest resource constraints.

In conclusion, thanks to identifying some weaknesses of existing instances, we are able to provide new and challenging datasets for MRCPSP/max. This is especially noticeable in mm100 sets which, with the exception of 3 instances for FDS, could not be solved in less than 600 s using the state-of-the-art exact methods for MRCPSP/max. We have provided an SMT formulation showing to be more efficient than other state-of-the-art approaches for heavily resource constrained instances. Interestingly, our approach takes an off-the-shelf SMT solver and, instead of using specialized propagators, it uses recent specialized encodings of PB constraints into SAT to deal with resource constraints. The resulting encoding uses only the IDL theory, to deal with generalized precedences. Left as future work are the incorporation of better relaxations on resource equations, as the ones in [17], for a better LB identification. Also other optimization approaches could be considered, as well as the use of a full SAT encoding.

Acknowledgments. Work supported by grants TIN2015-66293-R (MINECO/ FEDER, UE), MPCUdG2016/055 (UdG), and *Ayudas para Contratos Predoctorales 2016* (grant number BES-2016-076867, funded by MINECO and co-funded by FSE). We thank the authors of [12,17] for sharing with us their solvers.

---

[2] The solver used in the experiments, detailed results and the new instances are available at http://imae.udg.edu/recerca/LAP/.

References

1. Ansótegui, C., Bofill, M., Palahí, M., Suy, J., Villaret, M.: Satisfiability modulo theories: an efficient approach for the resource-constrained project scheduling problem. In: Proceedings of the Ninth Symposium on Abstraction, Reformulation, and Approximation (SARA), pp. 2–9. AAAI (2011)
2. Artigues, C., Demassey, S., Neron, E.: Resource-Constrained Project Scheduling: Models, Algorithms, Extensions and Applications. Wiley, Hoboken (2013)
3. Bofill, M., Coll, J., Suy, J., Villaret, M.: Solving the multi-mode resource-constrained project scheduling problem with SMT. In: 28th International Conference on Tools with Artificial Intelligence (ICTAI), pp. 239–246. IEEE (2016)
4. Bofill, M., Coll, J., Suy, J., Villaret, M.: Compact MDDs for pseudo-Boolean constraints with at-most-one relations in resource-constrained scheduling problems. In: International Joint Conference on Artificial Intelligence (IJCAI) (2017, to appear)
5. Brucker, P., Drexl, A., Mhring, R., Neumann, K., Pesch, E.: Resource-constrained project scheduling: notation, classification, models, and methods. Eur. J. Oper. Res. **112**(1), 3–41 (1999)
6. Dutertre, B., de Moura, L.: The yices SMT solver. Technical report, Computer Science Laboratory, SRI International (2006). http://yices.csl.sri.com
7. Eén, N., Sorensson, N.: Translating pseudo-Boolean constraints into SAT. J. Satisfiability Boolean Model. Comput. **2**, 1–26 (2006)
8. Herroelen, W., Demeulemeester, E., Reyck, B.: A classification scheme for project scheduling. In: Weglarz, J. (ed.) Project Scheduling. International Series in Operations Research & Management Science, vol. 14, pp. 1–26. Springer, New York (1999). doi:10.1007/978-1-4615-5533-9_1
9. Kolisch, R., Sprecher, A.: PSPLIB - a project scheduling problem library. Eur. J. Oper. Res. **96**(1), 205–216 (1997)
10. de Moura, L., Bjørner, N.: Z3: an efficient SMT solver. In: Ramakrishnan, C.R., Rehof, J. (eds.) TACAS 2008. LNCS, vol. 4963, pp. 337–340. Springer, Heidelberg (2008). doi:10.1007/978-3-540-78800-3_24
11. Pritsker, A.A.B., Waiters, L.J., Wolfe, P.M.: Multiproject scheduling with limited resources: a zero-one programming approach. Manag. Sci. **16**, 93–108 (1969)
12. Schnell, A., Hartl, R.F.: On the efficient modeling and solution of the multi-mode resource-constrained project scheduling problem with generalized precedence relations. OR Spectr. **38**(2), 283–303 (2016)
13. Schutt, A., Feydy, T., Stuckey, P.J., Wallace, M.G.: Solving the resource constrained project scheduling problem with generalized precedences by lazy clause generation. CoRR abs/1009.0347 (2010). http://arxiv.org/abs/1009.0347
14. Schwindt, C.: Generation of resource constrained project scheduling problems subject to temporal constraints. Inst. für Wirtschaftstheorie und Operations-Research (1998)
15. Sebastiani, R., Tomasi, S.: Optimization in SMT with $\mathcal{LA}(\mathbb{Q})$ cost functions. In: Gramlich, B., Miller, D., Sattler, U. (eds.) IJCAR 2012. LNCS, vol. 7364, pp. 484–498. Springer, Heidelberg (2012). doi:10.1007/978-3-642-31365-3_38
16. Szeredi, R., Schutt, A.: Modelling and solving multi-mode resource-constrained project scheduling. In: Rueher, M. (ed.) CP 2016. LNCS, vol. 9892, pp. 483–492. Springer, Cham (2016). doi:10.1007/978-3-319-44953-1_31
17. Vilím, P., Laborie, P., Shaw, P.: Failure-directed search for constraint-based scheduling. In: Michel, L. (ed.) CPAIOR 2015. LNCS, vol. 9075, pp. 437–453. Springer, Cham (2015). doi:10.1007/978-3-319-18008-3_30

Conjunctions of Among Constraints

Víctor Dalmau[(✉)]

Department of Information and Communication Technologies,
Universitat Pompeu Fabra, Barcelona, Spain
vdalmau@gmail.com

Abstract. Many existing global constraints can be encoded as a conjunction of among constraints. An among constraint holds if the number of the variables in its scope whose value belongs to a prespecified set, which we call its range, is within some given bounds. It is known that domain filtering algorithms can benefit from reasoning about the interaction of among constraints so that values can be filtered out taking into consideration several among constraints simultaneously. The present paper embarks into a systematic investigation on the circumstances under which it is possible to obtain efficient and complete domain filtering algorithms for conjunctions of among constraints. We start by observing that restrictions on both the scope and the range of the among constraints are necessary to obtain meaningful results. Then, we derive a domain flow-based filtering algorithm and present several applications. In particular, it is shown that the algorithm unifies and generalizes several previous existing results.

1 Introduction

Global constraints play a major role in constraint programming. Very informally, a global constraint is a constraint, or perhaps more precisely, a family of constraints, which is versatile enough to be able to express restrictions that are encountered often in practice. For example, one of the most widely used global constraints is the 'All different' constraint, $\text{ALLDIFF}(S)$ where $S = \{x_1, \ldots, x_n\}$ is a set of variables, which specifies that the values assigned to the variables in S must be all pairwise different. This sort of restriction arises naturally in many areas, such as for example scheduling problems, where the variables $x_1, \ldots, x_n$ could represent n activities that must be assigned different times of a common resource.

Besides is usefulness in simplifying the modeling or programming task, global constraints also improve greatly the efficiently of propagation-search based solvers. This type of solver performs a tree search that constructs partial assignments and enforces some sort of propagation or local consistency that prunes the space search. Different forms of consistency, including (singleton) bounds consistency, (singleton, generalized) arc-consistency, path consistency and many others, can be used in the propagation phase. One of the most commonly used

© Springer International Publishing AG 2017
J.C. Beck (Ed.): CP 2017, LNCS 10416, pp. 80–96, 2017.
DOI: 10.1007/978-3-319-66158-2_6

forms of local consistency is *domain consistency*, also called generalized arc-consistency. A domain consistency algorithm keeps, for every variable v, a list, $L(v)$, of feasible values, which is updated, by removing a value d from it, when some constraint in the problem guarantees that v cannot take value d in any solution. One of the key reasons of the success of global constraints is that they enable the use of efficient filtering algorithms specifically tailored for them.

Several constraints studied in the literature including ALLDIFF[19], GCC[22], SYMMETRIC-GCC [16], SEQUENCE [5], GLOBALSEQUENCING [29], ORDEREDDISTRIBUTE [23], and CARDINALITYMATRIX [28] can be decomposed as the conjunction of a simpler family of constraints, called among constraints [5]. An among constraint has the form AMONG(S, R, min, max) where S is again a set of variables called the *scope*, R is a subset of the possible values, called the *range*, and min, max are integers. This constraint specifies that the number of variables in S that take a value in R must be in the range $\{min, \ldots, max\}$. For example, the constraint ALLDIFF(S) can be expressed as the conjunction of constraints AMONG$(S, \{d_1\}, 0, 1), \ldots,$ AMONG$(S, \{d_k\}, 0, 1)$ where $d_1, \ldots, d_k$ are the set of all feasible values for the variables in S.

Besides encoding more complex global cardinality constraints, conjunctions of among constraints, (CAC), appear in many problems, such as Sudoku or latin squares. In consequence, CACs have been previously studied [9,27,33], specially the particular case of conjunctions of ALLDIFF constraints [2,3,8,12,17,18,20]. Although deciding the satisfiability of an arbitrary conjunction of among constraints is NP-complete [27] this body of work shows that sometimes there are benefits in reasoning about the interaction between the among constraints. Hence, it is important to understand under which circumstances among constraints can be combined in order to endow CSP solvers with the ability to propagate taking into consideration several among constraints simultaneously. The aim of the present paper is to contribute to this line of research. To this end we first observe that restrictions on *both* the scope and the range of the among constraints are necessary to obtain meaningful results. Then we embark in a systematic study of which such restrictions guarantee efficient propagation algorithms. In particular, we introduce a general condition such that every CAC satisfying it admits an efficient and complete domain filtering algorithm. This condition basically expresses that the matrix of a system of linear equations encoding the CAC instance belongs to a particular class of totally unimodular matrices known as network matrices. This allows to reformulate the domain filtering problem in terms of flows in a network graph and apply the methodology derived by Régin [25,26]. The algorithm thus obtained, although simple, unifies and generalizes existing domain filtering algorithms for several global constraints, including ALLDIFF, GCC, SEQUENCE, SYMMETRIC-GCC, ORDEREDDISTRIBUTE as well as for other problems expressed as conjunctions of among constraints in [24,27]. A nice feature of our approach is that it abstracts out the construction of the network flow problem, so that when exploring a new CAC one might leave out the usually messy details of the design of the network graph and reason purely in combinatorial terms.

Several filtering methods have been obtained by decomposing a global constraint into a combination of among constraints. For example the first polynomial-time filtering algorithm for the SEQUENCE constraint [33] is obtained explicitly in this way. However, there have been very few attempts to determine systematically which particular conjunctions of among constraints allow efficient filtering algorithms. The seminal paper in this direction is [26] which identifies several combinations of among and GCC constraints that admit a complete and efficient domain filtering algorithm (see Sect. 6 for more details). Our approach is more general as it subsumes the tractable cases introduced in [26]. Another closely related work is [24] where two tractable combinations of boolean CACs, called TFO and 3FO, are identified. The approach in [24] differs from ours in two aspects: it deals with optimization problem and also considers restrictions on the *min* and *max* parameters of the among constraints while we only consider restrictions on the scope and range. A different family of CACs has been investigated in [9] although the work in [9] focuses in bound consistency instead of domain consistency.

Other approaches to the design of filtering algorithms for combinations (but not necessarily conjunctions) of global (but not necessarily among) constraints are described in [4,6,7]. The method introduced in [4] deals with logical combinations of some primitive constraints but differs substantially from ours in the sense that it cannot capture a single among constraint. The work reported in [6,7] does not guarantee tractability.

Several proofs are omitted due to space restrictions. They can be found in the full version [10].

2   Preliminaries

A *conjunction of among constraints*, (CAC) is a tuple $(V, D, L, \mathcal{C})$ where V is a finite set whose elements are called *variables*, D is a finite set called *domain*, $L : V \to 2^D$ is a mapping that sends every variable v to a subset of D, which we call its *list*, and $\mathcal{C}$ is a finite set of *constraints* where a constraint is an expression of the form AMONG(S, R, min, max) where $S \subseteq V$ is called the *scope* of the constraint, $R \subseteq D$ is called *range* of the constraint, and min, max are integers satisfying $0 \le min \le max \le |S|$.

A *solution* of $(V, D, L, \mathcal{C})$ is a mapping $s : V \to D$ such that $s(v) \in L(v)$ for every variable $v \in V$ and $min \le |\{v \in S \mid s(v) \in R\}| \le max$ for every constraint AMONG(S, R, min, max) in $\mathcal{C}$.

Example 1 (GCC and ALLDIFF constraints).   The global cardinality constraint[1], GCC [22] corresponds to instances $(V, D, L, \mathcal{C})$ where all the constraints have the form AMONG$(V, \{d\}, min, max)$ with $d \in D$. The ALLDIFF constraint is the particular case obtained when, additionally, $min = 0$ and $max = 1$.

---

[1] We want to stress here that a global constraint is not a single constraint but, in fact, a family of them.

Let $I = (V, D, L, C)$ be a CAC. We say that a value $d \in D$ is *supported* for a variable $v \in V$ if there is a solution s of I with $s(v) = d$. In this paper we focus in the following computational problem, which we will call *domain filtering*: given a CAC, compute the set of all the non supported values for each of its variables.

This definition is motivated by the following scenario: think of (V, D, L, C) as defining a constraint which is part of a CSP instance that is being solved by a search-propagation algorithm that enforces domain consistency. Assume that at any stage of the execution of the algorithm, L encodes the actual feasible values for each variable in V. Then, the domain filtering problem is basically the task of identifying all the values that need to be pruned by considering the constraint encoded by (V, D, L, C).

3   Network Hypergraphs

An *hypergraph* H is a tuple, $(V(H), E(H))$, where $V(H)$ is a finite set whose elements are called *nodes* and $E(H)$ is set whose elements are subsets of $V(H)$, called hyperedges. An hypergraph is *totally unimodular* if its incidence matrix M is totally unimodular, that is, if every square submatrix of M has determinant 0, $+1$, or -1. In this paper we are concerned with a subset of totally unimodular hypergraphs called network hypergraphs. In order to define network hypergraph we need to introduce a few definitions.

An *oriented tree* T is any directed tree obtained by orienting the edges of an undirected tree. A *path* p in T is any sequence $x_1, e_1, x_2, \ldots, e_{n-1}, x_n$ where $x_1, \ldots, x_n$ are different vertices of T, $e_1, \ldots, e_{n-1}$ are edges in T and for every $1 \leq i < n$, either $e_i = (x_i, x_{i+1})$ or $e_i = (x_{i+1}, x_i)$. The *polarity* of an edge $e \in E(T)$ wrt. p is defined to be $+1$ (or positive) if $e = (x_i, x_{i+1})$ for some $1 \leq i < n$, -1 (or negative) if $e = (x_{i+1}, x_i)$ for some $1 \leq i < n$, and 0 if e does not appear in p. A path p has positive (resp. negative) polarity if all its edges have positive (resp. negative) polarity. Paths with positive polarity are also called directed paths. Since an oriented tree does not contain symmetric edges, we might represent a path by giving only its sequence of nodes $x_1, \ldots, x_n$.

We say that an oriented tree T *defines* an hypergraph H if we can associate to every hyperedge $h \in E(H)$ an edge $e_h \in E(T)$ and to every node $v \in V(H)$ a directed path p_v in T such that for every $v \in V(H)$ and $h \in E(V)$, $v \in h$ if and only if e_h belongs to p_v. We say that an H is a *network* hypergraph if there is an oriented tree that defines it.

Example 2. The hypergraph H with variable-set $\{v_1, \ldots, v_6\}$ and hyperedge-set $\{h_1, \ldots, h_5\}$ given in Fig. 1a is a network hypergraph as it is defined by tree T given in Fig. 1b where we have indicated, using labels on the edges, the edge in T associated to every hyperedge in H. We associate to every variable v_i, $1 \leq i \leq 6$ the directed path $s_{(i-1 \mod 2)}, r, t_{(i-1 \mod 3)}$ in T. It can be readily checked that, under this assignment, T defines H.

Sometimes, it will be convenient to assume that the tree T defining H is minimal in the sense that no tree with fewer nodes defines H. Minimal trees have

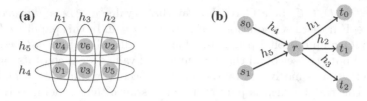

Fig. 1. (a): Hypergraph H, (b): oriented tree T

the nice property that every edge e in T is associated with some hyperedge of H. Indeed, assume that some edge $e = (x, y)$ is not associated to any hyperedge in H, then one could find a smaller tree T defining H by *contracting* edge e, that is, by merging x and y into a new node z that has as in-neighbors the union of all in-neighbors of x and y and, as out-neighbors, the union of all out-neighbors of x and y.

Since the vast majority of the trees defined in this paper will be oriented we shall usually drop 'oriented'. So, unless, otherwise explicitly stated, a tree is always an oriented tree. Finally, we note that one can decide whether a given hypergraph is a network hypergraph in time $O(e^3 v^2)$ where e is the number of hyperedges and v is the number of nodes (see Chap. 20 in [30] for example).

4   Restricting Only the Scope or the Range

It has been shown by Régin [27] that the domain filtering problem for CACs is NP-hard. Still, efficient algorithms are known for some particular cases. It seems natural to start by asking which tractable subcases of the problem can be explained by considering *only* the scopes of the constraints. This question has a close similarity to the study of the so-called structural restrictions of the CSP (see, for example [14] for a survey) and, not surprisingly, it can be solved by applying results developed there. Indeed, it follows easily from a result of Färnquivst and Jonsson [11] that, modulo some mild technical assumptions, if one allows arbitrary ranges in constraints, then the domain filtering problem is solvable in polynomial time if and only if the hypergraph of the scopes of the constraints has bounded tree-width (see the full version for precise statement and the proof). This result, although delineates exactly the border between tractability and intractability, turns out to be not very useful in explaining the tractability of global constraints. This is due to the fact that global cardinality constraints defined by conjunctions of among constraints usually have constraints with large scopes and the cardinality of the scope in a constraint is a lower bound on the tree-width of its scope hypergraph.

One can also turn the attention to the range of constraints and inquiry whether there are tractable subcases of the problem that can be explained *only* by the range of the constraints. Here, again the response is not too useful. Indeed, it is very easy to show (see again the full version) that as soon as we allow some non-trivial range R (that is some range different than the empty set and than the whole domain) and arbitrary scopes in the among constraints, then the domain filtering problem becomes NP-complete.

In view of this state of affairs it is meaningful to consider families of conjunctions of among constraints that are obtained by restricting *simultaneously* the scope and the range of the constraints occurring in them. This is done in the next section.

5   A Flow-Based Algorithm

Let $I = (V, D, L, \mathcal{C})$ be conjunction of among constraints. We will deal first with the case in which D is a boolean, say $D = \{0, 1\}$. Hence, we can assume that every constraint $\textsc{Among}(S, R, min, max)$ in $\mathcal{C}$ satisfies $R = \{1\}$ since, if $R = \{0\}$ it can be reformulated as $\textsc{Among}(S, \{1\}, |S| - max, |S| - min)$. We also assume that $L(v) = \{0, 1\}$ for every $v \in V$ since if $L(v) \neq \{0, 1\}$ we could obtain easily an equivalent instance without variable v.

It is easy to construct a system of linear equations whose feasible integer solutions encode the solutions of I. Let $v_1, \dots, v_n$ be the variables of I and let $C_j = (S_j, \{1\}, min_j, max_j), j = 1, \dots, m$, be its constraints. The system has variables $x_i (1 \leq i \leq n)$, $y_j (1 \leq j \leq m)$ and the following equations:

$$y_j + \sum_{v_i \in S_j} x_i = max_j \qquad\qquad j = 1, \dots, m$$

$$0 \leq y_j \leq max_j - min_j \qquad\qquad j = 1, \dots, m$$

$$0 \leq x_i \leq 1 \qquad\qquad i = 1, \dots, n$$

which we express in matrix form as

$$Mz = a$$

$$0 \leq z \leq c$$

with $z^T = (x_1, \dots, x_n, y_1, \dots, y_m)$ (see Example 3).

If M is totally unimodular then one can perform domain filtering in polynomial time. Indeed, for every $v_i \in V$ and $d \in L(v)$, we might decide whether d is a supported value for v as follows: add equation $x_i = d$ to the system and decide whether there exists a feasible solution of its linear relaxation using a LP solver. It follows from total unimodularity (see Theorem 19.1 in [30] for example) that such a feasible solution exists if and only if d is a support for v.

However, this approach implies invoking $O(n)$ times a LP solver, which might be too expensive to be practical, since, in addition, a propagation-based algorithm might call a domain filtering algorithm many times during its execution. To overcome this difficulty we shall require further conditions on the matrix M. To this end, we define the hypergraph associated to instance I to be the hypergraph H with $V(H) = V$ and $E(H) = \{S_j \mid 1 \leq j \leq m\}$.

Now, assume that H is a network hypergraph defined by a tree T. In this case, one can use specific and more efficient methods like the network simplex algorithm (see for example [1]) instead of a general purpose LP solver. However, it is still possible to do better (and avoid the $O(n)$ calls to the network simplex algorithm) by transforming it into a maximum flow problem. This idea has been used in [21] to obtain a domain filtering algorithm for the SEQUENCE constraint. More precisely, [21] deals with the particular case of network matrices defined by a directed path. Our approach draws upon [21] and generalizes it to network matrices defined by arbitrary trees. This is done as follows.

Let P be the incidence matrix of T. That is, let $t_1, \ldots, t_{m+1}$ be an arbitrary ordering of nodes in T and define P to be the $((m+1) \times m)$-matrix where $P_{i,j}$ is $+1$ if edge e_{S_j} starts at t_i, -1 if e_{S_j} ends at t_i, and 0 otherwise.

Let r be a m-ary (column) vector and let p be a path in T. We say that r is the *indicator vector* of p if for every $j = 1, \ldots, m$, r_j is the polarity of e_{S_j} wrt. p. The next observation follows directly from the definitions.

Observation 1. *Let p be a path in T and let r be its indicator vector. Then, the ith entry, $(Pr)_i$, of Pr, is $+1$ if t_i is the first node in p, -1 if t_i is the last node in p, and 0 otherwise.*

The next two lemmas follow directly from the previous observation.

Lemma 1. *P has full rank.*

Proof. Let P' be the $(m \times m)$ matrix obtained by removing the last row (corresponding to vertex t_{m+1}) and consider the $(m \times m)$-matrix Q such that for every $i = 1, \ldots, m$, the ith column of Q, which we shall denote as $Q_{*,i}$, is the indicator vector of the unique path in T starting at t_i and ending at t_{m+1}. It follows from Observation 1 that $P'Q$ is the identity matrix. ∎

Then, since P has full rank we can obtain an equivalent system $PMz = Pa$ by multiplying both sides of $Mz = a$ by P. Let $N = PM$ and $b = Pa$ (see Example 3).

Lemma 2. *In every column of N one entry is $+1$, one entry is -1, and all the other entries are 0.*

Proof. It is only necessary to show that every column, $M_{*,k}$, $k = 1, \ldots, m+n$ of M is the indicator vector of some directed path in T. If the variable corresponding to the k-column is x_i for some $1 \leq i \leq n$ then by construction $M_{*,k}$ is the indicator column of the path associated to v_i in T. Otherwise, if the variable corresponding to column k is y_j for some $1 \leq j \leq m$ then $M_{*,k}$ is the indicator vector of the directed path containing only edge e_{S_j}. ∎

Hence, matrix N is the incidence matrix of a directed graph G. Note that, by definition, G contains an edge $e_k, k = 1, \ldots, m+n$ for each variable z_k and a node $u_j, j = 1, \ldots, m+1$ for each row in N (that is, for every node in T). Define the capacity of every edge e_k to be c_k. Then, feasible solutions of the system correspond precisely to flows where every node u_j has a supply/demand specified by b_j (more precisely, node u_j has a supply of b_j units if $b_j > 0$ and a demand of $-b_j$ units if $b_j < 0$). It is well known that this problem can be reduced to the (standard) maximum flow problem by adding new source and sink nodes

s, t and edges from s to u_j with capacity b_j whenever $b_j > 0$ and from u_j to t with capacity $-b_j$ whenever $b_j < 0$.

It follows from this construction that for every $1 \leq i \leq n$ and every $d = \{0, 1\}$, there is a solution s of I with $s(v_i) = d$ if and only if there is a *saturating* flow (that is, a flow where all the edges leaving s or entering t are at full capacity) such that the edge associated to x_i carries d units of flow. Régin [25, 26] has shown that this later condition can be tested *simultaneously* for all $1 \leq i \leq n$ and $d \in \{0, 1\}$ by finding a maximal flow and computing the strongly connected components of its residual graph. Finding a maximal flow of a network with integral capacities can be done in time $O(min(v^{2/3}, e^{1/2})e \log(v^2/e) \log u)$ using Goldberg and Rao's algorithm [13] where v is the number of vertices, e is the number of edges, and u is the maximum capacity of an edge. Computing the strongly connected components of the residual graph takes $O(v + e)$ time using Tarjan's algorithm [31]. By construction, the network derived by our algorithm satisfies $e \leq n + 2m$ and $v \leq m + 1$. Furthermore, it is not difficult to see that $u \leq mn$. Indeed, note that the capacity of any edge is either some entry, c_i, of vector c or the absolute value of some entry, b_i, of vector b. It follows directly from the definition of c, that all its entries are at most n. As for b, the claim follows from the fact that $b = Pa$ where, by construction, a has m entries where every entry is in the range $\{1, \ldots, n\}$ and every entry in P is in $\{-1, 0, 1\}$.

Hence, if we define $f(n, m)$ to be $min(m^{2/3}, (n + m)^{1/2})(n + m) \log(m^2/(n + m)) \log mn$ we have:

Lemma 3. *There is a domain filtering algorithm for conjunctions of boolean among constraints whose associated hypergraph is a network hypergraph, which runs in time $O(f(n, m))$ where n is the number of variables and m is the number of constraints, assuming the instance is presented as a network flow problem.*

There are minor variants (leading to the same asymptotic complexity) obtained by modifying the treatment of the slack variables. Here we will discuss two of them. In the first variant one encodes a constraint $C_j = (S_j, \{1\}, min_j, max_j)$ with the equation $-y_j + \sum_{v_i \in S_j} x_i = 0$ where y_j satisfies $min_j \leq y_j \leq max_j$. Under this encoding, our approach produces a network problem where, instead of having nodes with specified supply or demand, we have edges with minimum demand. In the second variant, used in [21], one encodes a constraint C_j with two equations $y_j + \sum_{v_i \in S_j} x_i = max_j$, and $-z_j + \sum_{v_i \in S_j} x_i = min_j$ where y_j and z_j are new slack variables satisfying $0 \leq y_j, z_j$. This encoding produces a network that has m more nodes and edges.

When analyzing the time complexity of domain filtering it is customary to report, additionally, the so-called time complexity 'down a branch of a search tree' which consists in the aggregate time complexity of successive calls to the algorithm, when at each new call, the list of some of the variables has been decreased (as in the execution of a propagation-search based solver). It was observed again by Régin [26] that, in this setting, it is not necessary to solve the flow problem from scratch at each call, leading to a considerable reduction in total time. Applying the scheme in [26], we obtain that the time complexity down a branch of a search tree of our algorithm is $O(n(n + m))$. We omit the details because they are fairly standard.

If the instance is not presented as a network flow problem then one would need to add the cost of transforming the instance into it. However this cost would be easily amortized as a domain filtering algorithm is invoked several times during the execution of a constraint solver. Furthermore, in practical scenarios, the conjunction of among constraints will encode a global constraint from a catalog of available global constraints. Hence, it is reasonable to assume that the formulation of the global constraint as a network flow problem can be precomputed.

Example 3. Let I be a boolean instance with variables $\{v_1, \ldots, v_6\}$ and constraints: $C_1 = \text{AMONG}(\{v_1, v_4\}, \{1\}, 0, 1)$, $C_2 = \text{AMONG}(\{v_2, v_5\}, \{1\}, 0, 1)$, $C_3 = \text{AMONG}(\{v_3, v_6\}, \{1\}, 0, 1)$, $C_4 = \text{AMONG}(\{v_1, v_3, v_5\}, \{1\}, 1, 1)$, $C_5 = \text{AMONG}(\{v_2, v_4, v_6\}, \{1\}, 1, 1)$.

The specific values for M, a and c of the ILP formulation of I are:

$$
M = \begin{array}{c} \\ C_1 \\ C_2 \\ C_3 \\ C_4 \\ C_5 \end{array}
\begin{array}{c} y_1\ y_2\ y_3\ x_1\ x_2\ x_3\ x_4\ x_5\ x_6 \\
\left(\begin{array}{ccccccccc}
1 & 0 & 0 & 1 & 0 & 0 & 1 & 0 & 0 \\
0 & 1 & 0 & 0 & 1 & 0 & 0 & 1 & 0 \\
0 & 0 & 1 & 0 & 0 & 1 & 0 & 0 & 1 \\
0 & 0 & 0 & 1 & 0 & 1 & 0 & 1 & 0 \\
0 & 0 & 0 & 0 & 1 & 0 & 1 & 0 & 1
\end{array}\right)
\end{array}
\qquad
a = \begin{pmatrix} 1 \\ 1 \\ 1 \\ 1 \\ 1 \end{pmatrix}
$$

$$
c^T = \quad (1\ 1\ 1\ 1\ 1\ 1\ 1\ 1\ 1)
$$

Note that since $min_i = max_i$ for $i = 4, 5$ we did not need to add the slack variables y_4 and y_5. Note that the hypergraph of instance I is precisely the hypergraph H in Example 2. In particular, h_i is the hypererdge corresponding to the scope of constraint C_i for $i = 1, \ldots, 5$. The matrix P obtained from the tree T defining H is:

$$
P = \begin{array}{c} \\ t_0 \\ t_1 \\ t_2 \\ s_0 \\ s_1 \\ r \end{array}
\begin{array}{c} h_1\ h_2\ h_3\ h_4\ h_5 \\
\left(\begin{array}{ccccc}
-1 & 0 & 0 & 0 & 0 \\
0 & -1 & 0 & 0 & 0 \\
0 & 0 & -1 & 0 & 0 \\
0 & 0 & 0 & 1 & 0 \\
0 & 0 & 0 & 0 & 1 \\
1 & 1 & 1 & -1 & -1
\end{array}\right)
\end{array}
$$

Multiplying M and a by P we obtain:

$$
N = PM = \begin{array}{c} \\ t_0 \\ t_1 \\ t_2 \\ s_0 \\ s_1 \\ r \end{array}
\begin{array}{c} y_1\ y_2\ y_3\ x_1\ x_2\ x_3\ x_4\ x_5\ x_6 \\
\left(\begin{array}{ccccccccc}
-1 & 0 & 0 & -1 & 0 & 0 & -1 & 0 & 0 \\
0 & -1 & 0 & 0 & -1 & 0 & 0 & -1 & 0 \\
0 & 0 & -1 & 0 & 0 & -1 & 0 & 0 & -1 \\
0 & 0 & 0 & 1 & 0 & 1 & 0 & 1 & 0 \\
0 & 0 & 0 & 0 & 1 & 0 & 1 & 0 & 1 \\
1 & 1 & 1 & 0 & 0 & 0 & 0 & 0 & 0
\end{array}\right)
\end{array}
\qquad
b = Pa = \begin{pmatrix} -1 \\ -1 \\ -1 \\ 1 \\ 1 \\ 1 \end{pmatrix}
$$

The feasible solutions of the previous LP correspond to the feasible flows of the network in Fig. 2a, where nodes s_0, s_1 and r have a supply of one unit of flow and nodes y_0, y_1, y_2 have a demand of one unit of flow. We also give the final network (Fig. 2b) obtained by applying the first variant discussed after Lemma 3. In this case the edges (r, s_i), $i = 0, 1$ have a demand of one unit of flow. Finally, Fig. 2c contains the result of transforming the network in Fig. 2a to a (standard) max flow problem. In all three networks all edges have capacity 1.

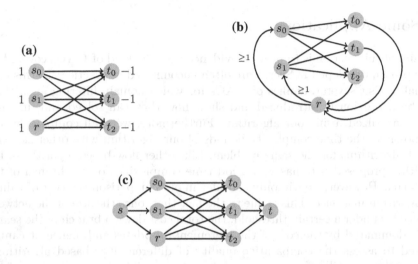

Fig. 2. (a): Network with supplies/demands, (b): network with edge demands, (c): standard flow network obtained from (2a)

This approach can be generalized to non-boolean domains via boolean encoding. The choice of boolean encoding might depend on the particular instance at hand but for concreteness we will fix one. The *canonical booleanization* (see Example 4) of a conjunction $I = (V, D, L, \mathcal{C})$ of among constraints with $|D| \geq 3$ is the boolean instance $(V \times D, \{0,1\}, L_b, \mathcal{C}_b)$ where $L_b(v, d) = \{0, 1\}$ if $d \in L(v)$ and $\{0\}$ otherwise, and $\mathcal{C}_b$ contains:

– AMONG$(S \times R, \{1\}, min, max)$ for every constraint AMONG$(S, R, min, max) \in \mathcal{C}$, and
– AMONG$(\{v\} \times D, \{1\}, 1, 1)$ for every variable $v \in V$. This family of constraints are called *non-empty assignment constraints*.

That is, the intended meaning of the encoding is that $(v, d) \in V \times D$ is true whenever v takes value d.

We define the hypergraph H associated to I to be the hypergraph associated to the canonical booleanization of I. That is, $V(H)$ is $V \times D$ and $E(H)$ contains hyperedge $\{(v, d) \mid v \in S, d \in R\} = S \times R$ for every constraint AMONG(S, R, min, max) in $\mathcal{C}$, and hyperedge $\{(v, d) \mid d \in D\} = \{v\} \times D$ for every variable $v \in V$. Thus, for arbitrary domains, we have:

Corollary 1. *There is a domain filtering algorithm for conjunctions $(V, D, L, \mathcal{C})$ of among constraints whose associated hypergraph is a network hypergraph, which runs in time $O(f(n, m))$ where $n = \sum_{v \in V} |L(v)|$ and $m = |\mathcal{C}| + |V|$, assuming the instance is presented as a network flow problem.*

Proof. It just follows from observing that the canonical booleanization of instance $(V, D, L, \mathcal{C})$ has at most $n = \sum_{v \in V} |L(v)|$ variables and at most $m = |\mathcal{C}| + |V|$ constraints.   ■

6   Some Applications

The aim of this section is to provide evidence that the kind of CACs covered by the approach developed in Sect. 5 are often encountered in practice. To this end, we shall revisit several families of CACs for which domain filtering algorithms have been previously introduced and show how they can solved and, in some cases generalized, using our algorithm. Furthermore, we shall compare, whenever possible, the time complexity bounds of our algorithm with other state-of-the-art algorithms for the same problem. Like other flow-based algorithms, the algorithm proposed here has very good time complexity down a branch of the search tree. However, we will only consider in our comparison the cost of calling the algorithm just once. This is due to the fact that, once the size of the network produced is under a certain threshold, the total cost down a branch of the search tree is dominated by the cost of the incremental updates and, hence, it cannot be used to assess the comparative quality of different flow-based algorithms. Furthermore, we will try to compare, whenever possible, the parameters of the obtained network flow problem (number of nodes, edges, capacities of the edges) instead of the actual running time since the latter is dependend on the choice of the max-flow algorithm. Somewhat surprisingly, in many of the cases, even if we did not attempt any fine-tuning, the network produced by the algorithm is essentially equivalent to the network produced by specific algorithms.

6.1   Disjoint Constraints

As a warm up we shall consider the GCC and ALLDIFF constraints. We have seen in Example 1 that both can be formulated as a conjunction $(V, D, L, \mathcal{C})$ of among constraints of the form $C_d = \text{AMONG}(V, \{d\}, min_d, max_d), d \in D$. Note that both have the same associated hypergraph H with node-set, $V \times D$, and edge set $E(H)$ containing hyperedge $h_v = \{v\} \times D$ for every $v \in V$, and hyperedge $h_d = V \times \{d\}$, for every $d \in D$.

It is not difficult to see (see Example 4) that H is defined by the tree T defined as follows: The node-set of T consists of $\{r\} \cup V \cup D$ where r is a new node. The edge-set of T contains an edge from every $v \in V$ to r (associated to h_v) and from r to every node $d \in D$ (associated to h_d). Consequently, both GCC and ALLDIFF are solvable by our algorithm. It can be seen that the network for the GCC using the tree T described above has $e = O(|V||D|)$ edges, $v = |V| + |D| + 3$

nodes, and the maximum capacity, u, of an edge is at most $|D||V|$. Hence, it follows that the total running time of our algorithm for the GCC constraint is $O(min(v^{2/3}, e^{1/2})e \log(v^2/e) \log u)$. Régin's algorithm [26] has a $O(|V|^2|D|)$ complexity which is better when $|V| \in O(|D|^{2/3})$ but the comparison between the two bounds is not very meaningful because it mainly reflects a different choice of max flow algorithm. Indeed, the network obtained using the first variant discussed after Lemma 3 is identical to the network constructed in [26]. In the particular case of the ALLDIFF constraint [25] shows how to produce a bipartite matching problem that can be solved using specialized algorithms, such as [15], leading to a total time complexity of $O(|V|^{5/2})$ which is better than ours.

Example 4. Consider the constraint $\text{ALLDIFF}(s_0, s_1)$ where the list of each variable contains the following three values: t_0, t_1, t_2. Then, $\text{ALLDIFF}(s_0, s_1)$ is encoded as a CAC I with the following constraints: $\text{AMONG}(\{s_0, s_1\}, \{t_0\}, 0, 1)$, $\text{AMONG}(\{s_0, s_1\}, \{t_1\}, 0, 1)$, $\text{AMONG}(\{s_0, s_1\}, \{t_2\}, 0, 1)$.

The canonical booleanization of I has variables $\{s_0, s_1\} \times \{t_0, t_1, t_2\}$ and constraints $C_1 = \text{AMONG}(\{s_0, s_1\} \times \{t_0\}, \{1\}, 0, 1)$, $C_2 = \text{AMONG}(\{s_0, s_1\} \times \{t_1\}, \{1\}, 0, 1)$, $C_3 = \text{AMONG}(\{s_0, s_1\} \times \{t_2\}, \{1\}, 0, 1)$, $C_4 = \text{AMONG}(\{s_0\} \times \{t_0, t_1, t_2\}, \{1\}, 1, 1)$, and $C_5 = \text{AMONG}(\{s_1\} \times \{t_0, t_1, t_2\}, \{1\}, 1, 1)$. Observe that this instance is, under the renaming $v_i \mapsto (s_{i-1 \mod 2}, t_{i-1 \mod 3})$, the same instance than we have considered previously in Example 3. The network flow problem that our algorithm constructs for this instance (see Fig. 2b in Example 3) is essentially identical to the one derived in [26]. Indeed, our network is obtained by merely merging the source and sink nodes of the network in [26].

A simple analysis reveals that the same approach can be generalized to instances $(V, D, L, \mathcal{C})$ satisfying the following *disjointedness* condition: for every pair of constraints $\text{AMONG}(S, R, min, max)$ and $\text{AMONG}(S', R', min', max')$ in $\mathcal{C}$, $(S \times R) \cap (S' \times R') = \emptyset$. The tractability of such instances was, to the best of our knowledge, not known before. The particular case in which $R \cap R' = \emptyset$ has been previously shown in [27] using a different approach. The proof given in [27] does not construct a flow problem nor gives run-time bounds so we omit a comparison.

6.2   Domains Consisting of Subsets

Consider the following generalization of our setting where in a CAC $(V, D, L, \mathcal{C})$ every variable v must be assigned to a subset of $L(v)$ (instead of a single element). In this case, the semantics of the among constraint need to be generalized as well. Instead, we will say a constraint $\text{AMONG}(S, R, min, max)$ is satisfied by a mapping $s : V \to 2^D$ if $min \leq \sum_{v \in S} |s(v) \cap R| \leq max$. To avoid confusion we shall refer to this variant of the among constraint as *set among* constraint.

For example, the SYMMETRIC-GCC constraint [16] is precisely a conjunction $(V, D, L, \mathcal{C})$ where in every constraint $\text{AMONG}(S, R, min, max)$, S or R is a singleton.

It is easy to reduce a conjunction of set among constraints $I = (V, D, L, \mathcal{C})$ to a conjunction of (ordinary, not set) boolean among constraints. Indeed, one only needs to remove the non-empty assignment constraints to the canonical booleanization of I. Consequently, if $(V, D, L, \mathcal{C})$ encodes a SYMMETRIC-GCC constraint then the resulting boolean instance has the same hypergraph than the canonical booleanization of the GCC constraint. It follows that the running time bounds of our algorithm for the SYMMETRIC-GCC constraint are exactly the same than for the GCC constraint (see Sect. 6.1) since the networks are essentially equivalent. The only differences are, possibly, in the capacities of the edges of the constructed network but it is still possible to bound them by $|V||D|$. Finally, we note that the network produced by our algorithm using the first variant discussed after Lemma 3 is identical the one given in [16].

6.3 The Sequence Constraint

The SEQUENCE constraint [5] corresponds to instances $(\{v_1, \ldots, v_n\}, D, L, \mathcal{C})$ with constraints $\text{AMONG}(\{v_i, \ldots, v_{i+k}\}, R, min, max), i = 1, \ldots, n - k$ for some fixed integers min, max, k, and fixed $R \subseteq D$. It is not difficult to see that the hypergraph of the canonical booleanization of the SEQUENCE constraint is *not* a network hypergraph. However, as shown in [21] one obtains an equivalent instance with a network hypergraph using a different encoding in which for every original variable $v_i \in V$, we have a boolean variable x_i which is intended to be true whenever v_i takes a value in R and false otherwise. Under this alternative encoding we obtain a boolean instance I which consists of constraints $\text{AMONG}(\{x_i, \ldots, x_{i+k}\}, \{1\}, min, max), i = 1, \ldots, n - k$. It is shown in [21] that the ILP formulation of this instance satisfies the so-called consecutive-ones property which implies that the hypergraph of I is defined by a tree T consisting of a single directed path. Indeed, the network flow obtained by our approach using such tree T is identical to the one derived in [21] if one encodes AMONG constraints using the second variant discussed after Lemma 3. Applying Lemma 3 and noting that, in the particular case of the SEQUENCE constraint, we have $m = O(n)$ we obtain the bound $O(n^{3/2} \log^2 n)$. By inspecting closely the proof of Lemma 3 this bound can be slightly improved (see full version) to $O(n^{3/2} \log n \log max)$ coinciding with the bound given in [21], which is not surprising since both networks are essentially equivalent. To the best of our knowledge $O(n^{3/2} \log n \log max)$ is the best bound among all complete domain filtering algorithms for the problem, jointly with the algorithm proposed in [32] which, with time complexity $O(n2^k)$, provides better bounds when $k \ll n$.

6.4 TFO model

The TFO model was introduced by Razgon et al. [24] as a generalization of several common global constraints. Formally, a TFO model is a triple (V, F_1, F_2) where V is a finite set of vertices and F_1 and F_2 are nonempty families of subsets of V such that two sets that belong to the same family are either disjoint or contained in each other. Each set Y in $F_1 \cup F_2$ is associated with two non-negative integers $min_Y, max_Y \leq |Y|$. A subset X of V is said to be *valid* if $min_Y \leq X \cap Y \leq max_Y$ for every $Y \in F_1 \cup F_2$. The task is to find the largest valid subset. Although the methods introduced in the present paper can be generalized to deal as well with optimization version we will consider only now the feasibility problem consisting in finding a valid subset (or report that none exists).

First, note that the existence of a valid subset in a TFO model can be formulated naturally as a satisfiability problem for a combination of among constraints. Indeed, there is a one-to-one correspondence between the valid subsets of (V, F_1, F_2) and the solutions of the instance $(V, \{0, 1\}, L, \mathcal{C})$ where $\mathcal{C}$ contains the constraints $\text{AMONG}(Y, \{1\}, min_Y, max_Y), Y \in F_1 \cup F_2$, and the list $L(v)$ of every variable $v \in V$ is $\{0, 1\}$. The hypergraph H associated to this instance is $(V, F_1 \cup F_2)$. It can be shown (see full version) that H is a network hypergraph and hence one can use our approach to decide the existence of a feasible solution of a TFO model. It turns out that the network introduced in [24] is identical to the network flow problem that would be obtained by our approach using the first variant described after Lemma 3. It is not meaningful to compare the running time of our algorithm with that of [24] since it deals with an optimization variant.

6.5 Conjunction of Among Constraints with Full Domain

Some global constraints studied in the literature correspond to conjunctions $(V, D, L, \mathcal{C})$ of among constraints where the scope of every constraint is the full set V of variables. This class contains, of course, the GCC constraint and also several others, since we do not require R to be a singleton. For example, the ORDEREDDISTRIBUTE constraint introduced by Petit and Régin [23] can be encoded as conjunction $(V, D, L, \mathcal{C})$ of among constraint where the domain D has some arbitrary (but fixed) ordering $d_1, \ldots, d_{|D|}$ and in every constraint $\text{AMONG}(S, R, min, max)$, $S = V$ and R is of the form $\{d_i, \ldots, d_{|D|}\}$.

We shall show that the hypergraph of the conjunction of among constraints defining ORDEREDDISTRIBUTE is a network hypergraph. Indeed, with some extra work we have managed to completely characterize all CAC instances containing only constraints with full scope that have an associated network hypergraph.

Theorem 1. *Let $I = (V, D, L, \mathcal{C})$ be a conjunction of among constraints with $|D| \geq 3$ such that the scope of each constraint is V. Then, the following are equivalent:*

1. *The hypergraph of the canonical booleanization of I is a network hypergraph.*
2. *For every pair of constraints in $\mathcal{C}$, their ranges are disjoint or one of them contained in the other.*

In the particular case of ORDEREDDISTRIBUTE constraint the network obtained by our approach using the first variant described after Lemma 3 is identical to the network introduced in ([23], Sect. V.A). However, in this case our algorithm seems far from optimal. In particular, a complete filtering algorithm with time complexity $O(|V| + |D|)$ is given also in [23].

6.6   Adding New Among Constraints to a GCC constraint

Let (V, D, L, C) be a conjunction of among constraints encoding the GCC constraint (see Example 1) and assume that we are interested in adding several new among constraints to it. In general, we might end up with a hard instance but depending on the shape of the new constraints we might perhaps still preserve tractability. Which among constraint can we safely add? This question has been addressed by Régin [27]. In particular, [27] shows that the domain filtering problem is still tractable whenever:

(a) every new constraint added has scope V and, furthermore, the ranges of every pair of new constraints are disjoint, or

(b) every new constraint added has range D and, furthermore, the scopes of every pair of new constraints are disjoint.

We can explore this question by inquiring which families of constraints can be added to an instance (V, D, L, C) encoding GCC such that its associate hypergraph is still a network hypergraph. Somewhat surprisingly we can solve completely this question (see Theorem 2). This is due to the fact that the presence of the global cardinality constraint restricts very much the shape of the tree defining the hypergraph of the instance.

Theorem 2. *Let $I = (V, D, L, C)$ be a conjunction of among constraints containing a global cardinality constraint with scope V with $|D| \geq 3$. Then the following are equivalent:*

1. *The hypergraph of the canonical booleanization of I is a network hypergraph.*
2. *In every constraint in C, the scope is a singleton or V, or the range is a singleton or D. Furthermore, for every pair $\mathrm{AMONG}(S_1, R_1, min_1, max_1)$, $\mathrm{AMONG}(S_2, R_2, min_2, max_2)$ of constraints in C the following two conditions hold:*
   (a) *If $S_1 = S_2 = V$ or $S_1 = S_2 = \{v\}$ for some $v \in V$ then R_1 and R_2 are disjoint or one of them is contained in the other.*
   (b) *If $R_1 = R_2 = D$ or $R_1 = R_2 = \{d\}$ for some $d \in D$ then S_1 and S_2 are disjoint or one of them is contained in the other.*

Note that the previous theorem covers cases (a) and (b) from [27] described at the beginning of this section. The network produced in [27] for the case (a) is identical to the one derived by our approach using the first variant described after Lemma 3. For the case (b) [27] does not construct a flow problem nor gives run-time bounds so we omit a comparison.

Acknowledgments. The author would like to thank the anonymous referees for many useful comments. This work was supported by the MEIC under grant TIN2016-76573-C2-1-P and the MECD under grant PRX16/00266.

References

1. Ahuja, R.K., Magnanti, T.L., Orlin, J.B.: Network Flows - Theory. Algorithms and Applications. Prentice Hall, Upper Saddle River (1993)
2. Appa, G., Magos, D., Mourtos, I.: LP relaxations of multiple all-different predicates. In: Régin, J.-C., Rueher, M. (eds.) CPAIOR 2004. LNCS, vol. 3011, pp. 364–369. Springer, Heidelberg (2004). doi:10.1007/978-3-540-24664-0_25
3. Appa, G., Magos, D., Mourtos, I.: On the system of two all-different predicates. Inf. Process. Lett. **94**(3), 99–105 (2005)
4. Bacchus, F., Walsh, T.: Propagating logical combinations of constraints. In: Proceedings of IJCAI 2005, pp. 35–40 (2005)
5. Beldiceanu, N., Contejean, E.: Introducing global constraints in chip. Math. Comput. Modell. **12**, 97–123 (1994)
6. Bessiere, C., Hebrard, E., Hnich, B., Kiziltan, Z., Toby Walsh, S.: A useful special case of the CARDPATH constraint. In: Proceedings of ECAI 2008, pp. 475–479 (2008)
7. Bessiere, C., Hebrard, E., Hnich, B., Kiziltan, Z., Walsh, T.: Range and roots: two common patterns for specifying and propagating counting and occurrence constraints. Artif. Intell. **173**(11), 1054–1078 (2009)
8. Bessiere, C., Katsirelos, G., Narodytska, N., Quimper, C.-G., Walsh, T.: Propagating conjunctions of alldifferent constraints. In: Proceedings of AAAI 2010 (2010)
9. Chabert, G., Demassey, S.: The conjunction of interval among constraints. In: Beldiceanu, N., Jussien, N., Pinson, É. (eds.) CPAIOR 2012. LNCS, vol. 7298, pp. 113–128. Springer, Heidelberg (2012). doi:10.1007/978-3-642-29828-8_8
10. Dalmau, V.: Conjunctions of among constraints. Technical report, eprint arXiv:1706.05059 (2017)
11. Färnqvist, T., Jonsson, P.: Bounded tree-width and CSP-related problems. In: Tokuyama, T. (ed.) ISAAC 2007. LNCS, vol. 4835, pp. 632–643. Springer, Heidelberg (2007). doi:10.1007/978-3-540-77120-3_55
12. Fellows, M.R., Friedrich, T., Hermelin, D., Narodytska, N., Rosamond, F.A.: Constraint satisfaction problems: convexity makes alldifferent constraints tractable. Theor. Comput. Sci. **472**, 81–89 (2013)
13. Goldberg, A.V., Rao, S.: Beyond the flow decomposition barrier. J. ACM **45**(5), 783–797 (1998)
14. Gottlob, G., Leone, N., Scarcello, F.: A comparison of structural CSP decomposition methods. Artif. Intell. **124**(2), 243–282 (2000)
15. Hopcroft, J.E., Karp, R.M.: An $n^{5/2}$ algorithm for maximum matchings in bipartite graphs. SIAM J. Comput. **2**(4), 225–231 (1973)
16. Kocjan, W., Kreuger, P.: Filtering methods for symmetric cardinality constraint. In: Régin, J.-C., Rueher, M. (eds.) CPAIOR 2004. LNCS, vol. 3011, pp. 200–208. Springer, Heidelberg (2004). doi:10.1007/978-3-540-24664-0_14
17. Kutz, M., Elbassioni, K.M., Katriel, I., Mahajan, M.: Simultaneous matchings: hardness and approximation. J. Comput. Syst. Sci. **74**(5), 884–897 (2008)
18. Lardeux, F., Monfroy, E., Saubion, F.: Interleaved alldifferent constraints: CSP vs. SAT approaches. In: Dochev, D., Pistore, M., Traverso, P. (eds.) AIMSA 2008. LNCS (LNAI), vol. 5253, pp. 380–384. Springer, Heidelberg (2008). doi:10.1007/978-3-540-85776-1_34
19. Laurière, J.-L.: A language and a program for stating and solving combinatorial problems. Artif. Intell. **10**(1), 29–127 (1978)

20. Magos, D., Mourtos, I., Appa, G.: A polyhedral approach to the alldifferent system. Math. Program. **132**(1–2), 209–260 (2012)
21. Maher, M., Narodytska, N., Quimper, C.-G., Walsh, T.: Flow-based propagators for the SEQUENCE and related global constraints. In: Stuckey, P.J. (ed.) CP 2008. LNCS, vol. 5202, pp. 159–174. Springer, Heidelberg (2008). doi:10.1007/978-3-540-85958-1_11
22. Oplobedu, A., Marcovitch, J., Toubier, Y.: CHARME: Un langage industriel de programmation par contraintes, illustré par une application chez renault. In: Proceedings of 9th International Workshop on Expert Systems and their Applications, pp. 55–70 (1989)
23. Petit, T., Régin, J.-C.: The ordered distribute constraint. Int. J. Artif. Intell. Tools **20**(4), 617–637 (2011)
24. Razgon, I., O'Sullivan, B., Provan, G.: Generalizing global constraints based on network flows. In: Fages, F., Rossi, F., Soliman, S. (eds.) CSCLP 2007. LNCS (LNAI), vol. 5129, pp. 127–141. Springer, Heidelberg (2008). doi:10.1007/978-3-540-89812-2_9
25. Régin, J.-C.: A filtering algorithm for constraints of difference in CSPs. In: Proceedings of AAAI 1994, pp. 362–367 (1994)
26. Régin, J.-C.: Generalized arc consistency for global cardinality constraint. In: Proceedings of AAAI 1996, pp. 209–215 (1996)
27. Régin, J.-C.: Combination of among and cardinality constraints. In: Barták, R., Milano, M. (eds.) CPAIOR 2005. LNCS, vol. 3524, pp. 288–303. Springer, Heidelberg (2005). doi:10.1007/11493853_22
28. Régin, J.-C., Gomes, C.P.: The cardinality matrix constraint. In: Proceedings of CP 2004, pp. 572–587 (2004)
29. Régin, J.-C., Puget, J.-F.: A filtering algorithm for global sequencing constraints. In: Smolka, G. (ed.) CP 1997. LNCS, vol. 1330, pp. 32–46. Springer, Heidelberg (1997). doi:10.1007/BFb0017428
30. Schrijver, A.: Theory of Linear and Integer Programming. Wiley, Hoboken (1998)
31. Tarjan, R.E.: Depth-first search and linear graph algorithms. SIAM J. Comput. **1**(2), 146–160 (1972)
32. Hoeve, W.-J., Pesant, G., Rousseau, L.-M., Sabharwal, A.: Revisiting the sequence constraint. In: Benhamou, F. (ed.) CP 2006. LNCS, vol. 4204, pp. 620–634. Springer, Heidelberg (2006). doi:10.1007/11889205_44
33. Jan van Hoeve, W., Pesant, G., Rousseau, L.-M., Sabharwal, A.: New filtering algorithms for combinations of among constraints. Constraints **14**(2), 273–292 (2009)

Clique Cuts in Weighted Constraint Satisfaction

Simon de Givry and George Katsirelos[✉]

MIAT, UR-875, INRA, 31320 Castanet Tolosan, France
simon.de-givry@inra.fr, gkatsi@gmail.com

Abstract. In integer programming, cut generation is crucial for improving the tightness of the linear relaxation of the problem. This is relevant for weighted constraint satisfaction problems (WCSPs) in which we use approximate dual feasible solutions to produce lower bounds during search. Here, we investigate using one class of cuts in WCSP: clique cuts. We show that clique cuts are likely to trigger suboptimal behavior in the specialized algorithms that are used in WCSP for generating dual bounds and show how these problems can be corrected. At the same time, the additional structure present in WCSP allows us to slightly generalize these cuts. Finally, we show that cliques exist in instances from several benchmark families and that exploiting them can lead to substantial performance improvement.

1 Introduction

The performance of branch and bound algorithms depends crucially on the quality of the dual bound produced during search. One of the techniques used in Integer Linear Programming (ILP) to improve dual bounds is cut generation. These work by adding to the LP relaxation constraints that are entailed by the integer program but not by the linear program. These eliminate optimal non-integral solutions and hence improve the dual bounds. One such class of cuts is clique cuts [4]. These are quite powerful, as they strengthen the inference possible in the LP from $\sum_{i=1}^{n} x_i \geq n/2$ to $\sum_{i=1}^{n} x_i \geq n - 1$, so when they can be found, they can increase performance significantly.

In weighted constraint satisfaction, adding cuts remains under explored so far. Instead, research has focused on other local techniques such as on-the-fly variable elimination [24], soft arc consistency [8,9,16,25,27,28,31], dominance detection [14,26], or global techniques like mini-bucket elimination [21] and tree-decomposition based search [1,12,15,20,29].

These techniques are useful, but they are orthogonal to cut generation and can be further improved by it. However, adding cuts to dual bound reasoning in WCSP presents several challenges. The bounds used in WCSP are based on producing feasible—but often suboptimal—dual solutions of the LP relaxation. The algorithms that do so are weaker than LP solvers, but are significantly faster. In some extreme cases, a WCSP solver can solve an instance to optimality by search on several thousand nodes in shorter time than it would take to solve the linear relaxation of that instance. This speed comes at a price, as it is not

© Springer International Publishing AG 2017
J.C. Beck (Ed.): CP 2017, LNCS 10416, pp. 97–113, 2017.
DOI: 10.1007/978-3-319-66158-2_7

easy to extend these algorithms to handle arbitrary linear constraints and to do so efficiently. In fact, we show that clique cuts reveal the worst cases for these bounds and, if used without care, may lead to no improvement whatsoever in the dual bound. We develop, however, a set of techniques and heuristics that prove successful in applying clique cuts to some families of instances. These families were cases where WCSP solving was much worse than applying ILP. With our contributions, the performance gap is closed significantly.

2   Background

Integer Linear Programming. An ILP has the general form

$$\min c^T x$$
$$s.t. Ax \geq b$$
$$x \geq 0$$
$$and$$
$$x \in Z^n$$

where c, b are constant vectors, x is a vector of variables and A is a constant matrix, and all entries in c, b and A are integral. If we ignore the integrality constraint, the resulting problem is the linear relaxation of the original ILP. As a relaxation, any dual bound of the LP is a dual bound of the ILP.

The dual of an LP as given above is

$$\max b^T y$$
$$s.t. A^T y \leq c$$
$$y \geq 0$$

Any feasible solution of the dual gives a lower bound of the primal and the optima meet.

Solving an ILP exactly is typically done using branch and bound. At each node, the linear relaxation is solved and if the lower bound is greater than the cost of the incumbent, the node is closed. The linear relaxation can be strengthened using cuts, i.e., linear inequalities that are not entailed by the LP relaxation, but are entailed by the ILP. All linear combinations of inequalities are entailed by the LP. A particular method of deriving inequalities that are not entailed by the LP is strengthening, which derives $\sum a_i x_i \geq \lceil b \rceil$ from $\sum a_i x_i \geq b$ when b is not integral. A special case that we use here is deriving $\sum a_i x_i \geq \lceil b/c \rceil$ from $\sum c a_i x_i \geq b$.

Weighted CSP. A WCSP is a tuple $\langle \mathcal{X}, \mathcal{C}, k \rangle$ where $\mathcal{X}$ is a set of discrete variables with associated domain $D(X)$ for each variable $X \in \mathcal{X}$, $\mathcal{C}$ is a set of cost functions with associated scope $scope(c) \subseteq \mathcal{X}$ for each $c \in \mathcal{C}$ and k is a distinguished "top" cost. For simplicity, we assume a single cost function per scope and write c_S for the unique function with scope S. We also write c_i for the unary cost function with scope $\{X_i\}$ and c_{ij} for the binary cost function with scope $\{X_i, X_j\}$. A partial assignment τ is a mapping from each of a subset of $\mathcal{X}$ (denoted $scope(\tau)$) to one of the values in its domain. A complete assignment is a partial assignment such that $scope(\tau) = \mathcal{X}$. We write $\tau(S)$ for all partial assignments such that $scope(\tau) = scope(S)$. Given a cost function c_S and a partial assignment τ such that $S \subseteq scope(\tau)$, $c(\tau|_S) \geq 0$ gives the cost of τ for c. If $c(\tau|_S) = k$, we say that the constraint is violated. Given a complete assignment τ, its cost is $c(\tau) = \sum_{c_S \in \mathcal{C}} c(\tau|_S)$. The objective of solving a WCSP is to find an assignment that minimizes $c(\tau)$. It defines an NP-hard problem.

Dual (lower) bounds in WCSP are derived by applying equivalence preserving transformations (EPTs). Given two cost functions c_{S_1}, c_{S_2} with $S_1 \subset S_2$, a partial assignment $\tau \in \tau(S_1)$ and a value α, we can move cost α between c_{S_1} and c_{S_2} by updating $c_{S_1}(\tau) \leftarrow c_{S_1}(\tau) + \alpha$ and $c_{S_2}(\tau') \leftarrow c_{S_2}(\tau') - \alpha$ for all $\tau' \in \tau(S_2)$ such that $\tau'|_{S_1} = \tau$ if this operation leaves no negative costs. This operation preserves the global cost of all assignments, so it is called an EPT. If α is positive, this operation is called a projection, otherwise it is an extension. They are written, respectively $project(c_{S_2}, c_{S_1}, \tau, \alpha)$ and $extend(c_{S_1}, \tau, c_{S_2}, -\alpha)$. For convenience, it is usual to assume the presence of a cost function with nullary scope, $c_\varnothing$. Since this function has nullary scope, its cost is used in every assignment of the WCSP and its value is a lower bound for the cost of any assignment of the WCSP (remember that all costs are non-negative). For any cost function c with non-empty scope, if $\alpha = \min_{scope(\tau)=scope(c)} c(\tau)$, we can apply $project(c, c_\varnothing, \emptyset, \alpha)$.

The *extend* and *project* operations preserve equivalence even if α is chosen so that the operation creates negative costs. But it would violate the requirement that all costs are non-negative. This implies that a given EPT may be inadmissible because it creates negative costs, but is admissible as part of a set of EPTs, such that if they are all applied no negative costs remain.

Soft Consistencies and Linear Programming. There has been a long sequence of algorithms for computing a sequence of EPTs. For the results from the WCSP literature, an overview is given by Cooper et al. [9]. A parallel development of algorithms has happened under the name maximum a posteriori (MAP) inference for Markov random fields (MRF) [39], starting with [36]. In most of the literature, the extension and projection operations are limited so that one of the cost functions involved has arity 1. We will also mostly limit our attention to this case here. In any case, it has been shown that all these algorithms find a feasible dual solution of the following linear relaxation of a WCSP/MRF:

$$\min \sum_{c_S \in \mathcal{C}, \tau \in \tau(S)} c_S(\tau) * y_\tau$$

$$s.t.$$

$$y_\tau = \sum_{\tau' \in \tau(S_2), \tau'|_{S_1} = \tau} y_{\tau'} \qquad \forall c_{S_1}, c_{S_2} \in \mathcal{C} \mid S_1 \subset S_2, \tau \in \tau(S_1), |S_1| \geq 1$$

$$\sum_{\tau \in \tau(S)} y_\tau = 1 \qquad \forall c_S \in \mathcal{C}, |S| \geq 1$$

However, specialized algorithms aim to be faster than using a linear solver on the above problem. But solving this special form is as hard as solving an arbitrary LP [32], hence these specialized algorithms typically converge on suboptimal dual solutions. One particular condition to which some algorithms converge is virtual arc consistency:

Definition 1. *Given a WCSP C, let $hard(C)$ be the CSP which has the same set of variables and domains as C and for each $c_S \in C$ has a hard constraint $hard(c_S)$, which is satisfied by an assignment τ if and only if $c_S(\tau) = 0$. C is virtually arc consistent if and only if $hard(C)$ is arc consistent.*

Among those algorithms that converge on virtual arc consistency is, not so surprisingly, VAC [9], to which we refer further in this paper.

Finally, we define EDAC, which is the default level of soft consistency enforced in the TOULBAR2 solver.

Definition 2. *A WCSP C is node consistent (NC) if for every cost function $c \in C$ with $|scope(c)| = 1$, there exists $\tau \in \tau(c)$ with $c(\tau) = 0$ and for all $\tau \in \tau(c)$, $c_\varnothing + c(\tau) < k$.*

Definition 3. *A WCSP C is generalized arc consistent (GAC) if for all $c_S \in C$, $|S| > 1$, $\forall \tau \in \tau(S)$, $c_S(\tau) = k$ if $c_\varnothing + c_S(\tau) + \sum_{X_i \in S} c_i(\tau|_{\{i\}}) = k$ and for all $X_i \in S$, $\forall v \in D(X_i)$, $\exists \tau \in \tau(S)$ such that $\tau_{\{i\}} = v$ and $c_S(\tau) = 0$.*

Definition 4. *A binary WCSP C is existential arc consistent (EAC) if there exists a value $u \in D(X_i)$ for each $X_i \in \mathcal{X}$ such that $c_i(u) = 0$ and $\forall c_{ij} \in C$, $\exists v \in D(X_j)$ s.t. $c_{ij}(u,v) + c_j(v) = 0$. It is existential directional arc consistent (EDAC) if it is NC, GAC, EAC, and directional arc consistent (DAC), i.e., $\forall c_{ij} \in C$, $i < j$, $\forall u \in D(X_i)$, $\exists v \in D(X_j)$ s.t. $c_{ij}(u,v) + c_j(v) = 0$.*

EDAC has been extended to ternary cost functions in [35]. Different generalizations for arbitrary arities have been proposed in [2,27,28].

3   Clique Cuts

An important class of cuts used by MIP solvers are *clique* cuts [4]. Given a set S of 0/1 variables and the constraints

$$x_i + x_j \geq 1 \qquad\qquad \forall x_i, x_j \in S, i \neq j$$

we can derive

$$\sum_{x_i \in S} x_i \geq |S| - 1 \tag{1}$$

This is done as follows. Given any triplet of distinct $x_i, x_j, x_k \in S$, we sum the binary constraints involving these three variables to get $2x_i + 2x_j + 2x_k \geq 3 \equiv x_i + x_j + x_k \geq 3/2$, which we strengthen to $x_i + x_j + x_k \geq 2$. We repeat this with m-tuples of variables and the $(m-1)$-ary constraints from the previous step to generate m-ary constraints until $m = |S|$, which gives the above constraint. This is in contrast to the much weaker constraint which can be derived by linear combinations only (that is, the constraint that is entailed by the LP), which is $\sum_{x_i \in S} x_i \geq |S|/2$.

The reason for the name of these cuts comes from the fact that the set S corresponds to a clique in a graphical representation of binary constraints of the IP. Specifically, we construct a graph with a vertex for each $0/1$ variable of the IP and have edges between two vertices if their variables appear together in a binary constraint of the form $x + y \geq 1$. From every clique in this graph we can derive a clique cut, i.e., a constraint of the form (1).

We can generalize this construction to have a vertex also for $1 - x$ (equivalently $\overline{x}$) for every $0/1$ variable x of the IP, connected to the vertex x. Then, there exists an edge between x and $\overline{y}$ for every constraint $x + (1-y) \geq 1 \equiv x - y \geq 0$ and an edge between $\overline{x}$ and $\overline{y}$ for every constraint $(1-x) + (1-y) \geq 1 \equiv -x - y \geq -1$. In this case if we find a clique that contains both x and $\overline{x}$, the clique cut requires that we must set all other variables in the clique to 1:

$$x + (1 - x) + \sum_{y_i \in S} y_i \geq |S| + 1 \Rightarrow$$

$$1 + \sum_{y_i \in S} y_i \geq |S| + 1 \Rightarrow$$

$$\sum_{y_i \in S} y_i \geq |S| \Rightarrow$$

$$y_i = 1 \qquad\qquad\qquad \forall y_i \in S$$

If we find a clique that contains $x, \overline{x}, y, \overline{y}$, the problem is unsatisfiable.

3.1   Cliques in WCSPs

We apply this reasoning to get clique cuts in WCSPs. From a WCSP P with top k, we construct a graph $G(P)$ as follows: we have a vertex for v_{xi} every variable x and every value $i \in D(x)$. There exists an edge between two vertices $v_{pv(p)}$, $v_{qv(q)}$ if there exists a cost function c_{pq} such that $c_{pq}(v(p), v(q)) = k$[1]. This corresponds to the constraint $(1 - x_{pv(p)}) + (1 - x_{qv(q)}) \geq 1$ in the LP relaxation.

Now, given a clique S in $G(P)$, we can add the clique constraint $\sum_{v_{pv(p)} \in S}(1 - x_{pv(p)}) \geq |S| - 1$. In other words, the constraint requires at least $n - 1$ variables in the clique must get a value other than the one included in the clique.

Overlapping Cliques. In the ILP case, we mentioned that the graph construction can be generalized to have vertices for both x and $(1 - x)$ and an edge between

---

[1] This is the micro-structure of the WCSP, restricted only to binary tuples with infinite cost.

them. In the WCSP case we already have vertices for different values of the same variable. We can ensure that each set of vertices corresponding to values of the same variable forms a clique. Then, the general case for a clique is that it may contain several values from each variable involved. We assume in the rest of this paper that this is the case. Then, given such a clique S, $varsof(S) = \{X_p \mid \exists i.v_{pi} \in S\}$. For every $X_p \in varsof(S)$ we write $S(p) = \{i \mid v_{pi} \in S\}$. The constraint then requires that at least $|varsof(S)| - 1$ of the variables get a value outside their respective set $S(p)$.

Despite this generalization, in order to simplify presentation, we will assume that every variable in the clique is a binary variable and $S(p) = \{1\}$. All results and algorithms are valid for the general case, with the caveat that when we use $c_i(0)$ ($c_i(1)$) on the right hand side of an expression, it means $\min_{i \notin S(i)} c_i(v)$ ($\min_{i \in S(i)} c_i(v)$) and when it appears on the left hand side, it means for all $v \notin S(i), c_i(v)$ ($v \in S(i), c_i(v)$).

3.2 Propagating Clique Constraints

We can encode a clique constraint with a non-uniform layered automaton (meaning an automaton where the transition function may differ in each layer) with two states q_0, q_1. The initial state is q_0 and both are accepting. Suppose $|S| = n$ and the variables involved are $x_1, \ldots, x_n$. Then the transition function at layer i is

Q_{i-1}	X_i	Q_i
q_0	$j \notin S(i)$	q_0
q_0	$j \in S(i)$	q_1
q_1	$j \notin S(i)$	q_1
q_1	$j \in S(i)$	$not\ allowed$

We can encode the automaton using the usual ternary construction [33,34]. This ensures that EDAC deduces the optimal lower bound for each clique constraint, at least when viewed in isolation. As the constraint is invariant under reordering of the variables, we can use an ordering that agrees with the EDAC ordering and place each state variable Q_i in between the variables x_i and x_{i+1} in the EDAC ordering. This is sufficient to guarantee that EDAC propagates each clique constraint optimally [2].

However, this is not an attractive option in practice. The reformulation to automata introduces many variables for each clique constraint, something to which EDAC is quite sensitive. That is, depending on the vagaries of the algorithm, such as the order in which constraints of a variable are processed, it may derive a stronger or weaker lower bound, even though the formulation is locally optimal.

Combined with the issues that we describe later, this leads us to implement a specialized propagator for these constraints, which includes the reasoning performed by the softregular constraint [18] on this automaton. For completeness, we describe the propagation independent of the softregular constraint. This also sets the stage for the discussion of Sect. 3.3.

Propagating Clique Constraints. For each clique constraint clq, we store a single integer a_0 which summarizes the effect on the constraint of all extensions and projections we have performed. This quantity is the cost of assigning all variables to 0.

We define the following transformation, $u \rightarrow clq$ which performs a set of extensions and projections through the clique constraint, getting the maximum increase in $c_\varnothing$ and extending the minimum amount from unary costs in order to achieve this increase.

Let clq be the clique constraint, $l_0 = \sum c_i(0) - \max\{c_i(0)\}$, $r_0 = \max(\{c_i(0)\}\setminus \max\{c_i(0)\})$ (the second largest $c_i(0)$, possibly equal to the largest), $t = \sum c_i(0)$, $t_r = l_0 + r_0$ and $l_1 = \min\{t_r - l_0, \min c_i(1), a_0\}$. Then $u \rightarrow clq$ comprises the following operations:

- If the arity is 2 (resp. 1), project a_0 to the $(0,0)$ tuple (resp. 0 value) of the corresponding binary (resp. unary) cost function. Otherwise:
- Add $l_0 + l_1$ to $c_\varnothing$
- Set each unary cost $c_i(0)$ to $\max(0, c_i(0) - r_0)$
- Add $t_r - l_0 - l_1$ to a_0 (possibly reducing a_0)
- Add $t_r - l_0 - l_1 - \min(c_i(0), r_0)$ to $c_i(1)$ for all i.

Proposition 1. $u \rightarrow clq$ *is an EPT*

Proof. We ignore here any binary costs. Since these are left untouched, they would contribute the same cost to any assignment before and after the EPT.

Suppose $n - 1$ variables are assigned 0 and that $x_i = 1$. The cost of this assignment is $c_i(1) + \sum_{j \neq i} c_j(0)$. In the reformulated problem, the cost of the assignment is $l_0 + l_1$ from $c_\varnothing$, $c_i(1) + t_r - l_0 - l_1 - \min(c_i(0), r_0)$ from the unary cost of $x_i = 1$, and $\sum_{j \neq i} \max(0, c_j(0) - r_0)$ from the unary costs of $x_j = 0$ for $j \neq i$. This sums to $c_i(1) + t_r - \min(c_i(0), r_0) + \sum_{j \neq i} \max(0, c_j(0) - r_0)$.

If $\max c_i(0) = r_0$, $\min(c_i(0), r_0)$ simplifies to $c_i(0)$, $\max(0, c_j(0) - r_0)$ and $t_r = t$. Then the above sum is $c_i(1) + t - c_i(0)$ which is $c_i(1) + \sum_{j \neq i} c_i(0)$, as in the original problem.

If $\max c_i(0) > r_0$, there exists a unique x_k for which $c_k(0) = \max c_i(0)$. Then $\min(c_i(0), r_0)$ simplifies to $c_i(0)$ for all $i \neq k$ and to r_0 for $i = k$, while $\sum_{j \neq i} \max(0, c_j(0) - r_0)$ is 0 if $i = k$ and $c_k(0) - r_0$ if $i \neq k$. So if $i = k$ the sum is $c_i(1) + t_r - r_0 = c_i(1) + l_0 = c_i(1) + \sum_{j \neq i} c_j(0)$, as in the original problem. If $i \neq k$, the sum is $c_1(1) - c_i(0) + t_r + c_k(0) - r_0 = c_1(1) + l_0 + c_k(0) - c_i(0)$. As l_0 is the sum of all $c_j(0)$ except for $c_k(0)$, that works out to $c_1(1) + \sum_{j \neq i} c_i(0)$, equivalent to the cost in the original problem.

Finally, assume all n variables are assigned 0, the cost of the assignment is $t = \sum_i c_i(0) = l_0 + l_1 + t - l_0 - l_1 = c_\varnothing + a_0 + (t - t_r)$. We have $t - t_r = \max c_i(0) - r_0$, which is exactly the cost left in $c_k(0)$ if $\max c_i(0) > r_0$ and 0 otherwise. In either case the cost remains the same before and after the transformation, as long as a_0 is later projected to $c_\varnothing$. As all variables are assigned 0, the cost a_0 is indeed used by projecting to a lower arity cost function, so the cost in the reformulated problem is the same as in the original problem.

In general, we apply $u \to clq$ after node consistency has been enforced. Therefore, for each variable, either $c_i(0) = 0$ or $c_i(1) = 0$ and so if $\min c_i(1) > 0$, it has to be that $t - l_0 = \max c_i(0) = 0$. This means that either l_0 or l_1 will be non-zero. But unary costs change non-monotonically as other constraints move costs through the variable of the clique, so it is possible for both to occur at different times in the lifetime of the same constraint.

3.3   Issues with Virtual Arc Consistency

Using clique constraints to improve lower bounds computed by VAC or any algorithm that converges to an arc consistent state is problematic. This includes algorithms such as MPLP [17,37,38], TRW-S [23] and other algorithms from the MRF community. We will show that the problem is that all these algorithms can only improve the lower bound by *sequences* of EPTs, but it is required to use *sets* of EPTs to fully exploit clique constraints.

As a hard constraint, the clique constraint is redundant, not only logically, but also in terms of propagation. Indeed, suppose a clique S contains values from three variables x, y and z, that the pairwise constraints are arc consistent, and let $d \in D(y)$ such that $v_{yd} \in S$ and t_{xy}, t_{yz} be the supports of $y = d$ in the corresponding binary constraints. Since $t_{xy}[x]$ and $t_{yz}[z]$ are values that may not be part of the clique, they are consistent with each other in the clique constraint and hence $t_{xyz} = t_{xy} \cup t_{yz}$ is a support for all three variables simultaneously. We can extend this reasoning to an arbitrary number of variables, hence we can get global supports from pairwise supports.

The fact that clique constraints add no propagation strength to $hard(C)$ means that adding clique constraints after VAC propagation will have no effect. Indeed, VAC can only improve the lower bound as long as $hard(C)$ is arc inconsistent. Since the clique constraint is propagation redundant, if $hard(C)$ is arc consistent, it will remain so after adding the clique constraint.

Empirically, we have observed the same is often true after EDAC, even though it does not necessarily converge to a virtually arc consistent state. Moreover, even if the clique constraint exists before we enforce VAC, the fact that VAC does not have a unique fixpoint means that we cannot predict whether the fixpoint that it does reach will use the clique constraint optimally. Hence, we need to devise a method to exploit clique constraints in a virtually arc consistent problem.

Example 1. To begin, consider what happens in a problem with 3 Boolean variables in a clique, where all costs are 0 except $c_x(0) = c_y(0) = c_z(0) = 1$ and $c_{xy}(1, 1) = c_{yz}(1, 1) = c_{xz}(1, 1) = k$. The following series of EPTs makes the problem VAC with $c_\varnothing = 3/2$:

1. $extend(x, 0, xy, 1/2)$
2. $extend(x, 0, xz, 1/2)$
3. $extend(z, 0, yz, 1/2)$
4. $project(xy, y, 1, 1/2)$
5. $project(xz, z, 1, 1/2)$

6. $project(yz, y, 1, 1/2)$
7. $project0(y, 1)$
8. $project0(z, 1/2)$.

The reformulated problem has $c_{xy}(1,1) = c_{yz}(1,1) = c_{xz}(1,1) = k$ and $c_{xy}(0,0) = c_{yz}(0,0) = c_{xz}(0,0) = 1/2$. Note that actually running VAC would perform a more convoluted series of moves: in the first iteration, a conflict involves only two variables, say x and y. That allows it to move 1 unit of cost from $x = 0$ to $y = 1$, and to project cost 1 to $c_\emptyset$. In the next iteration, the conflict involves all 3 variables and it moves $1/2$ a unit of cost from $z = 0$ to $x = 1$ and from there to $y = 0$ and another $1/2$ a unit directly from $z = 0$ to $y = 1$, projecting another $1/2$ to $c_\emptyset$.

Since this reformulated problem is VAC, adding a clique constraint will do nothing, but solving the linear relaxation detects that we can improve the lower bound by the following set of operations[2]:

1. $extend(y, 1, xy, 1/2)$. After this, $c_y(1) = -1/2$
2. $extend(z, 1, xz, 1/2)$, $extend(z, 1, yz, 1/2)$. After these, $c_z(1) = -1$
3. $project(xy, x, 0, 1/2)$
4. $project(xz, x, 0, 1/2)$
5. $project(yz, y, 0, 1/2)$
6. $extend(x, 0, clq, 1)$
7. $extend(y, 0, clq, 1/2)$
8. $project0(clq, 1/2)$
9. $project(clq, y, 1, 1/2)$
10. $project(clq, z, 1, 1)$.

This leaves the problem with lower bound increased to 2 and the only non-zero costs are the hard tuples, which are unchanged throughout, and the tuple $c_{clq}(0,0,0) = 1$.   □

More generally, we can derive a specific method for propagating cliques in virtually arc consistent instances by extending our reasoning to binary costs. In the following, let $n = |S|$. Since $(1 - x_{i1}) = x_{i0}$ for binary domains, the constraint can be written either as $\sum_{x_{i1} \in S} x_{i0} \geq n - 1$ or equivalently $\sum_{x_{i1} \in S} x_{i1} \leq 1$. Thus, either $n - 1$ or n variables must be assigned 0, meaning either $\binom{n}{2}$ or $\binom{n-1}{2}$ binary $\langle 0, 0 \rangle$ tuples will be used. This means

$$\sum_{x_{i1}, x_{j1} \in S, i < j} y_{ij00} \geq \binom{n-1}{2} \tag{2}$$

These constraints can be further strengthened by observing that if we assign $x_i = 1$, then none of the binary tuples $c_{ij}(0,0)$ for $j \neq i$ are used. Thus we can

---

[2] In this case, triangle-based consistencies [31] achieve the same effect, but not in arbitrary arity cliques.

treat these binary tuples as blocks, one for each variable, at least $n-1$ of which have to be used. This is captured by the following system:

$$\sum_{x_{i1},x_{j1}\in S,i<j} y_{ij00} \geq \binom{n-1}{2}$$

$$\sum_{x_{i1}\in S} x_{i0} \geq n-1$$

$$x_{i0} \geq y_{ij00} \forall i,j \tag{3}$$

We can incorporate (3) into propagation of the clique constraint. Let $base_i = \sum_{j<k\in[1,n]\backslash\{i\}} c_{jk}(0,0)$ for all $i \in [1,n]$, $base = \min\{base_i \mid i \in [1,n]\}$ and $total = \sum_{j<k\in[1,n]} c_{jk}(0,0)$. We define the *binary-to-clique* transform, denoted as $b \rightarrow clq$ as the transformation which performs the following operations:

1. Add $base$ to $c_\varnothing$
2. Set $c_{ij}(0,0) = 0$ for every binary constraint ij in the clique.
3. Add cost $total - base$ to a_0
4. For every variable i, add cost $comp_i = base_i - base$ to $c_i(1)$.

Proposition 2. *The $b \rightarrow clq$ transform is an EPT.*

Proof. Consider an arbitrary feasible solution of the subproblem. Since it is feasible, at least $n-1$ variables are assigned 0, so we only need to consider the cases where n or $n-1$ variables are 0.

Assume all n variables are assigned 0. The cost of this assignment in the original problem is $total = \sum_{j<k\in[1,n]} c_{jk}(0,0)$. In the reformulated problem, the clique constraint entails cost $a_0 = total - base$, because we assign all zeroes. Together with the cost $base$ projected to $c_\varnothing$, it gives $total$ in the reformulated problem as well.

Assume exactly $n-1$ variables are assigned 0 and that the variable assigned to 1 is X_i. The cost of this assignment is $base_i = \sum_{j<k\in[1,n]\backslash\{i\}} c_{jk}(0,0)$. In the reformulated problem, the cost of the assignment is $base$ from $c_\varnothing$ and $base_i-base$ from $c_i(1)$, giving $base_i$, identical to the original problem.

This transform involves a higher-order transformation, so it is tempting to think that it is stronger than the linear relaxation. Unfortunately, this is not the case.

Proposition 3. *The $b \rightarrow clq$ transform can be expanded to a set of EPTs.*

Proof (Sketch). There exist already positive costs on the $(0,0)$ tuple of binary cost functions. If the 1 value of one unary cost function has positive cost, we can extend it to the binary cost function and project it to the 0 value on the other variable involved. We can do this even if no such positive cost exists and create a temporary negative cost. However, we can now apply the $u \rightarrow clq$ transform, because we have positive costs on the 0 value of the unary cost functions. This projects costs back to the 1 value of the unary cost functions involved, hence covering the deficit created in the first step.

Convergence. As was observed in [28], EDAC may not converge when cost functions of higher arity exist. VAC is better behaved, but it may converge only as the number of iteration grows to infinity. It is straightforward to see that the presence of clique constraints does not raise any such issues: every time either the $u \to clq$ or $b \to clq$ transform is applied, the cost $c_\varnothing$ is increased. As $c_\varnothing$ cannot be increased past the global optimum, this means that application of these rules converges. Moreover, $c_\varnothing$ is increased by an integer, so the number of iterations is also bounded by the cost of the global optimum.

3.4   Clique Selection and Ordering Heuristic

Finding Cliques. We implement detection of cliques as a preprocessing step. We construct a graph as described previously: there exists a vertex for each variable-value pair and an edge between two vertices if $c_\varnothing + c_p(i) + c_q(j) + c_{pq}(i,j) \geq k$ or if they represent two values of the same variable. We then use the Bron-Kerbosch algorithm [6] with degeneracy ordering to generate a set of cliques. In some cases, the number of cliques can be overwhelming, so we place a limit on the maximum number of cliques generated per top-level branch (which should roughly correspond to the number of cliques that contain a single vertex), as well as a global limit on the total number of cliques. We discard cliques S for which $|varsof(S)| < 3$, as cliques over 2 variables are simply subsets of binary cost functions and can propagate no better.

Selection and Ordering. The order of EPTs may have a large impact on the quality of the resulting lower bounds.

Example 2. Consider two cliques C_1, C_2 with scope $\{X_1, X_2, X_3\}$ and $\{X_2, X_3, X_4\}$, respectively, such that all variables have binary domains and the 1 value from each variable participates in the cliques and $c_i(0) = i$. If we propagate C_1 first, we project cost 3 to $c_\varnothing$ and update the unary costs $c_1(0) = c_2(0) = 0$, $c_3(0) = 1$ and $c_1(1) = 1$ leaving the rest unchanged. We then propagate C_2 and project 1 unit of cost to $c_\varnothing$ and update $c_3(0) = 0$, $c_4(0) = 3$ and $c_2(1) = 1$. The final cost of $c_\varnothing$ is 4.

On the other hand, if we propagate C_2 first, we project 5 units of cost to $c_\varnothing$, leaving $c_2(0) = c_3(0) = 0, c_4(0) = 1$ and $c_2(1) = 1$. After this, C_1 does not propagate. Hence, by propagating C_2 before C_1, we get a stronger lower bound.

The above problem does not, of course, manifest itself when actually solving the linear relaxation of the problem. It is, however, an inherent limitation of algorithms like EDAC, which have no flexibility to process constraints in a different order in different passes. As we explained above, a more flexible algorithm like VAC would not help either, as clique constraints are redundant with respect to propagation on the hard problem. Therefore, our only recourse is to select an order of propagation before performing the actual propagation. We can choose this order either before search or dynamically at each node of the search. Here, we chose to select the order just once before search begins.

We choose a greedy heuristic to select and order the initial clique constraint propagation. For that, we collect a bounded number of potential clique constraints, as described previously. We then use the classical Chvatal's set covering heuristic [7] to find the best clique which maximizes the product of current arity and current lower bound increase (by simulating the effect of its unary-to-clique transform). We repeat this selection process followed by the corresponding unary-to-clique transform until all the remaining cliques do not increase the lower bound or have all their variables covered by another already-selected clique. Ties are broken by a lexicographic ordering on the scope of the cliques. We keep also all cliques found of arity 3 because they are natively managed by the TOUL-BAR2 solver as ternary/triangle cost functions with dedicated *EDAC* soft arc consistency [31,35].

4   Related Work

The most related work is that of Atamtürk et al. [4], who explore adding clique cuts in MIP solvers, not only during preprocessing but also during search. As we descend the search tree, more tuples become effectively hard, creating more cliques. This remains a future direction for us.

Khemmoudj and Bennaceur [22] studied the use of binary cliques to improve lower bounds. These are used to obtain better approximations than EDAC to the optimum of the LP relaxation of a MaxCSP. In contrast to our work here, the strength of the LP relaxation is not improved.

Sontag et al. [38] strengthen the LP relaxation by adding higher order cost functions, which may eventually make the relaxation as strong as the polytope of the integer program, at the cost of potentially making the LP exponentially larger. Later, Sontag et al. [37] considered adding cuts that correspond to frustrated cycles in the graph, which is more efficient but less powerful.

Cliques can also be handled efficiently in settings outside of linear relaxations. For example, Narodytska and Bacchus [30] proposed a method for MaxSAT which applies max-resolution on cores extracted from the instance. This method is complete, i.e., will eventually produce the optimum of an instance. This method can perform equivalent reasoning to a clique cut in a polynomial number of steps and with cores that can be discovered in polynomial time.

On the subject of clique detection, Dixon [13], Ansotegui Gil [3] and Biere et al. [5] showed several techniques that may uncover cliques that are not explicitly present in the microstructure (called NAND graph in SAT when restricted to binary clauses), using both syntactic and semantic (propagation-based) information.

5   Experimental Results

We have implemented the clique generation and clique constraint propagation inside TOULBAR2 (version 0.9.8), an open-source WCSP solver in C++. Among the various benchmarks from [19] (available in LP, WCNF, WCSP, UAI, and

MiniZinc formats) where CPLEX reports that clique cuts applied, we chose four problem categories, combinatorial auctions *Auction/path*, *Auction/sched*, maximum clique *MaxClique*, and satellite management *SPOT5*, a total of 252 instances, having binary forbidden tuples and initial unary cost functions such that the unary-to-clique transform increases the lower bound in preprocessing. The first three categories have Boolean domains, whereas SPOT5 has maximum domain size of 4. For each category, we report in Table 1 the mean value of the size of the problem, the number of cliques found, the number of selected cliques among them and their arity, and the CPU time to find and select the cliques. We limit the maximum number of cliques found to 10,000 in order to control the computation time. The largest CPU time was 11.61 s for *MaxClique/c-fat200-5* ($n = 200$ variables, $e = 11,627$ cost functions, and 10,000 selected cliques of arity 3). The arity of selected cliques varies from 3 to 67 (*MaxClique/san1000*).

Table 1. Clique generation process: number of instances per benchmark category followed by maximum domain size (d), mean number of variables (n), cost functions (e), graph vertices, graph edges, cliques found, and selected cliques, followed by mean arity of selected cliques and CPU time in seconds to find and select cliques. A limit on the maximum number of cliques found was set to 10,000.

Problem	nb	d	n	e	Vertices	Edges	Cliques	Selected c.	Arity	Time
Auction/path	86	2	120.2	1,475.7	120.2	1,355.5	143.9	28.8	7.3	0.02
Auction/sched	84	2	159.7	5,759.9	159.7	5,600.2	822.5	3.6	44.9	0.27
MaxClique	62	2	484.3	50,092.8	484.3	49,608.5	8,372.7	325.2	3.5	2.87
SPOT5	20	4	385.1	6,603.3	761.0	9,411.3	5,888.1	127.3	4.1	1.71

We compare solving time to find and prove optimality for TOULBAR2 exploiting cliques (denoted as TOULBAR2^{clq}) against the original code without cliques (both using default options, including hybrid best-first search [1]), and against the CP solver GECODE[3], the MaxSAT solvers MAXHS 2.51 [10,11] and EVA 500a [30], and IBM-ILOG CPLEX 12.6.0.0 (using a direct encoding [19] and parameters EPAGAP, EPGAP, and EPINT set to zero to avoid premature stop). All computations were performed on a single core of AMD Opteron 6176 at 2.3 GHz and 8 GB of RAM with a 1-h CPU time limit[4]. In Table 2, we give the number of solved instances within a 1-h CPU time limit. Among 252 instances, CPLEX solved 224 instances, MAXHS 216, TOULBAR2^{clq} 213, EVA 208, TOULBAR2 205, and GECODE 137. For *Auction*, TOULBAR2^{clq} is more than two orders of magnitude faster than TOULBAR2, GECODE (which cannot solve 57 *Auction/path* instances in 1 h) and EVA on *Auction/path* (*cat_paths_60_170_0005* instance unsolved in 1 h). Still TOULBAR2^{clq} is one order of magnitude slower than CPLEX and MAXHS. On *MaxClique* (resp. *SPOT5*), TOULBAR2^{clq} solved 6

---

[3] Version 4.4.0, using free search.

[4] Using parameter *-pe parallel_smp 2* on a SUN Grid Engine to ensure half-load of the cores on the cluster.

Table 2. Number of solved instances and mean solving computation time in seconds, for TOULBAR2 solver without using cliques compared to TOULBAR2^{clq} exploiting cliques, CPLEX, MAXHS, EVA and GECODE. TOULBAR2^{clq} solving time does not take into account clique generation time (see Table 1). A CPU time limit of 1 h was used for unsolved instances for reporting mean solving times.

Problem	TOULBAR2		TOULBAR2^{clq}		CPLEX		MAXHS		EVA		GECODE	
	Solv.	Time	Solv.	Time	Solv.	Time	Solv.	Time	Solv.	Time	Solv.	Time
Auction/path	**86**	59	**86**	0.18	**86**	0.01	**86**	0.01	85	102	29	2614
Auction/sched	**84**	110	**84**	0.23	**84**	0.04	**84**	0.04	**84**	0.28	**84**	76
MaxClique	31	1871	37	1508	38	1533	**40**	1510	26	2268	24	2314
SPOT5	4	2884	6	2603	**16**	738	6	2577	13	1260	0	3600

(resp. 2) more instances than without using cliques. For example, TOULBAR2^{clq} solved *MaxClique/MANN_a45* in 57.9 s (taking 0.24 supplementary seconds to generate the 330 selected cliques) whereas without using cliques it could not finish in 1 h (MAXHS took 28 s, CPLEX 93 s, and EVA and GECODE could not solve in 1 h). *SPOT5/404* was solved in 6 s using 32 cliques and 88.7 s without using cliques (CPLEX took 0.02 s, EVA 0.07 s, MAXHS 8 s, and GECODE could not solve in 1 h).

Finally, we summarize the evolution of lower and upper bounds for each solver over all instances in Fig. 1. Specifically, for each instance I we normalize all costs as follows: the initial lower bound produced by TOULBAR2 is 0; the best – but potentially suboptimal – solution found by any solver is 1; the worst solution

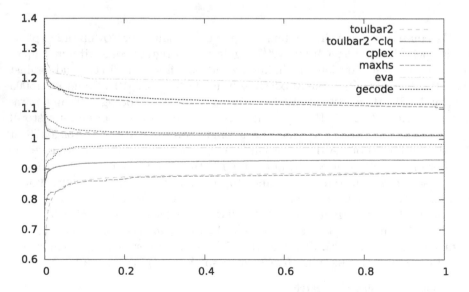

Fig. 1. Normalized lower and upper bounds on 252 instances as time passes.

is 2. This normalization is invariant to translation and scaling. Additionally, we simply normalize time from 0 to 1, corresponding to 1 h. A point $\langle x, y \rangle$ on the lower bound line for solver S in Fig. 1 means that after normalized runtime x, solver S has proved on average over all instances a normalized lower bound of y and similarly for the upper bound. We show both the upper and lower bound curves for all solvers evaluated here, except GECODE which produces no meaningful lower bound before it proves optimality. We observed that using cliques, it mainly improves lower bounds for TOULBAR2. For these benchmarks, CPLEX got the best lower bound curve and TOULBAR2^{clq} the best upper bound curve.

6   Conclusions

We have shown how the idea of clique cut originated from MIP can be exploited in the context of WCSPs. Using these cuts in WCSP is significantly more complicated than in integer programming, owing to a large degree to the fact that the fast specialized algorithms that are used in place of solving the linear relaxation have weaknesses which seem to be particularly exposed by clique constraints. To address these shortcomings, we provide two specific EPTs, unary-to-clique and binary-to-clique, which propagate isolated clique constraints optimally, even if costs are hidden in binary cost functions. We then gave an algorithm to greedily select and order a subset of potential cliques in preprocessing and do the propagation on these selected cliques during search. In an experimental evaluation, we have obtained large improvements over the existing complete solver TOULBAR2 on several benchmarks, significantly reducing the gap to state-of-the-art solver CPLEX.

Acknowledgements. This work has been partially funded by the french "Agence nationale de la Recherche", reference ANR-16-C40-0028. We are grateful to the Bioinfo Genotoul platform Toulouse Midi-Pyrenees for providing computing resources.

References

1. Allouche, D., de Givry, S., Katsirelos, G., Schiex, T., Zytnicki, M.: Anytime hybrid best-first search with tree decomposition for weighted CSP. In: Pesant, G. (ed.) CP 2015. LNCS, vol. 9255, pp. 12–29. Springer, Cham (2015). doi:10.1007/978-3-319-23219-5_2
2. Allouche, D., Bessière, C., Boizumault, P., de Givry, S., Gutierrez, P., Lee, J.H., Leung, K.L., Loudni, S., Métivier, J.P., Schiex, T., Wu, Y.: Tractability-preserving transformations of global cost functions. Artif. Intell. **238**, 166–189 (2016)
3. Anstegui Gil, C.: Complete SAT solvers for many-valued CNF formulas. Ph.D. thesis, University of Lleida (2004)
4. Atamtürk, A., Nemhauser, G.L., Savelsbergh, M.W.: Conflict graphs in solving integer programming problems. Eur. J. Oper. Res. **121**(1), 40–55 (2000)
5. Biere, A., Le Berre, D., Lonca, E., Manthey, N.: Detecting cardinality constraints in CNF. In: Sinz, C., Egly, U. (eds.) SAT 2014. LNCS, vol. 8561, pp. 285–301. Springer, Cham (2014). doi:10.1007/978-3-319-09284-3_22

6. Bron, C., Kerbosch, J.: Algorithm 457: finding all cliques of an undirected graph. Commun. ACM **16**(9), 575–577 (1973)
7. Chvatal, V.: A greedy heuristic for the set-covering problem. Math. Oper. Res. **4**(3), 233–235 (1979)
8. Cooper, M., de Givry, S., Sanchez, M., Schiex, T., Zytnicki, M.: Virtual arc consistency for weighted CSP. In: Proceedings of AAAI-2008, Chicago, IL (2008)
9. Cooper, M., de Givry, S., Sanchez, M., Schiex, T., Zytnicki, M., Werner, T.: Soft arc consistency revisited. Artif. Intell. **174**(7–8), 449–478 (2010)
10. Davies, J., Bacchus, F.: Solving MAXSAT by solving a sequence of simpler SAT instances. In: Lee, J. (ed.) CP 2011. LNCS, vol. 6876, pp. 225–239. Springer, Heidelberg (2011). doi:10.1007/978-3-642-23786-7_19
11. Davies, J., Bacchus, F.: Postponing optimization to speed up MAXSAT solving. In: Schulte, C. (ed.) CP 2013. LNCS, vol. 8124, pp. 247–262. Springer, Heidelberg (2013). doi:10.1007/978-3-642-40627-0_21
12. Dechter, R., Mateescu, R.: AND/OR search spaces for graphical models. Artif. Intell. **171**(2), 73–106 (2007)
13. Dixon, H.E.: Automating psuedo-Boolean inference within a DPLL framework. Ph.D. thesis, University of Oregon (2004)
14. de Givry, S., Prestwich, S.D., O'Sullivan, B.: Dead-end elimination for weighted CSP. In: Schulte, C. (ed.) CP 2013. LNCS, vol. 8124, pp. 263–272. Springer, Heidelberg (2013). doi:10.1007/978-3-642-40627-0_22
15. de Givry, S., Schiex, T., Verfaillie, G.: Exploiting tree decomposition and soft local consistency in weighted CSP. In: Proceedings of AAAI-2006, Boston, MA (2006)
16. de Givry, S., Zytnicki, M., Heras, F., Larrosa, J.: Existential arc consistency: getting closer to full arc consistency in weighted CSPs. In: Proceedings of IJCAI-2005, Edinburgh, Scotland, pp. 84–89 (2005)
17. Globerson, A., Jaakkola, T.: Fixing max-product: convergent message passing algorithms for MAP LP-relaxations. In: Proceedings of NIPS, Vancouver, Canada (2007)
18. van Hoeve, W.J., Pesant, G., Rousseau, L.: On global warming: flow-based soft global constraints. J. Heuristics **12**(4–5), 347–373 (2006)
19. Hurley, B., O'Sullivan, B., Allouche, D., Katsirelos, G., Schiex, T., Zytnicki, M., de Givry, S.: Multi-language evaluation of exact solvers in graphical model discrete optimization. Constraints **21**(3), 413–434 (2016)
20. Jégou, P., Terrioux, C.: Hybrid backtracking bounded by tree-decomposition of constraint networks. Artif. Intell. **146**(1), 43–75 (2003)
21. Kask, K., Dechter, R.: Branch and bound with mini-bucket heuristics. In: Proceedings of IJCAI-1999. vol. 99, pp. 426–433 (1999)
22. Khemmoudj, M.O.I., Bennaceur, H.: Clique inference process for solving Max-CSP. Eur. J. Oper. Res. **199**(3), 665–673 (2009)
23. Kolmogorov, V.: Convergent tree-reweighted message passing for energy minimization. IEEE Pattern Anal. Mach. Intell. **28**(10), 1568–1583 (2006)
24. Larrosa, J.: Boosting search with variable elimination. In: Dechter, R. (ed.) CP 2000. LNCS, vol. 1894, pp. 291–305. Springer, Heidelberg (2000). doi:10.1007/3-540-45349-0_22
25. Larrosa, J., Schiex, T.: In the quest of the best form of local consistency for weighted CSP. In: Proceedings of 18th IJCAI, Acapulco, Mexico, pp. 239–244 (2003)
26. Lecoutre, C., Roussel, O., Dehani, D.E.: WCSP integration of soft neighborhood substitutability. In: Milano, M. (ed.) CP 2012. LNCS, pp. 406–421. Springer, Heidelberg (2012). doi:10.1007/978-3-642-33558-7_31

27. Lee, J.H.M., Leung, K.L.: Consistency techniques for global cost functions in weighted constraint satisfaction. J. Artif. Intell. R. **43**, 257–292 (2012)
28. Lee, J.H., Leung, K.L.: A stronger consistency for soft global constraints in weighted constraint satisfaction. In: Proceedings of AAAI-2010, Atlanta, USA (2010)
29. Marinescu, R., Dechter, R.: AND/OR branch-and-bound for graphical models. In: Proceedings of IJCAI-2005, Edinburgh, Scotland, UK, pp. 224–229 (2005)
30. Narodytska, N., Bacchus, F.: Maximum satisfiability using core-guided MAXSAT resolution. In: Proceedings of AAAI-2014, Quebec City, Canada, pp. 2717–2723 (2014)
31. Nguyen, H., Bessiere, C., de Givry, S., Schiex, T.: Triangle-based consistencies for cost function networks. Constraints **22**(2), 230–264 (2016)
32. Prusa, D., Werner, T.: Universality of the local marginal polytope. In: Proceedings of IEEE Conference on Computer Vision and Pattern Recognition, pp. 1738–1743 (2013)
33. Quimper, C.-G., Walsh, T.: Global grammar constraints. In: Benhamou, F. (ed.) CP 2006. LNCS, vol. 4204, pp. 751–755. Springer, Heidelberg (2006). doi:10.1007/11889205_64
34. Quimper, C., Walsh, T.: Decompositions of grammar constraints. CoRR abs/0903.0470 (2009)
35. Sánchez, M., de Givry, S., Schiex, T.: Mendelian error detection in complex pedigrees using weighted constraint satisfaction techniques. Constraints **13**(1), 130–154 (2008)
36. Schlesinger, M.I.: Syntactic analysis of two-dimensional visual signals in noisy conditions. Kibernetika **4**, 113–130 (1976). (in Russian)
37. Sontag, D., Choe, D., Li, Y.: Efficiently searching for frustrated cycles in MAP inference. In: Proceedings of UAI, pp. 795–804 (2012)
38. Sontag, D., Meltzer, T., Globerson, A., Weiss, Y., Jaakkola, T.: Tightening LP relaxations for MAP using message-passing. In: Proceedings of UAI, pp. 503–510 (2008)
39. Werner, T.: A linear programming approach to max-sum problem: a review. IEEE Trans. Pattern Anal. Mach. Intell. **29**(7), 1165–1179 (2007). https://doi.org/10.1109/TPAMI.2007.1036

Arc Consistency via Linear Programming

Grigori German[✉], Olivier Briant, Hadrien Cambazard, and Vincent Jost

CNRS, Grenoble INP G-SCOP, University Grenoble Alpes, 38000 Grenoble, France
grigori.german@grenoble-inp.fr

Abstract. A typical technique in integer programming for filtering variables is known as *variable fixing*. The optimal dual solution of the linear relaxation can be used to detect some of the 0/1 variables that must be fixed to either 0 or 1 in any solution improving the best known, but this filtering is incomplete. A complete technique is proposed in this paper for satisfaction problems with an ideal integer programming formulation. We show, in this case, that the 0/1 variables taking the same value in all solutions can be identified by solving a single linear program with twice the number of the original variables. In other words, a complete variable fixing of the 0/1 variables can be performed for a small overhead. As a result, this technique can be used to design generic arc consistency algorithms. We believe it is particularly useful to quickly prototype arc consistency algorithms for numerous polynomial constraints and demonstrate it for the family of SEQUENCE global constraints.

Keywords: Constraint Programming · Linear Programming filtering · Arc consistency · Reduced-cost fixing

1 Introduction

Mixed integer programming (MIP) and Constraint Programming (CP) have benefited from each other increasingly in recent years due to the complementary strengths of the two frameworks. Many approaches have been proposed to combine their modeling and solving capabilities [1,2,6,18,21]. On one side, CP tailored algorithms for specific constraints take advantage of local combinatorial structures to reduce the search space efficiently. On the other, MIP techniques usually encompass the whole problem and typically compute lower/upper bounds of the objective function that propagation through the domains fails to derive. A typical integration of the two approaches is to use the linear relaxation of the entire problem in addition to the local consistencies enforced by the CP solver. The relaxation is used to perform filtering, in particular by providing a global bound of the objective but also by filtering the domains using a technique referred to as reduced-cost-based filtering [11,14]. Constraints can in turn provide specialized linear formulations and cutting planes.

As opposed to previous work, we investigate in this paper how linear programming (LP) can be used to filter individual constraints and in particular to

© Springer International Publishing AG 2017
J.C. Beck (Ed.): CP 2017, LNCS 10416, pp. 114–128, 2017.
DOI: 10.1007/978-3-319-66158-2_8

provide arc consistency algorithms. Let us suppose that an ideal linear formulation $\mathcal{F}$ over n variables is available for a global constraint. A formulation is referred to as ideal when it has the integrality property *i.e.* when the extreme points of the corresponding polytope are integer points. It is easy to come by such formulations for many global constraints [18] that include 0/1 variables typically encoding whether an original variable of the constraint is assigned to a given value of its domain. Since $\mathcal{F}$ is supposed ideal, a simple way to achieve arc consistency is to fix, in turn, each variable to each value and check the consistency by calling a linear solver. This is very similar to the failed literal test mentioned in [3] for achieving arc consistency with unit propagation over a SAT encoding of the constraint.

We show, however, that arc consistency can be achieved in this case by solving *a single linear program* with n additional variables and $2n$ additional constraints. The idea is to look for an interior point, of the convex hull of $\mathcal{F}$ maximizing the number of variables with a slack to their respective lower and upper bounds. Although this goes against the rationale explained above for integrating the two frameworks, we believe the advantages are twofold. First of all, since each solver only provides a handful of the existing constraints, it is particularly useful to quickly design arc consistency algorithms for many polynomial constraints. Secondly, it can provide a generic but competitive algorithm for constraints with a quadratic running time such as the GEN-SEQUENCE [13].

The linear relaxation has been used in the past for filtering and we now review several closely related works [1, 2, 18, 21] that propose frameworks for combining the linear relaxation, specialized cutting planes generation and filtering. An illustrative example is the work of [18] where each constraint is able to provide its linear relaxation so that a global relaxation of the entire problem is automatically derived from a CP model. Additionally, a constraint is able to add dedicated cutting planes during search, taking advantage of the corresponding combinatorial structure to build a stronger relaxation. The linear relaxations of common global constraints such as ELEMENT, ALLDIFFERENT, CIRCUIT and CUMULATIVE can be found in [12, 18] and relaxations of global constraints involving costs such as MINIMUMWEIGHTALLDIFFERENT or WEIGHTEDCIRCUIT are described in [10]. The linear relaxation is directly used for filtering by [1, 2, 10, 17, 18]. It can detect infeasibility, provide a bound for a cost variable and perform filtering using a technique referred to as reduced-cost based filtering [10, 11]. The latter is a specific case of cost-based filtering [9] that aims at filtering out values leading to non-improving solutions.

Section 2 summarizes the key notations. Section 3 reviews and explains reduced-cost based filtering in more details. The main result of this paper is presented Sect. 4 and its application to ALLDIFFERENT, GLOBALCARDINALITY and GEN-SEQUENCE constraints is described in Sect. 5. Finally experimental results are reported on three benchmarks in Sect. 6.

2   Notations

A constraint satisfaction problem is made of a set of variables, each with a given domain, *i.e.* a finite set of possible values, and a set of constraints specifying the allowed combinations of values for subset of variables. In the following, the variables, *e.g.* X_i, are denoted with upper case letters for the constraint programming models as opposed to the variables of linear programming models that are in lower case. $D(X_i) \subseteq \mathbb{Z}$ denotes the domain of X_i. A constraint C over a set of variables $\langle X_1, \ldots, X_n \rangle$ is defined by the allowed combinations of values (tuples) of its variables. Such tuples of values are also referred to as solutions of the constraint C. Given a constraint C with a scope $\langle X_1, \ldots, X_n \rangle$, a **support** for C is a tuple of values $\langle v_1, \ldots, v_n \rangle$ that is a solution of C and such that $v_i \in D(X_i)$ for all variable X_i in the scope of C. Consider a variable X_i in the scope of C, the domain $D(X_i)$ is said **arc consistent** for C if and only if all the values $v_j \in D(X_i)$ belong to a support for C. A constraint C is said arc consistent if and only if all its variables are arc consistent.

3   Traditional Filtering Using LP: Reduced-Cost Filtering

Let us first review how linear programming is traditionally used to perform filtering. Suppose we are dealing with a minimization problem. Cost-based filtering relies on a known upper bound $\overline{z}$ of the objective function which is usually the cost of the best feasible solution found so far. Since there is no need to consider solutions with a greater cost that $\overline{z}$, values of the domains that would necessarily lead to such non-improving solutions should be filtered. Linear reduced-costs provide valuable information to perform such reasonings. They are available from an optimal dual solution of the linear relaxation and give a minimum increase of the objective function. This increase can be used to detect if a bound of a variable leads to a non-improving solution. When applied to 0/1 variables, *i.e.* variables with the domain $\{0, 1\}$, any update of a bound leads to fixing a variable to 0 or 1. It has thus been known for a long time as *variable fixing*.

To our knowledge, the best account of this technique in the context of constraint programming is given in [11]. It is usually presented in textbooks on integer programming such as [14, 26] for 0/1 variables. We give a summary of this technique in the more general case of integer variables. Consider a linear programming formulation (P) where the feasible region is defined by a polytope $\mathcal{Q} = \{x \in \mathbb{R}^T \mid Ax \geq b, l \leq x \leq u\}$. Note that each variable x_t for all $t \in \{1, \ldots, T\}$, has a given lower and upper bound *i.e.* $x_t \in [l_t, u_t]$.

$$(P) \qquad z^* = \min\{cx : x \in \mathcal{Q}\}$$

Program (P) is typically the linear relaxation of an integer programming formulation identified for the whole problem or for a single constraint. Let α be the m dual variables of the constraints $Ax \geq b$. Moreover, let x^* be an optimal solution of (P) and α^* a set of optimal values of the α variables. The reduced cost

r_t of a variable x_t, with respect to α^*, is defined as $r_t = c_t - \alpha^* A_t$ where A_t is the t-th column of A. Note that the definition of r_t ignores the dual variables related to the lower and upper bounds since $x_t \leq u_t$ and $x_t \geq l_t$ are usually not added as constraints to the formulation of (P) but handled directly by the simplex algorithm (see [8] for more details). The reduced cost r_t is typically the quantity returned by linear programming solvers when the bounds are not explicitly stated as constraints in the model but directly as bounds of the variable's domains.

Reduced-cost-based filtering removes values necessarily leading to non-improving solutions *i.e.* solutions of cost greater than the known upper bound $\bar{z}$. The following rules can be used for filtering a variable x_t of (P) from an optimal dual solution.

Proposition 1 (Reduced cost filtering).

$$\text{If } r_t > 0 \text{ then } x_t \leq l_t + \frac{(\bar{z} - z^*)}{r_t} \text{ in any solution of cost less than } \bar{z} \quad (1)$$

$$\text{If } r_t < 0 \text{ then } x_t \geq u_t - \frac{(\bar{z} - z^*)}{-r_t} \text{ in any solution of cost less than } \bar{z} \quad (2)$$

Note that if x_t is originally an integer variable, the reasoning can be tightened as $x_t \leq l_t + \lfloor \frac{(\bar{z}-z^*)}{r_t} \rfloor$ and $x_t \geq u_t - \lfloor \frac{(\bar{z}-z^*)}{-r_t} \rfloor$.

The two rules are a direct consequence of traditional sensitivity analysis and the reader can refer to [11, 26] for more details.

The filtering obtained from a particular optimal dual solution is usually incomplete since r_t depends on the specific α^* found. In other words, considering several optimal dual solutions may provide more filtering. Let us go through a very simple example to illustrate this point. We consider a difference constraint $X_1 \neq X_2$ with $D(X_1) = \{1, 2\}$ and $D(X_2) = \{1\}$. Value 1 of $D(X_1)$ is thus expected to be filtered. A simple integer formulation of the feasible solutions can be written with 0/1 variables x_1, x_2, x_3 respectively encoding whether $X_1 = 1$, $X_1 = 2$ or $X_2 = 1$. The linear relaxation and its dual problem write as follows:

$$
\begin{array}{ll}
\min 0 & \max \sum_{i=1}^4 \alpha_i + \sum_{i=1}^3 \beta_i \\
\quad x_1 + x_2 \geq 1 & \quad \alpha_1 + \alpha_3 + \alpha_4 + \beta_1 \leq 0 \\
\quad x_3 \geq 1 & \quad \alpha_1 + \alpha_4 + \beta_2 \leq 0 \\
(P) \quad x_1 + x_3 \leq 1 & (D) \quad \alpha_2 + \alpha_3 + \beta_3 \leq 0 \\
\quad x_1 + x_2 \leq 1 & \quad \alpha_1, \alpha_2 \geq 0 \\
\quad x_1, x_2, x_3 \geq 0 & \quad \alpha_3, \alpha_4 \leq 0 \\
\quad x_1, x_2, x_3 \leq 1 & \quad \beta_1, \beta_2, \beta_3 \leq 0
\end{array}
$$

$x^* = (0, 1, 1)$ is an optimal solution of (P). It is easy to see that $\alpha^* = (0, 0, 0, 0)$ or $\alpha^* = (0, 1, -1, 0)$ are two vectors of optimal values for α (with $\beta^* = (0, 0, 0)$). The reduced cost of x_1 is $r_1 = 0 - \alpha_1^* - \alpha_3^* - \alpha_4^*$. In the first case, $r_1 = 0$ and none of the rules given in Proposition 1 is triggered. In the second case, $r_1 = 1$ and the first rule applies with $\bar{z} = 0$ enforcing $x_1 \leq 0$ as expected. In both cases $r_2 = r_3 = 0$, therefore x_2 and x_3 are not filtered.

Note that this drawback occurs even if the polytope $\mathcal{Q}$ has integer extreme points which is the case in the example. Moreover, in any case, minimizing 0 in (P) implies that $\alpha^* = 0$ is an optimal vector for α and the linear solver can always return it. To reduce this phenomenon, it is possible to use another objective function cx as long as all feasible solutions have the same cost (see Sect. 6 for an example).

An alternative approach to reduced costs is proposed in the next section to perform a complete variable fixing.

4 A New Generic Filtering Algorithm Based on LP

Let us briefly explain the general idea and its application to filtering global constraints before stating the result in detail.

Consider a polytope $\mathcal{Q} = \{x \in \mathbb{R}^T \mid Ax \geq b, l \leq x \leq u\}$ with integer extreme points and $l, u \in \mathbb{Z}^T$. We show in this section that a variable that is fixed to one of its bound (either l_t or u_t) in all extreme point of $\mathcal{Q}$ can be detected by solving a linear program with T additional variables and $2T$ additional constraints. The idea is to look for an interior point of $\mathcal{Q}$ maximizing the number of variables with a slack to their respective lower and upper bounds. When no slack is possible, the variable is proved to be fixed to the same value in all extreme points.

For numerous polynomial global constraints, it is possible to give an *ideal integer programming formulation* $\mathcal{F}$ of the solutions where 0/1 variables x_{ij} typically encode whether an integer variable X_i of the constraint's scope is assigned to value a v_j of its domain. The linear relaxation of $\mathcal{F}$ defines a polytope $\mathcal{Q}$ that represents the convex hull of the supports of the constraint. Each integer point in $\mathcal{Q}$ can be seen as a support of the constraint. The proposed technique identifies all 0/1 variables that are set to 0 in all extreme points. Since all interior points of $\mathcal{Q}$ are a convex combination of the extreme points, the same variables are thus also set to 0 for all interior points of $\mathcal{Q}$, i.e. the corresponding values do not belong to any support. Since *complete variable fixing* can be performed (all inconsistent values are removed), the proposed technique gives the arc consistent domains for the constraint.

The main result is stated as Theorem 1. The polytope $\mathcal{Q} = \{x \in \mathbb{R}^T \mid Ax \geq b, l \leq x \leq u\}$ is assumed to have integer extreme points and $l, u \in \mathbb{Z}^T$. Let us also denote by $\mathcal{S}$ the set of extreme points of $\mathcal{Q}$.

Theorem 1. *Let us define* $\varepsilon = \dfrac{1}{(T+1)}$, *and* (P') *the following linear program:*

$$\min\ z(x, e) = \sum_{t=1}^{T} e_t$$

$$
\begin{aligned}
s.t.\quad & x_t + e_t \geq l_t + \varepsilon & \forall t \in \{1, \ldots, T\} \\
& x_t - e_t \leq u_t - \varepsilon & \forall t \in \{1, \ldots, T\} \\
& e_t \geq 0 & \forall t \in \{1, \ldots, T\} \\
& x \in \mathcal{Q}
\end{aligned}
$$

For all $t \in \{1, \ldots, T\}$, all $\delta \in \{l_t, u_t\}$, and all optimal solution (x^*, e^*) of (P') we have:

$$x_t^* = \delta \qquad \Leftrightarrow \qquad \hat{x}_t = \delta \ \forall \hat{x} \in S$$

Note that a feasible solution of (P') with $e_t = 0$ indicates that variable x_t has a slack of at least ε to its lower and upper bound. Keep also in mind that the objective of (P') is to minimize the sum of the e_t which tend to create slack. We will show that any optimal solution of (P') actually maximizes the number of variables that can be unstuck from their bounds revealing all the x_t that are always instantiated to either l_t or u_t. We first make a simple observation:

Remark 1. *If (x^*, e^*) is an optimal solution of (P'), then $e^* = e(x^*)$ with $e : \mathbb{R}^T \to \mathbb{R}^T$ such that $e(x) = (e_1(x), \ldots, e_t(x), \ldots, e_T(x))$ and*

$$e_t(x) = \max\{0, \varepsilon + l_t - x_t, \varepsilon - u_t + x_t\} \quad \forall t \in \{1, \ldots, T\}$$

Proof: Each variable e_t only occurs in three constraints of (P'): $e_t \geq 0$, $e_t \geq \varepsilon + l_t - x_t$ and $e_t \geq \varepsilon - u_t + x_t$. Given a value of x^*, the minimum possible feasible value for e_t is thus $\max\{0, \varepsilon + l_t - x_t, \varepsilon - u_t + x_t\}$. $\qquad \square$

The optimal objective value of (P') is at least ε times the number of variables that are necessarily fixed to either l_t or u_t. The proof given below builds a feasible solution of (P') reaching this bound by setting $e_t(x) = 0$ for all other variables. Thus, any optimal solution highlights the fixed variables.

Proof of Theorem 1:
We denote by $T^l = \{t \in \{1, \ldots, T\} \mid \hat{x}_t = l_t, \ \forall \hat{x} \in S\}$ and $T^u = \{t \in \{1, \ldots, T\} \mid \hat{x}_t = u_t, \ \forall \hat{x} \in S\}$ the sets of indices referring to variables fixed respectively to their lower or upper bounds, in all extreme points of Q. As mentioned above, a valid lower bound (P') is $\varepsilon(|T^l| + |T^u|)$.

Let $(\hat{x}^0, \hat{x}^1, \hat{x}^2, \ldots, \hat{x}^T)$ be a series of extreme points of S defined as follows. $\hat{x}^0$ is chosen arbitrarily in S. Each $\hat{x}^t$ such that $t \notin T^l \cup T^u$ is chosen in S so that $\hat{x}_t^t \neq \hat{x}_t^0$. Finally, all remaining $\hat{x}^t$ are chosen arbitrarily in S.

Based on this series of points, we can define a feasible solution $(\bar{x}, e(\bar{x}))$ of (P') by considering $\bar{x}$ as the following convex combination $\bar{x} = \dfrac{1}{T+1} \displaystyle\sum_{t=0}^{T} \hat{x}^t$.

Firstly, note that $\bar{x}_t \in \{l_t, u_t\}$ if and only if $t \in T^l \cup T^u$. Indeed, $l_t \leq \hat{x}_t \leq u_t$ for all $\hat{x} \in S$, so we have $\bar{x}_t \in \{l_t, u_t\}$ if and only if $\hat{x}_t^{t'} = \bar{x}_t$ for all $t' \in \{0, \ldots, T\}$. Therefore, by construction, $\bar{x}_t \in \{l_t, u_t\}$ if and only if $\hat{x}_t = \hat{x}_t^0$ for all $\hat{x} \in S$, i.e. $t \in T^l \cup T^u$.

Secondly, all other $\bar{x}_t$ have a slack of at least ε. For all $t \notin T^l \cup T^u$, we have $\hat{x}_t^t \neq \hat{x}_t^0$, therefore $\max\{\hat{x}_t^t, \hat{x}_t^0\} \geq l_t + 1$ and $\min\{\hat{x}_t^t, \hat{x}_t^0\} \leq u_t - 1$ since extreme points of Q are integers. As a result:

$$\bar{x}_t \geq \frac{1}{T+1}(\max\{\hat{x}_t^t, \hat{x}_t^0\} + T\, l_t) \geq \frac{1}{T+1}(l_t + 1 + T\, l_t) = l_t + \varepsilon$$

$$\bar{x}_t \leq \frac{1}{T+1}(\min\{\hat{x}_t^t, \hat{x}_t^0\} + T\, u_t) \leq \frac{1}{T+1}(u_t - 1 + T\, u_t) = u_t - \varepsilon$$

Hence for all $t \in \{1, \ldots, T\}$, we have

$$e_t(\bar{x}) = \begin{cases} \varepsilon & \text{if } t \in T^l \cup T^u \\ 0 & \text{otherwise} \end{cases}$$

Thus $z(\bar{x}, e(\bar{x})) = \varepsilon(|T^l| + |T^u|)$ which proves that the solution $(\bar{x}, e(\bar{x}))$ is optimal. Any optimal solution x^* must therefore have a cost of $\varepsilon(|T^l| + |T^u|)$. Since $e_t(x^*) \geq 0$ for all t and $e_t(x^*) = \varepsilon$ for all $t \in T^l \cup T^u$, $e_t(x^*) = 0$ for all $t \notin T^l \cup T^u$.

Conclusion: for all optimal solutions (x^*, e^*) of (P'), all $t \in \{1, \ldots, T\}$ and all $\delta \in \{l_t, u_t\}$,

$$x_t^* = \delta \qquad \Leftrightarrow \qquad t \in T^l \cup T^u \qquad \Leftrightarrow \qquad \hat{x}_t = \delta \ \forall \hat{x} \in \mathcal{S}$$

$\square$

We now propose a simple application of this result to filtering global constraints. Consider a polynomial global constraint C over a scope $X = \langle X_1, \ldots, X_n \rangle$ of n integer variables with their respective domains $D(X_i) \in \mathbb{Z}$. The approach proposed to enforce arc consistency is summarized below:

LP-Based Filtering for Constraint C:
Inputs: A constraint C over the variables $\mathcal{X} = \langle X_1, \ldots, X_n \rangle$. An ideal integer formulation $\mathcal{F}$ of the solutions of C where a $0/1$ variable x_{ij} is present for all $X_i \in X, v_j \in D(X_i)$ to encode whether variable X_i takes value v_j.

Output: arc consistent domains $D(X_1), \ldots, D(X_n)$ for constraint C

1. Consider $\mathcal{Q}$ as the convex hull of $\mathcal{F}$ by simply relaxing the domain's constraint $x_{ij} \in \{0, 1\}$ into $x_{ij} \in [0, 1]$ for all x_{ij}
2. Find an optimal solution (x^*, e^*) of (P') as defined in Theorem 1
3. For each $X_i \in \mathcal{X}$ and each $v_j \in D(X_i)$, if $x_{ij}^* = 0$, remove value v_j from $D(X_i)$.

The procedure given above computes arc consistent domains as a direct consequence of Theorem 1: indeed $x_{ij}^* = 0$ means that $x_{ij} = 0$ for any solution of the LP, hence the corresponding value has to be removed. Furthermore, when $x_{ij}^* = 1$, $x_{ik}^* = 0$ for all $k \neq j$.

Corollary 1. *The procedure LP-based filtering is correct and establishes arc consistency for constraint C.*

Proof: Recall that the integer points of $\mathcal{Q}$ represent the supports of C. Since any interior point can be written as a convex combination of the extreme points, there exists at least one extreme point $\hat{x} \in \mathcal{S}$ such that $\hat{x}_{ij} = 1$ for any consistent value v_j of a $D(X_i)$. Similarly when all $\hat{x}_{ij} = 0$ for all $\hat{x} \in \mathcal{S}$, it is the case of all interior integer points. Keeping that in mind, we simply check that the procedure does not remove any consistent value (it is correct) and removes all inconsistent values (it is complete).

Correct: consider a consistent value v_j in $D(X_i)$. Since it belongs to a support, there exists at least one $\hat{x} \in \mathcal{S}$ such that $\hat{x}_{ij} = 1$. Therefore, $x_{ij}^* \neq 0$ according to Theorem 1 and value v_j is not removed by the proposed procedure.

Complete: Let us check that all remaining values belong to a support. Consider a value v_j of a domain $D(X_i)$ after the procedure has been called. Since v_j has not been filtered, $x_{ij}^* \neq 0$ implying by Theorem 1 that there exists at least one extreme point $\hat{x} \in \mathcal{S}$ such that $\hat{x}_{ij} = 1$. Therefore v_j belongs to the corresponding support. □

The complexity of LP-based filtering depends on the algorithm used to solve the LP. In practice, the simplex algorithm is known to have a number of iterations proportional to $m \log(n)$ [8] where n is the number of variables and m the number of constraints of the LP formulation.

5   Ideal Linear Formulations of Polynomial Global Constraints

We now provide an ideal integer programming formulation $\mathcal{F}$ for a number of polynomial global constraints. The fact that a variable X_i takes one and one value only of its domain ($X_i \in D(X_i)$) is typically expressed in $\mathcal{F}$ following [18]:

$$\left\{ \begin{array}{ll} \sum\limits_{v_j \in D(x_i)} x_{ij} = 1 & \\ x_{ij} \in \{0,1\} & \forall v_j \in D(X_i) \end{array} \right. \tag{3}$$

5.1   AllDifferent and GlobalCardinality

The ALLDIFFERENT$(X_1, \ldots, X_n)$ constraint [19] is satisfied when the variables $X_1, \ldots, X_n$ take different values. The formulation given below is the classical formulation for the matching problem and is known to have the integrality property:

$$\mathcal{F} = \left\{ \begin{array}{ll} \sum\limits_{i|v_j \in D(X_i)} x_{ij} \leq 1 & \forall v_j \in \bigcup_{i=1}^n D(X_i) \\ \sum\limits_{v_j \in D(x_i)} x_{ij} = 1 & \forall i \in \{1, \ldots, n\} \\ x_{ij} \in \{0,1\} & \forall i \in \{1, \ldots, n\}, \forall v_j \in D(X_i) \end{array} \right. \tag{4}$$

A related global constraint is the GLOBALCARDINALITY constraint [20]. It enforces the number of occurrences of each value v_j to be at least l_j and at most u_j in the set $X_1, \ldots, X_n$ of variables. The formulation $\mathcal{F}$ is the following:

$$\mathcal{F} = \left\{ \begin{array}{ll} \sum\limits_{i|v_j \in D(X_i)} x_{ij} \leq u_j & \forall v_j \in \bigcup_{i=1}^n D(X_i) \\ \sum\limits_{i|v_j \in D(X_i)} x_{ij} \geq l_j & \forall v_j \in \bigcup_{i=1}^n D(X_i) \\ \sum\limits_{v_j \in D(x_i)} x_{ij} = 1 & \forall i \in \{1, \ldots, n\} \\ x_{ij} \in \{0,1\} & \forall i \in \{1, \ldots, n\}, \forall v_j \in D(X_i) \end{array} \right. \tag{5}$$

$\mathcal{F}$ has the integrality property since the matrix can be seen as a specific case of a network flow matrix. Let us denote by $d = \max_{i=1}^{n} |D(X_i)|$, the maximum cardinality of the domains and $m = |\bigcup_{i=1}^{n} D(X_i)|$, the total number of distinct values. Both formulations given have $O(nd)$ variables and $O(n+m)$ constraints. Arc consistency can thus be established by solving the program (P') which has twice the number of variables and $O(nd + m)$ constraints. Recall that the dedicated algorithms for each constraint respectively runs in $O(n^{1.5}d)$ for the ALLDIFFERENT, $O(n^2m)$ for the GLOBALCARDINALITY and are incremental down a branch of the search tree.

5.2   The Family of Sequence Constraints

The SEQUENCE constraint restricts the number of occurrences of some given values in any sequence of k variables. It can be expressed as a conjunction of AMONG constraints and has been used for car sequencing [5] and nurse rostering [7]. More precisely, AMONG$(l, u, \langle X_1, \ldots, X_k \rangle, V)$ holds if and only if $l \leq |\{i|X_i \in V\}| \leq u$. In other words, at least l and at most u of the variables take their values in the set V. The SEQUENCE constraint can be defined as a conjunction of *sliding* AMONG constraints over k consecutive variables. SEQUENCE$(l, u, k, \langle X_1, \ldots, X_n \rangle, V)$ holds if and only if $\forall i \in \{1, \ldots, n - k + 1\}$ AMONG$(l, u, \langle X_i, \ldots, X_{i+k-1} \rangle, V)$ holds.

An incomplete filtering algorithm for SEQUENCE is proposed in [4]. Two arc consistency algorithms are later given in [24,25] with respective running times of $O(n^3)$ and $O(2^k n)$. Additionally, an encoding achieving arc consistency is presented in [7] and runs in $O(n^2 log(n))$ down a branch of a search tree. It is latter improved in [13] by using the fact that a natural integer programming formulation of the constraint has the consecutive ones property on the columns. This is used to build a network flow graph and derive an arc consistency algorithm. The complexity of this flow-based propagator to enforce arc consistency is $O(n((n-k)(u-l)+u))$ when using the Ford-Fulkerson algorithm for finding a maximum flow. The incremental cost when fixing a single variable is only $O(n)$ so that the algorithm runs in $O(n^2)$ down a branch of a search tree.

A generalization of SEQUENCE is known as GEN-SEQUENCE and allows different occurrences (l and u) and sizes (k) for an arbitrary set of m AMONG constraints over consecutive variables. GEN-SEQUENCE$(p_1, \ldots, p_m, \langle X_1, \ldots, X_n \rangle, V)$ holds if and only if $\forall 1 \leq i \leq m$, AMONG$(l_i, u_i, \langle X_{s_i}, \ldots, X_{s_i+k_i-1} \rangle, V)$ holds where $p_i = \{l_i, u_i, k_i, s_i\}$.

The GEN-SEQUENCE global constraint is defined in [24,25] and a $O(n^4)$ algorithm is proposed to enforce arc consistency. The consecutive one property does not hold in general for a GEN-SEQUENCE constraint. Although it may sometimes be possible to re-order the lines of the matrix to have the consecutive one property on the columns or to find an equivalent network matrix, [13] outlines that not all GEN-SEQUENCE constraints can be expressed as network flows. The flow-based algorithm for SEQUENCE can therefore not be reused in general. Nonetheless, the encoding of SEQUENCE proposed in [7] extends to GEN-SEQUENCE and runs in $O(nm + n^2 \log n)$ [13]. Finally, in [3], a filtering method based on unit propagation over a conjunctive normal form encoding of the constraint is proposed and achieves arc consistency in $O(mn^3)$.

Notice that the previous sequence constraints can be encoded with a simple boolean channeling, without hindering any filtering since the resulting constraint network is Berge-acyclic. Typically GEN-SEQUENCE$(p_1, \ldots, p_m, \langle X_1, \ldots, X_n \rangle, V)$ can be stated as:

$$\begin{cases} \text{GEN-SEQUENCE}(p_1, \ldots, p_m, \langle Y_1, \ldots, Y_n \rangle, 1) \\ \quad Y_i = 1 \Leftrightarrow X_i \in V, \forall i \in \{1, \ldots, n\} \\ \quad Y_i \in \{0, 1\}, \forall i \in \{1, \ldots, n\} \end{cases} \tag{6}$$

All previous studies thus focused on the restricted case where $V = \{1\}$. The integer linear formulation for GEN-SEQUENCE$(p_1, \ldots, p_m, \langle Y_1, \ldots, Y_n \rangle, 1)$ is the following:

$$\mathcal{F} = \begin{cases} \sum_{j=s_i}^{s_i+k_i-1} y_j \leq u_i & \forall i \in \{1, \ldots, m\} \\ \sum_{j=s_i}^{s_i+k_i-1} y_j \geq l_i & \forall i \in \{1, \ldots, m\} \\ y_j \in \{0, 1\} & \forall j \in \{1, \ldots, n\} \end{cases} \tag{7}$$

This formulation has the integrality property, as already mentioned in [13]. The linear program (P') solved by the proposed LP-based procedure to enforce arc consistency has $O(n)$ variables and $O(m + n)$ constraints. Recall that the best arc consistency algorithm for GEN-SEQUENCE is the encoding of [13] and runs in $O(nm + n^2 \log n)$.

6   Numerical Results

We carried out experiments on the ALLDIFFERENT and SEQUENCE global constraints. The first set of experiments compares the LP-based filtering to the dedicated filtering algorithm of ALLDIFFERENT (Sect. 6.1). It also evaluates in practice the power of reduced-cost based filtering. The second one (Sect. 6.2) compares two encodings of SEQUENCE to the LP-based filtering on random sequences alone following the experiments reported in [13]. Finally the last one evaluates the LP-based filtering on the Car-sequencing problem (Sect. 6.3).

The experiments were performed with Windows 8 on an Intel Core i5 @ 2.5 GHz with 12 GB of RAM. A memory limit of 4 GB of RAM was used. The indicators shown in the result tables are the average resolution time in seconds (CPU), the average number of nodes (N) and the average speed of the resolution in node per second (N/s). The constraint solver used is Choco 3.3 [16].

6.1   LP and Reduced-Cost Filtering for the AllDifferent constraint

The LP-based filtering is implemented for ALLDIFFERENT with the polytope given in Sect. 5.1. It is referred to as ALLDIFFERENTLPF (for LP Filtering) and compared with three other filtering algorithms: the Choco ALLDIFFERENT constraint, the decomposition into cliques of difference constraints (DEC) and the reduced-cost based filtering algorithm in addition to the decomposition (RCF). As mentioned in Sect. 3, when filtering via reduced costs, the use of an objective function of the form cx can increase the chances of having non null reduced costs. To perturb the dual, the objective function used is $\sum_{i=1}^{n} c_i(\sum_{j \in D(X_i)} x_{ij})$, where the c_i are randomly chosen in $[-10, 10]$ at each node. Note that this function guarantees that all feasible solutions have the same cost.

We solve the QuasiGroup Completion problem [15]. The problem is to fill a n by n matrix previously filled at $k\%$ with numbers from 1 to n such that on each line and on each column, each number appears only once. We compare four models and look for the number of nodes needed to find all solutions for small instances with a lexicographic branching heuristic. Table 1 shows the average results on 10 randomly generated instances for each size $n \in \{5, 10, 15\}$. These three classes of instances are respectively filled at 10, 40 and 50% to have instances solvable under 3600 s. One instance with $n = 15$ is solved within the time limit by Choco ALLDIFFERENT only and is thus not included in the results.

Table 1. QuasiGroup completion: filtering the ALLDIFFERENT constraint

n	5			10			15		
Filtering	CPU	N	N/s	CPU	N	N/s	CPU	N	N/s
Choco ALLDIFFERENT	0.1	3190.1	51453.2	1.2	51591.0	41405.3	5.1	87680.6	17244.9
ALLDIFFERENTLPF	2.7	3190.1	1184.6	110.2	51591.0	468.3	478.2	87680.6	183.4
RCF	0.9	3223.0	3679.2	40.7	57679.4	1416.6	292.6	159399.0	544.7
DEC	0.0	3285.6	547600.0	0.8	126613.2	150015.6	161.5	14338395.8	88805.2

As expected, ALLDIFFERENTLPF is slower than Choco ALLDIFFERENT that uses a dedicated algorithm. It is also on average three times slower than RCF. We can however see that it achieves arc consistency as it explores the same number of nodes than Choco ALLDIFFERENT. When $n = 15$, RCF explores approximately twice the number of nodes of Choco ALLDIFFERENT or ALLDIFFERENTLPF. RCF does not achieve arc consistency, yet filters more than the decomposition alone. DEC propagates small constraints and is faster than RCF but can explore up to a 100 times the number of nodes of RCF for these instances.

6.2   Filtering One Sequence Constraint

The LP-based filtering is implemented for SEQUENCE with the polytope given in Sect. 5.2. It is referred to as SEQUENCELPF and compared with an encoding of SEQUENCE: the PS encoding presented in [7] which achieves arc consistency using a decomposition based on partial sums with $O(nk^2)$ constraints. Following the experimentation of [13] we generated 20 instances of a single sequence for each combination of $n \in \{500, 1000, 2000, 3000, 4000, 5000\}$, $k \in \{5, 15, 50, 75\}$ and $\Delta = l - u \in \{1, 5\}$. We look for the first solution found with a heuristic that randomly chooses the variable and the value to branch on. Figure 1 shows the evolution of the resolution time (in s) with the size of the instances.

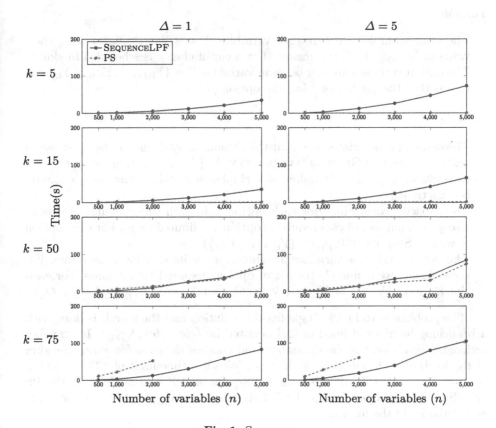

Fig. 1. Sequence

SEQUENCELPF is slower than the PS encoding on sequences with smaller k. However, when $k = 75$, PS runs out of memory due to the number of constraints, whereas SEQUENCELPF can solve the sequence. Most importantly, as we can see on Fig. 1, LP filtering scales better with the problem: it does not seem very sensitive to the value of k.

6.3 The Car-Sequencing Problem

We evaluate the performance of SEQUENCELPF on the Car-sequencing problem [23]. The goal is to schedule n cars partitioned in k classes: the demand of a class c is denoted by d_c. Each class requires a subset of options in a set of m options. For each option j, there must be at most p_j cars that require that option in each sub-sequence of size q_j. We consider the two first set of instances presented in [22]: Set 1 is composed of 70 feasible instances with 200 cars, Set 2 is composed of 4 feasible instances with 100 cars. All the instances are available in CSPLib. We use the model described by [22]:

Variables

- The class variables are n integer variables $\mathcal{X} = \{X_1, \ldots, X_n\}$ taking their value in $\{1, \ldots, k\}$. $X_i = c$ means that a car of class c is scheduled in slot i.
- The option variables are nm boolean variables $\mathcal{Y} = \{Y_{1,1}, \ldots, Y_{n,m}\}$. $Y_{i,j} = 1$ means that the car in slot i has the option j.

Constraints

- The demand constraint states that the demand of each class must be satisfied and is enforced by GLOBALCARDINALITY($\mathcal{X}, \{d_1, \ldots, d_k\}$), meaning that for each class $c \in [1, n]$ the number of variables in $\mathcal{X}$ taking the value c should be exactly d_c.
- The capacity constraints: for each option j, in each sub-sequence of cars of size q_j, the number of cars requiring option j is limited by p_j. For each option j, we set: SEQUENCE($0, p_j, q_j, \{Y_{1,j}, \ldots, Y_{n,j}\}$).
- The option and class variables are linked using implication constraints. For each class c, we define O_c the set of options required by the class. For each slot i, we set: $X_i = c \Rightarrow Y_{i,j} = 1 \ \forall j \in O_c$ and $X_i = c \Rightarrow Y_{i,j} = 0 \ \forall j \notin O_c$.

The problem is to find a single feasible solution and the search is done with a branching heuristic defined in [22] denoted by $\{class, lex, \delta, \leq_\Sigma\}$. It branches lexicographically on the class variables and chooses the class for which the sum of the loads of each option is the highest. We set a time limit of 1200 s. Table 2 compares a model with the filtering of SEQUENCELPF to a model with the PS encoding. The columns CST and VAR show the average number of constraints and variables of the model.

Table 2. Car-sequencing Sets 1 and 2

MODEL	Set 1 (70 instances)						Set 2 (4 instances)					
	CST	VAR	SOLV	CPU	N	N/s	CST	VAR	SOLV	CPU	N	N/s
SEQUENCELPF	1006	1212	100%	0.4	185.4	418.4	506	610	20.0%	900.0	603588.8	670.7
PS	11872	4786	100%	0.1	185.4	2948.9	5872	2384	20.0%	900.0	10666137.5	11850.8

The branching heuristic is very efficient for the first set of instances whereas only 1 instance out of the 4 of Set 2 is solved. The SEQUENCE constraints are not very big ($n \leq 200$ and $k \leq 5$), hence PS has no more than 12000 constraints and is faster than SEQUENCELPF. For this more complex problem, LP filtering is only 20 times slower than the encoding that achieves arc consistency while for QuasiGroup Completion, it is 100 times slower than the Choco ALLDIFFERENT constraint.

7   Conclusion and Future Work

Given a formulation of a constraint with the integrality property, we have shown that arc consistency can be achieved by solving a single linear program. We believe it is very useful to provide arc consistency algorithms for numerous polynomial constraints that are not available in solvers. Although it is unlikely to be competitive with dedicated and incremental filtering algorithms, a number of improvements have yet to be investigated. Firstly, the algorithm boils down to the search for an interior point in a polytope and there might be more efficient techniques, although maybe more difficult to implement, than the simplex algorithm for that purpose. Secondly, the result itself is more general than the specific usage done here to enforce arc consistency since it can be used to detect integer variables necessarily grounded to a bound of their domain. This raises the question whether more sparse LP formulations that does not necessarily introduce a variable per value of the original domains can be used. Finally, polynomial constraints with high running times often have a cost variable, for instance the MINIMUMWEIGHTALLDIFFERENT, so that a natural extension of this work is to handle an objective function.

References

1. Achterberg, T., Berthold, T., Koch, T., Wolter, K.: Constraint integer programming: a new approach to integrate CP and MIP. In: Perron, L., Trick, M.A. (eds.) CPAIOR 2008. LNCS, vol. 5015, pp. 6–20. Springer, Heidelberg (2008). doi:10.1007/978-3-540-68155-7_4
2. Aron, I., Hooker, J.N., Yunes, T.H.: SIMPL: a system for integrating optimization techniques. In: Régin, J.-C., Rueher, M. (eds.) CPAIOR 2004. LNCS, vol. 3011, pp. 21–36. Springer, Heidelberg (2004). doi:10.1007/978-3-540-24664-0_2
3. Bacchus, F.: GAC via unit propagation. In: Bessière, C. (ed.) CP 2007. LNCS, vol. 4741, pp. 133–147. Springer, Heidelberg (2007). doi:10.1007/978-3-540-74970-7_12
4. Beldiceanu, N., Carlsson, M.: Revisiting the cardinality operator and introducing the cardinality-pathconstraint family. In: Codognet, P. (ed.) ICLP 2001. LNCS, vol. 2237, pp. 59–73. Springer, Heidelberg (2001). doi:10.1007/3-540-45635-X_12
5. Beldiceanu, N., Contejean, E.: Introducing global constraints in chip. Math. Comput. Modell. **20**(12), 97–123 (1994)
6. Bockmayr, A., Kasper, T.: Branch and infer: a unifying framework for integer and finite domain constraint programming. INFORMS J. Comput. **10**(3), 287–300 (1998)
7. Brand, S., Narodytska, N., Quimper, C.-G., Stuckey, P., Walsh, T.: Encodings of the sequence constraint. In: Bessière, C. (ed.) CP 2007. LNCS, vol. 4741, pp. 210–224. Springer, Heidelberg (2007). doi:10.1007/978-3-540-74970-7_17
8. Chvátal, V.: Linear Programming. Series of Books in the Mathematical Sciences. W.H. Freeman, New York (1983)
9. Focacci, F., Lodi, A., Milano, M.: Cost-based domain filtering. In: Jaffar, J. (ed.) CP 1999. LNCS, vol. 1713, pp. 189–203. Springer, Heidelberg (1999). doi:10.1007/978-3-540-48085-3_14
10. Focacci, F., Lodi, A., Milano, M.: Embedding relaxations in global constraints for solving TSP and TSPTW. Ann. Math. Artif. Intell. **34**(4), 291–311 (2002)

11. Hooker, J.N.: Operations research methods in constraint programming (chap. 15). In: Rossi, F., van Beek, P., Walsh, T. (eds.) Handbook of Constraint Programming. Elsevier, Amsterdam (2006)

12. Hooker, J.N., Yan, H.: A relaxation of the cumulative constraint. In: van Hentenryck, P. (ed.) CP 2002. LNCS, vol. 2470, pp. 686–691. Springer, Heidelberg (2002). doi:10.1007/3-540-46135-3_46

13. Maher, M., Narodytska, N., Quimper, C.-G., Walsh, T.: Flow-based propagators for the SEQUENCE and related global constraints. In: Stuckey, P.J. (ed.) CP 2008. LNCS, vol. 5202, pp. 159–174. Springer, Heidelberg (2008). doi:10.1007/978-3-540-85958-1_11

14. George, L., Nemhauser, G.L., Wolsey, L.A.: Integer and Combinatorial Optimization. Wiley-Interscience, New York (1988)

15. Pesant, G.: CSPLib problem 067: Quasigroup completion. http://www.csplib.org/Problems/prob067

16. Prud'homme, C., Fages, J.-G., Lorca, X.: Choco Documentation. TASC, INRIA Rennes, LINA CNRS UMR 6241, COSLING S.A.S. (2016)

17. Refalo, P.: Tight cooperation and its application in piecewise linear optimization. In: Jaffar, J. (ed.) CP 1999. LNCS, vol. 1713, pp. 375–389. Springer, Heidelberg (1999). doi:10.1007/978-3-540-48085-3_27

18. Refalo, P.: Linear formulation of constraint programming models and hybrid solvers. In: Dechter, R. (ed.) CP 2000. LNCS, vol. 1894, pp. 369–383. Springer, Heidelberg (2000). doi:10.1007/3-540-45349-0_27

19. Régin, J.-C.: A filtering algorithm for constraints of difference in CSPs. In: AAAI, vol. 94, pp. 362–367 (1994)

20. Régin, J.-C., Puget, J.-F.: A filtering algorithm for global sequencing constraints. In: Smolka, G. (ed.) CP 1997. LNCS, vol. 1330, pp. 32–46. Springer, Heidelberg (1997). doi:10.1007/BFb0017428

21. Rodosek, R., Wallace, M.G., Hajian, M.T.: A new approach to integrating mixed integer programming and constraint logicprogramming. Ann. Oper. Res. **86**, 63–87 (1999)

22. Siala, M., Hebrard, E., Huguet, M.-J.: A study of constraint programming heuristics for the car-sequencing problem. Eng. Appl. Artif. Intell. **38**, 34–44 (2015)

23. Smith, B.: CSPLib problem 001: car sequencing. http://www.csplib.org/Problems/prob001

24. Van Hoeve, W.-J., Pesant, G., Rousseau, L.-M., Sabharwal, A.: Revisiting the sequence constraint. In: Benhamou, F. (ed.) CP 2006. LNCS, vol. 4204, pp. 620–634. Springer, Heidelberg (2006). doi:10.1007/11889205_44

25. Van Hoeve, W.-J., Pesant, G., Rousseau, L.-M., Sabharwal, A.: New filtering algorithms for combinations of among constraints. Constraints **14**(2), 273–292 (2009)

26. Wolsey, L.A.: Integer Programming. Wiley-Interscience, New York (1998)

Combining Nogoods in Restart-Based Search

Gael Glorian[✉], Frederic Boussemart, Jean-Marie Lagniez,
Christophe Lecoutre, and Bertrand Mazure

CRIL, CNRS, University Artois, F62300 Lens, France
{glorian,boussemart,lagniez,lecoutre,mazure}@cril.fr

Abstract. Nogood recording is a form of learning that has been shown useful for solving constraint satisfaction problems. One simple approach involves recording nogoods that are extracted from the rightmost branches of the successive trees built by a backtrack search algorithm with restarts. In this paper, we propose several mechanisms to reason with so-called increasing-nogoods that exactly correspond to the states reached at the end of each search run. Interestingly, some similarities that can be observed between increasing-nogoods allow us to propose new original ways of dynamically combining them in order to improve the overall filtering capability of the learning system. Our preliminary results show the practical interest of our approach.

Keywords: Learning · Increasing nogoods · Restarts · Filtering

1 Introduction

Nogood recording is a learning technique that has been applied to the Constraint Satisfaction Problem (CSP) in the 90's [2,4,14]. A classical nogood is defined as a partial instantiation that cannot be extended into a solution. Such nogoods have been cleverly exploited to manage explanations [5,7] of values that are deleted during search (when running constraint propagation). They have also been generalized [8] by incorporating both assigned variables (positive decisions) and refuted values (negative decisions). More recently, the practical interest of nogood recording has been revisited in the context of lazy clause generation [3].

Nogoods can also be effective in the context of a backtrack search algorithm that regularly triggers restarts. Indeed, just before restarting, a set of nogoods can be easily identified [10] on the rightmost branch of the search tree, which stands for the part of search space that has been explored during the last run. By recording these so-called nld-nogoods, we obtain the guarantee of never exploring the same subtrees, further making the approach complete. This restart-based learning mechanism has been extended to take into account symmetry breaking [11,13] and the increasing nature of nld-nogoods [12], called increasing-nogoods for this reason.

In this paper, we propose several mechanisms to combine increasing-nogoods, allowing us to increase their filtering capacity. By dynamically analyzing relevant

© Springer International Publishing AG 2017
J.C. Beck (Ed.): CP 2017, LNCS 10416, pp. 129–138, 2017.
DOI: 10.1007/978-3-319-66158-2_9

subsets of increasing-nogoods, especially from equivalence forms between decisions, we show that the search space can be more efficiently pruned. More specifically, we introduce three inference rules for deeper reasoning with increasing-nogoods.

2   Preliminaries

A constraint network P is a pair $(\mathcal{X}, \mathcal{C})$, where $\mathcal{X}$ is a finite set of variables and $\mathcal{C}$ a finite set of constraints. Each variable $x \in \mathcal{X}$ has a domain, denoted by $dom(x)$, which is the finite set of values a that can be assigned to x. Each constraint $c \in \mathcal{C}$ involves an ordered set of variables, called the scope of c and denoted by $scp(c)$. A constraint c is semantically defined by a relation, denoted by $rel(c)$, which is the set of tuples allowed by (variables of) c. Let $X \subseteq \mathcal{X}$ be a subset of variables, an instantiation I of X maps each variable $x \in X$ to a value in $dom(x)$; we note $I[x] = a$ and $vars(I) = X$. An instantiation I is *complete* iff $vars(I) = \mathcal{X}$, *partial* otherwise. A solution of P is a complete instantiation satisfying all constraints of P.

A nogood is an instantiation that cannot be extended to any solution. The benefit of recording nogoods is to avoid some form of thrashing, *i.e.* exploring the same unsatisfiable subtrees several times. There are two classical methods to identify and store nogoods: during search or at restarts. In this paper, we consider a complete backtrack search algorithm with binary branching and nogood recording from restarts [9]. Decisions taken during search are either positive (i.e., variable assignments such as $x = a$) or negative (i.e., value refutations such as $x \neq a$). A decision δ is checked to be positive or negative by simply writing $pos(\delta)$ and $neg(\delta)$, respectively. The variable involved in a decision δ is denoted by $var(\delta)$, whereas the value involved in a decision δ is denoted by $val(\delta)$. Binary branching means that at each search node a left branch labeled with a positive decision $x = a$ is developed first, and a right branch labeled with a negative decision $x \neq a$ is developed next. A decision δ is *satisfied* (resp., *falsified*) iff it holds (resp., does not hold) whatever is the value chosen in the current domain of $var(\delta)$. A decision that is not satisfied (resp., falsified) is said to be *unsatisfied* (resp., *unfalsified*).

3   Increasing Nogoods

So-called nld-nogoods [9] (negative last decision nogoods) can be extracted at each restart of a backtrack search algorithm. Let us assume that the sequence of labels all along the rightmost branch of a current search tree being developed is $\Sigma = \langle \delta_1, \ldots, \delta_m \rangle$, where each decision of Σ is either a positive or a negative decision. It is known that for any i such that $1 \leq i \leq m$ and $neg(\delta_i)$, the set $\{\delta_j : 1 \leq j < i \wedge pos(\delta_j)\} \cup \{\neg \delta_i\}$ is a reduced nld-nogood. Note that it only contains positive decisions (and so, is a standard nogood). From now on, for simplicity reasons, we simply call them nld-nogoods.

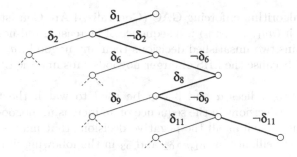

Fig. 1. The search tree at the end of a run.

Example 1. Let us consider the search tree depicted in Fig. 1, where the rightmost branch of the tree is $\Sigma = \{\delta_1, \neg \delta_2, \neg \delta_6, \delta_8, \neg \delta_9, \neg \delta_{11}\}$. The following (reduced) nld-nogoods can be extracted: $\{\delta_1, \delta_2\}$, $\{\delta_1, \delta_6\}$, $\{\delta_1, \delta_8, \delta_9\}$ and $\{\delta_1, \delta_8, \delta_{11}\}$.

As we can observe, there are some similarities between these nld-nogoods: they are said to be *increasing* [12,13]. An increasing-nogood compactly represents the full set of nld-nogoods that can be extracted from a branch. To obtain an increasing-nogood from a set of nld-nogoods, we have first to consider each nld-nogood under its *directed* form.

Example 2. Considering again the search tree in Fig. 1, let us assume the decisions of the last branch represent: $\Sigma = \langle x_2 = 1, x_3 \neq 0, x_4 \neq 1, x_5 = 2, x_1 \neq 1, x_6 \neq 2 \rangle$. The four nld-nogoods ng_0, ng_1, ng_2 and ng_3 are given below under their logical forms (middle) and directed forms (right):

$$ng_0 \equiv \neg(\qquad\qquad x_2 = 1 \wedge x_3 = 0) \equiv \qquad\qquad x_2 = 1 \Rightarrow x_3 \neq 0$$
$$ng_1 \equiv \neg(\qquad\qquad x_2 = 1 \wedge x_4 = 1) \equiv \qquad\qquad x_2 = 1 \Rightarrow x_4 \neq 1$$
$$ng_2 \equiv \neg(\ x_2 = 1 \wedge x_5 = 2 \wedge x_1 = 1) \equiv \quad x_2 = 1 \wedge x_5 = 2 \Rightarrow x_1 \neq 1$$
$$ng_3 \equiv \neg(\ x_2 = 1 \wedge x_5 = 2 \wedge x_6 = 2) \equiv \quad x_2 = 1 \wedge x_5 = 2 \Rightarrow x_6 \neq 2$$

In [12], the authors have shown that the set of directed nld-nogoods extracted from a branch are necessarily increasing, meaning that $\text{LHS}(ng_i) \subseteq \text{LHS}(ng_{i+1})$ where LHS designates the left hand side of the implication. This is illustrated on our example by:

$$ng_0 \equiv \qquad\qquad\qquad x_2 = 1 \Rightarrow x_3 \neq 0$$
$$ng_1 \equiv \qquad\qquad LHS(ng_0) \Rightarrow x_4 \neq 1$$
$$ng_2 \equiv \qquad LHS(ng_1) \wedge x_5 = 2 \Rightarrow x_1 \neq 1$$
$$ng_3 \equiv \qquad\qquad LHS(ng_2) \Rightarrow x_6 \neq 2$$

In practical terms, it means that it suffices to record the branch exactly as it is instead of extracting nld-nogoods independently. Another important observation is that each increasing-nogood can be viewed as a constraint, together with a filtering algorithm enforcing GAC (Generalized Arc Consistency). Interestingly enough, if $\langle ng_1, \ldots, ng_t \rangle$ is a sequence of increasing nld-nogoods, and if $LHS(ng_i)$ contains two unsatisfied decisions then any nogood ng_j with $j \geq i$ is necessarily GAC because the LHS of larger nogoods subsume the LHS of smaller ones.

Technically, two indices α and β can be used to watch the two leftmost unsatisfied positive decisions in the sequence of an increasing-nogood. These two watched decisions as well as all the negative decisions that may occur between them are under surveillance, as δ_1, $\neg \delta_2$ and δ_3 in the following illustration:

$$\Sigma = \langle \overbrace{\underset{\alpha}{\delta_1}, \neg \delta_2, \underset{\beta}{\delta_3}}^{\text{Watched}}, \delta_4, \neg \delta_5, \neg \delta_6 \rangle$$

For the sake of simplicity, we consider that for any increasing-nogood Σ, the decisions in Σ that are watched by the alpha and beta indices can be respectively accessed by using $\alpha(\Sigma)$ and $\beta(\Sigma)$. On our illustration, $\alpha(\Sigma)$ and $\beta(\Sigma)$ are respectively δ_1 and δ_3.

The filtering algorithm (called IncNG) associated with an increasing-nogood (constraint) is triggered in three cases:

1. a watched negative decision is falsified: $\alpha(\Sigma)$ must be forced to be falsified, and consequently all nogoods within the constraint are satisfied;
2. $\alpha(\Sigma)$ is satisfied: all negative decisions between α and β must be satisfied and we search for the next unsatisfied positive decision;
3. $\beta(\Sigma)$ is satisfied: we need to find the next unsatisfied positive decision.

4   Reasoning with Increasing Nogoods

When considered as constraints, it is quite natural that nld-nogoods and increasing-nogoods are solicited independently for filtering tasks. However, we show that it is possible to exploit the similarities that exist (rather frequently) between such nogoods. More specifically, we introduce in this section three rules for reasoning deeper with increasing-nogoods.

4.1   Reasoning with Watched Negative Decisions

By checking for each variable x and each increasing-nogood Σ that there exists a value in $dom(x)$ which is not involved in a negative decision for x between $\alpha(\Sigma)$ and $\beta(\Sigma)$, we have the guarantee of not missing some inferences from a simple reasoning on watched negative decisions.

Example 3. Consider the following increasing-nogood: $\Sigma = \langle x_2 = 1, x_3 \neq 2, x_3 \neq 4, x_5 = 3 \rangle$. Assume that we have $\alpha(\Sigma)$ and $\beta(\Sigma)$ being indices for $x_2 = 1$ and $x_5 = 3$, and all variables with the same domain $\{1, 2, 3, 4\}$. If x_2 is assigned to the value 1, then the values 2 and 4 can be removed from $dom(x_3)$. Of course, a conflict occurs if $dom(x_3)$ only contains these two values. However, this conflict could have been avoided (anticipated) by removing the value 1 from $dom(x_2)$ as soon as $dom(x_3)$ is reduced to $\{2, 4\}$. □

First, we introduce a function $\texttt{diffValues}(\Sigma, x_i)$ that returns for a given increasing-nogood Σ, the set of values present in a negative decision of Σ involving x_i and situated between $\alpha(\Sigma)$ and $\beta(\Sigma)$. We also introduce a function $\texttt{diffVars}(\Sigma)$ that returns the set of variables involved in a negative decision of Σ situated between $\alpha(\Sigma)$ and $\beta(\Sigma)$. For example, for $\Sigma = \langle x_2 = 1, x_3 \neq 2, x_3 \neq 4, x_5 = 3 \rangle$ with $\alpha(\Sigma)$ being $x_2 = 1$ and $\beta(\Sigma)$ being $x_5 = 3$, we have $\texttt{diffVars}(\Sigma) = \{x_3\}$ and $\texttt{diffValues}(\Sigma, x_3) = \{2, 4\}$. Algorithm 1 implements this way of reasoning, i.e. performs an inference by refuting the value involved in $\alpha(\Sigma)$, each time a conflict can be anticipated as discussed above. Even if increasing-nogoods are still reviewed independently (in turn), the filtering capability of the algorithm proposed in [12] is clearly improved if this simple procedure is systematically called. The worst-case time complexity of Algorithm 1 is $\mathcal{O}(nd)$ where n is the number of variables and d is the size of the largest domain. Indeed, we can precompute sets $\texttt{diffVars}(\Sigma)$ and $\texttt{diffValues}(\Sigma, x)$ by scanning the decisions in Σ whose size is $O(nd)$. With these precomputed sets, executing lines 1–2 is also in $O(nd)$.

Algorithm 1. checkNegativeDecisions(Σ : increasing-nogood)

---

1   **foreach** $x \in \texttt{diffVars}(\Sigma)$ **do**
2   **if** $dom(x) \subseteq \texttt{diffValues}(\Sigma, x)$ **then**
3   falsify $\alpha(\Sigma)$;

4.2 Combining Increasing Nogoods of Similar α

In this section, we extend the principle presented above to sets of increasing-nogoods. For this purpose, we partition the set of increasing-nogoods according to the decisions indexed by α: two increasing-nogoods Σ_i and Σ_j are in the same group iff $\alpha(\Sigma_i)$ is the same decision as $\alpha(\Sigma_j)$. Of course, it is therefore necessary to update the partition each time one α is modified (i.e., when filtering and backtracking). Despite that, reasoning about groups of increasing-nogoods allows us to improve the filtering capacity of the increasing-nogoods, and turns out to encompass the previous case.

Example 4. Let us consider the three following increasing-nogoods:

$$\Sigma_0 \equiv \quad \dots, \ x_6 \neq 2, \ \underbrace{x_2 = 1}_{\alpha}, \ x_1 \neq 3, \ x_3 \neq 1, \ \dots$$

$$\Sigma_1 \equiv \quad \dots, x_2 \neq 0, \ x_1 \neq 2, \ \underbrace{x_2 = 1}_{\alpha}, \ x_3 \neq 0, \ \dots$$

$$\Sigma_2 \equiv \quad \dots, \underbrace{x_2 = 1}_{\alpha}, \ x_3 \neq 2, \ x_6 \neq 1, \ x_8 \neq 3, \ \dots$$

and let us assume that all variables have the same domain $\{0,1,2,3\}$. On this example, we can observe that $x_2 = 1$ is the common α to this group of three increasing-nogoods. By looking at the negative decisions following these three occurrences of α (the precise values of β are not relevant for our illustration), we can collect $\{0,1,2\}$ as values involved in watched negative decisions for x_3 (they are necessarily put before β which is not represented here). This means that if x_2 is assigned the value 1 then the only remaining value in $dom(x_3)$ will be 3. On the other hand, if at a certain moment, the domain of x_3 does not contain anymore the value 3, it is absolutely necessary to prevent x_2 from being assigned the value 1. □

Algorithm 2. checkNegativeDecisions(Σs : set of increasing-nogoods)

　　Data: Increasing-nogoods in Σs have a common α
1　**foreach** $x \in \bigcup_{\Sigma \in \Sigma_s} \text{diffVars}(\Sigma)$ **do**
2　　　**if** $dom(x) \subseteq \bigcup_{\Sigma \in \Sigma_s} \text{diffValues}(\Sigma, x)$ **then**
3　　　　 falsify $\alpha(\Sigma)$;　　　　　　　　// Σ can be any increasing-nogood from Σs

---

Algorithm 2 is a generalization of Algorithm 1, by considering groups of increasing-nogoods instead of increasing-nogoods individually. Algorithm 2 has a worst-case time complexity in $\mathcal{O}(nd + g)$ where g is the sum of the size of the increasing-nogoods in Σs (we have $g = \sum_{\Sigma \in \Sigma_s} |\Sigma|$). Indeed, precomputing sets $\bigcup_{\Sigma \in \Sigma_s} \text{diffVars}(\Sigma)$ and $\bigcup_{\Sigma \in \Sigma_s} \text{diffValues}(\Sigma, x)$ can be performed in $O(g)$ by scanning every decision in the increasing-nogoods of Σs. With these precomputed sets, executing lines 1–2 is in $O(nd)$.

4.3　Combining Increasing Nogoods Using Pivots

We call *pivot* a variable x such that for any value $a \in dom(x)$ there exists an increasing-nogood Σ such that $\alpha(\Sigma)$ is the positive decision $x = a$; in that case, we say that Σ is a *support* of pivot x for a. Interestingly, once a pivot variable x is identified, it is possible to infer negative decisions that are shared by all supports of x. This is the principle of the algorithm we present after an illustration.

Algorithm 3. checkPivots(Σs : the full set of increasing-nogoods)

---

1 **foreach** $x \in \{var(\alpha(\Sigma)) : \Sigma \in \Sigma s\}$ **do**
2   **if** $dom(x) \subseteq \{val(\alpha(\Sigma)) : \Sigma \in \Sigma s \wedge var(\alpha(\Sigma)) = x\}$ **then**
3   **foreach** $\delta \in \bigcap\{\Sigma \in \Sigma s : var(\alpha(\Sigma)) = x\}$ **do**
4   satisfy δ;

Example 5. Let us consider the following three increasing-nogoods:

$$\Sigma_0 \equiv \qquad \dots,\ x_6 \neq 2,\ \underbrace{x_2 = 1}_{\alpha},\ x_1 \neq 0,\ x_3 \neq 1,\ \dots$$

$$\Sigma_1 \equiv \qquad \dots,\ x_7 \neq 0,\ x_1 \neq 2,\ \underbrace{x_2 = 0}_{\alpha},\ x_3 \neq 1,\ \dots$$

$$\Sigma_2 \equiv \qquad \dots,\ \underbrace{x_2 = 2}_{\alpha},\ x_3 \neq 1,\ x_6 \neq 1,\ x_8 \neq 2,\ \dots$$

and let us assume that all variables have the same domain $\{0, 1, 2\}$. On this example, we can see that x_2 is a pivot since all its possible values are involved in α of different increasing-nogoods. As $x_3 \neq 1$ is a negative decision that is watched in the three increasing-nogoods, we can deduce that x_3 must always be different from 1. $\square$

Algorithm 3 implements the use of pivot variables for making additional inferences. Line 1 only iterates over the variables that are involved in some α (i.e., α of some increasing-nogoods). Line 2 tests if the variable x is indeed a pivot. Line 3 iterates over the decisions that are shared by all supports of x. Each such decision must be forced to be satisfied. Note that an optimization consists in only checking that a decision is shared by some subsets of supports of x, the subsets with exactly one support of x for each value. Algorithm 3 has a worst-case time complexity in $\mathcal{O}(n^2 dp)$ where p is the number of increasing-nogoods.

5   Experiments

We have conducted an experimentation on a computer *Intel Xeon X5550* clocked at 2,67 GHz and equipped with 8 GB of RAM. Our initial benchmark was composed of all instances that were used during XCSP 2.1 solver competitions. We discarded the series of instances that were either too easy to solve (less than 1 s) or too hard to solve (more than 900 s) when employing MAC without nogood recording; this yielded a set composed of 3,744 instances. For our experiments, we have used the solver *rclCSP* introduced in [6]. The variable ordering heuristic is *dom/wdeg* [1] and the restart policy corresponds to a geometric series of first term 10 and common ratio 1.1. Given the complexity of Algorithm 3 and because Algorithm 1 is generalized by Algorithm 2, we chose to conduct our

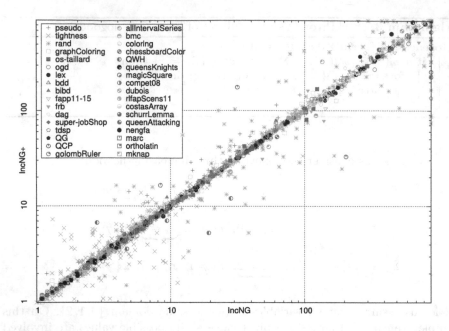

Fig. 2. Pairwise comparison (CPU time in seconds) of IncNG and IncNG+. Results obtained on the 3,744 instances used as benchmarks. The timeout to solve an instance is set to 900 s.

Table 1. Average results on 15 series of instances (timeout set to 900 s).

Series	#inst	NRR		IncNG		IncNG+	
		#sols	PAR10	#sols	PAR10	#sols	PAR10
costasArray	11	9	1,745	**9**	**1,686**	9	**1,686**
fapp11-15	55	42	2,162	42	2,157	**42**	**2,155**
frb45-21	10	9	1,138	**10**	**232**	10	282
nengfa	10	9	970	**9**	**954**	9	974
ogdVg	65	35	4,197	39	3,637	**40**	**3,511**
ortholatin	9	**4**	**5,045**	3	6,011	3	6,010
os-taillard-7	30	**19**	**3,353**	18	3,679	18	3,680
QCP-20	15	6	5,437	7	4,870	**8**	**4,289**
QWH-20	10	10	79	**10**	**47**	10	48
QWH-25	10	1	8,187	0	9,000	**2**	**7,291**
rand-2-50-23-fcd	50	5	8,146	12	6,953	**15**	**6,436**
rand-2-50-23	50	4	16,481	9	14,429	**9**	**13,899**
rand-3-24-24	50	14	11,605	16	10,557	**18**	**10,028**
rlfapScens11	12	11	852	**11**	**844**	11	852
super-jobShop	46	34	2,358	33	2,547	**35**	**2,358**

experiments with only Algorithm 2 (combining increasing-nogoods of similar α). For our comparison, we tested three methods: NRR (nogood recording from restarts as proposed in [12]), IncNG (managing increasing-nogoods as proposed in [10]) and IncNG+ (combining increasing-nogoods, our approach as introduced in Sect. 4.2). Figure 2 displays the overall results that we have obtained. The scatter plot shows that our approach (IncNG+) has usually a small overhead (when it turns out to be not very effective), and interestingly makes search a little bit more robust (see the dots on the right vertical line that corresponds to unsolved instances by IncNG). Table 1 shows a detailed comparison between the three methods on some series. The table contains the following information: the name of the series, the number of instances in the series (*#inst*) and for the three tested approaches, the number of solved instances (*#sols*) and the PAR10 score that is the average of the runtimes while considering 10 times the time-out (900) for unsolved instances. In general, our approach (IncNG+) solves at least as many instances as IncNG and sometimes more (see, for example, *super-jobShop*). When IncNG and IncNG+ both solve the same number of instances, we usually notice a slight loss for our approach due to the management of the partitions of the increasing-nogoods (for example, see *frb45-21*). But interestingly, there are series where this processing time is compensated by a better pruning of the search tree, with a substantial time saving as outcome (*rand-2-50-23*). Moreover, we observe that, in general, the more difficult the instance is, the more competitive our approach is, as illustrated by the *QWH-20* and *QWH-25* series which admit increasing sizes and complexities.

6 Conclusion

In this paper, we have introduced three general rules allowing us to reason with (groups of) increasing-nogoods. We have shown experimentally the practical interest of the second rule (which generalizes the first one): when it is effective, i.e., when inferences can be performed by reasoning on groups of increasing-nogoods, it saves computation time, and when it is not effective, the overhead is rather limited. We believe that some improvements are still possible, especially on the exploitation of pivot variables.

Acknowledgement. This work has been supported by the project CPER DATA from the "Hauts-de-France" Region.

References

1. Boussemart, F., Hemery, F., Lecoutre, C., Sais, L.: Boosting systematic search by weighting constraints. In: Proceedings of 16th European Conference on Artificial Intelligence, ECAI 2004, Including Prestigious Applicants of Intelligent Systems, PAIS 2004, Valencia, Spain, 22–27 August 2004, pp. 146–150. IOS Press (2004)
2. Dechter, R.: Enhancement schemes for constraint processing: backjumping, learning, and cutset decomposition. Artif. Intell. **41**(3), 273–312 (1990)

3. Feydy, T., Stuckey, P.J.: Lazy clause generation reengineered. In: Gent, I.P. (ed.) CP 2009. LNCS, vol. 5732, pp. 352–366. Springer, Heidelberg (2009). doi:10.1007/978-3-642-04244-7_29

4. Frost, D., Dechter, R.: Dead-end driven learning. In: Proceedings of 12th National Conference on Artificial Intelligence, Seattle, WA, USA, 31 July - 4 August 1994, vol. 1, pp. 294–300. AAAI Press/The MIT Press (1994)

5. Ginsberg, M.L.: Dynamic backtracking. J. Artif. Intell. Res. (JAIR) 1, 25–46 (1993)

6. Grégoire, É., Lagniez, J.-M., Mazure, B.: A CSP solver focusing on FAC variables. In: Lee, J. (ed.) CP 2011. LNCS, vol. 6876, pp. 493–507. Springer, Heidelberg (2011). doi:10.1007/978-3-642-23786-7_38

7. Jussien, N., Debruyne, R., Boizumault, P.: Maintaining arc-consistency within dynamic backtracking. In: Dechter, R. (ed.) CP 2000. LNCS, vol. 1894, pp. 249–261. Springer, Heidelberg (2000). doi:10.1007/3-540-45349-0_19

8. Katsirelos, G., Bacchus, F.: Unrestricted nogood recording in CSP search. In: Rossi, F. (ed.) CP 2003. LNCS, vol. 2833, pp. 873–877. Springer, Heidelberg (2003). doi:10.1007/978-3-540-45193-8_70

9. Lecoutre, C., Sais, L., Tabary, S., Vidal, V.: Nogood recording from restarts. In: IJCAI 2007, Proceedings of 20th International Joint Conference on Artificial Intelligence, Hyderabad, India, 6–12 January 2007, pp. 131–136 (2007)

10. Lecoutre, C., Sais, L., Tabary, S., Vidal, V.: Recording and minimizing nogoods from restarts. J. Satisf. Boolean Model. Comput. (JSAT) 1(3–4), 147–167 (2007)

11. Lecoutre, C., Tabary, S.: Symmetry-reinforced nogood recording from restarts. In: 11th International Workshop on Symmetry in Constraint Satisfaction Problems (SymCon 2011), Perugia, Italy, pp. 13–27 (2011)

12. Lee, J.H.M., Schulte, C., Zhu, Z.: Increasing nogoods in restart-based search. In: Proceedings of 30th AAAI Conference on Artificial Intelligence, 12–17 February 2016, Phoenix, Arizona, USA, pp. 3426–3433. AAAI Press (2016)

13. Lee, J.H.M., Zhu, Z.: An increasing-nogoods global constraint for symmetry breaking during search. In: O'Sullivan, B. (ed.) CP 2014. LNCS, vol. 8656, pp. 465–480. Springer, Cham (2014). doi:10.1007/978-3-319-10428-7_35

14. Schiex, T., Verfaillie, G.: Nogood recording for static and dynamic constraint satisfaction problems. Int. J. Artif. Intell. Tools 3(2), 187–208 (1994)

All or Nothing: Toward a Promise Problem Dichotomy for Constraint Problems

Lucy Ham and Marcel Jackson[(✉)]

Department of Mathematics and Statistics, La Trobe University,
Melbourne 3086, Australia
leham@students.latrobe.edu.au, m.g.jackson@latrobe.edu.au

Abstract. We show that intractability of the constraint satisfaction problem over a fixed finite constraint language can, in all known cases, be replaced by an infinite hierarchy of intractable promise problems of increasingly disparate promise conditions. The instances are guaranteed to either have no solutions at all, or to be k-robustly satisfiable (for any fixed k), meaning that every "reasonable" partial instantiation on k variables extends to a solution.

Keywords: Constraint satisfaction problem · Dichotomy · Robust satisfiability · Promise problem · Quasivariety · Universal horn class

1 Introduction

In the constraint satisfaction problem (CSP) we are given a domain A, a list of relations $\mathscr{R}$ on A and a finite set V of variables, in which various tuples of variables have been constrained by the relations in $\mathscr{R}$. The fundamental *satisfaction* question is to decide whether there is a function $\phi : V \to A$ such that $(\phi(v_1), \ldots, \phi(v_n)) \in r$ whenever $\langle (v_1, \ldots, v_n), r \rangle$ is a constraint (and $r \in \mathscr{R}$ is of arity n). Many computational problems are expressible in this framework, even in the nonuniform case, where the domain A and relations $\mathscr{R}$ are fixed. Such *fixed template* CSPs have received particular attention in theoretical investigations: examples include the SAT variants considered by Schaefer [40], graph homomorphism problems such as in the Hell-Nešetřil dichotomy [22] as well as list-homomorphism problems and conservative CSPs [9]. Feder and Vardi [14] generated particular attention on the theoretical analysis of computational complexity of fixed template CSPs, by tying the complexity of fixed finite template CSPs precisely to those complexities to be found in the largest logically definable class for which they were unable to prove that Ladner's Theorem holds. This motivated their famous *dichotomy conjecture*: is it true that a fixed finite template CSP is either solvable in polynomial time or is NP-complete?

M. Jackson—The second author was supported by ARC grants FT120100666 and DP1094578.

J.C. Beck (Ed.): CP 2017, LNCS 10416, pp. 139–156, 2017.
DOI: 10.1007/978-3-319-66158-2_10

A pivotal development in the efforts toward a possible proof of the dichotomy conjecture was the introduction of universal algebraic methods. This provided fresh tools to build tractable algorithms, and to build reductions for hardness, as well as an established mathematical landscape in which to formulate conjectures on complexity. The method is fundamental to Bulatov's classification of 3-element CSPs [8], of the Dichotomy Theorem for conservative CSPs [9], for homomorphism problems on digraphs without sources and sinks [6], in the classification of when a CSP is solvable by generalised Gaussian elimination [23], and of when a CSP is solvable by a local consistency check algorithm [5], among others. The *algebraic dichotomy conjecture* (ADC) of [11] refines the Feder-Vardi conjecture by speculating the precise boundary between P and NP, in terms of the presence of certain algebraic properties. The ADC has been verified in each of the aforementioned tractability classifications.

The present article shows that NP-completeness results obtained via the algebraic method also imply the NP-completeness of a strong promise problem. The NO instances are those for which there is no solution, but the YES instances are instances for which any "reasonable" partial assignment on k variables can extend to a solution. "Reasonable" here means subject to some finite set of local, necessary conditions. The All or Nothing Theorem (ANT) below proves the NP-completeness of this promise problem for any integer $k \geq 0$ and in any intractable CSP covered by the algebraic method. The promise conditions include satisfaction as a special case, and complement the promise condition on NO instances provided by the PCP Theorem [4] (at least ε proportion of the constraints must fail, for some $\varepsilon > 0$). We are also able to prove a dichotomy theorem by showing that for sufficiently large k, our promise problem is solvable in AC^0 if and only if the CSP is of bounded width (in the sense of Barto and Kozik [5]) and otherwise is hard for the complexity class $\text{Mod}_p(L)$ for some prime p.

A second contribution of the article is to connect the model-theoretic notion of *quasivariety* to the concept of *implied constraints*. Identifying various kinds of implied constraints is a central method employed in constraint solvers [36], and the proliferation of implied constraints is associated with phase transitions in randomly generated constraint problems [37]. We explain how the absence of implied constraints corresponds to membership in the quasivariety generated by the template. Intuitively, it seems quite unlikely that the problem of recognising "no implied constraints" can be approached using the algebraic method, because there is no obvious reduction between constraint languages $\mathscr{R}_1$ and $\mathscr{R}_2$ when $\mathscr{R}_1 \subsetneq \mathscr{R}_2$. Despite this intuition, the strength of the promise in the ANT enables us to show that whenever the algebraic method shows hardness of $CSP(\mathscr{R})$, then there is no polynomial time algorithm to distinguish constraint instances with no solution, from those that have no implied constraints with respect to $\mathscr{R}$. We can also use our bounded width dichotomy to obtain the most general nonfinite axiomatisability result known for finitely generated quasivarieties. A further important corollary is a promise problem extension of Hell and Nešetřil's well-known dichotomy for simple graphs [22].

More generally, the ability to extend all reasonable partial assignments holds potential for a wide range of applications, with recent examples including minimal networks [18], quantum mechanics [3], and semigroup theory [24].

2 Constraints and Implied Constraints

Since Feder and Vardi [14] it has been standard to reformulate the fixed template CSP over domain A and finite language $\mathscr{R}$ as a homomorphism problem between model-theoretic structures. The template is a relational structure $\mathbb{A} = \langle A, \mathscr{R}^A \rangle$ (with $\mathscr{R}$ a relational signature), as is the instance $(V, \mathscr{C})$ (where $\mathscr{C}$ is the list of constraints) where the variable set V is the universe, and with each $r \in \mathscr{R}$ being interpreted as the relation on V equal to the set of tuples constrained to r in the set $\mathscr{C}$. Thus each individual constraint $\langle (v_1, \ldots, v_n), r \rangle$ becomes a membership of a tuple $(v_1, \ldots, v_n)$ in the relation r^V on V. We refer to $(v_1, \ldots, v_n) \in r^V$ as a *hyperedge*. The *constraint satisfaction problem for* $\mathbb{A}$, which we denote by $\mathrm{CSP}(\mathbb{A})$ is the problem of deciding membership in the class of finite structures admitting homomorphism into $\mathbb{A}$. Throughout the article, $\mathbb{A}$ will be the default notation for a CSP template of signature $\mathscr{R}$ (both assumed to be finite) and $\mathbb{B}$ for a general (finite) $\mathscr{R}$-structure. We let arity($\mathscr{R}$) denote the maximal arity of any relation in $\mathscr{R}$.

A nonhyperedge $(v_1, \ldots, v_n) \notin r^B$, where $r \in \mathcal{R} \cup \{=\}$ (of arity n) satisfies the *separation condition* if there is a homomorphism $\phi : \mathbb{B} \to \mathbb{A}$ with $(\phi(v_1), \ldots, \phi(v_n)) \notin r^{\mathbb{A}}$. When a nonhyperedge fails the separation property we say that it is an *implied constraint*, as every homomorphism into $\mathbb{A}$ places it within the corresponding relation of $\mathbb{A}$; equivalently, $(v_1, \ldots, v_n)$ can be added to r^B without changing the set of possible solutions for $\mathbb{B}$ with respect to $\mathbb{A}$. We say that $\mathbb{B}$ *satisfies the separation condition* (w.r.t. $\mathbb{A}$), or *has no implied constraints* if no nonhyperedge is an implied constraint. Note that if $\mathbb{B}$ is a NO instance of $\mathrm{CSP}(\mathbb{A})$, then every nonhyperedge is implied (including equalities between distinct elements).

The separation condition for $\mathbb{B}$ is widely known to be equivalent to the property that $\mathbb{B}$ lies in the *quasivariety* generated by $\mathbb{A}$: the class of isomorphic copies of induced substructures of direct powers of $\mathbb{A}$; see Maltsev [34] and Gorbunov [17], but also [24, Theorem 2.1] and [26, §2.1,2.2] for the CSP interpretations and generalizations. The one-element total structure $\mathbf{1}_{\mathscr{R}}$, with no nonhyperedges, satisfies the separation condition vacuously. If we wish to exclude $\mathbf{1}_{\mathscr{R}}$ we arrive at the *universal Horn class* generated by $\mathbb{A}$ (which excludes the zeroth power from "direct powers"). We let $\mathsf{Q}(\mathbb{A})$ denote the quasivariety of $\mathbb{A}$ and $\mathsf{Q}^+(\mathbb{A})$ the universal Horn class of $\mathbb{A}$. Membership in $\mathsf{Q}(\mathbb{A})$ is the problem of deciding if an input has no implied constraints, which we denote by $\mathrm{CSP}_\infty(\mathbb{A})$. Membership in $\mathsf{Q}^+(\mathbb{A})$ is essentially the same as $\mathrm{CSP}_\infty(\mathbb{A})$ because $\mathsf{Q}(\mathbb{A})$ and $\mathsf{Q}^+(\mathbb{A})$ differ on at most the structure $\mathbf{1}_{\mathscr{R}}$.

> Problem: $\text{CSP}_\infty(\mathbb{A})$ (no implied constraints)
> Instance: a finite $\mathscr{R}$-structure $\mathbb{B}$.
> Question: for every nonhyperedge $(v_1, \ldots, v_n) \notin r^B$, is there a homomorphism into $\mathbb{B}$ taking $(v_1, \ldots, v_n) \notin r^B$ to a nonhyperedge $(a_1, \ldots, a_n) \notin r^A$ of $\mathbb{A}$?

The case of no implied *equalities* is considered in Ham [20,21], with a complete tractability classification in the case of Boolean constraint languages.

3   Primitive Positive Formulæ and Robust Satisfiability

Definition 1. *An* atomic formula *is an expression of the form* $(x_1, \ldots, x_n) \in r$ *for some* $r \in \mathscr{R}$ *or* $x = y$. *A* primitive positive formula (*abbreviated to* pp-formula) *is a formula obtained from a conjunction of atomic formulæ by existentially quantifying some variables. A pp-formula* $\phi(x_1, \ldots, x_n)$ *with free variables* $x_1, \ldots, x_n$ *defines an* n-*ary relation* r_ϕ, *which in any* $\mathscr{R}$-*structure* $\mathbb{C}$ *is interpreted as the solution set of* ϕ. *If* $\mathscr{F}$ *is a set of pp-formulæ, then* $\mathbb{C}_{\mathscr{F}}$ *denotes* $\langle C; \{r_\phi^C \mid \phi \in \mathscr{F}\} \rangle$.

We let $\text{pp}(\mathscr{R})$ be the set of all pp-formulæ (over some fixed countably infinite set of variables) in $\mathscr{R}$ and let $\text{pp}(\mathbb{C})$ denote the set $\{r_\phi^C \mid \phi \in \text{pp}(\mathscr{R})\}$ of all relations on C that are pp-definable from the fundamental relations of $\mathbb{C}$.

Let $\mathbb{A}, \mathbb{B}$ be $\mathscr{R}$-structures and $\mathscr{F} \subseteq \text{pp}(\mathscr{R})$. For a subset $S \subseteq B$, a function $\nu : S \to A$ is $\mathscr{F}$-*compatible* if it is a homomorphism from the substructure $\mathbb{S}$ of $\mathbb{B}_{\mathscr{F}}$ to $\mathbb{A}_{\mathscr{F}}$. A function $\nu : S \to A$ extends to a homomorphism precisely when it is $\text{pp}(\mathscr{R})$-compatible [24, Lemma 3.1], so restricting $\mathscr{F}$ to a fixed finite subset of $\text{pp}(\mathscr{R})$ is the natural local condition for extendability.

Definition 2 [21]. *Let* $\mathscr{F}$ *be a finite set of pp-formulæ in* $\mathscr{R}$ *and let* $\mathbb{A}$ *be a fixed finite* $\mathscr{R}$-*structure. For a finite* $\mathscr{R}$-*structure* $\mathbb{B}$, *we say that* $\mathbb{B}$ *is* $(k, \mathscr{F})$-*robustly satisfiable* (*with respect to* $\mathbb{A}$) *if* $\mathbb{B}$ *is a YES instance of* $\text{CSP}(\mathbb{A})$ *and for every* k-*element subset* S *of* B *and every* $\mathscr{F}$-*compatible assignment* $\nu : S \to A$, *there is a solution to* $\mathbb{B}$ *extending* ν. *The structure* $\mathbb{B}$ *is* $(\leq k, \mathscr{F})$-*robustly satisfiable if it is* $(\ell, \mathscr{F})$-*robustly satisfiable for every* $\ell \leq k$.

Note that $(0, \mathscr{F})$-robust satisfiability coincides with satisfiability. Intuitively, $(k, \mathscr{F})$-robust satisfiability is a very strong condition on an instance. For example, a graph is $(2, \mathscr{F})$-robustly 2-colorable if every $\mathscr{F}$-compatible 2-coloring of any 2 vertices extends to a full 2-coloring. It is an easy exercise to show that a $(2, \mathscr{F})$-robustly 2-colorable graph must have diameter at most m, where m is the number of variables appearing in $\mathscr{F}$.

In [7], the case of $(k, \varnothing)$-robust satisfiability is considered for SAT-related probems using the notation $\widehat{\mathbf{U}^k}$; this appears in the context of phase transitions and implied constraints. The concept of $(k, \varnothing)$-robust satisfiability is called k-*supersymmetric* in Gottlob [18], where it is used to show that there is no polynomial time solver for a minimal constraint network. If $\mathscr{P}$ denotes the conjunction-free pp-formulæ, then $(k, \mathscr{P})$-robust satisfiability is the "k-robust satisfiability" concept introduced in Abramsky, Gottlob and Kolaitis [3], where (for $k = 3$ in 3SAT) it is applied to show the intractability of detecting local hidden-variable models in quantum mechanics. Jackson [24] showed the NP-completeness of a promise problem form of $(2, \mathscr{P})$-robust satisfiability for positive 1-in-3SAT, and used it to solve a 20+ year old problem in semigroup theory [2, Problem 4], itself motivated by issues in formal languages. These examples involve very technical case-checking arguments. A more unified algebraic approach was very recently initiated by Ham [20,21], who classified the tractability of $(2, \mathscr{F})$-robust satisfiability (for some $\mathscr{F}$) in the case of Boolean constraint languages.

4 Primitive Positive Definability and Polymorphisms

When $\mathscr{R}$ is pp-definable from a set of relations $\mathscr{S}$ on a set A then there is logspace reduction from $\mathrm{CSP}(\langle A; \mathscr{R} \rangle)$ to $\mathrm{CSP}(\langle A; \mathscr{S} \rangle)$. This fundamental idea was primarily developed through the work of Cohen, Jeavons and others [27–31], though aspects appear in proof of Schaefer's original dichotomy for Boolean CSPs [40].

There is a well-known Galois correspondence between sets of relations on a set A and the sets of operations on A; see [16]. The link is via *polymorphisms*, which are homomorphisms from the direct product $\mathbb{A}^n$ to $\mathbb{A}$. In other words, for each relation $r \in \mathscr{R}$ (with arity k, say), if we are given an $k \times n$ matrix of entries from A, with each column being a k-tuple in r, then applying the polymorphism f to each row produces a k-tuple of outputs that also must lie in r. We let $\mathrm{Pol}(\mathbb{A})$ denote the family of all polymorphisms of the relational structure $\mathbb{A}$. For finite A we have $\mathrm{Pol}(\langle A; \mathscr{R} \rangle) \subseteq \mathrm{Pol}(\langle A; \mathscr{S} \rangle)$ if and only if $\mathscr{S} \subseteq \mathrm{pp}(\langle A, \mathscr{R} \rangle)$, so that pp-definability is captured by polymorphisms. We now list some conditions on polymorphisms that we will use; see an article such as [25] for a survey of other conditions that play a role in understanding the complexity of CSP complexity.

– An n-ary operation $w : A^n \to A$ on a set A is a *weak near unanimity* operation (or *WNU*) if it satisfies $w(x, x, \ldots, x) = x$ (idempotence) and $w(y, x, \ldots, x) = w(x, y, \ldots, x) = \cdots = w(x, x, \ldots, y)$ for all x, y. A weak near unanimity operation is *near unanimity* (NU) if it additionally satisfies $w(y, x, \ldots, x) = x$.
– If the condition of being idempotent is dropped, we refer to a *quasi WNU*, and a *quasi NU* respectively.

We mention that most algebraic approaches use the assumption that the template $\mathbb{A}$ is a *core*, meaning that it has no proper retracts. We now list a selection of pertinent results and conjectures that are expressed in the language of polymorphisms.

The fundamental conjecture on fixed template CSP complexity is the following refinement of Feder and Vardi's original.

Algebraic Dichotomy Conjecture (ADC) 3 [11]. *Let* $\mathbb{A}$ *be a finite core relational structure of finite relational signature. If* $\mathbb{A}$ *has a WNU polymorphism then* $\mathrm{CSP}(\mathbb{A})$ *is tractable. If* $\mathbb{A}$ *has no WNU polymorphism then* $\mathrm{CSP}(\mathbb{A})$ *is NP-complete.*

The final sentence in the conjecture is proved already in [11] (with the WNU condition we state established in [35]), with completeness with respect to first order reductions established in [33]. There are no counterexamples to the conjecture amongst known classifications, and recently several purported proofs have been claimed [10, 39, 42] (we do not assume these as verified).

In the following, *bounded width* corresponds to solvability by way of a local consistency check algorithm, while strict width is a restricted case of this, where every family of locally consistent partial solutions extends to a solution.

Theorem 4. *Let* $\mathbb{A}$ *be a finite core relational structure of finite relational signature.*

1. *(Feder and Vardi [14].)* $\mathrm{CSP}(\mathbb{A})$ *has strict width if and only if* $\mathbb{A}$ *has an NU polymorphism.*
2. *(Barto and Kozik [5].)* $\mathrm{CSP}(\mathbb{A})$ *has bounded width if and only if* $\mathbb{A}$ *has a 3-ary WNU* w_3 *and a 4-ary WNU* w_4 *such that* $w_3(y, x, x) = w_4(y, x, x, x)$ *holds for all* x, y.

5   Main Results

Recall that a *promise problem* consists of a pair of disjoint languages (Y, N). The question is conditional: given the promise that an instance lies in $Y \cup N$, decide if it lies in Y; see [19] for example.

The main result (ANT) concerns the following promise problem, which simultaneously extends $\mathrm{CSP}(\mathbb{A})$, $\mathrm{CSP}_\infty(\mathbb{A})$, robust-$\mathrm{CSP}(\mathbb{A})$ [3], $\mathrm{SEP}(\mathbb{A})$ [20, 21] and others. In the title line, k is a non-negative integer and $\mathscr{F}$ is a finite set of pp-formulæ in the signature of $\mathbb{A}$.

Promise problem: $(Y_{(k,\mathscr{F}),\mathsf{Q}}, N_{\mathrm{CSP}})$ for $\mathbb{A}$.

YES: $\mathbb{B}$ is $(k, \mathscr{F})$-robustly satisfiable with respect to $\mathbb{A}$ and has no implied constraints.
NO: $\mathbb{B}$ is a no instance of $\mathrm{CSP}(\mathbb{A})$.

We will let $Y_{(k,\mathscr{F})}$ denote the YES promise but where "no implied constraints" is omitted.

All or Nothing Theorem (ANT) 5. *Let* $\mathbb{A}$ *be a finite core relational structure in finite signature* $\mathscr{R}$.

1. *(Everything is easy.) If* $\mathrm{CSP}(\mathbb{A})$ *is tractable then so also is deciding both* $\mathrm{CSP}_\infty(\mathbb{A})$ *and* $(k, \mathscr{F})$-*robust satisfiability, for any* k *and any finite set* $\mathscr{F}$ *of pp-formulæ.*

2. (*Nothing is easy.*) If $\mathbb{A}$ has no WNU, then for all k there exists a finite set of pp-formulæ $\mathscr{F}$ such that $(Y_{(k,\mathscr{F}),\mathsf{Q}}, N_{\mathrm{CSP}})$ is NP-complete for $\mathbb{A}$ with respect to first order reductions.

Remark 6. *The ANT shows that the ADC is equivalent to the ostensibly far stronger dichotomy statement: either there is a WNU and (1) holds, or there is no WNU and $(Y_{(k,\mathscr{F}),\mathsf{Q}}, N_{\mathrm{CSP}})$ is NP-complete for some finite family of formulæ $\mathscr{F}$.*

As an example, the ANT shows that or all k there exists an $\mathscr{F}$ such that it is NP-hard to distinguish the $(k, \mathscr{F})$-robustly 3-colorable graphs from those that are not 3-colorable at all.

The following result gives a dichotomy within tractable complexity classes.

Theorem 7. *Let $\mathbb{A}$ be a finite core relational structure in finite signature $\mathscr{R}$.*

1. *If $\mathrm{CSP}(\mathbb{A})$ has bounded width, then there exists n such that for all $k \geq n$ and for all finite sets of pp-formulæ $\mathscr{F}$, the promise problem $(Y_{(k,\mathscr{F}),\mathsf{Q}}, N_{\mathrm{CSP}})$ lies in AC^0. (If $\mathrm{CSP}(\mathbb{A})$ has strict width, then the class of $(k, \mathscr{F})$-robustly satisfiable instances is itself first order definable.)*
2. *If $\mathrm{CSP}(\mathbb{A})$ does not have bounded width then for some prime number p and for all k there exists an $\mathscr{F}$ such that $(Y_{(k,\mathscr{F}),\mathsf{Q}}, N_{\mathrm{CSP}})$ is $\mathsf{Mod}_p(L)$-hard.*

Recall that the $\mathsf{Mod}_p(\mathsf{L})$ class contains L and hence properly contains AC^0; the precise relationship with NL is unknown. Thus Theorem 7 shows that in contrast to $\mathrm{CSP}(\mathbb{A})$ (see [1,33]), one cannot get L-completeness, nor NL-completeness for $(Y_{(k,\mathscr{F}),\mathsf{Q}}, N_{\mathrm{CSP}})$ over $\mathbb{A}$ unless there are unexpected collapses between L, NL and $\mathsf{Mod}_p(\mathsf{L})$ for various p. For example: while graph 2-colorability is L-complete, deciding $(k, \mathscr{F})$-robust 2-colorability is first-order when $k \geq 2$ (and for any $\mathscr{F}$).

The complexity of $\mathrm{CSP}(\mathbb{A})$ is determined by the core retract of $\mathbb{A}$, but this is not true for $(k, \mathscr{F})$-robust satisfiability and quasivariety membership; see [24] and [20,21] for example. The following results however apply regardless of whether $\mathbb{A}$ itself is a core.

Corollary 8. *Let $\mathbb{A}$ be finite relational structure of finite signature.*

- *If $\mathbb{A}$ has no quasi WNU polymorphism then $\mathrm{CSP}_\infty(\mathbb{A})$ is NP-complete,*
- *If the core retract of $\mathbb{A}$ fails to have bounded width then $\mathbb{Q}(\mathbb{A})$ is not finitely axiomatisable in first order logic, even amongst finite structures.*

The second statement is equivalent to the absence of quasi WNUs satisfying the conditions of Theorem 4(2). Similar statements to Corollary 8 hold for problems intermediate to $\mathrm{CSP}(\mathbb{A})$ and $\mathrm{CSP}_\infty(\mathbb{A})$, such as the $\mathrm{SEP}(\mathbb{A})$ of [21] and the problem of detecting if no variable is nontrivially forced to take a fixed value: variables with implicitly fixed values have been called the "backbone" or "frozen variables"; see [32] for example.

The following corollary simultaneously covers the original Hell-Nešetřil Dichotomy for simple graphs and a corresponding quasivariety dichotomy; again it does not assume cores.

Gap Dichotomy for Simple Graphs 9. *Let* $\mathbb{G}$ *be a finite simple graph.*

1. *If* $\mathbb{G}$ *is bipartite, then deciding* CSP($\mathbb{G}$) *and deciding membership in the quasivariety of* $\mathbb{G}$ *are both tractable.*
2. *Otherwise, the following promise problem is* NP-*complete with respect to first order reductions and for finite input graph* $\mathbb{H}$:
   Yes $\mathbb{H}$ *is in* $Q^{+}(\mathbb{G})$.
   No $\mathbb{H}$ *has no homomorphisms into* $\mathbb{G}$.

We also complete a line of investigation initiated by Beacham and Culberson [7], by identifying the threshold value for k in the intractability of $(k, \varnothing)$-robust satisfiability for nSAT; see Theorem 19 below.

To complete this section we give an overview of how the proof of the ANT develops across the remaining sections. Part (1) of ANT is a quite straightforward and is given in Sect. 12. The proof of ANT part (2) mimics the proof that CSP($\mathbb{A}$) is NP-complete when $\mathbb{A}$ has no WNU. Every step involves substantial difficulties in establishing that the promise $(Y_{(k,\mathscr{F}),Q}, N_{\text{CSP}})$ can be carried through for some suitably constructed $\mathscr{F}$. There are five main steps which are developed as separate sections once we have introduced some further preliminary development. The various stages of the proof are unified in Sect. 12, where an outline of the proof of Theorem 7 can also be found. Section 13 gives some ideas for future work, including an example demonstrating the limits to which the NO promise provided by the PCP Theorem can be incorporated in the ANT.

6   Preliminary Development: $\mathscr{F}$-Types and Claw Formulæ

We now establish some useful preliminary constructions relating to pp-formulæ and $(k, \mathscr{F})$-robustness. Throughout, $\mathbb{A}$ and $\mathscr{F} \subseteq \mathrm{pp}(\mathscr{R})$ are fixed and $\mathbb{B}$ is an input $\mathscr{R}$-structure; all are finite.

Let $x_1, \ldots, x_n$ denote the free variables in some pp-formula $\phi(x_1, \ldots, x_n) \in \mathscr{F}$ and let k be a nonnegative integer. For any function $\iota : \{x_1, \ldots, x_n\} \to \{x_1, \ldots, x_k\}$ we let $\phi^{\iota}(x_1, \ldots, x_k)$ denote the formula $\phi(\iota(x_1), \ldots, \iota(x_n))$. We let $\mathscr{F}_k$ denote the set of all formulæ obtained in this way. This is the standard way that high arity formulæ can produce lower arity ones, and the following is immediate.

Lemma 10. *Let* $\mathbb{A}$ *and* $\mathbb{B}$ *be* $\mathscr{R}$-*structures and consider a subset* $\{b_1, \ldots, b_k\}$ *of* B. *A function* $\nu \colon \{b_1, \ldots, b_k\} \to A$ *is* $\mathscr{F}$-*compatible if and only if for every* $\phi(x_1, \ldots, x_k) \in \mathscr{F}_k$, *if* $\mathbb{B} \models \phi(b_1, \ldots, b_k)$ *then* $\mathbb{A} \models \phi(\nu(b_1), \ldots, \nu(b_k))$.

The following gives a natural restriction of the model theoretic "k-type" to pp-formulæ.

Definition 11. *Let* $\mathscr{R}$ *be a finite relational signature and* $\mathscr{F}$ *a set of pp-formulæ in* $\mathscr{R}$.

1. *A* $(k, \mathscr{F})$-type *is any finite conjunction of distinct* k-ary formulæ in $\mathscr{F}_k$ over $x_1, \ldots, x_k$. *The set of all* $(k, \mathscr{F})$-types *is denoted by* $\mathrm{type}_k(\mathscr{F})$.

2. *The $(k, \mathscr{F})$-type of a tuple $\overrightarrow{b} \in B^k$ is the conjunction* $\displaystyle\bigwedge_{\substack{\phi(\overrightarrow{x}) \,\in\, \mathscr{F}_k, \\ \mathbb{B} \,\models\, \phi(\overrightarrow{b})}} \phi(\overrightarrow{x}).$

3. *For $\ell \leq k$ we let $\mathscr{F}|_\ell$ denote $\{\exists x_{\ell+1} \ldots \exists x_k \, \tau(x_1, \ldots, x_k) \mid \tau \in \mathrm{type}_k(\mathscr{F})\}$.*

The following follows immediately from Lemma 10 and the definition of $(k, \mathscr{F})$-types.

Lemma 12. *Let $\mathbb{A}$ and $\mathbb{B}$ be $\mathscr{R}$-structures and consider a subset $\{b_1, \ldots, b_k\}$ of B. A partial map $\nu \colon \{b_1, \ldots, b_k\} \to A$ is $\mathscr{F}$-compatible if and only if $\mathbb{A} \models \tau(\nu(b_1), \ldots, \nu(b_k))$, where τ is the $(k, \mathscr{F})$ type of $(b_1, \ldots, b_k)$.*

The next lemma has a straightforward proof. We assume that $|B| \geq k$, though minor amendment to the definition of $\mathscr{F}|_i$ can accommodate smaller $|B|$.

Lemma 13. *Let $\mathbb{A}$ and $\mathbb{B}$ be finite $\mathscr{R}$-structures and let $\mathscr{F}$ be a finite set of pp-formulæ in $\mathscr{R}$. If $\mathbb{B}$ is $(k, \mathscr{F})$-robustly satisfiable into $\mathbb{A}$ and $\ell \leq k$, then $\mathbb{B}$ is $(\ell, \mathscr{F}|_\ell)$-robustly satisfiable. In particular, if $\mathbb{B}$ is $(k, \mathscr{F})$-robustly satisfiable for some finite set of pp-formulæ $\mathscr{F}$, then $\mathbb{B}$ is $(\leq k, \bigcup_{0 \leq i \leq k} \mathscr{F}|_i)$-robustly satisfiable.*

Recall from Sect. 4 that when $\mathscr{R}$ is pp-definable from a set of relations $\mathscr{S}$ on a set A then there is logspace reduction from $\mathrm{CSP}(\langle A; \mathscr{R}^A \rangle)$ to $\mathrm{CSP}(\langle A; \mathscr{S}^A \rangle)$. Assume then that each relation symbol $r \in \mathscr{R}$ has been matched to some fixed defining $\mathscr{S}$-formula $\rho_r(x_1, \ldots, x_n)$ of the same arity n as r:

$$\exists y_1 \ldots \exists y_m \bigwedge_{1 \leq i \leq k} \alpha_i(x_{i,1}, \ldots, x_{i,n_i}, y_{i,1}, \ldots, y_{i,m_i}), \qquad (\dagger)$$

where each α_i is an atomic formula in $\mathscr{S} \cup \{=\}$, and $\bigcup_{1 \leq i \leq k}\{x_{i,1}, \ldots, x_{i,n_i}\} = \{x_1, \ldots, x_n\}$ and $\bigcup_{1 \leq i \leq k}\{y_{i,1}, \ldots, y_{i,m_i}\} = \{y_1, \ldots, y_m\}$. Let $\rho_r^\flat$ denote the underlying open formula obtained from ρ_r by removing quantifiers: variables of $\rho_r^\flat$ that are quantified in ρ_r will be called *existential variables* (or $\exists$-variables: the y_i in $\dagger$) and the other variables will be referred to as *open variables*.

Each pp-formula $\psi(x_1, \ldots, x_\ell)$ in the signature $\mathscr{R}$ is equivalent to a pp-formula $\psi^{\mathscr{S}}(x_1, \ldots, x_\ell)$ in the signature $\mathscr{S}$: replace each conjunct in ψ—an atomic formula $r(x_1, \ldots, x_n)$ in for some $r \in \mathscr{R}$—by the defining formula ρ_r as in $\dagger$, and then apply the usual logical rules for moving quantifiers to the front (including renaming quantified variables where necessary).

Definition 14. *Let $\mathscr{S}$ define $\mathscr{R}$ by pp-formulæ $\{\rho_r \mid r \in \mathscr{R}\} \subseteq \mathrm{pp}(\mathscr{S})$. Let k, ℓ be fixed non-negative integers and $\mathscr{F}$ a finite set of pp-formulæ in $\mathscr{R}$. A claw formula for $\mathscr{F}$ of arity k and bound ℓ is any pp-formula in $\mathscr{S}$ of the form constructed in the third step below:*

1. *(The talon.) Let γ denote any conjunction $\bigwedge_{1 \leq i \leq k'} \rho_{r_i}^\flat$, where $r_i \in \mathscr{R}$ and $k' \leq k$. We allow some identification between open variables, but not between existential variables.*

2. (*The* wrist.) *Let* σ *be an* $(\ell', \mathscr{F})$-*type in* $\mathscr{R}$ *for some* $\ell' \leq \ell$. *Some of the* ℓ' *free variables in* σ *may be identified with open variables in* γ, *but not with existential variables.*

3. (*The* claw.) *Existentially quantify all but* k *of the unquantified variables in the conjunction* $\gamma \wedge \sigma^{\mathscr{S}}$.

7 Step 1. Reflection

Definition 15. *Let* $\mathbb{A}$, k *and* $\mathscr{F}$ *be fixed. For an input* $\mathscr{R}$-*structure* $\mathbb{B}$, *let* $\mathbb{B}^{\downarrow}$ *be the result of adjoining all hyperedges to* $\mathbb{B}$ *that are implied by* $\mathscr{F}$-*compatible assignments from subsets of* $\mathbb{B}$ *on at most* k *elements. The structure* $\mathbb{B}^{\downarrow}$ *will be called the* 1-*step* $(k, \mathscr{F})$-*reflection of* $\mathbb{B}$.

Under the promise $(Y_{(k,\mathscr{F})}, N_{\mathrm{CSP}})$ it is possible to show that there is a first order query that defines $\mathbb{B}^{\downarrow}$. The details take some effort and we omit them. To achieve the main results with respect to polynomial time reductions however, simply observe that $\mathbb{B}^{\downarrow}$ can be constructed from $\mathbb{B}$ in polynomial time, because there are only polynomially many $\mathscr{F}$-compatible assignments from subsets of size at most k. Then all that is needed is the following lemma.

Lemma 16. *If* $\mathbb{B}$ *is* $(\leq k, \mathscr{F})$-*robustly satisfiable with respect to* $\mathbb{A}$, *and* $k \geq$ $\mathrm{arity}(\mathscr{R})$, *then* $\mathbb{B}^{\downarrow}$ *lies in the quasivariety of* $\mathbb{A}$ *and is also* $(\leq k, \mathscr{F})$-*robustly satisfiable. If* $\mathbb{B}$ *is a NO instance of* $\mathrm{CSP}(\mathbb{A})$ *then so also is* $\mathbb{B}^{\downarrow}$.

8 Step 2. Stability of Robustness over Primitive-Positive Reductions

We prove the following variant of the usual pp-reduction for CSPs.

Theorem 17. *Assume that* $\mathbb{A}_1 = \langle A, \mathscr{R}^A \rangle$ *and* $\mathbb{A}_2 = \langle A, \mathscr{S}^A \rangle$ *are two relational structures on the same finite set* A, *with* $\mathscr{R}^A \subseteq \mathrm{pp}(\mathbb{A}_2)$ *finite and* $\ell := \mathrm{arity}(\mathscr{R})$. *Let* $\mathscr{F}$ *be a finite set of pp-formulæ in the language of* $\mathscr{R}$. *Then, for any* k, *the standard pp-reduction of* $\mathrm{CSP}(\mathbb{A}_1)$ *to* $\mathrm{CSP}(\mathbb{A}_2)$ *takes* $(\leq k\ell, \mathscr{F})$-*robustly satisfiable instances of* $\mathrm{CSP}(\mathbb{A}_1)$ *to* $(k, \mathscr{G})$-*robustly satisfiable instances of* $\mathrm{CSP}(\mathbb{A}_2)$, *where* $\mathscr{G}$ *denotes the* k-*ary claw formulæ for* $\mathscr{F}$ *of bound* $k\ell$.

First briefly recall the precise nature of the "standard reduction" described in Theorem 17. Recall that each $r \in \mathscr{R}$ corresponds to an $\mathscr{S}$ formula ρ_r, as in (†). For an instance $\mathbb{B} = \langle B; \mathscr{R}^B \rangle$ of $\mathrm{CSP}(\mathbb{A}_1)$, an instance $\mathbb{B}^{\sharp}$ of $\mathrm{CSP}(\mathbb{A}_2)$ is constructed in the following way. For each hyperedge $(b_1, \ldots, b_n) \in r$ in $\mathbb{B}$ (and adopting the generic notation of †), new elements $c_1, \ldots, c_m$ are added to the universe of B, and the hyperedge $(b_1, \ldots, b_n) \in r$ is replaced by the hyperedges $\alpha_i(b_{i,1}, \ldots, b_{i,n_k}, c_{i,1}, \ldots, c_{i,m_i})$ for each $i = 1, \ldots, k$. Note that new elements $c_1, \ldots, c_m$ are introduced for every instance of a hyperedge. The new elements will be referred to as *existential elements* (or $\exists$-*elements*), and for any $D \subseteq B^{\sharp}$

we let $D_\exists$ denote $\exists$-elements in D. Elements of B will be referred to as *open elements*, and we write D_B for $D \cap B = D \backslash D_\exists$.

It is easy to see that there is a homomorphism from $\mathbb{B}$ to $\mathbb{A}_1$ if and only if there is one from $\mathbb{B}^\sharp$ to $\mathbb{A}_2$: this is the usual logspace CSP reduction, which is a first order reduction when none of the ρ_r formulæ involve equality [33]. Now assume that $\mathbb{B}$ is ($\leq k\ell, \mathscr{F}$)-robustly satisfiable with respect to $\mathbb{A}_1$ and consider a k-element subset $D \subseteq B^\sharp$, for which there is a $\mathscr{G}$-compatible assignment into A. The following arguments will refer back to the 3-step construction of claw formulæ in Definition 14.

Each $c \in D_\exists$ was introduced in replacing a hyperedge of $\mathbb{B}$ in signature $\mathscr{R}$ by a *family* of hyperedges in the signature $\mathscr{S}$, according to the pp-definition as in †. Each element of $D_\exists$ appears in at most one such family of $\mathscr{S}$-hyperedges, so the number of these, k', is at most $|D_\exists| \leq k$. Observe that these hyperedge families correspond to an interpretation of a conjunction γ of k' many formulæ as in step 1 of Definition 14: there is no identification of $\exists$-elements, but there may be of open elements. Each of these families involves at most ℓ open elements, so that at most $k' \times \ell$ open elements appear in these hyperedge families. Let O_B denote these elements. Because $k' + |D_B| \leq |D_\exists| + |D_B| = k$ and $|O_B| \leq k'\ell$, we have $|O_B \cup D_B| \leq k'\ell + |D_B| \leq k\ell$. Let σ denote the ($\ell', \mathscr{F}$)-type of $O_B \cup D_B$ in $\mathbb{B}$, as in the second step of Definition 14. (Here we treat $O_B \cup D_B$ as a tuple ordered in any fixed way.) Observe that some elements b of D_B may also lie in O_B, and we will assume then that the variable in σ corresponding to b has been identified with the variable in γ corresponding to b. Let U be the set of all unquantified variables in $\gamma \wedge \sigma^\mathscr{S}$ that do not correspond to elements of D. The claw formula $\exists U \; \gamma \wedge \sigma^\mathscr{S}$ is in $\mathscr{G}$ and is satisfied by $\mathbb{B}^\sharp$ at D (again, arbitrarily treated as a tuple). Hence $\exists U \; \gamma \wedge \sigma^\mathscr{S}$ is preserved by ν. In particular then, in $\mathbb{A}_2$ we can find values for the variables corresponding to the elements of O_B that witness the satisfaction of $\exists U \; \gamma \wedge \sigma^\mathscr{S}$ at $\nu(D)$. Let $\nu' : O_B \cup D \to A$ be the extension of ν obtained by giving elements of $O_B \backslash D_B$ these witnessing values. Because σ is the ($\ell', \mathscr{F}$)-type of $O_B \cup D_B$, it follows from Lemma 12 that $\nu'|_{O_B \cup D_B}$ is $\mathscr{F}$-compatible, so by the assumed ($\leq k\ell, \mathscr{F}$)-robust satisfiability of $\mathbb{B}$ it follows that $\nu'|_{O_B \cup D_B}$ extends to a homomorphism ν^+ from $\mathbb{B}$ to $\mathbb{A}_1$. By the usual pp-reduction, ν^+ extends to a homomorphism $\nu^\sharp$ from $\mathbb{B}^\sharp$ to $\mathbb{A}_2$. Now $\nu^\sharp$ agrees with ν on D_B, but also, we may assume that it agrees with ν on $D_\exists$, because the values given O_B by ν' (and hence $\nu^\sharp$) were such that γ held. Thus we have extended ν to a homomorphism, as required.

9   Step 3. $(k, \varnothing)$-Robustness of $(3k + 3)$SAT

Gottlob [18, Lemma 1] showed that the standard Yes/No decision problem 3SAT reduces to the promise $(Y_{(k,\varnothing)}, N_{\mathrm{CSP}})$ for $(3k+3)$SAT. For the sake of completeness of our sketch, we recall the basic idea. The construction is to replace in a 3SAT instance $\mathbb{B}$, each element b by $2k + 1$ copies $b_1, \ldots, b_{2k+1}$ and then each clause $(b \vee c \vee d)$ by all $\binom{2k+1}{k+1}^3$ clauses of the form $(b_{i_1} \vee \cdots \vee b_{i_{k+1}} \vee c_{i'_1} \vee \cdots \vee c_{i'_{k+1}} \vee d_{i''_1} \vee \cdots \vee d_{i''_{k+1}})$ where the i_j, i'_j, i''_j are from $\{1, \ldots, 2k+1\}$. No assignment on k elements covers all of the $k + 1$ copies of any element in a clause it appears, which enables the flexibility for such assignments to always extend to a solution, provided (and only when) $\mathbb{B}$ is a YES instance. We omit the details showing that this can be achieved via a first-order query.

10   Step 4. $(k, \mathscr{F})$-Robustness of 3SAT

We now establish the following theorem by reduction from the result in Step 3. Critically, the value of k is arbitrary, but the constraint language (3SAT) has fixed arity 3.

Theorem 18. *Fix any $k \geq 0$ and let $\mathscr{F}$ be the set of all claw formulæ for $\varnothing$ of arity k and with bound k. Then $(Y_{(k,\mathscr{F})}, N_{\mathrm{CSP}})$ for 3SAT is NP-complete via first order reductions.*

The usual reduction of nSAT to 3SAT (as in [15] for example) is an example of a pp-reduction, because the nSAT clause relation $(x_1 \vee \cdots \vee x_n)$ (where the x_i can be negated variables if need be) is equivalent to the following pp-formula over $n - 2$ clause relations of 3SAT:

$$\exists y_1 \ldots \exists y_{n-4} \, (x_1 \vee x_2 \vee y_1) \wedge \left(\bigwedge_{3 \leq i \leq n-2} (\neg y_{i-2} \vee x_i \vee y_{i-1}) \right) \wedge (\neg y_{n-3} \vee x_{n-1} \vee x_n)$$
$$(\ddagger)$$

As we are dealing with the standard pp-reduction, an instance $\mathbb{B}$ of $(3k + 3)$SAT is satisfiable if and only if the constructed instance $\mathbb{B}^\sharp$ of 3SAT is satisfiable.

Now assume that $\mathbb{B}$ is a $(k, \varnothing)$-robustly satisfiable instance of $(3k + 3)$SAT. Assume D is a k-set from $\mathbb{B}^\sharp$ and $\nu : D \to \{0, 1\}$ an $\mathscr{F}$-compatible partial assignment. As in the proof of Theorem 17, there are $k' \leq |D_\exists|$ different clause families involving elements from $D_\exists$; let F denote this set of families of clauses (each family arising by the replacement of a $(3k + 3)$SAT clause by the $3k + 1$ distinct 3SAT clauses). Let γ denote the conjunction of k' many pp-formulæ corresponding to these F: it is a conjunction of k' distinct copies of the underlying open formula of $\ddagger$, possibly with some of the open variables in different copies identified. Let U be the variables of γ that do not correspond to an element of D. Then $\exists U \, \gamma$ is a claw formula in the sense of Definition 14 because the only $(\ell, \varnothing)$-types (as detailed in step 2 of Definition 14) are empty formulæ. This formula $\exists U \, \gamma$ is obviously satisfied at D in $\mathbb{B}^\sharp$, so is preserved by ν. Now the proof deviates from Theorem 17. We show how to assign values to at most k of the remaining open elements of F such that any extension to a full solution on $\mathbb{B}$ extends to one for $\mathbb{B}^\sharp$ in a way consistent with the values given to $D_\exists$ by ν.

We introduce an arrow notation to help select the new open elements.

- Above the leftmost bracket of the clause family we place a right arrow $\mapsto$, and dually a $\hookleftarrow$ over the rightmost bracket.
- Place a left arrow $\hookleftarrow$ above a consecutive pair of brackets ")(" if the $\exists$-element immediately preceding it is given 0 by ν, and dually, $\mapsto$ if the $\exists$-element is assigned 1.

Let us say that two such arrows are *convergent* if they point toward one another. In order to extend ν to a solution, within each pair of convergent arrows, an open-literal to assign the value 1. We first give an example, consisting of a clause family, an assignment to some elements (say, $D_\exists = \{b_1, b_2, b_3\}$ and $D_B = \{a_1\}$) and the arrows placed as determined by the rules:

$$(\; a_1 \; \neg a_2 \; b_1 \;)(\; \neg b_1 \; a_3 \; b_2 \;)(\; \neg b_2 \; a_4 \; b_3 \;)(\; \neg b_3 \; a_5 \; a_6 \;)$$
$$(\; 0 \; \neg a_2 \; 0 \;)(\; 1 \; a_3 \; 1 \;)(\; 0 \; a_4 \; 0 \;)(\; 1 \; a_5 \; a_6 \;)$$
$$\overset{\rightarrow}{(} \quad \neg a_2 \quad)\overset{\leftarrow}{(} \quad a_3 \quad)\overset{\rightarrow}{(} \quad a_4 \quad)\overset{\leftarrow}{(} \quad a_5 \; a_6 \;)$$

By calling on witnesses to preservation of $\mathscr{F}$ by ν we can select open literals and values (here $\nu(a_4) = 1$ and $\nu(a_2) = 0$) that are consistent with the values assigned to $D_\exists$.

In the general case: because ν preserves the claw formula $\exists U \; \gamma$, the 2-element template for 3SAT has witnesses to all quantified variables. For each pair of convergent arrows under the assignment by ν for $D_\exists$, there is a witness to one of the open variables in γ taking the value 1; only one such witness is required for each pair of convergent arrows. Let E consist of the open elements in F corresponding to the selected witnesses, and extend ν to E by giving them the witness values. Note that $|E| \leq |D_\exists|$, so that $|E \cup D_B| \leq k$. Thus $\nu|_{E \cup D_B}$ extends to a solution for $\mathbb{B}$. This solution extends to a solution for $\mathbb{B}^\sharp$ in a way that is consistent with the values given elements of $D_\exists$ by ν.

By a variation of this argument and Sect. 9, we can also obtain the following theorem, which completes one line of investigation initiated by Beacham and Culberson [7].

Theorem 19. *Let $n > 2$ and consider the problem nSAT. If $k \geq n$ then deciding $(k, \varnothing)$-robust satisfiability is in AC^0. If $k < n$ then $(Y_{(k,\varnothing)}, N_{\mathrm{CSP}})$ is NP-complete.*

11 Step 5: Idempotence and the Algebraic Method

A key development in the algebraic method for CSP complexity was restriction to idempotent polymorphisms [11]. We now sketch how this works for the $(Y_{(k,\mathscr{F})}, N_{\mathrm{CSP}})$ promise.

Let $\mathscr{R}_{\mathrm{Con}}$ be the signature obtained by adding a unary relation symbol $\underline{a}$ for each element a of A, and let $\mathbb{A}_{\mathrm{Con}}$ denote the structure $\langle A; \mathscr{R}_{\mathrm{Con}} \rangle$, with $\underline{a}$ interpreted as $\{a\}$.

Theorem 20. *Let $\mathbb{A}$ be a core and $\mathscr{F}$ be a finite subset of $\mathrm{pp}(\mathscr{R}_{\mathrm{Con}})$. Then for any k, there exists a finite set $\mathscr{G}$ of pp-formulæ in the language of $\mathscr{R}$ such that the standard reduction from $\mathrm{CSP}(\mathbb{A}_{\mathrm{Con}})$ to $\mathrm{CSP}(\mathbb{A})$ takes $(k, \mathscr{F})$-robustly satisfiable instances of $\mathrm{CSP}(\mathbb{A}_{\mathrm{Con}})$ to the $(k, \mathscr{G})$-robustly satisfiable instances of $\mathrm{CSP}(\mathbb{A})$.*

Proof (Proof sketch). Let $\mathbb{B}$ be an instance of $\mathrm{CSP}(\mathbb{A}_{\mathrm{Con}})$. The standard reduction (first order by [33, Lemma 2.5]) involves adjoining a copy of $\mathbb{A}$ to the instance

$\mathbb{B}$, and replacing all hyperedges $b \in \underline{a}$ by identifying b with the adjoined copy of a; call this $\mathbb{B}^{\sharp}$. (A $(k, \mathscr{F})$-reflection, via the first order version of Lemma 16, can be used to circumvent some technical issues regarding identification of elements.) Our task is to show how to construct $\mathscr{G}$. Let diag($\mathbb{A}$) denote the *positive atomic diagram of* $\mathbb{A}$ on some set of variables $\{v_a \mid a \in A\}$; that is, the conjunction of all hyperedges of $\mathbb{A}$ (considered as atomic formulæ). We construct $\mathscr{G}$ by taking the conjunction of diag($\mathbb{A}$) with $\mathscr{F}$-types σ, and replacing each conjunct of the form $x \in \underline{a}$ in σ, by $x = v_a$.

Assume $\mathbb{B}$ is $(k, \mathscr{F})$-robustly satisfiable with respect to $\mathbb{A}_{\mathrm{Con}}$ and consider a $\mathscr{G}$-compatible assignment ν from some k-set in $\mathbb{B}^{\sharp}$. Because $\mathbb{A}$ is a core, there is an automorphism α of $\mathbb{A}$ mapping witnesses to diag($\mathbb{A}$) to their named location (that is, taking v_a to a). Then $\alpha \circ \nu$ is $\mathscr{F}$-compatible into $\mathbb{A}_{\mathrm{Con}}$, hence extends to a homomorphism ψ from $\mathbb{B}$. Then $\alpha^{-1} \circ \psi$ is a homomorphism from $\mathbb{B}^{\sharp}$ to $\mathbb{A}$ extending ν.   □

12   Proof of ANT, Corollaries and Theorem 7

Proof (Proof of ANT). For part (1), we extend an idea from [24]. Our proof will use only the assumption that CSP($\mathbb{A}_{\mathrm{Con}}$) is tractable. This is always true if $\mathbb{A}$ is a core with CSP($\mathbb{A}$) tractable. Now observe that an $\mathscr{F}$-compatible partial assignment $\nu : b_i \mapsto a_i$ from a subset $\{b_1, \ldots, b_k\}$ of an instance $\mathbb{B}$ into $\mathbb{A}$ extends to a solution if and only if the structure obtained from $\mathbb{B}$ by adjoining the constraints $\{b_i \in \{a_i\} \mid i = 1, \ldots, k\}$ is a YES instance of CSP($\mathbb{A}_{\mathrm{Con}}$). Thus after polynomially many calls on the tractable problem CSP($\mathbb{A}_{\mathrm{Con}}$), we can decide the $(k, \mathscr{F})$-robust satisfiability of $\mathbb{B}$. An almost identical argument will determine if $\mathbb{B}$ has no implied constraints, thus deciding $\mathrm{CSP}_{\infty}(\mathbb{A})$.

Now to prove ANT part (2). Let $\mathbf{A}$ denote the polymorphism algebra of $\mathbb{A}_{\mathrm{Con}}$. One of the fundamental consequences of the algebraic method is that if $\mathbf{A}$ has no WNU polymorphism, then the polymorphism algebra of 3SAT is a homomorphic image of a subalgebra of $\mathbf{A}$ (direct powers are not required; see [41, Prop 3.1]). For CSPs, these facts will give a first order reduction from 3SAT to some finite set of relations $\mathscr{S}^A$ in pp($\mathbb{A}_{\mathrm{Con}}$): see [33]. The first step of this reduction is to reduce through homomorphic preimages and subalgebras. Ham [21, Sect. 8] showed that these initial reductions also preserve the $(Y_{(\ell, \mathscr{F})}, N_{\mathrm{CSP}})$ promise, with only minor modification to $\mathscr{F}$. Combining this with Theorem 18 then Lemma 13 we find that for all ℓ there exists an $\mathscr{F}_2$ such that $(Y_{(\leq \ell, \mathscr{F}_2)}, N_{\mathrm{CSP}})$ is NP-complete for $\langle A, \mathscr{S}^A \rangle$. Then (using $\ell = arity(\mathscr{S}) \times k$) we can use Theorem 17 then Lemma 13 to find that for every k there exists $\mathscr{F}_3$ such that $(Y_{(\leq k, \mathscr{F}_3)}, N_{\mathrm{CSP}})$ is NP-complete for $\mathbb{A}_{\mathrm{Con}}$ with respect to first order reductions. By Theorem 20 the same is true for $\mathbb{A}$, with an amended compatibility condition $\mathscr{F}$ depending on k. Lemma 16 then extends the promise to $(Y_{(k, \mathscr{F}), \mathrm{Q}}, N_{\mathrm{CSP}})$, as required.   □

Proof (Proof of Corollary 8). Let $\mathbb{A}$ be a finite relational structure without a quasi WNU polymorphism. By Chen and Larose [13, Lemma 6.4] the core retract $\mathbb{A}^{\flat}$ of $\mathbb{A}$ has no WNU. Hence the ANT applies to $\mathbb{A}^{\flat}$. Now $\mathrm{Q}^{+}(\mathbb{A})$ contains $\mathrm{Q}^{+}(\mathbb{A}^{\flat})$, which contains the YES promise in the ANT and is disjoint from the NO promise. Hence membership in $Q^+(\mathbb{A})$ is NP-complete with respect to first order reductions, and hence is also not finitely axiomatisable in first order logic, even at the finite level. The same argument using Theorem 7(2) implies non-finite axiomatisability in the case that $\mathbb{A}^b$ does not have bounded width. $\square$

Proof (Proof of the Gap Dichotomy for Simple Graphs 9). If $\mathbb{G}$ is bipartite, then $\mathrm{CSP}(\mathbb{G})$ is tractable and so is deciding membership in $Q^+(\mathbb{G})$: there are only five distinct quasivarieties [12, 38]. Otherwise, $\mathbb{G}$ is not bipartite and so neither is its core retract. Hence $\mathbb{G}$ has no quasi WNU; see [6]. Then apply Corollary 8. $\square$

Proof (Proof of Theorem 7). Due to space constraints we give only a very brief overview of the method. A CSP has *bounded width* provided that there exists j such that the existence of a homomorphism from $\mathbb{B}$ to $\mathbb{A}$ is equivalent to a family of partial homomorphisms on all subsets of size at most $j+1$, with the family satisfying a compatibility condition, known as a $(j, j+1)$-*strategy*; see [5]. When $k > j$ and input $\mathbb{B}$ satisfies the $Y_{(k,\mathscr{F}),\mathrm{Q}}$ promise, there is an obvious choice for a $(j, j+1)$-strategy: the family of all maps that can extend to $\mathscr{F}$-compatible assignments on k points. The property that this family forms a $(j, j+1)$-strategy can be expressed as a first order sentence ξ. When $\mathrm{CSP}(\mathbb{A})$ has bounded width (so that NO instances do not have $(j, j+1)$-strategies) the sentence ξ must fail on instances satisfying the N_{CSP} promise, and must hold on those satisfying the $Y_{(k,\mathscr{F}),\mathrm{Q}}$ promise.

Now assume that $\mathbb{A}$ does not have bounded width. In this case, a direct analogue of the arguments of Sect. 12 lead back to a structure $\mathbb{C}$ (encoding ternary linear equations over an abelian group) whose CSP is $\mathrm{Mod}_p(\mathrm{L})$-complete; see proof of [33, Theorem 4.1]. The rest of the proof parallels that of the ANT 5, except that Sects. 9 and 10 are replaced by constructions concerning linear systems of equations.

13   Discussion and Extensions

We have shown in the ANT that the fundamental intractability result of [11] can be replaced by an unbounded hierarchy of intractable promise problems, and demonstrated in Theorem 7 a collapse in several intermediate complexity classes for these problems. We feel these results are just the beginning of new applications to ideas relating to the detection of more general implied constraints (as in [7]), minimal networks (as in [18]), as well as to other areas of mathematics and computer science, such as the quantum-theoretic applications in [3] and the semigroup-theoretic applications of [24]. Some further consequences of the ANT omitted from the present work include a substantial extension of the Ham's "Gap Trichotomy Theorem" [21] to the $(Y_{(k,\mathscr{F}),\mathrm{Q}}, N_{\mathrm{CSP}})$ promise.

Some specific new directions this work should be taken include the extension of ANT to noncore templates and to infinite templates, where a much wider array of important computational problems can be found. Another difficult question: can the promise supplied by the PCP Theorem be added as a restriction to N_{CSP} in the ANT? (We write $N_{\varepsilon\,\mathrm{CSP}}$ for this condition: ε proportion of the constraints must fail.) The answer is nearly yes, but not quite. It is quite routine to carry through the failure of a positive fraction of constraints through steps 1–5 of the proof of the ANT part (2), and through step 6 with more difficulty, thereby achieving the NP-completeness of $(Y_{(k,\mathscr{F})}, N_{\varepsilon}\,\mathrm{CSP})$ for core templates without a WNU. Surprisingly though, $N_{\varepsilon}\,\mathrm{CSP}$ does not in general survive reflection, as the following example demonstrates. Let 2^+ denote the template on $\{0,1\}$ with the fundamental ternary relation r of +1-in-3SAT and the 4-ary total relation $s := \{0,1\}^4$. This has no WNU, as +1-in-3SAT has no WNU, so the ANT part (2) and claims just made imply that both $(Y_{(k,\mathscr{F})}, N_{\varepsilon}\,\mathrm{CSP})$ and $(Y_{(k,\mathscr{F}),\mathsf{Q}}, N_{\mathrm{CSP}})$ are NP-complete. Yet $(Y_{(k,\mathscr{F}),\mathsf{Q}}, N_{\varepsilon}\,\mathrm{CSP})$ for 2^+ falls into AC^0! Indeed the first order property τ stating that s is total must hold on instances without implied constraints, and fail on any large enough instance $\mathbb{B}$ satisfying $N_{\varepsilon}\,\mathrm{CSP}$: the number of r-constraints is at most $|B|^3$ compared to the $|B|^4$-many s-constraints required by τ, and no s-constraint can fail into 2^+. For +1-in-3SAT itself we can show that $(Y_{(k,\mathscr{F}),\mathsf{Q}}, N_{\varepsilon}\,\mathrm{CSP})$ remains NP-complete.

References

1. Allender, E., Bauland, M., Immerman, N., Schnoor, H., Vollmer, H.: The complexity of satisfiability problems: refining Schaefer's Theorem. J. Comput. System Sci. **75**, 245–254 (2009)
2. Almeida, J.: Finite Semigroups and Universal Algebra. World Scientific, Singapore (1994)
3. Abramsky, S., Gottlob, G., Kolaitis, P.G.: Robust constraint satisfaction and local hidden variables in quantum mechanics. In: IJCAI 2013, pp. 440–446 (2013)
4. Arora, S., Safra, S.: Probabilistic checking of proofs: a new characterization of NP. J. ACM **45**(1), 70–122 (1998)
5. Barto, L., Kozik, M.: Constraint satisfaction problems of bounded width. In: Proceedings of FOCS 2009 (2009)
6. Barto, L., Kozik, M., Niven, T.: The CSP dichotomy holds for digraphs with no sources and no sinks (a positive answer to a conjecture of Bang-Jensen and Hell), SIAM J. Comput. **38**, 1782–1802 (2008/2009)
7. Beacham, A., Culberson, J.: On the complexity of unfrozen problems. Disc. Appl. Math. **153**, 3–24 (2005)
8. Bulatov, A.A.: A dichotomy theorem for constraint satisfaction problems on a 3-element set. J. ACM **53**, 66–120 (2006)
9. Bulatov, A.A.: Complexity of conservative constraint satisfaction problems. ACM Trans. Comput. Log. **12**(4), 24 (2011)
10. Bulatov, A.A.: A dichotomy theorem for nonuniform CSPs. arXiv:1703.03021v2
11. Bulatov, A.A., Jeavons, P.G., Krokhin, A.: Classifying the complexity of constraints using finite algebras. SIAM J. Comput. **34**(3), 720–742 (2005)
12. Caicedo, X.: Finitely axiomatizable quasivarieties of graphs. Algebra Univers. **34**, 314–321 (1995)
13. Chen, H., Larose, B.: Asking the metaquestions in constraint tractability. arxiv:1604.00932
14. Feder, T., Vardi, M.Y.: The computational structure of monotone monadic SNP and constraint satisfaction: a study through datalog and group theory. SIAM J. Comput. **28**(1), 57–104 (1998)

15. Garey, M.R., Johnson, D.S.: Computers and Intractability: A Guide to the Theory of NP-Completeness. W.H. Freeman & Co., New York (1979)
16. Geiger, D.: Closed systems of functions and predicates. Pacific J. Math. **27**, 95–100 (1968)
17. Gorbunov, V.A.: Algebraic Theory of Quasivarieties. Consultants Bureau, New York (1998)
18. Gottlob, G.: On minimal constraint networks. In: Lee, J. (ed.) CP 2011. LNCS, vol. 6876, pp. 325–339. Springer, Heidelberg (2011). doi:10.1007/978-3-642-23786-7_26
19. Goldreich, O.: On promise problems: a survey. In: Goldreich, O., Rosenberg, A.L., Selman, A.L. (eds.) Theoretical Computer Science. LNCS, vol. 3895, pp. 254–290. Springer, Heidelberg (2006). doi:10.1007/11685654_12
20. Ham, L.: A gap trichotomy theorem for Boolean constraint problems: extending Schaefer's theorem. In: ISAAC 2016, pp. 36: 1–36: 12 (2016)
21. Ham, L.: Gap theorems for robust satisfiability of constraint problems: Boolean CSPs and beyond. Theoret. Comp. Sci. **676**, 69–91 (2017)
22. Hell, P., Nešetřil, J.: On the complexity of H-colouring. J. Combin. Theory Ser. B **48**(1), 92–110 (1990)
23. Idziak, P., Markovic, P., McKenzie, R., Valeriote, M., Willard, R.: Tractability and learnability arising from algebras with few subpowers. SIAM J. Comput. **39**, 3023–3037 (2010)
24. Jackson, M.: Flexible constraint satisfiability and a problem in semigroup theory. arXiv:1512.03127
25. Jackson, M., Kowalski, T., Niven, T.: Digraph related constructions and the complexity of digraph homomorphism problems. Int. J. Algebra Comput. **26**, 1395–1433 (2016)
26. Jackson, M., Trotta, B.: Constraint satisfaction, irredundant axiomatisability and continuous colouring. Stud. Logica. **101**, 65–94 (2013)
27. Jeavons, P.: On the algebraic structure of combinatorial problems. Theor. Comput. Sci. **200**(1–2), 185–204 (1998)
28. Jeavons, P., Cohen, D.A., Martin, C.: Cooper.: constraints, consistency and closure. Artif. Intell. **101**(1–2), 251–265 (1998)
29. Jeavons, P., Cohen, D., Gyssens, M.: A unifying framework for tractable constraints. In: Montanari, U., Rossi, F. (eds.) CP 1995. LNCS, vol. 976, pp. 276–291. Springer, Heidelberg (1995). doi:10.1007/3-540-60299-2_17
30. Jeavons, P., Cohen, D.A., Gyssens, M.: Closure properties of constraints. J. ACM **44**, 527–548 (1997)
31. Jeavons, P., Cohen, D.A., Pearson, J.: Constraints and universal algebra. Ann. Math. Artif. Intell. **24**(1–4), 51–67 (1998)
32. Jonsson, P., Krokhin, A.: Recognizing frozen variables in constraint satisfaction problems. Theoret. Comp. Sci. **329**, 93–113 (2004)
33. Larose, B., Tesson, P.: Universal algebra and hardness results for constraint satisfaction problems. Theoret. Comput. Sci. **410**, 1629–1647 (2009)
34. Maltsev, A.I.: Algebraic Systems. Springer, Heidelberg (1973)
35. Maróti, M., McKenzie, R.: Existence theorems for weakly symmetric operations. Algebra Univers. **59**, 463–489 (2008)
36. Marques-Silva, J.P., Sakallah, K.A.: GRASP: a search algorithm for propositional satisfiability. IEEE Trans. Comput. **48**, 506–521 (1999)
37. Monasson, R., Zecchina, R., Kirkpatrick, S., Selman, B., Troyansky, L.: Determining computational complexity from characteristic phase transitions. Nature **400**, 133–137 (1998)

38. Nešetřil, J., Pultr, A.: On classes of relations and graphs determined by subobjects and factorobjects. Disc. Math. **22**, 287–300 (1978)
39. Rafiey, A., Kinne, J., Feder, T.: Dichotomy for digraph homomorphism problems, arXiv:1701.02409v2
40. Schaefer, T.J.: The complexity of satisfiability problems. In: STOC , pp. 216–226 (1978)
41. Valeriote, M.A.: A subalgebra intersection property for congruence distributive varieties. Canad. J. Math. **61**, 451–464 (2009)
42. Zhuk, D.: The proof of the CSP dichotomy conjecture, arXiv:1704.01914

Kernelization of Constraint Satisfaction Problems: A Study Through Universal Algebra

Victor Lagerkvist[1(✉)] and Magnus Wahlström[2]

[1] Institut für Algebra, TU Dresden, Dresden, Germany
victor.lagerqvist@tu-dresden.de
[2] Department of Computer Science, Royal Holloway, University of London, Egham,
Great Britain
magnus.wahlstrom@rhul.ac.uk

Abstract. A *kernelization* algorithm for a computational problem is
a procedure which compresses an instance into an equivalent instance
whose size is bounded with respect to a complexity parameter. For the
constraint satisfaction problem (CSP), there exist many results concerning upper and lower bounds for kernelizability of specific problems, but
it is safe to say that we lack general methods to determine whether a
given problem admits a kernel of a particular size. In this paper, we take
an algebraic approach to the problem of characterizing the kernelization
limits of NP-hard CSP problems, parameterized by the number of variables. Our main focus is on problems admitting linear kernels, as has,
somewhat surprisingly, previously been shown to exist. We show that a
finite-domain CSP problem has a kernel with $O(n)$ constraints if it can
be embedded (via a domain extension) into a CSP which is preserved by
a Maltsev operation. This result utilise a variant of the simple algorithm
for Maltsev constraints. In the complementary direction, we give indication that the Maltsev condition might be a complete characterization
for Boolean CSPs with linear kernels, by showing that an algebraic condition that is shared by all problems with a Maltsev embedding is also
necessary for the existence of a linear kernel unless $NP \subseteq co\text{-}NP/poly$.

1 Introduction

Kernelization is a preprocessing technique based on reducing an instance of a
computationally hard problem in polynomial time to an equivalent instance, a
kernel, whose size is bounded by a function f with respect to a given complexity
parameter. The function f is referred to as the *size* of the kernel, and if the size
is polynomially bounded we say that the problem admits a *polynomial kernel*.
A classical example is VERTEX COVER, which admits a kernel with $2k$ vertices,
where k denotes the size of the cover [25]. Polynomial kernels are of great interest
in parameterized complexity, as well as carrying practical significance in speeding
up subsequent computations (e.g., the winning contribution in the 2016 PACE
challenge for FEEDBACK VERTEX SET used a novel kernelization step as a key
component (see https://pacechallenge.wordpress.com/).

© Springer International Publishing AG 2017
J.C. Beck (Ed.): CP 2017, LNCS 10416, pp. 157–171, 2017.
DOI: 10.1007/978-3-319-66158-2_11

When the complexity parameter is a size parameter, e.g., the number of variables n, then such a size reduction is also referred to as *sparsification* (although a sparsification is not always required to run in polynomial time). A prominent example is the famous *sparsification lemma* that underpins research into the Exponential Time Hypothesis [10], which shows that for every k there is a subexponential-time reduction from k-SAT on n variables to k-SAT on $O(n)$ clauses, and hence $\tilde{O}(n)$ bits in size. However, the super-polynomial running time is essential to this result. Dell and van Melkebeek [5] showed that k-SAT cannot be kernelized even down to size $O(n^{k-\varepsilon})$, and VERTEX COVER cannot be kernelized to size $O(n^{2-\varepsilon})$, for any $\varepsilon > 0$ unless the polynomial hierarchy collapses (in the sequel, we will make this assumption implicitly). These results suggest that in general, polynomial-time sparsification cannot give non-trivial size guarantees. The first result to the contrary was by Bart Jansen (unpublished until recently [12]), who observed that 1-IN-k-SAT admits a kernel with at most n constraints using Gaussian elimination. More surprisingly, Jansen and Pieterse [11] showed that the NOT-ALL-EQUAL k-SAT problem admits a kernel with $O(n^{k-1})$ constraints, improving on the trivial bound by a factor of n and settling an implicit open problem. In later research, they improved and generalized the method, and also showed that the bound of $O(n^{k-1})$ is tight [12]. These improved upper bounds are all based on rephrasing the SAT problem as a problem of low-degree polynomials, and exploiting linear dependence to eliminate superfluous constraints. Still, it is fair to say that we currently lack the tools for making a general analysis of the kernelizability of a generic SAT problem.

In this paper we take a step in this direction, by studying the kernelizability of the *constraint satisfaction problem* over a constraint language Γ (CSP(Γ)), parameterized by the number of variables n, which can be viewed as the problem of determining whether a set of constraints over Γ is satisfiable. Some notable examples of problems of this kind are k-colouring, k-SAT, 1-in-k-SAT, and not-all-equal-k-SAT. We will occasionally put a particular emphasis on the Boolean CSP problem and therefore denote this problem by SAT(Γ). Note that CSP(Γ) has a trivial polynomial kernel for any finite language Γ (produced by simply discarding duplicate constraints), but the question remains for which languages Γ we can improve upon this. Concretely, our question in this paper is for which languages Γ the problem CSP(Γ) admits a kernel of $O(n^c)$ constraints, for some $c \geq 1$, with a particular focus on linear kernels ($c = 1$).

The Algebraic Approach in Parameterized and Fine-Grained Complexity. For any language Γ, the classical complexity of CSP(Γ) (i.e., whether CSP(Γ) is in P) is determined by the existence of certain algebraic invariants of Γ known as *polymorphisms* [13]. This gave rise to the *algebraic approach* to characterizing the complexity of CSP(Γ) by studying algebraic properties. It has been conjectured that for every Γ, CSP(Γ) is either in P or NP-complete, and that the tractability of a CSP problem can be characterized by a finite list of polymorphisms [3]. Recently, several independent results appeared, claiming to settle this conjecture in the positive [1,26,27]. However, for purposes of parameterized and fine-grained complexity questions, looking at polymor-

phisms alone is too coarse. More technically, the polymorphisms of Γ characterize the expressive power of Γ up to *primitive positive definitions*, i.e., up to the
use of conjunctions, equality constraints, and existential quantification, whereas
for many questions a liberal use of existentially quantified local variables is not
allowed. In such cases, one may look at the expressive power under *quantifier-
free* primitive positive definitions (qfpp-definitions), allowing only conjunctions
and equality constraints. This expressive power is characterized by more fine-
grained algebraic invariants called *partial polymorphisms*. For example, there
are numerous dichotomy results for the complexity of *parameterized* $\text{SAT}(\Gamma)$
and $\text{CSP}(\Gamma)$ problems, both for so-called FPT algorithms and for kernelization
[17–19, 24], and in each of the cases listed, a dichotomy is given which is equivalent to requiring a finite list of partial polymorphisms of Γ. Similarly, Jonsson
et al. [16] showed that the exact running times of NP-hard $\text{SAT}(\Gamma)$ and $\text{CSP}(\Gamma)$
problems in terms of the number of variables n are characterized by the partial
polymorphisms of Γ. Unfortunately, studying properties of $\text{SAT}(\Gamma)$ and $\text{CSP}(\Gamma)$
for questions phrased in terms of the size parameter n is again more complicated
than for more permissive parameters k. For example, it is known that for every
finite set P of strictly partial polymorphisms, the number of relations invariant
under P is double-exponential in terms of the arity n (hence they cannot all be
described in a polynomial number of bits) [20, Lemma 35]. It can similarly be
shown that the existence of a polynomial kernel cannot be characterized by such
a finite set P. Instead, such a characterization must be given in another way (for
example, Lagerkvist et al. [22] provide a way to finitely characterize all partial
polymorphisms of a finite Boolean language Γ).

Our Results. We generalize and extend the results of Jansen and Pieterse [12]
in the case of linear kernels to a general recipe for NP-hard SAT and CSP
problems in terms of the existence of a *Maltsev embedding*, i.e., an embedding
of a language Γ into a tractable language Γ' on a larger domain with a *Maltsev
polymorphism*. We show that for any language Γ with a Maltsev embedding into
a finite domain, $\text{CSP}(\Gamma)$ has a kernel with $O(n)$ constraints. Attempting an
algebraic characterization, we also show an infinite family of *universal* partial
operations which are partial polymorphisms of every language Γ with a Maltsev
embedding, and show that these operations guarantee the existence of a Maltsev
embedding for Γ, albeit into a language with an infinite domain. Turning to
lower bounds against linear kernels, we show that the smallest of these universal
partial operations is also necessary, in the sense that for any Boolean language
Γ which is not invariant under this operation, $\text{SAT}(\Gamma)$ admits no kernel of size
$O(n^{2-\varepsilon})$ for any $\varepsilon > 0$. We conjecture that this can be completed into a tight
characterization – i.e., that for Boolean languages Γ, $\text{SAT}(\Gamma)$ admits a linear
kernel if and only if it is invariant under all universal partial Maltsev operations.

Generalizations for kernels of higher degree are possible, but have been omitted for reasons of length, and we refer the reader to the extended preprint [21].

2 Preliminaries

2.1 The Constraint Satisfaction Problem and Kernelization

A *relation* R over a set of values D is a subset of D^k for some $k \geq 0$, and we write $\mathrm{ar}(R) = k$ to denote the arity of R. A set of relations Γ is referred to as a *constraint language*. An instance (V, C) of the *constraint satisfaction problem* over a constraint language Γ over D (CSP(Γ)) is a set V of variables and a set C of constraint applications $R(v_1, \ldots, v_k)$ where $R \in \Gamma$, $\mathrm{ar}(R) = k$, and $v_1, \ldots, v_k \in V$. The question is whether there exists a function $f : V \to D$ such that $(f(v_1), \ldots, f(v_k)) \in R$ for each $R(v_1, \ldots, v_k)$ in C? If Γ is Boolean we denote CSP(Γ) by SAT(Γ), and we let BR denote the set of all Boolean relations. As an example, let $R_{1/3} = \{(0,0,1), (0,1,0), (1,0,0)\}$. Then SAT($\{R_{1/3}\}$) can be viewed as an alternative formulation of the 1-in-3-SAT problem restricted to instances consisting only of positive literals. More generally, if we let $R_{1/k} = \{(x_1, \ldots, x_k) \in \{0,1\}^k \mid x_1 + \ldots + x_k = 1\}$, then SAT($\{R_{1/k}\}$) is a natural formulation of 1-in-k-SAT without negation.

A *parameterized problem* is a subset of $\Sigma^* \times \mathbb{N}$ where Σ is a finite alphabet. Hence, each instance is associated with a natural number, called the *parameter*.

Definition 1. *A* kernelization algorithm, *or a* kernel, *for a parameterized problem* $L \subseteq \Sigma^* \times \mathbb{N}$ *is a polynomial-time algorithm which, given an instance* $(x, k) \in \Sigma^* \times \mathbb{N}$, *computes* $(x', k') \in \Sigma^* \times \mathbb{N}$ *such that (1)* $(x, k) \in L$ *if and only if* $(x', k') \in L$ *and (2)* $|x'| + k' \leq f(k)$ *for some function* f.

The function f in the above definition is sometimes called the *size* of the kernel. In this paper, we are mainly interested in the case where the parameter denotes the number of variables in a given CSP(Γ) instance.

2.2 Operations and Relations

An n-ary function $f : D^n \to D$ over a domain D is typically referred to as an *operation* on D, although we will sometimes use the terms function and operation interchangeably. We let $\mathrm{ar}(f) = n$ denote the arity of f. Similarly, an n-ary *partial operation* over a set D of values is a map $f : X \to D$, where $X \subseteq D^n$ is called the *domain* of f. Again, we let $\mathrm{ar}(f) = n$, and furthermore let $\mathrm{domain}(f) = X$. If f and g are n-ary partial operations with $\mathrm{domain}(g) \subseteq \mathrm{domain}(f)$ and $f(x_1, \ldots, x_n) = g(x_1, \ldots, x_n)$ for each $(x_1, \ldots, x_n) \in \mathrm{domain}(g)$, then g is said to be a *subfunction* of f.

Definition 2. *An n-ary partial operation f is a* partial polymorphism *of a k-ary relation R if, for every sequence* $t_1, \ldots, t_n \in R$, *either* $f(t_1, \ldots, t_n) \in R$ *or* $(t_1[i], \ldots, t_n[i]) \notin \mathrm{domain}(f)$ *for some* $1 \leq i \leq k$, *where* $f(t_1, \ldots, t_n) = (f(t_1[1], \ldots, t_n[1]), \ldots, f(t_1[k], \ldots, t_n[k]))$.

If f is total we simply say that f is a *polymorphism* of R, and in both cases we sometimes also say that f *preserves* R, or that R is *invariant* under f. For a constraint language Γ we then let $\mathrm{Pol}(\Gamma)$ and $\mathrm{pPol}(\Gamma)$ denote the set of operations and partial operations preserving every relation in Γ, respectively, and if F is a set of total or partial operations we let $\mathrm{Inv}(F)$ denote the set of all relations invariant under F. It is known that $\mathrm{Pol}(\Gamma)$ and $\mathrm{pPol}(\Gamma)$ are closed under composition of (partial) operations, i.e., if $f \circ g_1, \ldots, g_m(x_1, \ldots, x_n) = f(g_1(x_1, \ldots, x_n), \ldots, g_m(x_1, \ldots, x_n))$ is included in $\mathrm{Pol}(\Gamma)$ (respectively $\mathrm{pPol}(\Gamma)$) then $f(g_1(x_1, \ldots, x_n), \ldots, g_m(x_1, \ldots, x_n))$ is included in $\mathrm{Pol}(\Gamma)$ (respectively $\mathrm{pPol}(\Gamma)$) [23]. It is also known that $\mathrm{Pol}(\Gamma)$ and $\mathrm{pPol}(\Gamma)$ for each n and $i \leq n$ contain every *projection* $\pi_i^n(x_1, \ldots, x_i, \ldots, x_n) = x_i$. On the relational side, if every operation in F is total, then $\mathrm{Inv}(F)$ is closed under *primitive positive definitions* (pp-definitions) which are logical formulas consisting of existential quantification, conjunction, and equality constraints. In symbols, we say that a k-ary relation R has a pp-definition over a constraint language Γ over a domain D if $R(x_1, \ldots, x_k) \equiv \exists y_1, \ldots, y_{k'} . R_1(\mathbf{x_1}) \wedge \ldots \wedge R_m(\mathbf{x_m})$, where each $R_i \in \Gamma \cup \{\mathrm{Eq}\}$, $\mathrm{Eq} = \{(\mathrm{x}, \mathrm{x}) \mid \mathrm{x} \in D\}$ and each $\mathbf{x_i}$ is an $\mathrm{ar}(R_i)$-ary tuple of variables over $x_1, \ldots, x_k, y_1, \ldots, y_{k'}$. If F is a set of partial operations then $\mathrm{Inv}(F)$ is closed under *quantifier-free primitive positive definitions* (qfpp-definitions), i.e., pp-definitions that do not make use of existential quantification. As a shorthand, we let $[F] = \mathrm{Pol}(\mathrm{Inv}(F))$, $\langle \Gamma \rangle = \mathrm{Inv}(\mathrm{Pol}(\Gamma))$, and $\langle \Gamma \rangle_{\not\exists} = \mathrm{Inv}(\mathrm{pPol}(\Gamma))$. We then have the following *Galois connections* [8].

Theorem 3. *Let* Γ, Γ' *be constraint languages. Then (1)* $\Gamma \subseteq \langle \Gamma' \rangle_{\not\exists}$ *if and only if* $\mathrm{pPol}(\Gamma') \subseteq \mathrm{pPol}(\Gamma)$ *and (2)* $\Gamma \subseteq \langle \Gamma' \rangle$ *if and only if* $\mathrm{Pol}(\Gamma') \subseteq \mathrm{Pol}(\Gamma)$.

Jonsson et al. [16] proved the following theorem, showing that partial polymorphisms are indeed a refinement over total polymorphisms, since the latter are only guaranteed to provide polynomial-time many-one reductions [15].

Theorem 4. *If* Γ, Γ' *are finite languages and* $\mathrm{pPol}(\Gamma) \subseteq \mathrm{pPol}(\Gamma')$ *there exists a constant* c *and a polynomial-time reduction from* $\mathrm{CSP}(\Gamma')$ *to* $\mathrm{CSP}(\Gamma)$ *mapping* (V, C) *of* $\mathrm{CSP}(\Gamma')$ *to* (V', C') *of* $\mathrm{CSP}(\Gamma)$ *where* $|V'| \leq |V|$ *and* $|C'| \leq c|C|$.

Last, we will define a particular type of operation which is central to our algebraic approach. A *Maltsev operation* over $D \supseteq \{0, 1\}$ is a ternary operation ϕ which for all $x, y \in D$ satisfies the two identities $\phi(x, x, y) = y$ and $\phi(x, y, y) = x$. Before we can explain the powerful, structural properties of relations invariant under Maltsev operations, we need a few technical definitions from Bulatov and Dalmau [2]. If $t \in D^n$ is a tuple we let $t[i]$ denote the ith element in t and we let $\mathrm{pr}_{i_1, \ldots, i_{n'}}(t) = (t[i_1], \ldots, t[i_{n'}])$, $n' \leq n$, denote the *projection* of t on (not necessarily distinct) coordinates $i_1, \ldots, i_{n'} \in \{1, \ldots, n\}$. Similarly, if R is an n-ary relation we let $\mathrm{pr}_{i_1, \ldots, i_{n'}}(R) = \{\mathrm{pr}_{i_1, \ldots, i_{n'}}(t) \mid t \in R\}$. Let t, t' be two n-ary tuples over D. We say that (t, t') *witnesses* a tuple $(i, a, b) \in \{1, \ldots, n\} \times D^2$ if $\mathrm{pr}_{1, \ldots, i-1}(t) = \mathrm{pr}_{1, \ldots, i-1}(t')$, $t[i] = a$, and $t'[i] = b$. The *signature* $\mathrm{Sig}(R)$ of an n-ary relation R over D is then defined as

$$\{(i, a, b) \in \{1, \ldots, n\} \times D^2 \mid \exists t, t' \in R \text{ such that } (t, t') \text{ witnesses } (i, a, b)\},$$

and we say that $R' \subseteq R$ is a *representation* of R if $\mathrm{Sig}(R) = \mathrm{Sig}(R')$. If R' is a representation of R it is said to be *compact* if $|R'| \leq 2|\mathrm{Sig}(R)|$, and it is known that every relation invariant under a Maltsev operation admits a compact representation. Furthermore, we have the following theorem from Bulatov and Dalmau, where we let $\langle R \rangle_f$ denote the smallest superset of R invariant under f.

Theorem 5 ([2]). *Let ϕ be a Maltsev operation over a finite domain, $R \in \mathrm{Inv}(\{\phi\})$ a relation, and R' a representation of R. Then $\langle R' \rangle_\phi = R$.*

Hence, relations invariant under Maltsev operations are reconstructible from their compact representations.

3   Maltsev Embeddings and Kernels of Linear Size

In this section we give general upper bounds for kernelization of NP-hard CSP problems, utilising Maltsev operations. At this stage the connection between Maltsev operations, compact representations and tractability of Maltsev constraints might not be immediate. In a nutshell, the Maltsev algorithm [2] works as follows (where ϕ is a Maltsev operation over a finite set D). First, let $(V, \{C_1, \ldots, C_m\})$ be an instance of $\mathrm{CSP}(\mathrm{Inv}(\{\phi\}))$, and let S_0 be a compact representation of $D^{|V|}$. Second, for each $i \in \{1, \ldots, m\}$ compute a compact representation S_i of the solution space of the instance $(V, \{C_1, \ldots, C_i\})$ using S_{i-1}. Third, answer yes if $S_m \neq \emptyset$ and no otherwise. For a full description of the involved procedures we refer the reader to Bulatov and Dalmau [2] and Dyer and Richerby [6].

Example 6. We review two familiar special cases of this result. First, consider a linear equation $\sum_i \alpha_i x_i = b$, interpreted over a finite field $\mathbb{F}$. It is clear that the set of solutions to such an equation is invariant under $x_1 - x_2 + x_3$ (over $\mathbb{F}$), hence systems of linear equations are a special case of Maltsev constraints, and can in principle be solved by the Maltsev algorithm. Second, for a more general example, let $G = (D, \cdot)$ be a finite group, and let $s(x, y, z) = x \cdot (y^{-1}) \cdot z$ be the *coset generating operation* of G. Then s is Maltsev, hence $\mathrm{CSP}(\mathrm{Inv}(\{s\}))$ is tractable; this was shown by Feder and Vardi [7], but also follows from the Maltsev algorithm. In particular, if $G = (D, +)$ is an Abelian group where $|D|$ is prime, then $R \in \mathrm{Inv}(\{s\})$ if and only if R is the solution space of a system of linear equations modulo $|D|$ [14].

Since $\mathrm{CSP}(\Gamma)$ is tractable whenever Γ is preserved by a Maltsev operation, it might not be evident how the Maltsev algorithm can be used for constructing kernels for NP-hard CSPs. The basic idea is to embed Γ into a language $\hat{\Gamma}$ over a larger domain, which is preserved by a Maltsev operation. This allows us to use the advantageous properties of relations invariant under Maltsev operations, in order to compute a kernel for the original problem.

Definition 7. *A constraint language Γ over D admits an embedding over the constraint language $\hat{\Gamma}$ over $D' \supseteq D$ if there exists a bijection $h : \Gamma \to \hat{\Gamma}$ such that $\mathrm{ar}(h(R)) = \mathrm{ar}(R)$ and $h(R) \cap D^{\mathrm{ar}(R)} = R$ for every $R \in \Gamma$.*

If $\hat{\Gamma}$ is preserved by a Maltsev operation then we say that Γ admits a *Maltsev embedding*. We do not exclude the possibility that D' is infinite, but in this section we will only be concerned with finite domains, and therefore do not explicitly state this assumption. If the bijection h is efficiently computable and there exists a polynomial p such that $h(R)$ can be computed in $O(p(|R|))$ time for each $R \in \Gamma$, then we say that Γ admits a *polynomially bounded* embedding. In particular, an embedding over a finite domain of any finite Γ is polynomially bounded.

Example 8. Recall from Sect. 2 that $R_{1/3} = \{(0,0,1), (0,1,0), (1,0,0)\}$. We claim that $R_{1/3}$ has a Maltsev embedding over $\{0,1,2\}$. Let $\hat{R}_{1/3} = \{(x,y,z) \in \{0,1,2\}^3 \mid x+y+z = 1 \,(\mathrm{mod}\,3)\}$. Then $\hat{R}_{1/3} \cap \{0,1\}^3 = R_{1/3}$, and from Example 6 we recall that $\hat{R}_{1/3}$ is preserved by a Maltsev operation. Hence, $\hat{R}_{1/3}$ is indeed a Maltev embedding of $R_{1/3}$. More generally, for every k, $R_{1/k}$ has a Maltsev embedding into equations over a finite field of size at least k.

For a CSP(Γ) instance $I = (\{x_1, \ldots, x_n\}, C)$ we let Ψ_I be the relation $\{(g(x_1), \ldots, g(x_n)) \mid g \text{ satisfies } I\}$, and if ϕ is a Maltsev operation and $I = (V, \{C_1, \ldots, C_m\})$ an instance of CSP$(\mathrm{Inv}(\{\phi\}))$ we let $\mathrm{Seq}(I) = (S_0, S_1 \ldots, S_m)$ denote the compact representations of the relations $\Psi_{(V,\emptyset)}$, $\Psi_{(V,\{C_1\})}$, $\ldots$, $\Psi_{(V,\{C_1,\ldots,C_m\})}$ computed by the Maltsev algorithm. We remark that the ordering of the constraints in $\mathrm{Seq}(I)$ does not influence the upper bound for the kernel.

Definition 9. *Let ϕ be a Maltsev operation, p a polynomial and let $\Delta \subseteq \mathrm{Inv}(\{\phi\})$. We say that Δ and CSP(Δ) have chain length p if $|\{\langle S_i \rangle_\phi \mid i \in \{0, 1, \ldots, |C|\}\}| \leq p(|V|)$ for each instance $I = (V, C)$ of CSP(Δ), where $\mathrm{Seq}(I) = (S_0, S_1, \ldots, S_{|C|})$.*

We now have everything in place to define our kernelization algorithm.

Theorem 10. *Let Γ be a constraint language over D which admits a polynomially bounded Maltsev embedding $\hat{\Gamma}$ with chain length p. Then CSP(Γ) has a kernel with $O(p(|V|))$ constraints.*

Proof. Let $\phi \in \mathrm{Pol}(\hat{\Gamma})$ denote the Maltsev operation witnessing the embedding $\hat{\Gamma}$. Given an instance $I = (V, C)$ of CSP(Γ) we can obtain an instance $I' = (V, C')$ of CSP$(\hat{\Gamma})$ by replacing each constraint $R_i(\mathbf{x_i})$ in C by $\hat{R}_i(\mathbf{x_i})$. We arbitrarily order the constraints as $C' = (C_1, \ldots, C_m)$ where $m = |C'|$. We then iteratively compute the corresponding sequence $\mathrm{Seq}(I') = (S_0, S_1, \ldots, S_{|C'|})$. This can be done in polynomial time with respect to the size of I via the same procedure as the Maltsev algorithm. For each $i \in \{1, \ldots, m\}$ we then do the following.

1. Let the ith constraint be $C_i = \hat{R}_i(x_{i_1}, \ldots, x_{i_r})$ with $\mathrm{ar}(R_i) = r$.
2. For each $t \in S_{i-1}$ determine whether $\mathrm{pr}_{i_1, \ldots, i_r}(t) \in \hat{R}_i$.
3. If yes, then remove the constraint C_i, otherwise keep it.

This can be done in polynomial time with respect to the size of the instance I', since (1) $|S_{i-1}|$ is bounded by a polynomial in $|V|$ and (2) the test $\mathrm{pr}_{i_1,\ldots,i_r}(t) \in \hat{R}_i$ can naively be checked in linear time with respect to $|\hat{R}_i|$. We claim that the procedure outlined above will correctly detect whether the constraint C_i is redundant or not with respect to $\langle S_{i-1} \rangle_\phi$, i.e., whether $\langle S_{i-1} \rangle_\phi = \langle S_i \rangle_\phi$. First, observe that if there exists $t \in S_{i-1}$ such that $\mathrm{pr}_{i_1,\ldots,i_r}(t) \notin \hat{R}_i$, then the constraint is clearly not redundant. Hence, assume that $\mathrm{pr}_{i_1,\ldots,i_r}(t) \in \hat{R}_i$ for every $t \in S_{i-1}$. Then $S_{i-1} \subseteq \langle S_i \rangle_\phi$, hence also $\langle S_{i-1} \rangle_\phi \subseteq \langle S_i \rangle_\phi$. On the other hand, $\langle S_i \rangle_\phi \subseteq \langle S_{i-1} \rangle_\phi$ holds trivially. Therefore, equality must hold. Let $I'' = (V, C'')$ denote the resulting instance. Since $\mathrm{CSP}(\mathrm{Inv}(\{\phi\}))$ has chain length p it follows that (1) the sequence $\langle S_0 \rangle_\phi, \langle S_1 \rangle_\phi, \ldots, \langle S_{|C'|} \rangle_\phi$ contains at most $p(|V|)$ distinct elements, hence $|C''| \leq p(|V|)$, and (2) $\Psi_{I'} = \Psi_{I''}$. Clearly, it also holds that $\Psi_I = (\Psi_{I'} \cap \{0,1\}^{|V|}) = (\Psi_{I''} \cap \{0,1\}^{|V|})$. Hence, we can safely transform I'' to an instance I^* of $\mathrm{CSP}(\Gamma)$ by replacing each constraint $\hat{R}_i(\mathbf{x_i})$ with $R_i(\mathbf{x_i})$. Then I^* is an instance of $\mathrm{CSP}(\Gamma)$ with at most $p(|V|)$ constraints, such that $\Psi_I = \Psi_{I^*}$. In particular, I^* has a solution if and only if I has a solution. □

All that remains to be proven now is that there actually exist Maltsev embeddings with bounded chain length.

Theorem 11. $\mathrm{CSP}(\mathrm{Inv}(\{\phi\}))$ *has chain length* $O(|D||V|)$ *for every Maltsev operation* ϕ *over a finite* D.

Proof. Let $I = (V, C)$ be an instance of $\mathrm{CSP}(\mathrm{Inv}(\{\phi\}))$, with $|V| = n$ and $|C| = m$, and let $\mathrm{Seq}(I) = (S_0, S_1, \ldots, S_m)$ be the sequence of compact representations computed by the Maltsev algorithm. First, we claim that $\mathrm{Sig}(S_{i+1}) \subseteq \mathrm{Sig}(S_i)$ for every $i < m$. To see this, pick $(j, a, b) \in \mathrm{Sig}(S_i)$, where $j \in \{1, \ldots, |V|\}$ and $a, b \in D$. Then there exists $t, t' \in S_i$ such that (t, t') witnesses (j, a, b), i.e., $\mathrm{pr}_{1,\ldots,j-1}(t) = \mathrm{pr}_{1,\ldots,j-1}(t')$, and $t[j] = a$, $t'[j] = b$. Since $\langle S_{i-1} \rangle_\phi \supseteq \langle S_i \rangle_\phi \supseteq S_i$, it follows that $t, t' \in \langle S_{i-1} \rangle_\phi$, and hence also that $(j, a, b) \in \mathrm{Sig}(\langle S_{i-1} \rangle_\phi)$. But since S_{i-1} is a representation of $\langle S_{i-1} \rangle_\phi$, $\mathrm{Sig}(S_{i-1}) = \mathrm{Sig}(\langle S_{i-1} \rangle_\phi)$, from which we infer that $(j, a, b) \in \mathrm{Sig}(S_{i-1})$. Second, we claim that the sets $(j, a, b) \in \mathrm{Sig}(S_i)$ induce an equivalence relation on $\mathrm{pr}_j(\langle S_i \rangle_\phi)$ for every $i \leq m$, $j \leq n$[1]. Let $a \sim b$ hold if and only if $(j, a, b) \in \mathrm{Sig}(S_i)$. Note that $(j, a, a) \in \mathrm{Sig}(S_i)$ if and only if $a \in \mathrm{pr}_j(S_i)$, and that $(j, a, b) \notin \mathrm{Sig}(S_i)$ for any b if $a \notin \mathrm{pr}_j(S_i)$. Also note that $\sim$ is symmetric by its definition. It remains to show transitivity. Let $(j, a, b) \in \mathrm{Sig}(S_i)$ be witnessed by (t_a, t_b) and $(j, a, c) \in \mathrm{Sig}(S_i)$ be witnessed by (t'_a, t'_c). We claim that $t_c := \phi(t_a, t'_a, t'_c) \in S_i$ is a tuple such that (t_b, t_c) witnesses $(i, b, c) \in \mathrm{Sig}(S_i)$. Indeed, for every $i' < i$ we have $\phi(t_a[i'], t'_a[i'], t'_c[i']) = \phi(t_a[i'], t'_a[i'], t'_a[i']) = t_a[i']$, whereas $\phi(t_a[i'], t'_a[i'], t'_c[i']) = (a, a, c) = c$. Since $t_a[i'] = t_b[i']$ for every $i' < i$, it follows that (t_b, t_c) witnesses $(j, b, c) \in \mathrm{Sig}(S_i)$. Hence $\sim$ is an equivalence relation on $\mathrm{pr}_j(S_i)$. We wrap up the proof as follows. Note that if $\mathrm{Sig}(S_{i+1}) = \mathrm{Sig}(S_i)$, then $\langle S_i \rangle_\phi = \langle S_{i+1} \rangle_\phi$ since S_{i+1} is a compact representation of $\langle S_i \rangle_\phi$. Hence, we need to bound the number of times that

---

[1] This property is essentially folklore in universal algebra, and follows from the *rectangularity* property of relations invariant under Maltsev operations.

$\text{Sig}(S_{i+1}) \subset \text{Sig}(S_i)$ can hold. Now, whenever $\text{Sig}(S_{i+1}) \subset \text{Sig}(S_i)$, then either $\text{pr}_j(\langle S_i \rangle_\phi) \subset \text{pr}_j(\langle S_{i+1} \rangle_\phi)$ for some j, or the equivalence relation induced by tuples $(j, a, b) \in \text{Sig}(S_{i+1})$ is a refinement of that induced by tuples $(j, a, b) \in \text{Sig}(S_i)$ for some j. Both of these events can only occur $|D| - 1$ times for every position j (unless $S_m = \emptyset$). Hence the chain length is bounded by $2|V||D|$. $\square$

This bound can be slightly improved for a particular class of Maltsev operations. Recall from Example 6 that $s(x, y, z) = x \cdot y^{-1} \cdot z$ is the coset generating operation of a group $G = (D, \cdot)$.

Lemma 12. *Let $G = (D, \cdot)$ be a finite group and let s be its coset generating operation. Then* $\text{CSP}(\text{Inv}(\{s\}))$ *has chain length* $O(|V| \log |D|)$.

Proof. Let $I = (V, C)$ be an instance of $\text{CSP}(\text{Inv}(\{s\}))$, where $|V| = n$ and $|C| = m$. Let $\text{Seq}(I) = (S_0, S_1, \ldots, S_m)$ be the corresponding sequence. First observe that S_0 is a compact representation of D^n and that $(D^n, \cdot)$ is nothing else than the nth direct power of G. It is well-known that R is a coset of a subgroup of $(D^n, \cdot)$ if and only if s preserves R [4]. In particular, this implies that S_1 is a compact representation of a subgroup of $(D^n, \cdot)$, and more generally that each S_i is a compact representation of a subgroup of $\langle S_{i-1} \rangle_s$. Lagrange's theorem then reveals that $|\langle S_i \rangle_s|$ divides $|\langle S_{i-1} \rangle_s|$, which implies that the sequence $\langle S_0 \rangle_s, \langle S_1 \rangle_s, \ldots, \langle S_m \rangle_s$ contains at most $n \log_2 |D| + 1$ distinct elements. $\square$

Note that the bound $|V| \log |D|$ is in fact a bound on the length of a chain of subgroups of G^n; thus it can be further strengthened in certain cases. In particular, if $|D|$ is prime then the bound on chain length is simply $|V| + 1$ and the resulting kernel has at most $|V|$ constraints. Thus, Theorem 10 and Lemma 12 (via Example 8) give an alternate proof of the result that $\text{SAT}(\{R_{1/k}\})$ has a kernel with at most $|V|$ constraints. More generally, we get the following cases. First, if Γ can be represented via linear equations over a finite field, then $\text{CSP}(\Gamma)$ has a kernel with at most $|V|$ constraints. This closely mirrors the result of Jansen and Pieterse [12]. Second, if Γ can be embedded into cosets of a finite group over a set D, then $\text{CSP}(\Gamma)$ has a kernel of $O(|V| \log |D|)$ constraints, but not necessarily $|V|$ constraints (for example, $x = 0 \,(\text{mod}\, 2)$ and $x = 0 \,(\text{mod}\, 3)$ are independent over Z_6). Third, in the general case, where Γ has an embedding into a language on domain D with some arbitrary Maltsev polymorphism with no further structure implied, $\text{CSP}(\Gamma)$ has a kernel with $O(|V||D|)$ constraints. (More generally, for $|\Gamma|$ finite, we may use different Maltsev embeddings for different $R \in \Gamma$, and apply the above kernel to each relation R in turn, for a kernel of $O(|\Gamma||D||V|)$ constraints, where $|D|$ is the largest domain used in these embeddings.) Each case is more general than the previous: there are groups whose coset generating operations cannot be represented by Abelian groups (for example A_n, the group of all even permutations over $\{1, \ldots, n\}$ for $n \geq 3$), and it is known that a Maltsev operation ϕ over D is the coset generating operation of a group $(D, \cdot)$ if and only if $\phi(\phi(x, y, z), z, u) = \phi(x, y, u)$, $\phi(u, z, \phi(z, y, x)) = \phi(u, y, x)$ for all $x, y, z, u \in D$ [4]. Hence, any Maltsev operation which does not satisfy these two identities cannot be viewed as the coset generating operation of a group.

4   Partial Polymorphisms and Lower Bounds

We have seen that Maltsev embeddings provide an algebraic criterion for determining that a $\mathrm{CSP}(\Gamma)$ problem admits a kernel of a fixed size. In this section we develop a connection between the partial polymorphisms of a constraint language and the existence of a Maltsev embedding, and leverage these results in order to prove lower bound on kernelization for $\mathrm{SAT}(\Gamma)$. Let $f : D^k \to D$ be a k-ary operation over $D \supseteq \{0,1\}$. We can then associate a partial Boolean operation $f_{|\mathbb{B}}$ with f by restricting f to the Boolean arguments which also result in a Boolean value. In other words $\mathrm{domain}(f_{|\mathbb{B}}) = \{(x_1,\ldots,x_k) \in \{0,1\}^k \mid f(x_1,\ldots,x_k) \in \{0,1\}\}$, and $f_{|\mathbb{B}}(x_1,\ldots,x_k) = f(x_1,\ldots,x_k)$ for every $(x_1,\ldots,x_k) \in \mathrm{domain}(f_{|\mathbb{B}})$. We then characterize the partial polymorphisms of Boolean constraint languages admitting Maltsev embeddings as follows.

Theorem 13. *Let Γ be a Boolean constraint language, ϕ a Maltsev operation, and $\hat{\Gamma} = \{\langle R \rangle_\phi \mid R \in \Gamma\}$. Then $\hat{\Gamma}$ is a Maltsev embedding of Γ if and only if $f_{|\mathbb{B}} \in \mathrm{pPol}(\Gamma)$ for every $f \in \mathrm{Pol}(\hat{\Gamma})$.*

Proof. For the first direction, assume that $\hat{\Gamma}$ is a Maltsev embedding of Γ, and assume that there exists $R \in \Gamma$ and an n-ary $f \in \mathrm{Pol}(\hat{\Gamma})$ such that $f_{|\mathbb{B}}(t_1,\ldots,t_n) \notin R$ for $t_1,\ldots,t_n \in R$. By construction, $f_{|\mathbb{B}}(t_1,\ldots,t_n) = t$ is a Boolean tuple. But since $\hat{R} \cap \{0,1\}^{\mathrm{ar}(R)} = R$, this implies (1) that $t \notin \hat{R}$ and (2) that $f_{|\mathbb{B}}(t_1,\ldots,t_n) = f(t_1,\ldots,t_n) = t \notin \hat{R}$. Hence, f does not preserve $\hat{R}$ or $\hat{\Gamma}$, and we conclude that $f_{|\mathbb{B}} \in \mathrm{pPol}(\Gamma)$. For the other direction, assume that $\{f_{|\mathbb{B}} \mid f \in \mathrm{Pol}(\hat{\Gamma})\} \subseteq \mathrm{pPol}(\Gamma)$ but that there exists $\hat{R} \in \hat{\Gamma}$ such that $\hat{R} \cap \{0,1\}^{\mathrm{ar}(R)} \supset R$. Let $t \in \hat{R} \cap \{0,1\}^{\mathrm{ar}(R)} \setminus R$. By construction of $\hat{R}$ it follows that there exists an n-ary $f \in [\{\phi\}]$ and $t_1,\ldots,t_n \in R$ such that $f(t_1,\ldots,t_n) = t \notin R$. But then it follows that $f_{|\mathbb{B}}(t_1,\ldots,t_n)$ is defined as well, implying that $f_{|\mathbb{B}}(t_1,\ldots,t_n) \notin R$. This contradicts the assumption that $f_{|\mathbb{B}} \in \mathrm{pPol}(\Gamma)$ for every $f \in \mathrm{Pol}(\hat{\Gamma})$.   □

Hence, the existence of a Maltsev embedding can always be witnessed by the partial polymorphisms of a constraint language. We will now describe the partial operations that preserve every Boolean language with a Maltsev embedding. Therefore, say that f is a *universal partial Maltsev operation* if $f \in \mathrm{pPol}(\Gamma)$ for every Boolean Γ admitting a Maltsev embedding. Due to Theorem 13 this is tantamount to finding a Maltsev operation ϕ such that every Boolean language with a Maltsev embedding admits a Maltsev embedding over ϕ.

Definition 14. *Let the infinite domain D_∞ be recursively defined to contain 0, 1, and ternary tuples of the form (x,y,z) where $x,y,z \in D_\infty$, $x \neq y$, $y \neq z$. The Maltsev operation u over D_∞ is defined as $u(x,x,y) = y$, $u(x,y,y) = x$, and $u(x,y,z) = (x,y,z)$ otherwise.*

We will now prove that $q_{|\mathbb{B}}$ is a universal partial Maltsev operation if $q \in [\{u\}]$.

Theorem 15. *Let $q \in [\{u\}]$. Then $q_{|\mathbb{B}}$ is a universal partial Maltsev operation.*

Proof. We provide a sketch of the most important ideas. Let $q \in [\{u\}]$ be n-ary, and let Γ be a Boolean constraint language admitting a Maltsev embedding with respect to an operation ϕ. It is known that every operation in $[\{u\}]$ can be expressed as a term over u [9], and if we let p denote the operation defined by replacing each occurence of u in this term by ϕ we obtain an operation included in $[\{\phi\}]$. We then claim that $q_{|\mathbb{B}}$ can be obtained as a subfunction of $p_{|\mathbb{B}}$, which is sufficient to prove the result since $p_{|\mathbb{B}} \in \mathrm{pPol}(\Gamma)$ via Theorem 13 and since $\mathrm{pPol}(\Gamma)$ is known to be closed under taking subfunctions [23]. The intuition behind this step is that $q(x_1, \ldots, x_n)$ for $x_1, \ldots, x_n \in \{0, 1\}$ may only return a Boolean value through a sequence of Maltsev conditions, and since ϕ is also a Maltsev operation, it has to abide by these conditions as well. Formally, this can be proven straightforwardly through induction on the terms defining q and p. □

We may thus combine Theorem 13 and Theorem 15 to obtain a complete description of all universal partial Maltsev operations. Even though these proofs are purely algebraic we will shortly see that universal Maltsev operations have strong implications for kernelizability of SAT. For this purpose we define the *first partial Maltsev operation* ϕ_1 as $\phi_1(x, y, y) = x$ and $\phi_1(x, x, y) = y$ for all $x, y \in \{0, 1\}$, and observe that $\mathrm{domain}(\phi_1) = \{(0, 0, 0), (1, 1, 1), (0, 0, 1), (1, 1, 0), (1, 0, 0), (0, 1, 1)\}$. Via Theorem 15 it follows that ϕ_1 is equivalent to $u_{|\mathbb{B}}$, and is therefore a universal partial Maltsev operation. We will now prove that $\phi_1 \in \mathrm{pPol}(\Gamma)$ is in fact a necessary condition for the existence of a linear-sized kernel for SAT(Γ), modulo a standard complexity theoretical assumption. A pivotal part of this proof is that if $\phi_1 \notin \mathrm{pPol}(\Gamma)$, then Γ can qfpp-define a relation Φ_1, which can be used as a gadget in a reduction from the VERTEX COVER problem. This relation is defined as $\Phi_1(x_1, x_2, x_3, x_4, x_5, x_6) \equiv (x_1 \vee x_4) \wedge (x_1 \neq x_3) \wedge (x_2 \neq x_4) \wedge (x_5 = 0) \wedge (x_6 = 1)$. The following lemma shows a strong relationship between ϕ_1 and Φ_1.

Lemma 16. *If Γ is a Boolean constraint language such that $\langle \Gamma \rangle = BR$ and $\phi_1 \notin \mathrm{pPol}(\Gamma)$ then $\Phi_1 \in \langle \Gamma \rangle_{\not\exists}$.*

Proof. Before the proof we need two central observations. First, the assumption that $\langle \Gamma \rangle = BR$ is well-known to be equivalent to that $\mathrm{Pol}(\Gamma)$ consists only of projections. Second, Φ_1 consists of three tuples which can be ordered as s_1, s_2, s_3 in such a way that for every $s \in \mathrm{domain}(\phi_1)$ there exists $1 \leq i \leq 6$ such that $s = (s_1[i], s_2[i], s_3[i])$. Now, assume that $\langle \Gamma \rangle = BR$, $\phi_1 \notin \mathrm{pPol}(\Gamma)$, but that $\Phi_1 \notin \langle \Gamma \rangle_{\not\exists}$. Then there exists an n-ary $f \in \mathrm{pPol}(\Gamma)$ such that $f \notin \mathrm{pPol}(\{\Phi_1\})$, and $t_1, \ldots, t_n \in \Phi_1$ such that $f(t_1, \ldots, t_n) \notin \Phi_1$. Now consider the value $k = |\{t_1, \ldots, t_n\}|$, i.e., the number of distinct tuples in the sequence. If $n > k$ then it is known that there exists a closely related partial operation g of arity at most k such that $g \notin \mathrm{pPol}(\{\Phi_1\})$ [22], and we may therefore assume that $n = k \leq |\Phi_1| = 3$. Assume first that $1 \leq n \leq 2$. Then, for every $t \in \{0, 1\}^n$ there exists i such that $(t_1[i], \ldots, t_n[i]) = t$. But then f must be a total operation which is not a projection, which is impossible since we assumed that $\langle \Gamma \rangle = BR$.

Hence, it must be the case that $n = 3$, and that $\{t_1, t_2, t_3\} = \Phi_1$. Assume without loss of generality that $t_1 = s_1$, $t_2 = s_2$, $t_3 = s_3$, and note that this implies that $\text{domain}(f) = \text{domain}(\phi_1)$ (otherwise f can simply be described as a permutation of ϕ_1). First, we will show that $f(0,0,0) = 0$ and that $f(1,1,1) = 1$. Indeed, if $f(0,0,0) = 1$ or $f(1,1,1) = 0$, it is possible to define a unary total f' as $f'(x) = f(x,x,x)$ which is not a projection since either $f'(0) = 1$ or $f'(1) = 0$. Second, assume there exists $(x,y,z) \in \text{domain}(f)$, distinct from $(0,0,0)$ and $(1,1,1)$, such that $f(x,y,z) \neq \phi_1(x,y,z)$. Without loss of generality assume that $(x,y,z) = (a,a,b)$ for $a,b \in \{0,1\}$, and note that $f(a,a,b) = a$ since $\phi_1(a,a,b) = b$. If also $f(b,b,a) = a$ it is possible to define a binary total operation $f'(x,y) = f(x,x,y)$ which is not a projection, therefore we have that $f(b,b,a) = b$. We next consider the values taken by f on the tuples (b,a,a) and (a,b,b). If $f(b,a,a) = f(a,b,b)$ then we can again define a total, binary operation which is not a projection, therefore it must hold that $f(b,a,a) \neq f(a,b,b)$. However, regardless of whether $f(b,a,a) = b$ or $f(b,a,a) = a$, f must be a partial projection. This contradicts the assumption that $f \notin \text{pPol}(\{\Phi_1\})$, and we conclude that $\Phi_1 \in \langle \Gamma \rangle_{\not\exists}$.   □

We will shortly use Lemma 16 to give a reduction from the VERTEX COVER problem, since it is known that VERTEX COVER does not admit a kernel with $O(n^{2-\varepsilon})$ edges for any $\varepsilon > 0$, unless NP $\subseteq$ co-NP/poly [5]. For each n and k let $H_{n,k}$ denote the relation $\{(b_1, \ldots, b_n) \in \{0,1\}^n \mid b_1 + \ldots + b_n = k\}$.

Lemma 17. *Let Γ be a constraint language. If $\langle \Gamma \rangle = BR$ then Γ can pp-define $H_{n,k}$ with $O(n+k)$ constraints and $O(n+k)$ existentially quantified variables.*

Proof. We first observe that one can recursively design a circuit consisting of fan-in 2 gates which computes the sum of n input gates as follows. At the lowest level, we split the input gates into pairs and compute the sum for each pair, producing an output of 2 bits for each pair. At every level i above that, we join each pair of outputs from the previous level, of i bits each, into a single output of $i+1$ bits which computes their sum. This can be done with $O(i)$ gates by chaining full adders. Finally, at level $\lceil \log_2 n \rceil$, we will have computed the sum. The total number of gates will be $\sum_{i=1}^{\lceil \log_2 n \rceil} (\frac{n}{2^i}) \cdot O(i)$, which sums to $O(n)$. Let $z_1, \ldots, z_{\log_2 n}$ denote the output gates of this circuit. By a standard Tseytin transformation we then obtain an equisatisfiable 3-SAT instance with $O(n)$ clauses and $O(n)$ variables. For each $1 \leq i \leq \log_2 n$, add the unary constraint $(z_i = k_i)$, where k_i denotes the ith bit of k written in binary. Each such constraint can be pp-defined with $O(1)$ existentially quantified variables over Γ. We then pp-define each 3-SAT clause in order to obtain a pp-definition of R over Γ, which in total only requires $O(n)$ existentially quantified variables. This is possible since if $\langle \Gamma \rangle = BR$ then Γ can pp-define every Boolean relation.   □

Theorem 18. *Let Γ be a finite Boolean constraint language such that $\langle \Gamma \rangle = BR$ and $\phi_1 \notin \text{pPol}(\Gamma)$. Then $\text{SAT}(\Gamma)$ does not have a kernel of size $O(n^{2-\varepsilon})$ for any $\varepsilon > 0$, unless NP $\subseteq$ co-NP/poly.*

Proof. We will give a reduction from VERTEX COVER parameterized by the number of vertices to $SAT(\Gamma \cup \{\Phi_1\})$, which via Theorem 4 and Lemma 16 has a reduction to $SAT(\Gamma)$ which does not increase the number of variables. Let (V, E) be the input graph and let k denote the maximum size of the cover. First, introduce two variables x_v and x'_v for each $v \in V$, and one variable y_i for each $1 \leq i \leq k$. Furthermore, introduce two variables x and y. For each edge $\{u, v\} \in E$ introduce a constraint $\Phi_1(x_u, x'_v, x'_u, x_v, x, y)$, and note that this enforces the constraint $(x_u \vee x_v)$. Let $\exists z_1, \ldots, z_m.\phi(x_1, \ldots, x_{|V|}, y_1, \ldots, y_k, z_1, \ldots, z_m)$ denote the pp-definition of $H_{|V|+k,k}$ over Γ where $m \in O(k + |V|)$, and consisting of at most $O(k + |V|)$ constraints. Such a pp-definition must exist according to Lemma 17. Drop the existential quantifiers and add the constraints of $\phi(x_1, \ldots, x_{|V|}, y_1, \ldots, y_k, z_1, \ldots, z_m)$. Let (V', C) denote this instance of $SAT(\Gamma \cup \{\Phi_1\})$. Assume first that (V, E) has a vertex cover of size $k' \leq k$. We first assign x the value 0 and y the value 1. For each v in this cover assign x_v the value 1 and x'_v the value 0. For any vertex not included in the cover we use the opposite values. Then set $y_1, \ldots, y_{k-k'}$ to 1 and $y_{k-k'+1}, \ldots, y_k$ to 0. For the other direction, assume that (V', C) is satisfiable. For any x_v variable assigned 1 we then let v be part of the vertex cover. Since $x_1 + \ldots + x_{|V|} + y_1 + \ldots + y_k = k$, the resulting vertex cover is smaller than or equal to k. $\qquad\square$

For example, let $R^k = \{(b_1, \ldots, b_k) \in \{0, 1\}^k \mid b_1 + \ldots + b_k \in \{1, 2\} \ (\bmod\ 6)\}$ and let $P = \{R^k \mid k \geq 1\}$. The kernelization status of $SAT(P)$ was left open in Jansen and Pieterse [12], and while a precise upper bound seems difficult to obtain, we can easily prove that this problem does not admit a kernel of linear size, unless $NP \subseteq co\text{-}NP/poly$. Simply observe that $(0, 0, 1), (0, 1, 1), (0, 1, 0) \in R^3$ but $\phi_1((0, 0, 1), (0, 1, 1), (0, 1, 0)) = (0, 0, 0) \notin R^3$. The result then follows from Theorem 18. At this stage, it might be tempting to conjecture that $\phi_1 \in pPol(\Gamma)$ is also a sufficient condition for a Maltsev embedding. We can immediately rule this out by finding a relation R and a universal partial Maltsev operation ϕ such that R is invariant under ϕ_1 but not under ϕ. For example, let q be the 9-ary function defined by $u(u(x_1, x_2, x_3), u(x_4, x_5, x_6), u(x_7, x_8, x_9))$. Then we by computer experiments have verified that there exists a relation R of cardinality 9, invariant under ϕ_1 but not under $q_{|\mathbb{B}}$ [21].

5   Concluding Remarks and Future Research

We have studied kernelization properties of SAT and CSP with tools from universal algebra. We focused on problems with linear kernels, and showed that a CSP problem has a kernel with $O(n)$ constraints if it can be embedded into a CSP problem preserved by a Maltsev operation; thus extending previous results in this direction. On the other hand, we showed that a SAT problem not preserved by a partial Maltsev operation does not admit such a kernel, unless $NP \subseteq co\text{-}NP/poly$. This shows that the algebraic approach is viable for studying such fine-grained kernelizability questions. Our work opens several directions for future research.

A Dichotomy Theorem for Linear Kernels? Our results suggest a possible dichotomy theorem for the existence of linear kernels for SAT problems. However, two gaps remain towards such a result. On the one hand, we proved that if Γ is preserved by the universal partial Maltsev operations then it admits a Maltsev embedding over an infinite domain. However, the kernelization algorithm only works for finite domains. Does the existence of an infinite-domain Maltsev embedding for a finite language imply the existence of a Maltsev embedding over a finite domain? Alternatively, can the algorithms be adjusted to work for languages with infinite domains, since D_∞ is finitely generated in a simple way? On the other hand, we only have necessity results for ϕ_1 out of an infinite set of conditions for the positive results. Is it true that every universal partial Maltsev operation is a partial polymorphism of every language with a linear kernel, or do there exist SAT problems with linear kernels that do not admit Maltsev embeddings?

The Algebraic CSP Dichotomy Conjecture. Several solutions to the CSP dichotomy conjecture have been announced [1,26,27]. If correct, these algorithms solve $CSP(\Gamma)$ in polynomial time whenever Γ is preserved by a *Taylor term*. One can then define the concept of a Taylor embedding, which raises the question of whether the proposed algorithms can be modified to construct polynomial kernels. More generally, when can an operation f such that $CSP(Inv(\{f\}))$ is tractable be used to construct improved kernels? On the one hand, one can prove that k-*edge operations*, which are generalized Maltsev operations, can be used to construct kernels with $O(n^{k-1})$ constraints via a variant of the *few subpowers algorithm*. On the other hand, it is known that relations invariant under *semilattice operations* can be described as generalized Horn formulas, but it is not evident how this property could be useful in a kernelization procedure.

Acknowledgements. We thank the anonymous reviewers for several helpful suggestions. The first author is supported by the DFG-funded project "Homogene Strukturen, Bedingungserfüllungsprobleme, und topologische Klone" (Project number 622397).

References

1. Bulatov, A.: A dichotomy theorem for nonuniform CSPs. CoRR, abs/1703.03021 (2017)
2. Bulatov, A., Dalmau, V.: A simple algorithm for Mal'tsev constraints. SICOMP **36**(1), 16–27 (2006)
3. Bulatov, A., Jeavons, P., Krokhin, A.: Classifying the complexity of constraints using finite algebras. SICOMP **34**(3), 720–742 (2005)
4. Dalmau, V., Jeavons, P.: Learnability of quantified formulas. TCS **306**(1–3), 485–511 (2003)
5. Dell, H., van Melkebeek, D.: Satisfiability allows no nontrivial sparsification unless the polynomial-time hierarchy collapses. J. ACM **61**(4), 23:1–23:27 (2014)
6. Dyer, M., Richerby, D.: An effective dichotomy for the counting constraint satisfaction problem. SICOMP **42**(3), 1245–1274 (2013)

7. Feder, T., Vardi, M.: The computational structure of monotone monadic SNP and constraint satisfaction: a study through datalog and group theory. SICOMP **28**(1), 57–104 (1998)
8. Geiger, D.: Closed systems of functions and predicates. Pac. J. Math. **27**(1), 95–100 (1968)
9. Goldstern, M., Pinsker, M.: A survey of clones on infinite sets. Algebra Univers. **59**(3), 365–403 (2008)
10. Impagliazzo, R., Paturi, R., Zane, F.: Which problems have strongly exponential complexity? J. Comput. Syst. Sci. **63**, 512–530 (2001)
11. Jansen, B.M.P., Pieterse, A.: Sparsification upper and lower bounds for graphs problems and not-all-equal SAT. In: Proceedings of IPEC 2015, Patras, Greece (2015)
12. Jansen, B.M.P., Pieterse, A.: Optimal sparsification for some binary CSPs using low-degree polynomials. In: Proceedings of MFCS 2016, vol. 58, pp. 71:1–71:14 (2016)
13. Jeavons, P.: On the algebraic structure of combinatorial problems. TCS **200**, 185–204 (1998)
14. Jeavons, P., Cohen, D., Gyssens, M.: A unifying framework for tractable constraints. In: Montanari, U., Rossi, F. (eds.) CP 1995. LNCS, vol. 976, pp. 276–291. Springer, Heidelberg (1995). doi:10.1007/3-540-60299-2_17
15. Jeavons, P., Cohen, D., Gyssens, M.: Closure properties of constraints. JACM **44**(4), 527–548 (1997)
16. Jonsson, P., Lagerkvist, V., Nordh, G., Zanuttini, B.: Strong partial clones and the time complexity of SAT problems. JCSS **84**, 52–78 (2017)
17. Kratsch, S., Marx, D., Wahlström, M.: Parameterized complexity and kernelizability of max ones and exact ones problems. TOCT **8**(1), 1 (2016)
18. Kratsch, S., Wahlström, M.: Preprocessing of min ones problems: a dichotomy. In: Abramsky, S., Gavoille, C., Kirchner, C., Meyer auf der Heide, F., Spirakis, P.G. (eds.) ICALP 2010. LNCS, vol. 6198, pp. 653–665. Springer, Heidelberg (2010). doi:10.1007/978-3-642-14165-2_55
19. Krokhin, A.A., Marx, D.: On the hardness of losing weight. ACM Trans. Algorithms **8**(2), 19 (2012)
20. Lagerkvist, V., Wahlström, M.: The power of primitive positive definitions with polynomially many variables. JLC **27**, 1465–1488 (2016)
21. Lagerkvist, V., Wahlström, M.: Kernelization of constraint satisfaction problems: a study through universal algebra. ArXiv e-prints, June 2017
22. Lagerkvist, V., Wahlström, M., Zanuttini, B.: Bounded bases of strong partial clones. In: Proceedings of ISMVL 2015 (2015)
23. Lau, D.: Function Algebras on Finite Sets: Basic Course on Many-Valued Logic and Clone Theory (Springer Monographs in Mathematics). Springer, New York (2006)
24. Marx, D.: Parameterized complexity of constraint satisfaction problems. Comput. Complex. **14**(2), 153–183 (2005)
25. Nemhauser, G.L., Trotter, L.E.: Vertex packings: structural properties and algorithms. Math. Program. **8**(1), 232–248 (1975)
26. Rafiey, A., Kinne, J., Feder, T.: Dichotomy for digraph homomorphism problems. CoRR, abs/1701.02409 (2017)
27. Zhuk, D.: The proof of CSP dichotomy conjecture. CoRR, abs/1704.01914 (2017)

Defining and Evaluating Heuristics
for the Compilation of Constraint Networks

Jean-Marie Lagniez, Pierre Marquis[(✉)], and Anastasia Paparrizou

CRIL, U. Artois & CNRS, Lens, France
{Lagniez,Marquis,Paparrizou}@cril.fr

Abstract. Several branching heuristics for compiling in a top-down fashion finite-domain constraint networks into multi-valued decision diagrams (MDD) or decomposable multi-valued decision graphs (MDDG) are empirically evaluated, using the cn2mddg compiler. This MDDG compiler has been enriched with various additional branching rules. These rules can be gathered into two families, the one consisting of heuristics for the satisfaction problem (which are suited to compiling networks into MDD representations) and the family of heuristics favoring decompositions (which are relevant when the MDDG language is targeted). Our empirical investigation on a large dataset shows the value of decomposability (targeting MDDG allows for compiling many more instances and leads to much smaller compiled representations). The well-known (Dom/Wdeg) heuristics appears as the best choice for compiling networks into MDD. When MDDG is the target, a new rule, based on a dynamic, yet parsimonious use of hypergraph partitioning for the decomposition purpose turns out to be the best option. As expected, the best heuristics for the satisfaction problem perform better than the best heuristics favoring decompositions when MDD is targeted, and the converse is the case when MDDG is targeted.

Keywords: Knowledge compilation · Top-down compiler · Heuristics

1 Introduction

The objective of this work is to evaluate several branching heuristics (both existing ones but also new ones) which are candidates for compiling in a top-down fashion finite-domain constraint networks into decision diagrams. Two target languages are considered: the language MDD of multi-valued (deterministic) decision diagrams, and its superset, the language MDDG of decomposable multi-valued decision graphs. The significance of those two compilation languages comes from the fact that they support many useful queries in polynomial time. For instance, it is possible to determine in polynomial time whether an MDD representation (or an MDDG representation) is consistent or not, and even to count in polynomial time its number of solutions, or more generally to compute in polynomial time the number of (possibly weighted) solutions compatible with a given (partial) instantiation. It is also possible to enumerate with a polynomial delay all the solutions. Answering such queries is fundamental in a number of applications like

© Springer International Publishing AG 2017
J.C. Beck (Ed.): CP 2017, LNCS 10416, pp. 172–188, 2017.
DOI: 10.1007/978-3-319-66158-2_12

product configuration (see e.g., [1]), where looking for a feasible product given the user choices amounts to decide the consistency of a representation conditioned by the instantiation encoding the user choices, or probabilistic inference in Bayesian networks (computing the probability of a piece of evidence amounts to a weighted model counting query, see e.g., [3]). However, all those queries are NP-hard when the input is a constraint network.

In the Boolean case, the MDD language corresponds to the language FBDD of free binary decision diagrams [13], while MDDG corresponds to the language Decision-DNNF [12,22]. Roughly, every internal node in an MDD representation is a decision node associated with a variable of the input constraint network and having as many children as the number of elements in the domain of the variable. In MDDG representations, internal nodes can also be decomposable ∧-nodes, i.e., conjunctions of representations based on pairwise disjoint sets of variables. Despite the increase of generality obtained by accepting non-Boolean domains, the key tractable queries and transformations offered by Decision-DNNF are also offered by MDDG and MDD. Note that MDD offers some transformations that MDDG does not, and this is why this subset of MDDG is interesting in its own right. For instance, from an MDD representation of the feasible products of a configuration problem, it is possible to enumerate with a polynomial delay all the full instantiations corresponding to the non-feasible products (while this is impossible from an MDDG representation unless P = NP).

In order to generate MDDG and MDD representations, one takes advantage of the cn2mddg compiler [17], see http://www.cril.fr/KC/mddg.html. As in the Boolean case [16], an MDDG representation of a constraint network can be generated by recording the trace of a solver (in the Boolean case, a SAT solver and here, a CSP solver). Accordingly, cn2mddg is a top-down constraint network compiler, based on a CSP solver. It exploits constraint propagation and conflict analysis to guide the search. It also benefits from a specific caching technique and it detects universal constraints during the search in order to perform additional simplifications. Though cn2mddg was primarily based on a specific branching heuristics relying on betweenness centrality for promoting decompositions, we have implemented in it a number of additional heuristics for the sake of further evaluations and comparisons.

We have considered two groups of heuristics. The first one is composed of six heuristics for the consistency issue, namely the *Dom/Wdeg* heuristics (Dom/Wdeg) [15], and its by-products (the *Dom* heuristics (Dom), and the *Wdeg* heuristics (Wdeg)), the *impact-based* heuristics (IBS) [23], the *activity-based* heuristics (ABS) [20], and the *conflict-ordering* search heuristics (COS) [18]. All those heuristics were already known.

The second one gathers heuristics for promoting the generation of decomposable ∧-nodes. It is composed of seven heuristics. Three of them are *static heuristics*, meaning that they are used for generating a decomposition tree (dTree) of the input network in a preliminary step, before the compilation phase. The three static heuristics considered for computing a dTree are *Min Degree* (dTree-MD), *Min Fill* (dTree-MF), *Hypergraph Partitioning* (dTree-HP). The four remaining heuristics are *dynamic ones*, which means that each selected variable is computed during the search from the current network (thus, not prior to the search). Two of them, namely *Closeness Centrality* (CC) and *Betweenness Centrality* (BC), are based on a notion of centrality of the variables in the primal graph of the constraint network, one on the *Hypergraph Partitioning* (HP) of the dual hypergraph of the network, and the remaining one aims at computing a *Cut Set* (CS) of the primal graph. Actually, we have also considered for each of those four heuristics (H) a *parsimonious variant* (H-P) of it, meaning that a branching variable is not computed at each decision step using the heuristics, but instead, one computes a set of variables (containing all the variables that are ranked first by the heuristics) and uses all those variables successively for branching until the set becomes empty. All those heuristics except the one based on betweenness centrality (computed at each decision step) are new in the sense that they have not been tested so far, even if they are based on ingredients which are not brand new.

The branching heuristics from the two groups have been implemented in cn2mddg, parameterized in such a way that the compiler computes either MDDG representations (its default mode) or MDD representations (which can be done easily, by freezing the detection of disjoint components). For evaluating and comparing their relative performances in generating MDD representations and MDDG representations, cn2mddg (either in MDDG mode or in MDD mode), equipped with each of the thirteen heuristics under consideration has been run on 546 benchmarks corresponding to several families of instances.

Our experiments show the value of decomposability (targeting MDDG allows for compiling many more instances and leads to much smaller compiled representations). The well-known (Dom/Wdeg) heuristics appears as the best choice for compiling networks into MDD. When MDDG is the target, the new rule (HP-P), based on a dynamic, yet parsimonious use of hypergraph partitioning for the decomposition purpose turns out to be the best option. As expected, the best heuristics for the satisfaction problem perform better than the best heuristics favoring decompositions when MDD is targeted and the converse is the case when MDDG is targeted.

Previous work on AND/OR search has shown the impact of variable ordering on performance. AND/OR search is a framework for solving optimization tasks in graphical models by detecting independencies in the model (decompositions). Marinescu and Dechter [19] have shown that combining static and dynamic variable orderings with problem decomposition principles results in exponential savings. Variable orderings based on decompositions are also important for compiling constraint networks in other graphical representations. Narodytska and Walsh [21] proposed heuristics to reduce the time and space requirements for compiling configuration problems into BDDs. These heuristics are based on the distinctive clustered and hierarchical structure of the constraint graphs and are used for a bottom-up compilation.

2 Formal Preliminaries

A *finite-domain constraint network (CN)* is a triple $\mathcal{N} = (\mathcal{X}, \mathcal{D}, \mathcal{C})$ consisting of a set $\mathcal{X} = \{X_1, \cdots, X_n\}$ of *variables*, a set $\mathcal{D} = \{D_1, \cdots, D_n\}$ of *domains*, and a set $\mathcal{C} = \{C_1, \cdots, C_m\}$ of *constraints*. Each domain D_i is a finite set containing the possible values of X_i. Each constraint C_j characterizes the combinations of values satisfying it. Formally, $C_j = (S_j, R_j)$, where $S_j = \{X_{j_1}, \cdots, X_{j_k}\}$ is a subset of variables from $\mathcal{X}$, called the *scope* of C_j, and R_j is a predicate over the Cartesian product $D_{j_1} \times \cdots \times D_{j_k}$, called the *relation* of C_j. R_j can be represented extensionally by the list of its satisfying tuples (or dually, by the list of its forbidden tuples), or intensionally by an oracle, i.e., a mapping from $D_{j_1} \times \cdots \times D_{j_k}$ to $\{0, 1\}$ which is supposed to be computable in time polynomial in its input size. The *arity* of a constraint is given by the size of its scope. Constraints of arity 2 are called *binary* and constraints of arity greater than 2 are called *non-binary*.

Example 1. Let $\mathcal{N}$ be the CN given by four variables X_1, X_2, X_3, and X_4, each of them being defined on the same domain $\{0, 1, 2\}$, and three constraints C_1, C_2, and C_3, specified by the following mathematical statements:

- $C_1 = (X_1 \neq X_2)$;
- $C_2 = (X_2 = 0) \lor (X_2 = 1) \lor (X_2 = X_3 + X_4 + 1)$;
- $C_3 = (X_3 > X_4)$.

Given a subset S of variables from $\mathcal{X}$, a *(decision) state* $\boldsymbol{s}$ over S is a mapping that associates with each variable X_i in S a subset $\boldsymbol{s}(X_i)$ of values in D_i. In what follows, states are often noted as union of elementary assignments, i.e., sets of the form $\{\langle X_i, x_j \rangle\}$, where $x_j \in \boldsymbol{s}(X_i)$. *scope*$(\boldsymbol{s})$ denotes the set S of variables over which $\boldsymbol{s}$ is defined. A state $\boldsymbol{s}$ is *partial* if *scope*$(\boldsymbol{s})$ is a proper subset of $\mathcal{X}$; otherwise, $\boldsymbol{s}$ is called a *full* state. A variable X_i in *scope*$(\boldsymbol{s})$ is *instantiated* if $\boldsymbol{s}(X_i)$ is a singleton set. The set of instantiated variables in $\boldsymbol{s}$ is noted *single*$(\boldsymbol{s})$. As usual, a state $\boldsymbol{s}$ is called an *instantiation* when all its variables are instantiated, i.e., *scope*$(\boldsymbol{s})$ = *single*$(\boldsymbol{s})$.

For a state $\boldsymbol{s}$ and a set of variables $T \subseteq$ *scope*$(\boldsymbol{s})$, $\boldsymbol{s}[T]$ denotes the *restriction* of $\boldsymbol{s}$ to T, i.e., $\boldsymbol{s}[T]$ is the set $\{\langle X_i, x_j \rangle \in \boldsymbol{s} \mid X_i \in T\}$. An instantiation $\boldsymbol{s}$ *satisfies* a contraint $C_j = (S_j, R_j)$ if $S_j \subseteq$ *scope*$(\boldsymbol{s})$ and $R_j(x_{j_1}, \ldots, x_{j_k}) = 1$, where $\forall l \in 1, \ldots, k$, $\langle X_{j_l}, x_{j_l} \rangle \in \boldsymbol{s}[S_j]$. A *solution* of a CN $\mathcal{N} = (\mathcal{X}, \mathcal{D}, \mathcal{C})$ is a full instantiation $\boldsymbol{s}$ satisfying all constraints C_j in $\mathcal{C}$. For example, $\boldsymbol{s} = \{\langle X_1, 1 \rangle, \langle X_2, 0 \rangle, \langle X_3, 1 \rangle, \langle X_4, 0 \rangle\}$ is a solution of the CN given at Example 1.

Given a CN $\mathcal{N} = (\mathcal{X}, \mathcal{D}, \mathcal{C})$ and a state $\boldsymbol{s}$ over a subset of $\mathcal{X}$, the *conditioning* $\mathcal{N} \mid \boldsymbol{s}$ of $\mathcal{N}$ by $\boldsymbol{s}$ is the CN $(\mathcal{X}', \mathcal{D}', \mathcal{C}')$ defined as follows: $\mathcal{X}' = \mathcal{X} \setminus$ *single*$(\boldsymbol{s})$; with each domain D_i in $\mathcal{D}$, one associates the domain $D_i' \in \mathcal{D}'$, where $D_i' = D_i$ if $X_i \notin$ *scope*$(\boldsymbol{s})$ and $D_i' = \boldsymbol{s}(D_i)$ otherwise; finally, with each constraint $C_j = (S_j, R_j)$ in $\mathcal{C}$, one associates the constraint $C_j' = (S_j', R_j')$ in $\mathcal{C}'$, where $S_j' = S_j \setminus$ *single*$(\boldsymbol{s})$ and R_j' is the restriction of R_j to S_j'.

The *primal graph* of a CN $\mathcal{N} = (\mathcal{X}, \mathcal{D}, \mathcal{C})$ is the undirected graph $PG(\mathcal{N})$ with vertex set $\mathcal{X}$ and edge set $\mathcal{E}$, such that $\{X_p, X_q\} \in \mathcal{E}$ if and only if $\{X_p, X_q\}$

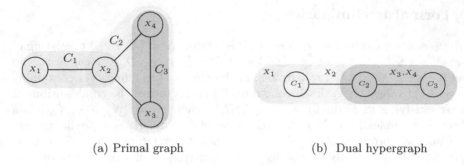

(a) Primal graph   (b) Dual hypergraph

Fig. 1. Graph representations of the CN given at Example 1.

is a subset of the scope S_j of some constraint C_j in $\mathcal{C}$. For instance, the primal graph of the CN given at Example 1 is depicted on Fig. 1a.

The *dual hypergraph* of a CN $\mathcal{N} = (\mathcal{X}, \mathcal{D}, \mathcal{C})$ is the undirected hypergraph $DH(\mathcal{N})$ with vertex set $\mathcal{C}$ and hyperedge set $\mathcal{H}$, such that $\mathcal{H} = \{H \subseteq \mathcal{C} \mid \exists X_i \in \mathcal{X} \text{ s.t. } \forall C_j \in \mathcal{C}, C_j \in H \text{ iff } X_i \in S_j\}$. Thus, every hyperedge corresponds precisely to a variable $\mathcal{X}$ which belongs to the scopes of all the constraints of the hyperedge, but does not belong to the scope of any other constraint of the input network. For instance, the dual hypergraph of the CN given at Example 1 is depicted on Fig. 1b. For this example, every hyperedge contains two constraints only, but of course this is not the case in general.

Let us now introduce a few definitions suited to the target languages considered for compiling CNs.

Definition 1 (MDG). *Given a finite set $\mathcal{X}$ of finite-domain variables, the (read-once) MDG language over $\mathcal{X}$ is the set of all single-rooted directed acyclic graphs Δ, where leaf nodes are labelled by $\top$ (true) or $\bot$ (false), and every internal node is either a $\wedge$-node $N = \wedge(N_1, \ldots, N_i)$ or a decision node N associated with variable $X_i \in \mathcal{X}$, i.e., a deterministic $\vee$-node $N = \vee(N_1, \ldots, N_j)$ such that $D_i = \{x_{i_1}, \ldots, x_{i_j}\}$ and the arc from N to N_k $(k \in 1, \ldots, j)$ is labelled by the elementary assignment $\{\langle X_i, x_{i_k}\rangle\}$. The paths of Δ must satisfy the read-once property: for every path from the root of Δ to a $\top$ leaf node, and for any $X_i \in \mathcal{X}$, no more than one arc can be labelled by an elementary assignment over X_i.*

For every node N in an MDG representation Δ, $Var(N)$ is defined inductively as follows:

- if N is a leaf node, then $Var(N) = \emptyset$;
- if N is a $\wedge$-node $N = \wedge(N_1, \ldots, N_i)$, then $Var(N) = \bigcup_{k=1}^{i} Var(N_i)$;
- if N is a decision node $N = \vee(N_1, \ldots, N_j)$ associated with variable X, then $Var(N) = \{X\} \cup \bigcup_{k=1}^{j} Var(N_k)$.

Let $\mathbf{s}$ be a full instantiation over $\mathcal{X}$ and let Δ be a MDG representation over $\mathcal{X}$, rooted at node N. Let $eval(N, \mathbf{s})$ be the MDG representation without any decision node, defined inductively by:

- if N is a leaf node, then $eval(N, \mathbf{s}) = N$;
- if N is a $\wedge$-node $N = \wedge(N_1, \ldots, N_i)$, then $eval(N, \mathbf{s}) = \wedge(eval(N_1, \mathbf{s}), \ldots, eval(N_i, \mathbf{s}))$;
- if N is a decision node $N = \vee(N_1, \ldots, N_j)$ associated with variable X_i, then $eval(N, \mathbf{s}) = eval(N_k, \mathbf{s})$, where $\langle X_i, x_{i_k} \rangle \in \mathbf{s}$.

$\mathbf{s}$ is a *solution* of Δ if and only $eval(N, \mathbf{s})$ evaluates to true.

The language MDDG we are interested in is the subset of MDG consisting of *decomposable* representations, those where the children of any $\wedge$-node do not share any variable. MDD is the subset of it, containing the representations without $\wedge$-nodes.

Definition 2 (MDDG, MDD). *Given a finite set $\mathcal{X}$ of finite-domain variables:*

- *the MDDG language over $\mathcal{X}$ is the subset of MDG representations Δ, where each $\wedge$-node $N = \wedge(N_1, \ldots, N_i)$ is decomposable, i.e., $\forall k, l \in 1, \ldots, i$, if $k \neq l$, then $Var(N_k) \cap Var(N_l) = \emptyset$.*
- *the MDD language over $\mathcal{X}$ is the subset of MDDG representations containing no $\wedge$-nodes.*

MDD can also be viewed as a restriction of the language of non-deterministic multi-valued decision diagrams considered in [2].

3   Heuristics for Compiling CNs

In this section, we briefly describe the branching heuristics which have been considered in our top-down constraint network compiler which targets the MDDG language (but can also be downsized to target the MDD language).

3.1   Heuristics Targeting the MDD Language

Dom/Wdeg. With (Dom/Wdeg), one selects a variable with minimum ratio of current domain size to weighted degree [15]. Each variable is associated with a weighted degree (Wdeg), which is the sum of the weights over all constraints involving the variable and at least another (unassigned) variable. A weight, initially set to one, is given to each constraint and each time a constraint causes a domain wipeout its weight is incremented by one. It is a generic state-of-the-art heuristics and the interesting is that it is adaptive, with the expectation to focus on the hard part(s) of the problem.

Dom. (Dom) is a first by-product of dom/wdeg. With (Dom), variables are ordered by considering their current domain size (the smallest cardinalities first).

Wdeg. (Wdeg) is a second by-product of (Dom/Wdeg). With (Wdeg), variables are ordered by considering their weighted degrees (the largest values first). The Wdeg score is close to the VSADS score used in model counters and compilers for propositional CNF formulae [24].

Impact-Based Search (IBS). With (IBS), one selects a variable with the highest impact, where impact measures the importance of a variable in reducing the search space [23]. An estimation of the size of the search space $S(P)$ is the product of every variable domain size:

$$S(P) = \prod_{x \in X} |D_x|$$

The impact of a variable assignment at a decision node k is computed by the ratio of the search space reduction as:

$$I(x = a) = 1 - \frac{S(P^k)}{S(P^{k-1})}$$

Note that if $x = a$ leads to a failure, then $I(x = a) = 1$, which is the maximum impact as $S(P^k) = 0$. It is easy to see that this heuristics can be used for value selection as well. For variable selection, the average impact is preferred, computed over the remaining values in its domain divided by its current domain size. For more accurate results, a forgetting strategy is used in order to give less importance to past variable assignments.

Activity-Based Search (ABS). With this heuristics, one selects a variable with the highest activity, where activity is measured by the times the domain of each variable is reduced during the search [20]. This heuristics is motivated by the key role of propagation in constraint programming and relies on a decaying sum to forget the oldest statistics progressively. The activities are initialized by making random probing in the search space.

More formally, the activity $A(x)$, of each variable x is updated at each node k of the search tree regardless of the outcome (success or failure) by the following two rules:

i. $A^k(x) = A^{k-1}(x) * \gamma$, where $0 \leq \gamma \leq 1$, $|D_x^k| > 1$ and $D_x^k = D_x^{k-1}$
ii. $A^k(x) = A^{k-1}(x) + 1$, where $D_x^k \subset D_x^{k-1}$

Conflict-Ordering Search (COS). This heuristics is considered more as a repairing mechanism than as a heuristics, that can be combined with any (underlying) variable ordering heuristics (e.g., (Dom/Wdeg)) [18]. When the solver needs to backtrack, the last conflicting variables, recorded during search, are selected in priority until they are all instantiated without causing any failure. Otherwise, in normal mode, the variable ordering heuristics is the one that decides the next variable.

3.2   Heuristics Targeting the MDDG Language

We have also considered eleven heuristics targeting the MDDG language, thus promoting the generation of decomposable ∧-nodes. The rationale for it is the gain in succinctness which mainly results in the generated representation when such nodes are allowed. Indeed, consider a constraint network $\mathcal{N}$ with x variables each of them having a domain of size $d > 1$ (for simplicity reasons). Suppose that every MDD representation of $\mathcal{N}$ has a size which is a fraction k ($0 < k \leq 1$) of the search space of all instantiations explored for generating it (which implies that the corresponding compilation time will be at least as high). Suppose now that a decomposition $(\mathcal{X}_1, \mathcal{X}_2, \mathcal{X}_3)$ of the set $\mathcal{X}$ of variables of $\mathcal{N}$ has been found. Such a decomposition is a tripartition of $\mathcal{X}$ such that for every assignment $\boldsymbol{x_1}$ over $\mathcal{X}_1$, the conditioned network $\mathcal{N} \mid \boldsymbol{x_1}$ has (at least) two disjoint components, one over the variables of $\mathcal{X}_2$ and one over the variables of $\mathcal{X}_3$. Then a decomposable $\wedge$-node can be generated. With $|\mathcal{X}_i| = x_i$ ($i \in \{1, 2, 3\}$), the size of the resulting MDDG representation of $\mathcal{N}$ will be at most $d^{x_1} \times (k \times d^{x_2} + k \times d^{x_3})$, which is always strictly smaller than the size $k \times d^{x_1 + x_2 + x_3}$ of the MDD representation of $\mathcal{N}$, unless $x_2 = x_3 = 1$ (in which case the decomposition is trivial). One can also easily check that the size of the resulting MDDG representation of $\mathcal{N}$ is as small as the decomposition is balanced, i.e., as x_2 and x_3 are close. More formally, $x_2^* = \lfloor \frac{x_2 + x_3}{2} \rfloor$ and $x_3^* = \lceil \frac{x_2 + x_3}{2} \rceil$ minimize the value of $d^{x_2} + d^{x_3}$ when the sum $x_2 + x_3$ is fixed (which is the case here whenever $\mathcal{X}_1$ has been set since $x_1 + x_2 + x_3 = x$). Accordingly, finding a "good" decomposition $(\mathcal{X}_1, \mathcal{X}_2, \mathcal{X}_3)$, i.e., a decomposition leading to an MDDG representation of "small" size) amounts to minimizing x_1 while making x_2 and x_3 as close as possible. It turns out that those two objectives can be antagonistic so that a trade-off must be looked for.

Static Heuristics. Static heuristics consist in generating a decomposition tree (dTree) prior to the compilation, which will be used to make precise the branching variables to be considered at each step. This approach is similar to the one used by the C2D compiler for propositional CNF formulae, which targets the Decision-DNNF language [8,9], see http://reasoning.cs.ucla.edu/c2d/. Roughly, a dTree for a constraint network is a full binary tree which induces a recursive decomposition of the network (for more details, see e.g., [10]).

The three static heuristics considered for computing a dTree are *Min Degree (dTree-MD)*, *Min Fill* (dTree-MF), and *Hypergraph Partitioning* (dTree-HP).

Min Degree (dTree-MD). The (dTree-MD) heuristics is used for generating a dTree in a bottom-up way and is driven by a variable elimination ordering: the variables are ordered by increasing degrees in the primal graph of the input network. The generation of the dTree starts with the leaves (each of them being associated with the scope of a constraint of the input network), and the initial forest (set of trees) to be dealt with is composed of those leaves. Then the variables of the network are considered according to the elimination ordering, and each time a variable is picked up, a tree is computed so that all the trees of the current forest which contain this variable are children of the resulting tree. Once all the variables have been processed, a forest of dTrees is generated (it consists of a single dTree when the primal graph of the network considered at start is connected).

Min Fill (dTree-MF). The (dTree-MF) heuristics is also used for generating a dTree in a bottom-up way and is driven by a variable elimination ordering (which is different from the one corresponding to the *Min Degree* heuristics): the variables are ordered by increasing numbers of non-connected neighbours in the primal graph of the input network. Then the generation of the dTree is made as in the case of *Min Degree* but considering the *Min Fill* elimination ordering instead.

Hypergraph Partitioning (dTree-HP). The hypergraph partitioning heuristics considers the dual hypergraph of the input network and generates a dTree for it in a top-down way. It looks for a subset of the set of hyperedges of the current network containing as few elements as possible such that removing them leads to a hypergraph containing (at least) two disjoint components having sizes as close as possible. When the variables corresponding to the selected hyperedges are instantiated (whatever the way they are assigned) it is guaranteed that the current network conditioned by the corresponding assignment has at least two disjoint components, so that a decomposable ∧-node can be generated in the compiled form. This set of variables is the cut set of the root of the dTree, and then the hypergraph partitioning approach proceeds recursively considering the disjoint hypergraphs which are generated by removing from the current hypergraph the hyperedges which have been selected at the previous step. One takes advantage of the partitioner PaToH – Partitioning Tools Hypergraph, v. 3.2 (http://bmi.osu.edu/~umit/software.html) [7] to do the job.

As explained above, hypergraph partitioning goes further than minimal cutting by taking account of the sizes of the subgraphs which are generated, which must be balanced, i.e., their difference in terms of numbers of vertices must be below a preset bound. This comes with a significant complexity increase since determining whether there exists a hypergraph partition of $DH(\mathcal{N})$ corresponding to a decomposition $(\mathcal{X}_1, \mathcal{X}_2, \mathcal{X}_3)$ of $\mathcal{N}$ such that $\#(\mathcal{X}_1) \leq c$ and $|\#(\mathcal{X}_2) - \#(\mathcal{X}_3)| \leq d$ (where c and d are two given bounds) is NP-complete. This departs deeply from the other heuristics considered in the paper which can be computed in polynomial time. In our experiments, we looked for 2-partitionings (i.e., one does not try to split the given hypergraph into more than two disjoint components) and used the default setting of PaToH.

Dynamic Heuristics. The four (pairs of) remaining heuristics we have considered are *dynamic ones*: each selected variable is computed during the search from the current network (thus, not prior to the search, from the input network), such that the propagations resulting from the previous variable assignments and leading to simplify the network, are taken into account. This can have a huge impact on the compilation process (both on the compilation times and on the sizes of the compiled forms).

We have considered two heuristics based on a notion of centrality (*Betweenness Centrality* (BC) vs. *Closeness Centrality* (CC)), which favor the decomposition of the current CN into components of balanced sizes by targeting the variables which are in some sense the central ones in its primal graph. We have also considered a heuristics based on *Hypergraph Partitioning* (HP) but this time it is not in the objective of generating first a dTree from the dual hypergraph of the input network but considering instead the dual hypergraph of the current network. Finally, we have also taken into account a *Cut Set* heuristics (CS), which focuses on identifying variables to be instantiated in order to split the current network into (at least) two disjoint components (whatever their sizes).

Each of those four heuristics (H) may point out several branching variables as the best ones. When they are several best variables, they ordered following their decreasing (Dom/Wdeg) score. When (H) is used in its default mode, the first variable w.r.t. this ordering is selected. When (H) is run in a parsimonious mode (H-P), the score of each variable from the current set of best candidates is not re-computed after each variable assignment but all the variables from this set are successively considered as branching variables up to exhaustion.

Closeness Centrality (CC). Closeness centrality is a measure of the centrality of a node in a graph [4]. Given a node X_i in a graph (here, the primal graph of the current CN in which the nodes can be identified as with the variables labelling them), the score $cc(X_i)$ is calculated as the sum of the lengths of the shortest paths between X_i and all other vertices in the graph. Thus the more central a vertex is, the closer it is to all other vertices. Formally:

$$cc(X_i) = \frac{1}{\Sigma_{X_j \neq X_i} d(X_i, X_j)}$$

where $d(X_i, X_j)$ is the geodesic distance between nodes X_i and X_j which belong to the same connected component of the graph. We also assume that the set of vertices of the primal graph of the component of the current CN which contains X_i is not a singleton, so that $\Sigma_{X_j \neq X_i} d(X_i, X_j) \neq 0$ (indeed in the remaining case, there is no option: X_i must be chosen in the component). The computation of the values of all $cc(X_i)$ when X_i varies in the set of vertices of the primal graph of the current CN can be done in time polynomial in the size of this graph, via a repeated use of breadth-first search from X_i or using Floyd-Warshall algorithm. We took advantage of the code of Floyd-Warshall algorithm from Boost Graph Library http://www.boost.org/doc/libs/1_64_0/libs/graph/doc/ for implementing *cc*. Clearly enough, by instantiating first the most central variables, the objective is to find out a decomposition $(\mathcal{X}_1, \mathcal{X}_2, \mathcal{X}_3)$ of the set of variables of the current network for which the cardinalities of $|\mathcal{X}_2|$ and $|\mathcal{X}_3|$ are close, however the number of variables in $\mathcal{X}_1$ can be large and one does not try to minimize it.

Betweenness Centrality (BC). Betweenness centrality is another measure of the centrality of a node in a graph [5,6]. The score $bc(X_i)$ is equal to the number of shortest paths from all nodes to all others that pass through X_i. Formally,

$$bc(X_i) = \Sigma_{X_j \neq X_i \neq X_k} \frac{\sigma_{X_i}(X_j, X_k)}{\sigma(X_j, X_k)}$$

where X_i, X_j, X_k are nodes of the same connected component of the given network, $\sigma(X_j, X_k)$ is the number of shortest paths from X_j to X_k, and $\sigma_{X_i}(X_j, X_k)$

are the number of those paths passing through X_i. Thus, for the CN $\mathcal{N}$ given at Example 1, X_2 is the unique variable maximizing the value of bc.

Interestingly, computing the betweenness centralities of all nodes in $(\mathcal{X}, \mathcal{E})$ can be done in time $\mathcal{O}(n.m)$, where n is the cardinality of $\mathcal{X}$ and m is the cardinality of $\mathcal{E}$. In practice, the computation of $bc(X_i)$ for each node X_i of the primal graph $(\mathcal{X}, \mathcal{E})$ of a CN is efficient enough so that it can be computed dynamically, i.e., for each network encountered during the compilation process. Again, we took advantage of the implementation of betweenness centrality available in the Boost Graph Library.

The rationale for instantiating first the most central variables (as to (BC)) is the same as the one for (CC), i.e., to split the network into two parts of similar sizes. (BC) can also be seen as an alternative of community structure [14], where instead of constructing communities by adding the strongest edges to an initially empty vertex set, it constructs them by progressively removing edges from the original graph. The community method detects which edges are most central to communities, while betweenness finds those edges that are most "between" communities. The community structure is thus more relevant for the bottom-up approaches to knowledge compilation like the one presented in [21].

Hypergraph Partitioning (HP). The approach is the same as the one described above, except that one does not compute a full dTree of the given hypergraph, but only its root. Note that when the current hypergraph has several disjoint components, the cut set returned by PaToH is empty so that no branching variable is defined. This is harmless when MDDG is targeted since this problem cannot happen in this case (a decomposable ∧-node would have been introduced before a branching variable is sought). However, this is still problematic in the case MDD is targeted. To deal with it, we switch to the (Dom/Wdeg) heuristics when such a pathological situation occurs.

Cut Set (CS). A (2-way) cut of a (undirected) graph is a partition of its vertices into two, non-empty sets. The corresponding cut set is the set of all edges between the two sets of vertices. A minimal cut set is a cut set of minimal size. Removing all the edges of the cut set of a graph leads to split it into two disjoint components. A minimal cut set of a graph $(\mathcal{X}, \mathcal{E})$ can be computed in time $\mathcal{O}(n.m^2)$ where n is the number of vertices of $\mathcal{X}$ and m is the number of edges in $\mathcal{E}$ [11]. Once a cut set of the primal graph of the current CN has been computed, one selects one variable per edge in the cut set. By construction, eliminating the variables of the resulting set in the primal graph is enough for ensuring that the resulting graph contains two disjoint components.

Using the cut set heuristics on the primal graph of the current constraint network, one computes decompositions $(\mathcal{X}_1, \mathcal{X}_2, \mathcal{X}_3)$ of the set of variables of the current network. One does not take care at all of balancing $|\mathcal{X}_2|$ and $|\mathcal{X}_3|$. In our implementation, we exploited the Stoer/Wagner algorithm [25] for min-cut available in the Boost Graph Library. As for (HP), if $\mathcal{X}_1$ turns out to be empty and MDD is targeted, then we switch to the (Dom/Wdeg) heuristics.

4 Empirical Evaluation

Setup. We have considered 546 CNs from 15 data sets.[1] Those data sets correspond to several families of problems, including configuration problems, graph coloring, scheduling problems, frequency allocation problems. For some instances, the constraints are represented extensionally, by the list of satisfying tuples or by the list of forbidden tuples; for other instances, they are given in intension. Each instance has been compiled using cn2mddg equipped with the various heuristics we focused on.

Our experiments have been conducted on a Quadcore Intel XEON X5550 with 32GiB of memory. A time limit of 1800 s for the off-line compilation phase (including the dTree generation when relevant) and a total amount of 8GiB of memory for storing the resulting compiled representation have been considered for each instance.

Results. We have first evaluated a random heuristics serving as base line for compiling the 546 benchmarks (for this heuristics, the decision variables are selected at random under a uniform distribution). Based on the random heuristics, cn2mddg has been able to compile 271 instances when MDDG was targeted and 238 instances when MDD was targeted.

We have then evaluated all the heuristic methods discussed before. In Fig. 2, we report their performances on a cactus plot where the x axis represents the number of "solved" (i.e., compiled) instances (numbers are displayed in the legend) and the y axis the CPU time needed per method, in logarithmic scale. Dotted lines correspond to MDD representations and solid to MDDG representations.

The general picture is that compiling constraint networks into MDDG allows to solve constantly more instances than when compiling to MDD, independently of the heuristics involved. The performance shift between the best approaches (ABS) and (Dom/Wdeg) targeting MDD and the one targeting MDDG (HP-P) is equal to 133–134 instances (over 546), which is significant. In more detail, when MDD is targeted, (ABS) solves 258 instances in the time and memory given and (Dom/Wdeg) solved 257 instances; the number of instances solved by the virtual best solver induced by the set of heuristics is 388. Its performance shift with (ABS) and/or (Dom/Wdeg) is thus large. When MDD is targeted, heuristics having instances solved by them and only them are (ABS), (IBS) and (dTree-MF). When MDDG is targeted, the best heuristics is (HP-P) with 391 instances solved; in this case, the number of instances solved by the virtual best solver induced by the set of heuristics is 404. Its performance shift with (HP-P) is thus quite limited (13 instances only). When MDDG is targeted, heuristics having instances solved by them and only them are (HP-P), (BC), (CC) and (dTree-MF).

When the target is MDD, as expected, heuristics that target to MDD work better than heuristics targeting MDDG. However, the performance shift is not that huge

---

[1] From www.cril.fr/~lecoutre/benchmarks.html, http://github.com/MiniZinc/minizinc
-benchmarks, and www.itu.dk/research/cla/externals/clib/.

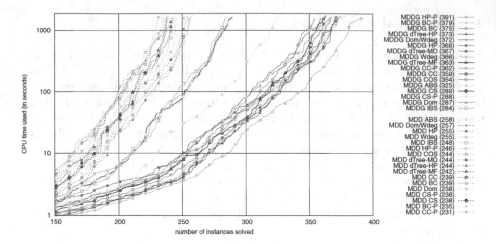

Fig. 2. Number of instances "solved" per method as the time allowed increases.

(it amounts to 27 instances, the worst heuristics for MDD being (CC-P) with 231 instances solved). Globally speaking, heuristics from the first group (the search-based ones) appear as slightly better than the other heuristics but the difference is not tremendous. Thus, one can also observe that the (HP) heuristics performs quite well when MDD is targeted, with 255 instances solved, despite the fact that it is designed for promoting decompositions. Contrastingly, some heuristics performed quite bad, actually as bad as the random heuristics ((Dom), (CS), (CS-P)) or even worse than it ((BC-P) and (CC-P)).

When MDDG is targeted, heuristics that take into account the graph structure are more suited. However, in that case, the performances shifts between the heuristics from the two groups and between the heuristics from the second group are important. Thus, the worst heuristics for MDDG is (IBS) which solves only 284 instances, showing a shift of 107 instances with (HP-P). Other search heuristics like (IBS), (Dom) and (ABS) also behave very poorly. Nevertheless, all the heuristics which have been considered performed better than the random one (with a shift at least equal to 13 instances). Search heuristics that take into account structural information, namely the degree like (Wdeg) are more efficient. It is interesting to note that the efficiency of (Dom/Wdeg) is mainly due to its second by-product (Wdeg), since when we separately try (Dom) and (Wdeg), there are 79 instances of difference in favor of (Wdeg). Such a difference does not appear when MDD is the target, demonstrating that structure plays a critical role for producing meaningful decompositions. Focusing now on the heuristics designed for promoting decompositions, significant shifts can also be observed, (CS-P) and (CS) with 288 and 289 instances solved, respectively, being far below the other heuristics (the next worst is (CC) with 359 instances solved). The best members from this second group are (HP-P), and then (BC-P), (BC), and (dTree-HP). Compared to (CS-P), (CS), (dTree-MD) and (dTree-MF), this suggests that the fact that the generated decompositions are balanced has a major impact, which is more significant than the fact that the decomposition is computed *ex ante* via the generation of a dTree, or on the fly. However, this does not explain the deceiving performances of (CC-P) and (CC) which try as well to split the network in two balanced parts by cutting it "in the middle".

Let us now provide a more detailed, pairwise comparison of the performances of the various heuristics. Tables 1 and 2 report (respectively) the dominance matrices of the heuristics under consideration for the two target languages. Each dominance matrix M gives for every pair of heuristics A, B, the number $M[A, B]$ of instances solved by A and not by B. Thus, when $M[A, B] = 0$ and $M[B, A] \neq 0$, A is strictly dominated by B.

Table 1. Dominance matrix (MDDG).

A/B	Dom/ Wdeg	Dom	Wdeg	IBS	ABS	COS	dTree-MD	dTree-MF	dTree-HP	CC-P	BC-P	HP-P	CS-P	CC	BC	HP	CS
Dom/Wdeg	-	85	8	89	48	19	7	12	1	27	17	5	85	30	20	5	84
Dom	0	-	0	11	3	0	1	4	0	12	8	0	16	13	9	0	16
Wdeg	2	79	-	83	46	15	8	14	3	27	17	5	82	29	19	4	81
IBS	1	8	1	-	3	2	2	5	1	14	9	0	16	15	9	1	16
ABS	1	41	5	44	-	5	6	10	2	20	15	5	45	22	17	5	44
COS	1	67	3	72	34	-	6	13	2	21	15	5	69	23	17	4	68
dTree-MD	2	81	9	85	48	19	-	10	0	24	14	3	81	27	17	4	80
dTree-MF	3	80	11	84	48	22	6	-	1	20	11	2	78	23	14	6	77
dTree-HP	2	86	10	90	50	21	6	11	-	27	17	5	86	30	20	5	85
CC-P	17	87	23	92	57	29	19	19	16	-	1	3	77	3	5	19	76
BC-P	24	100	30	104	69	40	26	27	23	18	-	4	93	21	4	26	92
HP-P	24	104	30	107	71	42	27	30	23	32	16	-	103	35	19	27	102
CS-P	1	17	4	20	8	3	2	3	1	3	2	0	-	5	4	1	0
CC	17	85	22	90	56	28	19	19	16	0	1	3	76	-	3	19	75
BC	23	97	28	100	67	38	25	26	22	18	0	3	91	19	-	25	90
HP	1	81	6	85	48	18	5	11	0	25	15	4	81	28	18	-	80
CS	1	18	4	21	8	3	2	3	1	3	2	0	1	5	4	1	-

For space reasons, we cannot detail the results obtained about the compilation times and the sizes of the compiled forms for the various heuristics. The benchmarks used and a detailed comparison in terms of compilation times and sizes of the compiled forms, reported in a number of scatter plots, of all the heuristics used can be found at http://www.cril.fr/KC/mddg.html. Roughly speaking, it turns out that the best heuristics in terms of number of instances solved are also the best heuristics when those two measures are considered instead (the intuition being that one obtains less time-out and/or memory-out precisely because the compilation times and the sizes of compiled forms are shorter ones). As expected, targeting MDDG leads to more compact representations than when MDD is targeted instead. The scatter plots on Fig. 3 illustrate it by focusing on the best heuristics for MDD (ABS) and (Dom/Wdeg) and the best heuristics for MDDG (HP-P). Each dot represents an instance. The size (in number of arcs) of the resulting compiled form, using the compiler cn2mddg

Table 2. Dominance matrix (MDD).

A/B	Dom/ Dom/	Dom	Wdeg	IBS	ABS	COS	dTree-MD	dTree-MF	dTree-HP	CC-P	BC-P	HP-P	CS-P	CC	BC	HP	CS
Dom/Wdeg	-	19	4	10	2	1	13	16	13	26	22	11	19	19	19	2	19
Dom	0	-	0	3	0	0	0	3	0	11	7	0	12	11	7	0	12
Wdeg	2	17	-	11	2	1	12	15	12	24	20	9	20	17	16	4	20
IBS	1	13	4	-	1	1	10	13	10	22	18	8	15	16	16	1	15
ABS	3	20	5	11	-	2	15	18	15	28	24	13	21	19	19	4	21
COS	2	20	4	11	2	-	14	17	14	27	23	12	20	19	19	3	20
dTree-MD	0	6	1	6	1	0	-	3	0	13	9	1	15	14	10	0	15
dTree-MF	1	7	2	7	2	1	1	-	1	11	7	2	13	12	8	1	13
dTree-HP	0	6	1	6	1	0	0	3	-	13	9	1	15	14	10	0	15
CC-P	0	4	0	5	1	0	0	0	0	-	0	0	3	1	1	0	3
BC-P	0	4	0	5	1	0	0	0	0	4	-	0	6	5	1	0	6
HP-P	0	8	0	6	1	0	3	6	3	15	11	-	15	14	10	0	15
CS-P	0	12	3	5	1	0	9	9	9	10	9	7	-	4	7	0	0
CC	1	12	1	7	0	0	9	9	9	9	9	7	5	-	4	1	5
BC	1	8	0	7	0	0	5	5	5	9	5	3	8	4	-	1	8
HP	0	17	4	8	1	0	11	14	11	24	20	9	17	17	17	-	17
CS	0	12	3	5	1	0	9	9	9	10	9	7	0	4	7	0	-

(a) (ABS) vs. (HP-P)   (b) (Dom/Wdeg) vs. (HP-P)

Fig. 3. Comparing the sizes of the MDD representations.

equipped with the heuristics corresponding to the x-axis (resp. y-axis) is given by its x-coordinate (resp. y-coordinate). Logarithmic scales are used for both coordinates. These empirical results clearly illustrates the value of the notion of decomposition in the compilation process from the practical side.

Again, due to space limitations, we cannot provide a differential analysis of the heuristics performances depending on the family of benchmarks. However, the family chosen has a clear impact on the performances. For instance, when MDDG is targeted, (HP-P) proved better than (Dom/Wdeg) for compiling Bayesian networks and (Dom/Wdeg) proved better than (HP-P) for instances of the frequency allocation problem (FAP) which are difficult to decompose.

5   Conclusion

In this work, we have evaluated several branching heuristics for the top-down compilation of constraint networks into the MDD language and into the MDDG language, through the use of the cn2mddg compiler. Our evaluation on a large dataset demonstrated that the decomposability of the constraint graph allows to compile many more instances and offers much smaller compiled representations. In particular, when compiling networks into MDD, the (Dom/Wdeg) heuristics proved to be the best, while for MDDG representations, a new heuristics (HP-P) based on dynamic, yet parsimonious hypergraph partitioning was the best performer.

References

1. Amilhastre, J., Fargier, H., Marquis, P.: Consistency restoration and explanations in dynamic CSPs application to configuration. Artif. Intell. **135**(1–2), 199–234 (2002)
2. Amilhastre, J., Fargier, H., Niveau, A., Pralet, C.: Compiling CSPs: a complexity map of (non-deterministic) multivalued decision diagrams. Int. J. Artif. Intell. Tools **23**(4) (2014)
3. Bart, A., Koriche, F., Lagniez, J.M., Marquis, P.: An improved CNF encoding scheme for probabilistic inference. In: Proceedings of ECAI 2016, pp. 613–621 (2016)
4. Bavelas, A.: Communication patterns in task-oriented groups. J. Acoust. Soc. Am. **22**(6), 725–730 (1950)
5. Brandes, U.: A faster algorithm for betweenness centrality. J. Math. Soc. **25**(2), 163–177 (2001)
6. Brandes, U.: On variants of shortest-path betweenness centrality and their generic computation. Soc. Netw. **30**(2), 136–145 (2008)
7. Catalyürek, U., Aykanat, C.: PaToH (Partitioning Tool for Hypergraphs), pp. 1479–1487. Encyclopedia of Parallel Computing (2011)
8. Darwiche, A.: Decomposable negation normal form. J. ACM **48**(4), 608–647 (2001)
9. Darwiche, A.: New advances in compiling CNF into decomposable negation normal form. In: Proceedings of ECAI 2004, pp. 328–332 (2004)
10. Darwiche, A., Hopkins, M.: Using recursive decomposition to construct elimination orders, jointrees, and dtrees. In: Benferhat, S., Besnard, P. (eds.) ECSQARU 2001. LNCS (LNAI), vol. 2143, pp. 180–191. Springer, Heidelberg (2001). doi:10.1007/3-540-44652-4_17
11. Edmonds, J., Karp, R.M.: Theoretical improvements in algorithmic efficiency for network flow problems. J. ACM **19**(2), 248–264 (1972). http://doi.acm.org/10.1145/321694.321699
12. Fargier, H., Marquis, P.: On the use of partially ordered decision graphs in knowledge compilation and quantified Boolean formulae. In: Proceedings of AAAI 2006, pp. 42–47 (2006)
13. Gergov, J., Meinel, C.: Efficient analysis and manipulation of OBDDs can be extended to FBDDs. IEEE Trans. Comput. **43**(10), 1197–1209 (1994)
14. Girvan, M., Newman, M.E.J.: Community structure in social and biological networks. Proc. Natl. Acad. Sci. **99**(12), 7821–7826 (2002)

15. Hemery, F., Lecoutre, C., Sais, L.: Boosting systematic search by weighting constraints. In: Proceedings of ECAI 2004, pp. 146–150 (2004)
16. Huang, J., Darwiche, A.: The language of search. J. Artif. Intell. Res. **29**, 191–219 (2007)
17. Koriche, F., Lagniez, J.M., Marquis, P., Thomas, S.: Compiling constraint networks into multivalued decomposable decision graphs. In: Proceedings of IJCAI 2015, pp. 332–338 (2015)
18. Lecoutre, C., Sais, L., Tabary, S., Vidal, V.: Reasoning from last conflict(s) in constraint programming. Artif. Intell. **173**(18), 1592–1614 (2009)
19. Marinescu, R., Dechter, R.: Dynamic orderings for AND/OR branch-and-bound search in graphical models. In: Proceedings of ECAI 2006, pp. 138–142 (2006)
20. Michel, L., Hentenryck, P.: Activity-based search for black-box constraint programming solvers. In: Beldiceanu, N., Jussien, N., Pinson, É. (eds.) CPAIOR 2012. LNCS, vol. 7298, pp. 228–243. Springer, Heidelberg (2012). doi:10.1007/978-3-642-29828-8_15
21. Narodytska, N., Walsh, T.: Constraint and variable ordering heuristics for compiling configuration problems. In: Proceedings of IJCAI 2007, pp. 149–154 (2007)
22. Oztok, U., Darwiche, A.: On compiling CNF into decision-DNNF. In: O'Sullivan, B. (ed.) CP 2014. LNCS, vol. 8656, pp. 42–57. Springer, Cham (2014). doi:10.1007/978-3-319-10428-7_7
23. Refalo, P.: Impact-based search strategies for constraint programming. In: Wallace, M. (ed.) CP 2004. LNCS, vol. 3258, pp. 557–571. Springer, Heidelberg (2004). doi:10.1007/978-3-540-30201-8_41
24. Sang, T., Beame, P., Kautz, H.A.: Performing Bayesian inference by weighted model counting. In: Proceedings of AAAI 2005, pp. 475–482 (2005)
25. Stoer, M., Wagner, F.: A simple min-cut algorithm. J. ACM **44**(4), 585–591 (1997)

A Tolerant Algebraic Side-Channel Attack on AES Using CP

Fanghui Liu, Waldemar Cruz, Chujiao Ma, Greg Johnson,
and Laurent Michel[✉]

Computer Science and Engineering Department, School of Engineering,
University of Connecticut, Storrs, CT 06269-4155, USA
{fanghui.liu,waldemar.cruz,chujiao.ma,greg.johnson,
laurent.michel}@uconn.edu

Abstract. AES is a mainstream block cipher used in many protocols and whose resilience against attack is essential for cybersecurity. In [14], Oren and Wool discuss a Tolerant Algebraic Side-Channel Analysis (TASCA) and show how to use optimization technology to exploit side-channel information and mount a computational attack against AES. This paper revisits the results and posits that Constraint Programming is a strong contender and a potent optimization solution. It extends bit-vector solving as introduced in [8], develops a CP and an IP model and compares them with the original Pseudo-Boolean formulation. The empirical results establish that CP can deliver solutions with orders of magnitude improvement in both run time and memory usage, traits that are essential to potential adoption by cryptographers.

Keywords: Algebraic Side-Channel Attack · AES · Cryptography · Block cipher · Constraint programming · Optimization

1 Introduction

Tolerant Algebraic Side-Channel Analysis (TASCA) attack is a combination of algebraic and side-channel attacks with possible errors taken into account. A cryptographic algorithm takes a plaintext and a secret cipher key as input. Through rounds of permutations and combinations, it outputs the encrypted message also known as ciphertext. Algebraic cryptanalysis attacks first model the target algorithm as a system of equations then solve for the secret key. Since solving such system of equations is generally prohibitively expensive computationally, side-channel information is added [17]. In side-channel analysis (SCA), data leaked during the encryption or decryption of the algorithm, such as power consumption, are related to the intermediate values internal to the algorithm. With such knowledge and given either plaintext or ciphertext, the attacker can utilize a divide-and-conquer strategy and use statistical analysis to recover bytes of the key [6]. However, such attack is dependent on the accuracy of the data gathered. To reduce the sensitivity of SCA to noise from the measurement or

© Springer International Publishing AG 2017
J.C. Beck (Ed.): CP 2017, LNCS 10416, pp. 189–205, 2017.
DOI: 10.1007/978-3-319-66158-2_13

decoding errors, a large number of samples may be needed. By combining algebraic cryptanalysis and SCA, fewer samples are needed and the attack can also succeed in unknown plaintext/ciphertext scenarios [16].

The success of the standard "Algebraic Side-channel Attacks" (ASCA), which include no attempt at being fault-tolerant regarding power trace measurements, depends on the accuracy of the side-channel data, i.e., the Hamming weights. Modification to ASCA to account for noise/error tolerance were introduced in [10]. One approach, introduced in [17], represents the Hamming weights as equations that accept any value from a set of several possible values instead of just a single possible value and relies on SAT solvers such as CryptoMiniSAT [18]. When multiple solutions are possible, the non-optimizing SAT solver arbitrarily chooses a solution and terminates, thus the success of the attack is still dependent on the error rate rather than the set size. For TASCA as presented by Oren and Wool [13], an optimizing solver with a goal function to minimize the amount of modeled noise is used. This transforms the problem from a satisfiability problem to an optimization problem and drastically increases the probability of successfully recovering the key. The key was recovered with error rate up to 20% using SCIP with a Pseudo-Boolean formulation. The error rate refers to the probability that the Hamming weight of a leak is incorrect. The key is counted as recovered or correct if four or less bytes of the sixteen byte key are wrong.

However, the solving time of the optimizer was inferior to that of the SAT solver and grew in proportion to the set size [14]. The optimizer used to perform TASCA required significant memory to represent the entire AES encryption in equation form [12]. The purpose of this paper is to revisit the TASCA attack with a Constraint Programming (CP) model over bit-vectors and explore its capabilities relative to IP solvers reported in [13]. CP was used in [5], to model a differential crypt-analysis attack which differs significantly from attacks based on side-channel data and relies on a multi-stage relaxation technique. CP was also used in [15] to design better S-boxes.

Section 2 of the paper introduces the modeling of AES and side-channel information. Section 3 presents TASCA and an IP approach. Section 4 introduces CP aspects. Section 5 discusses the CP implementation. Section 6 compares the IP and CP approaches. Section 7 concludes the paper.

2   Modeling AES and Side-Channel Information

AES is a symmetric block cipher that supports block lengths of 128, 192 and 256 bits. The version used in this paper is AES-128 and consists of 10 rounds. The 128-bit cipher key is first expanded into 11 round keys via key expansion [4]. During the encryption, the plaintext is separated into blocks of 16 bytes, where each block is represented by bytes $p_0..p_{15}$ in Fig. 1. The block is arranged as a 4×4 matrix and combined with an initial round key. It then goes through 9 rounds consisting of four subrounds, then one last round without MixColumns to produce the ciphertext c_0 to c_{15}. The subrounds in Fig. 2 are:

- **SubBytes:** substitute each byte of the state using an S-box.
- **ShiftRows:** shift each of the four rows 0, 1, 2 and 3 bytes to the left.
- **MixColumns:** multiply each column with a matrix of constants (MC).
- **AddRoundKey:** XOR the state with the round key from the key expansion.

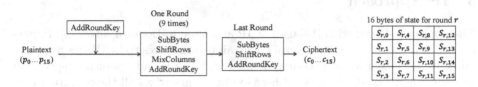

Fig. 1. AES flowchart.

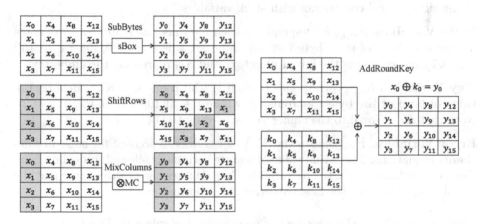

Fig. 2. AES subround operations, with x being the input state.

For more details on the structure of AES, please refer to [4]. This analysis considers a simulated implementation of the AES-128 as device under test (DUT). It assumes that the key expansion is done in advance and no leaks from the process are available. This corresponds to a more challenging scenario since it was shown in [7] that side-channel leakages from an 8-bit microcontroller implementation of AES during key expansion are sufficient to recover the complete key. Numerous research has shown that side-channel leakages occur regardless of whether the implementation of AES is for 8-bit microcontroller [16], 32-bit CPU [1], or an FPGA [19], hence we do not focus on the hardware aspect.

The purpose of this paper is to investigate the potential that constraint programming offers. For the sake of comparison, the attack model is the same as that of [13]. Namely, we assume that the attacker has access to a DUT (device under test) that emits a measurable power consumption trace, during encryption, from which the attacker is able to learn about the internal state of the DUT during the cryptographic operations and quantified as Hamming weights. The TASCA attack examines one round of AES only. The plaintext, AES model, and the Hamming weights with error taken into account are modeled as sets of constraints. The constraints are given to the solver, which solves for the key.

3 IP Approach

The approach described in [14] uses SCIP as a Pseudo-Boolean solver [2]. In particular, it does not linearize several constraints and leaves to SCIP the responsibility to produce a linear encoding or deal with the Pseudo-Boolean aspects on its own. The IP model considered here explicitly linearizes all the constructions.

Decision Variables. The encoding of the AES algorithm uses three types of variables to represent the state of the algorithm, the round keys and a relaxation of the side-channel constraints with slack variables.

State Variables. $S_{sr,i,j}$ corresponds to each intermediate state. $S_{sr,i,j}$ denotes the value of bit j of state byte i at subround sr, where $sr \in [0, 40]$, $i \in [0, 15]$, $j \in [0, 7]$. S_0 represents the initial plaintext and S_{40} represents the ciphertext.

Key Variables. $K_{r,i,j}$ corresponds to each 128-bit round key. $K_{r,i,j}$ denotes the value of bit j of key byte i at round r, where $r \in [0, 10]$, $i \in [0, 15]$, $j \in [0, 7]$. Note that K_0 is equal to the cipher key according to [4].

Error Variables. $E_{sr,i}$ relaxes the SCA constraints to account for noise in side-channel equations. The actual value of a state variable is allowed to deviate from the measured value by ± 1. The variable $E_{sr,0}$ denotes the positive error and the variable $E_{sr,1}$ denotes the negative error for subround sr.

Product Encodings. AES uses a set of non-linear algebraic equations that contain multiplication and exclusive OR. Encoding an exclusive OR $O = A \oplus B$ where O, $A, B \in \{0, 1\}$, is done with

$$A + B - 2AB - O = 0$$

which is linearized with the elimination of the product AB in favor of an auxiliary Boolean variable P and the inequalities $A \geq P$ (1), $B \geq P$ (2) and $A + B - P \leq 1$ (3). Indeed, $P = 1$ forces both A and B to be 1 whereas $A = B = 1$ forces $P = 1$ through (3). If only one of A or B is 1, by either (1) or (2), then $P = 0$. This can be generalized to a product of Boolean variables $\prod_{i=1}^{n} b_i$ with the set of inequalities

$$\forall i \in 1..n : b_i \geq P$$
$$\sum_{i=1}^{n} b_i - P \leq n - 1$$

This approach is also reported in [2].

XTime. AES implementations use a byte-level operation to further *mix* bits within a single byte. $\mathtt{xtime} : \{0,1\}^8 \rightarrow \{0,1\}^8$ is a function transforming an 8-bit input sequence x into an 8-bit output sequence y, i.e., $y = \mathtt{xtime}(x)$ specified as:

$$y = \begin{cases} (x << 1) & \Leftrightarrow x_7 = 0 \\ (x << 1) \oplus \mathtt{0x1b} & \Leftrightarrow x_7 = 1 \end{cases}$$

that applies a left shift to the 8-bits and subsequent conditional bitwise XOR with value $\mathtt{0x1b}$ if the most significant bit is 1. The operation is described in Sect. 4.2.1 of the FIPS specification of AES [11].

AES Constraints. Once the basic operations are linearized, we can then model the subrounds of AES as well as the side-channel constraints and error variables.

- **AddRoundKey** is a straightforward XOR operation. It takes in a state $S_{sr,i,j}$ and a round key $K_{r,i,j}$, then performs an XOR operation to translate $S_{sr,i,j}$ to $S_{sr+1,i,j}$.
- **SubBytes** is a non-linear byte-wise substitution. The mapping of the Sub-Bytes permutation $\pi : \{0,1\}^8 \Rightarrow \{0,1\}^8$ is defined by a look-up table $S_{r+1,i} = \pi[S_{r,i}]$, where the permutation π maps a group of 8 bits to another group of 8 bits. This operation transforms a state variable $S_{sr,i,[0..7]}$ to $S_{sr+1,i,[0..7]}$. The permutation π is represented by the following truth table:

x7	x6	x5	x4	x3	x2	x1	x0	y7	y6	y5	y4	y3	y2	y1	y0
0	0	0	0	0	0	0	0	1	0	0	1	1	1	0	0
0	0	0	0	0	0	0	1	0	1	1	1	1	1	0	0
.	.	.	.	.	.	.	.	.	.	.	.	.	.	.	.
1	1	1	1	1	1	1	1	1	1	1	0	1	0	0	1

The constraint $y = \pi[\mathcal{I}(x)]$ (where $\mathcal{I}$ is a function that gives the integer interpretation of its bit-vector argument) states that the value of y is the xth entry of the look-up table for π. For instance, row 0 of the table maps the bit-string 00000000 to the bit-string 01100011. Namely, when the bits of x are equal to 00000000, represented by the conjunction of literals $\bar{x}_7\bar{x}_6\bar{x}_5\bar{x}_4\bar{x}_3\bar{x}_2\bar{x}_1\bar{x}_0$, then the bits of y must take the value 10011100 implying that the conjunction of literals $\bar{y}_7 y_6 y_5 \bar{y}_4 \bar{y}_3 \bar{y}_2 y_1 y_0$ must also be true. Therefore, the non-linear Boolean equation

$$\bar{x}_7\bar{x}_6\bar{x}_5\bar{x}_4\bar{x}_3\bar{x}_2\bar{x}_1\bar{x}_0 - \bar{y}_7 y_6 y_5 \bar{y}_4 \bar{y}_3 \bar{y}_2 y_1 y_0 = 0$$

encodes the first row of the π lookup table and 256 such equations capture the entire table.

Naturally, each n-ary product is linearized with the encoding from the previous sub-section. Overall, this requires $9 \cdot 2 \cdot 256$ linear inequalities and equations for each occurrence of SubBytes in AES.

- **ShiftRows** is a logical circular shift on the state variables. Because there are no changes in the values, it does not leak any side-channel information. Therefore, ShiftRows is combined with MixColumns. To combine them, the state variables are shifted based on the rules for ShiftRows and then passed to MixColumns. For example, the ShiftRows operation for state 2 rearranges the entries in each row of the 4×4 state matrix S_2 as follows

$$[S_{2,0}, S_{2,5}, S_{2,10}, S_{2,15}] = ShiftRows([S_{2,0}, S_{2,1}, S_{2,2}, S_{2,3}])$$
$$[S_{2,4}, S_{2,9}, S_{2,14}, S_{2,3}] = ShiftRows([S_{2,4}, S_{2,5}, S_{2,6}, S_{2,7}])$$
$$[S_{2,8}, S_{2,13}, S_{2,2}, S_{2,7}] = ShiftRows([S_{2,8}, S_{2,9}, S_{2,10}, S_{2,11}])$$
$$[S_{2,12}, S_{2,1}, S_{2,6}, S_{2,11}] = ShiftRows([S_{2,12}, S_{2,13}, S_{2,14}, S_{2,15}])$$

- **MixColumns** is a more complex operation that applies to a column of the state matrix at a time. At a high-level it can be represented directly with:

$$[S_{4,0}, S_{4,1}, S_{4,2}, S_{4,3}] = MixColumns([S_{2,0}, S_{2,5}, S_{2,10}, S_{2,15}])$$
$$[S_{4,4}, S_{4,5}, S_{4,6}, S_{4,7}] = MixColumns([S_{2,4}, S_{2,9}, S_{2,14}, S_{2,3}])$$
$$[S_{4,8}, S_{4,9}, S_{4,10}, S_{4,11}] = MixColumns([S_{2,8}, S_{2,13}, S_{2,2}, S_{2,7}])$$
$$[S_{4,12}, S_{4,13}, S_{4,14}, S_{4,15}] = MixColumns([S_{2,12}, S_{2,1}, S_{2,6}, S_{2,11}])$$

The 32-bit operation is repeated 4 times, once for every column. For an 8-bit processor, the transformation has an efficient implementation using 8-bit words [14], which is used by the IP model. The following shows the four equations for one output column $[o_0, o_1, o_2, o_3]$ based on an input column $[a_0, a_1, a_2, a_3]$:

$$o_k = \left(\bigoplus_{i=0}^{3} a_i \right) \oplus \texttt{xtime}(a_k \oplus a_{(k+1) \bmod 4}) \oplus a_k \ \forall k \in 0..3$$

in which $\texttt{xtime}$ is the byte level operation described earlier. Suppose x is an 8-bit input sequence and y is an 8-bit output sequence, to linearize the xtime operation, the following encoding is applied

$$y_{i+1} = x_i \oplus x_7 \ \forall i \in \{0, 2, 3\}$$
$$y_{(i+1) \bmod 8} = x_i \ \forall i \in \{1, 4, 5, 6, 7\}$$

- **Key Expansion** is an invertible key derivation function that maps a given cipher key to a series of round keys. The key expansion derives the next round key by applying a series of XOR operations to the current round key with round constants RC and a series of SubBytes substitutions. The following is an example of the derivation of the second round key:

$$K_{1,0} = SubBytes(K_{0,13}) \oplus K_{0,0} \oplus RC_0 \tag{1}$$
$$K_{1,1} = SubBytes(K_{0,14}) \oplus K_{0,1} \tag{2}$$
$$K_{1,2} = SubBytes(K_{0,15}) \oplus K_{0,2} \tag{3}$$
$$K_{1,3} = SubBytes(K_{0,12}) \oplus K_{0,3} \tag{4}$$
$$\forall i \in \{0, ..., 11\} \ K_{1,i+4} = K_{1,i} \oplus K_{0,i+4} \tag{5}$$

– **Side-Channel Constraints** model the Hamming weights (number of bits with value of 1 for each byte) of state. To handle measurement errors, each equation uses slack variables. Namely, let $M_{sr,i}$ denote the Hamming weight for state $S_{sr,i}$. A *tolerant* side-channel constraint for byte i of subround sr becomes $\sum_{j=0}^{7} S_{sr,i,j} + E_{sr,i}^{+} - E_{sr,i}^{-} = M_{sr,i}$ where the slacks $E_{sr,i}^{+}$ and $E_{sr,i}^{-}$ relax the Hamming weight requirement from the power trace.
– **Objective Function** minimizes the total measurement of error and is written as

$$\text{min} : \sum E_{sr,i}^{+} + \sum E_{sr,i}^{-}$$

4   Constraint Programming

A Constraint Satisfaction Problem (CSP) is a tuple $\langle X, D, C \rangle$ where X is a finite set of variables, C is a set of constraints and D is a set of domains for the variables, i.e., $\forall x \in X$, $D(x)$ denotes the domain of variable x.

4.1   Bit-Vector Domains

A bit-vector $b_{[k]}$ is a sequence of $k \geq 1$ bits and b_i ($i \in 0..k-1$) is the ith bit. For instance, 100 is a bit-vector of length 3 with its least significant bit (b_0) equal to 0 and most significant bit (b_2) equal to 1. This paper assumes that bit-vectors are of length 8 and we slightly abuse notation using b to refer to $b_{[8]}$. Bit-vectors are denoted by the letters b, l, and u. The narrative adopts the bit-vectors from [8] in which the domain is a pair $\langle l, u \rangle$ of bit-sequences such that $l_i \leq u_i$ ($0 \leq i < k$). The bit-vector domain represents the set of bit-sequences

$$\{b \mid l \leq b \leq u \land \forall i \in 0..k-1 : l_i = u_i \Rightarrow b_i = l_i\}$$

The OBJECTIVECP solver offers an implementation for bit-vectors of arbitrary length. Emerging implementations have been introduced in the literature including one built on top of MiniSat [21] as well as another CP implementation [3].

4.2   Bit-Vector Constraints

The bit-vector library in [8] supports constraints for arithmetic, relational, logical and structural bit-vector manipulations. A significant number of propagators run in $\mathcal{O}(1)$ time. While extensive, the library misses two bit-vector constraints to neatly express the TASCA model. Those are a generalization of the element constraint over bit-vector arrays and the count constraint to determine the Hamming weight of a bit-vector. Both extensions are described next.

Element Constraint. The element constraint for finite-domains solvers appeared first in [20]. The constraint is defined as element(x,t,y) and states that $y = t[x]$, namely, y is constrained to be the xth entry of table t. This extension considers the constraint $y_{[k]} = t_{[k]}[x_{[n]}]$ where $x_{[n]}$ is an n-bit long bit-vector representing an unsigned integer between 0 and $2^n - 1$, $t_{[k]}$ is a 0-based table of bit-vectors of length k containing at most $2^n - 1$ entries and $y_{[k]}$ is a bit-vector of length k. Given the natural interpretation function $\mathcal{I}(x) = \sum_{i=0}^{n-1} x_i \cdot 2^i$, the semantics of the constraint is simply $y_{[k]} = t_{[k]}[\mathcal{I}(x_{[n]})]$.

The propagation algorithm uses an auxiliary function $IC(p, y)$ which, given a bit-sequence from $D(x)$, computes the set of incompatible bits. Its definition is

$$IC(p, y) = \{i \in 0..k-1 | y.l_i = y.u_i\} \cap \{i \in 0..k-1 | t[\mathcal{I}(p)]_i < y.l_i \vee y.u_i < t[\mathcal{I}(p)]_i\}$$

Namely, it is the set of bits fixed in y and disagreeing with the pth entry of table t. With IC, one can define the subset of $D(x)$ of *compatible* values with y, i.e., $I = \{p \in D(x) | IC(p, y) \neq \emptyset\}$. Four arrays can maintain the number of supports for 0 and 1 in x and y and are defined as

$$\forall v \in \{0,1\}, \ \forall i \in 0..k-1 : s_i^y(v) = |\{p \in I \mid t[\mathcal{I}(p)]_i = v\}|$$
$$\forall v \in \{0,1\}, \ \forall i \in 0..n-1 : s_i^x(v) = |\{p \in I \mid p_i = v\}|$$

The propagation algorithm triggered whenever $D(x)$ or $D(y)$ changes is:

---

procedure Propagate(y, t, x)   ▷ When $D(x)/D(y)$ changes
   for all $p \in I$ **do**
      if $IC(p, y) \neq \emptyset$ **then**   ▷ the index bit-sequence is incompatible.
         $I \leftarrow I \setminus \{p\}$
         if $|I| = 1$ **then** $D(x) \leftarrow k$, $y \leftarrow t[k]$ with $k \in I$
         if $|I| = 0$ **then** fail()
         update($p, t[p], s^y, s^x$)   ▷ Supports are decreased.
   for all $i \in 0..n-1 \mid x.l_i < x.u_i$ **do**   ▷ Consider free bits in x.
      if $s_i^x(0) = 0$ **then** $x.l_i \leftarrow 1$
      if $s_i^x(1) = 0$ **then** $x.u_i \leftarrow 0$
   for all $i \in 0..k-1 \mid y.l_i < y.u_i$ **do**   ▷ Consider free bits in y.
      if $s_i^y(0) = 0$ **then** $y.l_i \leftarrow 1$
      if $s_i^y(1) = 0$ **then** $y.u_i \leftarrow 0$

The algorithm runs in $\mathcal{O}(|D(x)| \cdot \max(n, k))$ as it must update the supports for each incompatible bit-sequence in I. Both s^y and s^x are reversible.

Bit-Vector Count Constraint. The count(b,x) constraint is a special case of the **cardinality** global constraint for integer variables. It is defined on one bit-vector variable b with $D(b) = \langle l, u \rangle$ and one integer variable x with a finite domain $D(x) = [L..U]$. It enforces $x = \sum_{i=0}^{n-1}(b_i = 1)$. Note that $M = \sum_{i=0}^{n-1} l_i$ is the number of mandatory bits (already set at 1) while $P = \sum_{i=0}^{n-1}(l \oplus u)_i$ is the number of possible bits (free bits that could go to 1). Both quantities can be computed in $\mathcal{O}(1)$ with the assembly-level popcount instruction. Whenever b changes or the bounds of x change, one can run the following constant time algorithm to enforce bit consistency.

---

procedure PROPAGATE($x = [L..U], b = \langle l, u \rangle$)   ▷ When $D(x)/D(b)$ changes
$\quad M = \sum_{i=0}^{n-1} l_i, \, P = \sum_{i=0}^{n-1} (l \oplus u)_i$ ▷ Runs in $\mathcal{O}(1)$
$\quad$ **if** $L > M + P \lor U < M$ **then** fail()
$\quad L \leftarrow \max(L, M), U \leftarrow \min(U, M + P)$
$\quad$ **if** $L = M + P$ **then** $D(x) \leftarrow \{L\}$, $D(b) \leftarrow \langle u, u \rangle$
$\quad$ **else if** $U = M$ **then** $D(x) \leftarrow \{U\}$, $D(b) \leftarrow \langle l, l \rangle$
$\quad$ **else**
$\quad\quad D(x) \leftarrow [L..U]$

---

5 CP Approach

The IP section showed that it is possible to produce a linear encoding of AES
with side-channel constraints. However, the encoding is cumbersome due to bit-
blasting and the heavy cost of modeling non-linear constructions such as the
S-Box. By using bit-vectors with a constraint programming model, a more direct
and natural formulation becomes possible. Bit-vectors [8] are the corner stone
of this formulation.

Variables/Bit-Vector. Each state variable is represented by 16 8-bit wide bit-
vectors. In the IP model, a byte would be represented by 8 Boolean variables
$S_{sr,i,j} \in \{0, 1\}$. In the CP model, it is represented by a single bit-vector $S_{sr,i} \in \{0, 1\}^8$. All the AES subrounds transformations operate over 8-bit values.

Constraints. The CP model is particularly attractive as the formulation does
not require any cumbersome encoding and is now lightweight direct expression
of the AES transformations.

- **AddRoundKey** is implemented as a bitwise XOR operation between two
  8-bit bit-vector variables. In the IP model, the linearization of an XOR operation requires several constraints. In the CP model, the XOR operation only
  requires one constraint per byte: $S_{sr,i} \oplus K_{r,i} = S_{sr+1,i}, \forall i = \{0, \cdots, 15\}$.
- **SubBytes** is implemented as an *element constraint* over bit-vectors. The
  element constraint $z = c[\mathcal{I}(x)]$ takes in an array c of (constant) bit-vectors
  and requires z to be equal to the $\mathcal{I}(x)^{th}$ entry of the array c. An array of 256
  constant bit-vectors model the full substitution box.
- **ShiftRows & MixColumns** are combined just like for the IP model. As
  before, the 8-bit efficient `MixColumns` implementation is used in the CP encoding. Bit-vector constraints are used to capture XOR as well as `xtime` resulting
  in the same non-linear Boolean constraints

$$\beta_k = \texttt{xtime}(a_k \oplus a_{(k+1) \bmod 4}) \qquad\qquad \forall\, k \in 0..3$$

$$o_k = \left(\bigoplus_{i=0}^{3} a_i \right) \oplus \beta_k \oplus a_k \qquad\qquad \forall\, k \in 0..3$$

meant to connect the input bytes of `MixColumns` to its output bytes. However, *it is no longer necessary to linearize these equations as the bit-vector solver natively supports all the operators.*

- **Key Expansion** It uses the Eqs. (1)–(5) but sidesteps any linear encoding in favor of a direct native expression based on bit-vectors and using bit-vector elements (for the S-Boxes) and XOR constraints.
- **Side-Channel Constraints** are created using the `count(b,x)` constraint on bit-vectors. The formulation is linear and therefore identical to the one used in the IP model:

$$count(S_{sr,i}) + E^+_{sr,i} - E^-_{sr,i} = M_{sr,i}$$

  The bit-vector constraint `count(b, c)` requires c to be the Hamming weight of the bit-vector b. Namely, $c = |\{k \in 0 \cdots n | b_{|k} = 1\}|$.
- **Objective Function** is the total number of errors, same as in the IP model:

$$\min : \sum E^+_{sr,i} + \sum E^-_{sr,i}$$

Search. The search is a major component in the optimized TASCA and focuses on the side-channel information. For each of the 16 bytes of the state variables appearing in each of the 40 stages, a Hamming weight is generated as the side-channel information. A candidate value $v \in D(S_{r,i})$ has a *Hamming weight* $H(v) = \sum_{b \in 0..7}(v_{|b} = 1)$ capturing the number of bits at 1 in v. TASCA imposes that $D(S_{r,i})$ be restricted to values v for which

$$-1 \leq H(v) - M_{r,i} \leq 1$$

i.e., the discrepancy between the measurement $M_{r,i}$ and the value v does not exceed ± 1. The CP model minimizes an objective that captures the total number of deviations (errors) that may be observed for the Hamming weights throughout the 40 replications of AES's state. Branching on value assignments that yield the least amount of errors in the objective will be most effective to get high-quality solutions early on. With the objective equal to the sum of the errors, we have

$$\min \sum_{j=0}^{41} E(j, \sigma(S_j)) \text{ where } E(j, v) = \begin{cases} 0 & \text{if } H(v) = M_j \\ 1 & \text{if } H(v) = M_j \pm 1 \end{cases}$$

and σ is the current value assignment. It is tempting to guide the search process with a custom procedure that takes advantage of the objective function and the semantics of AES. Such a procedure would first consider assignments that have the least impact on the objective function, and therefore on the errors. Unsurprisingly, this implies that it may be advantageous to branch not on individual bits, but instead on entire bit-vectors to quickly get good bounds on the number of errors induced by an assignment.

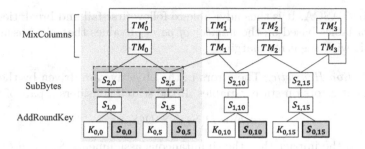

Fig. 3. Circuit for bytes $\{0, 1, 2, 3\}$.

A Circuit View. Consider a single round of the AES algorithm containing all 4 subrounds, AddRoundKey, SubBytes, and MixColumns/ShiftRows. Figure 3 illustrates the subrounds for the first column of the state matrix. The structure is repeated four times for the entire state. The column $[S_{0,0}S_{0,5}S_{0,10}S_{0,15}]^T$ of the state and the column $[K_{0,0}K_{0,5}K_{0,10}K_{0,15}]^T$ of the sub-round key form the inputs at the bottom of the Figure. In round 1, $[S_{0,0}S_{0,5}S_{0,10}S_{0,15}]^T$ are known since they represent bytes of the plaintext. AddRoundKey and SubBytes apply bijective transformations once the state is known. Consequently, as soon as $S_{2,0}, S_{1,0}$ or $K_{0,0}$ is fixed, the others are fixed by propagation (This is also true for $S_{2,5}, S_{1,5}$ and $K_{0,5}$). The dashed box on the far left that contains $S_{2,0}$ and $S_{2,5}$ highlights the inputs to an exclusive OR that yields the temporary value TM_0. The output of the circuit TM_0' is the first byte of the output, i.e., state $S_{4,0}$. The vertical light-gray column on the far left is a description of the relations defining byte 0 of the output. The evaluation of TM_0 rests upon the availability of values for both $S_{2,0}$ and $S_{2,5}$. Observe that if, for instance, $S_{2,5}$ is fixed, one only needs to fix $S_{2,10}$ to get propagation up and down and fix both $K_{0,10}$ and TM_1'.

Variable Selection Heuristic. Branching on TM_0 will only trigger propagation "up" as the exclusive OR will not be able to push information "down". Similarly, branching on $S_{2,0}$ or even $K_{0,0}$ will have a limited propagation given the bijective nature of the relations in the bottom "legs". However, simultaneously assigning both variables in the dashed box will trigger propagation up *and* down.

This is a key insight into the variable selection strategy. The search should branch on *pairs of variables* that trigger propagation within an entire gray box. In Fig. 3 this implies four columns for a total of 16 pairs of variables to consider for the first branching decision (recall that this structure is replicated 4 times). Note how the four columns are topped by TM_0', TM_1', TM_2' and TM_3'. Once a pair is selected, the search should create another pair by reusing one of the two variables from the first pair. For instance, if $\langle S_{2,0}, S_{2,5}\rangle$ was selected, it is tempting to consider the two pairs $\langle S_{2,5}, S_{2,10}\rangle$ and $\langle S_{2,15}, S_{2,0}\rangle$ as the domains of TM_1 and TM_3 were reduced by the first choice.

Finally, not all pairs of values drawn from the domains of $S_{2,0}$ and $S_{2,5}$ are compatible. Some of these pairs may induce errors that exceed the ± 1 margin dictated by TASCA. It is thus advisable to follow first-fail and break ties among pairs of variables based on the *number of pairs* of values that yield assignments compatible with the error margins.

Value Selection Heuristic. The errors in the objective are driven by the sum of measurement errors on state variables. If the search considers a pair of values

$$\langle a, b \rangle \in D(S_{2,0}) \times D(S_{2,5})$$

it can assess the impact that the simultaneous assignments $S_{2,0} = a \wedge S_{2,5} = b$ would have on the errors at the state variables in the leftmost gray column. This assessment is an under-approximation of the true error induced by the assignments. Indeed TM_1' can expose errors caused by the choice of value b for $S_{2,5}$, but that falls outside the gray column and is therefore ignored. A sensible value selection heuristic considers pairs of values and assess their quality with a scoring function C. Given a pair of values $\langle a, b \rangle$, the scoring is

$$C(\langle a, b \rangle) = C_{leg}(a, [S_{2,0}, S_{1,0}, K_{0,0}]) + C_{leg}(b, [S_{2,5}, S_{1,5}, K_{0,5}]) + C_{mc}(a \oplus b, [TM_0'])$$

The functions C_{leg} and C_{mc} model the errors attributable to a leg in the gray box, or the top of a gray box (the MixColumns operation) and $a \oplus b$ denotes the value inferred for TM_0 based on the connecting XOR constraint. C_{leg} and C_{mc} measure the differences between the value of the state variable and the expected Hamming weight. Given a pair of variables and the scoring function C, the value heuristic enumerate pairs that contribute the least to the objective function.

Optimality Pruning. Since the total contribution of errors due to Hamming weights accumulate as the search dives deeper, the total error can be used to further prune value-pairs whose contribution would bring the total beyond the total error for the incumbent solution.

Procedure. It is helpful to define *VarPairs* as the set of pairs of variables that the search will be branching on. Those correspond to the variables in the dashed boxes in Fig. 3 and are consecutive variables in the state matrix that `MixColumns` operates on. Additionally, let $Col(\langle X, Y \rangle)$ denote the state variables connected to X and Y in that same column (vertical gray highlight). Given a pair of variable $\langle X, Y \rangle$ it is possible to define the domain

$$D(\langle X, Y \rangle) = \left\{ \langle a, b \rangle \in D(X) \times D(Y) : -1 \leq \max_{z \in Col(\langle X, Y \rangle)} C(a, b) \leq 1 \right\}$$

as the set of value pairs from $D(X) \times D(Y)$ with error levels compatible with the Hamming weights. Let $B(X)$ return 0 or 1 based on whether X is bound.

6   Experimental Setup

The CP solver used in the experiments is OBJECTIVE-CP [9] which combines modeling and search including user-defined search. The IP approach relies on

---

1: $VarPairs \leftarrow \{\langle S_{2,0}, S_{2,5}\rangle, \langle S_{2,5}, S_{2,10}\rangle, \cdots\}$ ▷ all 16 pairs for the round
2: **for all** $\langle X, Y\rangle \in VarPairs$ **orderedBy** $(-(B(X) + B(Y)), |D(\langle X, Y\rangle)|)$ **do**
3:     **try all** $\langle a, b\rangle \in D(\langle X, Y\rangle)$ **orderedBy** $C(a, b)$ **tiebreak** $-(E(a) + E(b))$
4:         **post** $S_i = a \wedge S_j = b$
5:     **end tryall**

---

the Gurobi Solver (6.5.2) while SCIP 4.0 is used for both for IP and for Pseudo-Boolean formulations. The experiments ran on a 16-core Intel Xeon E52640 at 2.40 Ghz with 16 MB cache and Ubuntu 16.04 LTS. Two categories of instances were generated, **Structured** instances and **Random** instances. For random instances, the plaintexts are chosen uniformly in $\{0, 1\}^{128}$ and the cipher keys are fixed. For structured instances, the plaintexts are generated by picking a subset of 16 bytes from ascii text and the cipher keys are fixed.

Each instance contains a known plaintext and 100 Hamming weight leaks that correspond to the first four subrounds. For each instance, a 10% error rate is applied to the 100 Hamming weight leaks, the Hamming weights of 10 randomly chosen indices are modified by ± 1 (set size k = 3 [14]). To investigate the performance of all solvers, the hardest 3-Set TASCA benchmark instance from the original paper is used. To compare the average performance, 50 structured instances and 50 random instances are generated and solved by solvers separately. All instances are given 10 min limit. All the solvers run in sequential[1].

The Original TASCA Instance. The behavior on the original hardest TASCA instance (`aes_8bit_tasca_3set`) was captured for Objective-CP with custom and Fail-First searches. Gurobi uses an IP model while SCIP uses an IP model and Pseudo-Boolean model. Each solver was given 10 min time limit. Figure 4 shows the run time and comparative memory usage vs. the number of known key bytes. The CP approach delivers two orders of magnitude improvement over the recent SCIP (4.0) and Gurobi solvers. Note that the difference is far more dramatic if one compares to the published results that relied on SCIP 1.2. The custom CP search has two orders of magnitude advantage over the generic search when less than 7 known keys fixed.

CP Performance. Figure 5 shows that random instances display a better performance than structured instances, especially with few known key bytes. When solving for 16 bytes, on average, the random instances perform under 10 s, and the structured instances require around 80 s.

Comparing General Instances. Tables 1 and 2 compare all solvers in terms of solving times from 5 known key bytes down to 0 known key bytes. Figures 6 and 7 show the quartile plots of ratios $\frac{T_{Gurobi}(i)}{T_{CP}(i)}$ and $\frac{T_{SCIP}(i)}{T_{CP}(i)}$ over the 50 instances.

---

[1] Parallel search was also tested both for Gurobi and OBJECTIVE-CP but both solvers failed to gain speedups from parallelization.

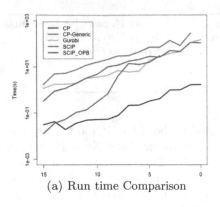 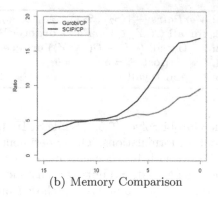

(a) Run time Comparison   (b) Memory Comparison

Fig. 4. Comparative TASCA performance on original instance.

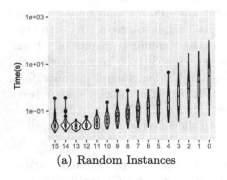 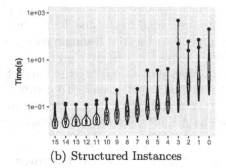

(a) Random Instances   (b) Structured Instances

Fig. 5. CP performance with x-axis being the number of known key bytes.

Table 1. Random instances   **Table 2.** Structured instances

	CP		Gurobi		SCIP	
KB	μ_T	σ_T	μ_T	σ_T	μ_T	σ_T
5	0:24	0:23	26:95	23:82	64:3	22:88
4	0:46	0:70	37:90	32:51	92:6	46:65
3	1:05	1:97	89:36	72:56	180:3	115:86
2	2:04	4:32	122:71	107:51	232:4	108:71
1	4:16	7:12	191:04	115:03	377:9	137:20
0	7:43	15:63	256:93	127:10	355:0	153:11

	CP		Gurobi		SCIP	
KB	μ_T	σ_T	μ_T	σ_T	μ_T	σ_T
5	0:24	0:51	27:51	22:48	63:5	17:31
4	0:32	0:60	41:18	29:92	87:3	46:80
3	11:43	67:87	100:60	71:28	191:2	109:97
2	3:39	9:56	138:58	79:76	296:5	138:33
1	4:87	11:55	228:29	106:90	348:4	108:90
0	13:49	34:52	247:84	120:89	384:0	98:05

Figure 6 shows that the CP solver can solve (on average) instances approximately 50 times faster than a state-of-the-art IP solver, peaking at 100 times faster than Gurobi when solving for all 16 keys. According to Fig. 7, for the majority of instances CP has more than 2 orders of magnitude advantage over SCIP. While SCIP cannot solve the key for most of the instances within 10 min when 0 key bytes are provided.

Memory Consumption. The CP approach for TASCA offers an extremely lightweight memory footprint. The original SCIP model is restricted to 5

sub-rounds due to the high memory usage. The memory consumption of the CP solver is significantly lower as indicated in Fig. 4. This is attributed to modeling with bit-vectors and avoiding costly linearizations. While the CP model has approximately 1000 variables and 600 constraints, the IP model has around 13000 variables and 10700 constraints. As Fig. 4 shows, the memory consumption for Gurobi is above 5 times that of CP, and the ratio increases to around 10 when solving for 16 keys. The memory usage ratio of (SCIP:CP) is higher with SCIP consuming 15 times more memory when solving for all 16 keys.

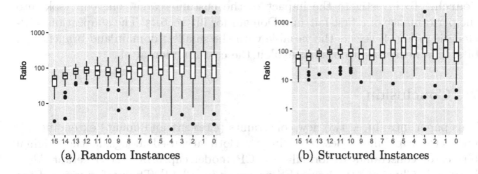

(a) Random Instances   (b) Structured Instances

Fig. 6. Performance comparison with Gurobi.

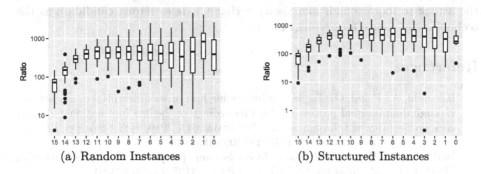

(a) Random Instances   (b) Structured Instances

Fig. 7. Performance comparison with SCIP.

Error Tolerance. A key advantage of CP is to not only solve the optimization problem, but also enumerate all optimal solutions. This is inherently valuable as each global optimum is a *candidate* key called a *partial correct solution* in [14]. A brute force approach can test the recovered keys to find the correct one. Naturally, a large number of incorrect bytes in the candidate keys make the brute force approach untractable. The size of the candidate pool is a critical indicator of the true potential of TASCA.

This experiment used the CP model to enumerate all global optima to assess the distribution of candidate pool sizes for 100 clear and cipher texts. The majority of instances provided a solution pool that was less than 100 solutions, and each solution pool contained the correct key. Figure 8 gives a histogram of the candidate pool sizes. 25 instances had pools with 0–10 keys while 50% of them have a pool size under 50.

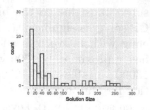

Fig. 8. Pool histogram

Search. To investigate the impact of the custom search, one can look into the performance of a first-fail/min-Domain for bit-vectors. The adaptation (CP-Generic in Fig. 4) selects the variable with the smallest domain and branches on bits. It is weaker as it does not exploit the cost function at all.

7   Conclusion

This paper introduced two new constraints (`element` and `count`) extending the bit-vector capabilities described in [8]. Both are used alongside the original bit-vector constraints to formulate a CP model capturing the Tolerant Algebraic Side-Channel Attack on AES proposed in [13,14]. The performance of the CP approach is empirically evaluated against the original model with the most recent SCIP and Gurobi versions. The CP approach delivers orders of magnitude improvement in run time and memory usage opening the door to scaling the attack to more rounds and showing that CP is as strong candidate as the technology of choice for cryptographers.

References

1. Barenghi, A., Pelosi, G., Teglia, Y.: Improving first order differential power attacks through digital signal processing. In: Proceedings of the 3rd International Conference on Security of Information and Networks, SIN 2010, NY, USA, pp. 124–133 (2010). http://doi.acm.org/10.1145/1854099.1854126
2. Berthold, T., Heinz, S., Pfetsch, M.E.: Solving pseudo-Boolean problems with SCIP. Technical report 08–12, ZIB, Takustr.7, 14195, Berlin (2008)
3. Chihani, Z., Marre, B., Bobot, F., Bardin, S.: Sharpening constraint programming approaches for bit-vector theory. In: Salvagnin, D., Lombardi, M. (eds.) CPAIOR 2017. LNCS, vol. 10335, pp. 3–20. Springer, Cham (2017). doi:10.1007/978-3-319-59776-8_1
4. Daemen, J., Rijmen, V.: The Design of Rijndael: AES - The Advanced Encryption Standard. Springer Science & Business Media, Berlin (2013)
5. Gerault, D., Minier, M., Solnon, C.: Constraint programming models for chosen key differential cryptanalysis. In: Rueher, M. (ed.) CP 2016. LNCS, vol. 9892, pp. 584–601. Springer, Cham (2016). doi:10.1007/978-3-319-44953-1_37
6. Kocher, P., Jaffe, J., Jun, B.: Differential power analysis. In: Wiener, M. (ed.) CRYPTO 1999. LNCS, vol. 1666, pp. 388–397. Springer, Heidelberg (1999). doi:10.1007/3-540-48405-1_25

7. Mangard, S.: A simple power-analysis (SPA) attack on implementations of the AES key expansion. In: Lee, P.J., Lim, C.H. (eds.) ICISC 2002. LNCS, vol. 2587, pp. 343–358. Springer, Heidelberg (2003). doi:10.1007/3-540-36552-4_24

8. Michel, L.D., Van Hentenryck, P.: Constraint satisfaction over bit-vectors. In: Milano, M. (ed.) CP 2012. LNCS, pp. 527–543. Springer, Heidelberg (2012). doi:10.1007/978-3-642-33558-7_39

9. Michel, L., Van Hentenryck, P.: A microkernel architecture for constraint programming. Constraints **22**(2), 107–151 (2017). http://dx.doi.org/10.1007/s10601-016-9242-1

10. Mohamed, M.S.E., Bulygin, S., Zohner, M., Heuser, A., Walter, M., Buchmann, J.: Improved algebraic side-channel attack on AES. J. Cryptographic Eng. **3**(3), 139–156 (2013). http://dx.doi.org/10.1007/s13389-013-0059-1

11. NIST: Federal information processing standards publication (FIPS 197). Advanced Encryption Standard (AES) (2001)

12. Oren, Y., Renauld, M., Standaert, F.-X., Wool, A.: Algebraic side-channel attacks beyond the hamming weight leakage model. In: Prouff, E., Schaumont, P. (eds.) CHES 2012. LNCS, vol. 7428, pp. 140–154. Springer, Heidelberg (2012). doi:10.1007/978-3-642-33027-8_9

13. Oren, Y., Wool, A.: Tolerant algebraic side-channel analysis of AES. IACR Cryptology ePrint Archive, Report 2012/092 (2012). http://iss.oy.ne.ro/TASCA-eprint

14. Oren, Y., Wool, A.: Side-channel cryptographic attacks using pseudo-Boolean optimization. Constraints **21**(4), 616–645 (2016). http://dx.doi.org/10.1007/s10601-015-9237-3

15. Ramamoorthy, V., Silaghi, M.C., Matsui, T., Hirayama, K., Yokoo, M.: The design of cryptographic S-boxes using CSPs. In: Lee, J. (ed.) CP 2011. LNCS, vol. 6876, pp. 54–68. Springer, Heidelberg (2011). doi:10.1007/978-3-642-23786-7_7

16. Renauld, M., Standaert, F.-X.: Algebraic side-channel attacks. In: Bao, F., Yung, M., Lin, D., Jing, J. (eds.) Inscrypt 2009. LNCS, vol. 6151, pp. 393–410. Springer, Heidelberg (2010). doi:10.1007/978-3-642-16342-5_29

17. Renauld, M., Standaert, F.-X., Veyrat-Charvillon, N.: Algebraic side-channel attacks on the AES: why time also matters in DPA. In: Clavier, C., Gaj, K. (eds.) CHES 2009. LNCS, vol. 5747, pp. 97–111. Springer, Heidelberg (2009). doi:10.1007/978-3-642-04138-9_8

18. Soos, M., Nohl, K., Castelluccia, C.: Extending SAT solvers to cryptographic problems. In: Kullmann, O. (ed.) SAT 2009. LNCS, vol. 5584, pp. 244–257. Springer, Heidelberg (2009). doi:10.1007/978-3-642-02777-2_24

19. Standaert, O.X., Peeters, E., Rouvroy, G., Quisquater, J.J.: An overview of power analysis attacks against field programmable gate arrays. Proc. IEEE **94**(2), 383–394 (2006)

20. Van Hentenryck, P., Carillon, J.P.: Generality versus specificity: an experience with AI and or techniques. In: 7th AAAI National Conference on Artificial Intelligence, AAAI 1988, pp. 660–664. AAAI Press (1988). http://dl.acm.org/citation.cfm?id=2887965.2888082

21. Wang, W., Søndergaard, H., Stuckey, P.J.: A bit-vector solver with word-level propagation. In: Quimper, C.-G. (ed.) CPAIOR 2016. LNCS, vol. 9676, pp. 374–391. Springer, Cham (2016). doi:10.1007/978-3-319-33954-2_27

On Maximum Weight Clique Algorithms, and How They Are Evaluated

Ciaran McCreesh, Patrick Prosser, Kyle Simpson, and James Trimble[⊠]

University of Glasgow, Glasgow, Scotland
j.trimble.1@research.gla.ac.uk

Abstract. Maximum weight clique and maximum weight independent set solvers are often benchmarked using maximum clique problem instances, with weights allocated to vertices by taking the vertex number mod 200 plus 1. For constraint programming approaches, this rule has clear implications, favouring weight-based rather than degree-based heuristics. We show that similar implications hold for dedicated algorithms, and that additionally, weight distributions affect whether certain inference rules are cost-effective. We look at other families of benchmark instances for the maximum weight clique problem, coming from winner determination problems, graph colouring, and error-correcting codes, and introduce two new families of instances, based upon kidney exchange and the Research Excellence Framework. In each case the weights carry much more interesting structure, and do not in any way resemble the 200 rule. We make these instances available in the hopes of improving the quality of future experiments.

1 Introduction

This paper does not present a better algorithm for the maximum weight clique problem. Instead, it argues that due to questionable benchmarking practices, we do not know what the state of the art for maximum weight clique algorithms is. This is unfortunate, because the problem is widely researched.

Of particular interest to us is using a maximum weight clique algorithm to solve certain kinds of constraint optimisation problem. The reduction of constraint *satisfaction* problems to finding a clique in a corresponding *microstructure* graph is primarily studied for its theoretical properties [13–16,28]. For problems with a special objective function, a reduction to the maximum clique problem which preserves the objective value is possible—indeed, recent experimental work shows that this encoding, rather than conventional constraint programming, is the best *practical* approach for solving the maximum common subgraph problem when vertex or edge labels are present [39]. To tackle other problems this way (such as graph edit distance problems with complex scoring schemes),

C. McCreesh, P. Prosser and J. Trimble—This work was supported by the Engineering and Physical Sciences Research Council [grant numbers EP/K503058/1, EP/M508056/1 and EP/P026842/1].

J.C. Beck (Ed.): CP 2017, LNCS 10416, pp. 206–225, 2017.
DOI: 10.1007/978-3-319-66158-2_14

we would like to be able to relax the restrictions on the objective function, by reducing to the maximum weight clique problem instead.

This paper also does not attempt to demonstrate that this is a viable approach. Although preliminary experiments suggest that this technique should not be dismissed out of hand, we believe that its chances of success would be much-improved by a change in how maximum weight clique algorithms are designed and evaluated. We therefore begin with a brief review of maximum weight clique algorithms. We then look at the set of instances usually used for benchmarking, focussing in particular upon a widespread practice of allocating weights to unweighted graphs. We show how this has affected the design of heuristics and other filtering rules. We finish by looking at other problem instances, where weights have real-world meanings and the vertices often have special structure; we make all of these instances available for other experimenters to use.

2   Maximum Weight Clique Algorithms

Given a graph G where each vertex v has an integer weight $w(v)$, the *maximum weight clique problem* is to find a subset of vertices of maximum sum of weights, such that every vertex in the subset is adjacent to every other in the subset; note that the *maximum weight independent set* and *minimum weighted vertex cover* problems are essentially equivalent. The problem also comes in an edge-weighted variant, which we do not discuss in this paper. We write $N(v)$ for the set of vertices adjacent to v (that is, its *neighbourhood*), the *degree* of a vertex is the cardinality of its neighbourhood, and the *density* of a graph is the proportion of potential edges which are present. We use C in our descriptions of algorithms to denote the current clique that is being built during search; P denotes the set of candidate (potential) vertices that may be added to this clique.

Cliquer [43] finds a maximum clique and an earlier paper [42] presents a sketch of how Cliquer can find a maximum weight clique. Cliquer is essentially a Russian Doll search [60], finding a maximum weight clique in iteration i with an initial clique $C = \{i\}$ and a candidate set of vertices to choose from $P = N(i) \cap \{0, \ldots, i - 1\}$. The weight of this clique is then stored in an array element $c[i]$. In the case of unweighted maximum clique, $c[i] = c[i - 1]$ if we cannot unseat the incumbent using vertices $\{0, \ldots, i\}$. If we can unseat the incumbent using vertices $\{0, \ldots, i\}$ then we must have added one more vertex, that vertex is i and $c[i] = c[i - 1] + 1$. In the case of maximum weight clique, $c[i]$ is a weight, and $c[i] > c[i - 1]$ if and only if we unseat the incumbent using vertices $\{0, \ldots, i\}$. Obviously, $0 \leq c[i] - c[i - 1] \leq w(i)$. When a vertex v is selected from P to add to C, $c[v]$ can be used to prune the search. Prior to adding vertex v to C, we know that the best possible clique that can be found using vertices $\{0, \ldots, v\}$ is $c[v]$, so if the weight of C plus $c[v]$ does not exceed the weight of the incumbent then search can be abandoned. The search is also pruned if the total weight of $C \cup P$ is no greater than the incumbent.

Kumlander's algorithm [32] is an enhancement of Cliquer. At the top of search the vertices of the graph are coloured, giving colour classes C_1 to C_k. In each colour class, vertices are then sorted by weight in ascending order. We now have a Cliquer-like search with iterations 1 to k, where iteration i finds the heaviest clique using vertices in the colour classes $C_1 \cup \ldots \cup C_i$ and the weight is recorded in $c[i]$ (as in Cliquer). This can then be used as a bound (as before) along with a *colouring upper bound*, that is the sum of the maximum weights in each of the colour classes $C_1 \ldots C_i$, i.e. $\sum_{j=1}^{i} \max\{w(v) : v \in C_j \cap P\}$.

MWCLQ [22] uses MaxSAT reasoning to tighten an upper bound given by vertex colouring. At each search node, the vertices of G are first partitioned into independent sets $C_1 \cup \ldots \cup C_n$. This allows conversion to a literal-weighted MaxSAT encoding: a variable x_i is created for every vertex v_i, having $w(x_i) = w(v_i)$, and a hard clause $\overline{x}_i \vee \overline{x}_j$ is posted for each pair of vertices v_i, v_j which are not adjacent. For each independent set $C_i = \{v_1, \ldots, v_l\}$, a soft clause is then added where literals are weighted, $c_i = (x_1, w(x_1)) \vee \cdots \vee (x_l, w(x_l))$, where $w(c)$ is the maximum literal weight within that clause and literals in a clause are sorted by weight in non-increasing order. The upper bound starts as the sum of all original soft clause weights. By applying unit propagation on an instance, the algorithm identifies sets S of conflicting soft clauses and for each, the accompanying set S_{topk} where the k highest-weight literals are failed. Defining $w(S) = \min\{w(c) : c \in S\}$ and $k(t)$ as the count of failed high-weight literals in t, the upper bound is then reduced by $\min(w(S), \min_{t \in S_{topk}}(w(t) - w(x_{k(t)+1})))$. Further such sets (and bounds reductions) are identified by iteratively splitting and transforming the soft clauses in S and S_{topk} to obtain new unit clauses.

Tavares [2,59] introduces a new heuristic colouring algorithm for calculating an upper bound, *BITCOLOR*, in which each vertex may appear in more than one colour class. Each colour class has an associated weight, and the colouring has the property that the weight of a vertex v equals the sum of the weights of the colour classes to which v belongs. If we produce a colouring of the candidate set P in this way, and let $\mathrm{UB}(P)$ be the sum of colour-class weights, it can be shown that $\mathrm{UB}(P) + \sum_{v \in C} w(v)$ is a valid upper bound on the maximum clique weight. In practice, this approach produces a tighter upper bound than simple colouring. Tavares uses BITCOLOR in three algorithms for maximum weight clique: a Cliquer-style Russian dolls algorithm, an algorithm which branches on vertices in reverse colouring order, and a resolution search algorithm [12].

OTClique [55] enhances the Russian dolls approach of Cliquer by precomputing, using dynamic programming, the maximum-weight clique in each of a large set of induced subgraphs of G. The precomputed values are used to quickly calculate a bound that is tighter than the naïve sum-of-vertex-weights bound used by Cliquer.

WLMC is an exact algorithm which is designed for large, sparse graphs, but also performs well on the relatively small, dense graphs that are the focus of this paper. In a preprocessing step—which is performed at the top of search and also after every possible choice of first vertex—WLMC uses the method of Carraghan and Pardalos [11] to produce a vertex ordering and an initial incumbent clique. The preprocessing step then reduces the size of the graph by deleting any vertex v such that the incumbent clique has weight greater than or equal to w(v) plus the sum of weights of v's neighbours. At each node of the branch-and-bound search, WLMC uses MaxSAT reasoning to find a set of vertices on which it is unnecessary to branch[1].

Other approaches. The problem has also been tackled using mathematical programming [25,27,61], and is the subject of ongoing research for inexact (heuristic) solutions [3,4,10,20,27,29,41,62,65,66]. Finally, sometimes alternative constraints or objectives are considered [6,34,54].

3   Current Practices in Benchmarking

For the maximum (unweighted) clique problem, experimenters are blessed with a suite of instances from the second DIMACS implementation challenge [30]. These are all relatively small, dense graphs. Most instances fit into one of three classes:

- Graphs which encode a problem from another domain. The "c-fat" family encode a problem involving fault diagnosis for distributed systems [5]. The "hamming" and "johnson" graphs model problems from coding theory [7]. The "keller" instances encode a geometric conjecture [17], and the MANN family is made from clique formulations of the Steiner triple problem [36]. In each of these cases, the size of the solution has a real-world interpretation (and sometimes the vertices contained therein also convey meaning).
- Randomly-generated graphs. The "C" and "DSJC" instances are simple random graphs of varying orders and densities. The "p_hat" family are also random graphs, but with an unusually large degree spread [23,56].
- Random graphs with large hidden solutions. The "brock" family of instances [9] are an attempt at camouflaging a known clique in a quasi-random graph for cryptographic purposes, in a way resistant to heuristic attacks. The "gen" and "san(r)" instances use a different technique for hiding a large clique of known size in a graph [50,51]: again, they are an attempt to create challenging instances with a known and unusually large optimal solution.

The instances from the first set are valuable because of their applications. Meanwhile, the randomly generated instances are useful because they provide a challenge: although being able to solve crafted hard instances is not the primary goal of developing clique algorithms, working on these instances has led to better solvers. For example, Depolli *et al.* [18] use a maximum clique algorithm

---

[1] The existing implementation of WLMC does not support the large weights that appear in many of the instances in this paper. Therefore, we could not include this program in our experimental evaluation.

for new instances from a biochemistry application, and note that although their instances are reasonably easy for a modern algorithm, they are challenging for earlier algorithms that predate experiments on these instances; a similar conclusion holds for clique-based solvers for maximum common subgraph problems [39].

For the weighted problem, standard practice is to take these same instances, and to follow a convention usually ascribed to Pullan [45]:

"Instances were converted to MVWC instances (the DIMACS-VW benchmark) by allocating weight, for vertex i, of $i \bmod 200 + 1$".

Incidentally, Mannino and Stefanutti [37] had used a similar convention previously, using modulo 10 rather than 200. Pullan justifies this rule and choice of constant as follows:

"This technique allows future investigators to simply replicate the experiments performed in this study. The constant 200 in the weight calculation was determined after a number of experiments showed that the generated problems appeared to be reasonably difficult for PLS (clearly, allocating weights in the range $1, \ldots, k$ results in an MC instance when $k = 1$ while, intuitively, it is reasonable to expect that as k increases, the difficulty in solving the instance will, in general, increase)."

This rule, together with a similar rule for allocating weights to edges for the edge-weighted variant of the problem, is very widely used [2–4, 20–22, 25, 27, 29, 32, 34, 41–43, 45, 55, 61, 62, 65, 66, and many more], often as the only way of evaluating a solver. It has also recently been adopted for large sparse graphs [10, 20, 29, 62], and for benchmarking the minimum weight dominating set problem [63], often as the only way of evaluating a solver. It has also recently been adopted for large sparse graphs [10, 20, 29, 62], and for benchmarking the minimum weight dominating set problem [63].

4   Experimental Setup

Our experiments are performed on a machine with dual Intel Xeon E5-2697A v4 CPUs and 512 GBytes RAM, running Ubuntu 16.04. We used the latest version of Cliquer (1.21), released in 2010, downloaded from the author's website. The source code for MWCLQ and OTClique was provided by these programs' authors. We modified each program by changing every occurrence of `int` to a 64-bit integer type, in order to accommodate the large weights that occur in some classes of instance. This change has a measurable effect on the runtime of the programs, particularly for OTClique.

Tavares' programs were not available when we ran our experiments. We have therefore written two programs using a Tavares-style colouring for use in our experiments, one which uses Russian Dolls and one which branches in an order based on colouring. We call these programs TR and TC, respectively. In both

Tavares' description and our implementations, bitsets are used to perform the colouring step efficiently.

All five programs are implemented in C, and were compiled using GCC 5.4.0 at optimisation level -O3. We set the OTClique parameter l to 20. Finally, we implemented a constraint programming model in Java, using the Choco solver version 3.3.3.

5   Does Weight Allocation Affect Algorithm Design?

The maximum weight clique problem has an obvious constraint programming model: we have a 0/1 variable for each vertex, a constraint for each non-adjacent pair of vertices prohibiting the two vertices from being set to 1 simultaneously, and an objective to maximise the weighted sum over all variables. But what about variable-ordering heuristics? For the unweighted maximum clique problem, we might use the degree of the vertex corresponding to each variable. In the weighted case, we could look either at degree, or at weight.

When weights are chosen to be between 1 and 200, we would expect weights to be much more important than degree: selecting a vertex of high weight would affect the solution more than selecting several vertices of low weight. Thus it seems likely that a variable-ordering heuristic which considered weights would be best. On the other hand, if we selected weights to be between 190 and 200, perhaps degree would matter more instead?

Figure 1 confirms this suspicion. We look at random graphs with 70 vertices and density 0.6. We assign weights sequentially, starting at $x + 1$ and wrapping back to $x + 1$ when we exceed 200. Thus, on the far left of the plot, we have weights ranging from 1 to 70, in the middle from 101 to 170, near the right from 180 to 200 (with weights repeated), and on the far right, every weight is 200. For the y-axis, we plot the mean search effort from a sample of 100 runs for our Choco model, using ascending and descending degree or weight as static variable-ordering heuristics, and preferring 1 over 0 as a value-ordering heuristic. Because weights are allocated sequentially, we randomly permute the order of vertices before solving to avoid using weight unintentionally as a tie-breaking heuristic. The results show that when weights are chosen to be between 1 and 70, it is indeed best to select a weight-based variable ordering heuristic. However, once weights are chosen to be between 50 and 119, it becomes much more useful to use degree-based heuristics.

The plot also shows the effects of using impact-based search [46]. These results are nearly as good as the degree-based heuristics, but do not beat tailored heuristics. Impact-based search is also unable to mimic weight-based heuristics in the parts of the parameter-space where weights are more informative, since impact is unaware of the effect of domain deletions upon the objective function. We also plot domain over weighted degree [8], which fares less well.

What about other densities? Most of the DIMACS instances are dense—does this affect algorithm design too? In the top left plot of Fig. 2, we vary both the graph density and weight range, and use colour to show which heuristic has best

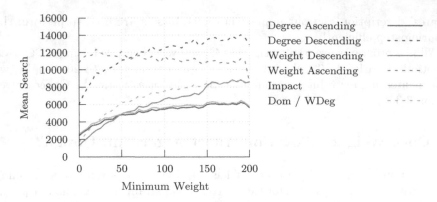

Fig. 1. A comparison of heuristics using a Choco model, over random graphs with 70 vertices, density 0.6, and sequential weights starting from x and never exceeding 200.

average performance at each point. The results show that when the minimum weight is low (such as when using the 200 rule), we should favour descending weight heuristics, but otherwise we should favour ascending degree.

This basic constraint programming approach is not performance-competitive, but the relative simplicity of the algorithm makes it easy to experiment with. What about the algorithms introduced in Sect. 2, which use more complex branching strategies and inference? Figure 2 also compares the run times of five dedicated algorithms with modified vertex orderings, working with 100 vertex graphs. (Note that several of the algorithms have branching strategies that are influenced by, but not identical to, the order of vertices in the graph.) The focus of this paper is not on explaining these results in great detail: we are simply demonstrating that the 200 rule has likely had an effect on algorithm design[2]. The default orderings for most of these algorithms are primarily weight-based, which appears to be a good choice when weight ranges are large (there are many dark and light blue points towards the bottom of these plots), but perhaps not otherwise (the plots are not monochromatic, and there are large orange and/or yellow areas in each plot).

Are heuristics the only design choice affected by the benchmark instances? Figure 3 compares our Tavares-style algorithm TC with a similar algorithm which uses a simple colour bound rather than the Tavares-style multi-colouring. The plots show mean search effort and runtimes for random 200-vertex instances of density 0.6. On the instances with a wide weight range such as 5–200, the

---

[2] It is interesting to note that MWCLQ often resembles Choco but with a higher threshold for switching away from weights, except that sometimes it is worth switching to descending degree as well as ascending degree, and that the three Russian Dolls algorithms exhibit similar heuristic behaviour to each other. We do not understand how ordering heuristics should work with Russian Dolls search, and suggest that this could be a good avenue for future research—for example, perhaps it is better to use different heuristics for different dolls?.

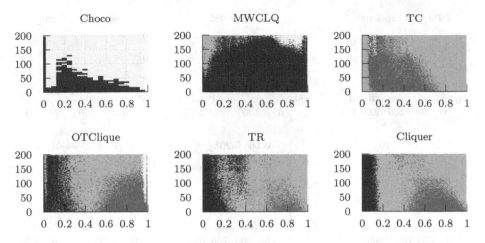

Fig. 2. Which heuristic is best when varying density (x-axis) and weight range (y-axis), choosing from descending/ascending weight (dark/medium blue), descending/ascending degree (orange/yellow), or no difference (white). Graphs have 70 vertices for Choco, and 100 for the other algorithms (which also have higher plotting densities). No plot is monochromatic, showing that heuristics are affected by density and weight ranges. (Color figure online)

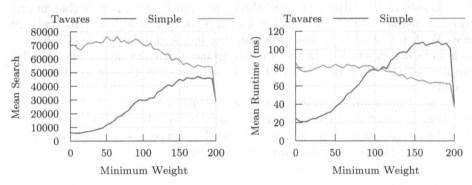

Fig. 3. Comparing TC with a simpler algorithm which does not use the Tavares multi-colouring rule, on 200 vertex graphs with density 0.6, and different weight ranges.

Tavares-style algorithm is the clear winner. When the minimum weight is greater than 100, the simple algorithm is faster; although it visits more search nodes, this is outweighed by the shorter time per node of the simpler algorithm. This shows that the practical benefits of Tavares' more complex bound are also heavily dependent upon how weights are allocated.

6   Other Families of Problem Instances

Having questioned the 200 rule and the use of unweighted DIMACS instances, we now discuss five families of instances which we hope will lead to better

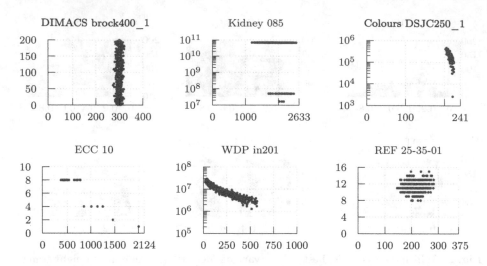

Fig. 4. Weight (y-axis) plotted against degree (x-axis) for an example instance from each different graph class. Note some plots use a log scale for the weight, and do not start at zero.

experiments in the future. Three of these are from existing papers (but in one case the instances are hard to find online in a convenient format), and two are new. We bring all of these instances together in the standard DIMACS format to help future experimenters, and we will update this collection as new families are uncovered[3]. Note that many of these instances require support for 64-bit weights.

Figure 4 plots weight versus degree for one instance from each of these families. We also plot "brock400_1" from the DIMACS set using the 200 rule: observe that degree gives almost no information for this instance, compared to weight. We return to this figure as we introduce each family.

6.1 Kidney Exchange

Kidney-exchange schemes exist in several countries to increase the number of transplants from living donors to patients with end-stage renal disease [24,35,. 47]. A patient enters the scheme along with a friend or family member who is willing to donate to that patient but unable to do so due to blood or tissue incompatibility. These two participants form a *donor-patient pair*. From the pool of donor-patient pairs, the scheme administrator periodically arranges *exchanges*, each of which involves two or more donor-patient pairs. In a two-way exchange, the donor of the first pair gives a kidney to the patient of the second pair, and the donor of the second pair gives a kidney to the patient of the first. In three-way and larger exchanges, kidneys are donated between the donor-patient pairs cyclically.

---

[3] https://doi.org/10.5281/zenodo.816293.

In addition, many schemes benefit from *altruistic donors*, who enter the scheme without a paired patient, and may initiate a chain of donations. For the optimisation problem, we may view an altruistic donor as a donor paired with a "dummy patient" who is compatible with any donor.

Each feasible exchange is given a score reflecting its desirability. This may, for example, take into account the size of the exchange, the time that patients have been waiting for a transplant, and the probability that the transplants will be successful. Typically, the scheme administrator carries out a *matching run* at fixed intervals, with the goal of maximising the sum of exchange scores. A popular approach to solving this optimisation problem is integer programming using the *cycle formulation*, in which we have one binary variable for each feasible exchange, and a constraint for each participant in the scheme ensuring that he or she is involved in at most one selected exchange [1, 48]. We propose that this optimisation problem may, alternatively, be solved by reduction to maximum-weight clique. Each vertex is an exchange, whose weight is its score. Two exchanges are adjacent if and only if they have no participants in common.

To create maximum weight clique instances, we used kidney instances by Dickerson [19], available on PrefLib [38], originally from a widely-used generator due to Saidman *et al.* [49] (real instances cannot be made public due to medical confidentiality). The weighting scheme and exchange size cap we used are based on the system used in the UK's National Living Donor Kidney Sharing Scheme (NLDKSS) [35]. The NLDKSS has a maximum exchange size of three, and has five objectives, ranked hierarchically. The first objective is optimised; subject to this being at its optimal value, the second objective is optimised, and so on.

The primary objective is to maximise the number of *effective two-way exchanges*: exchanges that either consist of only two donor-patient pairs, or which contain (as part of a larger exchange) two donor-patient pairs who could by themselves form a two-way exchange. This provides robustness: part of a larger exchange may still proceed even if the full exchange does not (for example, due to illness). The second objective is to maximise the total number of transplants. The third objective is to minimise the number of three-way exchanges. The fourth objective is to maximise the number of *back-arcs* in three-way exchanges; these are compatibilities between donor-patient pairs in the reverse direction of the exchange. The final objective is to optimise the *weight* of the exchange, which is a value based on factors including the number of previous matching runs that patients have been in, and the level of compatibility between donors and patients in planned transplants.

To create these instances, we used the first four of these objectives, combining them into a single long integer using the formula $x = 2^{36}x_1 + 2^{24}x_2 + 2^{12}x_3 + x_4$ where x_i is an exchange's score for the ith objective. We use a simple transformation to convert the third objective from a minimisation to a maximisation. This method of combining scores in order to perform a single optimisation is not practical using IP solvers because, as Manlove and O'Malley [35] observe, the resulting weights would be too large for IP solvers. By contrast, all of our maximum-weight clique solvers can use 64-bit weights without loss of precision.

(Ideally, we would like to use even larger weights, to include the fifth ranking criterion.) Note that due to the extreme ranges of weights requiring the use of a log scale, Fig. 4 does not clearly show the variation between weights.

6.2 Colouring Instances

In branch and bound graph colouring algorithms such as Held *et al.* [26] the *fractional chromatic number* $\chi_f(G)$ acts as a useful upper bound. This can be found according to an integer programming formulation introduced by Mehrotra and Trick [40]: the model starts with a subset of the required variables, which is extended if a *maximum weight independent set* (MWIS) of weight at least 1 can be found within the original graph. The weights themselves are the *dual price* of including that vertex in the model according to an independent set formulation, multiplied by some factor *scalef* to achieve integer values. As a result, these graphs feature very large weights to have sufficient resolution to encode small fractions of *scalef*.

The instances we include are due to Held *et al.* Each instance arises during colouring of a corresponding DIMACS instance; many of these are the last such MWIS instance encountered during search, and represent that problem's bottleneck.

6.3 Error-Correcting Codes

Östergård [42] describes the following problem from coding theory. Let a length n, a distance d, a weight w, and a permutation group G be given. The objective is to find a maximum-cardinality set C of codes (binary vectors) of length n, such that each element in S has Hamming weight w; each pair of elements in C is at least Hamming distance d apart; and for every permutation $\sigma \in G$ and every code $c \in C$, we have that $\sigma(c) \in C$. Östergård shows how this problem may be reduced to maximum weight clique by partitioning the set of all binary vectors of length n and weight w into orbits under the permutation group G, and creating a vertex for each orbit satisfying the condition that no two members of the orbit are less than Hamming distance d apart. The weight of each vertex equals the size of the corresponding orbit, and two vertices are adjacent if and only if all pairs of members of the two orbits are at least distance d apart.

The fifteen instances presented by Östergård are no longer readily available. We have written a program to reconstruct the instances. For the instance ECC10 shown in Fig. 4 the weights range from one to eight, and are roughly inversely correlated with degree; in other instances, the weights go as high as eighty.

6.4 The Winner Determination Problem

In a combinatorial auction, bidders are allowed to bid on sets of items rather than just single items. For example, at a furniture auction, agent A might bid for four dining chairs and a table, rather than bid for each chair and the table separately. Another bidder, agent B, might bid for the same table and a sideboard, whilst agent C bids for the sideboard and a set of crockery. Agent B's bid is incompatible with that of A (they want the same table) and that of C (they want the same sideboard), but A's bid is compatible with C's (there is no intersection on the items of interest).

Finding an allocation of items to bidders that maximizes the auctioneer's revenue is called the winner determination problem (WDP) [44,52,53]. A problem instance can be represented as a weighted graph. A vertex v in the graph corresponds to a bid, the weight of v is the value of that bid, and an edge exists between a pair of vertices (u, v) if the corresponding bids have no items in common (i.e. they are compatible with each other). Consequently, a maximum weight clique corresponds to an optimal allocation.

WDP instances, available via cspLib [44] and originally created by Lau and Goh [33], have been used as a benchmark suite by Fang *et al.* [22] and Wu and Hao [64] for comparing one maximum weight clique algorithm against another. But what do these instances look like? Figure 4, instance WDP in201, shows that high weight vertices have low degree, and light weight vertices have high degree. This is not surprising: a high value bid is typically a bid for many items and is incompatible with many other bids, and corresponds to a heavy vertex of low degree. Consequently, when used to compare algorithms, we might find that an ordering on decreasing weight will perform much the same as an ordering of increasing degree.

6.5   The Research Excellence Framework

In 2016, Her Majesty's Government proposed that in the next Research Excellence Framework (REF2021) academics would be allowed to submit exactly four publications over a given interval of time (typically 4 years)[4]. In a university, in each unit of assessment (typically a department or school) each member of staff would submit six publications and of those six publications management would select four. Papers are assigned rankings in the range 4 to 1, with 4 being "internationally excellent". Therefore, for each member of staff, there would be C_4^6 possible selections, where each selection would have a combined ranking in the range 4 to 16. At most one of these 4-selections would be allowed for each member of staff, and no publication could be counted more than once (that is, co-authors within the same unit of assessment cannot both submit a shared publication).

This has strong similarities to a winner determination problem: we must find an allocation of items (sets of four publications) to bidders (academic staff) that maximizes the auctioneer's (unit of assessment's) revenue (combined rankings).

Realistic instances were generated for departments with n members of academic staff producing m publications. A random number of papers were generated, each with a ranking in the range 2 to 4, with a specified distribution based on historical data[5]. For each member of staff 6 papers were randomly selected, and

---

[4] However, in July 2016 Lord Nicholas Stern suggested greater flexibility be allowed.
[5] Being a *"research-led institution"* no papers with a ranking of 1 are allowed.

that member of staff was then considered an author. This was then represented as a weighted graph. The graph has $15 \cdot n$ vertices (there are 15 ways for each author to choose 4 publications from 6) with weights in the range 8 to 16. The 15 vertices associated with an author form an independent set (at most one of the author's 4-selections can be selected).

As the number of publications to choose from increases, the likelihood of any pair of 4-selections having a publication in common falls, so bids become more compatible and the resultant graph has more edges (is denser), and this tends to increase the difficulty of the problem. For example with $n = 20$ and $m = 50$ graphs have 300 vertices and average density 0.67, and with $n = 20$ and $m = 30$ we again have 300 vertices and density is 0.52 on average. These graphs have a maximum (unweighted) clique of size no more than $\min(n, m/4)$. There is a small range of weights (8 to 16) and in any instance there is a small variation in degree (see instance REF 25-35-01 in Fig. 4).

6.6 Experiments

In Figs. 5, 6, 7, 8, 9 and 10 we plot, for each algorithm, the cumulative number of instances which can be solved in under a certain amount of time, for these different families of instances. The dark thick line in each plot shows that algorithm's default vertex ordering, and the light lines show ascending and descending weight and degree orderings. To interpret these plots, select a preferred timeout along the x-axis, and then select the line with highest y-value to determine the best-performing algorithm for that choice of timeout.

Although we did not intend to carry out an algorithmic beauty contest, these results support the simple conclusion that our implementation of Tavares' (little known) colour-ordering algorithm is consistently the best solver, and that the default heuristic we picked for it (decreasing degree order) is nearly always the best. This is a surprise. We were hoping to end this paper by stressing the importance of tailoring heuristics and solvers on a family by family basis, perhaps suggesting algorithm portfolios, but instead we have identified a clear winner.

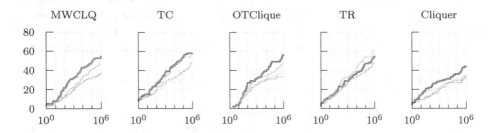

Fig. 5. Cumulative plots for DIMACS instances, with weights in range 1–200 added using the standard scheme. The dark thick line is the default heuristic for each solver, and the thin light lines show ascending and descending degree and weight heuristics. The x-axis is runtime in milliseconds, and the y-axis plots the cumulative number of instances which can be solved (individually) in time less than or equal to x.

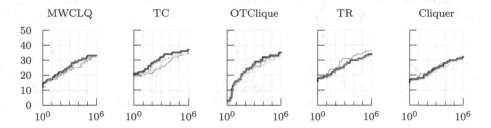

Fig. 6. Cumulative plots for kidney instances.

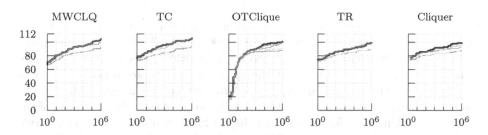

Fig. 7. Cumulative plots for colouring instances.

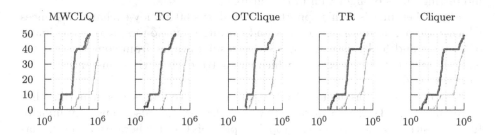

Fig. 8. Cumulative plots for winner determination problem instances.

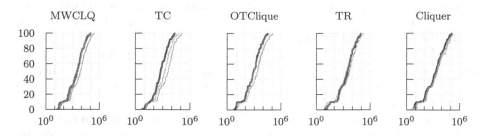

Fig. 9. Cumulative plots for REF instances.

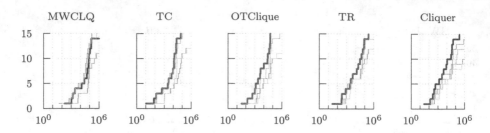

Fig. 10. Cumulative plots for error correcting code instances.

7 Conclusion

Despite our experiments suggesting a single winning algorithm, we believe our new sets of instances are valuable. The 200-weighted DIMACS benchmark instances are often cited as being good "real-world" tests for the maximum weight clique problem. This is not the case: some of these instances are real-world tests for the maximum clique problem, but adding weights destroys the real-world meaning of the results. Additionally, most of these instances are of the "crafted, challenging" (for unweighted clique) kind, and again, adding 200-rule weights destroys these properties. The other families we discuss in this paper are somewhat better in this respect, and if they replace the 200-weighted DIMACS instances as the standard for benchmarking, they may open the way up more interesting kinds of algorithm in the future.

Figure 6 emphasises this opportunity. It shows 50 kidney-exchange instances which have a minimum of 16 pairs and no altruistic donors, and a maximum of 64 pairs and 6 altruistic donors. These results are far from competitive with leading integer program solvers, which can solve each of these instances in less than a second.

Our discussion has focussed on weights. However, it is worth noting that for many of the DIMACS instances, vertex degrees are also unusually unhelpful. The situation shown in the top left plot of Fig. 4 where each vertex has similar degree is common, and for some instance families, the degrees are deliberately constructed to be misleading. In contrast, the vertices in our instances were not crafted with hostile intent, and they often carry a certain amount of structure. This is particularly true with microstructure-like encodings, where vertices from any given variable always form an independent set, and where we know that the graph may always be coloured in a particular way. Now that we have families of instances that have interesting, realistic structure, perhaps subsequent algorithms can be tailored to exploit these properties (such as treating the first branching vertex specially [58]), and it may also be worth considering preprocessing techniques [57].

We hope to extend our collections of instances and algorithms in the future, and perhaps this will make these results more interesting and inspiring. We are also interested in real instances for the edge-weighted variant of the problem, which suffers similarly from an arbitrary weight allocation rule.

We note in passing that all of these instances are dense, despite being "real-world" instances. It is important to distinguish between solving graph problems on graphs which directly *represent* real-world phenomena (which are often sparse and have power-law degree structures), with solving problems which *encode* the solution to a problem. Graphs of the latter kind may very well be dense. This is true even when the question being answered is regarding a sparse graph: for example, when solving the maximum common subgraph problem via reduction to clique, the encoding of two sparse graphs gives a dense graph [39]. Similarly, microstructure graphs for non-trivial problems are usually reasonably dense.

Finally, we observed (in Fig. 2) an anomaly with respect to the variable-ordering heuristics used by Russian Doll algorithms. Clearly, this deserves more attention.

Acknowledgments. The REF instance generator was joint work with David Manlove. We are grateful to David for this, and for helpful discussions on kidney exchange.

References

1. Abraham, D.J., Blum, A., Sandholm, T.: Clearing algorithms for barter exchange markets: enabling nationwide kidney exchanges. In: MacKie-Mason, J.K., Parkes, D.C., Resnick, P. (eds.) Proceedings 8th ACM Conference on Electronic Commerce (EC-2007), 11–15 June 2007, San Diego, California, USA, pp. 295–304. ACM (2007). http://doi.acm.org/10.1145/1250910.1250954
2. Araujo Tavares, W.: Algoritmos exatos para problema da clique maxima ponderada. Ph.D. thesis, Universidade federal do Ceará (2016). http://www.theses.fr/2016AVIG0211
3. Baz, D.E., Hifi, M., Wu, L., Shi, X.: A parallel ant colony optimization for the maximum-weight clique problem. In: 2016 IEEE International Parallel and Distributed Processing Symposium Workshops, IPDPS Workshops 2016, 23–27 May 2016, Chicago, IL, USA, pp. 796–800. IEEE Computer Society (2016). doi:10.1109/IPDPSW.2016.111
4. Benlic, U., Hao, J.: Breakout local search for maximum clique problems. Comput. OR **40**(1), 192–206 (2013). doi:10.1016/j.cor.2012.06.002
5. Berman, P., Pelc, A.: Distributed probabilistic fault diagnosis for multiprocessor systems. In: Proceedings of the 20th International Symposium on Fault-Tolerant Computing, FTCS 1990, 26–28 June 1990, Newcastle Upon Tyne, UK, pp. 340–346. IEEE Computer Society (1990). doi:10.1109/FTCS.1990.89383
6. Boginski, V., Butenko, S., Shirokikh, O., Trukhanov, S., Gil-Lafuente, J.: A network-based data mining approach to portfolio selection via weighted clique relaxations. Ann. OR **216**(1), 23–34 (2014). doi:10.1007/s10479-013-1395-3
7. Bomze, I.M., Budinich, M., Pardalos, P.M., Pelillo, M.: The maximum clique problem. In: Du, D.Z., Pardalos, P.M. (eds.) Handbook of Combinatorial Optimization, pp. 1–74. Springer, Boston (1999). doi:10.1007/978-1-4757-3023-4_1
8. Boussemart, F., Hemery, F., Lecoutre, C., Sais, L.: Boosting systematic search by weighting constraints. In: de Mántaras, R.L., Saitta, L. (eds.) Proceedings of the 16th Eureopean Conference on Artificial Intelligence, ECAI 2004, Including Prestigious Applicants of Intelligent Systems, PAIS 2004, 22–27 August 2004, Valencia, Spain, pp. 146–150. IOS Press (2004)

9. Brockington, M., Culberson, J.C.: Camouflaging independent sets in quasi-random graphs. In: Johnson and Trick [31], pp. 75–88. http://dimacs.rutgers.edu/Volumes/Vol26.html

10. Cai, S., Lin, J.: Fast solving maximum weight clique problem in massive graphs. In: Kambhampati, S. (ed.) Proceedings of the Twenty-Fifth International Joint Conference on Artificial Intelligence, IJCAI 2016, 9–15 July 2016, New York, NY, USA, pp. 568–574. IJCAI/AAAI Press (2016). http://www.ijcai.org/Abstract/16/087

11. Carraghan, R., Pardalos, P.M.: An exact algorithm for the maximum clique problem. Oper. Res. Lett. **9**, 375–382 (1990)

12. Chvátal, V.: Resolution search. Discrete Appl. Math. **73**(1), 81–99 (1997). doi:10.1016/S0166-218X(96)00003-0

13. Cohen, D.A., Cooper, M.C., Creed, P., Marx, D., Salamon, A.Z.: The tractability of CSP classes defined by forbidden patterns. J. Artif. Intell. Res. (JAIR) **45**, 47–78 (2012). doi:10.1613/jair.3651

14. Cohen, D.A., Jeavons, P., Jefferson, C., Petrie, K.E., Smith, B.M.: Symmetry definitions for constraint satisfaction problems. Constraints **11**(2–3), 115–137 (2006). doi:10.1007/s10601-006-8059-8

15. Cooper, M.C., Jeavons, P.G., Salamon, A.Z.: Generalizing constraint satisfaction on trees: hybrid tractability and variable elimination. Artif. Intell. **174**(9–10), 570–584 (2010). doi:10.1016/j.artint.2010.03.002

16. Cooper, M.C., Zivny, S.: Hybrid tractable classes of constraint problems. In: Krokhin, A.A., Zivny, S. (eds.) The Constraint Satisfaction Problem: Complexity and Approximability, Dagstuhl Follow-Ups, vol. 7, pp. 113–135. Schloss Dagstuhl-Leibniz-Zentrum fuer Informatik (2017). doi:10.4230/DFU.Vol7.15301.4

17. Debroni, J., Eblen, J.D., Langston, M.A., Myrvold, W., Shor, P.W., Weerapurage, D.: A complete resolution of the Keller maximum clique problem. In: Randall, D. (ed.) Proceedings of the Twenty-Second Annual ACM-SIAM Symposium on Discrete Algorithms, SODA 2011, 23–25 January 2011, San Francisco, California, USA, pp. 129–135. SIAM (2011). doi: 10.1137/1.9781611973082.11

18. Depolli, M., Konc, J., Rozman, K., Trobec, R., Janezic, D.: Exact parallel maximum clique algorithm for general and protein graphs. J. Chem. Inf. Model. **53**(9), 2217–2228 (2013). doi:10.1021/ci4002525

19. Dickerson, J.P., Procaccia, A.D., Sandholm, T.: Optimizing kidney exchange with transplant chains: theory and reality. In: van der Hoek, W., Padgham, L., Conitzer, V., Winikoff, M. (eds.) International Conference on Autonomous Agents and Multiagent Systems, AAMAS 2012, IFAAMAS, 4–8 June 2012, Valencia, Spain, vol. 3, pp. 711–718 (2012). http://dl.acm.org/citation.cfm?id=2343798

20. Fan, Y., Li, C., Ma, Z., Wen, L., Sattar, A., Su, K.: Local search for maximum vertex weight clique on large sparse graphs with efficient data structures. In: Kang, B.H., Bai, Q. (eds.) AI 2016. LNCS, vol. 9992, pp. 255–267. Springer, Cham (2016). doi:10.1007/978-3-319-50127-7_21

21. Fang, Z., Li, C., Qiao, K., Feng, X., Xu, K.: Solving maximum weight clique using maximum satisfiability reasoning. In: Schaub, T., Friedrich, G., ÓSullivan, B. (eds.) ECAI 2014–21st European Conference on Artificial Intelligence, 18–22 August 2014, Prague, Czech Republic - Including Prestigious Applications of Intelligent Systems (PAIS) 2014. Frontiers in Artificial Intelligence and Applications, vol. 263, pp. 303–308. IOS Press (2014). doi:10.3233/978-1-61499-419-0-303

22. Fang, Z., Li, C., Xu, K.: An exact algorithm based on maxsat reasoning for the maximum weight clique problem. J. Artif. Intell. Res. (JAIR) **55**, 799–833 (2016). doi:10.1613/jair.4953

23. Gendreau, M., Soriano, P., Salvail, L.: Solving the maximum clique problem using a tabu search approach. Ann. OR **41**(4), 385–403 (1993). doi:10.1007/BF02023002

24. Glorie, K., Haase-Kromwijk, B., van de Klundert, J., Wagelmans, A., Weimar, W.: Allocation and matching in kidney exchange programs. Transpl. Int. **27**(4), 333–343 (2014)

25. Gouveia, L., Martins, P.: Solving the maximum edge-weight clique problem in sparse graphs with compact formulations. EURO J. Comput. Optim. **3**(1), 1–30 (2015). doi:10.1007/s13675-014-0028-1

26. Held, S., Cook, W.J., Sewell, E.C.: Maximum-weight stable sets and safe lower bounds for graph coloring. Math. Program. Comput. **4**(4), 363–381 (2012). doi:10. 1007/s12532-012-0042-3

27. Hosseinian, S., Fontes, D., Butenko, S.: A quadratic approach to the maximum edge weight clique problem. In: XIII Global Optimization Workshop GOW 2016, pp. 125–128 (2016)

28. Jégou, P.: Decomposition of domains based on the micro-structure of finite constraint-satisfaction problems. In: Fikes, R., Lehnert, W.G. (eds.) Proceedings of the 11th National Conference on Artificial Intelligence, 11–15 July 1993, Washington, DC, USA, pp. 731–736. AAAI Press/The MIT Press (1993). http://www. aaai.org/Library/AAAI/1993/aaai93-109.php

29. Jiang, H., Li, C., Manyà, F.: An exact algorithm for the maximum weight clique problem in large graphs. In: Singh, S.P., Markovitch, S. (eds.) Proceedings of the Thirty-First AAAI Conference on Artificial Intelligence, 4–9 February 2017, San Francisco, California, USA, pp. 830–838. AAAI Press (2017). http://aaai.org/ocs/ index.php/AAAI/AAAI17/paper/view/14370

30. Johnson, D.S., Trick, M.A.: Introduction to the second DIMACS challenge: cliques, coloring, and satisfiability. In: Cliques, Coloring, and Satisfiability, Proceedings of a DIMACS Workshop, 11–13 October 1993, New Brunswick, New Jersey, USA, [31], pp. 1–10. http://dimacs.rutgers.edu/Volumes/Vol26.html

31. Johnson, D.S., Trick, M.A. (eds.): Cliques, coloring, and satisfiability. In: Proceedings of a DIMACS Workshop, DIMACS/AMS, 11–13 October 1993, New Brunswick, New Jersey, USA. DIMACS Series in Discrete Mathematics and Theoretical Computer Science, vol. 26 (1996). http://dimacs.rutgers.edu/Volumes/ Vol26.html

32. Kumlander, D.: On importance of a special sorting in the maximum-weight clique algorithm based on colour classes. In: An, L.T.H., Bouvry, P., Tao, P.D. (eds.) Modelling, Computation and Optimization in Information Systems and Management Sciences, MCO 2008. Communications in Computer and Information Science, vol. 14, pp. 165–174. Springer, Heidelberg (2008). doi:10.1007/978-3-540-87477-5-18

33. Lau, H.C., Goh, Y.G.: An intelligent brokering system to support multi-agent web-based 4th-party logistics. In: 14th IEEE International Conference on Tools with Artificial Intelligence (ICTAI), 4–6 November 2002, Washington, DC, USA, p. 154. IEEE Computer Society (2002). doi:10.1109/TAI.2002.1180800

34. Malladi, K.T., Mitrovic-Minic, S., Punnen, A.P.: Clustered maximum weight clique problem: algorithms and empirical analysis. Comput. Oper. Res. **85**, 113–128 (2017). http://www.sciencedirect.com/science/article/pii/S0305054817300837

35. Manlove, D.F., O'Malley, G.: Paired and altruistic kidney donation in the UK: algorithms and experimentation. ACM J. Exper. Algorithmics **19**(1) (2014). http:// doi.acm.org/10.1145/2670129

36. Mannino, C., Sassano, A.: Solving hard set covering problems. Oper. Res. Lett. **18**(1), 1–5 (1995). doi:10.1016/0167-6377(95)00034-H

37. Mannino, C., Stefanutti, E.: An augmentation algorithm for the maximum weighted stable set problem. Comput. Opt. Appl. **14**(3), 367–381 (1999). doi:10. 1023/A:1026456624746
38. Mattei, N., Walsh, T.: Preflib: a library of preference data. In: Perny, P., Pirlot, M., Tsoukiàs, A. (eds.) ADT2013, vol. 8176, pp. 259–270. Springer, Heidelberg (2013). doi:10.1007/978-3-642-41575-3_20. http://www.preflib.org
39. McCreesh, C., Ndiaye, S.N., Prosser, P., Solnon, C.: Clique and constraint models for maximum common (connected) subgraph problems. In: Rueher, M. (ed.) CP 2016. LNCS, vol. 9892, pp. 350–368. Springer, Cham (2016). doi:10.1007/ 978-3-319-44953-1_23
40. Mehrotra, A., Trick, M.A.: A column generation approach for graph coloring. INFORMS J. Comput. **8**(4), 344–354 (1996). doi:10.1287/ijoc.8.4.344
41. Nogueira, B., Pinheiro, R.G.S., Subramanian, A.: A hybrid iterated local search heuristic for the maximum weight independent set problem. Optim. Lett. 1–17 (2017). doi:10.1007/s11590-017-1128-7
42. Östergård, P.R.J.: A new algorithm for the maximum-weight clique problem. Nord. J. Comput. **8**(4), 424–436 (2001). http://www.cs.helsinki.fi/njc/References/ ostergard2001:424.html
43. Östergård, P.R.J.: A fast algorithm for the maximum clique problem. Discrete Appl. Math. **120**(1–3), 197–207 (2002). doi:10.1016/S0166-218X(01)00290-6
44. Prosser, P.: CSPLib problem 063: Winner determination problem (combinatorial auction)
45. Pullan, W.J.: Approximating the maximum vertex/edge weighted clique using local search. J. Heuristics **14**(2), 117–134 (2008). doi:10.1007/s10732-007-9026-2
46. Refalo, P.: Impact-based search strategies for constraint programming. In: Wallace, M. (ed.) CP 2004. LNCS, vol. 3258, pp. 557–571. Springer, Heidelberg (2004). doi:10.1007/978-3-540-30201-8_41
47. Roth, A.E., Sönmez, T., Ünver, M.U.: Kidney exchange. Q. J. Econ. **119**(2), 457 (2004). doi:10.1162/0033553041382157
48. Roth, A.E., Sönmez, T., Ünver, M.U.: Efficient kidney exchange: coincidence of wants in markets with compatibility-based preferences. Am. Econ. Rev. **97**(3), 828–851 (2007). http://www.aeaweb.org/articles?id=10.1257/aer.97.3.828
49. Saidman, S.L., Roth, A.E., Sonmez, T., Unver, M.U., Delmonico, F.L.: Increasing the opportunity of live kidney donation by matching for two- and three-way exchanges. Transplantation **81**(5), 773–782 (2006)
50. Sanchis, L.A.: Test case construction for the vertex cover problem. In: Dean, N., Shannon, G.E. (eds.) Computational Support for Discrete Mathematics, Proceedings of a DIMACS Workshop, 12–14 March 1992, Piscataway, New Jersey, USA. DIMACS Series in Discrete Mathematics and Theoretical Computer Science, DIMACS/AMS, vol. 15, pp. 315–326 (1992). http://dimacs.rutgers.edu/Volumes/ Vol15.html
51. Sanchis, L.A.: Generating hard and diverse test sets for NP-hard graph problems. Discrete Appl. Math. **58**(1), 35–66 (1995). doi:10.1016/0166-218X(93)E0140-T
52. Sandholm, T., Suri, S.: BOB: improved winner determination in combinatorial auctions and generalizations. Artif. Intell. **145**(1–2), 33–58 (2003). doi:10.1016/ S0004-3702(03)00015-8
53. Sandholm, T., Suri, S., Gilpin, A., Levine, D.: CABOB: a fast optimal algorithm for winner determination in combinatorial auctions. Manag. Sci. **51**(3), 374–390 (2005). doi:10.1287/mnsc.1040.0336
54. Sethuraman, S., Butenko, S.: The maximum ratio clique problem. Comput. Manag. Sci. **12**(1), 197–218 (2015). doi:10.1007/s10287-013-0197-z

55. Shimizu, S., Yamaguchi, K., Saitoh, T., Masuda, S.: Fast maximum weight clique extraction algorithm: optimal tables for branch-and-bound. Discrete Appl. Math. **223**, 120–134 (2017). http://www.sciencedirect.com/science/article/pii/S0166218X1730063X

56. Soriano, P., Gendreau, M.: Tabu search algorithms for the maximum clique problem. In: Johnson and Trick [31], pp. 221–244. http://dimacs.rutgers.edu/Volumes/Vol26.html

57. Strash, D.: On the power of simple reductions for the maximum independent set problem. In: Dinh, T.N., Thai, M.T. (eds.) COCOON 2016. LNCS, vol. 9797, pp. 345–356. Springer, Cham (2016). doi:10.1007/978-3-319-42634-1_28

58. Suters, W.H., Abu-Khzam, F.N., Zhang, Y., Symons, C.T., Samatova, N.F., Langston, M.A.: A new approach and faster exact methods for the maximum common subgraph problem. In: Wang, L. (ed.) COCOON 2005. LNCS, vol. 3595, pp. 717–727. Springer, Heidelberg (2005). doi:10.1007/11533719_73

59. Tavares, W.A., Neto, M.B.C., Rodrigues, C.D., Michelon, P.: Um algoritmo de branch and bound para o problema da clique máxima ponderada. In: Proceedings of XLVII SBPO, vol. 1 (2015)

60. Verfaillie, G., Lemaître, M., Schiex, T.: Russian doll search for solving constraint optimization problems. In: Clancey, W.J., Weld, D.S. (eds.) Proceedings of the Thirteenth National Conference on Artificial Intelligence and Eighth Innovative Applications of Artificial Intelligence Conference, AAAI 1996, IAAI 1996, 4–8 August 1996, Portland, Oregon, vol. 1, pp. 181–187. AAAI Press/The MIT Press (1996). http://www.aaai.org/Library/AAAI/1996/aaai96-027.php

61. Wang, Y., Hao, J., Glover, F., Lü, Z., Wu, Q.: Solving the maximum vertex weight clique problem via binary quadratic programming. J. Comb. Optim. **32**(2), 531–549 (2016)

62. Wang, Y., Cai, S., Yin, M.: Two efficient local search algorithms for maximum weight clique problem. In: Schuurmans, D., Wellman, M.P. (eds.) Proceedings of the Thirtieth AAAI Conference on Artificial Intelligence, 12–17 February 2016, Phoenix, Arizona, USA, pp. 805–811. AAAI Press (2016). http://www.aaai.org/ocs/index.php/AAAI/AAAI16/paper/view/11915

63. Wang, Y., Cai, S., Yin, M.: Local search for minimum weight dominating set with two-level configuration checking and frequency based scoring function. J. Artif. Intell. Res. (JAIR) **58**, 267–295 (2017). doi:10.1613/jair.5205

64. Wu, Q., Hao, J.: Solving the winner determination problem via a weighted maximum clique heuristic. Expert Syst. Appl. **42**(1), 355–365 (2015). doi:10.1016/j.eswa.2014.07.027

65. Wu, Q., Hao, J., Glover, F.: Multi-neighborhood tabu search for the maximum weight clique problem. Ann. OR **196**(1), 611–634 (2012). doi:10.1007/s10479-012-1124-3

66. Zhou, Y., Hao, J., Goëffon, A.: PUSH: a generalized operator for the maximum vertex weight clique problem. Eur. J. Oper. Res. **257**(1), 41–54 (2017). doi:10.1016/j.ejor.2016.07.056

MDDs: Sampling and Probability Constraints

Guillaume Perez and Jean-Charles Régin[(✉)]

Université Nice-Sophia Antipolis, I3S UMR 7271, CNRS, Sophia Antipolis, France
guillaume.perez06@gmail.com, jcregin@gmail.com

Abstract. We propose to combine two successful techniques of Artificial Intelligence: sampling and Multi-valued Decision Diagrams (MDDs). Sampling, and notably Markov sampling, is often used to generate data resembling to a corpus. However, this generation has usually to respect some additional constraints, for instance to avoid plagiarism or to respect some rules of the application domain. We propose to represent the corpus dependencies and these side constraints by an MDD and to develop some algorithms for sampling the solutions of an MDD while respecting some probabilities or a Markov chain. In that way, we obtain a generic method which avoids the development of ad-hoc algorithms for each application as it is currently the case. In addition, we introduce new constraints for controlling the probabilities of the solutions that are sampled. We experiments our method on a real life application: the geomodeling of a petroleum reservoir, and on the generation of French alexandrines. The obtained results show the advantage and the efficiency of our approach.

1 Introduction

Multi-valued decision diagrams (MDDs) are a compressing data structure defined over a set of variables and used to store a set of tuples of values. They are implemented in almost all constraint programming solvers and have been increasingly used to build models [1,3,5,8–10,21,23]. They can be constructed in several ways, from tables, automata, dynamic programming, etc.; or defined by combining two or more MDDs thanks to operators like intersection, union, or difference. They have a high compression efficiency. For instance, an MDD having 14,000 nodes and 600,000 arcs and representing 10^{90} tuples has been used to solve a music synchronization problem [23].

For solving some automatic generation problems, sampling from a knowledge data set is used to generate new data. Often, some additional control constraints must be satisfied. One approach is to generate a vast amount of sequences for little cost, and keep the satisfactory ones. However, this does not work well when constraints are complex and difficult to satisfy. Thus, some works have investigated to integrate the control constraints into the stochastic process.

For instance, in text generation, a Markov chain, which is a random process with a probability depending only on the last state (or a fixed number of them), is defined from a corpus [11,16,18]. In this case, a state can represent a word, and such a process will generate sequences of words, or phrases. It can be modeled as

© Springer International Publishing AG 2017
J.C. Beck (Ed.): CP 2017, LNCS 10416, pp. 226–242, 2017.
DOI: 10.1007/978-3-319-66158-2_15

a directed graph, encoding the dependency between the previous state and the next state. Then, a random walk, i.e. a walk in this graph where the probability for choosing each successor has been given by the Markov model, will correspond to a new phrase. Such a walk corresponds to a sampling of the solution set while respecting the probabilities given by the Markov chain. This process generates sequences imitating the statistical properties of the corpus. Then, the goal is to be able to incorporate some side constraints defining the type of phrases we would like to obtain. For example, we may want to only produce sequences of words that contain no subsequence belonging to the corpus or longer than a given threshold, in order to limit plagiarism [16].

Such Markov models have long been used to generate music in the style of a composer [7,13,16]. The techniques of Markov constraints have been introduced to deal precisely with the issue of generating sequences from a Markov model estimated from a corpus, that also satisfy non Markovian, user defined properties [2,14,15,24].

Hence, there is a real need for being able to sample some solutions while satisfying some other constraints.

The idea of this paper is to represent the corpus dependencies and the additional constraints by an MDD and develop sampling algorithms dealing with the solution set represented by this MDD.

Recently Papadopoulos *et al.* have designed a new algorithm which can be applied to a **regular** constraint [17]. However, the paper is complex because it is a direct adaption of the powerful and general belief propagation algorithm and requires the definition of a regular constraint. In this paper, we propose a conceptually simpler method defined on a more general data structure (the MDD), which may represent any regular constraint, but also different constraints. In addition, we show how to apply it for any kind of samplings and not only on Markov samplings. Thus, instead of developing ad-hoc algorithms or forcing the use of regular constraints, we propose a more general approach that could be used for a large range of problems provided that we have enough memory for representing the MDD.

However, combining samplings and MDDs is not an easy task. Consider, for instance, that we have a very simple MDD (Fig. 1) involving only two variables

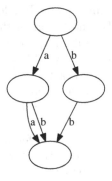

Fig. 1. A simple MDD.

x_1 and x_2 whose values are a and b and that it represents the three solutions $S = \{((x_1, a), (x_2, a)), ((x_1, a), (x_2, b)), ((x_1, b), (x_2, b))\}$. Assume that we want to sample uniformly the solution set. In other words, we want to randomly select one solution with an equal probability for each solution. This can easily be done by randomly selecting a solution in S. Since there are 3 solutions, any solution has a probability of $1/3$ to be selected. The issue with MDDs is that they compress the solution set, so picking a solution with a uniform probability is not straightforward. For instance, if we randomly select the first value of the first variable and if we randomly select the value of the second variable then the selection is not uniform, because we are going to select more often the solution $((x_1, b), (x_2, b))$ than the others. This problem can be solved by computing the local probabilities of selecting a value according to the probabilities of the solutions containing that value.

Furthermore, we study the case where the probabilities of values are not the same and we consider Markov sampling, that is sampling where instead of considering the probability of selecting one value, we consider the probability of selecting a sequence of values.

In addition, it is sometimes interesting to define some constraints on the sampling. For instance, the problem of generating the sequences with the maximum probability in the Markov chain estimated from the corpus satisfying other constraints has been studied by Pachet and Roy [14]. Hence we propose some constraints for imposing that the probabilities of the solutions belong to a given range of probabilities.

This paper is mainly a paper about modeling and the advantage of having general methods for dealing with different kinds of problems occurring in Artificial Intelligence. As an example of this advantage, we apply our method to the transformation of classical texts written in French into alexandrine texts. This means that we try to express the same idea as the original text with the same style but by using only sentences having twelve syllables. The generation of the text in the same style as an author uses a Markov chain that is extracted from the corpus. An MDD is defined from the corpus and ensures that each sentence will have exactly twelve syllables. Then, probabilities implementing the Markov chain are associated with arcs of the MDD, and a random walk procedure is used for sampling the solutions. Thus, the model of this problem is conceptually simple and easy to implement. In addition, thanks to the existence of efficient propagators for MDDs and the algorithms we propose for computing local probabilities, it gives good results in practice.

We also test our approach on a real world application mainly involving convolutions which are expressed by knapsack constraints (i.e. $\sum \alpha_i x_i$) in which the probability of a value to be taken by a variable is defined by a probability mass function. In addition outliers are not allowed. We show how solutions can be efficiently sampled.

Note that the problem we consider is different from the work of Morin and Quimper on the Markov transition constraint which proposes to compute the distribution of the states of a Markov chain [12].

The paper is organized as follows. First we recall some definitions about probability distribution, Markov chain and MDDs and their use in constraint programming. Then, we propose some algorithms for sampling the solution set of an MDD while respecting the probabilities given by a distribution, that can be a probability mass function or a Markov chain. Next, we introduce two constraints ensuring that any solution of an MDD associated with some probability distribution belongs to a given probability interval. Afterwards, we present some experiments on the geomodelling of a petroleum reservoir and on the generation of French alexandrines based on the famous La Fontaine's fables. Finally we conclude.

2   Preliminaries

2.1   Probability Distribution

We consider that the probability distribution is given by a probability mass function (PMF), which is a probability density function for a discrete random variable. The PMF gives for each value v, the probability $P(v)$ that v is taken:

Given a discrete random variable Y taking values in $Y = \{v_1, \ldots v_m\}$ its probability mass function P: $Y \rightarrow [0,1]$ is defined as $P(v_i) = Pr[Y = v_i]$ and satisfies the following condition: $P(v_i) \geq 0$ and $\sum_{i=1}^{m} P(v_i) = 1$.

Property 1. *Let f_P be a PMF and consider $\{x_i\}$ a set of n discrete integer variables independent from a probabilistic point of view and associated with f_P that specifies probabilities for their values. Then, the probability of an assignment of all the variables (i.e. a tuple) is equal to the product of the probabilities of the assigned values. That is $\forall i = 1..n$, $\forall a_i \in D(x_i)$ $P(a_1, a_2, \ldots, a_n) = P(a_1)P(a_2)...P(a_n)$.*

2.2   Markov Chain

A Markov chain[1] is a stochastic process, where the probability for state X_i, a random variable, depends only on the last state X_{i-1}. A Markov chain produces sequence $X_1, \ldots, X_n$ with a probability $P(X_1)P(X_2|X_1) \ldots P(X_n|X_{n-1})$.

Property 2. *Let P_M be a Markov chain and consider a set of n discrete integer variables associated with P_M that specifies probabilities for their values. Then, $\forall i = 1..n$, $\forall a_i \in D(x_i)$ $P(a_1, a_2, \ldots, a_n) = P(a_1)P(a_2|a_1) \ldots P(a_n|a_{n-1})$.*

Several methods can be used to estimate the Markov chain from a corpus, like the maximum likelihood estimation [11]. This paper is independent from such methods and considers that the Markov chain is given.

---

[1] Order k Markov chains have a longer memory: the Markov property states that $P(X_i|X_1, \ldots, X_{i-1}) = P(X_i|X_{i-k}, \ldots, X_{i-1})$. They are equivalent to order 1 Markov chains on an alphabet composed of k-grams, and therefore we assume only order 1 Markov chains [17].

Sampling a Markov chain can be simply and efficiently done by a random walk (i.e. a path consisting of a succession of random steps) driven by the distribution of the Markov chain. If we need to build a finite sequence of length k, then we perform a random walk of k iterations using the given distribution.

\	a	b	Tuple	Probability
a	0.9	0.1	aa	0.54
b	0.1	0.9	ab	0.06
			ba	0.04
			bb	0.36

Fig. 2. Markov chain for two variables. The starting probabilities are 0.6 for a and 0.4 for b.

Example. Consider M, the Markov chain in Fig. 2 and an initial probability of 0.6 for a and 0.4 for b. If we apply M on two variables x_1 and x_2, then the probability of the tuple (a, a) is $P(x_1, a)P((x_2, a)|(x_1, a)) = 0.6 \times 0.9 = 0.54$. The probabilities of the four possible tuples are given in Fig. 2. The sum of the probabilities is equal to 1.

2.3   Multi-valued Decision Diagram (MDD)

An MDD is a data-structure representing discrete functions. It is a multiple-valued extension of BDDs [4]. An MDD, as used in CP [1,3,5,8–10,20,23], is a rooted directed acyclic graph (DAG) used to represent some multi-valued function $f: \{0 \ldots d - 1\}^n \rightarrow \{true, false\}$. Given the n input variables, the DAG representation is designed to contain $n+1$ layers of nodes, such that each variable is represented at a specific layer of the graph. Each node on a given layer has at most d outgoing arcs to nodes in the next layer. Each arc is labeled by its corresponding integer. The arc (u, v, a) is from node u to node v and labeled by a. All outgoing arcs of the layer n reach tt, the true terminal node (the false terminal node is typically omitted). There is an equivalence between $f(a_1, \ldots, a_n) = true$ and the existence of a path from the root node to the true terminal node whose arcs are labeled $a_1, \ldots, a_n$. The number of nodes of an MDD is denoted by V, the number of edges by E and d is the largest domain size of the input variables.

MDD of a Constraint. Let C be a constraint defined on $X(C)$. The MDD associated with C, denoted by MDD(C), is an MDD which models the set of tuples satisfying C. MDD(C) is defined on $X(C)$, such that layer i corresponds to the variable x_i and the labels of arcs of the layer i correspond to values of x_i, and a path of MDD(C) where a_i is the label of layer i corresponds to a tuple $(a_1, \ldots, a_n)$ on $X(C)$.

Consistency with MDD(C). An arc (u, v, a) at layer i is valid iff $a \in D(x_i)$. A path is valid iff all its arcs are valid. The value $a \in D(x_i)$ is consistent with

MDD(C) iff there is a valid path in MDD(C) from the root node to tt which contains an arc at layer i labeled by a.

MDD Propagator. An MDD propagator associated with a constraint C is an algorithm which removes some inconsistent values of $X(C)$. It establishes arc consistency of C if and only if it removes all inconsistent values with MDD(C). This means that it ensures that there is a valid path from the root to the true terminal node in MDD(C) if and only if the corresponding tuple is allowed by C and valid.

Cost-MDD. A cost-MDD is an MDD whose arcs have an additional information: the cost c of the arc. That is, an arc is a 4-uplet $e = (u, v, a, c)$, where u is the head, v the tail, a the label and c the cost. Let M be a cost-MDD and p be a path of M. The cost of p is denoted by $\gamma(p)$ and is equal to the sum of the costs of the arcs it contains.

Cost-MDD of a Constraint [6,8]. Let C be a constraint and f_C be a function associating a cost with each value of each variable of $X(C)$. The cost-MDD of C and f_C is denoted by cost-MDD(C, f_C) and is MDD(C) whose the cost of an arc labeled by a at layer i is $f_C(x_i, a)$.

3 Sampling and MDD

We aim at sampling the solution set of an MDD while respecting the probabilities given by a distribution, that can be a PMF or a Markov chain.

Let M be an MDD whose n variables are associated with a distribution that specifies the probabilities of their values. For sampling the solutions of M, we propose to associate with each arc a probability, such that a simple random walk from the root node to tt according to these probabilities will sample the solution set of M while respecting the probabilities of the distribution of M.

First, we consider that the distribution of M is given by a PMF and that the variables of M are independent from a statistical point of view. Then, we will consider that we have a Markov chain for determining the probability of a value to be selected.

3.1 PMF and Independent Variables

If the distribution associated with M is defined by a PMF f_P and if the variables of M are independent from a statistical point of view, then we propose to associate with each arc e a probability $P(e)$. From Property 1 we know that the probability of a solution $(a_1, \ldots, a_n)$ must be equal to $\Pi_{i=1}^{n} P(a_i)$.

We could be tempted to define $P(e)$ as the value of $f_P(label(e))$ where $label(e)$ is the label (i.e. value) associated with e. However, this is not exact because the MDD usually does not contain all possible combinations of values as solutions.

For instance, consider the example of Fig. 1 with a uniform distribution. If all probabilities are equivalent then each solution must be able to be selected with the same probability, which is $1/3$ since there are three solutions (a, a), (a, b) and (b, b). Now, if we do a random walk considering that the probability of each arc is $1/2$ then we will choose with a probability $1/2$ the solution (b, b) which is incorrect. The problem stems from the fact that the probabilities of the higher layers are not determined according to the probabilities of solutions that they can reach while it should be the case. The choice (x_1, a) allows to reach 2 solutions and (x_1, b) one. So, with a uniform distribution the probability of choosing a for x_1 should be $2/3$ while that of choosing b should be $1/3$.

Definition 1. *The partial solutions that can be reached from a node n in an MDD are defined by the paths from n to tt.*

In order to compute the correct values, we compute for each node n the sum of the original probabilities of the partial solutions that we can reach from n. Then, we renormalize these values in order to have these sums equal to 1 for each node. For instance, for the node reached by traversing the first arc labeled by a in Fig. 1, the sum of the original probabilities is $1/2 + 1/2 = 1$, so the original probabilities are still valid. However, for the node reached by traversing the arc from the root and labeled by b, the sum of the original probabilities is $1/2$, so half of the combinations are lost. This probability is no longer valid and new values must be computed.

The sum of the original probabilities of the partial solutions that can be reached from a node is defined as follows:

Property 3. *Let M be an MDD defined on X and f_P a PMF associated with M. Let n be any of node of the MDD and A be any partial instantiation of X reaching node n. The sum of the original probabilities of the partial solutions that can be reached from n is $v(n) = \sum_{s \in S(n)} P(s|A)$, where $S(n)$ is the set of partial solutions that we can reach from n and $P(s|A)$ is the probability of s under condition A. The probability of any arc $e = (n', n, a)$ is defined by $P(e) = f_P(a) \times v(n)$.*

Proof. By induction from tt. Assume this is true at layer $i + 1$. Let n' be a node of layer i, n a node in layer $i + 1$ and $e = (n', n, a)$ an arc. We have $P(e) = f_P(a) \times v(n)$, that is $P(e) = f_P(a) \times \sum_{s \in S(n)} P(s|A)$, where A is any partial instantiation reaching node n. So for node n' we have:

$v(n') = \sum_{e \in \omega^+(n')} P(e)$, where $\omega^+(n')$ is the set of outgoing arcs of n'

$v(n') = \sum_{e \in \omega^+(n')} f_P(label(e)) \times \sum_{s \in S(n)} P(s|A)$. Note that A is any partial instantiation reaching node n, so it can go through e. So we have

$v(n') = \sum_{s \in S(n')} P(s|A')$ where A' is any partial instantiation reaching node n'. □

The correct probabilities can be computed by a bottom-up algorithm followed by a top-down algorithm. First, we consider the second to last layer and we define the probability P of an arc labeled by a as $f_P(a)$. Then, we directly apply Property 3 from the bottom of the MDD to the top: once the layer $i+1$ is determined, we compute for each node n' of the layer i the value $v(n') = \sum_{e \in \omega^+(n')} P(e) = \sum_{e \in \omega^+(n')} f_P(label(e)) \times v(n)$. Once the bottom-up part is finished, we normalize the computed values P in order to have $v(n) = 1$ for each node n. We use a simple top-down procedure for computing these values. Figure 3 details this process. The left graph simply contains the probability of the arc labels. The middle graph shows the bottom-up procedure. For instance, we can see that the right arc outgoing from the source has a probability equal to $1/2 \times 1/2 = 1/4$. Thus a normalization is needed for the root because the sum of the probabilities of the outgoing arcs is $1/2 + 1/4 = 3/4 < 1$. The right graph is obtained after normalization.

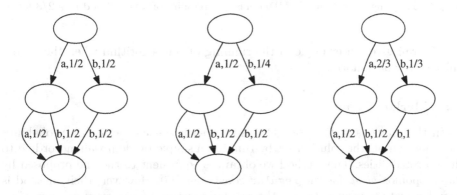

Fig. 3. Sampling from a simple MDD. The probability of a and b are $1/2$.

Note that the normalization consists of computing the probability according to the sum of the probabilities. If $P(e)$ is the current value for the arc $e = (u, v, a)$ and T is the sum of the probability of the outgoing arcs from u, then the probability of e becomes $P(e)/T$.

This step can be avoided in practice by computing such normalized values only when needed.

Algorithm COMPUTEMDDPROBABILITIES can be described as follows:

1. Set $v(tt) = 1$; For each node $v \neq tt$, in a Breadth First Search (BFS) in bottom-up fashion:
   (a) Compute $v(n)$ the sum of the original probabilities of the outgoing arcs of n.
   (b) Define the probability of each incoming arc e of n labeled by a as $P(e) = f_P(a) \times v(n)$.
2. For each node in a BFS top-down fashion, normalize the probabilities of the outgoing arcs.

During this algorithm, each sum is calculated once for each node during the bottom up processing, and the normalization is performed once for each arc. The final complexity is $O(|E| + |V|)$.

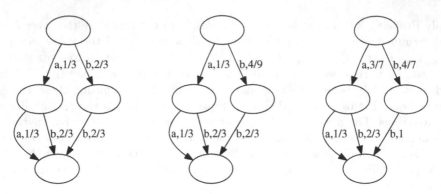

Fig. 4. Sampling from a simple MDD. The probability of a is $1/3$ and it is $2/3$ for b.

Figure 4 gives an example of the running of this algorithm when the probabilities are not uniform.

3.2   Markov Chain

As in the previous section, our goal is to associate each arc with a probability and then sample the solution set by running a simple random walk according to these probabilities. The method we obtain is equivalent to the one proposed by Papadopoulos *et al.* for the **regular** constraint [17]. However, their method is complex and the propagation of the regular constraint costs more memory than the one of an MDD [20]. We claim that our method is conceptually simpler.

It is more difficult to apply a Markov chain than a PMF because in a Markov chain the probability of selecting a value depends on the previous selected value, that is, probabilities must be defined in order to satisfy Property 2. More precisely, in an MDD, a node can have many incoming arcs, and these different incoming arcs can have different labels. Since the Markov probability depends on the previous value, the outgoing arcs of that node may have different probabilities depending on which was the incoming arc label. Thus, for an arc e, we need to have several probability values depending on the previous arc that has been used.

There are two possible ways to deal with a Markov chain. Either we transform the MDD by duplicating nodes in order to be able to apply an algorithm similar as COMPUTEMDDPROBABILITIES or we directly deal with the original MDD and we design a new algorithm.

Duplication of Nodes. We can note that the matrix of the Markov chain represents a compression of nodes. Thus, if we duplicate each node according to its incoming arcs then we obtain a new MDD for which the probabilities become independent. More precisely, for each node n we split the node n in as many nodes as there are different values incoming. This means that each node n has only incoming arcs having the same label, and so only one value a incoming.

Thus, the probability of each outgoing arc of the duplicated nodes of n can be determined directly by the Markov matrix.

For instance, consider the probabilities of Fig. 2 and that we have a node n with two incoming arcs: one labeled by a an the other labeled by b; and with two outgoing arcs: one labeled by a an the other labeled by b (Fig. 5). The node n is split into two nodes n_a and n_b. Node n_a has only incoming arcs labeled by a, and n_b has only incoming arcs labeled by b ((c) in Fig. 5). In this case, we can define the probabilities as if we had independent variables. The probability of the arc (n_a, x, a), is defined by $P(a|a) = 0.9$, the probability of the arc (n_a, x, b) is $P(b|a) = 0.1$, the probability of the arc (n_b, x, a) is $P(a|b) = 0.1$, the probability of the arc (n_b, x, b) is $P(b|b) = 0.9$. Figure 5 shows the duplication of a node. Note that when the node x will be split into two nodes x_a and x_b, then each of them will have two incoming arcs having the same label, a for x_a and b for x_b ((d) in Fig. 5).

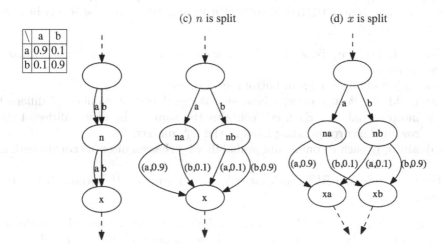

Fig. 5. Duplication of a node and computation of probabilities.

Let $P_C(e)$ be the computed probability of any edge e computed by the duplication process. We can establish a Property similar as Property 3.

Property 4. *Let M be an MDD defined on X and P_C a probability associated with each arc. Let n be any of node of the MDD and A be any partial instantiation of X reaching node n. The sum of the original probabilities of the partial solutions that can be reached from n is $v(n) = \sum_{s \in S(n)} P(s|A)$, where $S(n)$ is the set of partial solutions that we can reach from n and $P(s|A)$ is the probability of s under condition A. The probability of any arc $e = (n', n, a)$ is defined by $P(e) = P_C(e) \times v(n)$.*

Proof. Similar as for Property 3. □

From this property we can design an algorithm similar as COM-PUTEMDDPROBABILITIES by using $P_C(e)$ instead of $f_P(label(e))$ for each arc e. The drawback of this method is that it can multiply the number of nodes by at most d, the greatest cardinality domain of variables and also increases the number of edges which slowdowns the propagators. The next section presents another method avoiding this duplication.

A New Algorithm. In order to deal with the fact that the probability of an outgoing arc depends on the label of the incoming arc without duplicating nodes, we associate each node with a probability matrix whose row depends on the incoming arc label. We denote these matrices by P_M^n for the node n. For efficiency, we only have one vector by incoming value instead of the full matrix, and each vector contains only the probability of the possible outgoing arcs labels. Then, the same reasoning as previously can be applied. We just need to adapt the previous algorithm by using matrices instead of duplicating nodes:

Algorithm COMPUTEMDDMARKOVPROBABILITIES can be described as follows:

1. For each node n, build the P_M^n matrix by copying the initial Markov probabilities.
2. For each node n, in BFS in bottom-up fashion:
   (a) Build the vector $vv(n)$ whose size is equal to the number of different incoming labels[2]. Each cell contains the sum of the probabilities of the row of the corresponding label in the P_M^n matrix.
   (b) Multiply each incoming arc probability by the cell of $vv(n)$ corresponding to its label.
3. For each node in a BFS top-bottom fashion, normalize the probability of the outgoing arcs.

Example. Consider the MDD of Fig. 6a, if we reuse the Markov distribution of Fig. 2 and apply the step 1 of the method, we obtain the MDD in Fig. 6b.

Now from the MDD in Fig. 6b, we perform step 2, first (step 2.a) we process the sum of the outgoing probabilities for each node. For example for node 5 its probability is $0.1 + 0.9 = 1$ and for node 3 the sum is 0.9. For these two nodes the sum does not depend on the incoming arc label because there is only one. This is not the case for node 4 which has a sum of 0.1 for the incoming arc labeled by a and 0.9 for the incoming arc labeled by b. Now we apply step 2.b: we multiply the probability of the incoming arcs by the sum associated to their label in their destination node. Consider the arc from node 1 to node 3 and labeled by a, its probability was 0.9 and the sum of probabilities in its destination node is 0.9, then its new probability is 0.81. The arc from node 1 to node 4 is labeled by b; its probability was 0.1. For node 4, the sum is 0.9 for the incoming arc labeled by b, so the new probability of the $(1, 4, b)$ is $0.1 \times 0.9 = 0.09$. The MDD in Fig. 7a is labeled with the resulting global probabilities.

---

[2] $vv(n)$ represents a vector of $v(n)$.

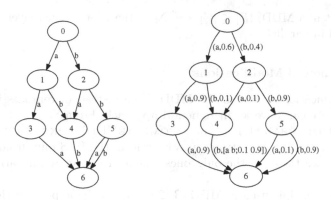

Fig. 6. (a) left: an MDD. (b) right: the MDD whose arcs have their probability set thanks to the Markov distribution.

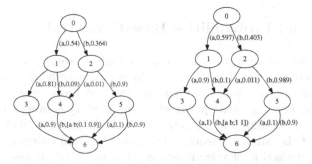

Fig. 7. (a) left: MDD from Fig. 6b whose arcs probability has been multiplied by the sum of the probabilities of the outgoing arcs from their destination node. (b) right: the MDD with renormalized probabilities.

Finally, from the MDD in Fig. 7a, we normalize the outgoing arc probability of each node (step 3). For the root node 0, the outgoing probabilities sum is $0.54 + 0.364 = 0.904$. For its arc labeled by a and directed to node 1, the probability become $0.54/0.904 = 0.597$, this value has been rounded to 3 digits for readability. For its arc labeled by b and directed to node 2, the probability becomes $0.364/0.904 = 0.403$ (rounded). Thus, the outgoing sum of probabilities emanating from node 0 becomes $0.597+0.403 = 1$. The MDD from Fig. 7b shows the normalized probabilities.

Complexities. The complexities of COMPUTEMDDMARKOVPROBABILITIES algorithm are the following. The number of matrices is $|V|$, in the worst case the number of columns and rows is d, so the global memory complexity is $O(|V| \times d^2)$. The complexity of each of the operations of this method are all linear over the matrices, so the overall time complexity is $O(|V| \times d^2)$. Since the number of columns of the matrix of a node is equal to the number of outgoing arcs of this node, a more realistic complexity for space and time is $O(|V| + |E| \times d)$, knowing that in a MDD, $|E| \leq |V| \times d$. Note that, for a given layer, nodes can be processed in parallel.

3.3   Incremental Modifications

If some modifications occur in the MDD, then instead of reprocessing all the probabilities we can have an incremental approach. From Step 2 of algorithms COMPUTEMDDPROBABILITIES or COMPUTEMDDMARKOVPROBABILITIES, which performs a BFS in bottom-up, we perform the BFS only from the modified nodes since they are the only ones that can trigger modifications of the probabilities.

The reset principle used in MDD4R [20] can also be applied in this case. In other words, when there is less remaining arcs than deleted arcs, it is worthwhile to recompute from scratch the values.

4   MDDs and Probabilities Based Constraints

For some reasons, like security or for avoiding outliers, some paths of MDDs can be unwanted, because they have only very little chance to be selected or because they contain almost only values having the strongest probability to be selected. In other words, we accept only paths whose probability is in a certain interval.

We define constraints for this purpose. One, named the MDDProbability, considered that the MDD is associated with a PMF and independent variables and the other, named MDDMarkovProcess, that the MDD is associated with a Markov chain.

Definition 2. *Given M an MDD defined on $X = \{x_1, x_2, \ldots, x_n\}$ that are independent from a probabilistic point of view and associated with f_P a probability mass function , P_{min} a minimum probability and P_{max} a maximum probability. The constraint MDDProbability$(X, f_P, M, P_{min}, P_{max})$ ensures that every allowed tuple $(a_1, a_2, \ldots a_n)$ is a solution of the MDD and satisfies $P_{min} \leq \Pi_{i=1}^{n} f_P(a_1) \leq P_{max}$.*

This constraint can be easily transformed into a cost-MDD constraint. The cost associated with an arc labeled by a is $\log(f_P(a))$, and the logarithms of P_{min} and P_{max} are considered for dealing with a sum instead of a product[3]. Thus, any cost-MDD propagator can be used [22].

Definition 3. *Given M an MDD defined on $X = \{x_1, x_2, \ldots, x_n\}$ and associated with P a Markov chain, P_{min} a minimum probability and P_{max} a maximum probability. The constraint MDDMarkovProcess$(X, P, M, P_{min}, P_{max})$ ensures that every allowed tuple $(a_1, a_2, \ldots a_n)$ is a solution of the MDD and satisfies $P_{min} \leq P(a_1)P(a_2|a_1)\ldots P(a_n|a_{n-1}) \leq P_{max}$.*

---

[3] We can also directly deal with products if we modify the costMDD propagator accordingly.

As we have seen, with a Markov chain, the probability for selecting an arc depends on the previous selected arc. Thus, each arc of the MDD is associated with several probabilities. So we cannot directly use a cost-MDD propagator as for the MDDProbability constraint. However, if we accept to duplicate the nodes as proposed in the previous section then we can immediately transforms the constraint into a simple cost-MDD constraint by considering logarithms of probabilities and any cost-MDD propagator can be used. Since the number of time a node can be duplicated is bounded by d, the overall complexity of this transformation is $O(d \times (|V| + |E|))$.

5   Evaluation

The experiments were run on a macbook pro (2013) Intel core i7 2.3 GHz with 8 GB of memory. The constraint solver used is or-tools. MDD4R [20] is used as MDD propagator and cost-MDD4R as cost-MDD propagator [22].

5.1   PMF Constraint and Sampling

The data come from a real life application: the geomodeling of a petroleum reservoir [19]. The problem is quite complex and we consider here only a sub-part. Given a seismic image we want to find the velocities. Velocities values are represented by a probability mass function (PMF) on the model space. Velocities are discrete values of variables. For each cell c_{ij} of the reservoir, the seismic image gives a value s_{ij} from which we define a sum constraint $C_{ij}: \sum_{k=1}^{22} \alpha_k log(x_{i-11+k-1} j) = s_{ij} \pm \epsilon$, where α_k are defined from the given seismic wavelet. Locally, that is, for each sum, we have to avoid outliers w.r.t. the PMF for the velocities. The problem is huge (millions of variables) so we consider here only a very small part.

We recall that the MDD of the constraint $\sum_{x_i \in X} f(x_i) \in I$, with $I = [a, b]$ is denoted by $MDD(\Sigma_{f,I}(X))$ and defined as follows. For the layer i, there are as many nodes as there are values of $\sum_{k=1}^{i} f(x_k)$. Each node is associated with such a value. A node n_p at layer i associated with value v_p is linked to a node n_q at layer $i + 1$ associated with value v_q if and only if $v_q = v_p + f(a_i)$ with $a_i \in D(x_i)$. Then, only values v of the layer $|X|$ with $a \leq v \leq b$ are linked to tt. The reduction operation is applied after the definition and delete invalid nodes [21]. The construction can be accelerated by removing states that are greater than b or that will not permit to reach a.

Each constraint C_{ij} is represented by $MDD(\Sigma_{f,I}(X))$ where $f(x_i) = \alpha_i x_i$ and I is the tight interval representing $[s_{ij} - \epsilon, s_{ij} + \epsilon]$. Outliers are avoided thanks to an MDDProbability constraint defined from the PMF for the velocities. P_{min} is defined by selecting only values having the 10% smaller probabilities, P_{max} is defined by selecting only values having the 10% greater probabilities. This constraint is represented by a cost-MDD constraint, as explained in the Sect. 4. Then, we intersect it with $MDD(\Sigma_{f,I}(X))$.

We consider 20 definitions of C_{ij}. We repeat the experiments 20 times and take the mean of the results.

For each constraint C_{ij}, the resulting MDD has in average 116,848 nodes and 1,239,220 edges. More than 320 s are needed to compute it. Only 8 ms are required by COMPUTEMDDPROBABILITIES algorithm in average. When a modification occurs the time to recompute the values are between a negligible value when the modifications are close to the root of the MDD and 8 ms when another part is modified.

For sampling 100,000 solutions we need 169 ms with the rand() function and 207 ms with the Mersenne-Twister random engine in conjunction with the uniform generator of the C++ standard library. Note that the time spends within the rand() function is 15 ms, whereas it is 82 ms with the second function. Therefore, the sampling procedures require less than 3 times the time spent in the random function.

5.2   Markov Chain and Sampling

We evaluate our method for generating French alexandrines. That is, sentences containing exactly twelve syllables. The goal is to transform an existing text into a text having the same meaning but using only alexandrines. From the corpus we define a Markov chain and an MDD representing the sentences having the right number of syllables. The sampling procedure we define generates solutions of the MDD associated with the Markov chain, that is, sentences hopefully resembling those of the corpus and having exactly 12 syllables. This model is simple and easy to implement. Note that we are not able to model this problem with any other technique, even the one proposed by Papadopoulos *et al*, because we need to deal only with sentences having 12 syllables and we do not know how to integrate this constraint into their model.

First, we use a corpus defined by one of the famous La Fontaine's fables. Here is the result we obtain for the fable: La grenouille qui veut se faire aussi grosse que le boeuf (The Frog and the Ox). We have underlined the syllables that must be pronounced when it is unclear:

La grenouille veut se faire aussi grosse que le bœuf

Grands seigneurs Tout bourgeois veut bâtir comme un Bœuf
Plus sages Tout marquis veut bâtir comme un œuf
Pour égaler l'animal en tout M'y voila
Voici donc Point du tout comme les grands seigneurs
Chétive Pécore S'enfla si bien qu'elle creva
Seigneurs Tout petit prince a des ambassadeurs

The generation of the MDD with the correct probabilities, that is just before the random walk, can be performed in negligible computational time.

We also considered a larger corpus: "A la recherche du temps perdu" of Proust, which contains more than 10,000 words. In this case, the results are less pertinent and some more work must be done about the meaning of the sentences.

However, the method is efficient in term of computing performance because only 2 s are needed to create the MDD with the correct probabilities.

6 Conclusion

We have presented two methods for sampling MDDs, one using a probability mass function and another one using a Markov chain. These methods require the definition of probabilities for each arc and we have given algorithms for performing this task. We have also proposed propagators for constraining these probabilities. Thanks to these algorithms and MDD propagators we can easily model and implement complex problems of automatic music or text generations having good performances in practice. We have experimented our method on a real life application: the geomodeling of a petroleum reservoir and on the problem of the transformation of French texts into alexandrines. We have shown how it is easy to define the model and to generate solutions.

Acknowledgments. This research is conducted within the Flow Machines project which received funding from the European Research Council under the European Unions Seventh Framework Programme (FP/2007–2013)/ERC Grant Agreement no. 291156. We would like to thank F. Pachet and P. Roy, who gave us the idea of this article.

References

1. Andersen, H.R., Hadzic, T., Hooker, J.N., Tiedemann, P.: A constraint store based on multivalued decision diagrams. In: Bessière, C. (ed.) CP 2007. LNCS, vol. 4741, pp. 118–132. Springer, Heidelberg (2007). doi:10.1007/978-3-540-74970-7_11
2. Barbieri, G., Pachet, F., Roy, P., Esposti, M.D.: Markov constraints for generating lyrics with style. In: ECAI 2012–20th European Conference on Artificial Intelligence, pp. 115–120 (2012)
3. Bergman, D., Hoeve, W.-J., Hooker, J.N.: Manipulating MDD relaxations for combinatorial optimization. In: Achterberg, T., Beck, J.C. (eds.) CPAIOR 2011. LNCS, vol. 6697, pp. 20–35. Springer, Heidelberg (2011). doi:10.1007/978-3-642-21311-3_5
4. Bryant, R.E.: Graph-based algorithms for Boolean function manipulation. IEEE Trans. Comput. **35**(8), 677–691 (1986)
5. Cheng, K.C.K., Yap, R.H.C.: An MDD-based generalized arc consistency algorithm for positive and negative table constraints and some global constraints. Constraints **15**(2), 265–304 (2010)
6. Demassey, S., Pesant, G., Rousseau, L.-M.: A cost-regular based hybrid column generation approach. Constraints **11**(4), 315–333 (2006)
7. Brooks, F., Hopkings, A., Neumann, P., Wright, W.: An experiment in musical composition. **3**(6), 175–182 (1957)
8. Gange, G., Stuckey, P.J., Hentenryck, P.: Explaining propagators for edge-valued decision diagrams. In: Schulte, C. (ed.) CP 2013. LNCS, vol. 8124, pp. 340–355. Springer, Heidelberg (2013). doi:10.1007/978-3-642-40627-0_28

9. Hadzic, T., Hooker, J.N., ÓSullivan, B., Tiedemann, P.: Approximate compilation of constraints into multivalued decision diagrams. In: Stuckey, P.J. (ed.) CP 2008. LNCS, vol. 5202, pp. 448–462. Springer, Heidelberg (2008). doi:10.1007/978-3-540-85958-1_30

10. Hoda, S., Hoeve, W.-J., Hooker, J.N.: A systematic approach to MDD-based constraint programming. In: Cohen, D. (ed.) CP 2010. LNCS, vol. 6308, pp. 266–280. Springer, Heidelberg (2010). doi:10.1007/978-3-642-15396-9_23

11. Jurafsky, D., Martin, J.H.: Speech and Language Processing. Pearson, London (2014)

12. Morin, M., Quimper, C.-G.: The Markov transition constraint. In: Simonis, H. (ed.) CPAIOR 2014. LNCS, vol. 8451, pp. 405–421. Springer, Cham (2014). doi:10.1007/978-3-319-07046-9_29

13. Nierhaus, G.: Algorithmic Composition: Paradigms of Automated Music Generation. Springer, Heidelberg (2009)

14. Pachet, F., Roy, P.: Markov constraints: steerable generation of Markov sequences. Constraints 16(2), 148–172 (2011)

15. Pachet, F., Roy, P., Barbieri, G.: Finite-length Markov processes with constraints. IJCAI 2011, 635–642 (2011)

16. Papadopoulos, A., Roy, P., Pachet, F.: Avoiding plagiarism in Markov sequence generation. In: Proceeding of the Twenty-Eight AAAI Conference on Artificial Intelligence, pp. 2731–2737 (2014)

17. Papadopoulos, A., Pachet, F., Roy, P., Sakellariou, J.: Exact sampling for regular and Markov constraints with belief propagation. In: Pesant, G. (ed.) CP 2015. LNCS, vol. 9255, pp. 341–350. Springer, Cham (2015). doi:10.1007/978-3-319-23219-5_24

18. Papadopoulos, A., Roy, P., Régin, J.-C., Pachet, F.: Generating all possible palindromes from Ngram corpora. In: Proceedings of the Twenty-Fourth International Joint Conference on Artificial Intelligence, IJCAI 2015, 25–31 July 2015, Buenos Aires, Argentina, pp. 2489–2495 (2015)

19. Pennington, W.D.: Reservoir geophysics 66(1) (2001)

20. Perez, G., Régin, J.-C.: Improving GAC-4 for table and MDD constraints. In: ÓSullivan, B. (ed.) CP 2014. LNCS, vol. 8656, pp. 606–621. Springer, Cham (2014). doi:10.1007/978-3-319-10428-7_44

21. Perez, G., Régin, J.-C.: Efficient operations on MDDs for building constraint programming models. In: International Joint Conference on Artificial Intelligence, IJCAI 2015, Argentina, pp. 374–380 (2015)

22. Perez, G., Régin, J.-C.: Soft and cost MDD propagators. In: The Thirty-First AAAI Conference on Artificial Intelligence AAAI 2017 (2017)

23. Roy, P., Perez, G., Régin, J.-C., Papadopoulos, A., Pachet, F., Marchini, M.: Enforcing structure on temporal sequences: the Allen constraint. In: Rueher, M. (ed.) CP 2016. LNCS, vol. 9892, pp. 786–801. Springer, Cham (2016). doi:10.1007/978-3-319-44953-1_49

24. Roy, P., Pachet, F.: Enforcing meter in finite-length Markov sequences. In: AAAI 2013 (2013)

An Incomplete Constraint-Based System for Scheduling with Renewable Resources

Cédric Pralet$^{(\boxtimes)}$

ONERA – The French Aerospace Lab, 31055 Toulouse, France
cedric.pralet@onera.fr

Abstract. In this paper, we introduce a new framework for managing several kinds of renewable resources, including disjunctive resources, cumulative resources, and resources with setup times. In this framework, we use a list scheduling approach in which a priority order between activities must be determined to solve resource usage conflicts. In this context, we define a new differentiable constraint-based local search invariant which transforms a priority order into a full schedule and which incrementally maintains this schedule in case of change in the order. On top of that, we use multiple neighborhoods and search strategies, and we get new best upper bounds on several scheduling benchmarks.

1 Introduction

In scheduling, renewable resources are resources which are consumed during the execution of activities and released in the same amount at the end of these activities. Such resources are present in most scheduling problems, and various types of renewable resources were extensively studied in the literature, such as *disjunctive resources*, which can perform only one activity at a time, *disjunctive resources with setup times*, which can require some time between activities successively realized by the resource, or *cumulative resources*, which can perform several activities in parallel up to a given capacity. These three kinds of renewable resources are respectively present in Job Shop Scheduling Problems (JSSPs [30]), Job Shop Scheduling Problems with Sequence-Dependent Setup Times (SDST-JSSPs [2]), and Resource Constrained Project Scheduling Problems (RCPSPs [8]).

In the constraint programming community, specific global constraints were defined to efficiently deal with renewable resources, like the *disjunctive* and *cumulative* constraints [1,10], together with efficient propagators based on edge-finding [11,36], timetable edge-finding [37], or on mechanisms to deal with setup times [34]. Following these developments, constraint programming is nowadays one of the best systematic approach for solving scheduling problems with renewable resources [19,21,32].

In parallel, several incomplete search techniques were developed in the scheduling community to quickly produce good-quality solutions on large instances. One of these is *list scheduling*. It manipulates a priority list between

© Springer International Publishing AG 2017
J.C. Beck (Ed.): CP 2017, LNCS 10416, pp. 243–261, 2017.
DOI: 10.1007/978-3-319-66158-2_16

activities, and at each step considers the next activity in the list and inserts it at the earliest possible time without delaying activities already placed in the schedule (so-called *serial schedule generation scheme*). For RCPSP, such a list scheduling approach is used for heuristic search [23,24] but also for designing local search [12] or genetic algorithms [20].

In this paper, we propose a combination between list scheduling and constraint programming. More specifically, we combine list scheduling with constraint-based local search [35], *with the goal of being able to deal with various types of renewable resources (disjunctive or cumulative, with or without setup times).* To get such a combination, we define a new constraint-based scheduling system composed of three layers: (1) an *incremental evaluation layer*, used for estimating very quickly the impact of local modifications on a given priority list, (2) a *neighborhood layer*, containing a catalog of neighborhoods usable for updating priority lists, and (3) a *search strategy layer*, which implements various techniques for escaping local minima and plateaus.

The paper is organized as follows. Sections 2–3 present the new framework considered and a lazy schedule generation scheme for this framework. Sections 4–5 detail the first layer mentioned above. Sections 6–7 give a brief overview of the second and third layers, and Sect. 8 shows the performance of the approach on standard benchmarks. In our current implementation, the techniques proposed are actually applied to a wider class of problems involving release and due dates for activities, time-dependent processing times, resource availability windows, and choices on the resources used by activities. We present here a simplified version for readability issues.

2   RCPSP with Sequence-Dependent Setup Times (SDST-RCPSP)

To simultaneously cover cumulative resources and disjunctive resources with setup times, we introduce a new framework called *Resource Constrained Project Scheduling Problem with Sequence-Dependent Setup Times* (SDST-RCPSP). The more complex part of this unifying framework (cumulative resources with setup times) can also be useful in practice. For instance, in manufacturing, a painting machine might be able to simultaneously paint several items with the same color, while requiring a setup time to change the color used by the machine when needed. In space applications, satellites can be equipped with several communication channels allowing them to transmit several data files in parallel to a given ground reception station, while requiring a setup time to change the pointing of the satellite to download data to another station.

Formally, an SDST-RCPSP is defined by a set of renewable resources $\mathcal{R}$ and by a set of activities $\mathcal{A}$ to be realized. Each resource $r \in \mathcal{R}$ has a maximum capacity K_r (equal to 1 for disjunctive resources), a set of possible running modes $\mathcal{M}_r$ (reduced to a singleton for resources without setup times), and an initial running mode $m_{0,r} \in \mathcal{M}_r$. For each pair of distinct resource modes $m, m' \in \mathcal{M}_r$, a setup time $\Delta_{r,m,m'}$ is introduced to represent the duration required by resource r to make a transition from mode m to mode m'.

Each activity $a \in \mathcal{A}$ has a duration (or processing time) p_a and consumes a set of resources $\mathcal{R}_a \subseteq \mathcal{R}$. We assume that $p_a > 0$ when activity a consumes at least one resource. With each activity a and each resource $r \in \mathcal{R}_a$ are associated the quantity of resource $q_{a,r} \in [1..K_r]$ consumed by a over r, and the resource mode $m_{a,r} \in \mathcal{M}_r$ required for realizing a. In the following, for every resource r, we denote by $\mathcal{A}_r$ the set of activities a which consume r (i.e. such that $r \in \mathcal{R}_a$). Activities are also subject to an acyclic set of *project precedence constraints* $\mathcal{P} \subseteq \mathcal{A} \times \mathcal{A}$, which contains pairs of activities (a,b) such that b cannot start before the end of a.

A *solution* to an SDST-RCPSP assigns a start time $\sigma_a \in \mathbb{N}$ to each activity $a \in \mathcal{A}$. The end time of a is then given by $\sigma_a + p_a$. A solution is said to be *consistent* when constraints in Eqs. 1 to 4 hold. These constraints impose that all project precedences must be satisfied (Eq. 1), that the capacity of resources must never be exceeded (Eq. 2), that there must be a sufficient setup time between activities requiring distinct resource modes (Eq. 3), and a sufficient setup time with regards to the initial modes (Eq. 4).

$$\forall (a,b) \in \mathcal{P}, \ \sigma_a + p_a \leq \sigma_b \tag{1}$$

$$\forall r \in \mathcal{R}, \forall a \in \mathcal{A}_r, \ \textstyle\sum_{b \in \mathcal{A}_r \mid \sigma_b \leq \sigma_a < \sigma_b + p_b} q_{b,r} \leq K_r \tag{2}$$

$$\forall r \in \mathcal{R}, \forall a,b \in \mathcal{A}_r \ \text{s.t.} \ m_{a,r} \neq m_{b,r},$$
$$(\sigma_a \geq \sigma_b + p_b + \Delta_{r,m_{b,r},m_{a,r}}) \vee (\sigma_b \geq \sigma_a + p_a + \Delta_{r,m_{a,r},m_{b,r}}) \tag{3}$$

$$\forall r \in \mathcal{R}, \forall a \in \mathcal{A}_r \ \text{s.t.} \ m_{a,r} \neq m_{0,r}, \ (\sigma_a \geq \Delta_{r,m_{0,r},m_{a,r}}) \tag{4}$$

A solution is said to be *optimal* iff it minimizes the makespan, defined as the end time of the last activity ($\max_{a \in \mathcal{A}}(\sigma_a + p_a)$).

Precedence Graphs. Another way of defining a solution schedule is the standard concept of *precedence graph*. Such a graph contains nodes labeled by activities, and arcs $a \rightarrow b$ labeled by the duration of a (see Fig. 1, upper part). These arcs correspond to precedence constraints "b can start only once a is finished". Each arc $a \rightarrow b$ corresponds either to a *project precedence* $(a,b) \in \mathcal{P}$ given in the initial specification (dotted lines in Fig. 1), or to a *resource precedence* posted to prevent resources from being overused (continuous lines in Fig. 1). A precedence graph must be acyclic, and it also contains two dummy activities of null duration called the *source node s* and the *sink node t*, which respectively represent the start and the end of the schedule. The precedence graph G contains arcs $s \rightarrow a$ and $a \rightarrow t$ that guarantee that the source and the sink activities respectively precede and follow every activity in $\mathcal{A}$. In the case of SDST-RCPSP, the precedence graph also contains setup activities $setup_{a,r}$ for changing the current running mode m of a resource r just before realizing an activity a requiring another running mode $m_{a,r} \neq m$. The duration of $setup_{a,r}$ is then given by $\Delta_{r,m,m_{a,r}}$.

From this, it is possible to compute, for every activity a, the length of the longest path from the source node to a in G, denoted by $d_{s,a}$, and the length of the longest path from a to the sink node, denoted by $d_{a,t}$. These distances are given inside each activity node in Fig. 1 (e.g., $d_{s,D} = 4$ and $d_{D,t} = 7$). Then, the earliest start time of a is given by $est_a = d_{s,a}$, the makespan mk of the schedule corresponds to the distance $d_{s,t}$ from the source to the sink, and the latest start time of a is given by $lst_a = mk - d_{a,t}$. The resulting earliest and latest time schedules are given in Fig. 1 (middle part). An activity is said to be *critical* iff $est_a = lst_a$, and its temporal flexibility is given by $mk - (d_{s,a} + d_{a,t})$. In Fig. 1, activities A, C, and E are critical.

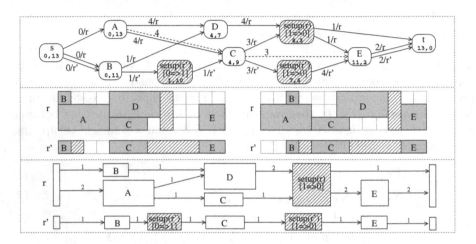

Fig. 1. Precedence graph (top), resulting earliest schedule (middle left) and latest schedule (middle right), and flow-network (bottom) for a SDST-RCPSP where $\mathcal{R} = \{r, r'\}$, $\mathcal{A} = \{A, B, C, D, E\}$, $\mathcal{P} = \{(A, C), (C, E)\}$, $p_A = p_D = 4$, $p_B = 1$, $p_E = 2$, $p_C = 3$, $\mathcal{R}_A = \mathcal{R}_D = \{r\}$, $\mathcal{R}_B = \mathcal{R}_C = \mathcal{R}_E = \{r, r'\}$, $K_r = 3$, $K_{r'} = 1$, $q_{A,r} = q_{D,r} = q_{E,r} = 2$, $q_{B,r} = q_{C,r} = 1$, $q_{x,r'} = 1$, $\mathcal{M}_r = \mathcal{M}_{r'} = \{0, 1\}$, $m_{0,r} = m_{A,r} = m_{B,r} = m_{C,r} = m_{D,r} = 1$, $m_{E,r} = 0$, $\Delta_{r,0,1} = \Delta_{r,1,0} = 1$, $m_{0,r'} = m_{B,r'} = m_{E,r'} = 0$, $m_{C,r'} = 1$, $\Delta_{r',0,1} = 1$, $\Delta_{r',1,0} = 4$. In the precedence graph, an arc $a \to b$ label by r/x is used when the duration of a is x and the precedence link is introduced because of resource r. The priority list used is $[A, B, C, D, E]$.

To deal with cumulative resources, as in [4], it is also possible to represent, for each resource precedence $a \to b$, the number of resource units $\phi^r_{a \to b}$ released at the end of activity a and used by activity b. These resource transfers are represented in a flow-network (see Fig. 1, bottom part). To get a consistent flow-network, the sum of all resource flows associated with a resource r and which respectively point to and come out of an activity node a must be equal to the amount of resource $q_{a,r}$ required by a. For setup activities, the sum of these flows must be equal to the total capacity of the resource (K_r), to guarantee that every resource has a unique running mode at any time. Last, all resource flows $\phi^r_{a \to b}$ used in the flow-network must be consistent with the resource modes associated with activities. This means for instance that there cannot be a direct flow $\phi^r_{a \to b}$ between two activities $a, b \in \mathcal{A}_r$ such that $m_{a,r} \neq m_{b,r}$.

3   A Lazy Precedence Graph Generation Scheme

As said in the introduction, instead of directly searching for activity start times σ_a or for precedence graphs, we use a list scheduling approach in which we search for a priority list $\mathcal{O} = [a_1, \ldots, a_n]$ containing all activities in $\mathcal{A}$. This priority list is then transformed into a precedence graph through a serial schedule generation scheme. In the following, we manipulate only *consistent priority lists* $\mathcal{O}$, which are such that for every project precedence (a, b) in $\mathcal{P}$, activity a is placed before activity b in $\mathcal{O}$. Also, we denote by $\mathcal{O}_r$ the sequence of successive activities which consume r. On the example of Fig. 1, $\mathcal{O} = [A, B, C, D, E]$, $\mathcal{O}_r = [A, B, C, D, E]$ and $\mathcal{O}_{r'} = [B, C, E]$.

The generation scheme that we use starts from a precedence graph containing only the source node. At each step, it considers the next activity a in priority list $\mathcal{O}$ and tries to insert it into the current schedule at the earliest possible time σ_a so that starting a at time σ_a is feasible in terms of project precedences and resource consumptions, and that activities already placed in the schedule are not delayed. The generation scheme also computes the so-called *pending resource flows* obtained for r after the insertion of a, denoted by $\Phi_{a,r}$. Formally, these pending resource flows correspond to a list

$$\Phi_{a,r} = [(\phi_1, z_1), \ldots, (\phi_h, z_h)]$$

such that for every $i \in [1..h]$, a resource amount ϕ_i is released at the end of activity z_i and is available for future activities. Figure 2 (upper part) gives the set of pending flows over resource r obtained after the insertion of each activity. For example, the set of pending flows after the insertion of C would be $\Phi_{C,r} = [(1, B), (1, A), (1, C)]$, and the set of pending flows after the insertion of E would be $\Phi_{E,r} = [(1, setup_{r,1,0}), (2, E)]$. *In the following, we assume that the pending flows are ordered by increasing release time, that is $\sigma_{z_1} + p_{z_1} \le \ldots \le \sigma_{z_h} + p_{z_h}$.*

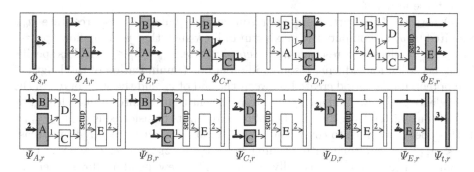

Fig. 2. Forward pending flows (upper part) and backward pending flows (bottom part)

We now come back to the generation process. For every resource r, the initial list of pending flows after source activity s is $\Phi_{s,r} = [(K_r, s)]$, since initially, the whole capacity of the resource is available. Then, let a be the next activity to consider in the priority list. For each resource $r \in \mathcal{R}_a$, let $\Phi_{prev_{a,r},r} = [(\phi_1, z_1), \ldots, (\phi_h, z_h)]$ be the pending flows obtained after the insertion of the activity $prev_{a,r}$ that immediately precedes a on r. If the resource mode $m_{prev_{a,r},r}$ associated with this predecessor is distinct from the resource mode $m_{a,r}$ required by a, a setup is needed before a and the earliest time $\sigma_{a,r}$ at which r can support the realization of a is given by Eq. 5. Otherwise, $\sigma_{a,r}$ corresponds to the earliest time at which resource level $q_{a,r}$ can be made available according to the pending resource flows in $\Phi_{prev_{a,r},r}$ (Eq. 6).

$$\sigma_{a,r} \leftarrow \sigma_{z_h} + p_{z_h} + \Delta_{r,m_{prev_{a,r},r},m_{a,r}} \text{ if } m_{prev_{a,r},r} \neq m_{a,r} \tag{5}$$

$$\sigma_{z_k} + p_{z_k} \text{ otherwise, with } k = \min\{k' \in [1..h] \mid \textstyle\sum_{i\in[1..k']} \phi_i \geq q_{a,r}\} \tag{6}$$

From these elements, the start time of a produced by the schedule generation scheme is given by the maximum between the end time of project predecessors of a, and the earliest start times of resource consumptions associated with a:

$$\sigma_a \leftarrow \max(\max_{(b,a)\in\mathcal{P}}(\sigma_b + p_b), \max_{r\in\mathcal{R}_a}\sigma_{a,r}) \tag{7}$$

In Fig. 1, the earliest start time of C on resource r is $\sigma_{C,r} = 1$ (the single resource unit required for C can be available just after the end of activity B), its earliest start time on r' is $\sigma_{C,r'} = 2$ (requirement to change the resource running mode), and its earliest start time according to its project predecessors equals 4 (end time of A). As a result, the earliest start time of C is $\sigma_C \leftarrow \max(4, 1, 2) = 4$.

The activity a considered is then truly introduced in the precedence graph. If the running mode of resource r before the introduction of a is distinct from the resource mode $m_{a,r}$ required by a, then a setup activity $setup_{a,r}$ is added to the precedence graph, together with one arc $z_i \rightarrow setup_{a,r}$ with flow ϕ_i for each pending flow (ϕ_i, z_i) in $\Phi_{prev_{a,r},r}$, and one arc $setup_{a,r} \rightarrow a$ with flow $q_{a,r}$ pointing to a. The new pending flows after the introduction of a are obtained from these updates.

Otherwise, if the current state of resource r is equal to the resource state $m_{a,r}$ required by a, we compute the maximum index k such that pending flow (ϕ_k, z_k) in $\Phi_{prev_{a,r},r}$ releases its resource units before σ_a (i.e. $\sigma_{z_k} + p_{z_k} \leq \sigma_a$). This specific index is chosen for reducing resource idle periods (minimization of the length of the potential idle period created between the end of z_k and the start of a). In the precedence graph, we add arc $z_k \rightarrow a$ to represent that resource units released by z_k are transmitted to a. The number of resource units transferred is $\min(\phi_k, q_{a,r})$. If this does not suffice to cover the total amount of resource required by a (case $\phi_k < q_{a,r}$), we continue with pending flows $(\phi_{k-1}, z_{k-1}), (\phi_{k-2}, z_{k-2}) \ldots$ until resource consumption $q_{a,r}$ is fully covered. The new pending flows after the introduction of a are obtained from these updates. For resource r in Fig. 1, activity C, for which $\sigma_C = 4$, consumes resource units released by A, while activity D consumes resource units released by both A and B.

These operations are successively realized for each activity in priority list $\mathcal{O}$. The generation scheme defined is *lazy* because it only considers pending flows, which are located at the end of the resource usage profile. It does not exploit potential valleys present in this profile. In terms of scheduling, this entails that the generation scheme does not necessarily generate *active schedules*, which means that some activities might be started earlier without delaying other activities. Nevertheless, for every active schedule S, there exists a priority list which generates this schedule (it suffices to order activities by increasing start times in S). This implies that for regular performance measures such as makespan minimization, there exists a priority list inducing an optimal schedule. Also, all schedules produced are *semi-active*, meaning that for every activity a, there is no date $t < \sigma_a$ such that all start times $t, t+1, \ldots, \sigma_a$ lead to a consistent schedule.

One advantage of the lazy schedule generation scheme proposed is its low time complexity compared to serial schedule generation schemes used for RCPSP, which maintain a global resource usage profile. Indeed, each insertion into a given resource has a *worst-case* time complexity which is linear in M, the maximum number of resource consumptions that might be performed in parallel on r, instead of a complexity linear in the number of activities in $\mathcal{A}$. In particular, for a disjunctive resource, our lazy schedule generation scheme takes a constant time for each consumption insertion.

4   Incremental Schedule Maintenance Techniques

The techniques defined in the previous section can be used to automatically derive a full schedule from a priority list $\mathcal{O}$. To progressively get better solutions, a strategy is to perform local search in the space of priority lists, for instance by changing the position of one activity in the list, by swapping the positions of two activities, or by moving a block of successive activities.

This is were constraint programming comes into play, since we use Constraint-Based Local Search (CBLS [35]) for realizing these updates efficiently. As in standard constraint programming, CBLS models are defined by decision variables, constraints, and criteria. One distinctive feature is that in CBLS, all decision variables are assigned when searching for a solution. The search space is then explored by performing local moves which reassign some decision variables, and it is explored more freely than in tree search with backtracking. One specificity of CBLS models is that they manipulate *invariants*, which are one-way constraints $x \leftarrow exp$ where x is a variable and exp is a functional expression of other variables of the problem, such as $x \leftarrow sum(i \in [1..N])\, y_i$. During local moves, these invariants are efficiently maintained thanks to specific procedures that incrementally reevaluate the output of invariants (left part) in case of change in their inputs (right part).

Invariant Inputs. To achieve our goal, we introduce a new CBLS invariant which takes as an input a priority list $\mathcal{O}$. In our CBLS solver, priority list $\mathcal{O}$ is implemented based on the data structure defined in [7] for encoding total orders. This data structure maintains the predecessor and the successor of each element a in the list. It also assigns to a an integer $tag_a \in [0..2^{31} - 1]$ such that if a is located before b in the list, then $tag_a < tag_b$. When inserting an element c between two successive elements a and b, the tag of c is set to $\frac{tag_a + tag_b}{2}$ if $tag_a + 1 < tag_b$, and otherwise operations are used to retag elements so as to allow some space between tag_a and tag_b.

The invariant introduced also takes as an input one boolean parameter u_a for each activity $a \in \mathcal{A}$, specifying whether resource consumptions associated with a are activated or not. When u_a takes value false, only the impact of project precedence constraints is taken into account for a, and in a full schedule, u_a must take value true for all activities. Adding such inputs to the invariant can bring a better view of the promising insertion positions for an activity (see Sect. 5).

Invariant Outputs. The invariant built maintains the precedence graph and returns, for each activity a, the distance from the source node to a ($d_{s,a}$, equal to σ_a), and the distance from a to the sink node ($d_{a,t}$). It also maintains, for each resource r, the sequence $\mathcal{O}_r$ of activities which successively use r. Each sequence $\mathcal{O}_r$ is represented as a linked list defining the predecessor $prev_{a,r}$ and the successor $next_{a,r}$ of a in $\mathcal{O}_r$. For incremental computation reasons, the invariant also maintains, for each activity a and each resource $r \in \mathcal{R}_a$, the pending resource flows $\Phi_{a,r}$ after the insertion of a. For each resource and each activity, the space complexity required to record these flows is $O(M)$, with M the maximum number of resource consumptions that might be performed in parallel on a resource. In the end, the *lazySGS* invariant defined takes the form:

$$(\{d_{s,a} \mid a \in \mathcal{A}\}, \{d_{a,t} \mid a \in \mathcal{A}\}, \{\mathcal{O}_r \mid r \in \mathcal{R}\}) \leftarrow lazySGS(\mathcal{O}, \{u_a \mid a \in \mathcal{A}\}) \quad (8)$$

Even if such an invariant is rather *large*, it can be used in a CBLS model containing other invariants. For example, output variables $d_{s,a}$ and $d_{a,t}$ are usually used as inputs for invariants that incrementally compute the temporal flexibility of each activity ($flex_a \leftarrow d_{s,t} - (d_{s,a} + d_{a,t})$), and then for computing the set of critical activities using a set invariant ($CriticalSet \leftarrow \{a \in \mathcal{A} \mid flex_a = 0\}$). Also, when considering extensions of the invariant in which there is a choice on the resources used by activities, the associated resource allocation decisions can be connected to other invariants managing other constraints such as limitations on non-renewable resources.

Incremental Evaluation. The incremental evaluation function of the CBLS invariant corresponds to Algorithm 1. This algorithm is inspired by the incremental longest paths maintenance algorithms introduced in [22]. The main difficulty is that in our case, we need more than incrementally computing distances in a precedence graph: we also need to incrementally manage the construction of the precedence graph itself, which depends on the pending resource flows successively obtained.

Two kinds of changes must be considered: (1) moves of some activities in the priority list; (2) activation/deactivation of resource consumptions for some activities. To handle these changes, the first step consists in updating sequences of resource consumptions applied to resources ($\mathcal{O}_r$). All activities for which changes occurred are removed from sequences $\mathcal{O}_r$, and all activities whose consumptions are still activated ($u_a = true$) are sorted by increasing tag and reintroduced in $\mathcal{O}_r$ via a single forward traversal. These updates are realized by a call to UPDATEORDERS at line 1.

For performing forward revisions, Algorithm 1 uses several data structures:

- a *global revision queue* Q^{rev} containing activities for which some revisions must be made; to ensure that revisions are realized in a topological order with relation to the precedence graph, activities are ordered in Q^{rev} as their tags in $\mathcal{O}$;
- for every activity a, a *precedence revision set* $\mathcal{P}_a^{rev}$ containing activities b such that (b, a) is a precedence in $\mathcal{P}$ and the end time of b has been updated;
- for every activity a, a *resource revision set* $\mathcal{R}_a^{rev}$ containing resources $r \in \mathcal{R}_a$ such that the impact of the consumption of resource r by a might not be up-to-date.

Data structures Q^{rev}, $\mathcal{P}_a^{rev}$, $\mathcal{R}_a^{rev}$ are initialized at line 2 by a call to INITREVISIONS. The latter adds to revision queue Q^{rev} all activities a which have been activated or deactivated since the last evaluation of the invariant. For these activities, all resources in $\mathcal{R}_a$ are added to revision set $\mathcal{R}_a^{rev}$. For every other activity a and every resource $r \in \mathcal{R}_a$, if activity a has a new predecessor in $\mathcal{O}_r$, then a is added to Q^{rev} and resource r is added to $\mathcal{R}_a^{rev}$. Last, $\mathcal{P}_a^{rev}$ is initially empty for every activity a.

After these initialization steps, while there are some revisions left in Q^{rev}, the revisions associated with the activity a that has the lowest tag in ordering $\mathcal{O}$ are considered (line 4). After that, the algorithm computes the contribution ρ^{old} of all elements in $\mathcal{P}_a^{rev}$ and $\mathcal{R}_a^{rev}$ to the previous value of the start time of a, denoted by σ_a^{old} (line 5). The computations performed take into account the value of u_a at the last evaluation of the invariant (value u_a^{old}). Then, the algorithm computes the new contribution ρ of these same elements to the new value of the start time of a (lines 6–7), by using function EARLIESTSTART which recomputes terms $\sigma_{a,r}$ seen in Eqs. 5 and 6.

If the new contribution ρ is greater than or equal to the current value of σ_a, this means that temporal constraints have been strengthened since the last evaluation of the invariant, hence the new value of σ_a is ρ (line 8). Otherwise, if the old contribution ρ^{old} was a support for σ_a^{old}, then σ_a is recomputed from scratch (lines 9–10). Otherwise, there is a weakening of the temporal constraints associated with the elements revised, but these elements did not support the previous value of σ_a, hence σ_a is up-to-date.

If the new value obtained for σ_a is distinct from its old value σ_a^{old}, revisions are triggered for project successors of a (lines 11–15). Last, resource consumptions associated with a and which require a revision are handled, by recomputing the pending flows after the insertion of a for these resources (function APPLY), and by triggering new resource revisions over successor activities in case of a change in these pending flows (lines 16–21). Note that it is more likely to get at some point an equality between two lists of pending resource flows than between two full resource usage profiles.

Algorithm 1. Incremental revision procedure for the CBLS invariant introduced

---

1 $\{\mathcal{O}_r \mid r \in \mathcal{R}\} \leftarrow$ UPDATEORDERS()

2 $(Q^{rev}, \{\mathcal{P}_a^{rev} \mid a \in \mathcal{A}\}, \{\mathcal{R}_a^{rev} \mid a \in \mathcal{A}\}) \leftarrow$ INITREVISIONS()

3 **while** $Q^{rev} \neq \emptyset$ **do**

4     $a \leftarrow$ EXTRACTMIN(Q^{rev})

5     $\rho^{old} \leftarrow \max_{b \in \mathcal{P}_a^{rev}}(\sigma_b^{old} + p_b)$; **if** u_a^{old} **then** $\rho^{old} \leftarrow \max(\rho^{old}, \max_{r \in \mathcal{R}_a^{rev}} \sigma_{a,r})$

6   **if** u_a **then** $\sigma_{a,r} \leftarrow$ EARLIESTSTART($a, r, \Phi_{prev_{a,r},r}, m_{prev_{a,r},r}$) for each $r \in \mathcal{R}_a^{rev}$

7     $\rho \leftarrow \max_{b \in \mathcal{P}_a^{rev}}(\sigma_b + p_b)$; **if** u_a **then** $\rho \leftarrow \max(\rho, \max_{r \in \mathcal{R}_a^{rev}} \sigma_{a,r})$

8   **if** $\rho \geq \sigma_a$ **then** $\sigma_a \leftarrow \rho$

9   **else if** $\rho^{old} = \sigma_a^{old}$ **then**

10        $\sigma_a \leftarrow \max_{(b,a) \in \mathcal{P}}(\sigma_b + p_b)$; **if** u_a **then** $\sigma_a \leftarrow \max(\sigma_a, \max_{r \in \mathcal{R}_a} \sigma_{a,r})$

11   **if** $\sigma_a \neq \sigma_a^{old}$ **then**

12   **foreach** $(a,b) \in \mathcal{P}$ **do**

13   **if** $b \notin Q^{rev}$ **then** ADD($\langle b, tag_b \rangle, Q^{rev}$)

14   ADD($a, \mathcal{P}_b^{rev}$)

15   **if** u_a **then** $\mathcal{R}_a^{rev} \leftarrow \mathcal{R}_a$

16   **if** u_a **then**

17   **foreach** $r \in \mathcal{R}_a^{rev}$ **do**

18           $\Phi^{old} \leftarrow \Phi_{a,r}$; $\Phi_{a,r} \leftarrow$ APPLY($a, r, \sigma_a, \Phi_{prev_{a,r},r}, m_{prev_{a,r},r}$)

19   **if** $(next_{a,r} \neq t) \wedge ((\Phi_{a,r} \neq \Phi^{old}) \vee (\exists(\phi,z) \in \Phi_{a,r} \ s.t. \ \sigma_z \neq \sigma_z^{old}))$ **then**

20   **if** $next_{a,r} \notin Q^{rev}$ **then** ADD($\langle next_{a,r}, tag_{next_{a,r}} \rangle, Q^{rev}$)

21   ADD($r, \mathcal{R}_{next_{a,r}}^{rev}$)

Backward revisions, which compute distances from every activity to the sink node, are performed similarly. The main differences are that revisions are triggered in the direction of predecessors of activities, and that backward revisions do not update the precedence graph but just follow the graph produced by the forward revisions. After all forward and backward revisions, the old values which need to be recorded are updated following the changes made ($u_a^{old} \leftarrow u_a$, $\sigma_a^{old} \leftarrow \sigma_a$, $d_{a,t}^{old} \leftarrow d_{a,t}$).

As a last remark, note that the CBLS invariant obtained can deal with planning horizons containing many time-steps, since the complexity of the algorithm defined does not depend on the number of possible values of temporal distances. Further analyses would be required to get, as in [22], the complexity of the incremental evaluation function in terms of changes in the inputs and outputs of the invariant. Finally, it would be possible to define another version of the invariant taking as inputs directly individual priority orders $\mathcal{O}_r$ over each resource, provided that these orders are compatible (invariant $(\{d_{s,a} \mid a \in \mathcal{A}\}, \{d_{a,t} \mid a \in \mathcal{A}\}) \leftarrow lazySGS'(\{\mathcal{O}_r \mid r \in \mathcal{R}\}, \{u_a \mid a \in \mathcal{A}\}))$.

5 Differentiability of the Invariant

In addition to incremental reevaluation issues, one key feature of invariants in CBLS is their *differentiability* [35], which allows to quickly *estimate* the quality of local moves instead of fully evaluating them. To get such a differentiability, in addition to the set of *forward* pending flows $\Phi_{a,r}$ obtained after the insertion of an activity a, the invariant also maintains a list of *backward* pending flows $\Psi_{a,r} = [(\psi_1, z_1), \ldots, (\psi_h, z_h)]$ obtained just before each activity a. This list contains pairs (ψ_i, z_i) such that activity z_i waits for a flow ψ_i which is released by activities placed before a in $\mathcal{O}_r$. See the bottom part of Fig. 2 for an illustration. On this figure, the backward pending flows for C and B on r are $\Psi_{C,r} = [(C, 1), (D, 2)]$ and $\Psi_{B,r} = [(B, 1), (C, 1), (D, 1)]$. Intuitively, $\Phi_{a,r}$ and $\Psi_{a,r}$ describe the resource usage frontiers obtained when cutting the flow-network respectively just after and just before the realization of a.

Let us now detail the techniques used for quickly estimating the quality of the possible reinsertions of an activity a in priority list $\mathcal{O}$. To do this efficiently, we first deactivate resource consumptions for a, to take into account only project precedences for a, and we evaluate the new schedule using the incremental evaluation function of the CBLS invariant. Then, from the current activity list $\mathcal{O}$, it is possible to compute all relevant insertion positions for a that do not lead to a precedence cycle. More precisely, we traverse the ancestors of a in the current precedence graph to find the greatest-tag ancestor activity b_1 such that there is a common resource used by a and b_1 ($\mathcal{R}_a \cap \mathcal{R}_{b_1} \neq \emptyset$). We also traverse the descendants of a to find the lowest-tag descendant activity b_2 such that there is a common resource used by a and b_2. Reinserting a in priority list $\mathcal{O}$ anywhere between b_1 and b_2 is consistent from the project precedences point of view. However, not all insertion positions deserve to be tested. It suffices to test the insertion of a just after activities c which share a common resource with a.

Let c be a relevant insertion position in $\mathcal{O}$ located between b_1 and b_2. To estimate the quality of the schedule obtained by inserting a just after c, we first simulate the insertion of a over resources $r \in \mathcal{R}_a$. For each resource $r \in \mathcal{R}_a$, we compute the activities $\hat{prev}_{a,r}$ and $\hat{next}_{a,r}$ that would immediately precede and follow a on r if a is inserted at the chosen position. We then simulate the merging of the forward pending flows in $\Phi_{\hat{prev}_{a,r},r}$, followed by a, followed by the backward pending flows in $\Psi_{\hat{next}_{a,r},r}$. To simulate this merging, we compute a set of additional resource flows and nodes that allow to get a consistent flow network (see Fig. 3). The quality of the resulting schedule is then estimated by *the contribution to the makespan associated with the part of the schedule modified by the insertion of a*, or in other words by the length of the longest source-to-sink path which traverses flow arcs added for merging the forward and backward pending flows (longest path through the gray area in Fig. 3).

In the case of disjunctive resources, the evaluation adopted generalizes classical formulas used in job shop scheduling and gives the exact value of the length of the longest source-to-sink path through a in the schedule that would be obtained [25]. In the case of cumulative resources, it gives a more global evaluation. To efficiently compute the length of such longest paths, we use the (already known) distances $d_{s,b}$ from the source node to activities b contained in the forward pending flows, and the (already known) distances $d_{b,t}$ from activities b contained in the backward pending flows to the sink node.

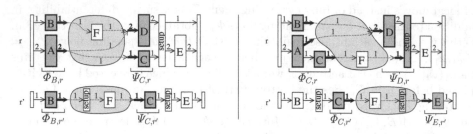

Fig. 3. Flow merging for estimating the quality of the insertion of activity F into the schedule given in Fig. 1: simulation of the insertion between B and C (left), and between C and D (right)

More formally, we first compute the earliest start time $\hat{\sigma}_a$ of a for the chosen insertion position. Next, for each resource $r \in \mathcal{R}_a$, we compute the new pending flows $\hat{\Phi}_{a,r}$ that would be obtained just after the insertion of a, and for each pending flow (ϕ, z) in $\hat{\Phi}_{a,r}$ we denote by $\hat{\sigma}_z$ the earliest start time obtained for z.

If no setup is required between a and $nêxt_{a,r}$ (case $m_{a,r} = m_{nêxt_{a,r},r}$), we compute a set of flows Γ which allow to merge forward pending flows in $\hat{\Phi}_{a,r}$ with backward pending flows in $\Psi_{nêxt_{a,r},r}$. To define Γ, flows (ψ, z) in $\Psi_{nêxt_{a,r},r}$ are ordered by increasing activity-tag, and for each of them we deliver from $\hat{\Phi}_{a,r}$ the resource flow ψ required by z, by using the same principles as in the schedule generation scheme of Sect. 3. For resource r, the quality of the insertion of a just after c in $\mathcal{O}$ is estimated by:

$$length_{a,c,r} = \max_{x \xrightarrow{\phi} y \in \Gamma} (\hat{\sigma}_x + p_x + d_{y,t}) \tag{9}$$

Otherwise (case $m_{a,r} \neq m_{nêxt_{a,r},r}$), a setup operation is required just after a, as for the insertion of F on resource r' on the right part of Fig. 3. In this case, the estimation associated with the insertion of a just after c in $\mathcal{O}$ is:

$$length_{a,c,r} = \max_{(\phi,z) \in \hat{\Phi}_{a,r}} (\hat{\sigma}_z + p_z) + \Delta_{r,m_{a,r},m_{nêxt_{a,r},r}} + \max_{(\psi,z) \in \Psi_{nêxt_{a,r},r}} d_{z,t} \tag{10}$$

Finally, the global estimation associated with the insertion of a just after c in $\mathcal{O}$ is:

$$length_{a,c} = \max((\max_{r \in \mathcal{R}_a} length_{a,c,r}), (\hat{\sigma}_a + p_a + \max_{(a,b) \in \mathcal{P}} d_{b,t})) \tag{11}$$

This estimation takes into account not only the estimation of the quality of the insertion over each resource, but also the project successors of the activity inserted. Compared to [4], which uses resource flows for guiding the way an activity should be inserted into an RCPSP schedule, testing the insertion at one position in our case has a lower complexity (complexity $O(Mm + P)$ with m the number of resources, M the maximum number of activities that might be performed in parallel over a resource, and P the maximum number of project predecessors and successors for an activity, instead of complexity $O(nm)$ with n the number of activities in $\mathcal{A}$).

6   Generic Local Search Neighborhoods

As said in the introduction, the other parts of our scheduling system are described with fewer details. Several neighborhoods are used when searching for good priority lists $\mathcal{O}$:

- REINSERTION: reinserts a given activity a at one of the best positions in the schedule according to the estimations provided by the differentiation techniques of the previous section. When choosing a particular insertion position for a, updates are automatically made in $\mathcal{O}$ to guarantee that all ancestors (resp. descendants) of a in the precedence graph are still located on the left (resp. on the right) of a in $\mathcal{O}$.
- REINSERTIONDEEP: considers an activity a and tries to reinsert a at the best possible position. To enlarge the set of candidate insertion positions, it also moves project ancestors of a at their leftmost position in the order and project descendants of a at their rightmost position, without increasing the current makespan.
- OR-OPT: consists in searching for a better positioning for a block B of k successive activities. It is inspired by the *or-opt-k* neighborhood [5] used for traveling salesman problems (TSPs), and it can be useful for scheduling with setup times for SDST-RCPSP. The candidate reinsertion positions for B are efficiently explored by adapting the differentiation techniques seen previously.
- 2-OPT: considers a set of successive activities $a_i, a_{i+1}, \ldots, a_j$ and tries to realize them in the reverse order. Such a neighborhood is inspired by the *2-opt* moves [13] used for TSPs and can be useful for scheduling with setup times. It is efficiently explored by extending the differentiation techniques seen in the previous section.
- SWAP: considers two activities a, b such that swapping a and b does not create a precedence cycle. Using this neighborhood is more expensive since for evaluating a swap, we explicitly make it and compute its exact effect.
- REORDERBYDISTTOSINK: transforms the current priority list $\mathcal{O}$ by ordering elements by decreasing distance to the sink node. For RCPSP, such a strategy is known as the *Forward-Backward Improvement* algorithm [33].

7   Search Strategy

The last part of our CBLS scheduling system allows search strategies to be defined. For space limitation reasons, we only give a brief overview of the strategy used for the experiments. One of the main component of this strategy is Variable Neighborhood Search (VNS [27]), which successively considers the neighborhoods defined and applies these neighborhoods to critical activities until reaching a locally optimal solution.

Then, tabu search is used to escape local optima and plateaus [14]. More precisely, we maintain a tabu list which contains *forbidden makespan values*, and schedules that have a forbidden (real or estimated) makespan cannot be selected anymore during VNS. Once VNS converges to a new locally optimal (and non-tabu) solution, the makespan of this solution is added to the tabu list, the oldest tabu makespan is removed if some place is needed, and VNS is launched again. Tabu search ends when the makespan of the current solution has not been improved during a certain number of VNS applications.

Another technique is used for focusing search on the bottleneck of the problem: (a) if after tabu search the current makespan is not strictly better than the best makespan known, we randomly choose one non-critical activity a, deactivate its resource consumptions ($u_a = false$), clear the tabu list, and apply tabu search again on the problem containing one less activity; (b) otherwise, we select an activity whose resource consumptions have been deactivated, reactivate these consumptions, clear the tabu list, and apply tabu search again on the problem containing one more activity. Each time a full schedule with a better makespan is found, it is recorded as the best solution. We call the obtained metaheuristic *Tabu Search with Repair* (**TSR**).

Last, to avoid spending too much time on conflict resolution rather than on search over full schedules, each non-critical activity can be removed at most once during each call to TSR. When no activity is candidate for being removed from the schedule, we select $x\%$ of the activities (x = 10% in the experiments), randomly move each of them in priority list $\mathcal{O}$ while preserving the satisfaction of all project precedences, and call TSR on the resulting schedule. From time to time, we also perform restarts. In the end, by combining tabu search and perturbation, we get a metaheuristic that we call *Iterated Tabu Search with Repair* (**ITSR**). It combines local and global search, and makes a trade-off between search intensification and diversification.

8   Experiments

Experiments are performed on clusters composed of 20 Intel Xeon 2.6 GHz processors with a shared memory of 65 GB of RAM. Each run used for solving one problem instance is performed on a single processor (no parallel solving). For each cluster, the 20 CBLS models together with the 20 search engines easily fit onto the available memory. The invariant defined is integrated into the *InCELL* CBLS library [31].

SDST-JSSP.   Table 1 summarizes the results obtained on SDST-JSSP instances [9]. Each instance contains n jobs and m resources, leading to nm activities to schedule. For ITSR, the results presented are obtained based on

10 runs, each run having a time limit of 1 h. All neighborhoods defined in Sect. 6 are used except for the swap neighborhood, and the or-opt-k neighborhood is limited to $k = 2$. The length of the tabu list which contains makespan values is set to 15, and for tabu search the number of iterations allowed without improvements is set to 15 as well. ITSR is compared against techniques of the state-of-the-art: the branch-and-bound method defined in [3], the method defined in [6] which adapts to SDST-JSSP the shifting bottleneck procedure used for JSSP, two methods based on genetic algorithms hybridized with local search [16] and tabu search [17], and the CP approach defined in [18]. In the table, these methods are respectively referred to as AF08, BSV08, GVV08, GVV09, and GH10. The latter uses the same 1 h time limit as ITSR on a similar processor. As in [18], the bold font in the table is used for best makespan values, underlined values are used when the solver proves optimality, and values marked with a star denote new best upper bounds found by ITSR, which have also been verified using a separate checker. *Globally, ITSR finds 6 new upper bounds and for all other instances it always finds the best known upper bound.* Table 1 also shows the impact of the incremental computation techniques and of the differentiability of the *lazySGS* invariant. Compared to ITSR, the version which fully evaluates local moves instead of using differentiability (column ITSR-D in Table 1) typically performs 10 times less local moves, while the version which uses neither differentiability nor incremental evaluations (column ITSR-D-I in Table 1), typically performs 20 times less local moves. The solutions produced after one hour by these versions have higher makespans on average.

MJSSP. Figure 4a summarizes results obtained for the Multi-Capacity Job Shop Scheduling Problem (MJSSP). This problem involves cumulative resources and activities which all consume one resource unit. As in [26,29], which introduces *iterative flattening* (a precedence constraint posting approach), the instances considered are derived from standard JSSP instances by duplicating jobs and increasing the capacity of resources accordingly. The advantage of doing this is that results over the initial JSSP instances provide lower and upper bounds for the MJSSP instances [28] (column NA96 in Fig. 4a). The instances selected are the 21 instances for which [26] provides new best upper bounds (column MV04 in Fig. 4a). In [26], these bounds are presented as the best ones found during the work on iterative flattening (no computation time specified). These bounds have been improved in [15], which uses randomized large neighborhood search (column GLN05). *Over the 21 instances, ITSR manages to find 12 best upper bounds compared to MV04, but only one compared to GLN05.* To get these results, the or-opt and 2-opt neighborhoods are deactivated because they are less relevant without setup times, and the length of the tabu list is set to 2. As for SDST-JSSP, Fig. 4a shows that deactivating the differentiability and the incremental computation capabilities of the *lazySGS* invariant degrades the average quality of the solutions produced.

RCPSP. Figure 4b reports the results obtained on RCPSP instances j60, j90, and j120 (www.om-db.wi.tum.de/psplib), which respectively contain 60, 90, and 120

Table 1. Experiments for SDST-JSSP: comparison with the state-of-the-art for makespan minimization (best & mean values over 10 runs, with a 1 h time limit for each run)

Instance	#jobs x #res	AF08 Best	BSV08 Best	GVV08 Best	Avg	GVV09 Best	Avg	GH10 Best	Avg	ITSR Best	Avg	ITSR-D Best	Avg	ITSR-D-I Best	Avg
t2-ps06	15 x 5	**1009**	1018	1026	1026			**1009**	1009.0	**1009**	1009.2	**1009**	1009.7	**1009**	1010.3
t2-ps07	15 x 5	**970**	1003	**970**	971			**970**	970.0	**970**	970.0	**970**	970.0	**970**	970.0
t2-ps08	15 x 5	**963**	975	**963**	966			**963**	963.0	**963**	963.0	**963**	963.0	**963**	963.0
t2-ps09	15 x 5	1061	**1060**	**1060**	1060			**1060**	1060.0	**1060**	1060.0	**1060**	1060.0	**1060**	1060.0
t2-ps10	15 x 5	**1018**	1018	**1018**	1018			**1018**	1018.0	**1018**	1018.0	**1018**	1018.0	**1018**	1018.0
t2-ps11	20 x 5	1494	1470	**1438**	1439	**1438**	1441	1443	1463.8	**1437***	1437.6	1438	1442.1	1438	1441.9
t2-ps12	20 x 5	1381	1305	**1269**	1291	**1269**	1277	**1269**	1322.2	**1269**	1269.0	**1269**	1270.8	**1269**	1271.1
t2-ps13	20 x 5	1457	1439	1406	1415	1415	1416	1415	1428.8	**1404***	1406.4	1406	1414.1	1406	1414.5
t2-ps14	20 x 5	1483	1485	**1452**	1489	**1452**	1489	**1452**	1470.5	**1452**	1452.0	**1452**	1452.0	**1452**	1452.0
t2-ps15	20 x 5	1661	1527	1485	1502	1485	1496	1486	1495.8	**1479***	1485.9	1485	1492.9	1485	1496.0
t2-pss06	15 x 5		1126					**1114**	1114.0	**1114**	1117.7	**1114**	1120.6	**1114**	1122.0
t2-pss07	15 x 5		1075					**1070**	1070.0	**1070**	1070.0	**1070**	1070.0	**1070**	1070.0
t2-pss08	15 x 5		1087					**1072**	1073.0	**1072**	1073.1	**1072**	1073.6	**1072**	1074.2
t2-pss09	15 x 5		1181					**1161**	1161.0	**1161**	1161.0	**1161**	1161.0	**1161**	1161.0
t2-pss10	15 x 5		1121					**1118**	1118.0	**1118**	1118.0	**1118**	1118.0	**1118**	1118.0
t2-pss11	20 x 5		1442					1412	1425.9	**1409***	1412.4	1412	1417.6	1414	1420.6
t2-pss12	20 x 5		1290			1258	1266	1269	1287.6	**1257***	1260.5	1258	1266.7	1260	1269.3
t2-pss13	20 x 5		1398			**1361**	1379	1365	1388.0	**1361**	1364.7	**1361**	1371.9	1367	1373.9
t2-pss14	20 x 5		1453					**1452**	1453.0	**1452**	1452.0	**1452**	1452.0	**1452**	1452.0
t2-pss15	20 x 5		1435					1417	1427.4	**1410***	1411.1	**1410***	1421.1	1417	1421.8

Instance	#jobs x #res	NA96 lb / ub	MV04 Best	GLN05 Best	ITSR Best	Avg	ITSR-D Best	Avg	ITSR-D-I Best	Avg
la04-2	20 x 5	572/590	577	**576**	**576**	576.0	584	587.3	582	588.7
la04-3	30 x 5	570/590	584	**573**	577	579.7	594	603.7	596	605.0
la16-2	20 x 10	888/935	929	**925**	933	938.4	932	936.6	930	937.6
la16-3	30 x 10	717/935	927	**918**	939	947.1	941	951.9	945	955.3
la17-2	20 x 10	750/765	756	**755**	756	757.8	760	763.3	760	763.4
la17-3	30 x 10	646/765	761	**755**	763	765.7	773	777.7	772	779.0
la18-2	20 x 10	783/844	818	**811**	814	818.1	828	833.7	832	836.6
la18-3	30 x 10	663/844	813	**808**	818	825.4	854	861.4	840	858.4
la19-2	20 x 10	730/840	803	795	**792***	799.9	818	826.2	822	827.5
la19-3	30 x 10	617/840	801	**787**	799	802.5	841	848.5	842	852.3
la20-2	20 x 10	829/902	864	**859**	**859**	867.2	872	878.9	871	878.6
la20-3	30 x 10	756/902	863	**854**	862	871.5	879	894.5	888	901.4
la24-2	30 x 10	704/935	932	**903**	911	917.9	966	976.6	963	975.3
la24-3	45 x 10	704/935	929	**898**	917	923.2	1000	1010.8	1000	1014.4
la25-3	45 x 10	723/977	965	**945**	977	983.2	1027	1037.6	1035	1043.9
la38-2	30 x 15	943/1196	1185	**1175**	1180	1188.6	1249	1260.8	1223	1261.5
la38-3	45 x 15	943/1196	1195	**1168**	1197	1204.9	1286	1304.6	1297	1311.3
ft10-2	20 x 10	835/930	913	**891**	906	908.9	921	931.3	922	934.2
ft10-3	30 x 10	655/930	912	**879**	915	921.8	946	957.7	955	962.3
ft20-2	40 x 5	1165/1165	1186	1182	*1172*	1175.0	1195	1215.6	1213	1226.0
ft20-3	60 x 5	387/1165	1205	*1179*	1182	1190.1	1239	1254.8	1220	1255.7

(a)   (b)

Fig. 4. Results on cumulative resources: (a) results on MJSSP instances (best & mean values over 10 runs with a 1 h time limit per run); (b) results on RCPSP instances j60 (top), j90 (middle), and j120 (bottom); representation of the mean number of instances (y-axis) for which the deviation percentage with regards to the best known upper bound is less than the value on the x-axis (average results over 5 runs with a 30 min time limit per run)

activities to schedule. This time, the length of the tabu list is set to 4 as well as the maximum number of iterations without improvement for tabu search. Figure 4b gives the distribution of the relative distance between the average makespan found by ITSR after 30 minutes and the best makespan reported in the PSPLIB. For example, for j60, the graph expresses that among the 450 instances of j60 considered, the distance to the best solution known is 0% for approximately 375 instances, it is $\leq 1\%$ for approximately 400 instances, and so on. Globally, for j60 (resp. j90 and j120), ITSR finds schedules which are within 5% (resp. 7% and 9%) from the best upper bounds. Figure 4b shows the degradation of the search efficiency when deactivating the differentiation techniques (ITSR-D), and then both differentiability and incremental computations (ITSR-D-I).

9   Conclusion and Future Work

In this paper, we gave a global view of CBLS techniques adapted to SDST-RCPSP, with a focus on a new CBLS invariant capable of dealing with a large class of renewable resources. With regards to existing CBLS systems, this invariant is rather large in the sense that it does not decompose the management of renewable resources into several smaller invariants which are then dynamically ordered in the graph of invariants. We believe that this allows us to get more powerful differentiation techniques, however additional experiments should be performed to confirm this point. For future work, it would be useful to extend the invariant introduced to take into account resources with time-varying availability profiles, activities with time-varying resource consumptions, or maximum distance constraints between activities, and to get more insight into the contribution of each component of the search strategy defined.

References

1. Aggoun, A., Beldiceanu, N.: Extending CHIP in order to solve complex scheduling and placement problems. Math. Comput. Modell. **17**(7), 57–73 (1993)
2. Allahverdi, A., Ng, C., Cheng, T., Kovalyov, M.Y.: A survey of scheduling problems with setup times or costs. Eur. J. Oper. Res. **187**(3), 985–1032 (2008)
3. Artigues, C., Feillet, D.: A branch and bound method for the job-shop problem with sequence-dependent setup times. Ann. Oper. Res. **159**(1), 135–159 (2008)
4. Artigues, C., Michelon, P., Reusser, S.: Insertion techniques for static and dynamic resource constrained project scheduling. Eur. J. Oper. Res. **149**(2), 249–267 (2003)
5. Babin, G., Deneault, S., Laporte, G.: Improvements to the Or-opt heuristic for the symmetric traveling salesman problem. J. Oper. Res. Soc. **58**, 402–407 (2007)
6. Balas, E., Simonetti, N., Vazacopoulos, A.: Job shop scheduling with setup times, deadlines and precedence constraints. J. Sched. **11**(4), 253–262 (2008)
7. Bender, M.A., Cole, R., Demaine, E.D., Farach-Colton, M., Zito, J.: Two simplified algorithms for maintaining order in a list. In: Möhring, R., Raman, R. (eds.) ESA 2002. LNCS, vol. 2461, pp. 152–164. Springer, Heidelberg (2002). doi:10.1007/3-540-45749-6_17

8. Brucker, P., Drexl, A., Möring, R., Neumann, K., Pesch, E.: Resource-constrained project scheduling: notation, classification, models, and methods. Eur. J. Oper. Res. **112**(1), 3–41 (1999)
9. Brucker, P., Thiele, O.: A branch and bound method for the general-shop problem with sequence dependent setup-times. Oper. Res. Spekt. **18**(3), 145–161 (1996)
10. Carlier, J.: The one machine sequencing problem. Eur. J. Oper. Res. **11**, 42–47 (1982)
11. Carlier, J., Pinson, E.: Adjustment of heads and tails for the job-shop problem. Eur. J. Oper. Res. **78**, 146–161 (1994)
12. Croce, F.D.: Generalized pairwise interchanges and machine scheduling. Eur. J. Oper. Res. **83**(2), 310–319 (1995)
13. Croes, G.A.: A method for solving traveling salesman problems. Oper. Res. **6**, 791–812 (1958)
14. Glover, F., Laguna, M.: Tabu search. In: Modern Heuristic Techniques for Combinatorial Problems, pp. 70–141. Blackwell Scientific Publishing (1993)
15. Godard, D., Laborie, P., Nuijten, W.: Randomized large neighborhood search for cumulative scheduling. In: Proceedings of the 15th International Conference on Automated Planning and Scheduling (ICAPS 2005), pp. 81–89 (2005)
16. González, M.A., Vela, C.R., Varela, R.: A new hybrid genetic algorithm for the job shop scheduling problem with setup times. In: Proceedings of the 18th International Conference on Automated Planning and Scheduling (ICAPS 2008), pp. 116–123 (2008)
17. González, M.A., Vela, C.R., Varela, R.: Genetic algorithm combined with tabu search for the job shop scheduling problem with setup times. In: Mira, J., Ferrández, J.M., Álvarez, J.R., Paz, F., Toledo, F.J. (eds.) IWINAC 2009. LNCS, vol. 5601, pp. 265–274. Springer, Heidelberg (2009). doi:10.1007/978-3-642-02264-7_28
18. Grimes, D., Hebrard, E.: Job shop scheduling with setup times and maximal time-lags: a simple constraint programming approach. In: Lodi, A., Milano, M., Toth, P. (eds.) CPAIOR 2010. LNCS, vol. 6140, pp. 147–161. Springer, Heidelberg (2010). doi:10.1007/978-3-642-13520-0_19
19. Grimes, D., Hebrard, E.: Solving variants of the job shop scheduling problem through conflict-directed search. INFORMS J. Comput. **27**(2), 268–284 (2015)
20. Hartmann, S.: A competitive genetic algorithm for resource-constrained project scheduling. Naval Res. Logist. **45**, 733–750 (1997)
21. Ilog: IBM ILOG CPLEX and CpOptimizer. http://www-03.ibm.com/software/products/
22. Katriel, I., Michel, L., Van Hentenryck, P.: Maintaining longest paths incrementally. Constraints **10**(2), 159–183 (2005)
23. Kolisch, R., Hartmann, S.: Heuristic algorithms for solving the resource-constrained project scheduling problem: classification and computational analysis. In: Handbook on Recent Advances in Project Scheduling, pp. 147–178. Kluwer Academic Publishers, Dordrecht (1999)
24. Kolisch, R., Hartmann, S.: Experimental investigation of heuristics for resource-constrained project scheduling: an update. Eur. J. Oper. Res. **174**(1), 23–37 (2006)
25. Mastrolilli, M., Gambardella, L.: Effective neighborhood functions for the flexible job shop problem. J. Sched. **3**(1), 3–20 (2000)
26. Michel, L., Van Hentenryck, P.: Iterative relaxations for iterative flattening in cumulative scheduling. In: Proceedings of the 14th International Conference on Automated Planning and Scheduling (ICAPS 2004), pp. 200–208 (2004)

27. Mladenovíc, N., Hansen, P.: Variable neighborhood search. Comput. Oper. Res. **24**(11), 1097–1100 (1997)
28. Nuijten, W.P.M., Aarts, E.H.L.: A computational study of constraint satisfaction for multiple capacitated job shop scheduling. Eur. J. Oper. Res. **90**(2), 269–284 (1996)
29. Oddi, A., Cesta, A., Policella, N., Smith, S.F.: Iterative flattening search for resource constrained scheduling. J. Intell. Manuf. **21**, 17–30 (2010)
30. Pinedo, M.: Scheduling: Theory, Algorithms, and Systems. Springer, New York (2012). doi:10.1007/978-1-4614-2361-4
31. Pralet, C., Verfaillie, G.: Dynamic online planning and scheduling using a static invariant-based evaluation model. In: Proceedings of the 23rd International Conference on Automated Planning and Scheduling (ICAPS 2013), pp. 171–179 (2013)
32. Schutt, A., Feydy, T., Stuckey, P.J., Wallace, M.G.: Solving the resource constrained project scheduling problem with generalized precedences by lazy clause generation. CoRR abs/1009.0347 (2010)
33. Valls, V., Ballestín, F., Quintanilla, S.: Justification and RCPSP: a technique that pays. Eur. J. Oper. Res. **165**(2), 375–386 (2005)
34. Van Cauwelaert, S., Dejemeppe, C., Monette, J.N., Schaus, P.: Efficient filtering for the unary resource with family-based transition times. In: Proceedings of the 22nd International Conference on Principles and Practice of Constraint Programming (CP 2016), pp. 520–535 (2016)
35. Van Hentenryck, P., Michel, L.: Constraint-Based Local Search. MIT Press, Cambridge (2005)
36. Vilím, P.: Edge finding filtering algorithm for discrete cumulative resources in O(k n log n). In: Proceedings of the 15th International Conference on Principles and Practice of Constraint Programming (CP 2009), pp. 802–816 (2009)
37. Vilím, P.: Timetable edge finding filtering algorithm for discrete cumulative resources. In: Achterberg, T., Beck, J.C. (eds.) CPAIOR 2011. LNCS, vol. 6697, pp. 230–245. Springer, Heidelberg (2011). doi:10.1007/978-3-642-21311-3_22

Rotation-Based Formulation for Stable Matching

Mohamed Siala$^{(\boxtimes)}$ and Barry O'Sullivan

Insight Centre for Data Analytics, Department of Computer Science, University
College Cork, Cork, Ireland
{mohamed.siala,barry.osullivan}@insight-centre.org

Abstract. We introduce new CP models for the many-to-many stable
matching problem. We use the notion of rotation to give a novel encoding
that is linear in the input size of the problem. We give extra filtering rules
to maintain arc consistency in quadratic time. Our experimental study
on hard instances of sex-equal and balanced stable matching shows the
efficiency of one of our propositions as compared with the state-of-the-art
constraint programming approach.

1 Introduction

In two-sided stable matching problems the objective is to assign some agents
to other agents based on their preferences [14]. The classic exemplar of such
problems is the well known *stable marriage (SM)* problem, first introduced by
Gale and Shapley [6]. In SM the two sets of agents are called men and women.
Each man has a preference list over the women and vice versa. The purpose
is to find a *matching* where each man (respectively woman) is associated to at
most one woman (respectively man) that respects a criterion called *stability*.
A matching M in this context is stable if any pair $\langle m, w \rangle$ (where m is a man and
w is a woman) that does not belong to M satisfies the property that m prefers
his partner in M to w or w prefers her partner in M to m.

This family of problems has gained considerable attention as it has a wide
range of applications such as assigning doctors to hospitals, students to college,
and in kidney exchange problems. The stable marriage problem itself can be
solved in $O(n^2)$ time [6] where n is the maximum number of men/women. This
is also true for the general case of many-to-many stable matching; the complexity
$O(n^2)$ is given in the proof of Theorem 1 in [1]. However, when facing real world
situations the problem often considers additional optimality criteria. In many
cases, the problem becomes intractable and specialized algorithms for solving
the standard version are usually hard to adapt. The use of a modular approach
such as constraint programming is very beneficial to tackle such cases.

Many constraint programming approaches exist in the literature for stable
matching problems. Examples of these concern stable marriage [7,21,22], hospital residents (HR) [13,20], many-to-many stable matching [3], and stable roommates [17]. Despite the fact that many-to-many stable matching generalizes HR
and SM, it has not gained as much attention as SM and HR in the constraint programming community. In this paper, we follow this line of research by proposing

© Springer International Publishing AG 2017
J.C. Beck (Ed.): CP 2017, LNCS 10416, pp. 262–277, 2017.
DOI: 10.1007/978-3-319-66158-2_17

an effective and efficient model for all three variants of stable matching: one-to-one, many-to-one, and many-to-many. Our propositions are based on a powerful structure called rotations. The latter has been used to model the stable roommates problem in [9] (p. 194) and [4,5].

We leverage some known properties related to rotations in order to propose a novel SAT formulation of the general case of many-to-many stable matching. We show that unit propagation on this formula ensures the existence of a particular solution. Next, we use this property to give an algorithm that maintains arc consistency if one considers many-to-many stable matching as a (global) constraint. The overall complexity for arc consistency is $O(L^2)$ time where L is total input size of all preference lists. Our experimental study on hard instances of sex-equal and balanced stable matching show that our approach outperforms the state-of-the-art constraint programming approach [20].

The remainder of this paper is organized as follows. In Sect. 2 we give a brief overview of constraint programming. We present the stable matching problem in Sect. 3 as well as various concepts related to rotations. In Sect. 4 we propose a novel formulation of stable matching based on the notion of rotation. We show in Sect. 5 some additional pruning rules and show that arc consistency can be maintained in $O(L^2)$ worst case time complexity. Lastly, in Sect. 6 we present an empirical experimental study on two hard variants of stable matching and show that one of our new models outperforms the state-of-the-art constraint programming approach in the literature.

2 Constraint Programming

We provide a short formal background related to constraint programming. Let $\mathcal{X}$ be a set of integer variables. A *domain* for $\mathcal{X}$, denoted by $\mathcal{D}$, is a mapping from variables to finite sets of integers. For each variable x, we call $\mathcal{D}(x)$ the *domain of the variable* x. A variable is called assigned when $\mathcal{D}(x) = \{v\}$. In this case, we say that v is assigned to x and that x is set to v. A variable is unassigned if it is not assigned. A *constraint* C defined over $[x_1, \ldots, x_k]$ ($k \in \mathbb{N}^*$) is a finite subset of $\mathbb{Z}^k$. The sequence $[x_1, \ldots, x_k]$ is the *scope* of C (denoted by $\mathcal{X}(C)$) and k is called the *arity* of C. A *support* for C in a domain $\mathcal{D}$ is a k-tuple τ such that $\tau \in C$ and $\tau[i] \in \mathcal{D}(x_i)$ for all $i \in [1, \ldots, k]$. Let $x_i \in \mathcal{X}(C)$ and $v \in \mathcal{D}(x_i)$. We say that the assignment of v to x has a support for C in $\mathcal{D}$ iff there exists a support τ for C in $\mathcal{D}$ such that $\tau[i] = v$. The constraint C is *arc consistent* (AC) in $\mathcal{D}$ iff $\forall i \in [1, \ldots, k]$, $\forall v \in \mathcal{D}(x_i)$, the assignment of v to x_i has a support in $\mathcal{D}$. A *filtering algorithm* (or *propagator*) for a constraint C takes as input a domain $\mathcal{D}$ and returns either $\emptyset$ if there is no support for C in $\mathcal{D}$ (i.e., failure) or a domain $\mathcal{D}'$ such that any support for C in $\mathcal{D}$ is a support for C in $\mathcal{D}'$, $\forall x \in \mathcal{X}(C)$, $\mathcal{D}'(x) \subseteq \mathcal{D}(x)$, and $\forall x \notin \mathcal{X}(C)$, $\mathcal{D}'(x) = \mathcal{D}(x)$. A *Boolean* variable has an initial domain equal to $\{0, 1\}$ (0 is considered as *false* and 1 as *true*). A clause is a disjunction of literals where a literal is a Boolean variable or its negation. Clauses are usually filtered with an algorithm called unit propagation [16].

Let $\mathcal{X}$ be a set of variables, $\mathcal{D}$ be a domain, and $\mathcal{C}$ be a set of constraints defined over subsets of $\mathcal{X}$. The *constraint satisfaction problem* (*CSP*) is the question of deciding if an $|\mathcal{X}|$−tuple of integers τ exists such that the projection of τ on the scope of every constraint $C \in \mathcal{C}$ is a support for C in $\mathcal{D}$. We consider in this paper classical backtracking algorithms to solve CSPs by using filtering algorithms at every node of the search tree [19].

3   Stable Matching

We consider the general case of the many-to-many stable matching problem. We follow the standard way of introducing this problem by naming the two sets of agents as *workers* and *firms* [14]. We use a notation similar to that of [3].

Let $n_F, n_W \in \mathbb{N}^*$, $F = \{f_1, f_2, \ldots, f_{n_F}\}$ be a set of firms, $W = \{w_1, w_2, \ldots, w_{n_W}\}$ be a set of workers, and $n = \max\{n_F, n_W\}$. Every firm f_i has a list, P_{f_i}, of workers given in a strict order of preference (i.e., no ties). The preference list of a worker w_i is similarly defined. We denote by $P_W = \{P_{w_i} \mid i \in [1, n_W]\}$ the set of preferences of workers, and by $P_F = \{P_{f_j} \mid j \in [1, n_F]\}$ the set of preferences of firms. We use L to denote the sum of the sizes of the preference lists. Note that the size of the input problem is $O(L)$. Therefore we shall give all our complexity results with respect to L.

For every firm f_j (respectively, worker w_i), we denote by q_{f_j} (respectively, q_{w_i}) its quota. We denote by $q_W = \{q_{w_i} \mid i \in [1, n_W]\}$ the set of quota for workers, and by $q_F = \{q_{f_j} \mid j \in [1, n_F]\}$ the set of quotas for firms. We use the notation $w_i \succ_{f_k} w_j$ when a firm f_k prefers worker w_i to worker w_j. The operator $\succ_{w_k}$ is defined similarly for any worker w_k.

A pair $\langle w_i, f_j \rangle$ is said to be *acceptable* if $w_i \in P_{f_j}$ and $f_j \in P_{w_i}$. A *matching* M is a set of acceptable pairs. Let $M(w_i) = \{f_j \mid \langle w_i, f_j \rangle \in M\}$, and $M(f_j) = \{w_i \mid \langle w_i, f_j \rangle \in M\}$. A worker w_i (respectively, firm f_j) is said to be *under-assigned* in M if $|M(w_i)| < q_{w_i}$ (respectively, $|M(f_j)| < q_{f_j}$). We define for every worker w_i, $last_M(w_i)$ as the least preferred firm for w_i in $M(w_i)$ if $M(w_i) \neq \emptyset$. For every firm f_j, $last_M(f_j)$ is similarly defined. A pair $\langle w_i, f_j \rangle \notin M$ is said to be *blocking* M if it is acceptable such that the following two conditions are true:

- w_i is under-assigned in M or $\exists f_k \in M(w_i)$ and $f_j \succ_{w_i} f_k$.
- f_j is under-assigned in M or $\exists w_l \in M(f_j)$ and $w_i \succ_{f_j} w_l$.

Definition 1 (Stability). *A matching M is* (pairwise) stable *if $\forall w_i \in W$, $|M(w_i)| \leq q_{w_i}$, $\forall f_j \in F$, $|M(f_j)| \leq q_{f_j}$, and there is no blocking pair for M.*

An instance of the many-to-many stable matching problem is defined by the tuple $\langle W, F, P_W, P_F, q_W, q_F \rangle$. The problem is to find a stable matching if one exists.

A pair $\langle w_i, f_j \rangle$ is *stable* if there exists a stable matching M containing $\langle w_i, f_j \rangle$, *unstable* otherwise. A pair $\langle w_i, f_j \rangle$ is *fixed* if it is included in all stable matchings.

Let M, M' be two stable matchings. A worker w_i prefers M no worse than M' (denoted by $M \succeq_{w_i} M'$) if (1) $M(w_i) = M'(w_i)$ or (2) $|M(w_i)| \geq |M'(w_i)|$

and $last_M(w_i) \succ_{w_i} last_{M'}(w_i)$. It should be noted that every worker (respectively, firm) is assigned to the same number of firms (respectively, workers) in every stable matching [1]. So the condition $|M(w_i)| \geq |M'(w_i)|$ is always true in the case of many-to-many stable matching. Let M, M' be two different stable matchings. We say that M dominates M' (denoted by $M \succeq_W M'$) if $M \succeq_{w_i} M'$ for every worker w_i. This is called the worker-oriented dominance relation. The *firm-oriented dominance relation ($\succeq_F$)* is similarly defined for firms.

The authors of [1] showed that a stable matching always exists and can be found in $O(n^2)$ time. More precisely, the complexity of finding a stable matching is $O(L)$. Moreover, they showed that there always exist worker-optimal and firm-optimal stable matchings (with respect to $\succeq_W$ and $\succeq_F$). We denote these two matchings by M_0 and M_z, respectively.

Example 1 (An instance of many-to-many stable matching (from [3]). Consider the example where $n_W = 5, n_F = 5$, and for all $1 \leq i, j \leq 5$, $q_{w_i} = q_{f_j} = 2$. The preference lists for workers and firms are given in Table 1.

Table 1. Example of preference lists

$P_{w_1} = [f_1, f_2, f_3, f_4, f_5]$	$P_{f_1} = [w_3, w_2, w_4, w_5, w_1]$
$P_{w_2} = [f_2, f_3, f_4, f_5, f_1]$	$P_{f_2} = [w_2, w_3, w_5, w_4, w_1]$
$P_{w_3} = [f_3, f_4, f_5, f_1, f_2]$	$P_{f_3} = [w_4, w_5, w_2, w_1, w_3]$
$P_{w_4} = [f_4, f_5, f_1, f_2, f_3]$	$P_{f_4} = [w_1, w_5, w_3, w_2, w_4]$
$P_{w_5} = [f_5, f_1, f_2, f_3, f_4]$	$P_{f_5} = [w_4, w_1, w_2, w_3, w_5]$

There exist seven stable matchings for this instance:

- $M_0 = \{\langle w_1, f_1 \rangle, \langle w_1, f_2 \rangle, \langle w_2, f_2 \rangle, \langle w_2, f_3 \rangle, \langle w_3, f_3 \rangle, \langle w_3, f_4 \rangle, \langle w_4, f_4 \rangle,$
  $\langle w_4, f_5 \rangle, \langle w_5, f_5 \rangle, \langle w_5, f_1 \rangle\}$
- $M_1 = \{\langle w_1, f_1 \rangle, \langle w_1, f_3 \rangle, \langle w_2, f_2 \rangle, \langle w_2, f_3 \rangle, \langle w_3, f_5 \rangle, \langle w_3, f_4 \rangle, \langle w_4, f_4 \rangle, \langle w_4, f_5 \rangle,$
  $\langle w_5, f_2 \rangle, \langle w_5, f_1 \rangle\}$
- $M_2 = \{\langle w_1, f_4 \rangle, \langle w_1, f_3 \rangle, \langle w_2, f_2 \rangle, \langle w_2, f_3 \rangle, \langle w_3, f_5 \rangle, \langle w_3, f_4 \rangle, \langle w_4, f_1 \rangle, \langle w_4, f_5 \rangle,$
  $\langle w_5, f_2 \rangle, \langle w_5, f_1 \rangle\}$
- $M_3 = \{\langle w_1, f_4 \rangle, \langle w_1, f_5 \rangle, \langle w_2, f_2 \rangle, \langle w_2, f_3 \rangle, \langle w_3, f_1 \rangle, \langle w_3, f_4 \rangle, \langle w_4, f_1 \rangle, \langle w_4, f_5 \rangle,$
  $\langle w_5, f_2 \rangle, \langle w_5, f_3 \rangle\}$
- $M_4 = \{\langle w_1, f_4 \rangle, \langle w_1, f_5 \rangle, \langle w_2, f_2 \rangle, \langle w_2, f_3 \rangle, \langle w_3, f_1 \rangle, \langle w_3, f_2 \rangle, \langle w_4, f_1 \rangle, \langle w_4, f_5 \rangle,$
  $\langle w_5, f_4 \rangle, \langle w_5, f_3 \rangle\}$
- $M_5 = \{\langle w_1, f_4 \rangle, \langle w_1, f_5 \rangle, \langle w_2, f_2 \rangle, \langle w_2, f_1 \rangle, \langle w_3, f_1 \rangle, \langle w_3, f_4 \rangle, \langle w_4, f_3 \rangle, \langle w_4, f_5 \rangle,$
  $\langle w_5, f_2 \rangle, \langle w_5, f_3 \rangle\}$
- $M_z = M_6 = \{\langle w_1, f_4 \rangle, \langle w_1, f_5 \rangle, \langle w_2, f_2 \rangle, \langle w_2, f_1 \rangle, \langle w_3, f_1 \rangle, \langle w_3, f_2 \rangle, \langle w_4, f_3 \rangle,$
  $\langle w_4, f_5 \rangle, \langle w_5, f_4 \rangle, \langle w_5, f_3 \rangle\}$

In this instance, $\langle w_1, f_1 \rangle$ is a stable pair since $\langle w_1, f_1 \rangle \in M_0$ and $\langle w_2, f_4 \rangle$ is not stable since it is not included in any stable matching. Regarding the dominance relation, we have $M_1 \succeq_W M_2$, and $M_2 \succeq_W M_3$, Using transitivity, we obtain $M_1 \succeq_W M_3$, Note that M_4 and M_5 are incomparable.   □

In the following, we introduce a central notion in this paper called rotation. Consider the matching M_0 from the instance given in Example 1 and the list of pairs $\rho_0 = [\langle w_1, f_2 \rangle, \langle w_5, f_5 \rangle, \langle w_3, f_3 \rangle]$. Notice that every pair in ρ_0 is part of M_0. Consider now the operation of shifting the firms in a cyclic way as follows: f_2 is paired with w_5, f_5 is paired with w_3, and f_3 is paired with w_1. This operation changes M_0 to M_1. In this case, we say ρ_0 is a rotation.

Formally, for any stable matching $M \neq M_z$ and any worker w_i such that $M(w_i) \neq \emptyset$, we define $r_M(w_i)$ to be the most preferred firm f_j for w_i such that $w_i \succ_{f_j} last_M(f_j)$ and $\langle w_i, f_j \rangle \notin M$. In other words, given $\langle w_i, f_j \rangle \notin M$, $r_M(w_i)$ is a firm that is the most preferred firm to w_i such that it prefers w_i to her worst assigned partner in M.

Definition 2 (Rotation [2]). *A rotation ρ is an ordered list of pairs $[\langle w_{i_0}, f_{j_0} \rangle, \langle w_{i_1}, f_{j_1} \rangle, \ldots, \langle w_{i_{t-1}}, f_{j_{t-1}} \rangle]$ such that $t \in [2, \min(n_W, n_F)]$, $i_k \in [1, n_W]$, $j_k \in [1, n_F]$ for all $0 \leq k < t$ and there exists a stable matching M where $\langle w_{i_k}, f_{j_k} \rangle \in M$, $w_{i_k} = last_M(f_{j_k})$, and $f_{j_k} = r_M(w_{i_{k+1} \mod t})$ for all $0 \leq k < t$. In this case we say that ρ is* exposed *in M.*

Let ρ be a rotation exposed in a stable matching M. The operation of *eliminating* a rotation ρ from M consists of removing each pair $\langle w_{i_k}, f_{j_k} \rangle \in \rho$ from M, then adding $\langle w_{i_{k+1} \mod t}, f_{j_k} \rangle$. The new set of pairs, denoted by $M_{/\rho}$ constitutes a stable matching that is dominated (w.r.t. workers) by M [3,8]. We say that ρ *produces* $\langle w_i, f_j \rangle$ if $\langle w_i, f_j \rangle \in M_{/\rho} \setminus M$.

The following three lemmas are either known in the literature [3] or are a direct consequence of [3].

Lemma 1. *In every stable matching $M \neq M_z$, there exists (at least) a rotation that can be exposed in M.*

Lemma 2. *Every stable matching $M \neq M_0$ can be obtained by iteratively eliminating some rotations, without repetition, starting from M_0.*

Lemma 3. *Any succession of eliminations leading from M_0 to M_z contains all the possible rotations (without repetition).*

We say that a rotation ρ_1 precedes another rotation ρ_2 (denoted by $\rho_1 \prec\prec \rho_2$) if ρ_1 is exposed before ρ_2 in every succession of eliminations leading from M_0 to M_z. Note that this precedence relation is transitive and partial. That is, $\rho_1 \prec\prec \rho_2 \wedge \rho_2 \prec\prec \rho_3$, implies $\rho_1 \prec\prec \rho_3$, and there might exist two rotations ρ_1, and ρ_2 where neither $\rho_1 \prec\prec \rho_2$ nor $\rho_2 \prec\prec \rho_1$.

Example 2 (Rotation precedence). In the previous example we have $\rho_0 \prec\prec \rho_1$, $\rho_1 \prec\prec \rho_2$, $\rho_2 \prec\prec \rho_3$, $\rho_2 \prec\prec \rho_4$. By transitivity we obtain $\rho_0 \prec\prec \rho_4$. Note that in this example neither $\rho_3 \prec\prec \rho_4$ nor $\rho_4 \prec\prec \rho_3$.   □

Let R be the set of all rotations. The precedence relation $\prec\prec$ with R forms the *rotation poset* Π_R. Let $G = (V_G, A_G)$ be the directed graph corresponding to the rotation poset. That is, every vertex corresponds to a rotation, and there is an arc $(\rho_j, \rho_i) \in A_G$ iff $\rho_j \prec\prec \rho_i$. The construction of R and G can be performed in $O(L)$ time [3]. For each rotation $\rho_i \in R$, we denote by $N^-(\rho_i)$ the set of rotations having an outgoing edge towards ρ_i, i.e., these rotations dominate ρ_i. We introduce below the notion of closed subset and a very important theorem.

Definition 3 (Closed subset). *A subset of rotations $S \subseteq V_G$ is closed iff $\forall \rho_i \in S$, $\forall \rho_j \in V_G$, if $\rho_j \prec\prec \rho_i$, then $\rho_j \in S$.*

Theorem 1 (From [2]). *There is a one-to-one correspondence between closed subsets and stable matchings.*

The solution corresponding to a closed subset S is obtained by eliminating all the rotations in S starting from M_0 while respecting the order of precedence between the rotations. Recall from Lemma 2 that every stable matching $M \neq M_0$ can be obtained by iteratively eliminating some rotations, without any repetition, starting from M_0. The closed subset corresponding to a stable matching M is indeed the set of rotations in any succession of eliminations of rotations leading to M. Notice that M_0 corresponds to the empty set and that M_z is the set of all rotations.

We denote by Δ the set of stable pairs. Let $\langle w_i, f_j \rangle$ be a stable pair. There exists a unique rotation containing $\langle w_i, f_j \rangle$ if $\langle w_i, f_j \rangle \notin M_z$ [3]. We denote this rotation by $\rho_{e_{ij}}$. Similarly, $\forall \langle w_i, f_j \rangle \in \Delta \setminus M_0$ there exists a unique rotation ρ such that eliminating ρ produces $\langle w_i, f_j \rangle$. We denote by $\rho_{p_{ij}}$ the rotation that produces the stable pair $\langle w_i, f_j \rangle \in \Delta \setminus M_0$. Notice that it is always the case that $\rho_{p_{ij}} \prec\prec \rho_{e_{ij}}$ for any stable pair that is not part of $M_0 \cup M_z$.

Example 3 (The rotations $\rho_{e_{ij}}$ and $\rho_{p_{ij}}$). For the previous example, we have $\rho_{e_{23}} = \rho_4$, and $\rho_{p_{31}} = \rho_2$ since ρ_2 produces the pair $\langle w_3, f_1 \rangle$.   □

Lastly, we denote by FP the set of fixed pairs, SP is the set of stable pairs that are not fixed, and NSP is the set of non stable pairs. Note that $\langle w_i, f_j \rangle \in FP$ iff $\langle w_i, f_j \rangle \in M_0 \cap M_z$. These three sets can be constructed in $O(L)$ time [3].

4   A Rotation-Based Formulation

We first show that the problem of finding a stable matching can be formulated as a SAT formula using properties from rotations. Next, we show that for any input domain $\mathcal{D}$, if unit propagation is performed without failure, then there exists necessarily a solution in $\mathcal{D}$. Recall that there exists an algorithm (called the Extended Gale-Shapley algorithm) to find a solution to the many-to-many stable matching that runs in $O(L)$ time [1,3]. However, using a CP formulation such as the one that we propose in this section is very beneficial when dealing with NP-Hard variants of the problem.

In out model, a preprocessing step is performed to compute M_0, M_z, SP, FP, NSP, the graph posed, $\rho_{e_{ij}}$ for all $\langle w_i, f_j \rangle \in SP \setminus M_z$, and $\rho_{p_{ij}}$ for all $\langle w_i, f_j \rangle \in SP \setminus M_0$. This preprocessing is done in $O(L)$ time [3].

4.1 A SAT Encoding

We introduce for each pair $\langle w_i, f_j \rangle$ a Boolean variable $x_{i,j}$. The latter is set to true iff $\langle w_i, f_j \rangle$ is part of the stable matching. Moreover, we use for each rotation ρ_k a Boolean variable r_k (called *rotation variable*) to indicate whether the rotation ρ_k is in the closed subset that corresponds to the solution.

Observe first that for all $\langle w_i, f_j \rangle \in FP$, $x_{i,j}$ has to be true, and for all $\langle w_i, f_j \rangle \in NSP$, $x_{i,j}$ has to be false.

We present three lemmas that are mandatory for the soundness and completeness of the SAT formula. Let M be a stable matching and S its closed subset (Theorem 1).

Lemma 4. $\forall \langle w_i, f_j \rangle \in SP \cap M_0 : \langle w_i, f_j \rangle \in M$ iff $\rho_{e_{ij}} \notin S$.

Proof. $\Rightarrow$ Suppose that $\rho_{e_{ij}} \in S$. Let *Sequence* be an ordered list of the rotations in S such that exposing the rotations of S starting from M_0 leads to M. For all $a \in [1, |S|]$, we define M'_a to be the stable matching corresponding the closed subset $S'_a = \{Sequence[k] \mid k \in [1, a]\}$. We also use M'_0 to denote the particular case of M_0 and $S'_0 = \emptyset$. Notice that $M'_{|S|} = M$ and $S'_{|S|} = S$. Let $a \in [1, |S|]$ such that $Sequence[a] = \rho_{e_{ij}}$. We know that exposing the rotation $\rho_{e_{ij}}$ from S'_{a-1} moves worker w_i to a partner that is worse than f_i. For any matching M'_b where $b \in [a, |S|]$, w_i either has the same partners in M'_{b-1} or is assigned a new partner that is worse than f_i. Hence $\langle w_i, f_j \rangle$ cannot be part of $M'_{|S|} = M$.

$\Leftarrow$ $\langle w_i, f_j \rangle$ must be part of the solution since it is part of M_0 and $\rho_{e_{ij}} \notin S$. $\square$

Lemma 5. $\forall \langle w_i, f_j \rangle \in SP \cap M_z : \langle w_i, f_j \rangle \in M$ iff $\rho_{p_{ij}} \in S$.

Proof. $\Rightarrow$ Suppose that $\rho_{p_{ij}} \notin S$. The pair $\langle w_i, f_j \rangle$ cannot be produced when eliminating rotations in S since $\rho_{p_{ij}}$ is unique. Therefore $\rho_{p_{ij}} \in S$.

$\Leftarrow$ Suppose that $\rho_{p_{ij}} \in S$. The pair $\langle w_i, f_j \rangle$ must be part of the solution since $\rho_{p_{ij}} \in S$ and it can never be eliminated by any rotation since $\langle w_i, f_j \rangle \in M_z$. $\square$

Lemma 6. $\forall \langle w_i, f_j \rangle \in SP \setminus (M_0 \cup M_z) : \langle w_i, f_j \rangle \in M$ iff $\rho_{p_{ij}} \in S \wedge \rho_{e_{ij}} \notin S$.

Proof. $\Rightarrow$ Suppose that $\langle w_i, f_j \rangle$ is part of M.

- If $\rho_{p_{ij}} \notin S$, then $\langle w_i, f_j \rangle$ can never be produced when eliminating rotations in S. Therefore $\rho_{p_{ij}} \in S$.
- If $\rho_{e_{ij}} \in S$, similarly to the proof of Lemma 4, we can show that the pair $\langle w_i, f_j \rangle$ cannot be part of the solution.

$\Leftarrow$ Suppose that $\rho_{p_{ij}} \in S$ and $\rho_{e_{ij}} \notin S$. The pair $\langle w_i, f_j \rangle$ must be part of the solution since it is produced by $\rho_{p_{ij}}$ and not eliminated since $\rho_{e_{ij}} \notin S$. $\square$

Using Lemmas 4, 5 and 6, we can formulate the problem of finding a stable matching as follows.

$$\forall \rho_i \in R, \forall \rho_j \in N^-(\rho_i) : \ \neg r_i \vee r_j \tag{1}$$

$$\forall \langle w_i, f_j \rangle \in SP \cap M_0 : \neg x_{i,j} \vee \neg r_{e_{ij}} \; ; \; x_{i,j} \vee r_{e_{ij}} \tag{2}$$

$$\forall \langle w_i, f_j \rangle \in SP \cap M_z : \neg x_{i,j} \vee r_{p_{ij}} \; ; \; x_{i,j} \vee \neg r_{p_{ij}} \tag{3}$$

$$\forall \langle w_i, f_j \rangle \in SP \setminus (M_0 \cup M_z) : \neg x_{i,j} \vee r_{p_{ij}} \; ; \; \neg x_{i,j} \vee \neg r_{e_{ij}} \; ; \; x_{i,j} \vee \neg r_{p_{ij}} \vee r_{e_{ij}} \tag{4}$$

$$\forall \langle w_i, f_j \rangle \in FP : x_{i,j} \tag{5}$$

$$\forall \langle w_i, f_j \rangle \in NSP : \neg x_{i,j} \tag{6}$$

We denote this formula by Γ. Clauses 1 make sure that the set of rotation variables that are set to true corresponds to a closed subset. Clauses 2, 3, and 4 correspond (respectively) to Lemmas 4, 5, and 6. Lastly, Clauses 5 and 6 handle the particular cases of fixed and non stable pairs (respectively). Observe that each clause is of size at most 3. Moreover, since the the number of edges in the graph poset is bounded by $O(L)$ [3], then the size of this formula is $O(L)$.

The only CP formulation for the case of many-to-many stable matching was proposed in [3]. It is a straightforward generalization of the CSP model proposed for the hospital/residents problem in [13]. The authors use q_{w_i} variables per worker, and q_{f_j} variables per firm. The variables related to a worker w_i represent the rank of the firm assigned at each position (out of the q_{w_i} available positions). A similar set of variables is used for firms. The model contains $|W| \times (\sum_i q_{w_i} + |F| \times (1 + \sum_j q_{f_j} \times (2 + \sum_i (q_{w_i} - 1))))$ constraints related to workers. Likewise, $|F| \times (\sum_j q_{f_j} + |W| \times (1 + \sum_i q_{w_i} \times (2 + \sum_j (q_{f_j} - 1))))$ constraints are used for firms.

4.2 Properties Related to Unit Propagation

In the following, we show that once unit propagation is performed without failure then there exists necessarily a solution.

Suppose that $\mathcal{D}$ is a domain where unit propagation has been performed without failure. Let S_1 be the set of rotation variables that are set to 1.

Lemma 7. S_1 is a closed subset.

Proof. Let ρ_i be a rotation in S_1 and let ρ_j be rotation such that $\rho_j \prec\prec \rho_i$. Unit propagation on Clauses 1 enforces r_j to be true. Therefore $\rho_j \in S_1$. Hence S_1 is a closed subset. □

Let M_1 be the stable matching corresponding to S_1 (Theorem 1). We show that M_1 is part of the solution space in $\mathcal{D}$.

Lemma 8. For any $x_{i,j}$ that is set to 1, $\langle w_i, f_j \rangle \in M_1$.

Proof. The case where $\langle w_i, f_j \rangle$ is a fixed pair or non stable is trivial. Take a non-fixed stable pair $\langle w_i, f_j \rangle$ and suppose that $\mathcal{D}(x_{i,j}) = \{1\}$. There are three cases to distinguish.

1. $\langle w_i, f_j \rangle \in SP \cap M_0$: Unit propagation on Clauses 2 enforces $r_{e_{ij}}$ to be false. Therefore, $\rho_{e_{ij}} \notin S_1$. Hence by Lemma 4 we obtain: $\langle w_i, f_j \rangle \in M_1$.

2. $\langle w_i, f_j \rangle \in SP \cap M_z$: Unit propagation on Clauses 3 enforces $r_{p_{ij}}$ to be true. Therefore, $\rho_{p_{ij}} \in S_1$. Hence by Lemma 5 we obtain: $\langle w_i, f_j \rangle \in M_1$.
3. $\langle w_i, f_j \rangle \in SP \setminus (M_0 \cup M_z)$: Unit propagation on Clauses 4 enforces $r_{p_{ij}}$ to be true and $r_{e_{ij}}$ to be false. Therefore, $\rho_{p_{ij}} \in S_1$, $\rho_{e_{ij}} \notin S_1$. Hence by Lemma 6 we obtain: $\langle w_i, f_j \rangle \in M_1$. $\qquad \square$

Lemma 9. *For any $x_{i,j}$ that is set to 0, $\langle w_i, f_j \rangle \notin M_1$.*

Proof. The case where $\langle w_i, f_j \rangle$ is a fixed pair or non-stable is trivial. Take a non-fixed stable pair $\langle w_i, f_j \rangle$ and suppose that $\mathcal{D}(x_{i,j}) = \{0\}$. There are three cases to distinguish.

1. $\langle w_i, f_j \rangle \in SP \cap M_0$: Unit propagation on Clauses 2 enforces $r_{e_{ij}}$ to be true. Therefore, $\rho_{e_{ij}} \in S_1$. Hence by Lemma 4 we obtain: $\langle w_i, f_j \rangle \notin M_1$.
2. $\langle w_i, f_j \rangle \in SP \cap M_z$: Unit propagation on Clauses 3 enforces $r_{p_{ij}}$ to be false. Therefore, $\rho_{p_{ij}} \notin S_1$. Hence by Lemma 5 we obtain: $\langle w_i, f_j \rangle \notin M_1$.
3. $\langle w_i, f_j \rangle \in SP \setminus (M_0 \cup M_z)$: We distinguish two cases:
   (a) $\mathcal{D}(\rho_{p_{ij}}) \neq \{1\}$: In this case $\rho_{p_{ij}} \notin S_1$ hence by Lemma 6 we obtain: $\langle w_i, f_j \rangle \notin M_1$
   (b) $\mathcal{D}(\rho_{p_{ij}}) = \{1\}$: In this case, unit propagation on Clauses 4 enforces $r_{e_{ij}}$ to be true. Therefore, $\rho_{e_{ij}} \in S_1$. Hence by Lemma 6 we obtain: $\langle w_i, f_j \rangle \notin M_1$. $\qquad \square$

Recall that Γ denotes the SAT formula defined in Sect. 4.1.

Theorem 2. *Let $\mathcal{D}$ be a domain such that unit propagation is performed without failure on Γ. There exists at least a solution in $\mathcal{D}$ that satisfies Γ.*

Proof. We show that M_1 corresponds to a solution under $\mathcal{D}$. To do so, one needs to set every unassigned variable to a particular value. We propose the following assignment. Let $x_{i,j}$ be an unassigned variable. Note that $\langle w_i, f_i \rangle$ has to be part of SP.

1. If $\langle w_i, f_j \rangle \in SP \cap M_0$: $x_{i,j}$ is set to 1 if $\rho_{e_{ij}} \notin S_1$; and 0 otherwise.
2. If $\langle w_i, f_j \rangle \in SP \cap M_z$: $x_{i,j}$ is set to 1 if $\rho_{p_{ij}} \in S_1$; and 0 otherwise.
3. If $\langle w_i, f_j \rangle \in SP \setminus (M_0 \cup M_z)$: $x_{i,j}$ is set to 1 if $\rho_{p_{ij}} \in S_1 \wedge \rho_{e_{ij}} \notin S_1$; and 0 otherwise.

This assignment corresponds to a solution as a consequence of Lemmas 4, 5, 6, 8, and 9. Therefore, once unit propagation is established without failure, we know that there exists at least one solution. $\qquad \square$

5   Arc Consistency

We propose in this section a procedure to filter more of the search space. We assume in the rest of this section that I is a stable matching instance defined by $\langle W, F, P_W, P_F, q_W, q_F \rangle$ using the same notations introduced in Sect. 3.

Let $\mathcal{X}(M2M) = \{x_{1,1}, \ldots x_{n_W,n_F}, r_1, \ldots r_{|R|}\}$ be the set of Boolean variables defined in Sect. 4.1. We define the many-to-many stable matching constraint as $M2M(I, \mathcal{X}(M2M))$. Given a complete assignment of the variables in $\mathcal{X}(M2M)$, this constraint is satisfied iff the set M of pairs corresponding to Boolean variables $x_{i,j}$ that are set to 1 is a solution to I and the set of rotations corresponding to Boolean variables r_k that are set to 1 is the closed subset corresponding to M.

Example 4 shows an instance with a particular domain where unit propagation on Γ is not enough to establish arc consistency on the $M2M$ constraint.

Example 4 (Missing Support). Consider the example where $n_W = 4, n_F = 4$, and for all $1 \le i, j \le 4$, $q_{w_i} = q_{f_j} = 1$. The preference lists for workers and firms are given in Table 2.

Table 2. Preference lists

$P_{w_1} = [f_3, f_2, f_4, f_1]$	$P_{f_1} = [w_1, w_2, w_4, w_3]$
$P_{w_2} = [f_2, f_4, f_1, f_3]$	$P_{f_2} = [w_3, w_1, w_2, w_4]$
$P_{w_3} = [f_4, f_1, f_3, f_2]$	$P_{f_3} = [w_2, w_3, w_4, w_1]$
$P_{w_4} = [f_1, f_2, f_3, f_4]$	$P_{f_4} = [w_4, w_1, w_2, w_3]$

Consider the domain such that all the variables are unassigned except for $x_{1,4}$, $x_{3,1}$, $x_{3,3}$, $x_{4,2}$, and $x_{4,3}$ where the value 0 is assigned to each of these variables. Unit propagation on the encoding Γ of this instance does not trigger a failure. It also does not change the domain of $x_{2,1}$ (i.e., $\{0,1\}$). However, the assignment of 1 to $x_{2,1}$ does not have a support in $\mathcal{D}$ for $M2M$.   □

In the following, we assume that unit propagation is established on an input domain $\mathcal{D}$ and that it propagated the clauses without finding a failure. In the rest of this section, we use the term 'support' to say 'support for $M2M(I, \mathcal{X}(M2M))$'. We shall use unit propagation to find a support for any assignment using the property we showed in Theorem 2.

In order to construct supports, we need to introduce the following two lemmas.

Lemma 10. *For any rotation ρ_i where $\mathcal{D}(r_i) = \{0,1\}$, assigning 1 to r_i has a support.*

Proof. Consider the set of rotations $S = S_1 \cup \{r_j \mid r_j \prec\prec r_i\}$. Clearly S is a closed subset (Lemma 7). Let M be the corresponding stable matching of S. We show that M corresponds to a valid support.

By construction, we have any variable $x_{i,j}$ set to 1 is part of M and any variable set to 0 is not. Consider now the rotation variables. Recall that S_1 is the set of rotation variables that are set to 1. Observe that $\{r_j | r_j \prec\prec r_i\}$ can only contain rotations that are unassigned because otherwise, unit propagation would assign 0 to r_i. In our support, every rotation variable whose rotation is in $\{r_j | r_j \prec\prec r_i\}$ is set to 1. Consider $x_{i,j}$ an unassigned variable. We set $x_{i,j}$ as follows

1. If $\langle w_i, f_j \rangle \in SP \cap M_0 : x_{i,j}$ is set to 1 if $\rho_{e_{ij}} \notin S$; and 0 otherwise.
2. If $\langle w_i, f_j \rangle \in SP \cap M_z : x_{i,j}$ is set to 1 if $\rho_{p_{ij}} \in S$; and 0 otherwise.
3. If $\langle w_i, f_j \rangle \in SP \setminus (M_0 \cup M_z) : x_{i,j}$ is set to 1 if $\rho_{p_{ij}} \in S \wedge \rho_{e_{ij}} \notin S$; and 0 otherwise.

This assignment corresponds by construction to M as a consequence of Lemmas 4, 5, 6, 8, and 9.   □

Lemma 11. *For any rotation ρ_i where $\mathcal{D}(r_i) = \{0, 1\}$, assigning 0 to r_i has a support.*

Proof. Recall that S_1 is the set of rotation variables that are set to 1 and that M_1 is its corresponding stable matching. By construction, we can show that M_1 corresponds to a support.   □

Consider now an unassigned variable $x_{i,j}$. Notice that $\langle w_i, f_i \rangle \in SP$. Lemma 12 show that there is always a support for 0.

Lemma 12. *For any unassigned variable $x_{i,j}$, assigning 0 to $x_{i,j}$ has a support.*

Proof. We distinguish three cases:

1. $\langle w_i, f_j \rangle \in SP \cap M_0$: Observe that $\rho_{e_{ij}}$ is unassigned. We know by Lemma 10 that assigning 1 to $r_{e_{ij}}$ has a support. In this support 0 is assigned to $x_{i,j}$.
2. $\langle w_i, f_j \rangle \in SP \cap M_z$: In this case $\rho_{p_{ij}}$ is unassigned. We know by Lemma 11 that assigning 0 to $r_{p_{ij}}$ has a support. In this support, 0 is assigned to $x_{i,j}$.
3. $\langle w_i, f_j \rangle \in SP \setminus (M_0 \cup M_z)$: Note that 0 cannot be assigned to $\rho_{p_{ij}}$ because otherwise $x_{i,j}$ would be set to 0. We distinguish two cases:
   (a) $\rho_{p_{ij}}$ is set to 1: In this case $\rho_{e_{ij}}$ is unassigned (otherwise $x_{i,j}$ would be assigned). We know by Lemma 10 that assigning 1 to $r_{e_{ij}}$ has a support. In this support 0 is assigned to $x_{i,j}$.
   (b) $\rho_{p_{ij}}$ is unassigned: We know by Lemma 11 that assigning 0 to $r_{p_{ij}}$ has a support. In this support 0 is assigned to $x_{i,j}$.   □

In the case of finding supports when assigning 1 to $x_{i,j}$, there are three cases. These cases are detailed in Lemmas 13, 14, and 15.

Lemma 13. *If $\langle w_i, f_j \rangle \in SP \cap M_0$, then assigning 1 to $x_{i,j}$ has a support.*

Proof. In this case $\rho_{e_{ij}}$ is unassigned. We know by Lemma 11 that assigning 0 to $r_{e_{ij}}$ has a support. In this support 1 is assigned to $x_{i,j}$.   □

Lemma 14. *If $\langle w_i, f_j \rangle \in SP \cap M_z$, then assigning 1 to $x_{i,j}$ has a support.*

Proof. In this case $\rho_{p_{ij}}$ is unassigned. We know by Lemma 10 that assigning 1 to $r_{p_{ij}}$ has a support. In this support $x_{i,j}$ is set to 1.   □

Let $\mathcal{D}^1_{x_{i,j}}$ be the domain identical to $\mathcal{D}$ except for $\mathcal{D}(x_{i,j}) = \{1\}$.

Lemma 15. *If $\langle w_i, f_j \rangle \in SP \setminus M_0 \cup M_z$, then*

- *If $\mathcal{D}(r_{p_{ij}}) = \{1\}$, then assigning 1 to $x_{i,j}$ has a support.*
- *Otherwise, we have $\mathcal{D}(r_{p_{ij}}) = \{0,1\}$ and assigning 1 to $x_{i,j}$ has a support iff unit propagation on $\mathcal{D}^1_{x_{i,j}}$ does not fail.*

Proof. For the first case, we can argue that $\rho_{e_{ij}}$ is unassigned (otherwise $x_{i,j}$ would be assigned). By Lemma 11, we have a support if we set $r_{e_{ij}}$ to 0. In this support $x_{i,j}$ is set to 1.

For the second case, we have necessarily $\mathcal{D}(r_{p_{ij}}) = \{0,1\}$ (otherwise $x_{i,j}$ would be assigned) and it is easy to see that there exists a support iff unit propagation does not fail on $\mathcal{D}^1_{x_{i,j}}$ by Theorem 2.   □

We summarize all the properties of the previous lemmas in Algorithm 1. This algorithm shows a pseudo-code to maintain arc consistency on $M2M(I, \mathcal{X}(M2M))$. In this algorithm, UP$(\mathcal{D})$ is the output domain after performing unit propagation on a domain $\mathcal{D}$. The output of UP$(\mathcal{D})$ is $\emptyset$ iff a failure is found.

Algorithm 1. Arc Consistency for $M2M(I, \mathcal{X}(M2M))$

---

1 $\mathcal{D} \leftarrow$ UP$(\mathcal{D})$;
   if $\mathcal{D} \neq \emptyset$ **then**
2   **foreach** $\langle w_i, f_j \rangle \in SP \wedge \langle w_i, f_j \rangle \notin M_0 \cup M_z \wedge \mathcal{D}(r_{p_{ij}}) = \{0,1\}$ **do**
3        $\mathcal{D}' \leftarrow$ UP$(\mathcal{D}^1_{x_{i,j}})$;
      if $\mathcal{D}' = \emptyset$ **then**
4          $\mathcal{D}(x_{i,j}) = \{0\}$;

   return $\mathcal{D}$

Suppose that $\mathcal{D}$ is a domain where unit propagation is established without failure. First, for any variable that is set to a value v, the assignment of v to this variable has a support in $\mathcal{D}$ since there exists necessarily a solution (Theorem 2). Second, we know that any assignment of any rotation variable has a support in $\mathcal{D}$ by Lemmas 10 and 11. Also, the assignment of 0 to any unassigned variable $x_{i,j}$ has a support (Lemma 12). Lastly, by Lemmas 13, 14, and 15, we know that we need to check supports only for the assignment of 1 to some particular unassigned variables $x_{i,j}$. These variables correspond to the pairs of the set $\Psi = \{\langle w_i, f_j \rangle | \langle w_i, f_j \rangle \in SP \wedge \langle w_i, f_j \rangle \notin M_0 \cup M_z \wedge \mathcal{D}(r_{p_{ij}}) = \{0,1\}\}$ (Lemma 15).

Algorithm 1 first performs unit propagation on the input domain $\mathcal{D}$ in Line 1. If a failure is not found, we loop over the pairs in Ψ in Line 2 and call unit propagation on the new domain $\mathcal{D}^1_{x_{i,j}}$ in Line 3 for each $\langle w_i, f_j \rangle \in \Psi$. If this call results in failure then $x_{i,j}$ does not have a support for the value 1. In this case, such a variable is set to 0 in Line 4.

We discuss now the complexity of Algorithm 1. Observe first that since the SAT formula contains only clauses of size at most 3, and since the number of clauses is $O(L)$, then unit propagation takes $O(L)$ time. Notice that by using the two-watched literal procedure [16], there is no data structure to update between the different calls. Lastly, observe that the number of calls to unit propagation in Line 3 is bounded by the number of unassigned variables. Therefore the worst-case time complexity to maintain arc consistency is $O(U_x \times L)$ where U_x is the number of unassigned $x_{i,j}$ variables. Therefore the overall complexity is $O(L^2)$.

6   Experimental Results

In the absence of known hard problems for many to many stable matching, we propose to evaluate our approach on two NP-hard variants of stable marriage called sex-equal stable matching and balanced stable matching [14]. Let M be a stable marriage. Let C_M^m (respectively C_M^w) be the sum of the ranks of each man's partner (respectively woman's partner). In balanced stable matching, the problem is to find a stable matching M with the minimum value of $max\{C_M^m, C_M^w\}$. In sex-equal stable matching, the problem is to find a stable matching M with the minimum value of $|C_M^m - C_M^w|$ [14]. Modeling these problems in constraint programming is straightforward by using an integer variable X_i for each man m_i whose domain represents the rank of the partner of m_i.

We implemented our two propositions in the Mistral-2.0 [10] solver (denoted by fr for the first formulation and ac for the arc consistency algorithm) and we compare them against the bound ($\mathcal{D}$) consistency algorithm of [20] implemented in the same solver (denoted by bc). We restrict the search strategy to branch on the sequence $[X_1, \ldots, X_n]$ since it is sufficient to decide the problem. We used four different heuristics: a lexicographic branching (lx) with random value selection (rd); lx with random min/max value selection (mn); activity based search (as) [15]; and impact-based search (is) [18]. We use geometric restarts and we run 5 randomization seeds. There is a time cutoff of 15 min for each model on each instance.

We first run all the configurations on purely random instances with complete preference lists of size up to 500×500 and observed that these instances are extremely easy to solve for all configurations without valuable outcome. We therefore propose to use a new benchmark of hard instances.

Irving and Leather [12] described a family of stable marriage instances, where the number of solutions for stable matching grows exponentially. In this family, the number of stable matchings $g(n)$ for an instance of size $n \times n$ respects the recursive formula $g(n) \geq 2 \times g(n/2)^2$, and $g(1) = 1$, where n, the number of men, is of the form 2^k. To give an idea of the exponential explosion, when $n = 16$, the number of solutions is 195472, and when $n = 32$, the number of solutions is 104310534400. We generate instances of sizes $n \in \{32, 64, 128, 256\}$ as follows. For each size, we generate the instance as in [12], then swap $\alpha\%$ of n random pairs from the preference lists of men. We apply the same swapping procedure for woman. We generated 50 instances for each size with $\alpha = 10$, $\alpha = 20$, and $\alpha = 30$. This gives us a total of 600 instances available in http://siala.github.io/sm/sm.zip.

In the following figures we represent every configuration by "A-B" where A$\in \{fr, ac, bc\}$ is the constraint model for stability and B $\in \{lx\text{-}rd, lx\text{-}mn, as, is\}$

is the search strategy. In Figs. 1a and 2b we give the cactus plots of proving optimality for these instances on the two problems. That is, after a given CPU time in seconds (y-axis), we give the percentage of instances proved to optimality for each configuration on the x-axis. In Figs. 1a and 2b we study the quality of solutions by plotting the normalized objective value of the best solution found by the configuration h (x-axis) after a given time in seconds (y-axis) [11]. Let $h(I)$ be the objective value of the best solution found using model h on instance I and $lb(I)$ (resp. $ub(I)$) the lowest (resp. highest) objective value found by any model on I. We use a normalized score in the interval $[0, 1]$: $score(h, I) = \frac{ub(I)-h(I)+1}{ub(I)-lb(I)+1}$. The value of $score(h, I)$ is equal to 1 if h has found the best solution for this instance among all models, decreases as $h(I)$ gets further from the optimal objective value, and is equal to 0 if and only if h did not find any solution for I. Note that for fr and ac the CPU time in all there figures includes the $O(L)$ preprocessing step that we mentioned at the beginning of Sect. 4.

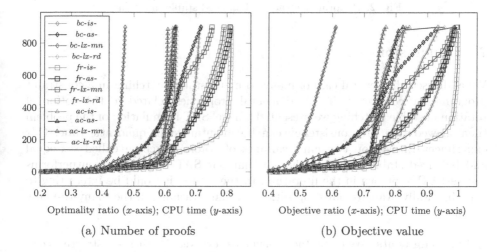

(a) Number of proofs   (b) Objective value

Fig. 1. Performance cactus, sex equal stable matching

These figures show that the arc consistency model (ac) does not pay off as it considerably slows down the speed of exploration. It should be noted that between bc and ac there is no clear winner. The SAT formulation (fr), on the other hand, outperforms both bc and ac using any search strategy. This is true for both finding proofs of optimality and finding the best objective values. In fact, fr clearly finds better solutions faster than any other approach.

Lastly, we note that the best search strategy for sex-equal stable matching is, surprisingly, the one branching lexicographically using a random value selection (Figs. 1a and b). For the case of balanced stable matching, clearly impact-based search is the best choice for finding proofs (Fig. 2a) whereas activity based search finds better solutions (Fig. 2b).

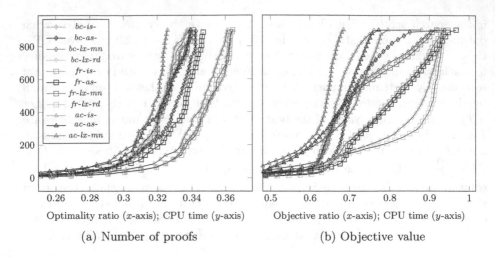

Optimality ratio (x-axis); CPU time (y-axis)   Objective ratio (x-axis); CPU time (y-axis)

(a) Number of proofs   (b) Objective value

Fig. 2. Performance cactus, balanced stable matching

7   Conclusion

We addressed the general case of many-to-many stable matching in a constraint programming context. Using fundamental properties related to the notion of rotation in stable matching we presented a novel SAT formulation of the problem then showed that arc consistency can be maintained in quadratic time. Our experimental study on two hard variants of stable matching called sex-equal and balanced stable matching showed that our SAT formulation outperforms the best CP approach in the literature. In the future, it would be interesting to experimentally evaluate our propositions on hard variants in the many-to-many setting.

Acknowledgments. We thank the anonymous reviewers for their constructive comments that helped to improve the presentation of the paper.

We thank Begum Genc for generating the instances.

This publication has emanated from research conducted with the financial support of Science Foundation Ireland (SFI) under Grant Number SFI/12/RC/2289.

References

1. Baïou, M., Balinski, M.: Many-to-many matching: stable polyandrous polygamy (or polygamous polyandry). Discrete Appl. Math. **101**(1–3), 1–12 (2000)
2. Bansal, V., Agrawal, A., Malhotra, V.S.: Polynomial time algorithm for an optimal stable assignment with multiple partners. Theor. Comput. Sci. **379**(3), 317–328 (2007)
3. Eirinakis, P., Magos, D., Mourtos, I., Miliotis, P.: Finding all stable pairs and solutions to the many-to-many stable matching problem. INFORMS J. Comput. **24**(2), 245–259 (2012)

4. Feder, T.: A new fixed point approach for stable networks and stable marriages. J. Comput. Syst. Sci. **45**(2), 233–284 (1992)
5. Fleiner, T., Irving, R.W., Manlove, D.: Efficient algorithms for generalized stable marriage and roommates problems. Theor. Comput. Sci. **381**(1–3), 162–176 (2007)
6. Gale, D., Shapley, L.S.: College admissions and the stability of marriage. Am. Math. Mon. **69**(1), 9–15 (1962)
7. Gent, I.P., Irving, R.W., Manlove, D.F., Prosser, P., Smith, B.M.: A constraint programming approach to the stable marriage problem. In: Walsh, T. (ed.) CP 2001. LNCS, vol. 2239, pp. 225–239. Springer, Heidelberg (2001). doi:10.1007/3-540-45578-7_16
8. Gusfield, D., Irving, R.W.: The Stable Marriage Problem - Structure and Algorithms. Foundations of Computing Series. MIT Press, Cambridge (1989)
9. Gusfield, D., Irving, R.W.: The Stable Marriage Problem: Structure and Algorithms. MIT Press, Cambridge (1989)
10. Hebrard, E.: Mistral, a constraint satisfaction library. In: Proceedings of the CP 2008 Third International CSP Solvers Competition, pp. 31–40 (2008)
11. Hebrard, E., Siala, M.: Explanation-based weighted degree. In: Salvagnin, D., Lombardi, M. (eds.) CPAIOR 2017. LNCS, vol. 10335, pp. 167–175. Springer, Cham (2017). doi:10.1007/978-3-319-59776-8_13
12. Irving, R.W., Leather, P.: The complexity of counting stable marriages. SIAM J. Comput. **15**(3), 655–667 (1986)
13. Manlove, D.F., O'Malley, G., Prosser, P., Unsworth, C.: A constraint programming approach to the hospitals/residents problem. In: Hentenryck, P., Wolsey, L. (eds.) CPAIOR 2007. LNCS, vol. 4510, pp. 155–170. Springer, Heidelberg (2007). doi:10.1007/978-3-540-72397-4_12
14. Manlove, D.F.: Algorithmics of Matching Under Preferences. Series on Theoretical Computer Science, vol. 2. WorldScientific, Singapore (2013)
15. Michel, L., Hentenryck, P.: Activity-based search for black-box constraint programming solvers. In: Beldiceanu, N., Jussien, N., Pinson, É. (eds.) CPAIOR 2012. LNCS, vol. 7298, pp. 228–243. Springer, Heidelberg (2012). doi:10.1007/978-3-642-29828-8_15
16. Moskewicz, M.W., Madigan, C.F., Zhao, Y., Zhang, L., Malik, S.: Chaff: engineering an efficient SAT solver. In: Proceedings of the 38th Design Automation Conference, DAC 2001, Las Vegas, NV, USA, 18–22 June 2001, pp. 530–535 (2001)
17. Prosser, P.: Stable roommates and constraint programming. In: Simonis, H. (ed.) CPAIOR 2014. LNCS, vol. 8451, pp. 15–28. Springer, Cham (2014). doi:10.1007/978-3-319-07046-9_2
18. Refalo, P.: Impact-based search strategies for constraint programming. In: Wallace, M. (ed.) CP 2004. LNCS, vol. 3258, pp. 557–571. Springer, Heidelberg (2004). doi:10.1007/978-3-540-30201-8_41
19. Rossi, F., van Beek, P., Walsh, T. (eds.): Handbook of Constraint Programming. Foundations of Artificial Intelligence, vol. 2. Elsevier (2006)
20. Siala, M., O'Sullivan, B.: Revisiting two-sided stability constraints. In: Quimper, C.-G. (ed.) CPAIOR 2016. LNCS, vol. 9676, pp. 342–357. Springer, Cham (2016). doi:10.1007/978-3-319-33954-2_25
21. Unsworth, C., Prosser, P.: A specialised binary constraint for the stable marriage problem. In: Zucker, J.-D., Saitta, L. (eds.) SARA 2005. LNCS (LNAI), vol. 3607, pp. 218–233. Springer, Heidelberg (2005). doi:10.1007/11527862_16
22. Unsworth, C., Prosser, P.: An n-ary constraint for the stable marriage problem. CoRR, abs/1308.0183 (2013)

Preference Elicitation for DCOPs

Atena M. Tabakhi[1]([envelope]), Tiep Le[2], Ferdinando Fioretto[3], and William Yeoh[1]

[1] Department of Computer Science and Engineering,
Washington University in St. Louis, St. Louis, USA
{amtabakhi,wyeoh}@wustl.edu
[2] Department of Computer Science, New Mexico State University,
Las Cruces, USA
tile@cs.nmsu.edu
[3] Department of Industrial and Operations Engineering,
University of Michigan, Ann Arbor, USA
fioretto@umich.edu

Abstract. *Distributed Constraint Optimization Problems* (DCOPs) offer a powerful approach for the description and resolution of cooperative multi-agent problems. In this model, a group of agents coordinate their actions to optimize a global objective function, taking into account their preferences or constraints. A core limitation of this model is the assumption that the preferences of all agents or the costs of all constraints are specified a priori. Unfortunately, this assumption does not hold in a number of application domains where preferences or constraints must be elicited from the users. One of such domains is the *Smart Home Device Scheduling* (SHDS) problem. Motivated by this limitation, we make the following contributions in this paper: **(1)** We propose a general model for preference elicitation in DCOPs; **(2)** We propose several heuristics to elicit preferences in DCOPs; and **(3)** We empirically evaluate the effect of these heuristics on random binary DCOPs as well as SHDS problems.

Keywords: Distributed Constraint Optimization · Smart homes · Preference elicitation

1 Introduction

The importance of constraint optimization is outlined by the impact of its application in a range of *Weighted Constraint Satisfaction Problems* (WCSPs), also known as *Constraint Optimization Problems* (COPs), such as supply chain management [34] and roster scheduling [1]. When resources are distributed among a set of autonomous agents and communication among the agents are

This research is partially supported by NSF grant 1345232. The views and conclusions contained in this document are those of the authors and should not be interpreted as representing the official policies, either expressed or implied, of the sponsoring organizations, agencies, or the U.S. government.

© Springer International Publishing AG 2017
J.C. Beck (Ed.): CP 2017, LNCS 10416, pp. 278–296, 2017.
DOI: 10.1007/978-3-319-66158-2_18

restricted, COPs take the form of *Distributed Constraint Optimization Problems* (DCOPs) [10,27,33,45]. In this context, agents coordinate their value assignments to minimize the overall sum of resulting constraint costs. DCOPs are suitable to model problems that are distributed in nature and where a collection of agents attempts to optimize a global objective within the confines of localized communication. They have been employed to model various distributed optimization problems, such as meeting scheduling [44,46], sensor networks [9], coalition formation [40], and smart grids [14,24].

The field of DCOP has matured significantly over the past decade since its inception [27]. DCOP researchers have proposed a wide variety of solution approaches, from complete approaches that use distributed search-based techniques [27,28,44] to distributed inference-based techniques [33,41]. There is also a significant body of work on incomplete methods that can be similarly categorized into local search-based methods [9,23], inference-based techniques [41], and sampling-based methods [11,29,31]. Researchers have also proposed the use of other off-the-shelf solvers such as logic programming solvers [21,22] and mixed-integer programming solvers [17].

One of the core limitations of all these approaches is that they assume that the constraint costs in a DCOP are known a priori. Unfortunately, in some application domains, these costs are only known after they are queried or elicited from experts or users in the domain. One such application is the *Smart Home Device Scheduling* (SHDS) problem [13]. In this problem, agents have to coordinate with each other to schedule smart devices (e.g., smart thermostats, smart light-bulbs, smart washers, etc.) distributed across a network of smart homes, where the goal is to schedule them in such a way that optimizes the preferences of occupants in those homes subject to a larger constraint that the peak energy demand in the network does not exceed an energy utility defined limit. Through the introduction of a number of smart devices in the commercial market, they are starting to become ubiquitous in today's very interconnected environment, consistent with the Internet-of-Things paradigm [25]. Therefore, we suspect that this SHDS problem will become more important in the future.

DCOPs are a natural framework to represent this problem as each home can be represented as an agent and the preferences of occupants can be represented as constraints. Furthermore, due to privacy reasons, it is preferred that the preferences of each occupant are not revealed to other occupants. The DCOP formulation allows the preservation of such privacy since agents are only aware of constraints that they are involved in. We further describe this motivating application and its mapping to DCOPs in more detail in Sect. 3.

A priori knowledge on the constraint costs is infeasible in our motivating SHDS application. A key challenge is thus in the elicitation of user preferences to populate the constraint cost tables. Due to the infeasibility of eliciting preferences to populate *all* preferences, in this paper, we introduce the *preference elicitation problem* for DCOPs, which studies *how to select a subset of k cost tables to elicit from each agent* with the goal of choosing those having a large impact on the overall solution quality. We propose several methods to select this subset of cost tables to elicit, based on the notion of *partial orderings*. Additionally, we extend the SHDS problem to allow for the encoding and elicitation of soft preferences, and evaluate our methods on this extended SHDS problem as well as on random graphs to show generality. Our results illustrate the effectiveness of our approach in contrast to a baseline evaluator that randomly selects cost tables to elicit. While the description of our solution focuses on DCOPs, our approach is also suitable to solve WCSPs.

2 Background

WCSP: A *Weighted Constraint Satisfaction Problem* (WCSP) [20,36] is a tuple $\mathcal{P} = \langle \mathcal{X}, \mathcal{D}, \mathcal{F} \rangle$, where $\mathcal{X} = \{x_1, \ldots, x_n\}$ is a finite set of variables, $\mathcal{D} = \{D_1, \ldots, D_n\}$ is a set of finite domains for the variables in $\mathcal{X}$, with D_i being the set of possible values for the variable x_i, and $\mathcal{F}$ is a set of *weighted constraints* (or *cost tables*). A weighted constraint $f_i \in \mathcal{F}$ is a function, $f_i : \times_{x_j \in \mathbf{x}^{f_i}} D_j \rightarrow \mathbb{R}_0^+ \cup \{\bot\}$, where $\mathbf{x}^{f_i} \subseteq \mathcal{X}$ is the set of variables relevant to f_i, referred to as the *scope* of f_i, and $\bot$ is a special element used to denote that a given combination of value assignments is not allowed. A *solution* $\mathbf{x}$ is a value assignment to a set of variables $X_{\mathbf{x}} \subseteq \mathcal{X}$ that is consistent with the variables' domains. The cost $\mathbf{F}_{\mathcal{P}}(\mathbf{x}) = \sum_{f \in \mathcal{F}, \mathbf{x}^f \subseteq X_{\mathbf{x}}} f(\mathbf{x})$ is the sum of the costs of all the applicable cost functions in $\mathbf{x}$. A solution $\mathbf{x}$ is said *complete* if $X_{\mathbf{x}} = \mathcal{X}$. The goal is to find an optimal complete solution $\mathbf{x}^* = \operatorname{argmin}_{\mathbf{x}} \mathbf{F}_{\mathcal{P}}(\mathbf{x})$.

DCOP: When the elements of a WCSP are distributed among a set of autonomous agents, we refer to it as a *Distributed Constraint Optimization Problem* (DCOP) [27,33,45]. Formally, a DCOP is described by a tuple $\mathcal{P} = \langle \mathcal{X}, \mathcal{D}, \mathcal{F}, \mathcal{A}, \alpha \rangle$, where $\mathcal{X}$, $\mathcal{D}$, and $\mathcal{F}$ are the set of variables, their domains, and the set of cost functions, defined as in a classical WCSP, $\mathcal{A} = \{a_1, \ldots, a_p\}$ $(p \leq n)$ is a set of autonomous agents, and $\alpha : \mathcal{X} \rightarrow \mathcal{A}$ is a surjective function, from variables to agents, mapping the control of each variable $x \in \mathcal{X}$ to an agent $\alpha(x)$. The goal in a DCOP is to find a complete solution that minimizes its cost: $\mathbf{x}^* = \operatorname{argmin}_{\mathbf{x}} \mathbf{F}_{\mathcal{P}}(\mathbf{x})$. A DCOP can be described by a *constraint graph*, where the nodes correspond to the variables in the DCOP, and the edges connect pairs of variables in the scope of the same cost functions. Following [12], we introduce the following definitions:

Definition 1. *For each agent* $a_i \in \mathcal{A}$, $\mathbf{L}_i = \{x_j \in \mathcal{X} \mid \alpha(x_j) = a_i\}$ *is the set of its* local *variables.* $\mathbf{I}_i = \{x_j \in \mathbf{L}_i \mid \exists x_k \in \mathcal{X} \wedge \exists f_s \in \mathcal{F} : \alpha(x_k) \neq a_i \wedge \{x_j, x_k\} \subseteq \mathbf{x}^{f_s}\}$ *is the set of its* interface *variables.*

Definition 2. *For each agent* $a_i \in \mathcal{A}$, *its* local constraint graph $G_i = (\mathbf{L}_i, \mathcal{E}_{\mathcal{F}_i})$ *is a subgraph of the constraint graph, where* $\mathcal{F}_i = \{f_j \in \mathcal{F} \mid \mathbf{x}^{f_j} \subseteq \mathbf{L}_i\}$.

Figure 1(a) shows the constraint graph of a sample DCOP with 3 agents a_1, a_2, and a_3, where $\mathbf{L}_1 = \{x_1, x_2\}$, $\mathbf{L}_2 = \{x_3, x_4\}$, $\mathbf{L}_3 = \{x_5, x_6\}$, $\mathbf{I}_1 = \{x_2\}$, $\mathbf{I}_2 = \{x_4\}$, and $\mathbf{I}_3 = \{x_6\}$. The domains are $D_1 = \cdots = D_6 = \{0, 1\}$. Figure 1(b) shows the cost table of all constraints; all constraints have the same cost table for simplicity.

a_1	(x1)—(x2)
a_2	(x3)—(x4)
a_3	(x5)—(x6)

for $i < j$

x_i	x_j	Costs
0	0	20
0	1	8
1	0	10
1	1	3

for $i = 2, j = 4$

x_i	x_j	Costs
0	0	$\mathcal{N}(65,8)$
0	1	$\mathcal{N}(71,8)$
1	0	$\mathcal{N}(41,6)$
1	1	$\mathcal{N}(29,5)$

for $i = 2, j = 6$

x_i	x_j	Costs
0	0	$\mathcal{N}(10,6)$
0	1	$\mathcal{N}(9,5)$
1	0	$\mathcal{N}(8,5)$
1	1	$\mathcal{N}(19,6)$

(a) Constraint Graph   (b) Cost Table   (c) Uncertain Cost Tables

Fig. 1. Example DCOP and uncertain DCOP

3 Motivating Domain: Smart Home Device Scheduling Problem

We now provide a description of (a variant of) the *Smart Home Device Scheduling* (SHDS) problem [13]. An SHDS problem is composed of a neighborhood $\mathcal{H}$ of smart homes $h_i \in \mathcal{H}$ that are able to communicate with one another and whose energy demands are served by an energy provider. The energy prices are set according to a real-time pricing schema specified at regular intervals t within a finite time horizon H. We use $\mathbf{T} = \{1, \ldots, H\}$ to denote the set of time intervals and $\theta : \mathbf{T} \to \mathbb{R}^+$ to represent the price function associated with the pricing schema adopted, which expresses the cost per kWh of energy consumed by a consumer.

Within each smart home h_i there is a set of (smart) electric devices $\mathcal{Z}_i$ networked together and controlled by a home automation system. We assume all the devices are uninterruptible (i.e., they cannot be stopped once they are started) and use s_{z_j} and δ_{z_j} to denote, respectively, the start time and duration (expressed in multiples of time intervals) of device $z_j \in \mathcal{Z}_i$. The energy consumption of each device z_j is ρ_{z_j} kWh for each hour that it is *on*. It will not consume any energy if it is *off*. We use the indicator function $\phi^t_{z_j}$ to indicate the state of the device z_j at time step t:

$$\phi^t_{z_j} = \begin{cases} 1 \text{ if } & s_{z_j} \leq t \wedge s_{z_j} + \delta_{z_j} \geq t \\ 0 \text{ otherwise} \end{cases}$$

Additionally, the usage of a device z_j is characterized by a *cost*, representing the monetary expense to schedule z_j at a given time. The aggregated cost of the home h_i at time step t is denoted with C^t_i and expressed as:

$$C^t_i = E^t_i \cdot \theta(t) \tag{1}$$

where $E^t_i = \sum_{z_j \in \mathcal{Z}_i} \phi^t_{z_j} \cdot \rho_{z_j}$ is the aggregated energy consumed by home h_i at time step t.

The SHDS problem seeks a schedule for the devices of each home in the neighborhood in a coordinated fashion so as to minimize the monetary costs and, at the same time, ensure that user-defined scheduling constraints (called *active scheduling rules* in [13]) are satisfied. The SHDS problem is also subject to the following constraints:

$$1 \leq s_{z_j} \leq T - \delta_{z_j} \qquad \forall h_i \in \mathcal{H}, z_j \in \mathcal{Z}_i \qquad (2)$$

$$\sum_{t \in \mathbf{T}} \phi_{z_j}^t = \delta_{z_j} \qquad \forall h_i \in \mathcal{H}, z_j \in \mathcal{Z}_i \qquad (3)$$

$$\sum_{h_i \in \mathcal{H}} E_i^t \leq \ell^t \qquad \forall t \in \mathbf{T} \qquad (4)$$

where $\ell^t \in \mathbb{R}^+$ is the maximum allowed total energy consumed by all the homes in the neighborhood at time step t. This constraint is typically imposed by the energy provider and is adopted to guarantee reliable electricity delivery. Constraint (2) expresses the lower and upper bounds for the start time associated to the schedule of each device. Constraint (3) ensures the devices are scheduled and executed for exactly their duration time. Constraint (4) ensures the total amount of energy consumed by the homes in the neighborhood does not exceed the maximum allowed threshold.

3.1 DCOP Representation

Fioretto *et al.* introduced a mapping of the SHDS problem to a DCOP [13]. At a high level, each home $h_i \in \mathcal{H}$ is mapped to an autonomous agent in the DCOP. For each home, the start times s_{z_j}, indicator variables $\phi_{z_j}^t$, and aggregated energy in the home are mapped to DCOP variables, which are controlled by the agent for that home. Constraints (2) to (4) are enforced by the DCOP constraints. Finally, the objective function of the SHDS is expressed through agents' preferences.

4   Encoding and Eliciting Preferences in SHDS

The above SHDS problem thus far includes exclusively hard constraints and has no soft constraints (i.e., preferences for when devices are scheduled). Thus, we will describe in this section how to integrate such preferences as soft constraints into SHDS.

We consider the scenario in which a single home h_i may host multiple users $u \in \mathbf{U}_{h_i}$, with $\mathbf{U}_{h_i}$ denoting the set of users in h_i. In modeling agents' preferences, we introduce *discomfort values* $d_{z_j,u}^t \in \mathbb{R}_0^+$ describing the degree of dissatisfaction for a user u to schedule the device z_j at a given time step t. Note that the monetary cost is the same for all users while the degree of dissatisfaction is user dependent. Thus, to avoid conflicting users' decision over the control of the device, we assume that there is one user who has exclusive access to a device $z \in \mathcal{Z}_i$ at any point in time. In this paper, for each device $z_j \in \mathcal{Z}_i$ in home h_i and each time step t, we assume the likelihood for a user to gain exclusive access on a device z_j is expressed through a probability $Pr_{z_j}^t$ (i.e., $\forall u \in \mathbf{U}_{h_i}, Pr_{z_j}^t(u) \in [0,1]$ and $\sum_{u \in \mathbf{U}_{h_i}} Pr_{z_j}^t(u) = 1$). Additionally, we use $\mathbf{d}_i^t = \sum_{z_j \in \mathcal{Z}_i} \phi_{z_j}^t \cdot d_{z_j}^t$ to denote the aggregated discomfort in home h_i at time step t, where $d_{z_j}^t$ is the discomfort value of the user who has exclusive access to the device z_j at time step t.

We can update the SHDS objective to take into account the users' preferences in addition to minimizing the monetary costs. While this is a multi-objective problem, we combine the two objectives into a single one through the use of a weighted sum:

$$\text{minimize} \quad \sum_{t \in \mathbf{T}} \sum_{h_i \in \mathcal{H}} \alpha_c \cdot C_i^t + \alpha_u \cdot \mathbf{d}_i^t \tag{5}$$

where α_c and α_u are weights in the open interval $(0, 1) \subseteq \mathbb{R}$ such that $\alpha_c + \alpha_u = 1$.

While, in general, the real-time pricing schema θ that defines the cost per kWh of energy consumed and the energy consumption ρ_{z_j} of each device z_j are well-defined concepts and can be easily acquired or modeled, the preferences on the users' discomfort values $d_{z_j,u}^t$ on scheduling a device z_j at time step t are subjective and, thus, more difficult to model explicitly.

We foresee two approaches to acquire these preferences: *(1)* eliciting them directly from the users and *(2)* estimating them based on historical preferences or from preferences of similar users. While the former method will be more accurate and reliable, it is cumbersome for the user to enter their preference for every device z_j and every time step t of the problem. Therefore, in this paper, we assume that a combination of the two approaches will be used, where a subset of preferences will be elicited and the remaining preferences will be estimated from historical sources or similar users. We believe that this strategy is especially important in application domains such as the SHDS problem, where users' preferences may be learned over time, thus, ensuring a continuous elicitation process of the unknown users preferences.

5   Preference Elicitation in DCOPs

A key drawback of existing DCOP approaches is the underlying assumption of a *total* knowledge of the model, which is not the case for a number of applications involving users' preferences, including the SHDS problem. Due to the infeasibility of eliciting *all* users' preferences—and, thus, their associated complete cost tables—in this paper, we study how to choose a subset of k cost tables to elicit. We first cast this problem as an optimization problem, before describing our proposed techniques.

Let $\hat{\mathcal{P}} = \langle \mathcal{X}, \mathcal{D}, \hat{\mathcal{F}}, \mathcal{A}, \alpha \rangle$ denote a DCOP with partial knowledge on the cost tables in $\hat{\mathcal{F}}$. The constraints $\hat{\mathcal{F}} = \mathcal{F}_r \cup \mathcal{F}_u$ are composed of *revealed constraints* $\mathcal{F}_r$, whose cost tables are accurately revealed, and *uncertain constraints* $\mathcal{F}_u$, whose cost tables are unrevealed and must be either estimated from historical sources or elicited. We refer to this problem as the *uncertain DCOP*.

In this paper, we assume that the costs of the uncertain constraints are sampled from Normal distributions that can be estimated from historical sources.[1]

---

[1] Other forms of distributions can also be used, but our minimax regret heuristics require that the form of the distributions have the following property: The sum of two distributions has the same form as their individual distributions.

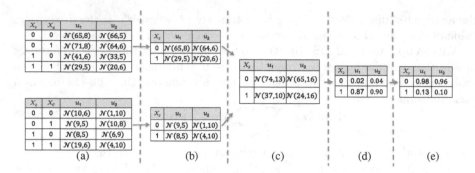

Fig. 2. Minimax regret example

Further, we assume that the distribution for each cost value is independent from the distribution of all other cost values. Figure 1(c) illustrates an uncertain cost table whose costs are modeled via random variables obeying Normal distributions, and u_1 and u_2 denote two distinct users that can control the associated device.

5.1 The Preference Elicitation Problem

The preference elicitation problem in DCOPs is formalized as follows: Given an oracle DCOP $\mathcal{P}$ and a value $k \in \mathbb{N}$, construct an uncertain DCOP $\hat{\mathcal{P}}$ that reveals only k constraints per agent (i.e., $|\mathcal{F}_r| = k \cdot |\mathcal{X}|$) and minimizes the error:

$$\epsilon_{\hat{\mathcal{P}}} = \mathbb{E}\left[\mathbf{F}_{\mathcal{P}}(\hat{\mathbf{x}}^*) - \mathbf{F}_{\mathcal{P}}(\mathbf{x}^*)\right] \tag{6}$$

where $\hat{\mathbf{x}}^*$ is the optimal solution for a *realization* of the uncertain DCOP $\hat{\mathcal{P}}$, and $\mathbf{x}^*$ is the optimal complete solution for the oracle DCOP $\mathcal{P}$. A realization of an uncertain DCOP $\hat{\mathcal{P}}$ is a DCOP (with no uncertainty), whose values for the cost tables are sampled from their corresponding Normal distributions. Note that the possible numbers of uncertain DCOPs that can be generated is $\binom{|\mathcal{F}|}{k \cdot |\mathcal{X}|}$. Since solving each DCOP is NP-hard [26], the preference elicitation problem is a particularly challenging one. Thus, we propose a number of heuristic methods to determine the subset of constraints to reveal, and to construct an uncertain problem $\hat{\mathcal{P}}$.

5.2 Preference Elicitation Heuristics

Let us first introduce a general concept of dominance between cost tables of uncertain constraints. Given two cost tables of uncertain constraints $f_{z_i}, f_{z_j} \in \mathcal{F}_u \subset \hat{\mathcal{F}}$, let $\succeq_\circ$ denote the dominance between the two cost tables according to partial ordering criteria $\circ$. In other words, $f_{z_i} \succeq_\circ f_{z_j}$ means that f_{z_i} *dominates* f_{z_j} according to criteria $\circ$. We now introduce the heuristic methods for different possible ordering criteria $\circ$.

Minimax Regret: *Minimax regret* is a well-known strategy that minimizes the maximum regret, and it is particularly suitable in a risk-neutral environment. At a high level, the minimax regret approach seeks to approximate and minimize the impact of the worst-case scenario. The idea of using minimax regret in our domain of interest is derived by the desire of taking into account the possible different outcomes occurring when eliciting the preferences of different users for a single device. Further, we assume that constraints that can be elicited are either unary or binary constraints. We leave to future work the extension to higher arity constraints. We now describe how to compute the regret for a single user u, and later how to combine the regrets across multiple users.

We use $Pr_{x_i}(d)$ to estimate the likelihood of an assignment $d \in D_i$ to a variable x_i:

$$Pr_{x_i,u}(d) = \Pi_{d' \in D_i \setminus \{d\}} Pr(\psi^d_{x_i,u} \leq \psi^{d'}_{x_i,u}) \tag{7}$$

where $\psi^d_{x_i,u}$ is the random variable representing the total cost incurred by x_i if it is assigned value d from its domain under user u. Then, the value

$$d^*_{x_i,u} = \underset{d}{\mathrm{argmax}} \; Pr_{x_i,u}(d) \tag{8}$$

with the largest probability is the one that is most likely to be assigned to x_i.

The probability $Pr(\psi^d_{x_i,u} \leq \psi^{d'}_{x_i,u})$ can be computed using:

$$Pr(\psi^d_{x_i,u} \leq \psi^{d'}_{x_i,u}) = \int_{c'=0}^{\infty} \int_{c=0}^{c'} Pr^d_{x_i,u}(c) Pr^{d'}_{x_i,u}(c') \, dc \, dc' \tag{9}$$

where $Pr^d_{x_i,u}$ is the *probability distribution function* (PDF) for random variable $\psi^d_{x_i,u}$. Unfortunately, the PDF $Pr^d_{x_i,u}$ is not explicitly defined in the uncertain DCOP. There are two challenges that one needs to address to obtain or estimate this PDF:

i. First, the total cost incurred by an agent is the summation of the costs over all constraints of that agent. Thus, the PDF for the total cost needs to be obtained by summing over the PDFs of all the individual constraint costs. Since we assume that these PDFs are all Normally distributed, one can efficiently construct the summed PDF, which is also a Normal distribution. Specifically, if $\mathcal{N}(\mu_i, \sigma_i^2)$ is the PDF for random variables c_i ($i = 1, 2$), then $\mathcal{N}(\mu_1 + \mu_2, \sigma_1^2 + \sigma_2^2)$ is the PDF for $c_1 + c_2$.

ii. Second, the cost associated to a variable for each constraint is not only dependent on its value but also on the value of the other variables constrained with it. In turn, the value of those variables depend on the variables that they are constrained with, and so on. As a result, estimating the true PDF requires the estimation of all the constraint costs in the entire DCOP. To simplify the computation process and introduce an independence between the costs of all variables, we propose the three following variants, each of which *estimates* the true PDF $Pr^{d,f}_{x_i,u}$ of a random variable $\psi^{d,f}_{x_i,u}$, representing the cost incurred by x_i from constraint f if assigned value d when its control is under user u:

- OPTIMISTIC: In this variant, the agent will optimistically choose the PDF with smallest mean among all the PDFs for all possible values of variables $x_j \in \mathbf{x}^f \setminus \{x_i\}$ in the scope of constraint f:

$$Pr_{x_i,u}^{d,f} = \mathcal{N}(\mu^*, \sigma_{\hat{d}}^2) \tag{10}$$

$$\mu^* = \min_{\hat{d} \in D_j} \mu_{\hat{d}} \tag{11}$$

where $\mathcal{N}(\mu_{\hat{d}}, \sigma_{\hat{d}}^2)$ is the PDF of the constraint cost if $x_i = d$ and $x_j = \hat{d}$ under user u. For example, in the uncertain cost tables in Fig. 1(c), the estimated PDF of the cost incurred for the choice $x_2 = 0$ from constraint f_{24} is $Pr_{x_2,u}^{0,f_{24}} = \mathcal{N}(65, 8^2)$, which *optimistically* assumes that x_4 will be assigned value 0 to minimize the incurred cost.

- PESSIMISTIC: In this variant, the agent chooses the PDF with largest mean among all the PDFs for all possible values of $x_j \in \mathbf{x}^f \setminus \{x_i\}$:

$$Pr_{x_i,u}^{d,f} = \mathcal{N}(\mu^*, \sigma_{\hat{d}}^2) \tag{12}$$

$$\mu^* = \max_{\hat{d} \in D_j} \mu_{\hat{d}} \tag{13}$$

In Fig. 1(c), the estimated PDF of the cost incurred by $x_2 = 0$ from constraint f_{24} is $Pr_{x_2,u}^{0,f_{24}} = \mathcal{N}(71, 8^2)$, which *pessimistically* assumes that x_4 will be assigned value 1 to maximize the incurred cost.

- EXPECTED: In this variant, the agent chooses the PDF with the "average" value of all the PDFs for all possible values of $x_j \in \mathbf{x}^f \setminus \{x_i\}$:

$$Pr_{x_i,u}^{d,f} = \mathcal{N}\left(\frac{1}{|D_j|} \sum_{\hat{d} \in D_j} \mu_{\hat{d}}, \frac{1}{|D_j|^2} \sum_{\hat{d} \in D_j} \sigma_{\hat{d}}^2 \right) \tag{14}$$

In Fig. 1(c), the estimated PDF of the cost incurred by $x_2 = 0$ from constraint f_{24} is $Pr_{x_2,u}^{0,f_{24}} = \mathcal{N}(68, 8^2)$, assuming that $x_4 = 0$ or $x_4 = 1$ with equal probability.

The *regret* $R_{x_i,u}^d$ of variable x_i being assigned value d is defined as:

$$R_{x_i,u}^d = 1 - Pr_{x_i,u}(d) \tag{15}$$

Each variable x_i will most likely be assigned the value $d_{x_i}^*$ with the smallest regret by definition (see Eqs. 7 and 8). We thus define the regret $R_{x_i,u}$ for each variable x_i to be the regret for this value:

$$R_{x_i,u} = R_{x_i,u}^{d_{x_i}^*} = \min_{d \in D_i} R_{x_i,u}^d \tag{16}$$

To generalize our approach to also handle multiple users in each house, where the PDFs differ across users, we take the maximum regret over all users u for each variable x_i and its value d before taking the minimum over all values. More precisely,

$$R_{x_i} = \min_{d \in D_i} \max_s R_{x_i, u}^d \tag{17}$$

Therefore, the minimax regret approach seeks to approximate the impact of the worst-case scenario. Finally, we define the regret R_{f_i} for a constraint f_i to be the absolute difference between the regrets of the variables in the scope of the function:

$$R_{f_i} = |R_{x_{i_1}} - R_{x_{i_2}}| \tag{18}$$

where $\mathbf{x}^{f_i} = \{x_{i_1}, x_{i_2}\}$.

While defining the regret to be the sum of the two variables' regrets may be more intuitive, our experimental results show that the above definition provides better results. Intuitively, if the regret of a variable x_i is large, then there is little confidence that it will take on value $d_{x_i}^*$ with the smallest regret because the PDFs for all its values are very similar and have significant overlaps. Thus, eliciting a constraint between two variables with large regrets will likely not help in improving the overall solution quality since the PDFs for all value combinations for that constraint are likely to be similar.

Similarly, if the regret of a variable is small, then there is a high confidence that it will be assigned value with the smallest regret because the PDFs for its values are sufficiently distinct that regardless of the actual realizations of the random variables (costs in the cost table), the value with the smallest regret will be the one with the smallest cost. Therefore, eliciting a constraint between two agents with small regrets will also not help. Therefore, we define the regret of a constraint to be the difference in the regrets of the variables in its scope (see Eq. 18).

If we order the constraints using the ordering criteria $\circ = MR[\cdot]$, that is, according to the minimax regret criterion, then, given two uncertain constraints $f_i, f_j \in \mathcal{F}_u$, we say that $f_i \succeq_{MR} f_j$ iff $MR[f_i] \geq MR[f_j]$, where $MR[f_j] = R_{f_j}$ is the regret as defined in Eq. 18.

Figure 2 illustrates a partial trace of this approach on the example DCOP of Fig. 1 with two users u_1 and u_2. Figure 2(a) shows the uncertain cost tables for constraint f_{24} between variables x_2 and x_4 and constraint f_{26} between variables x_2 and x_6. Figure 2(b) shows the estimated PDFs $Pr_{x_2, u}^{d, f}$ of the constraint costs incurred by variables x_2 from constraint f under user u if it takes on value d. In this trace, we use the "optimistic" variant of the algorithm, and the PDFs are estimated using Eqs. 10 and 11. Figure 2(c) shows estimated summed PDFs $Pr_{x_2, u}^{d, f}$ of the total constraint costs incurred by the agent, summed over all of its constraints. Here, we only sum the PDFs for the two constraints f_{24} and f_{26}. Figure 2(d) shows the probabilities $Pr_{x_2, u}(d)$ of $x_2 = d$ under user u, computed using Eq. 7, and Fig. 2(e) shows the regrets $R_{x_2, u}^d$, computed using Eq. 15. Thus, the regret R_{x_2} for x_2 is 0.13, computed using Eq. 17. Assume that the regret R_{x_1} of x_1 is 0.50. Then, the regret $R_{f_{12}}$ of constraint $f_{12} = |R_{x_1} - R_{x_2}| = |0.50 - 0.13| = 0.37$.

Maximum Standard Deviation: We now propose a different heuristic that makes use of the degree of uncertainty in the constraint costs $\chi_{f,u}^v$ for constraint f, value combination $v = \langle x_{i_1} = d_{i_1}, \ldots, x_{i_k} = d_{i_k} \rangle$, and user u, where $\mathbf{x}^f = \{x_{i_1}, \ldots, x_{i_k}\}$, $d_{i_1} \in D_{i_1}$, ..., and $d_{i_k} \in D_{i_k}$.

Assume that there is only a single user u. Then, using the same motivation described for the minimax regret heuristic, and assuming that *variables be assigned a different value for different constraints*, the value combination chosen for a constraint f will be the $v^* = \mathrm{argmin}_v \chi_{f,u}^v$ that has the smallest cost. Unfortunately, the actual constraint costs are not known and only their PDFs $\mathcal{N}(\mu_{f,u}^v, (\sigma_{f,u}^v)^2)$ are known.

Since the constraint costs' distribution means are known, we assume that the value combination chosen for a constraint f will be the value $v^* = \mathrm{argmin}_v \mu_{f,u}^v$ that has the smallest mean. The degree of uncertainty in the constraint cost for that constraint f is thus the standard deviation associated $\sigma_{f,u}^{v^*}$ with that value combination v^*.

To generalize this approach to multiple users, we take the maximum standard deviations over all users u. More precisely, the degree of uncertainty in the constraint costs for a constraint f is:

$$\sigma_f = \max_s \sigma_{f,u}^{v^*} \tag{19}$$

One can then use this maximum standard deviation criterion to order the constraints. In other words, if the ordering criteria $\circ = MS[\cdot]$ is done according to the maximum standard deviation criterion, then, given two unknown functions $f_i, f_j \in \mathcal{F}_u$, we say that $f_i \succeq_{MS} f_j$ iff $MS[f_i] \geq MS[f_j]$, where $MS[f_j] = \sigma_{f_j}$ is the maximum standard deviation as defined in Eq. 19.

6   Related Work

There is an extensive body of work on the topic of modeling preferences [16]. In particular, Rossi *et al.* discussed *conditional-preference networks* (CP-nets) for handling preferences [35], which provide a qualitative graphical representation of preferences reflecting the conditional dependence of the problem variables. Differently from CP-nets, our proposal focuses on the notion of *conditional additive independence* [3], which requires the utility of an outcome to be the sum of the "utilities" of the different variable values of the outcome.

In terms of preference elicitation, two major approaches are studied in the literature [6]: A Bayesian approach [4,7] and a minimax regret approach [5,15,42]. The former is typically adopted when the uncertainty can be quantified probabilistically, and preference elicitation is often formalized as a *partially-observable Markov decision process* (POMDP) [18] that assumes each query to a user is associated with a finite set of possible responses. In contrast, our proposal follows the minimax regret approach [5,15,42]. The proposed framework differs from other proposals in the literature in the following ways: We assume the unknown costs are sampled from a Normal distribution and compute the regret based on such distributions. In contrast, other minimax regret based methods have different assumptions. For example, Boutilier *et al.* assumes that a set of (hard) constraints together with a graphical utility model captures user preferences [5]. While the structure of the utility model is known, the parameters of this utility model are imprecise, given by upper and lower bounds. The notion of regret is computed based on those upper and lower bounds. Differently, Wang and Boutilier computes regrets under the assumption that constraints over unknown utility values are linear [42]. Finally, Gelain *et al.* computes regrets by taking the minimum among the known utilities associated to the projections of an assignment, that is, of the appropriated sub-tuples in the constraints [15].

Finally, preference elicitation has never been applied directly on DCOPs before. The closest DCOP-related problem is a class of DCOPs where agents have partial knowledge on the costs of their constraints and, therefore, they may discover the unknown costs via exploration [39,47]. In this context, agents must balance the coordinated *exploration* of the unknown environment and the *exploitation* of the known portion of the rewards, in order to optimize the global objective [37]. Another orthogonal related DCOP model is the problem where costs are sampled from probability distribution functions [30]. In such a problem, agents seek to minimize either the worst-case regret [43] or the expected regret [21].

7 Empirical Evaluation

We evaluate our preference elicitation framework on distributed random binary graphs and smart home device scheduling (SHDS) problems [13], where we compared our four heuristics–minimax regret with the three variants: optimistic (MR-O), pessimistic (MR-P), and expected (MR-E) and maximum standard deviation (MS)–against a random baseline (RD) that chooses the constraints to elicit randomly. All the problems are modeled and solved optimally on multiple computers with Intel Core i7-3770 CPU 3.40 GHz and 16 GB of RAM. We use MiniZinc [38], an off-the-shelf centralized CP solver, to solve all the DCOPs.

In our experiments the preference elicitation heuristics are evaluated in terms of the *normalized error* $\frac{\epsilon_{\hat{p}}}{F_{\mathcal{P}}(\mathbf{x}^*)}$, where $\epsilon_{\hat{p}}$ is the error as defined by Eq. 6. An accurate computation of this error requires us to generate all possible realizations for the uncertain DCOPs. Due to the complexity of such task, we create $m = 50$ realizations of the uncertain DCOPs and compute the error $\epsilon_{\hat{p}}$ in this reduced sampled space.

7.1 Random Graphs

We create 100 random graphs whose topologies are based on the Erdős and Rényi model [8] with the following parameters: $|\mathcal{X}| = 50$, $|\mathcal{A}| = 5$, and $|D_i| = 2$ for all variables $x_i \in \mathcal{X}$. Each agent a_i has $|\mathbf{L}_i| = 10$ local variables with density $p_1 = 0.8$ that produces $|\mathcal{F}_i| = 36$ local constraints per agent. These constraints are unknown (uncertain constraints) and we set two scenarios (called

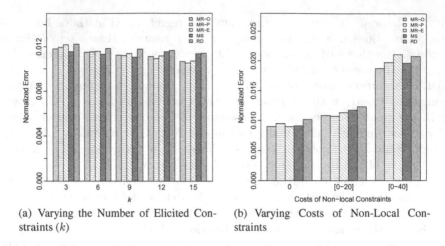

(a) Varying the Number of Elicited Constraints (k)

(b) Varying Costs of Non-Local Constraints

Fig. 3. Random graphs preference elicitation

users in Sect. 5) for all uncertain cost tables. All constraint costs are modeled as random variables following a Normal distribution $\mathcal{N}(\mu, \sigma^2)$, where σ is uniformly sampled from the range $[5, 10]$ and the means μ are uniformly sampled from one of the following six ranges $[5, 70]$, $[5, 80]$, $[5, 90]$, $[5, 100]$, $[5, 110]$, and $[5, 120]$. The different ranges are to introduce some heterogeneity into the constraints. We set the non-local constraints (i.e., inter-agent constraints) to be uncertain constraints as well, where we vary the mean μ of their Normal distributions to be from different distributions: $\mu = 0$; $\mu \in [0, 20]$; and $\mu \in [0, 40]$. Finally, we allow only local constraints (or preferences) to be elicited.

Figure 3(a) illustrates the normalized errors of our heuristics and that of the random baseline heuristic, where the mean μ of the non-local constraints are uniformly sampled from the range $[0, 20]$. We control k so that the number of the constraints per agent elicited from the oracle DCOP varies from 3 to 15 with increment of 3. We make the following observations:

- As the number of constraints (k) to elicit increases, the errors of the MR-P and MR-O heuristic decrease for all values of k as opposed to the random heuristic which is approximately the same for all values of k. The reason is, as we increase k, the random heuristic randomly selects k constraint to elicit with high likelihood of choosing the wrong constraints. However, since the regret-based heuristic (e.g., MR-P) takes into account the uncertain cost of the constraints it chooses those minimizing the regret.

- The MS heuristic performs slightly better than random heuristic. The reason is that MS orders the uncertain constraints by their degrees of uncertainty (i.e., σ) corresponding to the most likely value combinations to be assigned (i.e., the ones with the smallest μ). In contrast, the random heuristic chooses constraint randomly without taking into account the degree of uncertainty.

Table 1. Smart devices and their energy consumption (in kWh)

Dish-washer	Washer	Dryer	Hob	Oven	Microwave	Laptop	Desktop	Vacuum cleaner	Fridge	Electrical vehicle
0.75	1.20	2.50	3.00	5.00	1.70	0.10	0.30	1.20	0.30	3.50

- All regret-based heuristics outperform the baseline heuristic, especially for larger values of k, indicating that they are able to effectively take the regrets of the constraints into account.

Figure 3(b) illustrates the normalized errors for the random problems, where we vary the mean values μ of the non-local constraints, sampled from different distributions; we set $k = 15$ for all cases. The same trends observed above apply here. However, the normalized error increases as the range of the mean increases for all heuristics. The reason is because the magnitude in the error (when variables are assigned wrong value due to wrong guesses in the cost of the constraints) increases when the range increases. However, generally, the optimistic and pessimistic variants of the minimax regret heuristics still perform better in all three cases.

7.2   Smart Home Device Scheduling (SHDS) Problems

SHDS Problem Construction: We now describe how we construct SHDS problems. As the only uncertain element in the uncertain constraints are the discomfort values $d^t_{z_j,u}$ (defined in Sect. 3) for devices z_j, time steps t, and users u, we model these values as random variables following a Normal distribution (e.g., one could fit a Normal distribution to the historical data). As the distribution for one user may be different from the distribution for a different user in a home, for each user u, we generate a discomfort table composed of a Normal distribution $\mathcal{N}(\mu^t_{z_j,u}, (\sigma^t_{z_j,u})^2)$ for each device z_j and time step t. Each user u can gain the exclusive access to a device z_j with the probability $Pr^t_{z_j}$, and the Normal distribution of the discomfort of device z_j at time step t is the Normal distribution of the user that has exclusive access for that device and time step.

Next, let $\mathcal{P} = \langle \mathcal{X}, \mathcal{D}, \mathcal{F}, \mathcal{A}, \alpha \rangle$ denote the DCOP whose constraints $\mathcal{F}$ have *accurate* cost tables that depend only on external parameters and are easily obtained (e.g., price function θ and energy consumption of devices ρ_{z_j}) or they depend on user preferences that are accurately obtained through an oracle. Using the same process described above, we combine the discomfort tables for multiple users into a single aggregated discomfort table $\mathcal{U}$. Note that this aggregated discomfort table may be different than the one $\hat{\mathcal{U}}$ for the uncertain DCOP if there are multiple users. Then, the actual discomfort value $d^t_{z_j}$ for each device z_j and time step t is sampled from the Normal distribution $\mathcal{N}(\mu^t_{z_j}, (\sigma^t_{z_j})^2)$ for that device and time step in the aggregated discomfort table $\mathcal{U}$. We refer to this problem as the *oracle DCOP*. In summary, when a constraint is elicited, the actual discomfort values are retrieved from the oracle DCOP.

Experimental Setup: In our experiments, we consider $|\mathcal{H}| = 10$ homes, each controlling $|\mathcal{Z}_i| = 10$ smart devices, listed in Table 1 along with their energy consumption. We populate the set of smart devices $\mathcal{Z}_i$ of each home by randomly sampling 10 elements from $\mathcal{Z}$. Thus, a home might control multiple devices of the same type. We set a time horizon $H = 6$ with increments of 4 h. We use the same real-time pricing schema as proposed by Fioretto *et al.* [13], which is the one used by the Pacific Gas and Electric Company for their Californian consumers during peak summer months.[2]

To generate the discomfort table for each user, we assume that there is a weak correlation between the price of energy and the level of discomfort of the user. Specifically, we assume that users will prefer (i.e., they are more comfortable) using their devices when prices are low to save money. Therefore, the higher the price, the more uncomfortable the user will be at using the device at that time. Based on this assumption, for each home, user, and device, the mean μ^t at each time step t is an integer that is uniformly sampled from the range $[\max\{1, \theta(t) - 50\}, \theta(t) + 50]$, where $\theta(t)$ is the real-time pricing at time step t used by the Pacific Gas and Electric Company. Therefore, the range of the means differ across time steps but are the same for all devices as the discomfort level is primarily motivated by the pricing schema.

The weights α_c and α_u of the objective function defined in Eq. 5 are both set to 0.5. These settings are employed to create both an oracle DCOP and the corresponding uncertain DCOP, except that the values of the constraints of the uncertain DCOPs are not realized (i.e., they are distributions).

Finally, since all uncertain constraints in an SHDS problem are unary constraints, all three variants of the minimax regret heuristics are identical, and we use "MR" to label this heuristic.

Single User Experiments: In the first set of experiments, we set each home to have only one user. Figure 4(a) plots the error for our heuristics compared against the random baseline heuristic. The results are averaged over 100 randomly generated SHDS problem instances. We make the following observations:

- As expected, for all elicitation heuristics, the error decreases as the number of cost tables to elicit increases.
- Both the MR and MS heuristics consistently outperform the random heuristic for all values of k. Like the results in random graphs, the random heuristic has a higher likelihood of choosing the wrong constraint to elicit, while MR and MS choose better constraints.
- Interestingly, we observe that MS selects the constraints slightly better than MR, indicating that despite the fact that MS is a simpler heuristic, it is well-suited in problems with single users. The reason is that the key feature of MR—maximizing the regret over all users—is ignored when there is only one user.

---

[2] https://www.pge.com/en_US/business/rate-plans/rate-plans/peak-day-pricing/peak-day-pricing.page. Retrieved in November 2016.

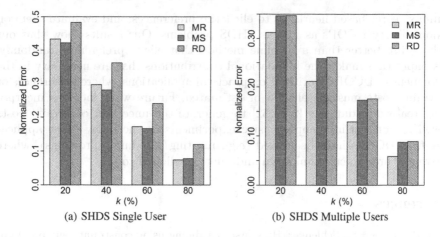

(a) SHDS Single User

(b) SHDS Multiple Users

Fig. 4. Smart homes device scheduling preference elicitation

Multiple User Experiments: In the second set of experiments, we set each home to have two users, where both users have equal likelihood of controlling the devices (i.e., $P_{u_1} = P_{u_2} = 0.5$). Figure 4(b) shows the results. The trends for this experiment is similar to that shown for Fig. 4(a), where our MR heuristic outperforms the random heuristic. However, the MS heuristic performs poorly in this experiment, with similar performance as the random baseline. In general, our results show that the regret-based method outperforms other heuristics in multiple users scenarios, as it takes into account the discomfort values of all users, orders the constraints to elicit based on their minimum regrets. Similar to random graph results, MR performs better in the scenarios that multiple users take control of the devices in a building.

Finally, the SHDS and random graph experiment results demonstrate that the regret-based elicitation heuristics achieve approximately 30% and 11% improvement over the baseline random heuristic in minimizing the error, respectively. The improvements in SHDS problems are larger than those in random graphs because variables are highly connected ($p_1 = 0.8$) in random graph problems. In contrast, variables in SHDS problems are mostly independent as they mostly have unary constraints. The higher dependency between variables in random graphs reduces the improvements of our heuristics over the baseline random heuristic.

8   Conclusions and Future Work

DCOPs have been used to model a number of multi-agent coordination problems including the smart home device scheduling (SHDS) problem. However, one of the key assumptions in DCOPs—that constraint costs are known a priori—do not apply to many applications including SHDS. Thus, in this paper, we propose the problem of preference (i.e., constraint cost) elicitation for DCOPs; introduce minimax regret based heuristics to elicit the preferences; and evaluate them on random binary DCOPs as well as SHDS problems. Our results show that our methods are better than a baseline method that elicits preferences randomly. This paper thus makes the foundational contributions that are necessary in the deployment of DCOP algorithms on practical applications, where preferences or constraint costs must be elicited or estimated. Future work includes incorporating real-world datasets [2,19,32] to generate the uncertain constraint costs as well as conducting comprehensive experiments in the many other applications that DCOPs have been used (e.g., meeting scheduling problems), where preferences are typically unknown and must be elicited too.

References

1. Abdennadher, S., Schlenker, H.: Nurse scheduling using constraint logic programming. In: Proceedings of the Conference on Innovative Applications of Artificial Intelligence (IAAI), pp. 838–843 (1999)
2. Anderson, B., Lin, S., Newing, A., Bahaj, A., James, P.: Electricity consumption and household characteristics: implications for census-taking in a smart metered future. Comput. Environ. Urban Syst. **63**, 58–67 (2017)
3. Bacchus, F., Grove, A.J.: Utility independence in a qualitative decision theory. In: Proceedings of the International Conference on Principles of Knowledge Representation and Reasoning (KR), pp. 542–552 (1996)
4. Boutilier, C.: A POMDP formulation of preference elicitation problems. In: Proceedings of the National Conference on Artificial Intelligence (AAAI), pp. 239–246 (2002)
5. Boutilier, C., Patrascu, R., Poupart, P., Schuurmans, D.: Regret-based utility elicitation in constraint-based decision problems. In: Proceedings of the International Joint Conference on Artificial Intelligence (IJCAI), pp. 929–934 (2005)
6. Braziunas, D.: Computational approaches to preference elicitation. Technical report (2006)
7. Braziunas, D., Boutilier, C.: Local utility elicitation in GAI models. In: Proceedings of the Conference in Uncertainty in Artificial Intelligence (UAI), pp. 42–49 (2005)
8. Erdös, P., Rényi, A.: On random graphs, I. Publ. Math. (Debr.) **6**, 290–297 (1959)
9. Farinelli, A., Rogers, A., Petcu, A., Jennings, N.: Decentralised coordination of low-power embedded devices using the Max-Sum algorithm. In: Proceedings of the International Conference on Autonomous Agents and Multiagent Systems (AAMAS), pp. 639–646 (2008)
10. Fioretto, F., Pontelli, E., Yeoh, W.: Distributed constraint optimization problems and applications: a survey. CoRR, abs/1602.06347 (2016)
11. Fioretto, F. Yeoh, W., Pontelli, E.: A dynamic programming-based MCMC framework for solving DCOPs with GPUs. In: Proceedings of Principles and Practice of Constraint Programming (CP), pp. 813–831 (2016)
12. Fioretto, F., Yeoh, W., Pontelli, E.: Multi-variable agent decomposition for DCOPs. In: Proceedings of the AAAI Conference on Artificial Intelligence (AAAI) (2016)
13. Fioretto, F., Yeoh, W., Pontelli, E.: A multiagent system approach to scheduling devices in smart homes. In: Proceedings of the International Conference on Autonomous Agents and Multiagent Systems (AAMAS), pp. 981–989 (2017)

14. Fioretto, F., Yeoh, W., Pontelli, E., Ma, Y., Ranade, S.: A DCOP approach to the economic dispatch with demand response. In: Proceedings of the International Conference on Autonomous Agents and Multiagent Systems (AAMAS), pp. 981–989 (2017)
15. Gelain, M., Pini, M.S., Rossi, F., Venable, K.B., Walsh, T.: Elicitation strategies for fuzzy constraint problems with missing preferences: algorithms and experimental studies. In: Stuckey, P.J. (ed.) CP 2008. LNCS, vol. 5202, pp. 402–417. Springer, Heidelberg (2008). doi:10.1007/978-3-540-85958-1_27
16. Goldsmith, J., Junker, U.: Preference handling for artificial intelligence. AI Mag. **29**(4), 9–12 (2008)
17. Hatano, D., Hirayama, K.: DeQED: an efficient divide-and-coordinate algorithm for DCOP. In: Proceedings of the International Joint Conference on Artificial Intelligence (IJCAI), pp. 566–572 (2013)
18. Kaelbling, L.P., Littman, M.L., Cassandra, A.R.: Planning and acting in partially observable stochastic domains. Artif. Intell. **101**(1–2), 99–134 (1998)
19. Kolter, J.Z., Johnson, M.J.: REDD: a public data set for energy disaggregation research. In: Proceedings of the Workshop on Data Mining Applications in Sustainability, pp. 59–62 (2011)
20. Larrosa, J.: Node and arc consistency in weighted CSP. In: Proceedings of the AAAI Conference on Artificial Intelligence (AAAI), pp. 48–53 (2002)
21. Le, T., Fioretto, F., Yeoh, W., Son, T.C., Pontelli, E.: ER-DCOPs: a framework for distributed constraint optimization with uncertainty in constraint utilities. In: Proceedings of the International Conference on Autonomous Agents and Multiagent Systems (AAMAS) (2016)
22. Le, T., Son, T.C., Pontelli, E., Yeoh, W.: Solving distributed constraint optimization problems with logic programming. In: Proceedings of the AAAI Conference on Artificial Intelligence (AAAI) (2015)
23. Maheswaran, R., Pearce, J., Tambe, M.: Distributed algorithms for DCOP: a graphical game-based approach. In: Proceedings of the International Conference on Parallel and Distributed Computing Systems (PDCS), pp. 432–439 (2004)
24. Miller, S., Ramchurn, S., Rogers, A.: Optimal decentralised dispatch of embedded generation in the smart grid. In: Proceedings of the International Conference on Autonomous Agents and Multiagent Systems (AAMAS), pp. 281–288 (2012)
25. Miorandi, D., Sicari, S., De Pellegrini, F., Chlamtac, I.: Internet of things: vision, applications and research challenges. Ad Hoc Netw. **10**(7), 1497–1516 (2012)
26. Modi, P.: Distributed constraint optimization for multiagent systems. Ph.D. thesis, University of Southern California, Los Angeles (United States) (2003)
27. Modi, P., Shen, W.-M., Tambe, M., Yokoo, M.: ADOPT: asynchronous distributed constraint optimization with quality guarantees. Artif. Intell. **161**(1–2), 149–180 (2005)
28. Netzer, A., Grubshtein, A., Meisels, A.: Concurrent forward bounding for distributed constraint optimization problems. Artif. Intell. **193**, 186–216 (2012)
29. Nguyen, D.T., Yeoh, W., Lau, H.C., Gibbs, D.: A memory-bounded sampling-based DCOP algorithm. In: Proceedings of the International Conference on Autonomous Agents and Multiagent Systems (AAMAS), pp. 167–174 (2013)
30. Nguyen, D.T., Yeoh, W., Lau, H.C., Zilberstein, S., Zhang, C.: Decentralized multi-agent reinforcement learning in average-reward dynamic DCOPs. In: Proceedings of the AAAI Conference on Artificial Intelligence (AAAI), pp. 1447–1455 (2014)
31. Ottens, B., Dimitrakakis, C., Faltings, B.: DUCT: an upper confidence bound approach to distributed constraint optimization problems. In: Proceedings of the AAAI Conference on Artificial Intelligence (AAAI), pp. 528–534 (2012)

32. Paatero, J.V., Lund, P.D.: A model for generating household electricity load profiles. Int. J. Energy Res. **30**(5), 273–290 (2006)
33. Petcu, A., Faltings, B.: A scalable method for multiagent constraint optimization. In: Proceedings of the International Joint Conference on Artificial Intelligence (IJCAI), pp. 1413–1420 (2005)
34. Rodrigues, L., Magatao, L.: Enhancing supply chain decisions using constraint programming: a case study. In: Proceedings of the Mexican International Conference on Artificial Intelligence (MICAI), pp. 1110–1121 (2007)
35. Rossi, F., Venable, K.B., Walsh, T.: Preferences in constraint satisfaction and optimization. AI Mag. **29**(4), 58–68 (2008)
36. Shapiro, L.G., Haralick, R.M.: Structural descriptions and inexact matching. IEEE Trans. Pattern Anal. Mach. Intell. **5**, 504–519 (1981)
37. Stranders, R., Delle Fave, F., Rogers, A., Jennings, N.: DCOPs and bandits: exploration and exploitation in decentralised coordination. In: Proceedings of the International Conference on Autonomous Agents and Multiagent Systems (AAMAS), pp. 289–297 (2012)
38. Stuckey, P.J., Becket, R., Brand, S., Brown, M., Feydy, T., Fischer, J., de la Banda, M.G., Marriott, K., Wallace, M.: The evolving world of MiniZinc. In: Constraint Modelling and Reformulation, pp. 156–170 (2007)
39. Taylor, M., Jain, M., Tandon, P., Yokoo, M., Tambe, M.: Distributed on-line multiagent optimization under uncertainty: balancing exploration and exploitation. Adv. Complex Syst. **14**(03), 471–528 (2011)
40. Ueda, S., Iwasaki, A., Yokoo, M.: Coalition structure generation based on distributed constraint optimization. In: Proceedings of the AAAI Conference on Artificial Intelligence (AAAI), pp. 197–203 (2010)
41. Vinyals, M., Rodríguez-Aguilar, J., Cerquides, J.: Constructing a unifying theory of dynamic programming DCOP algorithms via the generalized distributive law. J. Auton. Agents Multi-Agent Syst. **22**(3), 439–464 (2011)
42. Wang, T., Boutilier, C.: Incremental utility elicitation with the minimax regret decision criterion. In: Proceedings of the International Joint Conference on Artificial Intelligence (IJCAI), pp. 309–318 (2003)
43. Wu, F., Jennings, N.: Regret-based multi-agent coordination with uncertain task rewards. In: Proceedings of the AAAI Conference on Artificial Intelligence (AAAI), pp. 1492–1499 (2014)
44. Yeoh, W., Felner, A., Koenig, S.: BnB-ADOPT: an asynchronous branch-and-bound DCOP algorithm. J. Artif. Intell. Res. **38**, 85–133 (2010)
45. Yeoh, W., Yokoo, M.: Distributed problem solving. AI Mag. **33**(3), 53–65 (2012)
46. Zivan, R., Okamoto, S., Peled, H.: Explorative anytime local search for distributed constraint optimization. Artif. Intell. **212**, 1–26 (2014)
47. Zivan, R., Yedidsion, H., Okamoto, S., Glinton, R., Sycara, K.: Distributed constraint optimization for teams of mobile sensing agents. J. Auton. Agents Multi-Agent Syst. **29**(3), 495–536 (2015)

Extending Compact-Table to Basic Smart Tables

Hélène Verhaeghe[1]([⊠]), Christophe Lecoutre[2], Yves Deville[1], and Pierre Schaus[1]

[1] UCLouvain, ICTEAM, Place Sainte Barbe 2, 1348 Louvain-la-Neuve, Belgium
{helene.verhaeghe,yves.deville,pierre.schaus}@uclouvain.be
[2] CRIL-CNRS UMR 8188, Université d'Artois, 62307 Lens, France
lecoutre@cril.fr

Abstract. Table constraints are instrumental in modelling combinatorial problems with Constraint Programming. Recently, Compact-Table (CT) has been proposed and shown to be as an efficient filtering algorithm for table constraints, notably because of bitwise operations. CT has already been extended to handle non-ordinary tables, namely, short tables and/or negative tables. In this paper, we introduce another extension so as to deal with basic smart tables, which are tables containing universal values ($*$), restrictions on values ($\neq v$) bounds ($\leq v$ or $\geq v$) and sets ($\in S$). Such tables offer the user a better expressiveness and permit to deal efficiently with compressed tuples. Our experiments show a substantial speedup when compression is possible (and a very limited overhead otherwise).

Keywords: Table constraints · Filtering · Compression · Compact-table · Bitset

1 Introduction

Table constraints, also known as extension(al) constraints, express on sequences of variables the combinations of values that are allowed (*supports*) or forbidden (*conflicts*). Lots of efforts [1,5,7,9,13,16,17,19,24,28,30] have been made during the last decade to enhance the filtering process of such constraints, in order to establish the property known as Generalized Arc Consistency (GAC). Motivation behind this excitement comes from the fact that tables can theoretically encode any other kind of constraints. They are thus used in many application fields, as stated in the industry.

The last big improvement in this domain has been the introduction of Compact-Table [5], an algorithm that advantageously combines tabular reduction and bitwise operations (a related algorithm, independently proposed in the literature, is STRBit [30]). Quite interestingly, Compact-Table (CT) has been shown to be about one order of magnitude faster than the best algorithm(s) developed during the last decade.

Unfortunately, tables have a major drawback: the memory space required to store them, which may grow exponentially with the number of columns (arity). To address this issue, various compression techniques have already been studied.

© Springer International Publishing AG 2017
J.C. Beck (Ed.): CP 2017, LNCS 10416, pp. 297–307, 2017.
DOI: 10.1007/978-3-319-66158-2_19

Some are based on using particular data structures, like Multi-valued Decision Diagrams (MDDs) [4,24], tries [7] and Deterministic Finite Automata (DFA) [25]. Other approaches attempt to keep a table-like structure, which is made compact by reasoning on Cartesian products and some intentional forms of column restrictions, like short tuples [10], compressed tuples [11,31], sliced tables [8] and smart tables [21].

In this paper, we show how CT can be extended for basic smart tables, i.e., tables that may contain universal values ($*$), restrictions ($\neq$), bounds ($\leq$ and $\geq$) and sets ($\in S$).

2   Technical Background

A *constraint network* (CN) is composed of a set of n variables and a set of e constraints. Each *variable* x has an associated domain $(dom(x))$ that contains the finite set of values that can be assigned to it. Each *constraint* c involves an ordered set of variables, called the *scope* of c and denoted by $scp(c)$, and is semantically defined by a *relation* $(rel(c))$ which contains the set of tuples allowed for the variables involved in c. The *arity* of a constraint c is $|scp(c)|$. For simplicity, a variable-value pair (x, a) such that $x \in scp(c)$ and $a \in dom(x)$ is called a *value* (of c). A *table constraint* c is a constraint such that $rel(c)$ is explicitly defined by listing (in a table) the tuples that are allowed[1] by c.

Let $\tau = (a_1, \dots, a_r)$ be a tuple of values associated with an ordered set of variables $vars(\tau) = \{x_1, \dots, x_r\}$. The ith value of τ is denoted by $\tau[i]$ or $\tau[x_i]$, and τ is *valid* iff $\forall i \in 1..r, \tau[i] \in dom(x_i)$. τ is a *support* on a constraint c iff $vars(\tau) = scp(c)$ and τ is a valid tuple allowed by c. If τ is a support on a constraint c involving a variable x and such that $\tau[x] = a$, we say that τ is a *support for* the value (x, a) of c. Enforcing Generalized Arc Consistency (GAC) on a constraint c means removing all values without any support on c.

Over the years, there have been many developments about compact forms of tables. Ordinary tables contain *ordinary* or *ground* tuples, i.e., classical sequences of values as in $(1, 2, 0)$. Short tables can additionally contain *short* tuples, which are tuples involving the special symbol $*$ as in $(0, *, 2)$, and compressed tables can additionally contain *compressed* tuples, which are tuples involving sets of values as in $(0, \{1, 2\}, 3)$. Assuming that the tuples mentioned just above are associated with the ordered set of variables $\{x_1, x_2, x_3\}$, in $(0, *, 2)$, x_2 can take any value from its domain and in $(0, \{1, 2\}, 3)$, x_2 can take the value 1 or the value 2. Smart tables[2] are composed of *smart* tuples, which are tuples containing expressions (column constraints) of one the following forms: $*$, $<$op$>v$, $\in S$, $\notin S$, $<$op$>x_j$ and $<$op$>x_j + v$; v being a value, S a set of values, and $<$op$>$ an operator in $\{<, \leq, =, \neq, \geq, >\}$. Finally, a *basic* smart table is a restricted form of smart table where column constraints are unary, that is where smart tuples are of the form $*$, $<$op$>v$, $\in S$ and $\notin S$. For example, in the smart tuple $(\neq 1, 2, > 1)$, x_1 must be

---

[1] We only deal with positive forms of table constraints in this paper.

[2] For simplicity, we consider here a slightly simpler form of smart table constraints than in [21].

different from 1, x_2 must be equal to 2 ('$= 2$' being trivially simplified in '2') and x_3 must be greater than 1. In term of expressiveness, one can observe that basic smart tables are equivalent to compressed tables, meaning that any compressed tuple can be represented by a basic smart tuple (this is immediate), and any basic smart tuple can be represented by a compressed tuple. However, they allow more compact representation (having '$\neq 2$' is shorter than '$\{\ldots, 1, 3, 4, 5, \ldots\}$').

3   CT on Ordinary and Short Tables

This section briefly introduces Compact-Table (CT), a state-of-the-art filtering algorithm [5] initially introduced for enforcing GAC on positive (ordinary) table constraints. It first appeared in Or-Tools [22], the solver developed at Google, and is now implemented in OscaR [23], AbsCon and Choco [26]. CT benefits from well-established techniques: bitwise operations [2,18], residual supports [14,15,20], tabular reduction [13,16,28], reversible sparse sets [27] and resetting operations [24].

The core structure of CT, when applied to a constraint c, is a reversible sparse bitset, called currTable, responsible for keeping track of the current supports of c: the ith bit of currTable is set to 1 iff the ith tuple τ_i of the table of c is currently valid. It is updated by means of precomputed bitsets: for each value (x, a) of c, supports$[x, a]$ is the bitset that identifies the set of tuples that are initially supports of (x, a) on c.

Algorithm 1 presents a simplified version of CT, which consists of two main steps. First, updateTable(), iterates for each variable x involved in c over either the set Δ_x of values that have been removed from the domain of x since the last invocation of the algorithm (Line 5) or the current domain of x (Line 9), and use the appropriate bitsets supports to update currTable using bitwise operations (Lines 6 and 10). The test at Line 4 ensures minimizing the number of operations that must be performed during the update. Secondly, filterDomains(), iterates over every value (x, a) of c (Lines 13 and 14) and use the corresponding bitset supports$[x, a]$ to verify whether the value is still supported or not (Line 15). It should be noted that the bitwise operations on the bitsets currtable and mask are only performed on the active (i.e., non null) words of currtable.

In [29], it was shown how CT can be extended to (positive) short tables. CT* just requires the introduction of a second pool of bitsets: for each value (x, a) of c, supports*$[x, a]$ is the bitset that identifies the set of tuples τ that are *explicit* supports of (x, a), i.e., such that $\tau[x] = a$. This means that any occurrence of * in a short tuple implies that the corresponding bits are always set to 0 in the bitsets supports* (unlike bitsets supports). An illustration is given by Fig. 1 where the bits for τ_1 in bitsets supports and supports* are different because $\tau_1[y] = *$. CT* is obtained from CT by simply replacing Line 6 of Algorithm 1 with:

$$\text{mask} \leftarrow \text{mask} \mid \text{supports}^*[x, a]$$

	x	y	z
τ_1	$\neq a$	$*$	c
τ_2	c	$\leq b$	$\neq a$
τ_3	$< c$	b	$\neq b$
τ_4	$> b$	$\geq b$	$*$

(a) Table

	supports				supports*				supportsMin				supportsMax			
	τ_1	τ_2	τ_3	τ_4	τ_1	τ_2	τ_3	τ_4	τ_1	τ_2	τ_3	τ_4	τ_1	τ_2	τ_3	τ_4
(x,a)	0	0	1	0	0	0	0	0	1	1	1	1	1	0	1	0
(x,b)	1	0	1	0	0	0	0	0	1	1	1	1	1	0	1	0
(x,c)	1	1	0	1	0	1	0	0	1	1	0	1	1	1	1	1
(y,a)	1	1	0	0	0	0	0	0	1	1	1	1	1	1	0	0
(y,b)	1	1	1	1	0	0	1	0	1	1	1	1	1	1	1	1
(y,c)	1	0	0	1	0	0	0	0	1	0	0	1	1	1	1	1

...

(b) Bitsets

Fig. 1. Bitsets supports, supports*, supportsMin and supportsMax

The complexity of CT* is $\mathcal{O}(rd\frac{t}{w})$ where r denotes the arity, d the size of the largest domain, t the number of tuples and w the size of the computer words (e.g., $w = 64$).

Algorithm 1. Class ConstraintCT

```
 1  Method updateTable()
 2      foreach variable x ∈ scp do
 3          mask ← 0
 4          if |Δx| < |dom(x)| then                 // Incremental Update
 5              foreach value a ∈ Δx do
 6                  mask ← mask | supports[x,a]  // Use supports*[x,a] in CT*
 7              mask ← ~mask                        // ~: bitwise negation
 8          else                                    // Reset-based Update
 9              foreach value a ∈ dom(x) do
10                  mask ← mask | supports[x,a]       // | : bitwise or
11          currTable ← currTable & mask             // & : bitwise and

12  Method filterDomains()
13      foreach variable x ∈ scp do
14          foreach value a ∈ dom(x) do
15              if currTable & supports[x,a] = 0 then
16                  dom(x) ← dom(x) \ {a}

17  Method enforceGAC()
18      updateTable()
19      filterDomains()
```

---

4   CT on Basic Smart Tables

A basic smart table is composed of smart tuples containing expressions of the following forms: $*$, $<op>v$, $\in S$ and $\notin S$, with $<op> \in \{<, \leq, =, \neq, \geq, >\}$. As CT^* already covers the forms $*$ and $= v$, we need to further extend CT^* to handle the remaining forms of smart tuples in a basic smart table. The resulting algorithm is called CT^{bs}.

4.1   Handling $\neq v$

In reality, CT^* can already handle expressions of the form $\neq v$. We simply have to slightly extend the semantics of bitsets supports*. Any occurrence of $*$ or $\neq v$ in a tuple implies that the corresponding bits are set to 0 in supports* (unlike supports). The complexity is obviously unchanged. An example is depicted with $\tau_1[x]$ in Fig. 1. Correctness is proved by showing that the table is always properly updated. Let us cover all possible cases for an expression $\neq v$ at column x. When $|dom(x)| = 0$, the solver detect the failure through the variables. For the case of the reset-based update, as support precisely depicts the acceptance of values by tuples, this is necessarily correct. In the incremental update, we will necessary have $|dom(x)| \geq 2$. This comes from the structure of the algorithm when the full version, as described in [5], is used. In this case, the tuple is always support for the given variable and as the corresponding bits in supports* are set to 0 by construction, the tuple is not removed from currTable.

4.2   Handling $<op>v$, with $<op> \in \{<, \leq, \geq, >\}$

First, note that it is sufficient to focus on expressions of the form $\geq v$ and $\leq v$ since $> v$ and $< v$ are equivalent to $\geq v + 1$ and $\leq v - 1$. We first introduce two additional arrays of bitsets: supportsMin for $\leq v$ and supportsMax for $\geq v$. For each value (x, a) of c, the ith bit of supportsMin$[x, a]$ (resp., supportsMax$[x, a]$) is 1 iff $\tau_i[x]$ allows at least one value $\geq a$ (resp., $\leq a$). An example of the different bitsets for each of the operators $<, \leq, \geq, >$ can be found in Fig. 1. We assume the ordering $a < b < c$ on domains.

To handle $<op>v$, with $<op> \in \{\leq, \geq\}$, Lines 4–7 (incremental update) in Algorithm 1 must be replaced by the lines given in Algorithm 2. Note that min (resp. max) denotes the smallest (resp. largest) value of $dom(x)$, whereas minChanged() (resp. maxChanged()) is a method that return true when min (resp. max) have changed since the last call of the algorithm. Line 1 is slightly modified to compensate the overhead induced by the two operations. Because Lines 5–8 handle all the values that are less than and greater than min and max, we only consider at Line 3 the values $a \in \Delta_x$ such that $dom(x).\text{min} < a < dom(x).\text{max}$. Note that the semantics of supports* is unchanged: only *explicit* supports of (x, a) are considered, meaning that we have supports$^*[x, a] = 0$ when $\tau[x] = *$, $\tau[x] \neq b$, $\tau[x] \geq b$ or $\tau[x] \leq b$ for any value b.

Correctness is shown for $\leq v$, considering all cases at column x for tuple τ. The case $|dom(x)| = 0$ is as trivial as in the last section. For the case of the reset-based update, as supports precisely depicts the acceptance of values by tuples, this is necessarily correct. Finally in the incremental update (Algorithm 2), due to the constructions of the bitsets, i.e., the bit for τ in supports* (resp. supportsMax) is always set to 0 (resp. 1), updating depends only on supportsMin. By definition, if $dom(x).\min \leq v$, meaning still supported, the bit for τ in supportsMin is 1, keeping τ in currTable. If $dom(x).\min > v$, bit for τ is 0, removing τ. Time complexity remains $\mathcal{O}(rd\frac{t}{w})$.

Algorithm 2. Incremental Update for CT^{bs}

---

1 **if** $|\Delta_x| + 2 < |dom(x)|$ **then**
2   **foreach** *value* $a \in \Delta_x$ *such that* $dom(x).\min < a < dom(x).\max$ **do**
3   mask $\leftarrow$ mask | supports*$[x, a]$
4   mask $\leftarrow$ ˜mask
5   **if** $dom(x).\text{minChanged}()$ **then**
6   mask $\leftarrow$ mask & supportsMin$[x, dom(x).\min]$
7   **if** $dom(x).\text{maxChanged}()$ **then**
8   mask $\leftarrow$ mask & supportsMax$[x, dom(x).\max]$

4.3 Handling $\in S$ (and $\notin S$)

There is no easy way to handle expressions of the form $\in S$ using incremental update (on bitsets). We then propose to systematically execute reset-based update as they do in [30] for passing from STRbit to STRbit-C. More precisely, as soon as a variable is involved in an expression of the form $\in S$ in one of the tuples of the basic smart table, a reset-based update is forced in Algorithm 1 (Lines 9–10). We do not present the (rather immediate) code. Dealing with $\notin S$ can be conducted similarly.

5 Compression

Computing the smallest short table (compression with only $*$) is known to be NP-complete [10]. Not surprisingly computing the smallest basic smart table is also NP-complete. Indeed minimizing the size of a DNF formula (NP-complete [12]) can be reduced in polynomial time to minimizing the size of a basic smart table.

    We introduce a heuristic compression algorithm to generate a basic smart table from a given (ordinary) table. It focuses on column constraints of the form $\leq v$ and $\geq v$. Other forms can be obtained by post-processing: (i) expressions $\leq dom(x).\max$ or $\geq dom(x).\min$ can be replaced by $*$, and (ii) two tuples that are identical except on a column where we have respectively $\leq v-1$ and $\geq v+1$ can be merged by simply using $\neq v$. Expressions $\in S$ and $\notin S$ were not considered in this heuristic to avoid costly set operations.

The compression algorithm proceeds in r steps, r being the arity of the table. At each step, the algorithm handles two tables: the c-table (compressed table) and the r-table (residual table). The union of c-table and r-table is always equivalent to the initial table. At step i, each tuple of the c-table has exactly i column constraints of the form $\leq v$ or $\geq v$. When $i = 0$, c-table is the initial table and r-table is empty. After step r, the resulting table of the algorithm is the union of c-table and r-table. The computation at a given step is the following. From the tuples in c-table, several abstract tuples are generated, used to introduce new tuples with one more column constraint of the form $\leq v$ or $\geq v$. The new tuples that cover at least two tuples in c-table are gathered in a new c-table used in the next step. The uncovered tuples in c-table are added to r-table.

More formally, at a given step, we define an *abstract tuple* as a tuple taken from the current c-table with one of its literal value $x = a$ replaced by the symbol '?'. At step i, there are thus $(r - i) \cdot t_c$ possible abstract tuples, with t_c the size of c-table. An abstract tuple can be matched against so-called strictly compatible (resp. compatible) tuples. A basic smart tuple τ is strictly compatible (resp. compatible) with an abstract tuple ρ iff for each $1 \leq j \leq r$, the form of $\tau[j]$ is strictly compatible (resp. compatible) with the form of $\rho[j]$. Compatibility of forms is intuitive: a value v is compatible with the same value v and also with '?', the form $\leq v$ (resp. $\geq v$) is compatible with $\leq w$ (resp. $\geq w$) provided that $w \geq v$ (resp. $w \leq v$). Strict compatibility requires compatibility and $w = v$.

We denote by S_c^ρ (resp., S_{sc}^ρ) the sets of tuples from the current c-table that are compatible (resp., strictly compatible) with ρ, an abstract tuple. Note that the computation of these two sets can be done in $O(r.t_c)$ and that we have $S_{sc}^\rho \subset S_c^\rho$. Given $S_c^\rho = \{\tau_1, \ldots, \tau_k\}$, we denote by V^ρ set of values $\{\tau_1[j], \ldots, \tau_k[j]\}$ where j is the column index of ? in ρ. If, given the domain of x_j, a subset of V^ρ can be represented by $x_j \leq v$ (or $x_j \geq v$), then a new basic smart tuple ρ' is generated, where ρ' is the tuple ρ with ? replaced by $\leq v$ (or $\geq v$). The corresponding tuples in S_{sc} can be removed as they are covered by the new smart tuple. However, the tuples only present in S_c cannot be removed. In practice, a new basic smart tuple is only introduced if it ensures a reduction of the table (i.e., at least two tuples can be removed). As t_c is $O(t)$, the total complexity of the compression algorithm is $O(r^3 t^2)$.

Example. Let us consider the abstract tuple $\rho = (1, ?, \leq 1)$. In the following set of basic smart tuples $\{\tau_1 = (1, 0, \leq 1), \tau_2 = (1, 1, \leq 2), \tau_3 = (1, 2, \leq 1)\}$, the tuples τ_1 and τ_3 are strictly compatible with ρ, the tuple τ_2 is only compatible with ρ. The new smart tuple $(1, \leq 2, \leq 1)$ is then generated, allowing us to remove both τ_1 and τ_3. The tuple τ_2 is necessary to generate this new tuple, but cannot be removed from the table.

6 Experimental Results

We have selected from the XCSP3 website [3] the instances that exclusively contain positive table constraints. This benchmark includes a large variety of series.

Compression of Ordinary Tables into Basic Smart Tables. The compression ratio is defined as $\frac{t'}{t}$, where t and t' respectively denote the numbers of tuples in the initial and compressed tables. Using the algorithm described in Sect. 5, we obtain the results displayed in Fig. 2. As we expected, dense tables (i.e., tables with a high number of tuples compared to the Cartesian product of domains) lead to good compression. This can be observed in particular with the series PigeonsPlus that contains really dense instances (making them highly compressible), and also the series Renault that contains instances with a wide range of tables (many of them being well compressed). On the other hand, the series Kakuro contains very sparse tables that cannot be compressed at all.

Practical Efficiency. To assess the efficiency of CT^{bs}, notably the interest of using the different forms of expressions, tables have been compressed using our algorithm in three different related ways: (1) compression with $\leq$ and $\geq$, (2) compression with $\leq$ and $\geq$ followed by a post-processing to detect $*$ and $\neq$ and (3) compression with $\leq$ and $\geq$ followed by a transformation into set restrictions (e.g., $\leq v$ is written as $\{i : i \leq v\}$).

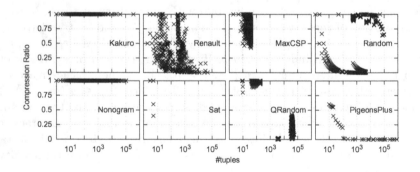

Fig. 2. Distribution of table compression ratios on 8 series of instances.

Figure 3 shows the performance profile [6] for CT^{bs} with these three related compression approaches and also for standard CT on uncompressed tables. A point (x, y) on the plot indicates the percentage of instances that can be solved within a time-limit that is at most x times the time taken by the best algorithm. The performance profile was based only on instances showing enough compression (rate ≤ 0.9) and requiring at least 2 s of solving time. With a timeout set to 10 min, only 60 instances matched out these criteria out of the 4,000 tested instances.

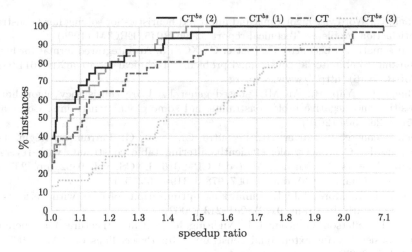

Fig. 3. Performance profile for CT^{bs} versus CT

Obtained results show that simple compression (1) brings a slight speedup compared to CT. Notice however that the computation time for an instance was reduced up to a factor of 7. Because post-processing (2) brought less than 3% of additional compression, it is not surprising that CT^{bs} with approaches (1) and (2) are close. As expected, handling tables with set restrictions only, approach (3), induces an overhead as no incremental updates can be performed. The overhead is however limited (at most a factor two). The computation time taken by Method `updateDomain()` in Algorithm 1 is not much reduced when using basic smart tables (mainly, because of the residue caching described in [5]). This explains why the observed speed-ups are not proportional to the compression ratios.

7   Conclusion

In this paper, we have shown how to extend CT^*, the Compact-Table algorithm devised for (ordinary and) short tables, to basic smart tables. The new algorithm CT^{bs} benefits from the highly optimized mechanisms of CT and can be attractively applied to expressive forms of tables involving natural conditions on values ($*$, $\neq v$, $\leq v$, $\geq v$ and $\in S$).

We have also proposed a heuristic algorithm to generate basic smart tables. Our experimental results show both the usefulness of this form of compression and the good behavior of CT^{bs} compared to CT.

References

1. Bessiere, C., Régin, J.C.: Arc consistency for general constraint networks: preliminary results. In: Proceedings of IJCAI 1997, pp. 398–404 (1997)

2. Bliek, C.: Wordwise algorithms and improved heuristics for solving hard constraint satisfaction problems. Technical report 12–96-R045, ERCIM (1996)
3. Boussemart, F., Lecoutre, C., Piette, C.: XCSP3: An integrated format for benchmarking combinatorial constrained problems. Technical report arXiv:1611.03398, CoRR (2016). http://www.xcsp.org
4. Cheng, K., Yap, R.: An MDD-based generalized arc consistency algorithm for positive and negative table constraints and some global constraints. Constraints 15(2), 265–304 (2010)
5. Demeulenaere, J., Hartert, R., Lecoutre, C., Perez, G., Perron, L., Régin, J.-C., Schaus, P.: Compact-table: efficiently filtering table constraints with reversible sparse bit-sets. In: Rueher, M. (ed.) CP 2016. LNCS, vol. 9892, pp. 207–223. Springer, Cham (2016). doi:10.1007/978-3-319-44953-1_14
6. Dolan, E.D., Moré, J.J.: Benchmarking optimization software with performance profiles. Math. Program. 91(2), 201–213 (2002)
7. Gent, I., Jefferson, C., Miguel, I., Nightingale, P.: Data structures for generalised arc consistency for extensional constraints. In: Proceedings of AAAI 2007, pp. 191–197 (2007)
8. Gharbi, N., Hemery, F., Lecoutre, C., Roussel, O.: Sliced table constraints: combining compression and tabular reduction. In: Simonis, H. (ed.) CPAIOR 2014. LNCS, vol. 8451, pp. 120–135. Springer, Cham (2014). doi:10.1007/978-3-319-07046-9_9
9. Mairy, J.B., Van Hentenryck, P., Deville, Y.: Optimal and efficient filtering algorithms for table constraints. Constraints 19(1), 77–120 (2014)
10. Jefferson, C., Nightingale, P.: Extending simple tabular reduction with short supports. In: Proceedings of IJCAI 2013, pp. 573–579 (2013)
11. Katsirelos, G., Walsh, T.: A compression algorithm for large arity extensional constraints. In: Bessière, C. (ed.) CP 2007. LNCS, vol. 4741, pp. 379–393. Springer, Heidelberg (2007). doi:10.1007/978-3-540-74970-7_28
12. Khot, S., Saket, R.: Hardness of minimizing and learning DNF expressions. In: IEEE 49th Annual IEEE Symposium on Foundations of Computer Science, FOCS 2008, pp. 231–240. IEEE (2008)
13. Lecoutre, C.: STR2: optimized simple tabular reduction for table constraints. Constraints 16(4), 341–371 (2011)
14. Lecoutre, C., Boussemart, F., Hemery, F.: Exploiting multidirectionality in coarse-grained arc consistency algorithms. In: Rossi, F. (ed.) CP 2003. LNCS, vol. 2833, pp. 480–494. Springer, Heidelberg (2003). doi:10.1007/978-3-540-45193-8_33
15. Lecoutre, C., Hemery, F.: A study of residual supports in arc consistency. In: Proceedings of IJCAI 2007, pp. 125–130 (2007)
16. Lecoutre, C., Likitvivatanavong, C., Yap, R.: STR3: a path-optimal filtering algorithm for table constraints. Artif. Intell. 220, 1–27 (2015)
17. Lecoutre, C., Szymanek, R.: Generalized arc consistency for positive table constraints. In: Benhamou, F. (ed.) CP 2006. LNCS, vol. 4204, pp. 284–298. Springer, Heidelberg (2006). doi:10.1007/11889205_22
18. Lecoutre, C., Vion, J.: Enforcing arc consistency using bitwise operations. Constraint Program. Lett. 2, 21–35 (2008)
19. Lhomme, O., Régin, J.C.: A fast arc consistency algorithm for n-ary constraints. In: Proceedings of AAAI 2005, pp. 405–410 (2005)
20. Likitvivatanavong, C., Zhang, Y., Bowen, J., Freuder, E.: Arc consistency in MAC: a new perspective. In: Proceedings of CPAI'04 Workshop held with CP 2004, pp. 93–107 (2004)

21. Mairy, J.-B., Deville, Y., Lecoutre, C.: The smart table constraint. In: Michel, L. (ed.) CPAIOR 2015. LNCS, vol. 9075, pp. 271–287. Springer, Cham (2015). doi:10. 1007/978-3-319-18008-3_19

22. van Omme, N., Perron, L., Furnon, V.: or-tools user's manual. Technical report, Google (2014). https://github.com/google/or-tools

23. OscaR Team: OscaR: Scala in OR (2012). https://bitbucket.org/oscarlib/oscar

24. Perez, G., Régin, J.-C.: Improving GAC-4 for table and MDD constraints. In: O'Sullivan, B. (ed.) CP 2014. LNCS, vol. 8656, pp. 606–621. Springer, Cham (2014). doi:10.1007/978-3-319-10428-7_44

25. Pesant, G.: A regular language membership constraint for finite sequences of variables. In: Wallace, M. (ed.) CP 2004. LNCS, vol. 3258, pp. 482–495. Springer, Heidelberg (2004). doi:10.1007/978-3-540-30201-8_36

26. Prud'homme, C., Fages, J.G., Lorca, X.: Choco3 documentation. TASC, INRIA Rennes, LINA CNRS UMR 6241 (2014)

27. de Saint-Marcq, V.L.C., Schaus, P., Solnon, C., Lecoutre, C.: Sparse-sets for domain implementation. In: (TRICS) Workshop on Techniques for Implementing Constraint Programming Systems (2013)

28. Ullmann, J.: Partition search for non-binary constraint satisfaction. Inf. Sci. **177**, 3639–3678 (2007)

29. Verhaeghe, H., Lecoutre, C., Schaus, P.: Extending compact-table to negative and short tables. In: Proceedings of AAAI 2017 (2017)

30. Wang, R., Xia, W., Yap, R., Li, Z.: Optimizing Simple Tabular Reduction with a bitwise representation. In: Proceedings of IJCAI 2016, pp. 787–795 (2016)

31. Xia, W., Yap, R.H.C.: Optimizing STR algorithms with tuple compression. In: Schulte, C. (ed.) CP 2013. LNCS, vol. 8124, pp. 724–732. Springer, Heidelberg (2013). doi:10.1007/978-3-642-40627-0_53

Constraint Programming Applied to the Multi-Skill Project Scheduling Problem

Kenneth D. Young[1], Thibaut Feydy[2], and Andreas Schutt[1,2(✉)]

[1] The University of Melbourne, Melbourne, Australia
kdyoung@student.unimelb.edu.au
[2] Decision Sciences, Data61, CSIRO, Melbourne, Australia
{thibaut.feydy,andreas.schutt}@data61.csiro.au

Abstract. The Multi-Skill Project Scheduling Problem is a variant of the well-studied Resource Constrained Project Scheduling Problem, in which the resources are assumed to be multi-skilled. Practical applications of this problem occur when the resources considered are a multi-skilled workforce or multi-purpose machines. This variant introduces a set of assignment decisions between the resources and activities, further to the usual scheduling decisions. This additional layer of complexity results in the problem becoming far more difficult to solve. We investigate different constraint programming models and searches tailored for solvers with nogood learning. These models and searches are then evaluated on instances available from the literature as well as newly generated ones. Using the best performing model and search, we are able to close at least 87 open instances from the literature.

1 Introduction

The Resource Constrained Project Scheduling Problem (RCPSP) is one of the most widely studied combinatorial optimization problems [11,27] and is the basic problem for the herein studied Multi-Skill Project Scheduling Problem (MSPSP). Solving RCPSP involves finding the optimal schedule to perform a set of non-preemptive activities which satisfies the given precedence relations. Limitations on the available resources are also imposed and must be respected throughout the project's duration. In most cases, the objective when solving this problem is to find the shortest possible duration of the project, which we call the makespan.

In RCPSP and many of its variants, it is assumed that each resource only has one capability or *skill*. However, when the resources considered are multi-skilled workers or multi-purpose machines then each could have a variety of skills. MSPSP is the extension of RCPSP, where each resource can have multiple skills. In addition, it is assumed that each resource can use only one skill at each time.

MSPSP consists of non-preemptive activities, which require time and skills for their execution, some precedence relations between pairs of activities, and renewable multi-skilled resources. We assume that all resources are unary, because multi-skilled resources are typical human resources in real scenarios. In a solution, all precedence relations are satisfied, no resource is overloaded at any time,

© Springer International Publishing AG 2017
J.C. Beck (Ed.): CP 2017, LNCS 10416, pp. 308–317, 2017.
DOI: 10.1007/978-3-319-66158-2_20

and no resource uses more than one skill at any time. In addition, each activity must be fully processed by the same resources from the start to the end. Naturally, a solution can be divided into the scheduling decisions, which assign each activity to a start time, and the assignment decisions, which assign each activity to one or more resources having the required skills. As noted in [7], adding these assignment decisions to the unit-skilled RCPSP presents a non-trivial layer of complexity. Hence, the MSPSP is NP-hard [3].

Example 1 (Adapted from [17]). The small example is made up of activities $A_0, A_1, \ldots, A_5$, for which A_0 (A_5) is a fictitious source (sink) activity having a zero duration, and resources $R_1, R_2, \ldots, R_4$ having one or more of the three skills S_1, S_2, and S_3. Table 1 defines the skills which each resource has mastered and Table 2 defines the skill requirements of each activity. Figure 1 shows the precedence relations which must be respected, with the processing times of each activity shown on the directed edges. Figure 2 contains a solution showing the start times of the non-fictitious activities, assignment of the resources to activities and skill contribution of each resource. This solution has a makespan of 8.   □

In the past two decades Constraint Programming (CP) has been used extensively to tackle a variety of combinatorial optimization problems. One of its successes lies in solving resource-constrained scheduling problems, which was further boosted since the introduction of Lazy Clause Generation (LCG) [19]. LCG is a CP solver with nogood learning facility from Boolean Satisfiability Solving (SAT). It has been shown to effectively prune the search space with conflict driven search to guide the solver [13, 20–24, 26]. In line with these works, we investigate the different CP models and searches for MSPSP using the MiniZinc modeling language and show the effectiveness of LCG by beating the state-of-the-art methods on instances from the literature and newly generated ones. When

Table 1. Resources' skills

	R_1	R_2	R_3	R_4
S_1	-	✓	✓	✓
S_2	✓	-	-	-
S_3	✓	✓	-	✓

Table 2. Activity skill requirement

	A_0	A_1	A_2	A_3	A_4	A_5
S_1	-	-	1	2	1	-
S_2	-	1	-	-	1	-
S_3	-	1	1	-	-	-

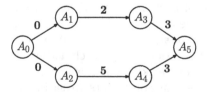

Fig. 1. Precedence graph

R_1	$A_1 : S_2$	$A_4 : S_2$
R_2	$A_1 : S_3$	$A_3 : S_1$
R_3	$A_2 : S_1$	$A_3 : S_1$
R_4	$A_2 : S_3$	$A_4 : S_1$

time 0   2   4   6   8

Fig. 2. Feasible schedule and assignment

using the best performing model and search, which utilizes a recent extension of the search facilities of MiniZinc, the LCG solver Chuffed [6] was able to close at least 87 open instances that were made available to us from the literature.

2   Literature Review

To the best of our knowledge, the first contribution to the literature concerning the MSPSP was by Néron [17] who proposed two lower bounds on the makespan. These bounds were later enhanced in [9]. [16] provides strong lower bounds using Lagrangian relaxation and column generation. Exact branch-and-bound methods were developed in [4,5] based on the reduction of the slack of one activity at each node. More recently, the authors of [14,15] propose an approach using column generation and a branch-and-price framework.

Correia et al. [8] propose a mixed-integer linear program for solving MSPSP. They strengthen their model by adding many valid cuts, which are computed in a pre-processing step. For evaluating their method, they created a data set which was inspired by the data generator for RCPSP in [12]. Building upon this work, [1] develop an instance generator for MSPSP and create a larger data set. Later the same author [2] proposed a constructive heuristic using a priority-based parallel scheduling scheme. Correia and Saldanha-da-Gama [7] design a generic modeling framework for the MSPSP as well as preprocessing and enhancement methods. Their framework provides the basis of our CP model in Sect. 4.

A closely related problem is the Multi-Mode RCPSP, in which each activity has one or more execution modes and a solver has to decide, in which mode an activity is processed. Recently all open instances from standard benchmark sets were closed by [26] using the solver Chuffed. As noted in [4], MSPSP can be reduced to Multi-Mode RCPSP by creating one execution mode for each possible activity-resource assignment potentially leading to a large number of modes.

3   Problem Definition

The MSPSP consists of *non-preemptive activities* $V = \{0, 1, \ldots, n+1\}$, *a set of precedence relations* $E \subseteq V \times V$, *unary renewable resources* $R = \{1, 2, \ldots, m\}$, and *skills* $S = \{1, 2, \ldots, l\}$. An activity $i \in V$ is characterized by a *duration* (processing time) $p_i \in \mathbb{N}^0$ and the *skill requirement* $sr_i^s \in \mathbb{N}^0$ for each skill $s \in S$, where $sr_i^s = 0$ means the skill s is not required. The activities 0 and $n+1$ are fictitious marking the start and the end of the project. Both activities have a zero duration and no skill requirements. A resource $r \in R$ is defined by the *skill mastery* $mast_r^s \in \{0, 1\}$ for each skill $s \in S$, where $mast_r^s = 1$ ($mast_r^s = 0$) means that the resource (does not) masters the skill. A solution assigns each activity i to not only a start time $s_i \in \mathbb{N}^0$, but also the resource-skill combinations $y_{ir}^s \in \{0, 1\}$, that are used for executing i, so that the following constraints are satisfied.

$$s_i + p_i \leq s_j \qquad\qquad \forall(i,j) \in E \qquad\qquad (1)$$

$$\sum_{r \in R} y_{ir}^s = sr_i^s \qquad\qquad \forall i \in V, \forall s \in S \qquad\qquad (2)$$

$$\sum_{s \in S} \sum_{i \in V: s_i \leq t < s_i + p_i} y_{ir}^s \leq 1 \quad \forall r \in R, \forall t \in \left\{0, 1, \dots, \sum_{i \in V} p_i\right\} \quad (3)$$

$$y_{ir}^s \leq mast_r^s \qquad\qquad \forall i \in V, \forall r \in R, \forall s \in S \qquad\qquad (4)$$

Constraint (1) models the precedence relations, (2) ensures that required resource-skills are assigned to each activity, (3) states that a unary resource only can use one skill at any time, and (4) ensures that a resource only contributes with a skill that it mastered. The goal is to find a solution that minimizes the makespan, *i.e.*, $\min_{i \in V} s_i + p_i$ or, simply, $\min s_{n+1}$.

4   Constraint Programming Model

The inherent similarities between the Multi-Skill and the Multi-Mode RCPSP motivated us to employ a similar approach to Szeredi and Schutt [26] using CP with the modeling language MiniZinc [18]. MiniZinc allows the user to define application-tailored search strategies that guide the solver's branching strategy when exploring the solution space which proved helpful for this problem.

4.1   Basic Variables and Constraints

As the previous section indicated, we have a start time variable s_i and resource-skill assignment variables y_{ir}^s for each activity i. They are linked with the input data via (2), (4), and the following ones, which ensures that a resource only contributes at most one skill for the execution of an activity.

$$\sum_{s \in S} y_{ir}^s \leq 1 \qquad\qquad \forall i \in V, \forall r \in R \qquad\qquad (5)$$

Precedence relations are modeled as stated in constraint (1). Note that all non-fictitious activities have at least one predecessor and one successor, whereas the fictitious activity 0 $(n+1)$ has no predecessor (successor). We set the start time $s_0 = 0$ and, thus, the start time of $n+1$ equals the project makespan if started immediately after the completion of all its predecessors.

4.2   Unary Resource Constraints

The modeling of the unary resource constraints (3) is an important ingredient for efficient solving of MSPSP. We investigate different model choices tailored to the LCG solver Chuffed and other CP solvers with nogood learning. Note that some model choices are clearly weaker for CP solvers without nogood learning, due to the weaker propagation strength, but for CP solvers with nogood learning the weaker propagation strength might be more than compensated by the stronger learning that is provided by the model choice.

Time-Indexed Decomposition. The straightforward way is to model it as a time-indexed decomposition, as stated in the constraint (3). Since the size of the decomposition also depends on the size of the planning horizon, the resulting model size quickly became prohibitive, which was supported by bad results in preliminary experiments. Thus, we did not consider this option further.

Global Constraints. The standard way in CP is to use the global constraint `disjunctive` or `cumulative`, if the first one is not supported by the solver, for each resource and re-using the variables y_{ir}^s for modeling the optionality of each activity on the resource. The solver `Chuffed` does not support the constraint `disjunctive` in the form that is needed for MSPSP. Thus, we model the resource constraints by `cumulative`.

$$\texttt{cumulative}((s_i)_{i \in V, s \in S}, (p_i)_{i \in V, s \in S}, (y_{ir}^s)_{i \in V, s \in S}, 1) \qquad \forall r \in R$$

The parameters of `cumulative` represent the start times of activities, the durations of activities, the resource requirements of activities, and the resource capacity. This resource model creates nl "optional" tasks with a variable resource requirement, which equals to 1 if the activity uses a skill of the resource; otherwise it is 0 and the activity is absent.

This model choice works better than the time-indexed decomposition, and can be further improved as the number of "optional" tasks created can be reduced to n by introducing auxiliary variables x_{ir} for each activity $i \in V$ and each resource $r \in R$. These variables are reflecting the fact whether a resource r contributes with any of its skills to the execution of the activity i.

$$y_{ir}^s \leq x_{ir} \qquad\qquad\qquad\qquad \forall i \in V, r \in R, s \in S \qquad (6)$$
$$\texttt{cumulative}((s_i)_{i \in V}, (p_i)_{i \in V}, (x_{ir})_{i \in V}, 1) \qquad \forall r \in R \qquad\qquad (7)$$

Constraint (6) links the auxiliary variables to the resource-skill assignment variables, whereas (7) uses these auxiliary variables as resource requirement ones in the constraint `cumulative`. Note that these cumulative constraints will perform the same propagation when using the variables y_{ir}^s, but the learning would be stronger, because it relaxes the exact skill contribution for the resources.

Order Constraints. Instead of using the `cumulative` constraint, which provides the strongest propagation on the start time variables and lets the solver learn about the time resource usage connection, we can use ordering constraints that enforce a non-overload of any resource. In order to reduce the number of those ordering constraints, we introduce the concept of *unrelated* activity pairs.

Let $clo(E)$ be the transitive closure of the set of the precedence relations in E, i.e., $\forall (i, j) \in clo(E), \exists (k_1, k_2), (k_2, k_3), \ldots, (k_o, k_{o+1}) \in E, o \geq 1$ with $k_1 = i$ and $k_{o+1} = j$. Two activities i and j are *unrelated* if $(i, j), (j, i) \notin clo(E)$. Let $U = \{(i, j) \mid i, j \in V, i < j, \{(i, j), (j, i)\} \cap clo(E) = \emptyset\}$ be the set of all unrelated activity pairs. Only the execution of unrelated activities can overlap in time; all others cannot due to the precedence relations. Even if unrelated activities run concurrently they might not cause a resource overload if different resources are assigned to them. For each pair of unrelated activities $(i, j) \in U$, we introduce an order variable $o_{ij} \in \{0, 1\}$, which takes the value 1 if their executions are overlapping and otherwise 0.

$$\neg o_{ij} \Leftrightarrow (s_i + p_i \leq s_j) \lor (s_j + p_j \leq s_i) \qquad \forall (i, j) \in U \qquad (8)$$

$$(x_{ir} \land x_{jr}) \Rightarrow \neg o_{ij} \qquad \forall (i, j) \in U, r \in R \qquad (9)$$

Constraint (8) uses an equality for linking the order variables to whether one of the activity runs before the other one. Constraint (9) enforces that, when two unrelated activities are assigned to the same resource, then their execution cannot overlap. We note that the order variables allow the solver to learn about the relative position of activities, which is stronger than the time-dependent learning that happens with the above described models. Thus, the stronger learning might compensate the weaker propagation.

Moreover, for some unrelated activities, we may know a priori that they cannot overlap due their combined requirements of a given skill exceeding the number of available resources mastering that skill. For such unrelated activities, we can replace the corresponding constraints in (8) and (9) by this conjunction.

$$(o_{ij} \Rightarrow s_i + p_i \leq s_j) \land (\neg o_{ij} \Rightarrow s_j + p_j \leq s_i)$$
$$\forall (i, j) \in U, \exists s \in S \colon sr_i^s + sr_j^s > \sum_{r \in R} mast_r^s \qquad (10)$$

4.3   Redundant Constraints

Similar to the valid inequalities presented in [8], we add redundant constraints to enhance our formulation. Since the assignment of activities to resources is unknown at the beginning of any search, the unary resource constraints will propagate poorly, potentially resulting in poor early search decisions, which are hard to recover for any solver. Two ways to allow more propagation earlier in the search is to relax the skill from the resources and the resource from the skills.

$$\texttt{cumulative}((s_i)_{i \in V}, (p_i)_{i \in V}, (sr_i^s)_{i \in V}, |\{r \in R \mid mast_r^s = 1\}|) \quad \forall s \in S \quad (11)$$

$$\texttt{cumulative}((s_i)_{i \in V}, (p_i)_{i \in V}, (\sum_{s \in S} sr_i^s)_{i \in V}, |R|) \qquad (12)$$

Constraint (11) ensures that at any time no more than the available number of resources having a particular skill are taken, whereas (12) states that at any time no more than the available resources can be used.

4.4   Search Procedures

Preliminary experiments showed that the basic search procedures in MiniZinc performed poorly, because of the disconnection of branching on the resource assignment and start time variables related to one activity. However, we made use of the new search facilities of MiniZinc `priority_search` [10], which is supported by `Chuffed`. We grouped the start time variables s_i and the resource-skill assignment variables y_{ir}^s by activities $i \in V$. The search procedures then branch over each activity group. For each group, the search assigns the smallest possible start time of s_i before assigning the maximal value of its corresponding resource skill variables y_{ir}^s in input order. We investigate three different branching orders of activity groups, which depend on the domain of their start time variables s_i. The first one `priority-sm` picks the group having the smallest value in the domain of s_i, the second one `priority-sml` the smallest largest value, and the last one `priority-ff` the smallest domain size.

5   Computational Experiments

All experiments were run on a PC with an Intel i7 2600 CPU 3.4 GHz and 8 GB of memory. The model was compiled to the `Chuffed` FlatZinc format using MiniZinc 2.1.2. We used `Chuffed` [6] from https://github.com/chuffed (branch: develop, commit 1f37fde). All experiments were run with a time limit of 10 min. We evaluated our models and search strategies on two data sets detailed in Table 3. This table provides the number of instances in each set and the range of n, l and m values. The last two columns give source and performance of the best known exact solution methods for each corresponding data set. Set 1 was proposed in [1,8], but we were unable to get in contact with the authors and access their data sets. Thus, we created a new data set 1′ using the same instance generator and parameters as they did. [15] selected a subset of 271 instances from the available 278 MSPSP instances in set 2. To the best of our knowledge, their exact method provides the best results for this subset. All data we used can be found in [28]. We run preliminary experiments on set 1′a to inform our decision on the best model formulation and search strategy.

Table 3. Data sets summary

Set	#instances	n	l	m	Best known results	
					Source	%optimal
1a	216	22	4	10–30	[8]	93.98
1b	216	42	4	20–60	[2]	2.31
2a	110	20–51	2–8	5–14	[15]	43.64
2b	77	32–62	9–15	5–19	[15]	66.20
2c	91	22–32	3–12	4–15	[15]	51.11

Table 4 presents a comparison between the different unary and redundant constraints, as well as the search strategies tested on set 1′a. In the table, we first define the basic, unary and redundant constraints used in the model and search strategy. We provide the mean number of nodes explored by the search,

Table 4. Model and search comparison – set $1'a$

Basic cons.	Unary cons.	Redundant cons.	Search	#nodes	%optimal	Runtime
(1–2),(4–6)	(7)	(11–12)	Default	370,174	100.00	10.23 s
(1–2),(4–6)	(8–9)	(11–12)	Default	97,085	100.00	2.73 s
(1–2),(4–6)	(8–10)	(11–12)	Default	54,282	100.00	1.30 s
(1–2),(4–6)	(8–10)	(11–12)	priority-ff	41,762	100.00	1.25 s
(1–2),(4–6)	(8–10)	(11–12)	priority-sml	20,786	100.00	0.68 s
(1–2),(4–6)	(8–10)	(11–12)	priority-sm	13,241	100.00	**0.51 s**
(1–2),(4–6)	(8–10)	(11)	priority-sm	847,879	85.19	94.81 s
(1–2),(4–6)	(8–10)	(12)	priority-sm	13,953	100.00	0.67 s

the percentage of instances optimally solved and the mean runtime. We used the time-tabling filtering as described in [22] for all cumulative constraints (7), (11–12) and, additionally, the time-tabling-edge-finding filtering [25] for (12). The first three rows show that modeling the unary constraints by (8–10) is superior to the other formulations, where Chuffed's default search is an activity-based search with restarts. The next three rows show the results for the search strategies from Sect. 4.4, which are alternated by the default search on restarts. The priority-based searches are superior to the default search and prioritizing the activity selection by the smallest possible start is the best. The last two rows reveal that the performance decays when one of the redundant constraints is removed, especially for (12). Constraints (8–12) using the priority-sm search were used for all remaining experiments.

Table 5 presents the results of testing the CP model on all remaining benchmark instances. We include the mean nodes explored, mean optimality gap for unsolved instances, optimal solutions found, mean runtime on solved instances and the mean runtime across all instances. The gap has been calculated using a naïve lower bound defined by the length of the critical path in the precedence graph. The results of set $2a$ make it clear that this is a very loose bound.

The CP model performed well on the instances of set 2 previously tackled by the literature as we see that all instances of set $2c$ have been solved to optimality in an average of 1.20 s. Previously, half of this set was unsolved.

Table 5. Benchmark results – set $1'b$ and set 2

Set	#nodes	%gap	#opt	%opt	Mean opt runtime	Mean runtime
$1'b$	7,584,577	49.32	27/216	12.50	77.12 s	534.64 s
$2a$	2,223,060	185.20	81/110	73.64	50.29 s	195.22 s
$2b$	816,068	22.42	63/77	81.82	16.88 s	122.90 s
$2c$	14,035	0.00	91/91	100.00	1.20 s	1.20 s

6   Conclusion

We have investigated different CP models and searches for the MSPSP and evaluated them on 710 instances. These models were tailored for CP solvers deploying nogood learning, for which the best trade-off between propagation and learning strength needs to be found. The key ingredients for a successful solution procedure was the combination of problem tailored search with the use of redundant resource constraints and learning on the order between activities by not modeling unary resources with global constraints. The CP solver `Chuffed` was able to optimally solve 67.3% of the instances with a time limit of 10 min. In total, we closed at least 87 open instances from the literature.

Acknowledgments. This work was partially supported by the Asian Office of Aerospace Research and Development grant 15-4016.

References

1. Almeida, B.F., Correia, I., Saldanha-da Gama, F.: An instance generator for the multi-skill resource-constrained project scheduling problem (2015). https://ciencias.ulisboa.pt/sites/default/files/fcul/unidinvestig/cmaf-cio/SGama.pdf. Accessed 26 Apr 2017
2. Almeida, B.F., Correia, I., Saldanha-da Gama, F.: Priority-based heuristics for the multi-skill resource constrained project scheduling problem. Expert Syst. Appl. **57**, 91–103 (2016)
3. Artigues, C., Demassey, S., Néron, E.: Resource-Constrained Project Scheduling: Models, Algorithms, Extensions and Applications. ISTE/Wiley, Hoboken (2008)
4. Bellenguez-Morineau, O.: Methods to solve multi-skill project scheduling problem. 4OR **6**(1), 85–88 (2008)
5. Bellenguez-Morineau, O., Néron, E.: A branch-and-bound method for solving multi-skill project scheduling problem. RAIRO - Oper. Res. **41**(2), 155–170 (2007)
6. Chu, G.G.: Improving Combinatorial Optimization. Ph.D. thesis, The University of Melbourne (2011). http://hdl.handle.net/11343/36679
7. Correia, I., Saldanha-da-Gama, F.: A modeling framework for project staffing and scheduling problems. In: Schwindt, C., Zimmermann, J. (eds.) Handbook on Project Management and Scheduling Vol.1. IHIS, pp. 547–564. Springer, Cham (2015). doi:10.1007/978-3-319-05443-8_25
8. Correia, I., Loureno, L.L., Saldanha-da Gama, F.: Project scheduling with flexible resources: formulation and inequalities. OR Spectr. **34**(3), 635–663 (2012)
9. Dhib, C., Kooli, A., Soukhal, A., Néron, E.: Lower bounds for a multi-skill project scheduling problem. In: Klatte, D., Lüthi, H.J., Schmedders, K. (eds.) OR 2011. ORP, pp. 471–476. Springer, Heidelberg (2012)
10. Feydy, T., Goldwaser, A., Schutt, A., Stuckey, P.J., Young, K.D.: Priority search with MiniZinc. In: ModRef 2017: The Sixteenth International Workshop on Constraint Modelling and Reformulation at CP2017 (2017)
11. Hartmann, S., Briskorn, D.: A survey of variants and extensions of the resource-constrained project scheduling problem, Working Paper Series 02/2008, Hamburg School of Business Administration (HSBA) (2008). https://ideas.repec.org/p/zbw/hsbawp/022008.html

12. Kolisch, R., Sprecher, A.: PSPLIB - a project scheduling problem library. Eur. J. Oper. Res. **96**(1), 205–216 (1997)
13. Kreter, S., Schutt, A., Stuckey, P.J.: Modeling and solving project scheduling with calendars. In: Pesant, G. (ed.) CP 2015. LNCS, vol. 9255, pp. 262–278. Springer, Cham (2015). doi:10.1007/978-3-319-23219-5_19
14. Montoya, C.: New methods for the multi-skills project scheduling problem. Ph.D. thesis, Ecole des Mines de Nantes (2012)
15. Montoya, C., Bellenguez-Morineau, O., Pinson, E., Rivreau, D.: Branch-and-price approach for the multi-skill project scheduling problem. Optim. Lett. **8**(5), 1721–1734 (2014)
16. Montoya, C., Bellenguez-Morineau, O., Pinson, E., Rivreau, D.: Integrated column generation and lagrangian relaxation approach for the multi-skill project scheduling problem. In: Schwindt, C., Zimmermann, J. (eds.) Handbook on Project Management and Scheduling Vol.1. IHIS, pp. 565–586. Springer, Cham (2015). doi:10.1007/978-3-319-05443-8_26
17. Néron, E.: Lower bounds for the multi-skill project scheduling problem. In: Proceedings of 8th International Workshop on Project Management and Scheduling, pp. 274–277 (2002)
18. Nethercote, N., Stuckey, P.J., Becket, R., Brand, S., Duck, G.J., Tack, G.: MiniZinc: towards a standard CP modelling language. In: Bessière, C. (ed.) CP 2007. LNCS, vol. 4741, pp. 529–543. Springer, Heidelberg (2007). doi:10.1007/978-3-540-74970-7_38
19. Ohrimenko, O., Stuckey, P.J., Codish, M.: Propagation via lazy clause generation. Constraints **14**(3), 357–391 (2009)
20. Schutt, A., Chu, G., Stuckey, P.J., Wallace, M.G.: Maximising the net present value for resource-constrained project scheduling. In: Beldiceanu, N., Jussien, N., Pinson, É. (eds.) Integration of AI and OR Techniques in Contraint Programming for Combinatorial Optimzation Problems, pp. 362–378. Springer, Heidelberg (2012)
21. Schutt, A., Feydy, T., Stuckey, P.J.: Scheduling optional tasks with explanation. In: Schulte, C. (ed.) CP 2013. LNCS, vol. 8124, pp. 628–644. Springer, Heidelberg (2013). doi:10.1007/978-3-642-40627-0_47
22. Schutt, A., Feydy, T., Stuckey, P.J., Wallace, M.G.: Explaining the cumulative propagator. Constraints **16**(3), 250–282 (2011)
23. Schutt, A., Feydy, T., Stuckey, P.J., Wallace, M.G.: Solving RCPSP/max by lazy clause generation. J. Sched. **16**(3), 273–289 (2013)
24. Schutt, A., Stuckey, P.J., Verden, A.R.: Optimal carpet cutting. In: Lee, J. (ed.) CP 2011. LNCS, vol. 6876, pp. 69–84. Springer, Heidelberg (2011). doi:10.1007/978-3-642-23786-7_8
25. Schutt, A., Feydy, T., Stuckey, P.J.: Explaining time-table-edge-finding propagation for the cumulative resource constraint. In: Gomes, C., Sellmann, M. (eds.) CPAIOR 2013. LNCS, vol. 7874, pp. 234–250. Springer, Heidelberg (2013). doi:10.1007/978-3-642-38171-3_16
26. Szeredi, R., Schutt, A.: Modelling and solving multi-mode resource-constrained project scheduling. In: Rueher, M. (ed.) CP 2016. LNCS, vol. 9892, pp. 483–492. Springer, Cham (2016). doi:10.1007/978-3-319-44953-1_31
27. Węglarz, J., Józefowska, J., Mika, M., Waligóra, G.: Project scheduling with finite or infinite number of activity processing modes a survey. Eur. J. Oper. Res. **208**(3), 177–205 (2011)
28. Young, K.D.: Multi-skill project scheduling problem instance library (2017). https://github.com/youngkd/MSPSP-InstLib

Application Track

An Optimization Model for 3D Pipe Routing with Flexibility Constraints

Gleb Belov[1]([✉]), Tobias Czauderna[1], Amel Dzaferovic[2],
Maria Garcia de la Banda[1], Michael Wybrow[1], and Mark Wallace[1]

[1] Faculty of Information Technology, Monash University, Melbourne, Australia
{gleb.belov,tobias.czauderna,maria.garciadelabanda,michael.wybrow,
mark.wallace}@monash.edu
[2] Woodside Energy Ltd., Perth, Australia
amel.dzaferovic@woodside.com.au

Abstract. Optimizing the layout of the equipment and connecting pipes that form a chemical plant is an important problem, where the aim is to minimize the total cost of the plant while ensuring its safety and correct operation. The complexity of this problem is such that it is still solved manually, taking multiple engineers several years to complete. Most research in this area focuses on the simpler subproblem of placing the equipment, while the approaches that take pipe routing into account are either based on heuristics or do not consider sufficiently realistic scenarios. Our work presents a new model of the pipe routing subproblem that integrates realistic requirements, such as flexibility constraints, and aims for optimality while solving the largest problem instance considered in the literature. The model is being developed in collaboration with Woodside Energy Ltd. for their Liquefied Natural Gas plants, and is implemented in the high-level modeling language MiniZinc. The use of MiniZinc has both reduced the amount of time required to develop the model, and allowed us to easily experiment with different solvers.

1 Introduction

A chemical process plant produces chemicals through the transformation or separation of materials. This is achieved as the materials pass through the different equipment in the plant via the required connecting pipes. Perhaps surprisingly, the cost of the pipes and associated support structures takes the largest share of the material cost for constructing such plants. This paper focuses on determining the 3D layout of the pipes [5] required to connect the equipment in a chemical plant. The aim is to obtain a layout that minimizes the cost of the pipes and their support structures, while satisfying the constraints needed to ensure the safety and proper functioning of the plant. The *pipe routing problem*, as we will refer to this application, occurs in many different industries, from water desalination to natural gas production. It is part of the more general *process plant layout problem* [9], which is in turn a special case of the *spatial packaging problem* [15].

While there has been some research in optimization methods to solve these problems (e.g., [9,12,15,19–21]), the approaches that consider more realistic pipe

© Springer International Publishing AG 2017
J.C. Beck (Ed.): CP 2017, LNCS 10416, pp. 321–337, 2017.
DOI: 10.1007/978-3-319-66158-2_21

routing scenarios are based on local/heuristic search methods. Realistic scenarios include not only a reasonably large amount of pipes, but also take into account features like pipe flexibility constraints (which deal with the ability for each pipe to cope with the stress inflicted by thermal expansion), and pipe supporting structures (such as pipe racks and other equipment). These features are crucial for the correct functioning and costing of the plant. Currently, producing layouts for such realistic scenarios is a manual task and may take multiple engineers several years to complete.

This paper presents a new model for the 3D pipe routing problem that aims for optimality and explicitly incorporates flexibility constraints and supporting structures, while solving the largest problem instance considered in the literature. The model, which has been developed in collaboration with engineers from Woodside Energy Ltd. for their Liquefied Natural Gas (LNG) plants, is implemented in MiniZinc [13], a high-level, solver-independent Constraint Programming (CP) modeling language for combinatorial optimization and satisfiability problems. This has allowed us to produce a high-level but quite realistic model relatively quickly, and then explore different solving approaches to this model: from constraint propagation solvers with a variety of search strategies, to mixed-integer linear programming (MIP) solvers with different translations for non-linear constraints.

Our aim is to build a tool that will allow plant designers and piping engineers at Woodside to explore and evaluate alternative layouts for realistic pipe routing scenarios and, thus, support them in designing a better plant in a shorter amount of time. Note that we are not attempting to algorithmically compute the final plant layout without human intervention, as we do not believe this is yet feasible. This is both due to the huge complexity of modern plants and to the amount of "undocumented" requirements that seem to exist mainly in the head of the plant and pipe layout engineers. Therefore, while it is important for our system to find a high quality layout solution to a realistic pipe routing scenario, it is also important to display the solution via a visual interface that allows engineers to interrogate the proposed solution, as well as guide the optimization process by requesting changes to various parameters and constraints. We have already performed the first steps towards such an interactive optimization system, in the form of a 3D visualization tool that is connected to our modeling system. This visualization tool enables engineers to explore the produced layout, and to evaluate and validate the proposed solution in a familiar way.

The rest of the paper is structured as follows. Section 2 discusses the related literature and past work on the general plant design problem. Section 3 describes the problem in greater detail and provides the decision variables, constraints and search strategies used in our MiniZinc model. Section 4 briefly describes how the non-linear constraints in the model are appropriately translated by MiniZinc for MIP and CP solvers. Section 5 presents a series of experiments aimed at evaluating the scalability and accuracy of our model. Section 6 describes the interactive 3D visualization tool we have developed for communicating and exploring the results provided by the model. Finally, Sect. 7 presents our conclusions and future work.

2 Literature Review

Solving the overall plant design problem requires finding 3D location coordinates for all the equipment and connecting pipes within a plant's volume (referred to as the *container cuboid*), in such a way as to minimize the total cost of the plant while, at the same time, ensuring its safety and correct functionality. For small problem instances, Sakti et al. [15] successfully apply an integrated approach for a satisfaction version of the simultaneous equipment and piping layout design problem, where the aim is to find any feasible solution that places the equipment and connects the pipes within the given container cuboid. In particular, they considered 10 equipment pieces and up to 15 pipes with 4 segments on average.

For larger, more realistic problem instances, and those where the goal is to find an optimal (or high quality) solution, their integrated approach does not scale. In these cases the problem is naturally divided into two phases [9]. The first phase aims at positioning the equipment, that is, obtaining the 3D location coordinates for each piece of equipment, while minimizing an approximate total cost of the plant. In this phase the focus is on the equipment, ensuring it is supported, safely positioned and can correctly function, while the cost of the pipes is approximated using rough measures, such as Manhattan distances. The second phase aims at determining an optimal layout of the pipes connecting the already positioned equipment. The focus this time is on the pipe routing, taking into consideration issues such as pipe stress and flexibility, the need to support the pipes and the cost associated to these supports. Theoretically, this separation into phases can lead to infeasibilities in the later phases, which can be made less probable using ample safety distances between equipment.

There has already been some research devoted to this problem. However, most of it (e.g., [19,20]) focuses only on the efficient modeling and solving of the first phase (equipment location), which is considerably simpler than the second. While there has been research that includes the second phase (e.g., [9,15]) or even focuses on it (e.g., [12,21]), the existing approaches do not satisfy Woodside's requirements.

On the one hand, the more realistic approaches, which take into account the simultaneous optimized routing of several pipes (including branching pipes and support placement), are based on heuristic algorithms (rather than complete search methods), such as the ant-colony evolutionary algorithms used by [7,12]. This is not our focus, as Woodside is more interested in pursuing complete approaches.

On the other hand, it is difficult to extend the approaches that rely on complete search methods to take into account some of the required constraints, particularly flexibility constraints. In its basic version, single-pipe routing can be modeled as a 3D rectilinear shortest path problem solvable by Dijkstra's algorithm. One of the most realistic of the complete approaches is that of Guirardello and Swaney [9], which provides a detailed MIP model for solving phase one and a general overview of a network-flow MIP model for solving phase two. This second MIP model relies on the construction of a reduced connection graph that limits the possible routes of the pipes. This is used to route pipes one by one, since they suggest that simultaneous routing of the pipes is too costly for a MIP model. While they do not give enough details regarding how the connection graph is constructed, an approach to construct such a connection graph is given, e.g., by de Berg et al. [6], who present a higher-dimensional rectilinear shortest path model that considers bend costs. A more hierarchical method using cuboid free space decomposition is given by Zhu and Latombe [21] and applied to pipe routing. Unfortunately, even if these methods are used, it is not clear how [9] performs sequential pipe routing when pipes interfere with each other ([9] talks about "some tuning by hand" which might be required for these cases). Further, none of these methods can be easily extended to take pipe flexibility constraints into account. In fact, we are not aware of any published results on a general pipe routing method that incorporates flexibility constraints. While [9] mentions the use of Guided Cantilever flexibility constraints in an iterative barrier method to eliminate over-stressed pipe solutions, extending their method to achieve this is not straightforward, and their work provides no information on how to do so.

3   Problem Description and Associated Model

As mentioned before, this paper focuses on the second phase of the process plant layout problem, where the equipment has already been positioned safely and correctly within the container cuboid, and the aim now is to determine the best routing for the pipes that connect the equipment (see the rightmost picture in Fig. 2 for a final solution to our full benchmark). While we have also implemented a MiniZinc model for solving phase one that provides the equipment locations, due to space considerations we will not provide details regarding this model.

As done in the literature, we limit our pipe routing approach to rectilinear axis-parallel routing; in particular, we constrain all bends to be 90°. This is acceptable as non-90° bends are extremely rare in real plants, and can be added later by a post-process step that looks for possible pipe simplifications based on the addition of such bends and non-axis-parallel segments. Further, we approximate pipe segments by cuboids with a square cross-section and ignore bend radius at this stage. This simplification is also acceptable since the resulting loss of space in a large process plant (versus, say, the pipe routing in a jet engine as performed in [15]) is negligible. In this form, the resulting model extends the one in [15] to account for supports and flexibility constraints, and is related to 3D orthogonal packing [14,17].

3.1   Input and Derived Data

The input data to the second phase includes the following:

- Dimensions of the container cuboid (integer values in millimeters) and an associated discretization parameter $K \geq 1$ for placement positions, where K is the size of the model's length unit in mm. The smaller the value of K, the higher the number of possible position points for placing pipes and, thus, the more difficult the problem to solve.
- Set $\mathcal{M}$ of equipment and their locations as provided by phase one, which models each equipment piece $m \in \mathcal{M}$ as a cuboid with given length, width and height $W^m \in \mathbb{Z}^3$ (depending on the chosen rotation), and returns its location as the 3D coordinates $X^m \in \mathbb{Z}^3$ of the corner with the smallest x, y and z values. The elements of $\mathcal{M}$ can be of several types, including vessels, heat exchangers, pumps, the pipe rack (a multi-platform structure that traverses the entire plant), and the source/sink points that connect the main pipe to parts of the plant not considered by the problem instance. Note that, in this second phase, the elements of $\mathcal{M}$ represent either support structures or obstacles to which some pipes are attached via "nozzles". Only some of these elements (mostly vessels and racks) can provide support to nearby pipes.
- Two pieces of information for each nozzle: its location modeled as the 3D coordinates of the center of the attachment area (i.e., the intersection of the equipment surface and the pipe axis), and its direction, modeled as a value from the set $\{0,\ldots,5\}$, where values $\{0,1,2\}$ indicate the nozzle has a positive direction along the x,y,z-axis, respectively, and values $\{3,4,5\}$ indicate a negative direction along the same axes.
- Set $\mathcal{P}$ of pipes and, for each pipe $p \in \mathcal{P}$, the following information: external diameter D_p (corresponding to the length of any side of the square cross-section), cost per length unit C_p^L, cost per bend C_p^B, maximal number of pipe segments allowed N_p^S, safety distance to equipment and other pipes s_p^M, thermal expansion unit e_p, elasticity modulus E_p (which is associated to p's material), and maximum stress allowed (stress capacity) S_p^A. We also input lower/upper pipe segment length bounds, where *segment length* denotes the distance between the segment's defining *nodes*, and a node is either a bend (assuming bend radius zero) or the pipe nozzle attachment point.
- Set $\mathcal{SZ}$ of cuboid support zones (i.e., areas that can support pipes), and for each $j \in \mathcal{SZ}$ the following information: its location as the smallest corner $X^j \in \mathbb{Z}^3$ and a length, width, and height vector $W^j \in \mathbb{Z}^3$ as above, together with the cost penalty C_{jp}^{SZ} for any pipe p that uses j. Note that some of these support zones are associated to equipment (mostly vessels), while some are associated to each of the platforms in the pipe rack. We currently have only two cost classes for support zones: $C_{jp}^{SZ} \in \{0, C_p^{SZ}\}$ with the zero value being the one given to the rack's platform support levels.

From the above input data, our model derives the following parameters for each pipe $p \in \mathcal{P}$:

- Set $\mathcal{S}_p$ of pipe segment indices $\mathcal{S}_p = \{1,\ldots,N_p^S\}$. Note that if the number n of segments of p is less than the maximum allowed ($n < N_p^S$), the length of all segments $j : n < j \leq N_p^S$ will be 0.

- Set $\mathcal{N}_p$ of *pipe node* indices: $\mathcal{N}_p = \{1, \ldots, N_p^S + 1\}$, where the first and last nodes are the nozzle attachment points (and thus, they are known as they were set in phase one), and the rest correspond to pipe bends. If $n < N_p^S$, the position of all nodes $j : n < j \leq N_p^S + 1$ will be the same.

3.2   Decision Variables

The solution to the problem is expressed in terms of the values of the decision variables representing for each pipe $p \in \mathcal{P}$, the pipe node positions $X_i^p \in \mathbb{Z}^3$ of each $i \in \mathcal{N}_p$, subject to the container cuboid. The objective function (and associated decision variable *obj*) is the sum of the cost penalties associated to the length, bends and supports of each pipe.

In addition to the decision variables associated to the solution representation and objective function, the following intermediate decision variables are used in our model to better express the required constraints and/or search strategy:

- Pipe segment existence flags $f\exists_i^p \in \{0, 1\}$, $p \in \mathcal{P}, i \in \mathcal{S}_p$, with $f\exists_i^p = 1$ *iff* segment i of pipe p has length greater than 0.
- Pipe segment directions $d_i^p \in \{0, \ldots, 5\}$, $p \in \mathcal{P}, i \in \mathcal{S}_p$ using the same convention as for nozzles, that is, values $\{0, 1, 2\}$ indicate segment i of pipe p has a non-negative direction along the x, y and z axes, respectively, while values $\{3, 4, 5\}$ indicate a non-positive direction along the same axes. Note that the values for the first and last nodes (the nozzles) are given as input.
- Absolute pipe segment lengths $L_{ix}^p, L_{iy}^p, L_{iz}^p$ for each pipe $p \in \mathcal{P}$ and segment $i \in \mathcal{S}_p$ along the axes x, y, z, respectively.
- Pipe segment cuboid hulls as the smallest-corner coordinates $\tilde{X}_i^p \in \mathbb{Z}^3$ and sizes $\tilde{W}_i^p \in \mathbb{Z}^3$ (ignoring bend radius), for each pipe $p \in \mathcal{P}$, similar to the notation for equipment and support zone cuboids.

3.3   Constraints

Symmetry Breaking Constraints. We want to avoid searching for symmetric solutions, that is, solutions where the flags $f\exists_i^p$ are either 0 at the end, or 0 at the beginning of a pipe. The following symmetry breaking constraints ensure only those with 0s at the end are considered solutions, by imposing a non-increasing order among the segment existence flags of every pipe:

$$f\exists_i^p \geq f\exists_{i+1}^p, \qquad p \in \mathcal{P}, i \in \mathcal{S}_p \setminus \{\max\{\mathcal{S}_p\}\} \tag{1}$$

Orthogonal Direction Change Constraints. They ensure consecutive segments of a pipe do not form a line by, imposing a change in pipe segment direction for any two consecutive segments of every pipe:

$$0 = f\exists_{i+1}^p \ \vee \ d_i^p \bmod 3 \neq d_{i+1}^p \bmod 3, \qquad p \in \mathcal{P}, i \in \mathcal{S}_p \setminus \{\max\{\mathcal{S}_p\}\} \tag{2}$$

Object Non-overlapping Constraints. They ensure the pipes do not overlap with other pipes and any other equipment (note that the non-overlapping among equipment has already been achieved in phase one) and take into account the required safety distances. They are implemented by disjunctions over the coordinate points at each of the three axes, similarly to [14,17]. For example, for ensuring non-overlapping between pipes and other equipment, the constraints are as follows:

$$\bigvee_{c=1}^{3}\left(\tilde{X}_{ic}^{p}+\tilde{W}_{ic}^{p}+s_{p}^{M}\leq X_{c}^{m}\vee X_{c}^{m}+W_{c}^{m}+s_{p}^{M}\leq\tilde{X}_{ic}^{p}\right),\; p\in\mathcal{P}, i\in\mathcal{S}_{p}, m\in\mathcal{M}\quad(3)$$

For subsets of objects with equal safety distances, we could have instead used the `diffn_k` global constraint [1]. However, among the solvers we tested only OR-Tools [8] has it, and it is not enough to express the support constraints.

Support Constraints. Each bend's base point (the intersection of the neighboring segments' base lines) is required to be in a provided support zone, whose cost penalty is the one added to the objective function. Supporting only bends can be an under approximation, as long segments might also need to be supported. However, this seems to happen very rarely (once in our biggest benchmark).

The placement of bends within a support zone is modeled as a reified form of condition (3), namely we denote this placement by a boolean variable b_{pij}^{SZ} for each pipe p, bend i and support zone j:

$$b_{pij}^{\mathrm{SZ}}\;\leftrightarrow\;\bigwedge_{c=1}^{3}\left(\tilde{X}_{ic}^{p}\geq X_{c}^{j}\wedge\tilde{X}_{ic}^{p}\leq X_{c}^{j}+W_{c}^{j}\right),\quad p\in\mathcal{P}, i\in\mathcal{S}_{p}\backslash\{1\}, j\in\mathcal{SZ}\quad(4)$$

To ensure each bend is placed within a valid support zone, we demand:

$$\exists_{j}b_{pij}^{\mathrm{SZ}},\qquad p\in\mathcal{P}, i\in\mathcal{S}_{p}\backslash\{1\}\qquad\qquad(5)$$

According to our cost assumption, namely $C_{jp}^{\mathrm{SZ}}\in\{0,C_{p}^{\mathrm{SZ}}\}\;\forall j\in\mathcal{SZ}, p\in\mathcal{P}$, with the zero cost belonging to (disjoint) rack support levels, we add to the objective function the following variables for each bend's support cost:

$$C_{pi}^{S}=C_{p}^{\mathrm{SZ}}f\exists_{i}^{p}\left(1-\sum_{j:C_{jp}^{\mathrm{SZ}}=0}b_{pij}^{\mathrm{SZ}}\right),\qquad p\in\mathcal{P}, i\in\mathcal{S}_{p}\backslash\{1\}\qquad(6)$$

Flexibility Constraints. Several approximate methods are described in [5], including the Guided Cantilever Method (GCM), which is the one implemented by our model, as it is reasonably accurate and not too complex. This method assumes that pipes are only fixed at the nozzle attachment points and their bends are rectilinear. The *thermal expansion* of pipe $p\in\mathcal{P}$ along axis x is defined as:

$$\Delta_{x}^{p}=L_{x}^{p}e_{p},\qquad\qquad(7)$$

where L_{x}^{p} is the x-distance between p's nozzles and e_{p} is the *unit thermal expansion* of p. The Δ_{y}^{p} and Δ_{z}^{p} are defined similarly.

According to the GCM, a segment $i \in \mathcal{S}_p$ with a non-zero length L in the y or z direction (with length $L = L_{iy}^p$ or $L = L_{iz}^p$, respectively), absorbs the following portion of the thermal expansion in the x-direction:

$$\delta_x = \frac{L^3}{\sum_i (L_{iy}^p)^3 + \sum_i (L_{iz}^p)^3} \Delta_x^p, \tag{8}$$

where δ_x is the segment's lateral deflection in the x-direction, Δ_x^p is the overall thermal expansion of p in the x-direction given by (7), and $\sum_i (L_{iy}^p)^3 + \sum_i (L_{iz}^p)^3$ is the sum of the cubed lengths of all pipe segments of p that are perpendicular to x. Similar equations can be written for the lateral deflections of a segment in the y- and z-directions.

Also, the deflection capacity $\bar{\delta}$ of a segment under the method's assumptions can be given as:

$$\bar{\delta} = \frac{48L^2 S_p^A}{E_p D_p}, \tag{9}$$

where S_p^A is the allowable stress range for p, L is the segment length, E_p is the modulus of elasticity associated to p's material, and D_p is the external diameter of pipe p, all in appropriate units. Finally, the expansion stress on p's segment is permissible when:

$$\max\{\delta_x, \delta_y, \delta_z\} \leq \bar{\delta}. \tag{10}$$

For δ_x this can be re-written as:

$$\kappa L \leq \sum_i (L_{iy}^p)^3 + \sum_i (L_{iz}^p)^3 \tag{11}$$

with $\kappa = \Delta_x^p E_p D_p / S_p^A / 48$. We add inequalities (11) and their y-, z-counterparts to the model. Value $\max\{\delta_x, \delta_y, \delta_z\}/\bar{\delta}$ is called the *stress ratio* of the segment.

3.4   Search Strategy

MiniZinc allows us to specify a custom search strategy for the current optimization model being solved. One of the advantages of constraint programming solvers is that they can benefit from such a custom search strategy, either following it strictly or interleaved with their own search strategy. Our MiniZinc code declared the following strategy for routing each pipe p: the search first explores the values of search pairs $(f\exists_i^p, d_i^p)$ in the order of the pipe's segments $i \in \mathcal{S}_p$. For each variable, the value selection strategy uses binary search (called `indomain_split` in MiniZinc), which first splits the domain of the variable around the integer mean of its lower and upper bounds, and then searches within the lower half, followed by the upper half. Search strategies are simply ignored by MIP solvers.

4   MiniZinc and Its Solver-Specific Redefinitions

We have implemented the above model in the MiniZinc language [13] and solved it using several CP and MIP solvers. This is despite constraints (6) and (11) being non-linear and, thus, not directly supported by MIP solvers. The MIP interface of MiniZinc [2] handles this by using an automatic solver-specific redefinition mechanism for constraints defined as *predicates* or *functions*: when the model is compiled for a target solver, the front-end looks for a solver-specific redefinition of each predicate or function used in the model. If none is provided, MiniZinc uses the default decomposition appearing in its standard library or forwards the constraint to the solver backend. For example, the standard library definition for the `pow` function of an integer variable, is as follows:

```
1  /** @group builtins.arithmetic Return \(\a x ^ {\a y}\) */
2  function var int: pow(var int: x, var int: y) =
3      let {
4        int: yy = if is_fixed(y) then fix(y) else -1 endif;
5      } in
6      if yy = 0 then 1
7      elseif yy = 1 then x else
8      let { var int: r;
9              constraint int_pow(x,y,r);
10     } in r
11     endif;
```

which calls predicate `int_pow(x,y,r)`. As no solver we tested handles this predicate, we defined it to represent x^3 as $x \cdot x \cdot x$ for CP solvers, and as $\sum_{v=lb(x)}^{ub(x)} v^3(x = v)$ for MIP ones. For example, the latter was achieved by adding the following MiniZinc code to the linearization library:

```
1  predicate int_pow(var int: x, var int: y, var int: r) =
2      let {
3        array[int, int] of int: x2y
4            = array2d(lb(x)..ub(x), lb(y)..ub(y),
5              [ pow(X, Y) | X in lb(x)..ub(x), Y in lb(y)..ub(y) ] )
6      } in
7        r == x2y[x, y];
```

5 Evaluation

We have performed several experiments aimed at evaluating the scalability and accuracy of our method. All these experiments were executed as a 1-thread process on an Intel(R) Core(TM) i7-4771 CPU @ 3.50 GHz. Figure 2 provides a view of solutions obtained for the largest benchmark.

5.1 Default Benchmark

All our benchmarks modify a default benchmark by considering either subsets of its $\mathcal{M}$ and $\mathcal{P}$ sets, or a different discretization parameter. This default benchmark models the *acid gas removal unit* of an existing LNG plant. Its *container cuboid* is sized $76 \times 40 \times 43$ m length by width by height, and its discretization parameter is $K = 200$ mm, which gives $381 \times 201 \times 216$ position points along axes x, y, z, respectively. It also has a set $\mathcal{P}$ with 27 pipes, with diameters D_p between 50 and 750 mm, and a set $\mathcal{M}$, with the following 17 equipment pieces already positioned by phase one:

- 4 column vessels, with heights between 17 and 40 m
- 1 horizontal drainage vessel, grounded, with a footprint of 7×2.5 m
- 3 heat exchanger groups and 2 individual heat exchangers. Two of the groups are fin-fan blocks of sizes $17 \times 15 \times 2.5$ m and $10 \times 15 \times 2.5$ m. The two individual exchangers are $21 \times 3 \times 3$ m each and the third group is $8 \times 2 \times 3$ m
- 4 pump groups, grounded, of sizes from $3 \times 1 \times 1.5$ to $15 \times 8 \times 1.5$ m
- a source point and a sink point connecting the current unit to other parts of the plant
- a pipe rack of size 13×13 m cross-section running through the container cuboid length-wise (see below for details on support areas).

The set $\mathcal{SZ}$ of support zones, where all pipe bends must be located, is as follows:

- 3 m zones around the 5 vessels, the 2 individual heat exchanges, one of the heat exchanger groups, and the multi-level pipe rack
- 50 cm thick "preferred levels" at heights 3, 6, 9, and 12 m in the pipe rack, corresponding to the established platforms in the rack
- 0–3 m layer above ground.

The above pipes, bends and support zones have the following costs:

- length cost C_p^L: \$25–\$400 per meter, depending on diameter and routing requirements
- bend cost C_p^B: twice the per-meter cost
- support cost C_{jp}^{SZ}: 10× the per-meter length cost of p for all bends except those located in the 0–3 m "ground level zone" or in the "preferred rack levels" at heights 3, 6, 9, and 12 m, which have no cost.

All benchmarks require a safety distance S_p^M of 75 mm between pipes and between pipes and equipment. The upper bound for the length of a nozzle segment is 6 m. The lower bound for the length of any segment is 2× the diameter. Finally, all benchmarks impose the following flexibility/stress capacity requirements on all pipes in the plant, which are taken from the example in Sect. 4.5 in [5]:

- $e = 0.078$ in/ft, corresponding to the temperature range from 70 to 480 °F
- $E = 29 \cdot 10^6$ psi
- $S^A = 21625$ psi.

While the costs and stress values given above are not the real ones, we use them to report on the approach (the actual values cannot be disclosed).

5.2   Overall Approach

The overall approach follows a single-pipe configuration strategy, where each pipe is routed in sequence and the resulting route becomes an obstacle for the next pipe to be routed. In other words, the first pipe is routed in the context of the equipment, plus other pipes' nozzle segments, as obstacles. The second pipe is routed in the same context plus the obstacles resulting from the segments of the first pipe routed, and so on. We experimented with several pipe orders, including widest pipe first (according to diameter), and largest-surface pipe first (according to a combined measure of approximated length × width). With a minimal advantage for the latter, we choose it for the experiments. As shown below, the loss in accuracy due to our single-pipe routing is small.

Each pipe starts with 10 as the maximal number N_p^S of segments, and increases it if infeasible. Also, each pipe is first routed without the GCM flexibility constraints, and subsequently re-routed with them if stress violations occur (i.e., if the stress is greater than that allowed by S_p^A). For example, Fig. 1 shows the result of routing pipes without GCM constraints (left), and the subsequent loops added to reduce the stress down to allowable levels (right).

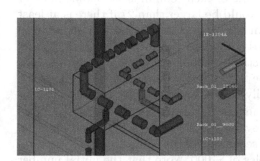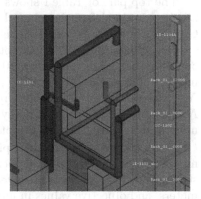

Fig. 1. Left: shows pipes routed without flexibility considerations, where dashed pipes indicate pipe segments with greater than the allowable stress S_p^A: the shorter the dashes the higher the stress. Right: shows re-routed pipes with extension loops to relieve stress.

5.3   Results for Solvers Gurobi, IBM ILOG CPLEX, Chuffed, Gecode, and OR-Tools

The wide choice of solver backends for MiniZinc allowed us to try several solvers, including the two state-of-the-art MIP solvers Gurobi 7.0.2 [10] and IBM ILOG CPLEX 12.7.1 [11]. In terms of CP solvers, we tried the following three: Chuffed [4], one of the best solvers to combine constraint propagation with (SAT-style) clause learning, compiled from the develop branch on [3]; OR-Tools FlatZinc 5.1.4045 [8], Google's fast and portable suite for combinatorial optimization; and Gecode 5.1.0 [16], one of the most popular CP solvers based on traditional constraint propagation. All these solvers have shown prominent performance in the annual MiniZinc Challenges [18].

For the MIP solvers it proved best to use the simple 'big-M' translation of logical constraints, cf. [2]. For CP solvers, OR-Tools did not produce any feasible solutions in the 1800-seconds time limit allowed per benchmark, while the best results for Chuffed were produced with option -f (free search).

Table 1. Results for routing default benchmark without flexibility constraints. The total objective value is different among solvers despite every pipe being routed optimally. This is due to the sequential routing approach

Solver	N^{calls}	$N^{\text{init}}_{\text{obst}}$	$N^{\text{last}}_{\text{obst}}$	N^{opt}	t^{min}	$t^{\text{max}}_{\text{opt}}$	t^{Total}	L	S	Obj	NB	NOS	r^{max}
CPLEX	27	68	152	27	3.1	14	**150**	741	982	1765	111	55	2809%
Gurobi	27	68	153	27	3.6	71	318	741	982	1769	114	60	2755%
Chuffed	27	68	153	27	0.5	565	2038	753	985	1757	114	61	2809%
Gecode	27	68	155	27	0.3	261	685	764	987	1758	116	59	2809%
indep.	27	16	16	27	1.0	3	49	758	1004	1660	106	48	2809%

The top part of Table 1 shows the results of sequentially routing each of the 27 pipes in the default benchmark without flexibility constraints using CPLEX, Gurobi, Chuffed and Gecode. The meaning of each column is as follows: number of routing instances N^{calls} attempted (would be larger than 27 if there is at least one pipe that needs more than 10 segments, or such a solution is hard to find); number of obstacles (i.e., already placed equipment or pipe segments) $N^{\text{init}}_{\text{obst}}$ and $N^{\text{last}}_{\text{obst}}$ for the first and last pipe, respectively; number of optimally solved pipes N^{opt} (would be less than 27 if any pipe timed-out); minimum solving time t^{min}; maximum solving time for an optimally solved instance $t^{\text{max}}_{\text{opt}}$; total time spent t^{Total}; total pipe length L, total pipe surface S, total objective value Obj; total number of bends NB; total number of overstressed segments NOS; and maximal stress ratio r^{max}. All times are given in seconds, lengths (surface area) in (square) meters, and objective values in multiples of \$1000. Note that the initial number of obstacles is larger than the cardinality of set $\mathcal{M}$, because the start and end nozzle segments of each pipe are also considered as obstacles for all other pipes. Also note that no pipe needed more than 10 segments (as $N^{\text{calls}} = 27$).

All four solvers produce similar solutions, solving each pipe optimally. Thus, differences in the total objective value are explained only by the sequentiality of the approach (pipe after pipe) and the possibly different optimal solutions found by each solver. Since CPLEX is the fastest overall, we select CPLEX to perform the initial routing of each pipe without GCM constraints.

The last line of Table 1 presents a computation (with CPLEX) where each pipe was routed independently, i.e., ignoring other pipes. The total objective value is 6.2% smaller than the worst one among sequential approaches, giving a lower bound on what can be achieved with simultaneous routing. Note that this lower bound is independent of the solver, as long as every pipe is optimal.

As shown in Table 1, the maximal stress ratio r^{max} is well above the allowed 100%, which is why flexibility constraints are needed. Table 2 shows the results for the full method, where we first try each pipe without flexibility constraints and only re-solve with GCM flexibility constraints if the stress is over the allowable limit. The first four rows of Table 2 show the results for the default benchmark, which is the same as that in Table 1 (27 pipes in $\mathcal{P}$, 17 elements in $\mathcal{M}$, and discretization parameter $K = 20\,\text{cm}$), and different solvers for GCM routing:

Table 2. Comparison of various configurations of routing with GCM constraints

Config		$|\mathcal{P}|$	N_{GCM}^{pipes}	N_{obst}^{init}	N_{obst}^{lst}	N^{opt}	t^{min}	t_{opt}^{max}	t^{Total}	L	S	Obj	NB	r^{max}
Default	CPLEX	27	22	72	181	22	8.0	925	2361	1002	1379	2089	144	98%
	Gurobi	27	21	72	180	18	13.9	1501	10^4	1010	1381	2048	143	99%
	Chuffed	27	23	72	181	23	1.1	441	**1476**	997	1362	2105	144	99%
	Gecode	27	23	72	181	23	0.9	1661	3711	1008	1378	2053	144	98%
indep.		27	22	16	16	22	2.8	691	1228	1008	1380	1999	133	99%
Decrease	$|\mathcal{M}| = 13$	17	13	44	108	13	1.5	420	925	656	889	1270	88	96%
	$|\mathcal{M}| = 9$	10	6	30	59	6	1.8	121	234	419	631	573	51	98%
	$|\mathcal{M}| = 5$	3	3	8	19	3	1.9	267	281	252	594	389	18	97%
Discr	CPLEX	27	22	68	175	21	12.2	328	**2414**	1009	1377	**1948**	144	98%
	Chuffed	27	23	68	182	20	4.1	823	8531	1016	1387	2218	151	99%

CPLEX, Gurobi, Chuffed and Gecode. The data shown is similar to that of Table 1, with the addition of N_{GCM}^{pipes}, which shows the number of pipes that needed to be re-routed with GCM constraints. Again, no pipes needed more than 10 segments (data not shown).

The next row shows, again, a computation where each pipe was routed (with CPLEX) independently of other pipes. The objective value improvement from the worst sequential approach is 5%. Interestingly, independent routing with the chosen objective function reduces the number of bends, but not the total length.

The next three rows show the results for decreasing numbers of equipment (and thus, of associated pipes) as follows: we first removed from the original $\mathcal{M}$ all but 1 heat exchanger, leaving 13 elements; then removed all pump groups, leaving 9; and finally removed all but the 2 main components of the plant unit plus the source, sink, and pipe rack. GCM routing was done with Chuffed. The results show that even with fewer obstacles, routing remains a hard problem, probably due to the GCM constraints.

The final two rows show the results for the default benchmark with a smaller discretization parameter: $K = 10$ cm. CPLEX had a timeout on one pipe but the total objective value is 6.7% smaller than its own with $K = 20$ cm and smaller than independent routing with $K = 20$ cm. This shows that finer discretization can lead to qualitatively better solutions for the current model.

6 Visualization of Layout Solutions

The solutions to our model are text-based geometric descriptions of the optimized plant layout. These descriptions list the positions and dimensions of the equipment and pipe racks (from phase one), and the routes for the pipes as a series of intermediate bend locations (from phase two). This format is not understood by humans as easily as an image of the 3D layout. Moreover, plant layout and piping engineers are nowadays used to working with technical drawings and interactive 3D CAD models that allow them to zoom in and out and rotate the model in three dimensions in order to fully understand it.

We have developed an interactive 3D visualization that enables engineers to explore the produced layout, and to evaluate and validate the proposed solution in a familiar way (see Fig. 2 for two examples). This visualization displays pipes in different colors, drawing them as cylinders (rather than cuboids) of the appropriate diameter with visual bends.

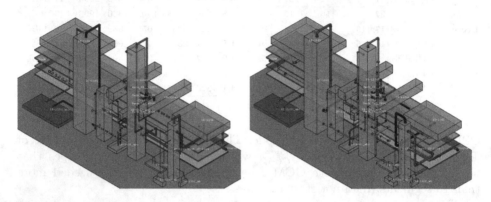

Fig. 2. Solutions for our largest problem instance with (right) and without (left) GCM constraints. The brown base represents the ground level. The cuboids represent equipment. The pipe rack is comprised of the four stacked plane-like cuboids that span most of the length of the volume. Pipes are depicted in different colors to differentiate them. (Color figure online)

Our visualization is displayed in a web-based 3D viewer, allowing the 3D model to be rotated, panned and zoomed. We attach additional metadata to objects in the 3D model, such as equipment and pipe IDs and pipe segment stress. Engineers can see this information by selecting these elements and viewing their properties. We also attach high-level information (such as parameters of the optimization model) to the base of the container cuboid, allowing this to be easily viewed. We have also experimented with using colors or dashes to visualize stress on pipe segments when it is over the allowable levels, although this is mainly for our own benefit while working on the optimization model, since our final solutions must always keep stress within allowable levels.

We have found that using 3D visualizations during our discussions with Woodside engineers not only leads to a more fluid discussion, but allows them to quickly identify layouts that look very different from what they would expect. Often this means there is some requirement we were not aware of. In this case a nice benefit of the problem being specified as a high-level constraint model is that these additional requirements can generally be encoded as constraints and easily added to the model. This has happened multiple times resulting in an iterative evolution of our MiniZinc model and the overall approach over time. On other occasions, the unexpected results did not violate any requirement. Instead, they challenged Woodside engineers to reconsider long-held process plant construction and layout conventions, where diverging from the status quo has the potential to result in significant cost savings.

While discussing various solutions and visualizations, Woodside engineers have identified the need to easily evaluate the quality of various solutions. This requires us to annotate the 3D models with information regarding pipe and support structure material costs, on a pipe-by-pipe basis. These annotations will allow engineers to compare multiple potential solutions and compare the costs of particular subsets of the plant.

We are currently extending our plant layout visualization in two ways. The first extension allows engineers at different physical locations to explore the same visualization in a collaborative way, as is required in, say, a video conference setting. This extension allows the interactions (e.g., rotations and pipe selections) made in one location to be seen at all other locations. The second extension will allow engineers to modify parameters and constraints, such as safety distances or location of some equipment and/or pipes, thus either triggering a re-optimization step (for the pipes connected to the moved equipment) or a re-evaluation step to obtain the new stress and length values for the moved pipes. While re-optimization steps might take hours (depending on how many pipes need to be re-computed), this can be a significant improvement from the current manual process. A comparison between the two layouts will be performed in both cases. The aim is for both extensions to eventually form part of an interactive optimization tool for plant layout that can be used directly by plant layout and piping engineers and be closely integrated into the design process.

7   Conclusions and Outlook

We have presented a MiniZinc model to solve the 3D pipe routing problem under thermal expansion requirements (modeled by the Guided Cantilever Method) and support constraints, thus addressing the major constraints of the practical problem to a certain degree of detail. Different MIP and CP solvers were used to solve this model for the largest benchmark presented in the literature, and the results used to fine-tune a solving method that was shown to achieve near-optimal results with reasonable efficiency. The flexibility constraints represent a major challenge in the current version, significantly increasing the optimal subproblem solving times (which are nevertheless a step forward compared to the manual design process). We plan to extend this model to include maintenance constraints and other units within the plant.

We have also developed a 3D visualization of solutions as a first step towards an interactive optimization system. While it already allows users to evaluate and validate a solution, the aim is to transform the current version into an interactive visual interface that allows users to make changes of the optimization model via direct manipulation of a 3D model. These actions would trigger the optimization software to compute a new solution, which is then visualized and compared to the previous one.

Acknowledgments. This research was funded by Woodside Energy Ltd. We thank all our Woodside collaborators, particularly Solomon Faka, for the many useful discussions, as well as for the enlightening visit to their LNG plant.

References

1. Beldiceanu, N., Carlsson, M., Demassey, S., Petit, T.: Global constraint catalogue: past, present and future. Constraints **12**(1), 21–62 (2007)
2. Belov, G., Stuckey, P.J., Tack, G., Wallace, M.: Improved linearization of constraint programming models. In: Rueher, M. (ed.) CP 2016. LNCS, vol. 9892, pp. 49–65. Springer, Cham (2016). doi:10.1007/978-3-319-44953-1_4
3. Chu, G., Stuckey, P.J., Schutt, A., Ehlers, T., Gange, G., Francis, K.: Chuffed – a lazy clause solver (2017). https://github.com/chuffed/chuffed. Accessed 23 Mar 2017
4. Chu, G.G.: Improving combinatorial optimization. Ph.D. thesis (2011)
5. M.W. Kellogg Company: Design of Piping Systems. Wiley series in Chemical Engineering. Wiley, Hoboken (1956)
6. de Berg, M., van Kreveld, M., Nilsson, B.J., Overmars, M.: Shortest path queries in rectilinear worlds. Int. J. Comput. Geom. Appl. **02**(03), 287–309 (1992)
7. Furuholmen, M., Glette, K., Hovin, M., Torresen, J.: Evolutionary approaches to the three-dimensional multi-pipe routing problem: a comparative study using direct encodings. In: Cowling, P., Merz, P. (eds.) EvoCOP 2010. LNCS, vol. 6022, pp. 71–82. Springer, Heidelberg (2010). doi:10.1007/978-3-642-12139-5_7
8. Google: Google optimization tools (2017). https://developers.google.com/optimization/
9. Guirardello, R., Swaney, R.E.: Optimization of process plant layout with pipe routing. Comput. Chem. Eng. **30**(1), 99–114 (2005)
10. Gurobi Optimization, Inc.: Gurobi Optimizer Reference Manual Version 7.0. Houston. Gurobi Optimization, Texas (2016)
11. IBM: IBM ILOG CPLEX Optimization Studio. CPLEX User's Manual (2017)
12. Jiang, W.-Y., Lin, Y., Chen, M., Yu, Y.-Y.: A co-evolutionary improved multi-ant colony optimization for ship multiple and branch pipe route design. Ocean Eng. **102**, 63–70 (2015)
13. Nethercote, N., Stuckey, P.J., Becket, R., Brand, S., Duck, G.J., Tack, G.: MiniZinc: towards a standard CP modelling language. In: Bessière, C. (ed.) CP 2007. LNCS, vol. 4741, pp. 529–543. Springer, Heidelberg (2007). doi:10.1007/978-3-540-74970-7_38
14. Padberg, M.: Packing small boxes into a big box. Math. Methods Oper. Res. **52**(1), 1–21 (2000)
15. Sakti, A., Zeidner, L., Hadzic, T., Rock, B.S., Quartarone, G.: Constraint programming approach for spatial packaging problem. In: Quimper, C.-G. (ed.) CPAIOR 2016. LNCS, vol. 9676, pp. 319–328. Springer, Cham (2016). doi:10.1007/978-3-319-33954-2_23
16. Schulte, C., Tack, G., Lagerkvist, M.Z.: Modeling and programming with Gecode (2017). www.gecode.org
17. Simonis, H., O'Sullivan, B.: Search strategies for rectangle packing. In: Stuckey, P.J. (ed.) CP 2008. LNCS, vol. 5202, pp. 52–66. Springer, Heidelberg (2008). doi:10.1007/978-3-540-85958-1_4
18. Stuckey, P.J., Becket, R., Fischer, J.: Philosophy of the MiniZinc challenge. Constraints **15**(3), 307–316 (2010)

19. Xu, G., Papageorgiou, L.G.: A construction-based approach to process plant layout using mixed-integer optimization. Ind. Eng. Chem. Res. **46**(1), 351–358 (2007)
20. Xu, G., Papageorgiou, L.G.: Process plant layout using an improvement-type algorithm. Chem. Eng. Res. Des. **87**(6), 780–788 (2009)
21. Zhu, D., Latombe, J.C.: Pipe routing-path planning (with many constraints). In: Proceedings of 1991 IEEE International Conference on Robotics and Automation, vol. 3, pp. 1940–1947 (1991)

Optimal Torpedo Scheduling

Adrian Goldwaser[1,3(✉)] and Andreas Schutt[2,3(✉)]

[1] The University of New South Wales, Sydney, Australia
adrian.goldwaser@gmail.com
[2] The University of Melbourne, Melbourne, Australia
[3] Decision Sciences, Data61, CSIRO, Melbourne, Australia
andreas.schutt@data61.csiro.au

Abstract. We consider the torpedo scheduling problem in steel production, which is concerned with the transport of hot metal from a blast furnace to an oxygen converter. A schedule must satisfy, amongst other considerations, resource capacity constraints along the path, the locations traversed and the sulfur level of the hot metal. The goal is first to minimize the number of torpedo cars used during the planning horizon and second to minimize the time spent desulfurizing the hot metal. We propose an exact solution method based on Logic-based Benders Decomposition using Mixed-Integer and Constraint Programming, which optimally solves and proves, for the first time, the optimality of all instances from the ACP Challenge 2016 within 20 min. In addition, we adapted our method to handle large-scale instances. This adaptation optimally solved all challenge instances within one minute and was able to solve instances of up to 100,000 hot metal pickups.

1 Introduction

Steel production is a complex process of sequential stages from raw materials to a final product in the form of, *e.g.*, wire plate coils. In the first stage, the iron making, raw materials are melted in a blast furnace. In the second stage, the steel making, the hot metal is loaded in torpedo cars, or *torpedoes*, transported to different locations for improving its quality, and finally brought to an oxygen converter, in which it is poured. Once at the oxygen converter, the hot metal is further refined before the last two stages of continuous casting and hot trip mill. This work focuses on the rotation of the torpedoes between the blast furnace and oxygen converter in the steel making stage. At the steel making area, there are a number of blast furnaces producing hot metal of different qualities. At certain times or *events*, the hot metal in the blast furnace has to be loaded into a torpedo. Then the torpedo moves on a rail network to different locations for improving the quality of the hot metal if needed. After that, the hot metal is transported to the oxygen converter and poured into it at a pre-defined event time. Now, the empty torpedo is available for the next pick up of hot metal.

We study the torpedo scheduling problem that was proposed by Schaus et al. [12] for the ACP Challenge 2016. This problem focuses on the assignment

© Springer International Publishing AG 2017
J.C. Beck (Ed.): CP 2017, LNCS 10416, pp. 338–353, 2017.
DOI: 10.1007/978-3-319-66158-2_22

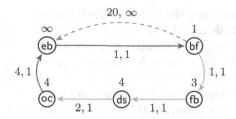

Fig. 1. The graph for Example 1.

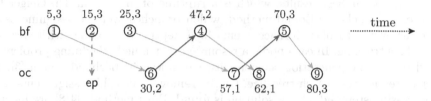

Fig. 2. A small example of a torpedo scheduling problem.

of blast furnace events to oxygen converter events and the scheduling problem of transporting the hot metal through different locations while satisfying all scheduling constraints and the quality constraint, sulfurization level, on the hot metal. Figure 1 shows the rail network considered. There are five different locations: blast furnace (bf), full buffer (fb), desulfurization station (ds), oxygen converter (oc), and empty buffer (eb). The full and empty buffers are waiting areas for full and empty torpedoes, whereas at the desulfurization station the sulfur level of the hot metal can be reduced by chemical processes. Each location has a torpedo capacity, which is shown above the node. Each link or edge has a minimal transition time and a torpedo capacity, which is shown next to the edge in the same order. The dashed edge from bf to eb represents the emergency pit, in which hot metal can be dumped if required. The objective is a lexicographical one, first to minimize the number of torpedoes and second to minimize the total time spent at the desulfurization station by torpedoes.

Example 1. Figure 2 shows a small problem with five blast furnace $1, 2, \ldots, 5$ and four converter events $6, 7, 8, 9$. Each event is specified by its due date and (maximal) sulfur level given above or below its node. We assume that the loading time at the blast furnace, the unloading time at the oxygen converter, and the time to desulfurize the hot metal by one level are 5 time units. The transition times between different locations and the torpedo capacities are shown in Fig. 1, *e.g.*, the blast furnace bf has torpedo capacity 1 and the emergency trip (dashed line) a transition time of 20 and no capacity limit. A solution is depicted by the arrows between the events, which shows the usage of three torpedoes. The first torpedo serves the events 1, 6, 4, 8, the second one only 2, and the third one 3, 7, 5, 9 in this order. For fulfilling the demands for oxygen converter events 6, 7, and 8, a total of 20 time units have to be spent for desulfurization.   □

To best of our knowledge, the torpedo scheduling problem we study was proposed at the ACP Challenge 2016. From ten teams, who took part, we are only aware of the publication of the winning team [8] and the third placed team [4]. Kletzander and Musliu [8] propose a two-stage simulated annealing approach. The first stage minimizes the number of torpedoes by tracking the maximal number of torpedoes simultaneously used at any one time, whereas the second stage minimizes the desulfurization time. They relax some constraints, but add penalty terms to their objective. One iteration of the method takes between four to ten minutes time for the ACP challenge instances. They run 50 iterations to get their best results, which is a runtime of more than 3 h. Geiger [4] proposes a Branch-and-Bound method, which branches over the assignments of converter events to blast furnace events in a depth first manner with chronological backtracking. In each node, a resource-constrained scheduling problem is solved by a serial generation scheme with variable neighborhood search [3]. In order to reduce the search tree size, Geiger removes infeasible assignments in a preprocessing step and after a solution is found. Both methods [4,8] are incomplete and thus cannot prove optimality of an instance, unless the lower bound of the objective is the optimal value.

Different aspects on the torpedo scheduling problem have been studied in the literature: the routing of torpedoes through the rail network while minimizing the transportation time of the hot metal [2,7,10], the molten iron allocation problem [13], the molten scheduling problem [6,9], and the locomotive scheduling problem [14]. All those works use different solution methods such as local search, mixed integer programming, and column generation, but none use Logic-based Benders Decomposition [5] and Constraint Programming (CP) as we do in the present paper.

We propose a Logic-Based Benders Decomposition [5] method, in which the assignment problem and the lexicographical objective is handled in the master problem. The remaining scheduling is partitioned, if possible, and solved minimizing the desulfurization time. The master problem is solved by a Mixed Integer Programming (MIP) solver whereas the scheduling problems by CP solver with Nogood Learning applying Lazy Clause Generation (LCG) [11]. In preprocessing, we simplify the problem by removing symmetries. To the best of our knowledge, our method is the first published complete method for the torpedo scheduling and proves the optimality of found solutions of the simulated annealing approach [8] for all ACP challenge instances, in an even shorter runtime. We modified our method for handling large scale problems, but with the price of losing optimality. The modified method improved the runtime by orders of magnitude and was able to solve instances with 100,000 events in 70 min.

2   Torpedo Scheduling

The torpedo scheduling problem consists of a set of *blast furnace events* $N = \{1, 2, \ldots, n\}$, a set of *(oxygen) converter events* $M = \{n + 1, n + 2, \ldots, n + m\}$, and a set of *locations* $L = \{\mathsf{bf}, \mathsf{fb}, \mathsf{ds}, \mathsf{oc}, \mathsf{eb}\}$ in the production plant. In addition, the *torpedo graph* $G = (L, P)$ is a directed graph which specifies the two possible traversals of the torpedoes through the plant. The *oxygen converter trip* delivers the hot metal to the converter and visits the locations in this order eb, bf, fb, ds, oc, and eb, whereas the *emergency pit trip* dumps the hot metal at the emergency pit and visits the locations in this order eb, bf, and eb. Thus, $P = \{(\text{eb}, \text{bf}), (\text{bf}, \text{eb}), (\text{bf}, \text{fb}), (\text{fb}, \text{ds}), (\text{ds}, \text{oc}), (\text{oc}, \text{eb})\}$.

Each location $l \in L$ has a *torpedo capacity* cap_l where $cap_{\text{bf}} = 1$ and $cap_{\text{ep}} = \infty$. We extend this notation for edges $p \in P$, which is cap_p. All edges $p \in P \setminus \{(\text{bf}, \text{eb})\}$ have a unit capacity $cap_p = 1$, whereas $cap_{(\text{bf},\text{eb})} = \infty$. A torpedo traversing the edge p requires a minimal *transition time* of tt_p.

Each blast furnace event $i \in N$ is characterized by a *due date*, $dd_i^{\text{bf}} \in \mathbb{N}^0$, at which hot metal is picked up by exactly one empty torpedo, and a *sulfur level*, $sul_i^{\text{bf}} \in \{1, 2, \dots, 5\}$, of the hot metal. Each oxygen converter event $j \in M$ has a *due date*, $dd_j^{\text{oc}} \in \mathbb{N}^0$, at which hot metal from exactly one full torpedo is poured into the converter, and a *maximal sulfur level*, $sul_j^{\text{oc}} \in \{1, 2, \dots, 5\}$, of the hot metal. Loading of a torpedo takes $dur^{\text{bf}} \in \mathbb{N}$ time periods at the blast furnace, while unloading takes $dur^{\text{oc}} \in \mathbb{N}$ time periods at the oxygen converter. Reducing the sulfur level of hot metal by one unit requires $dur^{\text{ds}} \in \mathbb{N}$ time periods at the desulfurization station.

Following [8], a *torpedo run* i is either a converter or emergency pit trip. In the former case, it is specified by variable departure times dep_i^l, variable arrival times arr_i^l for locations in $\{\text{eb}, \text{bf}, \text{fb}, \text{ds}, \text{oc}\}$, and the variable converter event $oc_i \in M$, that it serves. For the latter case, it is specified by the variable departure and variable arrival times for only the locations eb and bf. We denote by ep_i whether it is a converter trip $ep_i = 0$ or an emergency pit trip $ep_i = 1$.

Definition 1 (Torpedo Scheduling Problem). *A torpedo scheduling problem consists of a triplet $(N, M, G = (L, P))$. A solution $S = (1, 2, \dots, n)$ is a vector of n torpedo runs, in which the i-th run picks up the hot metal of the i-th blast furnace event, matches the blast furnace event to an oxygen converter event or an emergency pit trip, and assigns all corresponding arrival and departure times. A solution satisfies the capacity constraints on each location (1) and on each edge (2),*

$$\sum_{i \in S: arr_i^l \leq t < dep_i^l} 1 \leq cap_l \qquad \forall l \in L, \forall t \in \mathbb{N}^0 \qquad (1)$$

$$\sum_{i \in S: dep_i^l \leq t < arr_i^k} 1 \leq cap_{(l,k)} \qquad \forall (l, k) \in P, \forall t \in \mathbb{N}^0 \qquad (2)$$

the minimal transition times for oxygen converter (3) and emergency pit trips (4),

$$arr_i^k - dep_i^l \geq tt_{(l,k)} \qquad \forall i \in S : ep_i = 0, \forall (l, k) \in P \setminus \{(\textbf{bf}, \textbf{eb})\} \qquad (3)$$

$$arr_i^k - dep_i^l \geq tt_{(l,k)} \qquad \forall i \in S : ep_i = 1, \forall (l, k) \in \{(\textbf{eb}, \textbf{bf}), (\textbf{bf}, \textbf{eb})\} \qquad (4)$$

the loading constraints at the blast furnace (5), the unloading constraints (6), and the maximal sulfurization level (7) at the oxygen converter.

$$arr_i^{bf} \leq dd_i^{bf} \quad \wedge \quad dd_i^{bf} + dur^{bf} \leq dep_i^{bf} \qquad \forall i \in S \qquad (5)$$

$$arr_i^{oc} \leq dd_{oc_i}^{oc} \quad \wedge \quad dd_{oc_i}^{oc} + dur^{oc} \leq dep_i^{oc} \qquad \forall i \in S : ep_i = 0 \qquad (6)$$

$$sul_i - \left\lceil \frac{dep_i^{ds} - arr_i^{ds}}{dur^{ds}} \right\rceil \leq sul_{oc_i} \qquad \forall i \in S : ep_i = 0 \qquad (7)$$

All torpedoes, which are identical, are located at **eb** at time 0. Here, we are interested in a solution that minimizes two objective functions in lexicographic order. The primary objective (8) is to minimize the number of torpedoes used, which can be stated as minimizing the maximal number of "active" torpedo runs at any time [4, 8]. The secondary objective (9) is to minimize the total time spent at the desulfurization station.

$$\min \max_{t \in \mathbb{N}^0} |\{i \in S \mid dep_i^{eb} \leq t \wedge t < arr_i^{eb}\}| \qquad (8)$$

$$\min \sum_{i \in S : ep_i = 0} dep_i^{ds} - arr_i^{ds} \qquad (9)$$

Note that the solution does not provide an assignment of individual torpedoes to the torpedo runs. But such an assignment can be computed in polynomial time with respect to the number of torpedo runs using a stack for torpedoes in the empty buffer and a return queue of torpedoes sorted ascending by their arrival (return) times to the empty buffer. The algorithm would iterate over due dates of blast furnace events and the arrival times in the return queue in chronological order. Depending on the case, it either pops a torpedo from the stack and pushes it into the return queue or vice versa.

Moreover, as already observed in [4,8] the possible oxygen event matches for a blast furnace event can be reduced by simply calculating the minimal travel time including a minimal time for desulfurization from the blast furnace to the oxygen converter. We denote $X = \{(b, o) \in N \times M \mid dd_b^{bf} + tt_{(bf,fb)} + tt_{(fb,ds)} + tt_{(ds,oc)} + dur^{ds} \cdot \max(0, sul_b - sul_o) \leq dd_o^{oc}\}$ the set of possible matchings of blast furnace to oxygen converter events. In addition, they also observed that there is no reason to delay a departure of a torpedo from the blast furnace in the case of an emergency trip due to the uncapacitated path (bf, eb) and empty buffer. Thus, we can fix $dep_i^{bf} = dd_i^{bf} + dur^{bf}$ and $arr_i^{eb} = dep_i^{bf} + tt_{(bf,eb)}$ if the torpedo run i goes to the emergency pit.

Example 2. Given the example from Example 1. Then, $X = \{(1,6), (1,7), (1,8), (1,9), (2,7), (2,8), (2,9), (3,7), (3,8), (3,9), (4,8), (4,9), (5,9)\}$ and the departure times at bf respectively are 10, 20, 30, 52, and 75 for events 1, 2, 3, 4, and 5 if they go to the emergency pit. Note that only bf event 1 can deliver hot metal for event 6, we leave such simple reductions to the solver.

3   Preprocessing

Before solving the problem, we perform preprocessing steps in order to simplify the problem and setup the structure needed for our solution approach.

Algorithm 1. Computation of departure times from the oxygen converter.

Input: M an array of m oxygen converter events sorted in chronological order.

1 $j := M[1]$; $depOC[j] := dd_j^{oc} + dur^{oc}$; $arrEB[j] := depOC[j] + tt_{(oc,eb)}$;
2 **for** $jj := 2$ **to** m **do**
3      $j := M[jj]$;
4      $depOC[j] := \max(arrEB[M[jj-1]], dd_j^{oc} + dur^{oc})$;
5      $arrEB[j] := depOC[j] + tt_{(oc,eb)}$;

3.1 Departure Times from the Oxygen Converter

The empty buffer has unlimited capacity, this means that is it never suboptimal to get an empty torpedo there earlier rather than later as it can be reused earlier, it frees space at the oxygen converter earlier, and clears the path from the oxygen converter to the empty buffer earlier. Thus, an empty torpedo should leave the oxygen converter as early as possible, which is the latest time of the completion unloading the torpedo, *i.e.*, $dd_i^{oc} + dur^{oc}$, and the arrival time of the previous torpedo at the empty buffer from the oxygen converter, *i.e.*, arr_j^{eb}.

Since the due dates for the oxygen converter events are known a priori, the departure dates from the oxygen converter and the arrival times to the empty buffer can be computed in linear time with the respect to the number of those events, if the events are given in chronological order, as shown in Algorithm 1. Note that the order of torpedoes serving oxygen converter events remains unchanged by the algorithm. It is obvious that the following holds.

Proposition 1. *Algorithm 1 computes the earliest departure times for each oxygen converter event without changing the order of their corresponding earliest arrival times at the empty buffer and without creating an overload on the path between both locations.*

Example 3. Given the example from Example 1 from page 2. Then Algorithm 1 respectively computes departure times $35, 62, 67$, and 85 for the oxygen converter events $6, 7, 8$, and 9. □

3.2 Arrival Times at the Blast Furnace

A similar observation to the departure times at the oxygen converter can be seen for the arrival times at the blast furnace. Since the empty buffer is uncapacitated and the hot metal cannot be picked up before its due date, it is never suboptimal to get an empty torpedo there later than rather earlier.

Algorithm 2 is symmetric to Algorithm 1 for the arrival times at the blast furnace. It computes the times in reverse-chronological order of the blast furnace events. With similar arguments as in the oxygen converter case, the following claim holds.

Proposition 2. *Algorithm 2 computes the latest arrival time for each blast furnace event and their latest departure time from the empty buffer without creating an overload on the path between both locations.*

Algorithm 2. Computation of arrival times at the blast furnace.

Input: N an array of n blast furnace events sorted in chronological order.

1   $i := N[n]$; $arrBF[i] = dd_i^{bf}$; $depEB[i] := arrBF[i] - tt_{(eb,bf)}$;
2   **for** $ii := n - 1$ **down to** 1 **do**
3      $i := N[ii]$;
4      $arrBF[i] := \min(depEB[N[ii + 1]], dd_i^{bf})$;
5      $depEB[i] := arrBF[i] - tt_{(eb,bf)}$;

Algorithm 3. Computation of a backward matching.

Input: N an array of n blast furnace events sorted in chronological order.

Input: M an array of m oxygen converter events sorted in chronological order.

1   $dep := $ Alg. $1(M)$;
2   **for** $o = 1$ **to** m **do** $bm[o] := \infty$;
3   $bb := 1$; $oo := 1$;
4   **while** $bb \leq n$ **and** $oo \leq m$ **do**
5      $b := N[bb]$; $o := M[oo]$;
6   **if** $dep[o] + tt_{(oc,eb)} + tt_{(eb,bf)} \leq dd_b^{bf}$ **then** $bm[o] := b$; $bb{+}{+}$; $oo{+}{+}$;
7   **else** $bb{+}{+}$;

Note that for torpedo runs using the emergency pit, we can now fix its remaining departure and arrival times. Thus, we only have to decide which run is an emergency pit trip.

Example 4. Given the example from Example 1 from page 2. Then Algorithm 2 respectively computes arrival times 4, 14, 24, 46, and 69 for the events 1, 2, 3, 4, and 5. □

Since the blast furnace and oxygen converter events are independent of each other, hence it follows that an optimal solution exists, which has the same arrival and departure times for the corresponding events as computed in Algorithms 1 and 2. Thus, fixing the corresponding variables to those times removes symmetries from the problem.

3.3   Backward Matching

We introduce the concept of *backward matches*, i.e., matches from oxygen converter events to blast furnace events. The meaning of such a match is that a torpedo fulfilling the demand for the oxygen converter event $o \in M$ is used to serve the request for the blast furnace event $b \in N$. In other words, the torpedo used for o is *reused* for b.

Since the torpedoes are identical and each blast furnace event requires exactly one torpedo, it does not matter which empty torpedo serves the event if more than one can be at the blast furnace in time. Algorithm 3 computes a backward matching in linear time with respect to the number of blast furnace events, when these events and the oxygen converter events are already sorted. Let $bm : M \rightarrow$

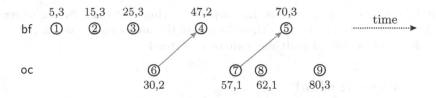

Fig. 3. The backward matching for Example 1.

$N \cup \{\infty\}$ denote the backward matching returned by Algorithm 3. Note that some of the last oxygen converter events can not be matched with any blast furnace event. We represent this case by a match to ∞.

Example 5. Given the example from Example 1 from page 2. Then Algorithm 3 computes the backward matching as shown by the arrows in Fig. 3, in which events 8 and 9 do not get a match.   □

Theorem 1. *Let* (N, M, G) *be a torpedo scheduling problem. Then there exists an optimal solution* S *using the backward matching computed by Algorithm 3 for the reuse of torpedoes.*

Proof. Let S' be an optimal solution. We can assume that S' uses the departure times at the oxygen converter computed by Algorithm 1. We will construct a solution S by swapping torpedoes in S'. Consider the first blast furnace event b_1 which uses a torpedo t_1 other than the assigned one t_2 in the backward matching **bm**. Without loss of generality, we assume that b_2 is the next blast furnace event that t_2 serves. Since b_1 is the earliest event that t_2 can serve after finishing its oxygen converter run, it holds $dd_{b_1}^{\text{bf}} \leq dd_{b_2}^{\text{bf}}$. Now, we distinguish regarding the origin of torpedo t_1. If the torpedo t_1 was never used before or returns from an emergency pit run then, clearly, we can swap the torpedoes for b_1 and b_2. If the torpedo t_1 returned from an oxygen converter trip then the departure time from the oxygen converter must be later than for t_2, otherwise S' would deviate earlier from the backward matching. Since Algorithm 3 matched t_2 with the earliest possible blast furnace event, it holds that $dd_{b_1}^{\text{bf}} \leq dd_{b_2}^{\text{bf}}$. Therefore, the torpedoes can be swapped.   □

Given a backward matching, it divides the blast furnace events into the set of *matched* events, i.e., $V = \text{bm}(M) \setminus \{\infty\}$, and the set of *unmatched* events, i.e., $U = N \setminus \text{bm}(M)$. Our solution method presented in the next section will extend this matching by matching torpedoes used for an emergency pit trip to unmatched events. As all departure and arrival times are known in the case of those trips, we reduce possible matchings to $R = \{(i, j) \in N \times U \mid arr_i^{\text{bf}} + dur^{\text{bf}} + tt_{(\text{bf,eb})} + tt_{(\text{eb,bf})} \leq arr_j^{\text{bf}}\}$.

Example 6. Given the example from Example 5. Then, $V = \{4, 5\}$ and $U = \{1, 2, 3\}$. The time cost for an emergency trip is from bf (including loading) back to it is $5 + 20 + 1 = 26$. Thus, no torpedo serving any blast furnace events would be able to return to bf in time for one of the unmatched one, *i.e.*, $R = \emptyset$. Therefore, the backward matching cannot be extended.

4   Solution Method

At first, we preprocess an instance for determining the various arrival and departure times, and the backward matching bm as described in the previous section. After that, we start the Benders decomposition, which alternates between solving the master and scheduling problems until an optimal solution is found. The master problem is formulated as a MIP, in which each oxygen converter event is assigned to a torpedo run, unmatched blast furnace events are matched with emergency pit trips, and the lexicographic objective of the problem is minimized. Then the remaining scheduling problem is split into smaller sub-problems using the optimal matching from the MIP solution. Each sub-problem is then solved as a constraint optimization problem minimizing the total time spent at the desulfurization station. If all sub-problems are feasible and the total time spent at the desulfurization station equals the corresponding lower bound in the MIP solution then we have found a globally optimal solution. If some sub-problems are not feasible, we compute minimal Benders cuts, add them to the MIP problem, and re-optimize the MIP. If some sub-problems require extra desulfurization time, we add optimality cuts, which forces the objective to take into account the extra desulfurization time, and re-optimize the MIP.[1] The optimality cuts can also make the MIP problem infeasible. In this case, it proves that the last found solution was the optimal one.

4.1   MIP Model

The MIP model tries to find the mapping of blast furnace events to converter events and reuse of torpedoes after emergency trips such that the number of torpedoes is minimized and for the minimal number of torpedoes, the lower bound on the desulfurization time is minimized.

Contrary to [4,8], the idea of counting the number of torpedoes used is not based on how many torpedoes are doing an emergency pit or an oxygen converter trip at the same time, but rather to model it via the reuse of torpedoes. The backward matching bm already provides the reuse of torpedoes used for oxygen converter trips. Solving the MIP model just extends this backward matching for torpedoes used for emergency trips.

Besides the binary variables ep_i from the torpedo run, the MIP model uses the following binary variables. For each $(i, o) \in X$, we create a variable $x_{io} \in \{0, 1\}$ expressing whether the torpedo run i serves the demand of the oxygen converter

---

[1] Note that this case never occurred for generated instances and was only tested on handcrafted instances.

event o. For each $(i, j) \in R$, the variables $r_{ij} \in \{0, 1\}$ models whether the torpedo from torpedo run i is reused for the blast furnace event j.

$$\min \quad dur^{ds} \cdot n \cdot obj_1 + obj_2 \tag{10}$$

$$\text{s.t.} \quad obj_1 = |U| - \sum_{(i,j) \in R} r_{ij} \tag{11}$$

$$obj_2 = \sum_{(i,o) \in X} x_{io} \cdot \max(0, sul_i - sul_o) \cdot dur^{ds} \tag{12}$$

$$\sum_{(i,o) \in X} x_{io} = 1 \qquad \forall o \in M \tag{13}$$

$$ep_i + \sum_{(i,o) \in X} x_{io} = 1 \qquad \forall i \in N \tag{14}$$

$$\sum_{i \in N} ep_i = N - M \tag{15}$$

$$\sum_{(i,j) \in R} r_{ij} \le ep_i \qquad \forall i \in N \tag{16}$$

$$\sum_{(i,j) \in R} r_{ij} \le 1 \qquad \forall j \in U \tag{17}$$

Constraint (10) states the objective of the MIP, which is split into two parts. The first part (11) models the minimization of the number of torpedoes, by maximizing the number of reused torpedoes for unmatched blast furnace events. We scale this objective by the product of number of blast furnace events and the duration for desulfurizing the hot metal by one sulfur level in order to account for the lexicographic problem objective. Constraint (13) ensures each oxygen converter event is matched by one blast furnace event, whereas (14) matches each blast furnace event to an oxygen converter event or emergency trip. Constraint (15) ensures that there are the right number of emergency pit trips. Constraint (16) models that a torpedo used for an emergency trip can be reused for an unmatched blast furnace event, whereas (17) ensures that at most one torpedo is reused for each unmatched blast furnace event. Note that the reuse of torpedoes for an oxygen converter trip is already determined by the backward matching **bm**, and thus can be left out of the model.

A MIP solution provides not only the matching of torpedo runs to oxygen converter events and the matching for the reuse of torpedoes, but also a lower bound on the desulfurization time, which is used as a quality measurement for the scheduling solution.

4.2   The CP Model

Once, we have a mapping of blast furnace to oxygen converter events, we can split the remaining scheduling problem into several smaller ones. The idea is to split the problem at those blast furnace events serving an oxygen converter event, that do not interfere with any previous torpedo runs serving oxygen converter events. Algorithm 4 computes all sub-problems.

Example 7. Given the example from Example 1 from page 2. Algorithm 4 will split the problem into three sub-problems as depicted in Fig. 4.

Algorithm 4. Computation of the sub-problems.

Input: N an array of n blast furnace events sorted in non-decreasing order of
        the due dates.
Input: oc a mapping from blast furnace events to oxygen converter events
        or ∞.

1 $latestDepDS := -\infty;\ A := \emptyset;\ B := \emptyset;$
2 **for** $ii := 2$ **to** n **do**
3   **if** $oc(N[ii]) = \infty$ **then continue** ;
4      $i := N[ii];\ earliestArrDS := dd_i^{\text{bf}} + tt_{(\text{bf},\text{fb})} + tt_{(\text{fb},\text{ds})};$
5   **if** $latestDepDS \leq earliestArrDS$ **then**
6          $B := B \cup \{A\};\ A := \{i\};$
7   **else** $A := A \cup \{i\};$
8      $latestDepDS := \max(latestDepDS, dd_{oc(i)}^{\text{oc}} - tt_{(\text{ds},\text{oc})});$
9 **return** $B;$

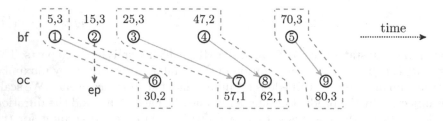

Fig. 4. Partition of the scheduling given the shown matching.

Each sub-problem is then modeled as a constraint optimization problem using
the same model, but restricted to torpedo runs in the sub-problem, as in Definition 1 on page 4 except for the torpedo capacity constraints and an additional
constraint enforcing an upper bound on the departure times at the blast furnace
for avoiding an overload (18).

$$\min \quad \sum_{i \in S'} dep_i^{\text{ds}} - arr_i^{\text{ds}}$$

$$s.t. \quad (3\text{--}7)$$

$$dep_i^{\text{bf}} \leq arr_{i+1}^{\text{bf}} \qquad\qquad \forall i \in S' \setminus \{n\} \quad (18)$$

$$\text{disjunctive}((dep_i^l)_{i \in S'}, (arr_i^k - dep_i^l)_{i \in S'}) \qquad \forall (k,l) \in P'$$

$$\text{cumulative}((arr_i^l)_{i \in S'}, (dep_i^l - arr_i^l)_{i \in S'}, (1)_{i \in S'}, cap_l) \quad \forall l \in L \setminus \{\text{bf}\}$$

where S' are the torpedo runs of the sub-problem, $P' = \{(\text{bf},\text{fb}), (\text{fb},\text{ds}),$
$(\text{ds},\text{oc})\}$, and disjunctive and cumulative are global constraints modeling unary
and non-unary resources. Note that for each torpedo run $i \in S$, Algorithms 1
and 2 provide the arrival times at bf and eb, and the departure times at oc
and eb. In the case of an emergency trip, we also know dep_i^{bf}. Due to these pre-
assigned times, the CP model only has to take care of oxygen converter trips
from the blast furnace to the converter.

We employ a sequential search over sub-searches, which represent a location or a path. Each sub-search branches over the duration of the torpedoes i used for the location l or the path (k, q), $i.e.$, $dep_i^l - arr_i^l$ and $arr_i^q - dep_i^k$. The most constrained duration variable is selected first and its smallest possible duration is assigned to it. The sub-searches are explored in this order ds, (fb, ds), (ds, oc), oc, (bf, fb), and bf. There are two important ingredients for this search. First, the first sub-search is objective driven, because it tries to minimize the duration spent at ds. Second, branching on the durations rather than on the departure or arrival times keeps the schedule flexible while providing some propagation on the departure and arrival time variables. Other searches tested, that did not follow both ingredients, were inferior.

Note that in order to avoid resolving sub-problems from scratch, we cached all sub-problems and their solution in a hash map.

4.3   Benders Cuts

The scheduling problem can have three possible outcomes. First, it is infeasible. Second, it is schedulable, but not with the lower bound on the desulfurization time from the MIP solution. Last, it is schedulable with the same desulfurization time. Only in the first two cases do we need to create Benders cuts in terms of the decision variables in the master problem. In the third case, the combined MIP and CP solution is an optimal solution of the entire problem.

We express the cuts in terms of the variables x_{io} from the MIP problem. Let oc be the mapping from the MIP restricted to the sub-problem and N' the blast furnace events in the sub-problem.

Infeasibility Cuts. The sub-problem is infeasible, which is a direct result of the mapping. Thus, $\sum_{i \in N'} x_{ioc(i)} < |N'|$ is valid cut, because it forces the MIP solver to choose a different oxygen converter event for at least one torpedo run.

To strengthen the cut, we rerun $|N'|$-times the CP model, but with a small modification. For each rerun, we remove one torpedo run including the matched oxygen converter event from the model. If the model is still infeasible then this run does not contribute to the infeasibility and we can leave it out; otherwise it contributes to the infeasibility and we reinsert it. The removals are performed in chronological order of the blast furnace events. At the end of the process, we obtain a minimal unsatisfiable set of torpedo runs $N'' \subseteq N'$ leading to the stronger cut $\sum_{i \in N''} x_{ioc(i)} < |N''|$, which is minimal too. In preliminary testing, this minimization resulted in an order of magnitude less MIP iterations.

We also investigated more general cuts by relaxing the conditions on the start time of a torpedo trip instead of removing it completely, but they were not beneficial for the overall runtime.

Optimality Cuts. The sub-problem is schedulable with minimal desulfurization time β, but the desulfurization time α from the MIP solution is smaller, $i.e.$, $\alpha < \beta$. In this case, we introduce a new binary variable b for the MIP model, add the term

$(\beta - \alpha) \cdot b$ to the objective (10), and add the constraint $\sum_{i \in N'} x_{ioc(i)} - (|N'| - 1) \leq b$ to the MIP model. The variable b takes value 1 if and only if the MIP uses the same mapping oc for the sub-problem. In that case, the added objective term accounts for the difference in the desulfurization time derived by the CP model. If the variable b takes value 0 then the MIP model is forced to take a different mapping due to the added constraint and the added objective term is zero.

4.4   Limited Forward Matchings

The size of the MIP model, *i.e.*, the number of constraints, variables, and the size of constraints, depends on the number of blast furnace and oxygen converter events. For example, the objective (12) has a quadratic size of $\mathcal{O}(nm)$. For large problems, the MIP model is so large that the MIP solver runs out of memory or is extremely slow. A way to reduce the size is to limit the oxygen converter events to which a blast furnace event can be matched, for example to the 10 next closest oxygen converter events which it can reach. The same is also done for the reuse of torpedoes after emergency trip events. Not only does this drastically shrink the model size, but also significantly speeds up the solving time. The drawback is that we cannot prove the optimality of the original problem and the optimal solution of this relaxed problem can be worse than the one from the original problem. However, from a practical point of view, it might be the preferred mode because a matching of a blast furnace to an oxygen converter event far in the future can be seen as not preferable or sub-optimal due to cooling of the molten metal. In the experiments, we show the sweet spot for the number of forward matchings.

5   Experiments

We conducted experiments on the ACP 2016 Challenge instances and created larger ones using the instance generator provided at the ACP Challenge website. All generated solutions were checked using the provided ACP solution checker. We grouped all instances in the test sets **small** having 15 instances with 30 to 500 blast furnace events, **comp** having 6 ACP challenge instances with 850 to 2500 blast furnace events, **medium** having 19 instances with 1000 to 3000 blast furnace events, and **large** having 3 instances with 10000 blast furnace events. All instances are available at https://github.com/AdGold/TorpedoSchedulingInstances. We ran all our experiments on a machine with an Intel(R) Core(TM) i7-5500U CPU at 2.40 GHz and 8 GB RAM unless otherwise stated. The solution was implemented in Python 3.5.2 interfacing Gurobi 7.0.1 using the Python library gurobipy. Gurobi was used for solving the MIP problem and Chuffed [1] for solving the CP problem. No runtime limit was imposed.

Unlimited Forward Matchings. Table 1 shows the results on the set **comp** for each instance. We list the number of torpedoes (#T), the desulfurization time spent at ds (Desulf), the total runtime (RT), the percentage of the total runtime that was used by the MIP solver (MT), the number of iterations (#I), the cache

Table 1. Detailed results on **comp**.

Inst	#T	Desulf	RT	MT	#I	CHR	#SP	SSR	S1	SAve	SMax
instance01	4	7695	41 s	88%	1	0%	10	100%	70%	264	316
instance02	4	5302	119 s	88%	1	0%	43	100%	67%	98	774
instance03	3	27150	415 s	97%	1	0%	583	100%	84%	17	123
instance04	3	10676	35 s	92%	1	0%	839	100%	84%	2	4
instance05	4	16308	575 s	97%	3	50%	1074	99%	83%	14	410
instance06	4	7755	1134 s	97%	2	48%	34	98%	62%	237	755

Table 2. Results on each test set excluding infeasible instances.

Inst	#T	Desulf	RT	MT	#I	CHR	#SP	SSR
small	3.5	362	2 s	51%	1.1	3%	26	99.2%
comp	3.7	12481	386 s	93%	1.5	16%	431	99.7%
medium	4.4	1224	484 s	95%	1.2	4%	170	99.6%
large	4.5	6481	124093 s	99%	2.0	31%	848	99.9%

hit rate for sub-problems (CHR), the number of total sub-problems stores (#SP), the success rate of sub-problems (SSR), *i.e.*, no cuts needed to be generated, and the percentage of sub-problems with size 1 (S1), the average size of sub-problems with size greater than 1 (SAvg), and the maximal size of sub-problems (SMax). All ACP challenge instances were optimally solved in less than 20 min, which is much quicker than the winning method presented in [8]. The results also reveal that the MIP solver used the majority of the runtime and the sub-problems had almost 100% success, which lead to a very low number of iterations. In addition, most sub-problems were small and only a few contained 100 s of blast furnace events. Note that the nogood learning solver Chuffed was essential for quickly solving the sub-problems. In particular on larger and infeasible ones, we had to terminate the process if using Gecode.

Table 2 presents the results on all test sets excluding infeasible instances. The table shows a subset of columns, but each entry is an average over the number of feasible instances. The results show a similar picture to the ACP challenge instances. However, for the large size instances in **large**, the runtime was more than 30 h and we needed to run it on a machine with extra RAM in order to be able to solve the MIP. The machine used was an Intel(R) Xeon(R) CPU E5-2660 at 2.60 GHz with 128 GB RAM which is not practical in most cases.

Limited Forward Matchings. Figures 5, 6 and 7 show the development of the optimality gap on the desulfurization time spent, of the percentage of instances optimally solved, and of the runtime when the limit on the forward matchings increases. Note that Fig. 7 uses logarithmic scale for the y-axis. The optimality gap on the desulfurization time spent converges quickly on each test set. Between

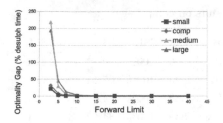

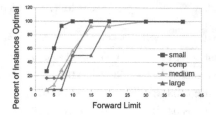

Fig. 5. Optimality gap in desulfurization time.

Fig. 6. Percent of instances solved optimally.

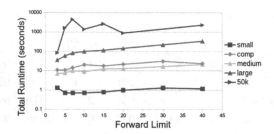

Fig. 7. Total runtime.

a limit of 30 and 40 the last optimal solution was found even on the test set **large**. The runtime could be reduced by orders of magnitude for medium and large scale problems, especially for large scale instances where the runtime was reduced to less than 10 min. All ACP challenge instances were solved in less than one minute, down from 20 min. In order to test the limit of our method, we created three instances with 50,000 and 100,000 blast furnace events, respectively. The average total runtime of the 50 k instances were below 20 min except for a limit of 7 as shown in Fig. 7. Interestingly, the same optimal solutions were generated with limits of at least 10. The 100 k instances were solved between 70 min and 3.5 h for a limit of 20.

6   Conclusion

We propose a logic-based Benders decomposition solution method for the industrial problem of torpedo scheduling in the steel production. The master problem was modeled as a MIP, which takes care of the assignment component of the problem and the lexicographical objective. The remaining scheduling problem was split into smaller sub-problems and solved by a CP solver with nogood learning. This solution method is the first exact one for the torpedo scheduling problem and is the first one, that could prove the optimality of all instances from the ACP 2016 Challenge in less than 20 min. Thus, it outperforms the previous state of the art. A limited version of our method, which cannot guarantee optimality, could reduce the runtime by an order of magnitude and was able to find optimal solutions very quickly for even larger instances that we created.

Acknowledgments. This work was partially supported by the Asian Office of Aerospace Research and Development grant 15-4016.

References

1. Chu, G.G.: Improving combinatorial optimization. Ph.D. thesis, The University of Melbourne (2011). http://hdl.handle.net/11343/36679
2. Deng, M., Inoue, A., Kawakami, S.: Optimal path planning for material and products transfer in steel works using ACO. In: The 2011 International Conference on Advanced Mechatronic Systems, pp. 47–50, August 2011
3. Geiger, M.J.: A multi-threaded local search algorithm and computer implementation for the multi-mode, resource-constrained multi-project scheduling problem. Eur. J. Oper. Res. **256**(3), 729–741 (2017)
4. Geiger, M.J.: Optimale Torpedo-Einsatzplanung – Analyse und Lösung eines Ablaufplanungsproblems der Stahlindustrie. In: Spengler, T., Fichtner, W., Geiger, M.J., Rommelfanger, H., Metzger, O. (eds.) Entscheidungsunterstützung in Theorie und Praxis, pp. 63–86. Springer, Wiesbaden (2017). doi:10.1007/978-3-658-17580-1_4
5. Hooker, J., Ottosson, G.: Logic-based benders decomposition. Math. Program. **96**(1), 33–60 (2003)
6. Huang, H., Chai, T., Luo, X., Zheng, B., Wang, H.: Two-stage method and application for molten iron scheduling problem between iron-making plants and steel-making plants. IFAC Proc. **44**(1), 9476–9481 (2011)
7. Kikuchi, J., Konishi, M., Imai, J.: Transfer planning of molten metals in steel works by decentralized agent. In: Memoirs of the Faculty of Engineering, vol. 42, pp. 60–70. Okayama University (2008)
8. Kletzander, L., Musliu, N.: A multi-stage simulated annealing algorithm for the torpedo scheduling problem. In: Salvagnin, D., Lombardi, M. (eds.) CPAIOR 2017. LNCS, vol. 10335, pp. 344–358. Springer, Cham (2017). doi:10.1007/978-3-319-59776-8_28
9. Li, J.Q., Pan, Q.K.P., Duan, P.Y.: An improved artificial bee colony algorithm for solving hybrid flexible flowshop with dynamic operation skipping. IEEE Trans. Cybern. **46**(6), 1311–1324 (2016)
10. Liu, Y.Y., Wang, G.S.: The mix integer programming model for torpedo car scheduling in iron and steel industry. In: International Conference on Computer Information Systems and Industrial Applications - CISIA 2015, pp. 731–734. Atlantis Press (2015)
11. Ohrimenko, O., Stuckey, P.J., Codish, M.: Propagation via lazy clause generation. Constraints **14**(3), 357–391 (2009)
12. Schaus, P., Dejemeppe, C., Mouthuy, S., Mouthuy, F.X., Allouche, D., Zytnicki, M., Pralet, C., Barnier, N.: The torpedo scheduling problem: description (2016). http://cp2016.a4cp.org/program/acp-challenge/problem.html. Accessed 28 April 2017
13. Tang, L., Wang, G., Liu, J.: A branch-and-price algorithm to solve the molten iron allocation problem in iron and steel industry. Comput. Oper. Res. **34**(10), 3001–3015 (2007)
14. Wang, G., Tang, L.: A column generation for locomotive scheduling problem in molten iron transportation. In: 2007 IEEE International Conference on Automation and Logistics, pp. 2227–2233, August 2007

Constraint Handling in Flight Planning

Anders Nicolai Knudsen, Marco Chiarandini[✉], and Kim S. Larsen

Department of Mathematics and Computer Science,
University of Southern Denmark, Campusvej 55, 5230 Odense M, Denmark
{andersnk,marco,kslarsen}@imada.sdu.dk

Abstract. Flight routes are paths in a network, the nodes of which represent waypoints in a 3D space. A common approach to route planning is first to calculate a cheapest path in a 2D space, and then to optimize the flight cost in the third dimension. We focus on the problem of finding a cheapest path through a network describing the 2D projection of the 3D waypoints. In European airspaces, traffic flow is handled by heavily constraining the flight network. The constraints can have very diverse structures, among them a generalization of the forbidden pairs type. They invalidate the FIFO property, commonly assumed in shortest path problems. We formalize the problem and provide a framework for the description, representation and propagation of the constraints in path finding algorithms, best-first, and A* search. In addition, we study a lazy approach to deal with the constraints. We conduct an experimental evaluation based on real-life data and conclude that our techniques for constraint propagation work best together with an iterative search approach, in which only constraints that are violated in previously found routes are introduced in the constraint set before the search is restarted.

1 Introduction

The Flight Planning Problem (FPP) aims at finding 3D paths for an aircraft in an airway network, minimizing the total cost determined by fuel consumption and flying time. The motivation is financial and environmental. Airway networks can be huge, due to the added dimension compared with road networks, and side constraints complicate the problem further. Most of the constraints are determined by a central control institution, e.g., Eurocontrol in Europe and FAA in USA, and change rapidly with time in order to take traffic conditions into account and to minimize the need for later changes by the institution itself. Therefore, the common practice is to determine the precise flight route only a few hours before take-off. For this to be feasible and bring any advantage, the route determination must be quite fast, say on the order of a few seconds. If necessary, the route can then be adjusted during the flight by real-time optimization, considering more up-to-date information. Over the last years, the strain on the European airspace has increased to a level where the network must be heavily

K.S. Larsen—was supported in part by the Danish Council for Independent Research, Natural Sciences, grant DFF-1323-00247.

J.C. Beck (Ed.): CP 2017, LNCS 10416, pp. 354–369, 2017.
DOI: 10.1007/978-3-319-66158-2_23

constrained to ensure safe flights. This also implies increased difficulty in finding cost-efficient routes respecting the constraints.

We focus on the problem of finding 2D routes in European airspaces in an off-line setting. Normally, waypoints are defined for one or more intervals of flight levels, but here we assume that flights are cruising at a given altitude. This version of the problem is relevant because a common approach to flight planning in industry is to decompose the problem into two subproblems: finding a 2D route and expanding it in the third dimension. For both problems, there are constraints to satisfy and costs to minimize. Moreover, costs are resource-dependent because they depend on the weather conditions, which vary with time, and on the weight of the aircraft. This latter depends on the fuel level, the initial amount of which is also a decision variable of the problem. We use an estimate based on a great circle distance for this initial amount of fuel and assume that its more precise determination is done during the vertical route optimization (see [1], for instance).

The classic shortest path problem has been the focus of considerable amounts of research for many years. For an extensive survey on recent advances, see [2]. However, many of the new advances rely on preprocessing techniques, most of which we deem inapplicable in the flight planning context, due to the impact of the constraints. The problem of finding cheapest flight routes with resource-dependent costs was studied in [3,4], and more recently in [5]. The latter focuses on a 2D version, presenting three A* algorithms with different heuristic functions. However, constraints are not taken into account in these works while they are the main focus of our work.

The constraints in the European airspaces come in three different forms: Conditional Routes (CDR), Route Availability Document (RAD), and Restricted Airspaces (RSA). These are all published by the European air traffic management institution, Eurocontrol. RAD constraints are the most general and challenging. They include local constraints affecting the availability of airways and airspaces at certain times, but they are primarily conditional types of constraints. For example, if the route comes from a given airway, then it can only continue through another airway. Or some airways can only be used if coming from, or arriving to, certain airspaces. Or flights between some locations are not allowed to fly over certain airways. Or short-haul and long-haul are segregated in congested zones. RAD constraints must be handled during the route construction or checked later. There are more than 16,000 of these constraints and they can be updated several times a day, although most of them remain unchanged for longer time periods.

Some of these constraints are generalizations of the *forbidden pairs* type, which make the problem at least as hard as the *path avoiding forbidden pairs* problem that was shown to be NP-hard [6]. Given a topological sorting of the nodes, restricting to certain structures of forbidden pairs makes the problem polynomially solvable [7]. However, none of these structures can be guaranteed in the European airspace network.

Our contribution is the design of a framework for the representation and propagation of RAD constraints during the search. We formalize the constraints and extend path finding algorithms, such as best-first and A* search [8], to handle them. In particular, we propose an ad hoc tree structure to represent the constraints and to check their satisfiability and implications and to simplify their structure during the search. Then, we study two lazy approaches to constraint propagation. In one approach, we postpone the expansion of partial paths that cannot be dominated due to the constraints, but that are less promising than others in terms of costs. We then reconsider them only if it becomes necessary. In the other approach, which is similar to a Bender's decomposition with nogood cuts, we ignore all constraints in an initial search. If the path found is feasible, then we found a solution. Otherwise, we include only those constraints that are violated by the current path and iterate the whole process until a feasible solution is found. Additionally, we consider an exact and a heuristic approach to removing active constraints during the search based on geographical considerations.

This work is in collaboration with an industrial partner. Their core business is in flight route planning. Many of their customers are owners of private planes who plan their flights shortly before departure. Once they have chosen a destination, they send a query for a route from some portable device and they expect an almost immediate answer. Hence, this company is interested in an algorithm that can solve the problem within a few seconds. The size of the network used by this company is approximately 100,000 nodes and 3,000,000 edges and we use these real life data to test our ideas.

2   The Constrained Horizontal Flight Planning Problem

The European airspace is a network of *waypoints* that can be traversed at different altitudes (*flight levels*). Waypoints are connected across different flight levels by *airways*. The overall network could be described as a layered digraph, with several nodes for each waypoint representing different flying altitudes and arcs connecting these nodes if they belong to different waypoints. We simplify the situation by only allowing flights at a single flight level. The flight level is chosen to be the best cruising altitude for the tested aircraft. Hence, we represent the European airspace as a 2D network formed by a directed graph $D = (V, A)$, where the nodes in V represent waypoints defined by latitude and longitude coordinates and the arcs in A represent feasible airways between the waypoints. Each arc has associated resource consumptions and costs. The resource consumption for flying through an arc $a \in A$ is defined by a pair $\tau_a = (\tau_a^x, \tau_a^t) \in \mathbb{R}_+^2$, where the superscripts x and t denote the fuel and time components of the consumption, respectively. The cost c_a is a function of the resource consumption, i.e., $c_a = f(\tau_a)$.[1] A 2D (*flying*) *route* is an (s, g)-path in D represented by n waypoints plus a departure node (source) s and an arrival node (goal) g, that is, $P = (s, v_1, \ldots, v_n, g)$,

---

[1] The total cost is calculated as a weighted sum of time and fuel consumed. In our specific case, we have used 3$ per gallon of fuel and 1000$ per hour.

with $s, v_i, g \in V$ for $i = 1..n$, $v_i v_{i+1} \in A$ for $i = 1 \ldots n - 1$, and $s v_1, v_n g \in A$. The cost of a route is defined as $c_P = c_{s v_1} + \sum_{i=1..n-1} c_{v_i v_{i+1}} + c_{v_n g}$.

The route must satisfy a set $\mathcal{C}$ of *constraints* imposed on the path. These constraints are of the following type: if a set of nodes or arcs A is visited then another set of nodes or arcs B must be avoided or visited. The visit or avoidance of the sets A and B can be further specified by restrictions on the order of the elements, on the time window, and on the flight level range (although the latter does not play a role in our 2D setting).

Definition 1 (Constrained Horizontal Flight Planning Problem). *Given a network $N = (V, A, \boldsymbol{\tau}, \mathbf{c})$, a departure node s, an arrival node g, and a set of side constraints $\mathcal{C}$, find an (s, g)-path P in D that satisfies all constraints in $\mathcal{C}$ and that minimizes the total cost, c_P.*

We use the abbreviation CHFPP for the above.

In most common shortest path problems, a property that usually holds is the First In First Out (FIFO) property. It states that a path P' reaching a node with a cost worse than another path P reaching the same node cannot become part of the final solution and can, therefore, be discarded. This property plays a fundamental role in the efficiency of both Dijkstra and A* algorithms.

However, this property does not hold in our CHFPP. Indeed, a path P' arriving at a node with a cost worse than another path P reaching the same node cannot be discarded, because if the conditions activated during the path P' are less stringent than those activated during the path P, then P' could still become part of the best route. Moreover, the performance on a given arc is influenced by the weight (which in turn depends on the fuel consumed up to that arc) and by the time at which the arc is traversed (due to the possibly changing weather conditions). These dependencies of the resource consumptions on the path up to a given point are reflected in the cost of the next arcs, which, therefore, cannot be statically determined. Therefore, because of these dependencies of the cost function, the FIFO property would not hold even if there were no constraints. However, our experiments (see Sect. 4) show that for the real cases studied, no optimal solution is missed by assuming the FIFO property on cost. Hence, to simplify the presentation, we will assume the FIFO property on cost, but emphasize that we will *not* assume the FIFO property on constraints.

2.1 Definition of RAD Constraints

RAD constraints are implications of two types: *forbidden* and *mandatory*. They consist of an *antecedent* expression p and a *consequent* expression q. The expressions are Boolean and contain identifiers of locations visited during the flight and relationships between these. A RAD constraint is *satisfied* when the antecedent is false or when an antecedent is true and the consequent is true (in the mandatory case) or false (in the forbidden case). Thus, the interpretation is $p \to q$ for the mandatory and $p \to \neg q$ for the forbidden case. On the other hand, a RAD constraint is *violated* if it is mandatory and $p \to q$ is false or if it is forbidden and $p \rightarrow \neg q$ is false. In Fig. 1, we have defined a grammar to specify all possible types of RAD constraints.

```
constraint : 'Forbidden:' ID 'Antecedent:' expr 'Consequent:' expr
           | 'Mandatory:' ID 'Antecedent:' expr 'Consequent:' expr
expr_list  : expr
           | expr expr_list
expr       : '(' AND expr expr_list ')'
           | '(' OR expr expr_list ')'
           | '(' SEQ expr expr_list ')'
           | '(' NOT expr ')'
           | term
           | term time
term       : point | airway | airspace | arrival | departure
point      : 'Point:' ID
           | 'Point:' ID 'FL:' FLIGHT_LEVELS
airway     : 'Airway: from' ID 'to' ID
           | 'Airway: from' ID 'to' ID 'FL:' FLIGHT_LEVELS
airspace   : 'Airspace:' ID 'FL' FLIGHT_LEVELS
departure  : 'Dep:' ID
arrival    : 'Arr:' ID
time       : 'Time:' date 'to' date '-' time 'to' time '-' WEEKDAYS
```

Fig. 1. Bison (yacc) grammar for RAD constraints

Expressions are written in prefix notation using non-binary operators. Besides well-known operators, there is SEQ (sequence) for which all operands must be satisfied in the same order as they are presented in the constraint. The terms represent the possible flight choices, such as waypoint, airways between them, airspaces, and departure/arrival airports. ID's are the identifiers of the respective terms. Note that flight levels, included in the grammar for completeness, are not relevant in this exposition. Terms that have a `time` associated with them, are only satisfied if they are visited within the specified time window. An example of a constraint can be seen to the left in Fig. 2.

```
Forbidden: ID xxxx
Antecedent: (AND (OR Airspace: eg Airspace: ee) (NOT
      Point: mohni))
Consequent: Point: petot FL: 0-200
Time: 03-07-16 to 20-12-16 - 08:00 to 16:00 - FrSaSu
```

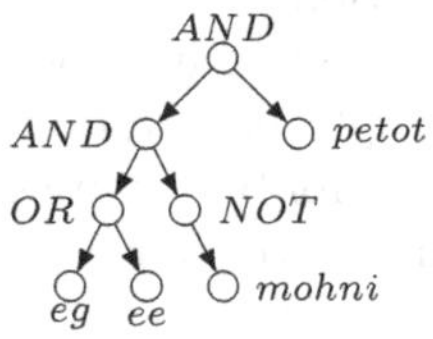

Fig. 2. Example of a forbidden constraint and its tree representation.

3   Path Finding Algorithms

In this section, we present path finding algorithms for solving the CHFPP to optimality. We consider classic best-first and A* search modified to take the constraints and the cost dependencies into account. Then, we introduce lazy approaches to deal with constraints, both during the search and after the search, leading to an iterated search process.

3.1   Handling the Constraints

In our path finding algorithms, constraints are checked while the route is constructed. Each RAD constraint is encoded in a tree data structure, where leaves are terms and internal nodes are operators (see Fig. 2, right). The truth values of the leaves are propagated up to the root, which must evaluate to false for the route to be feasible with respect to the corresponding constraint. Initially, all terms are in an unknown state. Then, if a term is resolved, the respective term is removed from the tree and the truth value is propagated upwards.

The set of constraints C is translated into a dictionary of constraint trees Γ with constraint identifiers as keys and the corresponding trees as values. For a constraint $\gamma \in \Gamma$, we let $\iota(\gamma)$ denote the constraint identifier and $T(\gamma)$ the corresponding tree. Then, for each node, $v \in V$, and each arc, $uv \in A$, we maintain a set of identifiers of the constraints that have those nodes or arcs, respectively, as leaves in the corresponding tree. We denote these sets E_v and E_{uv}, with $E_v = \{\iota(\gamma) \mid \gamma \in \Gamma, v \text{ appears in } \gamma\}$ and E_{uv} defined similarly.

Partial paths under construction are represented by labels. A *label* ℓ is associated with a node $\phi(\ell) = u \in V$ and contains information about a partial route from the departure node $s \in V$ to the node u. It is written as $\ell = (P_\ell, c_\ell, \Delta_\ell)$, where $P_\ell = (s, \ldots, u)$ is the path taken, c_ℓ is the cost of the path, and Δ_ℓ is the set of constraint trees of active constraints for the label ℓ. *Active constraints* are those where at least one term in the antecedent or consequent part has been determined, but where the complete satisfaction of the constraint has not yet been decided. Note that the constraint trees in Δ_ℓ are different from the initial ones in Γ because some terms may have been resolved and the tree consequently reduced. Formally, $\Delta_\ell = \{\rho(\gamma, P_\ell) \mid \iota(\gamma) \in E_u \cup E_{uv}, uv \in P_\ell\}$, where $\rho(\gamma, P_\ell)$ is the tree $T(\gamma)$ after propagation of the terms in P_ℓ. However, active constraints preserve the original identifiers, that is, $I(\Delta_\ell) = \{\iota(\gamma) \mid \gamma \in \Delta_\ell\} = \{\iota(\gamma) \in E_u \cup E_{uv}, uv \in P_\ell\}$. Depending on whether the term is negated or not, some locations can be advantageous or disadvantageous for a label to visit, opening up or restricting possibilities ahead. This can be determined for each term while building the constraints and active constraints are flagged as belonging to one of the two categories when a term is resolved.

All labels created are maintained in a structure Q, called the *open list*. The *expansion* of a label is the operation of extracting a label from Q and inserting a new label into Q for any node in D reachable by an outgoing arc from the node of the label under expansion. When a label ℓ with $\phi(\ell) = u \in V$ is expanded along an arc $uv \in A$, a new label $\ell' = ((s, \ldots, u, v), c_\ell + c_{uv}, \Delta_{\ell'})$ is created. The new set of constraint trees is obtained by copying the trees from Δ_ℓ, and the trees from Γ identified by E_v and E_{uv}. While performing these operations, the trees are reduced based on the satisfaction of u and/or uv. If the root of a constraint tree in $\Delta_{\ell'}$ evaluates to true, then the label ℓ' is deleted, because the corresponding route would be infeasible. On the other hand, if a root evaluates to false, then the corresponding constraint tree is resolved but is kept in $\Delta_{\ell'}$ to prevent re-evaluating it if, at a later stage, one of the terms that were logically deduced appears in the path. Formally, $\Delta_{\ell'} = \{\rho(\gamma, P_{\ell'}) \mid \gamma \in \Delta_\ell\} \cup \{\rho(\gamma, P_{\ell'}) \mid \iota(\gamma) \in E_{\phi(\ell')} \cup E_{\phi(\ell)\phi(\ell')}\}$.

For efficiency reasons, we use the following conservative approximation of logical implication between sets of constraints. We say that Δ_{ℓ_a} is *implied* by Δ_{ℓ_b} if no constraints in $I(\Delta_{\ell_b}) \setminus I(\Delta_{\ell_a})$ are marked as advantageous, no constraints in $I(\Delta_{\ell_a}) \setminus I(\Delta_{\ell_b})$ are marked as disadvantageous, and the trees of all constraints in $I(\Delta_{\ell_a}) \cap I(\Delta_{\ell_b})$ are identical, in the sense that they are isomorphic when regarded as ordered trees (sorted in the order of the input of the corresponding constraints).[2] Further, we hash a post-order traversal of the tree so that identity check is fast. The traversal is performed anew any time the tree is evaluated (during an expansion) and the hash function is recomputed and stored at the same time. If we hash to a 64 bit value, false positives are extremely unlikely.

A label ℓ_a is *dominated* by another label ℓ_b if $\phi(\ell_a) = \phi(\ell_b)$, $c_{\ell_a} > c_{\ell_b}$ and Δ_{ℓ_b} is implied by Δ_{ℓ_a}. A label that is dominated is removed from the open list and deleted. If $\phi(\ell_a) = \phi(\ell_b)$, $c_{\ell_a} > c_{\ell_b}$ but Δ_{ℓ_b} is *not* implied by Δ_{ℓ_a}, then we say that ℓ_a is *partly dominated* by ℓ_b. Partly dominated labels cannot be removed from the open list. As an example, consider the scenario in Fig. 3. Let $C_1 = ((a \vee b) \wedge c)$ be the only constraint relevant to the example. Let $\ell_1 = ((s, a, x), 3, \{C_1\})$, $\ell_2 = ((s, d, x), 4, \varnothing)$ and $\ell_3 = ((s, b, x), 2, \{C_1\})$ be the only three labels at x. The label ℓ_1 is dominated by ℓ_3 and can be discarded. The route is cheaper and Δ_{ℓ_3} is implied by Δ_{ℓ_1}, as they both contain only C_1. On the other hand, ℓ_2 is only partly dominated by ℓ_3, because Δ_{ℓ_3} is not implied by Δ_{ℓ_2}. Hence, ℓ_2 is not discarded. Indeed, ℓ_2 leads to the cheapest route to g, since ℓ_3 must avoid c while ℓ_2 does not have to.

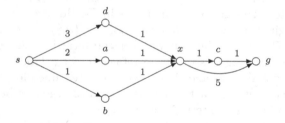

Fig. 3. An example where a partly dominated label leads to an optimal route. The labels are $\ell_1 = ((s, a, x), 3, \{C_1\})$, $\ell_2 = ((s, d, x), 4, \varnothing)$ and $\ell_3 = ((s, b, x), 2, \{C_1\})$.

3.2 Best-First and A* Algorithms

Our algorithms are based on classic best-first and A* algorithms. These algorithms expand labels from an open list $\mathcal{Q}$ until a path from source to goal is proven optimal. When we extract a label from the open list, we choose one of

---

[2] Note that a more accurate determination of subsumption between two trees, accurately reflecting semantic logical implication, would require solving a subgraph isomorphism that can be quite costly due to its NP-completeness.

smallest cost (best-first) or smallest sum of the cost of the label and a heuristic estimate of the cost from the corresponding node to the goal (A^*). The algorithm terminates when the goal g has been reached and the incumbent best path to g is cheaper than the cheapest label in Q. In best-first, the solution returned is optimal. In A^*, the solution returned is optimal if the heuristic is both admissible (the estimated cost must never overestimate the cost from a node to the goal) and consistent (for every node u, the estimated cost of reaching the goal must not be greater than the cost of getting to a successor v plus the estimated cost of reaching the goal from v). Consistency can be shown to be a stronger property as it also implies admissibility. As heuristic, we use the cheapest path determined by preprocessing the graph with a backward breadth-first search from the goal to all other nodes. The guarantee of admissibility and consistency of these estimates is obtained by disregarding the constraints and assuming a cost on each arc that is a lower bound of the corresponding costs. The lower bound can be computed by choosing the best weather conditions in the period between the departure time and an upper bound on the arrival time. ·

A baseline of the resulting path finding algorithm for solving CHFPP is given in Algorithm 1. The function FINDPATH takes the initial conditions of the aircraft as input, i.e., the initial fuel load τ_0^x, the departure time τ_0^t, a network $N = (D, \tau, c)$ built using information from the airspace, aircraft performance data, and weather conditions. Here, τ and c are intended as data tables. The time and fuel consumption for an arc is looked up in these tables using the inputs: (i) the fixed flight level, (ii) weight, (iii) international standard atmosphere deviation (i.e., temperature), (iv) wind component, and (v) cost index.[3] Inputs (ii), (iii), and (iv) depend on the partial path.

Differently from classic path finding algorithms, the algorithm in Algorithm 1 includes an extra comparison with respect to constraints for the domination of labels (lines 20–21 and 23). Under the FIFO assumption, it would be possible to determine a strict domination among labels and to add nodes of expanded labels to a closed list. As a consequence, at most one label per node would be expanded. However, in our case, domination of labels also needs to take constraints into account, for which the FIFO property does not hold. Thus, partly dominated labels cannot be discarded and the closed list becomes unnecessary. As a consequence, more than one label from a node can be expanded.

Finally, although D contains cycles and although, theoretically, the cycles could be profitable because of the time dependency of costs, labels are not allowed to expand to already visited vertices because routes with cycles would be impractical.

---

[3] The cost index is an efficiency ratio between the time-related cost and the fuel cost that airlines use to specify how to operate the aircraft, determining the speed of the aircraft. It is decided upon at a strategic level and cannot be changed during the planning phase.

```
 1  Function FINDPATH((τ₀ˣ, τ₀ᵗ), N = (D(V, A), τ, c), Γ)
 2  │   initialize the open list Q by inserting ℓₛ = ((s), 0, {})
 3  │   initialize ℓᵣ = ((), ∞, {})
 4  │   while Q is not empty do
 5  │   │   ℓ ← extract the cheapest label from Q
 6  │   │   if (cℓ > cℓᵣ) then break                              ▷ termination criterion
 7  │   │   if (φ(ℓ) = g) and (cℓ < cℓᵣ) then
 8  │   │   │   ℓᵣ ← ℓ
 9  │   │   │   continue
10  │   │   foreach node v such that uv in A do
11  │   │   │   ℓ' ← label at v expanded from ℓ
12  │   │   │   evaluate constraint trees in Δℓ'
13  │   │   │   if one or more constraints in Δℓ' are violated then
14  │   │   │   │   continue
15  │   │   │   INSERT(ℓ', Q)
16  │   return Pℓᵣ and cℓᵣ

17  Function INSERT(ℓ', Q)
18  │   foreach label ℓ ∈ Q with φ(ℓ) = φ(ℓ') do
19  │   │   if (cℓ > cℓ') then
20  │   │   │   if (Δℓ is implied by Δℓ') then                    ▷ ℓ is dominated
21  │   │   │   │   remove ℓ from Q
22  │   │   else if (c'ℓ > cℓ) then
23  │   │   │   if (Δℓ is implied by Δℓ') then return             ▷ ℓ' is dominated
24  │   insert ℓ' in Q
25  │   return
```

Algorithm 1. A general template for solving CHFPP

3.3   Lazy Expansion

In Algorithm 1, partly dominated labels are also added to Q, so only few labels can actually be dominated. To speed up the algorithm, we attempt a *lazy* approach to expansions by postponing the expansion of partly dominated labels. This is achieved as follows. At each node $v \in V$, we maintain a waiting list of labels, ω_v. Then, instead of adding partly dominated labels at a node v to Q, we add them to ω_v. The idea is that if all successors of the cheapest label at v, $\ell = ((s, \ldots, v), c_\ell, \Delta_\ell)$, are able to expand throughout the cheapest path from ϕ_ℓ to g without being affected by constraints in Δ_ℓ, then there is no label that was partly dominated by ℓ that would lead to a better route. However, if there is a successor ℓ' of ℓ that cannot expand to the next node in the cheapest path from ϕ_ℓ, then one of the labels in ω_v could potentially lead to a better route, and thus must be inserted into Q. This is done by *backtracking* through every node in the path of the label ℓ' and, at each node in $P_{\ell'}$, inserting into Q the cheapest label from the corresponding waiting list.

Backtracking is triggered whenever a label cannot be expanded to a reachable node because a constraint becomes violated and the path infeasible or whenever a label is dominated. An example of backtracking due to infeasibility was presented in Fig. 3. There the label l_2 is partly dominated by l_3 and hence set in w_x, but when l_3 fails to expand to c, l_3 is backtracked, resuming l_2, which is moved from w_x to $\mathcal{Q}$. For an example where backtracking is needed because of domination, consider the situation in Fig. 4 (left). The only relevant constraint is $C_2 = (a \wedge b)$. At the node x, we have the labels: $\ell_1 = ((s,a,x),2,\{C_2\})$ and $\ell_2 = ((s,x),3,\varnothing)$ with ℓ_2 in w_x because of being partly dominated by ℓ_1. At the node y, the labels are: $\ell_3 = ((s,a,x,y),4,\{C_1\})$ (which is the expansion of ℓ_1) and $\ell_4 = ((s,a,y),3,\{C_1\})$. When the label ℓ_3 is discovered to be dominated by ℓ_4, it cannot simply be removed because then we would lose the label ℓ_2 that, when expanded to y, becomes $\ell_5 = ((s,x,y),5,\varnothing)$, which is only partly dominated by ℓ_4. Hence, we need to backtrack ℓ_3 and include ℓ_2 in $\mathcal{Q}$.

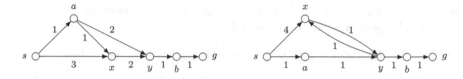

Fig. 4. Backtracking triggered by domination (left) and cycling (right)

When we backtrack a label ℓ to a given node u, we select the cheapest label ℓ' from the waiting list at u and add that to the open list. We only need to backtrack ℓ once, since backtracking ℓ' will trigger further moves to the open list, if it becomes necessary. Therefore, we associate a backtracking indicator with each label to prevent backtracking from the same label a second time.

Particular care must be devoted to potential cycles. Routes are not allowed to visit the same node twice, so the detection of cycling in D can also be the cause of a label not being expanded. Consider the situation in Fig. 4 (right). The only relevant constraint is $C_3 = (a \wedge b)$. The labels at x are $\ell_1 = ((s,a,y,x),3,\{C_3\})$ and $\ell_2 = ((s,x),4,\varnothing)$, with ℓ_2 in w_x because of being partly dominated by ℓ_1. Further, at b, we have the label $\ell_3 = ((s,a,y,b),3,\{C_3\})$. When we try to expand ℓ_1, we discover it cannot be expanded anywhere without creating a cycle. Then we consider ℓ_3, and discover that it has become infeasible. However, backtracking ℓ_3 does not allow us to resume ℓ_2 from w_x because we do not pass through x. Thus, ℓ_2 would never be added to $\mathcal{Q}$ and we would not find the one feasible route. Hence, backtracking must be triggered also when cycles are detected.

To handle this efficiently, we equip all labels ℓ with a dictionary, H, associating nodes with labels. The keys of such a dictionary are the nodes of the path P_ℓ, and the associated value, $H(u)$, is the label at u which is eventually expanded into ℓ. We use a small hash table and get expected constant time lookups. After the initialization of the dictionary, cycles can be detected in constant time by a look-up. Additionally, we let $\pi(\ell)$ denote the label associated with the predecessor of the last node in P_ℓ.

Further, it should be noted that, when backtracking is caused by domination or cycle detection, it can be delayed. Let ℓ' be a label that we need to backtrack and let ℓ be the *blocking* label, that is either the dominating label (domination case) or $H(u)$ if ℓ' is trying to expand to a node u that is already in $P_{\ell'}$. Let $B_{\ell'}$ denote the set of labels that would be added to Q, if ℓ' were to be backtracked immediately. Since the labels in $B_{\ell'}$ were all partly dominated predecessors of ℓ', any successor of those reaching $\phi(\ell')$ would be more expensive than ℓ' and thus they would be (partly) dominated by ℓ as well. Therefore, backtracking can be delayed until ℓ is backtracked, which would allow the successors of labels in $B_{\ell'}$ to reach farther than $\phi(\ell)$.

To implement delayed backtracking, we add to the information maintained with each label ℓ a list β_ℓ of labels that were blocked by ℓ at some point. Thus, whenever a label ℓ' is blocked by ℓ and should be backtracked, we do the following. If ℓ has already been backtracked, we backtrack ℓ' immediately, but otherwise, we delay and add ℓ' to β_ℓ instead. When ℓ itself is backtracked, besides $\pi(\ell)$, also all labels in β_ℓ are backtracked.

Theorem 1. *Algorithm 1 with lazy expansion returns optimal routes.*

Proof. The algorithm is derived from Algorithm 1, which is optimal, by adding lazy expansion. To show that lazy expansion maintains optimality, we need to show that all labels that are still in any waiting list when the algorithm terminates cannot be part of an optimal (s, g)-path.

Let ℓ be a label at $v \in V$, stored in ω_v when the algorithm terminates. Since ℓ is in ω_v, there must exist a label ℓ' which partly dominated ℓ, i.e., $c_{\ell'} < c_\ell$. Since ℓ is still in ω_v when the algorithm terminates, none of the expanded successors of ℓ' can have caused a backtrack. Thus, any possible path from v to any node in V originating from ℓ has also been explored by the expanded successors of ℓ' and is also cheaper. ∎

3.4   Further Elements: Lazy Constraints and Constraint Pruning

Lazy Constraints and Iterative Path Finding. An alternative approach is to ignore the constraints initially and to iterate the search, adding constraints only when they are actually violated in the route found. First, a route is found without considering any RAD constraints. Then, the route is checked against all constraints. If no constraints are violated, the route is valid and the algorithm terminates. Otherwise, if one or more constraints are violated, the constraints are added to the input data of the path finding algorithm and a new search is started. The advantage of this procedure is that it avoids considering many constraints that never turn out to be relevant for the optimal route.

Heuristic Constraint Pruning. Some active constraints may become very unlikely to be violated if their terms correspond to locations that are already passed by the label or far from the direct route between the current node of the label and the goal. Thus, whenever during expansion a label ℓ evaluates a constraint, we try to estimate heuristically whether it is still relevant or not. More precisely, we compare a lower bound and an upper bound for the length of a route from $\phi(\ell) \in V$ to the destination passing through the location $u \in V$ (or $uv \in A$) represented by the term. If the lower bound is larger than the upper bound, then we declare the term not satisfiable. Let $d(P)$ be the flying distance covered by a path P and $\gcd(u, v)$ be the great circle distance between two airway points u and v. The lower bound is given by $d(P_\ell) + \gcd(\phi(\ell), u) + \gcd(u, g)$. We use two different heuristic values for the upper bound. One is the *current result*: once the search finds any feasible (s, g)-path, with ℓ' being the final label, $d(P_{\ell'})$ is saved as the upper bound. If a better (s, g)-path is found, the bound is updated. The second heuristic is the *remaining distance*. It uses $d(P_\ell) + (1.3 \cdot \gcd(\phi(\ell), g))$ as the upper bound. The factor 1.3 was determined by observing the maximal deviation of historical routes from the great circle distance. This heuristic is disabled when close (i.e., within 20 nautical miles) to the departure or destination, as the procedures to exit and enter airports are unpredictable and can deviate considerably from great circle distances.

4   Experimental Results

Experimentally, we have compared different algorithms obtained from the combination of the elements presented in the previous sections. We consider computation time, number of labels expanded, and the quality of the routes.

We use real-life data provided by our industrial partner. This data consists of aircraft performance data, weather forecast data in standardized GRIB2 format, and a navigation database containing all the information for the graph. The graph consists of approximately 100,000 nodes and 3,000,000 edges. The aircraft performance data refers to one single aircraft and tests are run on the optimal cruising flight level of that aircraft, i.e., the one that yields the best cost on average weather conditions. The data for the weather forecast is given at intervals of three hours on specific grid points that may differ from the airspace waypoints. They are then interpolated both in space and time. A test instance is specified by a departure airport and time, and a destination airport. A set of 13 major airports in Europe was selected uniformly at random to pursue a uniform coverage of the constraints in the network. Among the 156 possible pairings, 14 were discarded because of short distance, resulting in 142 pairs that were used as queries. Great circle distances range from 317 to 1721 nautical miles. All algorithms were implemented in C# and the tests were conducted on a virtual machine in a cloud environment with an Intel Xeon E5-2673 processor at 2.40 GHz and with 7 GB RAM. To account for fluctuations in CPU time measurement, each algorithm was run 5 times on each instance and only the fastest was recorded. A preliminary comparison between A* and best-first unveiled that best-first is impracticable. Within a time limit of one minute, it terminates only in 11 instances against 103 of A*.

Assessment of the FIFO Assumption on Costs. We tested whether assuming the FIFO property on costs would lead to suboptimal results. Removing the FIFO property means that labels in the open list are never dominated. More specifically, we tested two versions of A*, one that assumes FIFO, and thus is as described in the previous sections, and one where the lines 18–23 of the INSERT function in Algorithm 1 are omitted. With a timeout of 10 min, A* without the FIFO assumption solved 78 out of 142 instances, and in these instances, all returned solutions were of the same cost as A*. Thus, we conclude that at least for our real-life setting, assuming the FIFO property on costs seems to be a good heuristic that does not affect the optimality of results. Henceforth, we continue to assess only algorithms that use this assumption.

Empirical Run-Time. We include in the run-time of the path finding algorithms both the time used for preprocessing (determining lower bounds for each vertex) and the time spent for actually performing the search. Initially, we compare the run-time time of 4 different algorithms: A*, A* with the upper bound heuristic to ignore constraints (A*UB), A* with the remaining distance heuristic (A*RD), and iterative A* (iA*). We use A* as a baseline algorithm and calculate the percentage deviation of running time per instance of the other algorithms with respect to A*. A scatter plot of the run-time percentage deviations from A* is shown in Fig. 5 (left column), where the x-axis represents different instances sorted by great circle distance between query airports. A time limit of one minute was used in these experiments. Within this time limit, A* did not terminate in 39 queries. These cases are detectable by the lack of points for some ordinate in the first panel in Fig. 5. There does not seem to be a correlation between the size of the instance and the non-termination of A*.

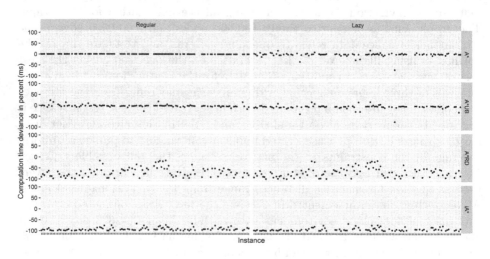

Fig. 5. Regular algorithms (left) and lazy expansion (right).

We observe that A*UB keeps returning optimal solutions (not shown), only results in minor runtime improvements compared to A*, and does not terminate in the same 39 instances. Separately, we observed that A* does little work after finding the first path to the goal, indicating that the heuristic cost value used for selection in all our A* algorithms must be very close to the exact value. Thus, since A*UB has an impact only after an (s, g)-path has been found, the space for improvement is small.

A*RD is considerably faster than A* and the number of instances unsolved within 1 min is reduced to 15. Unfortunately, the omission of constraints is sometimes too optimistic, leading to suboptimal routes due to the inaccurate domination of some labels. This happens in 11 out of the 142 instances where the solution quality was within 0.1–0.6% of the optimal solution. This effect can be controlled by increasing the 1.3 factor in the remaining distance heuristic, but this increases the running time. On the other hand, we never experienced that A*RD returned infeasible solutions (which could theoretically happen).

The winner of the comparison is by far iA*. The reduction in computation time with respect to A* is up to 99% in all instances going from running times of the order of seconds to running times of the order of milliseconds. It solves all cases where A* does not terminate, taking 12 s in the worst case (which is an extreme outlier in iA* running times). The number of iterations ranges between 1 and 10 and although the overall number of expanded labels can in some cases become comparable to that of A*, the reduction in computation time from not having to handle a large number of constraints is huge.

In the right column of Fig. 5 we assess the impact of the lazy expansion to all four algorithms. The deviations are still calculated with respect to the results of the baseline A* algorithm. The visual comparisons performed row-wise inform us that the lazy expansion improves the running time of the algorithms only in few cases. While in many instances there is a reduced number of label expansions, the overhead in run-time due to maintaining the waiting and backtracking lists is sometimes larger than the time saved.

Instance Complexity. In Fig. 6, we investigate the scaling of the algorithms with respect to instance size. We removed iA* from the analysis because its running times and number of constraints activated are too small (note that iA* is however using A* and hence it is implicitly represented). The plots are on a semi-logarithmic scale with the run-time expressed in milliseconds on the y-axis. In the left column, we show the dependency on the great circle distance between the departure and destination airports of the query. We observe that there is no pattern in the points, indicating that this distance is not a good predictor for the complexity of the search. In the right column, we show the dependency of the run-time on the number of constraints that became active during the search. The plots indicate that this can be an important regressor hinting at an exponential relationship. Unfortunately, the number of constraints that become active is known only after the search has taken place. An analysis of the correlation between great circle distance and number of constraints was found to be

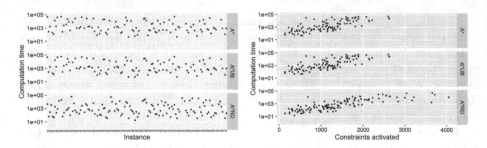

Fig. 6. Time complexity of the search as a function of distance (left) and as a function of constraints activated (right). The search is truncated at a time limit of one minute.

inconclusive, hinting at the fact that it is not the length of the route but rather the density of constraints in the area it crosses that is important.

5   Conclusions

We have studied constraint handling in path finding algorithms for 2D route planning. We formalized the structure of these constraints and represented them with an ad hoc tree structure that makes it efficient to gradually update constraints and eliminate terms that become irrelevant during the search. We showed that from a collection of 16,000 constraints arising in a real-life setting, up to 4,000 were activated during the search of the algorithms. We concluded that a combination of constraint handling during the search and iterating A*, introducing only relevant constraints, leads to significantly better running times than including all constraints from the beginning. We regarded this approach as a lazy constraint approach, but it can also be seen as a form of logic-based Benders decomposition driven by nogood cuts [9]. In our experiments, this approach reduced the running time of A* from a few seconds to a few milliseconds. We also investigated another type of lazy approach, where the label expansions in path finding algorithms is conducted lazily. However, our experimental evaluation indicated that in our specific real-life instances, the contribution of this technique is not as pronounced as the lazy constraint approach. The handling of constraints during the search was new for our industrial partner, who decided to implement our algorithms in their product, obtaining an increased robustness and considerable reductions in running times.

We have also approached the problem with a generic purpose solver via mixed integer programming. If costs are considered static, the model is a classic min cost flow model with additional constraints derived from the RAD constraints that break the total unimodular structure of the constraint matrix. Preliminary results showed that this approach is slow. The instances were solved on average in about 12 min on an 8 core machine using about 10 GB of memory. However, this approach cannot deal with the—here fundamental—resource dependency structure of the costs. We expect this to be an issue with SAT solvers as well.

Throughout we have assumed a static flight level chosen as the one with best average performance. As future work, we plan to include the vertical dimension in our flight planning. The size of the network grows dramatically, and this leads to entirely new challenges.

References

1. Knudsen, A.N., Chiarandini, M., Larsen, K.S.: Vertical optimization of resource dependent flight paths. In: Twentysecond European Conference on Artificial Intelligence (ECAI). Frontiers in Artificial Intelligence and Applications, vol. 285, pp. 639–645. IOS Press (2016)
2. Bast, H., Delling, D., Goldberg, A., Müller-Hannemann, M., Pajor, M., Sanders, T., Wagner, D., Werneck, R.F.: Route planning in transportation networks. Technical report. arXiv:1504.05140 [cs.DS] (2015)
3. Olivares, A., Soler, M., Staffetti, E.: Multiphase mixed-integer optimal control applied to 4D trajectory planning in air traffic management. In: Proceedings of the 3rd International Conference on Application and Theory of Automation in Command and Control Systems (ATACCS), pp. 85–94. ACM (2013)
4. de Jong, H.M.: Optimal Track Selection and 3-Dimensional Flight Planning: Theory and Practice of the Optimization Problem in air Navigation Under Space-Time Varying Meteorological Conditions. Staatsuitgeverij, Madison (1974)
5. Blanco, M., Borndörfer, R., Hoang, N.-D., Kaier, A., Schienle, A., Schlechte, T., Schlobach, S.: Solving time dependent shortest path problems on airway networks using super-optimal wind. In: 16th Workshop on Algorithmic Approaches for Transportation Modelling, Optimization, and Systems (ATMOS), pp. 12:1–12:15. Schloss Dagstuhl-Leibniz-Zentrum für Informatik (2016)
6. Yinnone, H.: On paths avoiding forbidden pairs of vertices in a graph. Discret. Appl. Math. **74**(1), 85–92 (1997)
7. Ková č, J.: Complexity of the path avoiding forbidden pairs problem revisited. Discret. Appl. Math. **161**(10–11), 1506–1512 (2013)
8. Hart, P.E., Nilsson, N.J., Raphael, B.: A formal basis for the heuristic determination of minimum cost paths. IEEE Trans. Syst. Sci. Cybern. **4**(2), 100–107 (1968)
9. Hooker, J.N., Ottosson, G.: Logic-based benders decomposition. Math. Program. **96**(1), 33–60 (2003)

NightSplitter: A Scheduling Tool to Optimize (Sub)group Activities

Tong Liu[1]([⊠]), Roberto Di Cosmo[2], Maurizio Gabbrielli[1], and Jacopo Mauro[3]

[1] DISI, University of Bologna, Bologna, Italy
{t.liu,maurizio.gabbrielli}@unibo.it
[2] INRIA and University Paris Diderot, Paris, France
roberto@dicosmo.org
[3] Department of Informatics, University of Oslo, Oslo, Norway
jacopom@ifi.uio.no

Abstract. Humans are social animals and usually organize activities in groups. However, they are often willing to split temporarily a bigger group in subgroups to enhance their preferences. In this work we present NightSplitter, an on-line tool that is able to plan movie and dinner activities for a group of users, possibly splitting them in subgroups to optimally satisfy their preferences. We first model and prove that this problem is NP-complete. We then use Constraint Programming (CP) or alternatively Simulated Annealing (SA) to solve it. Empirical results show the feasibility of the approach even for big cities where hundreds of users can select among hundreds of movies and thousand of restaurants.

1 Introduction

Nowadays, most of the city activities such as restaurants, cinemas, museums, theaters have complete and detailed information on web pages and offer a variety of online services and options for consulting programs, making reservations, buying tickets, etc. One of the main problems that the customer has to face in order to take advantage of this huge offer is to master the information overload which comes with it. For example, in Paris, our reference town for this work, there are more than 13500 restaurants and around 100 cinemas with 150 movies each night. Hence, the apparently simple task of organising a night out with a movie followed by a dinner can already turn into a serious planning exercise.

When there are several persons involved, e.g., a family or a group of friends, with different ideas, preferences, and needs, coordinating the activities of the group becomes significantly more complex. It is quite natural, in order to satisfy all the preferences of the members of a group, to take a pragmatic approach and split the group of persons into several sub-groups performing different activities, in order to enhance the individual satisfactions: some groups will watch the latest Hollywood blockbusters, while some others will prefer an indie movie, provided, of course, this can take place approximately at the same time, and in the same movie theater, or in movie theaters not too far apart.

ⓒ Springer International Publishing AG 2017
J.C. Beck (Ed.): CP 2017, LNCS 10416, pp. 370–386, 2017.
DOI: 10.1007/978-3-319-66158-2_24

And that's not all: one needs to take into account both time constraints (e.g., we need to be home before midnight) and spatial constraints (e.g., we do not have the car and we do not want to walk for one hour). The planning of a night out can therefore easily become a daunting task.

Recommender systems and planners provide tools that can help users to manage these difficulties by filtering information, suggesting solutions, predicting some needs and planning the activities. However, most of the existing tools focus on a single user, so they cannot be used when several users interact and participate in a group activity [7,17]. Tools considering group experiences exist [3,5,20] but they mainly focus on methods for aggregating preferences for a fixed group of users in order to optimize (some notions of) group satisfaction.

Only a few research papers [4,18] consider the problem of sub-group formation and group splitting, but they do not take into account time and space constraints or they impose the same subgroups for all the activities, thus forbidding the most interesting cases, like a group that splits into subgroups to see different movies, but then joins at the same restaurant.

In this work we present NightSplitter, an on-line tool that is able to plan movie and dinner activities for a group of users, possibly splitting them in subgroups to optimally satisfy their preferences. We first model this problem and prove that it is NP-complete. We then use Constraint Programming (CP) or alternatively Simulated Annealing (SA) to solve it. Empirical results, obtained on real data for the city of Paris, show the feasibility and scalability of the approach even when hundred of users can select among hundreds of movies and thousand of restaurants.

It is worth noticing that even though, for the sake of clarity and concreteness, in this paper we focus on the above mentioned activities, our approach is completely general and our tool can be easily adapted to any problem which has the following features: (1) there is a group of users who have to perform a sequence of n activities; (2) each user can express some preferences on these activities; (3) the group can be divided in several sub-groups, each one performing a different activity at a given time frame; (4) temporal and spacial constraints can be added on the different activities; (5) the aim of the tool is to optimize the overall satisfaction of all the users involved in the activities.

Structure of the Paper. In Sect. 2 we describe NightSplitter from the user perspective. In Sect. 3 we first formalize the problem solved by NightSplitter proving its NP-hardness while in Sect. 4 we present how CP and SA techniques are used to solve it. Section 5 presents the experiment results that validate the use of NightSplitter. Related work and conclusions are in Sects. 6 and 7 respectively.

2 NightSplitter

NightSplitter, the tool we have developed and that we present in this Section, is a web application for planning movie and restaurant activities in the city of

Paris. It may be used by a group of users and it can split them in subgroups
to optimally satisfy their preferences. The application uses real data for (currently) 13598 restaurants and 93 cinemas with 153 movies, which are stored in
a database and are constantly updated by a crawler embedded in the application. Using `NightSplitter`, an initial user dubbed *group initiator* can create
a "group event" for a certain date. The group initiator is able to tune several
parameters and constraints such as the number of possible subgroups, the size
of subgroups, the total time window for performing the activities, the maximal
time one is forced to wait between the activities. The group initiator can then
invite other members to participate to the group by sharing a reference link. The
invited member, by clicking on the link, is included automatically into the group
and will be able to express his/her preferences, possibly inviting other persons
to join the group.

As can be seen from Fig. 1 showing a screenshot of `NightSplitter`, by using
some simple menus each user can express preferences on movies and restaurants
in Paris. Social interaction among group members is possible, since each user
can see the preferences of others and can instantly see the results of updating or modifying his/her own preferences. The main interface is divided in two
parts: a dashboard for preferences and a digital map for showing the solutions.
In the preference dashboard (right side of Fig. 1), users can input their preferred
movie and restaurant names (or alternatively movie and cuisine categories). The
introduction of this information is facilitated by an autocomplete function that
suggest possible values. The expressed preference is represented by a tag with
color, where the tag shows the name of the preference and the color indicates
its scale: deep blue to signal a strong like, light blue for like, yellow for dislike,
red for strong dislike, and gray for neutral. On the top of the dashboard, there

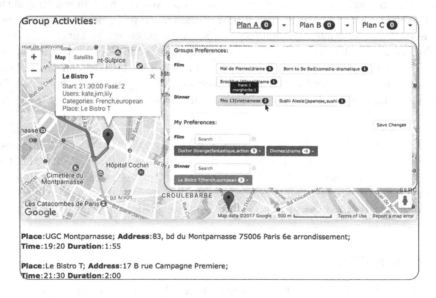

Fig. 1. NightSplitter screenshot (Color figure online)

is a summary of the group preferences, where in each tag, next to the activity name, there is an aggregated score. Each time a user enters or modifies a preference, the preference dashboard will be updated in real time and the system will start to compute a new solution.[1] The computation, as later detailed in Sect. 4, uses either a Constraint Programming or Simulated Annealing technique. The averages of the individual preferences and the public ratings of the selected activities are weighted and combined to form a unique evaluation metric to establish the quality of every solution (cf. Definition 6). The 3 solutions with highest aggregated preference are provided and displayed on-the-fly to the users, both in textual form and on the digital map. The text informs the user about their tentative scheduled activities while the map provides a global view of the subgroups activities with their cinema-restaurant paths. Given the different solution plans, group members have the option to like or dislike them by clicking "Plan A/B/C" as shown in the upper part of Fig. 1. Based on these votes the group initiator can finalize the decision and pick up the plan for the entire group.

The online version of `NightSplitter` is available at [29][2].

3 NightSplit

In this section we formalize the definition of the optimization problem solved by `NightSplitter` and dubbed `NightSplit`. The key elements of `NightSplit` are the users and the activities that users can perform. We therefore assume the following finite disjoint sets: $\mathcal{U}$ for users range over by $u_1, u_2, \ldots$, $\mathcal{A}_M$ and $\mathcal{A}_R$ for the movie and restaurant activities respectively. We will denote with $\mathcal{A} = \mathcal{A}_M \cup \mathcal{A}_R$ a generic activity ranged over by $a_1, a_2, \ldots$.

Activities have properties such as a possible starting time or the location where they are performed. The planning problem therefore needs to consider two dimensions: time and space. As far as the time is concerned, for `NightSplit` we consider only a fixed time window assuming that we want to plan all the activities within a given time range. In particular, for simplicity we use a discrete notion of time dividing the time window in time slots of fixed duration. Similarly, we discretize also the space by dividing it into a finite number of different locations. The granularity of the time and the space can be arbitrarily improved by reducing the duration of the time slot or considering smaller locations. In the following we denote with $TIME = \{1, \ldots, T_{max}\}$ and $Loc = \{1, \ldots, Loc_{max}\}$ the time slots and the locations where T_{max} and Loc_{max} are the number of time slots and the number of locations. In our examples, we consider 5 min as the time slot unit. We can therefore define the general properties of an activity as follows.

Definition 1 (Activity Proprieties). *Given a set of activities $\mathcal{A}$ we denote with:*

---

[1] Currently preferences are visible to all the users. However, mechanisms to hide the individual preferences such as differential privacy [8] are under consideration.

[2] We are developing the tool for commercial use.

- **startTime** the total function $\mathcal{A} \to TIME$ that associates to an activity its starting time slot (i.e., when the movie starts or when the restaurant opens),
- **endTime** the total function $\mathcal{A} \to TIME$ that associates to an activity its finishing time slot (i.e., when the movie ends or when the restaurant closes),
- **duration** the total function $\mathcal{A} \to TIME$ that associates to an activity the user's duration in time slots.
- **area** the total function $\mathcal{A} \to Loc$ that associates to an activity the location where it takes place.
- **publicRating** a complete function $\mathcal{A} \to \mathbb{N}$ that associates to an activity a possible rating.[3] Ratings are represented with natural numbers: the bigger the rating, the better the activity is considered.

With a slight abuse of notation, given an activity a and a property p we denote with $a.p$ (rather than with p(a)) the value of the propriety p for activity a.

Example 1. A restaurant activity $a \in \mathcal{A}_r$ might be characterized by $a.\texttt{startTime} = 228$, meaning that the restaurant opens at 19:00 (assuming a time slot of 5 min 228 corresponds to 19), $a.\texttt{endTime} = 276$, meaning that the restaurant closes at 23:00, $a.\texttt{duration} = 18$ meaning that the dinner will last 90 min, $a.\texttt{area} = 5$ meaning that the location is identified with id 5, and $a.\texttt{publicRating} = 3$ meaning that the public rating is 3.

As far as preferences are concerned, based on findings such as those reported in [23], we avoid using a very refined scale and we allow only 5 values: from -2 indicating a strong dislike to a $+2$ indicating a strong preference, and 0 indicating a neutral opinion. Formally user preferences are defined as follows.

Definition 2 (Activity Preferences). *Given a set of users $\mathcal{U}$ and a set of activities $\mathcal{A}$, an activity preference is a total function* $\textbf{pref} : \mathcal{U} \times \mathcal{A} \to \{-2, -1, 0, 1, 2\}$.

Since the user has to move between different locations, to properly define a valid plan we need a metric that evaluates the distance between different activities. We are only interested in the time to go from one activity to another. Hence, we abstract from physical details such as GPS coordinates and means of transportation and we simply consider a distance metric between locations which is given in terms of times slots (needed to go from one location to the other).

Definition 3 (Distance Metric). *Given a set of locations Loc and a set of time slots $TIME = \{1, \dots, T_{max}\}$ a distance metric is a total function* $\textbf{dist} : Loc \times Loc \to TIME$.

We are now ready to define what is a plan: a simple association of activities to the users.

---

[3] Specifically, the rating value of activity ranges from 0 to 5, where 0 means "no rating information is given".

Definition 4 (Plan). *Let us consider a set of users $\mathcal{U}$, two sets of activities $\mathcal{A}_M$ and $\mathcal{A}_R$ and a set of time slots $TIME$. A plan is a total function* $\mathtt{plan}$: $\mathcal{U} \rightarrow (\mathcal{A}_M \times TIME) \times (\mathcal{A}_R \times TIME)$ *that associates to a user a movie and restaurant activity with their beginning time slots.*

Example 2. A plan $\mathtt{plan}(u) = ((a_1, 108), (a_2, 138))$ means that to the user u is assigned the activity a_1 that starts at 9:00 and the activity a_2 at 11:30.

Not all the plans are valid: For instance a plan may schedule two overlapping activities for a user. For this reason, we introduce the notion of plan validity that captures the constraints that a feasible plan must possess.

Definition 5 (Plan Validity). *Given a positive integer* $\mathtt{maxGroupNum}$ *representing the maximal number of sub-groups allowed, a positive integer* $\mathtt{minCardinality}$ *representing the minimal size of a group, and a positive integer* $\mathtt{maxWait} \in TIME$ *representing the maximal waiting time between two activities, a plan* $\mathtt{plan}$ *is said valid iff:*

- *starting and ending time are satisfied. Formally, for each user $u \in \mathcal{U}$, if* $\mathtt{plan}(u) = ((a_m, t_m), (a_r, t_r))$ *then* $\mathtt{startTime}(a_m) \leq t_m \leq \mathtt{endTime}(a_m) - \mathtt{duration}(a_m)$ *and* $\mathtt{startTime}(a_r) \leq t_r \leq \mathtt{endTime}(a_r) - \mathtt{duration}(a_r)$;
- *activities do not overlap. Formally, $\forall u \in \mathcal{U}$, if $\mathtt{plan}(u) = ((a_m, t_m), (a_r, t_r))$ then $t_r \geq t_m + \mathtt{duration}(a_m) + \mathtt{dist}(\mathtt{area}(a_m), \mathtt{area}(a_r))$;*
- *activities are not too far apart. Formally, $\forall u \in \mathcal{U}$, if $\mathtt{plan}(u) = ((a_m, t_m), (a_r, t_r))$ then $t_r \leq t_m + \mathtt{duration}(a_m) + \mathtt{maxWait}$;*
- *the number of groups is limited by $\mathtt{maxGroupNum}$. Formally, $|\{(a_m, t_m) \mid \forall u \in \mathcal{U} . \mathtt{plan}(u) = ((a_m, t_m), (a_r, t_r))\}| \leq \mathtt{maxGroupNum}$ and $|\{(a_r, t_r) \mid \forall u \in \mathcal{U} . \mathtt{plan}(u) = ((a_m, t_m), (a_r, t_r))\}| \leq \mathtt{maxGroupNum}$;*
- *the cardinality of the group is bounded by $\mathtt{minCardinality}$. Formally, for all activities $a_m \in \mathcal{A}_m$, and time slots $t_m \in Time$ $|\{u \mid \forall u \in \mathcal{U} . \mathtt{plan}(u) = ((a_m, t_m), (a_r, t_r))\}|$ is 0 or greater or equal than $\mathtt{minCardinality}$. Similarly, for all activities $a_r \in \mathcal{A}_R$, and time slots $t_r, \in Time$ $|\{u \mid \forall u \in \mathcal{U} . \mathtt{plan}(u) = ((a_m, t_m), (a_r, t_r))|$ is 0 or greater or equal than $\mathtt{minCardinality}$.*

In order to simplify the presentation, given a plan $\mathtt{plan}(u) = ((a_1, t_1), (a_2, t_2))$ in the following we will use $\mathtt{plan}(u).a_m$ for denoting a_1, $\mathtt{plan}(u).a_r$ for a_2, $\mathtt{plan}(u).t_m$ for t_1, and $\mathtt{plan}(u).t_r$ for t_2 (m stands for movie, r for restaurant).

We are now ready to define the $\mathtt{NightSplit}$ optimization problem. Intuitively, the $\mathtt{NightSplit}$ goal is to find a valid plan that optimizes the individual activity preferences and the public activity preferences. Different criteria may be used to combine these preferences. $\mathtt{NightSplit}$ allows a great flexibility combining all these objectives into one by summing them according to some weights.

Definition 6 ($\mathtt{NightSplit}$). *Let η be a real number $\in [0,1]$ representing the weight associated to the individual activity preferences and the public preferences[4]. The $\mathtt{NightSplit}$ problem is to find the valid plan $\mathtt{plan}^*$ that maximizes*

---

[4] Public preferences are useful to break the ties when users have very general individual preferences (e.g., I like all the movies).

the following objective function.

$$obj(plan) = \eta \cdot sum_{act}(plan) + (1 - \eta) \cdot sum_{pub}(plan) \qquad (1)$$

where sum_{act} and sum_{pub} are the sum of the individual activities preferences and public preferences as define below:

$$sum_{act}(plan) = \sum_{u \in \mathcal{U}} (pref(u, plan(u).a_m) + pref(u, plan(u).a_r)) \qquad (2)$$

$$sum_{pub}(plan) = \sum_{u \in \mathcal{U}} (plan(u).a_m.publicRating + plan(u).a_r.publicRating)$$
$$(3)$$

As can be expected, even tough this formulation is rather simple, `NightSplit` is an NP-hard problem.

Theorem 1 (NP-hardness). *The `NightSplit` is NP-hard.*

Proof. To prove hardness, we reduce the NP-complete problem Perfect Expected Component Sum (PECS) [4] to the decision version of `NightSplit`, i.e., the problem to find whether there exists a valid plan such that the objective function `obj` is greater or equal than a given value.[5] An instance of PECS consists of a collection V of m-dimensional boolean vectors, i.e., $V \subset \{0,1\}^m$ and a number k. The problem is to determine whether there exists a disjoint partition of V into k subsets $V_1, \ldots, V_k$ such that $\sum_{i=1}^{k} \max_{1 \leq j \leq m}(\sum_{\bar{v} \in V_i} \bar{v}|j|) = |V|$.

Given an instance of PECS we map every vector $\bar{v}_i \in V$ as a user u_i having some preferences over m different movies. The intuition behind the hardness proof is to exploit the planning of the movie activities to find a solution for PECS. We assume that there is only one location, that the m movie activities start at the time slot 0 and end at time slot 1 with duration 1. Similarly, we assume that there are m different restaurant activities that start at time slot 1 and end at time slot 2 with duration 1. We set `maxGroupNum` to k, `minCardinality` to 1, `maxWait` to 1, and we assume that the function `dist` is the constant function 0. In this way all the movie activities are compatible with the restaurant activities and all the possible plans that have a maximal number of k groups are valid. We set the preferences of the movie activities to reflect the values of the vector $\bar{v}$. Formally, for all $1 \leq i \leq |V|$ and $1 \leq j \leq m$ we define $pref(u_i, a_j) = \bar{v}|j|$. We set to 0 instead the preferences for all the restaurant. We set the weight of the user preferences η to 1 while we discard the public preferences with $1 - \eta = 0$.

Based on the definition of `NightSplit`, it is easy to see that $\sum_{i=1}^{k} \max_{1 \leq j \leq m}$ $(\sum_{\bar{v} \in V_i} \bar{v}|j|) = |V|$ iff the `obj` of the `NightSplit` problem is equal to $|V|$. The partition induced on the users performed by `NightSplit` corresponds to the partition of V into the k set of vectors $V_1, \ldots, V_k$.   □

---

[5] The decision version of the problem requires the "greater or equal" operator. Similar to the theorem presented in [4], our theorem holds because the sum of the preferences is never greater than V.

3.1   Useful Extensions

While NightSplit is already NP-hard, there are some useful extensions of it that do not alter its complexity class and its nature. In the following we just comment on some of them that are considered in the online NightSplitter. For space reasons they are not formally defined here, however their definition is straightforward.

First observe that the notion of a valid plan can be further restricted considering additional constraints. For example, it may be useful to allow users to indicate that they are not available before or after a given time. Moreover, the minimal number of people required to form a group or the number of groups can vary depending on the activity (e.g., it may be the case that for going to the movie we accept to split the group in two while to eat in a restaurant we do not allow any split). Other useful extensions concern the definition of different kinds of user preferences. For instance, usually users like to hang out in certain locations and they want to minimize the traveling time between the activities, minimize the waiting time, start the activities as soon as possible, etc. All these preferences may be considered by adding further terms to the objective function that we optimize in NightSplit, possibly reducing its weight by an appropriate parameter. NightSplitter has been designed to be easily extensible and take into account new sources of user preferences or constraints. For instance, the preferences over some areas can can be easily defined in the profile menu of the user and then taken into account when generating the plans.

Finally, we could also relax the limit of two activities, considered in this paper, and we could extend our system to applications where more activities can be performed in sequence, especially in the tourism industry, following, e.g., [18, 25].

4   Solution Approaches

To solve the NightSplit problem we propose two different approaches. The first one relies on Constraint Programming (CP) and allows us, in principle, to obtain the optimal solution. The second approach uses Simulated Annealing (SA), a probabilistic local search procedure which, under certain conditions for its parameters, is known to find the optimal solution with a probability approaching one. In this section we briefly describe the CP and SA approaches, while we defer to Sect. 5 for their comparison.

4.1   NightSplit and Constraint Programming

Constraint Programming (CP) [24, 26] is a widely adopted approach for solving NP-hard problems. The CP paradigm enables to express complex relations in form of constraints to be satisfied. In particular a Constraint Satisfaction Problem (CSP) $\mathcal{P} = (\mathcal{X}, \mathcal{D}, \mathcal{C})$ consists of a finite set of variables $\mathcal{X}$, each of which associated with a domain $D_x \in \mathcal{D}$ of possible values that it could take, and a set of constraints $\mathcal{C}$ that defines all the admissible assignments of values to the variables [19]. Given a CSP the goal is normally to find a solution, i.e., an assignment to the variables that satisfies all the constraints of the problem. When an objective function needs to be minimized or maximized we deal instead with a Constraint Optimization Problems (COPs), i.e., a generalized CSP where the goal is not only to find a solution but among all possible solutions the one that maximizes or minimizes the objective function.

Clearly the NightSplit problem can be seen as a COP. For every user u we have introduced:

- a variable M_u representing the selection of the movie activity. The domain of this value is the finite domain of all the possible movie activities;
- a variable R_u representing the selection of the restaurant. The domain of this value is the finite domain of all the possible restaurant activities;
- two variables $S_{u,1}, S_{u,2}$ representing the beginning of the activities. The domain of these variables is the finite set of the possible time slots;
- two variables $G_{u,1}, G_{u,2}$ representing the subgroup to which user u belongs (for the first and second activity respectively). The domain of these variables depend on the maximal number of groups allowed for activity.

With these variables it is possible to state all the constraints as listed in Definition 5. For instance, the first constraint bounding the starting time of the activities might be expressed by stating that movie_start$[M_u] \leq S_{u,1}$ where movie_start is the array storing the movies starting time. This constraint is simply a disequality between two expressions: the first retrieves the concrete value from an array while the second is the variable $S_{u,1}$. Note that CP solvers can employ efficient techniques to handle this kind of equalities or disequalities (global constraints). Moreover, for this particular case, the constraint setting x as the value taken by the y-th value of the array is known as element constraint [24], which is often supported by constraint solvers that adopt ad-hoc propagation algorithms to speed up the search of solutions.

To model all the constraints we used MiniZinc [21], which is the de-facto language to define CSPs and COPs and is supported by a huge variety of constraint solvers. Since the majority of the solvers does not support real variables, we restrict the use of the preference weights η to rational numbers only. A detailed explanation of the MiniZinc model and all the constraints defined is outside the scope of this paper. For more information we invite the reader to consult [14].

Remark 1. Beside CP, we have also tried to encode the NightSplit to exploit Satisfiability-Modulo-Theories (SMT) solvers. SMT solving extends and improves upon SAT solving by introducing the possibility of stating constraints in some expressive theories, e.g., arithmetic or bit-vector expressions. While all the constraints of NightSplit can be encoded in SMT, we were not able to provide an encoding linear w.r.t. the number of activity locations. Indeed, differently to what happens in CP where the *element* constraint can be used [24], in the SMT case the encoding of the traveling time between two activities requires the introduction of a quadratic number of constraints w.r.t. the number of locations. Based on our test, since we had more than 300 locations, the addition of these quadratic number of constraints hindered the use of SMT solvers. For this reason, in Sect. 5, we will compare only the performances of the CP and SA approaches.

4.2 NightSplit and Simulated Annealing

Simulated Annealing (SA) [1] is a local search technique inspired by the annealing process in metallurgy. SA has been widely used for approximating the global optimum of a given function. Given an initial solution, random moves are made to produce new potential solutions. A new solution that improves the previous one is (usually) always accepted. Solutions that worsen the current solution are instead accepted with a probability that, like the temperature in the annealing process, is gradually decreasing. Accepting worse solutions is a fundamental property because it allows for a more extensive search for the optimal solution, possibly avoiding getting stuck in local optima.

Contrary to the CP technique described before, SA can not guarantee that the final solution obtained is optimal. However, for discrete and large search spaces, SA scales better and could produce (sub)optimal solution very quickly.

Among all the different implementations of SA available we rely on the re-implementaiton in PHP of the python SA module [22]. After some manual tuning, we have fixed the parameters to control the decreasing of the temperature and the number of iterations (50000). The temperature exponentially decreases as the algorithm progresses. As customary, a move causing a decrease in state energy (i.e., an improvement of the NightSplit objective function) was always accepted. Moves instead increasing the state energy (i.e., a worse solution) but within the bounds of the temperature are also accepted.

The initial solution is obtained by randomly generating the assignments from users to activities. To obtain instead a valid plan from a current solution we proceed as follows: (i) we randomly select movie activity assignments or restaurants activity assignments and modify them; (ii) we randomly select a subset of users U; (iii) we assign a new activity a to the selected users in U. This activity is randomly chosen among all the activities for which the aggregated preference of the U users is positive. Intuitively, this avoids selecting an activity that no user in U wants to perform; (iv) if the assigned activity is not compatible with other existing ones (e.g., if for user u we select a movie activity a that overlaps with his/her restaurant activity) we delete these activities; (v) for every user u that has no activity assigned we look at the activities assigned to other users, check if any of them is compatible with the updated activity and if so we assign this activity to the user u assuming that this does not violate the group constraints.

5 Empirical Experiment

In this section we describe the experiments performed in order to validate the scalability of NightSplitter and we discuss the results.

We have considered for the experiments real data from the city of Paris: The movies information - for 93 cinemas and currently 153 different movies (with 1950 projections a day) - is retrieved from Allociné [2, 10], restaurant data - for 13598 restaurants - from Tripadvisor [30]. OpenStreetMap and GoogleMaps were also used to identified 317 positions of metro stations: for each activity we considered its nearest metro station as its location. The data related to the preferences was collected from Movielens [12] and Yelp [31]. These datasets, originally defined for activities in the U.S., were converted for Paris activities. This was done by mapping the names of the Paris activities to the activity existing in the preference dataset while preserving the activity category and the public rating. After that, we randomly sampled 8,000 users for the restaurant activity and 5.300 users for the movies activity to use their individual preferences for the experiments. The statistics related to the activities and preference data are summarized in Table 1 where the last column indicates the average preferences of the users. Note that if a restaurant was open for two separate intervals (e.g., from 11 to 15 and from 19 to 23) this was captured by considering two separate activities.

Table 1. Summary statistics of the dataset

Activity type	Activities	Users	Avg. pref
Movies	1950	5300	6
Restaurants	17069	8000	2

Since the goal is to provide a responsive tool, for the experiments we fixed a timeout of 60 s taking the best solution found by the tested approach within this time frame. For each testing scenario we repeated the experiment 30 times. For every experiment we match the chosen number of user with random user from the dataset using their preferences. We allow the subgroups to be formed by at least 2 people, the time slot unit to be 5 min assuming that the duration for a dinner/lunch is 90 min. The experiments were run on an Ubuntu Intel Core 3.30 GHz machine with 8 GB of RAM.

We compared the performance of three different state-of-the-art CP solvers, namely Chuffed [6], Or-Tools [11], and HCSP [13],[6] and the SA method described in the previous section.

We first compare the three different CP solvers for different number of users, assuming to have only 2 subgroups and not taking into account the public ratings (i.e., $\eta = 1$). Figure 2 shows the average times needed by the solvers to find

---

[6] We selected these solvers based on the recent results of the MiniZinc Challenge 2016 [27]. In particular Or-Tools won a golden medal in the Fixed category and HCSP won a golden medal in Free and Parallel category. Chuffed was the second best solver of the entire Challenge after LCG-Glucose-free which is not publicly available. We would remark also that our problem instances have been submitted to the incoming MiniZinc Challenge 2017 [28].

the optimal value by varying the number of users, where the filled icons mean that the solver has proven the optimality of the solution for all the 30 repeated tests. Chuffed has always computed the optimal solution for values up to 9 users and it is the fastest among the three solvers. The Or-Tools cannot find the optimal solution within the timeout for more than 5 users, while the HCSP solver performs slightly better Or-Tools and occasionally it is still capable to prove optimal solution for up to 13 users. Similar results are obtained when increasing the number of subgroups or when public ratings are taken into account by lowering the value of the η parameter. Since Chuffed outperforms the other solvers in our application, in the following we show only the performance of this solver for the comparison with SA approach.

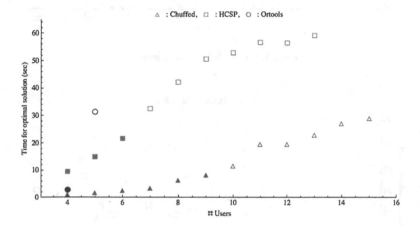

Fig. 2. CP Solvers comparison.

We compare the performance of Chuffed and SA in terms of quality of the solution for a number of users ranging between 4 and 40, assuming 2 subgroups could be formed, and the weight associated to the individual preference η to be 1. Figures 3 and 4 depicts respectively the average solution score and the average time needed to find the best solution for the 30 repeated tests (the green dot in Fig. 3 representing the number of tests such that CP proves solution optimality). The plots show that for a limited number of users SA is competitive with Chuffed, while for more than 15 users SA is definitely better. The advantage of the CP solution is that for less than 10 users the solutions are proven optimal while some SA solutions were suboptimal. From the plot it is however possible to see that the number of solutions that could be proven optimal in less than 60 s decreases at the increase of the number of users. With more than 20 users no solution was proven optimal. It is clearly visible that Chuffed is better only for a limited number of users while the SA is often able to find the best solution within the first 15 s.

We then compare the two approaches by varying the number of possible subgroups from 1 to 8. In Fig. 5 we present the plots obtained considering 32,

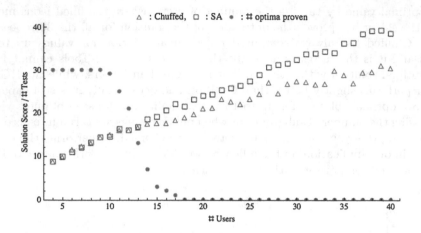

Fig. 3. CP vs SA comparison.

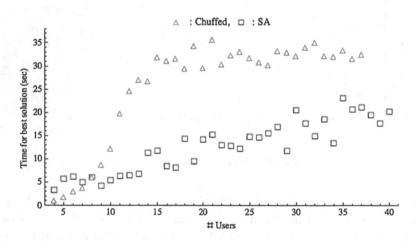

Fig. 4. Time to find the best solution.

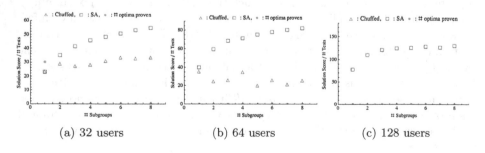

(a) 32 users   (b) 64 users   (c) 128 users

Fig. 5. Comparison of CP and SA varying the number of subgroups.

64, and 128 users. From the plots it can be seen that the CP technique is only suitable with few users and when no more than 2 subgroups can be formed. When the number of users increases or more than 2 groups can be formed the solutions provided by the CP solver within 60 s are worse than the ones produced by the SA. In our biggest scenario, considering 128 users, the SA is the only viable choice because unfortunately the CP solver is not even able to provide a single solution (hence the lack of points for Chuffed in Fig. 5c). We conduct experiments also varying the weights used to aggregate the individual and public preferences. In these cases there are no significant changes, except that the final score increases. Figure 6 shows for instance the performances of Chuffed and SA while varying the parameter η considering 32 users and 2 subgroups. In particular, Fig. 6a presents the average time when the best value is found while Fig. 6b presents the average score found after 60 s. As long as the user's preferences are accounted for (i.e., $\eta \neq 0$), it is immediately visible that with this amount of users the SA approach is better than Chuffed since SA is able to find better values in a short amount of time and Chuffed is not able to prove the optimality of the solutions within 60 s.

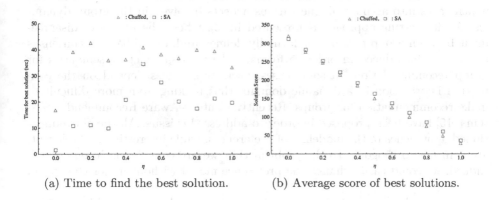

(a) Time to find the best solution.   (b) Average score of best solutions.

Fig. 6. Comparison of CP and SA varying η.

Summarizing, we may conclude that when considering two subgroups and few users the CP approach may be useful and even prove the optimality of the solution. For more subgroups and more users the SA approach is better. For those experiments where the optimality of the solutions was proven, the SA approach was able to propose competitive solutions. We conjecture that this holds also for big instances where we were not being able to prove the optima.

6 Related Work

The literature on recommender or planning systems is very large and we omit all the references to works which consider the case of a single user only, with the exception of [25], which uses CSP techniques for building a tourist recommendation and planning application. Concerning group recommender systems, [5] provide a survey on several existing approaches while [9] presents a recommender system for tourism based on the tastes of the users, their demographic classification and the places they have visited in former trips. More recently, the idea of group splitting has appeared in some papers. Notably [4] proposes an approach for forming groups of users in order to maximize satisfaction. The work [18] introduces the problem of group tour recommendation which includes the problem of forming tour groups whose members have similar interests. Differently from our case, all the above mentioned papers consider groups or sub-groups as fixed entities, which once are created cannot be modified. With our approach, instead, for each activity we have a different group formation, that is, we can have two users who are in the same group for the first activity (the movie) and are in different groups for the second one (the dinner). Moreover, the above papers focus on the theoretical aspects rather than presenting a tool.

There exist also several works which address the problem of group preference modeling and the definition of an appropriate notion of "group satisfaction" [16,20]. In general these are difficult tasks, since it is hard to find a definition which takes into account all the various aspects involved in the group dynamics. An interesting approach is presented in [3], where the notion of disagreement between group members is formally defined and, on its basis, a consensus function is introduced in order to formally define a satisfactory semantics for group recommendation. In some cases, users preferences depend on the contextual information in a dynamic domain, thus making even more difficult to make recommendation for groups. Recently Context-Aware Recommender Systems [15] have been proposed in order to address this issue. All the above mentioned approaches to the modeling of preferences, while interesting and relevant, are somehow orthogonal to the problem that we are considering in our paper. Indeed, we could easily change the preference model without major changes in our tool.

To conclude we would like to mention also the works conducted in [7,17,25] which present recommendation and planning systems targeting a single user only but are interesting for us since they consider models of generating itineraries (for touristic applications) which could be integrated with our tool.

7   Conclusions and Future Work

We have presented NightSplitter, an on-line tool that is able to plan movie and dinner activities for a group of users, possibly splitting them in subgroups to optimally satisfy their preferences. The tool is based on a formal model and two different technologies - Constraint Programming and Simulate Annealing - which can be easily adapted to other applications. The tests we have conducted show that our tool can be effectively used on real data for the city of Paris, with thousands of activities and hundred of users. The comparison between CP and the simulated annealing approach show that the latter can scale up to consider larger number of users, making our approach feasible also for quite different social applications.

We are now extending our work along several directions: First, we are considering a greater number of different activities and we are adding some more features such as, e.g., the selection of a preferred limited area for the activities (this is done by selecting an area on the map). Second, the recommendation semantics adopted in our model is aggregated preference: we are now exploring different notions of group recommendation semantics such as least misery, most pleasure, Borda count, etc. [20]. In particular we would like to see whether the semantics proposed in [4] with the related algorithms could improve our approach. Third, we would like to investigate techniques for group definition using social factors and group dynamics as those suggested in [16]. Fourth, we would like to explore possible improvements for the CP approach by using, e.g., linearizion of the constraints, column generation methods, or the use of presolve.

References

1. Aarts, E.H., Korst, J.H.: Simulated annealing. ISSUES 1, 16 (1988)
2. AlloCiné (2016). http://www.allocine.fr
3. Amer-Yahia, S., Roy, S.B., Chawlat, A., Das, G., Yu, C.: Group recommendation: semantics and efficiency. Proc. VLDB Endow. 2(1), 754–765 (2009)
4. Basu Roy, S., Lakshmanan, L.V., Liu, R.: From group recommendations to group formation. In: Proceedings of the 2015 ACM SIGMOD International Conference on Management of Data, pp. 1603–1616. ACM (2015)
5. Boratto, L., Carta, S.: State-of-the-art in group recommendation and new approaches for automatic identification of groups. In: Soro, A., Vargiu, E., Armano, G., Paddeu, G. (eds.) Information Retrieval and Mining in Distributed Environments, vol. 324, pp. 1–20. Springer, Heidelberg (2010). doi:10.1007/978-3-642-16089-9_1
6. Chu, G., de la Banda, M.G., Mears, C., Stuckey, P.J.: Symmetries and lazy clause generation. In: Proceedings of the 16th International Conference on Principles and Practice of Constraint Programming (CP 2010) Doctoral Programme, pp. 43–48 (2010)
7. Di Bitonto, P., Di Tria, F., Laterza, M., Roselli, T., Rossano, V., Tangorra, F.: A model for generating tourist itineraries. In: 2010 10th International Conference on Intelligent Systems Design and Applications, pp. 971–976. IEEE (2010)
8. Dwork, C.: Differential privacy: a survey of results. In: Agrawal, M., Du, D., Duan, Z., Li, A. (eds.) TAMC 2008. LNCS, vol. 4978, pp. 1–19. Springer, Heidelberg (2008). doi:10.1007/978-3-540-79228-4_1
9. Garcia, I., Sebastia, L., Onaindia, E.: On the design of individual and group recommender systems for tourism. Expert Syst. Appl. 38(6), 7683–7692 (2011)
10. Gauvin, E.: Allocine helper (2016). https://github.com/etienne-gauvin/api-allocine-helper
11. Google: Google or-tools (2016). https://developers.google.com/optimization/
12. Harper, F.M., Konstan, J.A.: The movielens datasets: history and context. ACM Trans. Interact. Intell. Syst. (TiiS) 5(4), 19 (2016)
13. Ivrii, A., Ryvchin, V., Strichman, O.: Mining backbone literals in incremental SAT. In: Heule, M., Weaver, S. (eds.) SAT 2015. LNCS, vol. 9340, pp. 88–103. Springer, Cham (2015). doi:10.1007/978-3-319-24318-4_8

14. Jacopo Mauro, T.L.: Minizinc model (2017). http://cs.unibo.it/t.liu/nightsplitter/mzn.html

15. Khoshkangini, R., Pini, M.S., Rossi, F.: A self-adaptive context-aware group recommender system. In: Adorni, G., Cagnoni, S., Gori, M., Maratea, M. (eds.) AI*IA 2016. LNCS, vol. 10037, pp. 250–265. Springer, Cham (2016). doi:10.1007/978-3-319-49130-1_19

16. Kompan, M., Bielikova, M.: Group recommendations: survey and perspectives. Comput. Inform. **33**(2), 446–476 (2014)

17. Le Berre, D., Marquis, P., Roussel, S.: Planning personalised museum visits. In: ICAPS (2013)

18. Lim, K.H., Chan, J., Leckie, C., Karunasekera, S.: Towards next generation touring: personalized group tours (2016)

19. Mackworth, A.K.: Consistency in networks of relations. Artif. Intell. **8**(1), 99–118 (1977)

20. Masthoff, J.: Group recommender systems: combining individual models. In: Ricci, F., Rokach, L., Shapira, B., Kantor, P.B. (eds.) Recommender Systems Handbook, pp. 677–702. Springer, Heidelberg (2011). doi:10.1007/978-0-387-85820-3_21

21. Nethercote, N., Stuckey, P.J., Becket, R., Brand, S., Duck, G.J., Tack, G.: MiniZinc: towards a standard CP modelling language. In: Bessière, C. (ed.) CP 2007. LNCS, vol. 4741, pp. 529–543. Springer, Heidelberg (2007). doi:10.1007/978-3-540-74970-7_38

22. Perry, M.: Python module for simulated annealing (2017). https://github.com/perrygeo/simanneal

23. Preston, C.C., Colman, A.M.: Optimal number of response categories in rating scales: reliability, validity, discriminating power, and respondent preferences. Acta psychol. **104**(1), 1–15 (2000)

24. Rossi, F., Van Beek, P., Walsh, T.: Handbook of Constraint Programming. Elsevier, Amsterdam (2006)

25. Sebastia, L., Garcia, I., Onaindia, E., Guzman, C.: E-Tourism: a tourist recommendation and planning application. Int. J. Artif. Intell. Tools **18**(5), 717–738 (2009). http://dx.doi.org/10.1142/S0218213009000378

26. Smith, B.M.: Modelling for constraint programming. In: Lecture Notes for the First International Summer School on Constraint Programming (2005)

27. Minizinc Team: Minizinc challenge 2016 (2016). http://www.minizinc.org/challenge2016/challenge.html

28. Minizinc Team: Minizinc challenge 2017 (2017). http://www.minizinc.org/challenge2017/challenge.html

29. Liu, T., Mauro, J., Di Cosmo, R., Gabbrielli, M.: Nightsplitter (2017). http://cs.unibo.it/t.liu/nightsplitter

30. TripAdvisor (2016). https://www.tripadvisor.com

31. Yelp: Yelp dataset challenge (2016). http://yelp.com/dataset_challenge/

Time-Aware Test Case Execution Scheduling for Cyber-Physical Systems

Morten Mossige[1,3], Arnaud Gotlieb[2], Helge Spieker[2(✉)], Hein Meling[3], and Mats Carlsson[4]

[1] ABB Robotics, Bryne, Norway
morten.mossige@uis.no
[2] Simula Research Laboratory, Lysaker, Norway
{arnaud,helge}@simula.no
[3] University of Stavanger, Stavanger, Norway
hein.meling@uis.no
[4] RISE SICS, Kista, Sweden
mats.carlsson@ri.se

Abstract. Testing cyber-physical systems involves the execution of test cases on target-machines equipped with the latest release of a software control system. When testing industrial robots, it is common that the target machines need to share some common resources, e.g., costly hardware devices, and so there is a need to schedule test case execution on the target machines, accounting for these shared resources. With a large number of such tests executed on a regular basis, this scheduling becomes difficult to manage manually. In fact, with manual test execution planning and scheduling, some robots may remain unoccupied for long periods of time and some test cases may not be executed.

This paper introduces TC-Sched, a time-aware method for automated test case execution scheduling. TC-Sched uses Constraint Programming to schedule tests to run on multiple machines constrained by the tests' access to shared resources, such as measurement or networking devices. The CP model is written in SICStus Prolog and uses the Cumulatives global constraint. Given a set of test cases, a set of machines, and a set of shared resources, TC-Sched produces an execution schedule where each test is executed once with minimal time between when a source code change is committed and the test results are reported to the developer. Experiments reveal that TC-Sched can schedule 500 test cases over 100 machines in less than 4 min for 99.5% of the instances. In addition, TC-Sched largely outperforms simpler methods based on a greedy algorithm and is suitable for deployment on industrial robot testing.

1 Introduction

Continuous integration (CI) aims to uncover defects in early stages of software development by frequently building, integrating, and testing software systems.

A. Gotlieb and H. Spieker—These authors are supported by the ResearchCouncil of Norway (RCN) through the research-based innovation center Certus, under the SFI programme.

© Springer International Publishing AG 2017
J.C. Beck (Ed.): CP 2017, LNCS 10416, pp. 387–404, 2017.
DOI: 10.1007/978-3-319-66158-2_25

When applied to the development of cyber-physical systems (CPS)[1], the process may include running integration test cases involving real hardware components on different machines or machines equipped with specific devices. In the last decade, CI has been recognized as an effective process to improve software quality at reasonable costs [13,14,27,35].

Different from traditional testing methods, running a test case in CI requires tight control over the *round-trip time*, that is, the time from when a source code change is committed until the success or failure of the build and test processes is reported back to the developer [15]. Admittedly, the easiest way to minimize the round-trip time is simply to execute as many tests as possible in the shortest amount of time. But the achievable parallelism is limited by the availability of scarce global resources, such as a costly measurement instrument or network device, and the compatible machines per test case, targeting different machine architecture and operating systems. These global resources are required in addition to the machine executing the test case and thereby require parallel adjustments of the schedule for multiple machines.

Thus, computing an optimal test schedule with minimal round-trip time is a challenging optimization problem. Since different test cases have different execution times and may use different global resources that are locked during execution, finding an optimal schedule manually is mostly impossible. Nevertheless, manual scheduling still is state-of-the-practice in many industrial applications, besides simple heuristics. In general, successful approaches to scheduling use techniques from Constraint Programming (CP) and Operations Research (OR), additionally metaheuristics are able to provide good solutions to certain scheduling problems. We discuss these approaches further in Sect. 2.

Informally, the optimal test scheduling problem (OTS) is to find an execution order and assignment of all test cases to machines. Each test case has to be executed once and no global resource can be used by two test cases at the same time. The objective is to minimize the overall test scheduling and test execution time. The assignment is constrained by the compatibility between test cases and machines, that is, each test case can only be executed on a subset of machines.

This paper introduces TC-Sched, a time-aware method to solve OTS. Using the CUMULATIVES [1,5] global constraint, we propose a cost-effective constraint optimization search technique. This method allows us to (1) automatically filter invalid test execution schedules, and (2) find among possible valid schedules, those that minimize the global test execution time (i.e., makespan). To the best of our knowledge, this is the first time the problem of optimal scheduling test suite execution is formalized and a fully automated solution is developed using constraint optimization techniques. TC-Sched has been developed and deployed together with ABB Robotics, Norway.

An extensive experimental evaluation is conducted over test suites from industrial software systems, namely an integrated control system for industrial robots and a product line of video-conferencing systems. The primary goal in this paper is to demonstrate the scalability of the proposed approach for CI

---

[1] CPS can simply be seen as communicating embedded software systems.

processes involving hundreds of test cases and tens of machines, which corresponds to a realistic development environment. Furthermore, we demonstrate the cost-effectiveness of integrating our approach within an actual CI process.

2   Existing Solutions and Related Work

Automated solutions to address the OTS problem are not yet common practice. In industrial settings, test engineers manually design the scheduling of test case execution by allocating executions to certain machines at a given time or following a given order. In practice, they manage the constraints as an aggregate and try to find the best compromise in terms of the time needed to execute the test cases. Keeping this process manual in CI is paradoxical, since every activity should, in principle, be automated.

Regression testing [28], i.e. the repeated testing of systems after changes were made, in CI covers a broad area of research works, including automatic test case generation [9], test suite prioritization and test suite reduction [14]. There, the idea of controlling the time taken by optimization processes in test suite prioritization is not new [12]. In test suite prioritization, [38] proposed to use time-aware genetic algorithms to optimize the order in which to execute the test cases. Zhang et al. further refined this approach in [39] by using integer linear programming. On-demand test suite reduction [17] also exploits integer linear programming for preserving the fault-detection capability of a test suite while performing test suite reduction. Cost-aware methods are also available for selecting minimal subsets of test cases covering a number of requirements [16,23]. All these approaches participate in a general effort to better control the time allocated to the optimization algorithms when they are used in CI processes. Note however that test suite execution scheduling is different to prioritization or reduction as it deals with the notion of scheduling in time the execution of all test cases, without paying attention to any prioritization or reduction.

Scheduling problems have been studied in other contexts for decades and an extensive body of research exists on resource-constrained approaches. The scheduling domain is divided into distinct areas such as process execution scheduling in operating systems and scheduling of workforces in a construction project. The scheduling problem of this paper belongs to a scheduling category named resource-constrained project scheduling problem (RCPSP; see [7,8,18] for an extensive overview). RCPSP is concerned with finding schedules for resource-consuming tasks with precedence constraints in a fixed time horizon, such that the makespan is minimized [18]. From the angle of RCPSP, global resources can be expressed as *renewable resources* which are available with exactly one unit per timestep and can therefore only be consumed by a single job per timestep.

RCPSP has been addressed by both exact methods [22,30,32,36], as well as heuristic methods [19,21]. Due to the vast amount of literature, we will focus on CP/OR-methods most closely related to the work of this paper. The clear trend in both CP and OR is to solve such problems with hybrid approaches, like, for instance, the work by Schutt et al. [29] or Beck et al. [3]. Furthermore, *disjunctive scheduling problems*, a subfamily of RCPSP addressing unary resources (in our terms global resources), have been effectively solved, e.g. by lazy clause generation [33].

RCPSP is considered to be a generalization of *machine scheduling problems* where *job shop scheduling* (JSS) is one of the best known [20]. JSS is the special case of RCPSP where each operation uses exactly one resource, and FJSS (*flexible job shop scheduling*) further extends JSS such that each operation can be processed on any machine from a given set. The FJSS is known to be NP-hard [4].

While OTS is closely related to FJSS, and efficient approaches to FJSS are known [6,31], there are some differences. First, in OTS, execution times are machine-independent. Second, each job in OTS consists of only one operation, while in FJSS one job can contain several operations, where there are precedences between the operations. Finally, some operations additionally require exclusive access to a global resource, preventing overlap with other operations.

3   Problem Modeling

This section contains a formal definition of the OTS problem for test suite execution on multiple machines with resource constraints. Based on this definition, we propose a constraint optimization model using CUMULATIVES global constraint.

3.1   Optimal Test Case Execution Scheduling

Optimal test case scheduling[2] (OTS) is an optimization problem $(\mathcal{T}, \mathcal{G}, \mathcal{M}, d, g, f)$, where $\mathcal{T}$ is a set of n test cases along with a function $d : \mathcal{T} \longrightarrow \mathbb{N}$ giving each test case a duration d_i; a set of global resources $\mathcal{G}$ along with a function $g : \mathcal{T} \longrightarrow 2^{\mathcal{G}}$ that describes which resources are used by each test case; and a set of machines $\mathcal{M}$ and a function $f : \mathcal{T} \longrightarrow 2^{\mathcal{M}}$ that assigns to each test case a subset of machines on which the test case can be executed. The function d is usually obtained by measuring the execution time of each test case in previous test campaigns and by over-approximating each duration to account for small variations between the different execution machines. OTS is the optimization problem of finding an execution ordering and assignment of all test cases to machines, such that each test case is executed once, no global resource is used by two test cases at the same time, and the overall test execution time, T_t, is minimized. We define T_t as the time needed to compute the schedule (T_s) plus the time needed to execute the schedule (C^*), $T_t = T_s + C^*$. Machine assignment and test case execution ordering can be described either by a time-discretized table containing a line per machine or a starting time for each test case and its assignment to a given machine.

The problem addressed in this paper aims to execute each test case once while minimizing the total duration of the execution of the test cases. That is, to find an assignment $a : \mathcal{T} \longrightarrow \mathcal{M}$ and an execution order for each machine to run its test cases.

---

[2] OTS was part of the Industrial Modelling Competition at CP 2015.

In its basic version, the OTS problem includes the following constraints:

Disjunctive Scheduling: Two test cases cannot be executed at the same time on a single machine.

Non-preemptive Scheduling: The execution of a test case cannot be temporarily interrupted to execute another test case on the same machine.

Non-shared Resources: When a test case uses a global resource, no other test case needing this resource can be executed at the same time.

Machine-Independent Execution Time: The execution time of a test case is assumed to be independent of the executing machine. This is reasonable for test cases in which the time is dominated by external physical factors such as a robot's motion, the opening of a valve, or sending an Ethernet frame. Such test cases typically have execution times that are uncorrelated with machine performance. In any case, a sufficient over-approximation will satisfy the assumption.

There are cases where OTS can be trivially solved, e.g. with only one machine executing all test cases in sequence. Indeed, the global execution time remains unchanged, whatever the execution order. Similarly, when there are no global resources and when test cases can be executed on any available machine, then simply allocating the longest test cases first to the available execution machine easily calculates a best-effort solution.

Example. Considering the test suite in Table 1, we present a small example. Let $\mathcal{T}$ be the test cases $\{1, \ldots, 10\}$, $\mathcal{G}$ be the global resources $\{1, 2\}$, and $\mathcal{M}$ be the machines $\{1, 2, 3\}$. The machines on which each test case in $\mathcal{T}$ can run is given in Table 1. This table can be extracted by analyzing the test scripts or querying the test management. By sharing the same resource 1, test cases $2, 3, 4$ cannot be executed at the same time, even if their execution is scheduled on different machines. Since test case 7 can only be executed on machine 1, test case 8 on machine 2, test case 9 on machine 3, and test case 10 on machines 1 or 3, we have to solve a complex scheduling problem. One possible *optimal* schedule is given in Fig. 1, where the time needed to execute the test campaign is $C^* = 11$. For this small problem the solving time, T_s, can be assumed to be very short, so the total execution time will be $T_t \approx C^*$.

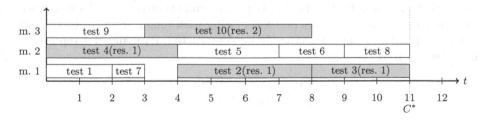

Fig. 1. An optimal solution to the scheduling problem given in Table 1. Test cases in light gray require exclusive access to a global resource

Table 1. Test suite for example.

Test	Duration	Executable on	Use of global resource
1	2	$1, 2, 3$	-
2	4	$1, 2, 3$	1
3	3	$1, 2, 3$	1
4	4	$1, 2, 3$	1
5	3	$1, 2, 3$	-
6	2	$1, 2, 3$	-
7	1	1	-
8	2	2	-
9	3	3	-
10	5	$1, 3$	2

3.2 The Cumulatives Global Constraint

The CUMULATIVES global constraint [5] is a powerful tool for modeling cumulative scheduling of multiple operations on multiple machines, where each operation can be set up to consume a given amount of a resources, and each machine can be set up to provide a given amount of resources.

CUMULATIVES($[O_1, \ldots, O_n], [c_1, \ldots, c_p]$)[3] constrains n operations on p machines such that the total resource consumption on each machine j does not exceed the given threshold c_j at any time [10]. An operation O_i is typically represented by a tuple $(S_i, d_i, E_i, r_i, M_i)$[4] where S_i (resp. E_i) is a variable that denotes the starting (resp. ending) instant of the operation, d_i is a constant representing the total duration of the operation, r_i is a constant representing the amount of resource used by the operation. S_i, E_i and M_i are bounded integer variables. S_i and E_i have the domains $est_i \ldots let_i$, where est_i denotes the operation's earliest starting time and let_i denotes its latest ending time and $let_i \geq est_i + d_i$. M_i is bounded by the number of machines available, that is $1, \ldots, p$. By reducing the domain of M_i it is possible to force a specific operation to be assigned to only a subset of the available machines, or even to one specific machine. It is worth noting that this formalization implicitly uses discrete time instants. Indeed, since est_i and let_i are integers, a function associating each time instant to the current executed operations can automatically be constructed. Formally, if h represents an instant in time, we have:

$$r_i^h = \begin{cases} r_i & \text{if } S_i \leq h < S_i + d_i \\ 0 & \text{otherwise} \end{cases}$$

---

[3] In [5] an additional third argument to CUMULATIVES, $Op \in \{\leq, \geq\}$ is defined. We omit it throughout our work and always set $Op = \leq$.

[4] Throughout the paper, lower-case characters are used to represent constants and upper-case characters are used to represent variables.

CUMULATIVES holds if and only if, for every operation O_i, $S_i + d_i = E_i$, and, for all machines k and instants h, $\sum_{i|M_i=k} r_i^h \leq c_k$. In fact, CUMULATIVES captures a disjunctive relation between different scenarios and applies deductive reasoning to the possible values in the domains of its variables. This constraint provides a cost-effective process for pruning the search space of some impossible schedules.

3.3 Modeling Test Case Execution Scheduling

This section shows how the CUMULATIVES constraint can be used to model a schedule. In this small example, we disregard the use of global resources, and the constraints that some operations can only be executed on a subset of the available machines, since that will be covered in Sect. 3.4. By the schedule in Fig. 1, we have ten operations $\mathcal{O} = \{O_1, \ldots, O_{10}\}$ and three available machines. By encoding the data from Table 1, we get $O_1 = (S_1, 2, E_1, 1, M_1)$, $O_2 = (S_2, 4, E_2, 1, M_2) \ldots$, $O_{10} = (S_{10}, 5, E_{10}, 1, M_{10})$, $c_1 = 1$, $c_2 = 1$, $c_3 = 1$. Note that each operation has a resource consumption of one and all three machines have a resource capacity of one. This implies that one machine can only execute one operation at a time. Here, a resource refers to an execution machine and not to a global resource.

3.4 Introducing Global Resources

As mentioned above, global resources corresponding to physical equipment such as valves, air sensors, measurement instruments, or network devices, have limited and exclusive access. To avoid concurrent access from two test cases, additional constraints are introduced. Note that global resources must not be confused with the resource consumption or resource bounds of operations and machines.

The CUMULATIVES constraint does not support native modelling of these global resources without additional, user-defined constraints. However, there are ways to model exclusive access to such global resources by means of further constraints. The naive approach to prevent two operations from overlapping is to consider constraints over the start and stop time of the operations. For instance, if O_1 and O_2 both require exclusive access to a global resource, then the constraint $E_1 \leq S_2 \vee E_2 \leq S_1$ can be added. A less naive approach is to use a DISJUNCTIVE($\mathcal{O}^k$) constraint per global resource k, where $\mathcal{O}^k$ is the set of tasks that require that global resource, and DISJUNCTIVE prevents any pair of tasks from overlapping.

Referring to the example in Fig. 2, there are ten operations to be scheduled on three machines, and two global resources, 1 and 2. The basic scheduling constraint is set up as explained in Sect. 3.3. Yet another way to model the global resources is to treat each resource as a new quasi-machine $1'$ corresponding to $c_{1'} = 1$ and $2'$ corresponding to $c_{2'} = 1$. For each operation requiring a global resource, we create a "mirrored" operation of the corresponding quasi-machine: $\mathcal{O}'_1 = \{O'_2, O'_3, O'_4\}$ and $\mathcal{O}'_2 = \{O'_{10}\}$. Finally, we can express the schedule with a single constraint: CUMULATIVES($\mathcal{O} \cup \mathcal{O}'_1 \cup \mathcal{O}'_2, [c_1, c_2, c_3, c_{1'}, c_{2'}]$). For each operation in $\mathcal{O}'_1$ and $\mathcal{O}'_2$ we also reuse the same domain variables for start-time, duration and end-time. The operation O_4 will be forced to have the same start-/end-time as O'_4, while they are scheduled on two different machines 2 and $1'$ (Fig. 2).

4   The TC-Sched Method

This section describes our method, TC-Sched, to solve the OTS problem. It is a *time-constrained cumulative scheduling technique*, as (1) it allows to keep fine-grained control over the time allocated to the constraint solving process (i.e., *time-constrained*), (2) it encodes exclusive resource use with constraints (i.e., *constraint-based*), and (3) it solves the problem by using the CUMULATIVES constraint. The TC-Sched method is composed of three elements, namely, the constraint model described in Sect. 4.1, the search procedure described in Sect. 4.2, and the time-constrained minimization process described in Sect. 4.3.

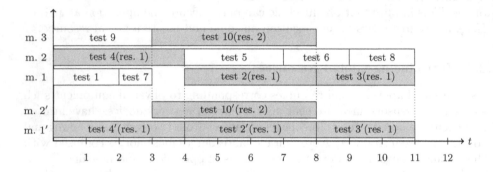

Fig. 2. Modeling global resources by creating quasi-machines and CUMULATIVES

4.1   Constraint Model

We encode the OTS problem with one CUMULATIVES($\mathcal{O}, \mathcal{C}$) constraint, one DISJUNCTIVE($\mathcal{O}^k$) constraint per global resource k, using the second scheme from Sect. 3.4, and a search procedure able to find an optimal schedule among many feasible schedules. Each test case i is encoded as an operation $(S_i, d_i, E_i, 1, M_i)$ as explained in Sect. 3.2. $\mathcal{O}$ is simply the array of all such operations and $\mathcal{C}$ is an array of 1s of length equal to the number of machines. Suppose that there are three execution machines numbered $1, 2$, and 3; then, to say that test i can be executed on any machine, we just add the domain constraint $M_i \in \{1, 2, 3\}$, whereas to say that test i can only be executed on machine 1, we replace M_i by 1. Finally, to complete the model, we introduce the variable *MakeSpan* representing the completion time of the entire schedule and seek to minimize it. *MakeSpan* is lower bounded by the ending time of each individual test case. The generic model is captured by:

$$\text{CUMULATIVES}(\mathcal{O}, \mathcal{C}) \wedge$$
$$\forall \text{global resource } k : \text{DISJUNCTIVE}(\mathcal{O}^k) \wedge$$
$$\forall 1 \leq i \leq n : M_i \in f(i) \wedge \tag{1}$$
$$\forall 1 \leq i \leq n : E_i \leq \textit{MakeSpan} \wedge$$
$$\text{LABEL}(\text{MINIMIZE}(\textit{MakeSpan}), [S_1, M_1, \ldots, S_n, M_n])$$

Note that the ending times depend functionally on the starting times. Thus, a solution to the OTS problem can be obtained by searching among the starting times and the assignment of test cases to execution machines.

4.2   Search Procedure

Our search procedure is called *test case duration splitting*, and is a branch-and-bound search that seeks to minimize the *Makespan*. The procedure makes two passes over the set of test cases. A key idea is to allocate the most demanding test cases first. To this end, the test cases are initially sorted by decreasing r_i where r_i is the number of global resources used by test case i, breaking ties by choosing the test case with the longest duration d_i.

In Phase 1, two actions are performed on each test case. First, in order to avoid a large branching factor in the choice of start time and to effectively fix the relative order among the tasks on the same machine or resource, we split the domain of the start variable, forcing an obligatory part of the corresponding task, as described in [34, Sect. 3.6]. Next, in order to balance the load on the machines, we choose machines in round-robin fashion. These two choices are of course backtrackable, to ensure completeness of the search procedure.

Note that at the end of Phase 1, the constraint system effectively forms a directed acyclic graph where every node is a task and every arc is a precedence constraint induced by the relative order. It is well known that such constraint systems can be solved without search by topologically sorting the start variables and assigning each of them to its minimal value. This is Phase 2 of the search.

In this procedure, the load-balancing component has shown to be particularly effective in a CI context and makes the first solution found a good compromise between solving and execution time of the schedule, which is one of the key factors in CI. Our preliminary experiments concluded, that the presented strategy provided the best compromise between cost and solution quality. Furthermore, we tried a more precise but costlier load-balancing scheme, but it did not significantly improve the quality. We also tried to sort the tests by decreasing $d_i \cdot (r_i + 1)$, which did not significantly improve the quality, either.

4.3   Time-Constrained Minimization

The third necessary ingredient of the TC-Sched method is to perform branch-and-bound search under a time contract. That is, to settle on the schedule with the shortest *MakeSpan* found when the time contract ends. When the number of test cases grows to be several hundred, finding a globally optimal schedule may become an intractable problem[5], but in practical applications it is often sufficient to find a "best-effort" solution. This leads to the important question to select the most appropriate contract of time for the minimization process, as the time used to optimize the schedule is not available to actually execute the schedule. We address this question in the experimental evaluation.

5   Implementation and Exploitation

This section details our implementation of the TC-Sched method and its insertion into CI. We implemented the TC-Sched method in SICStus Prolog [11]. The CUMULATIVES constraint is available as part of the `clpfd` library [10]. The `clpfd` library also provides an implementation of the time-constrained branch-and-bound with the option to express individual search strategy (see Sect. 4.2). Using `clpfd`, a generic constraint model for the TC-Sched method is designed, which takes an OTS problem as input and returns an (quasi-)optimal schedule (Fig. 3).

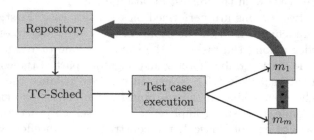

Fig. 3. Integration of TC-Sched into a CI process. The test case schedule solved by TC-Sched is transmitted for execution to the machines in the machine pool, $\mathcal{M}$. The results including actual test case durations are then feed back into the repository

Since TC-Sched is designed to run as part of a CI process, we describe how it can be integrated within the CI environment. Because CI environments change and test cases and agents are constantly added or removed, TC-Sched has to be provided with a list of test cases and available machines at runtime. Furthermore, an estimation of the test case durations on the available agents has to be provided. This can either be gathered from historical execution data and then (over-)estimated to account for differences in execution machines, or, for some kinds to test suites, they are fixed and can be precisely given [26], e.g. for robotic applications where the duration is determined by the movement of the robot.

A test campaign in a CI cycle is typically initiated upon a successful build of the software being tested. As a first step, all machines available for test execution are identified and updated with the newly built software. Then, TC-Sched

---

[5] The general cumulative scheduling problem is known to be NP-hard [2].

takes as input the test cases of the test campaign and the previous test case execution times from the storage repository. After TC-Sched calculated an optimal schedule, that schedule is handed over to a dedicated dispatch server which is responsible for distributing the test cases to the physical machines and the actual execution. Finally, after the test execution finished, the overall result of the test campaign is reported back to the users and the storage repository is updated with the latest test case execution times. Of course, minimizing the *round-trip time* leads to earlier notifications of the developers in case the software system fails and helps to improve the development cycle in CI.

6   Experimental Evaluation

This section presents our findings from the experimental evaluation of TC-Sched. To this end, we address the following three research questions:

RQ1: How does the first solution provided by TC-Sched compare with simpler scheduling methods in terms of schedule execution time? This research question states the crucial question of whether using complex constraint optimization is useful despite simpler approaches being available at almost no cost to implement.
RQ2: For TC-Sched, will an increased investment in the solving time in TC-Sched reduce the overall time of a CI cycle? This question is about finding the most appropriate trade-off between the solving time and the execution time of the test campaign in the proposed approach.
RQ3: In addition to random OTS problem instances, can TC-Sched efficiently and effectively handle industrial case studies? These cases can lead to structured problems which exhibit very different properties than random instances.

All experiments were performed on a 2.7 GHz Intel Core i7 processor with 16 GB RAM, running SICStus Prolog 4.3.5 on a Linux operating system.

6.1   Experimental Artifacts

To answer RQ1, we implemented two scheduling methods, referred to as the *random* method and the *greedy* method.

The *random* method works as follows: It first picks a test case at random and then picks a machine at random such that no resource constraint is violated. Finally, the test case is assigned the lowest possible starting time on the selected machine. The *greedy* method is more advanced. At first, it assigns test cases by decreasing resource demands. Afterwards, test cases without any resource demands are assigned to the remaining machines. For each assignment, the machine that can provide the earliest starting time is selected. Note that none of the two methods can backtrack to improve upon the initial solution.

The reason we have chosen to compare with these two methods is threefold: (1) As explained in Sect. 2, we are not aware of any previously published work related to test case execution scheduling, which means that there is no baseline to compare against; (2) From cooperation with our industrial partners, we know that this is, in the best case, the industrial state of the art (i.e., non-optimal schedules computed manually); (3) We manually checked the results on simple schedules and found them to be satisfactory, so they are a suitable comparison.

To answer our research questions, we have considered randomly generated benchmarks and industrial case studies. Although there are benchmark test suites for both JSS and FJSS, e.g., [37] or [4], they cannot be used as a comparison baseline. Furthermore, as our method approaches testing applications, a thorough evaluation on data from the target domain is justifiable.

We generated a benchmark library containing 840 OTS instances[6]. The library is structured by data collected from three different real-world test suites, provided by our industrial partners: a test suite for video conferencing systems (VCS) [24], a test suite for integrated painting systems (IPS) [26], and a test suite for a mobile application called *TV-everywhere*.

VCS is a test suite for testing commercial video conferencing systems, developed by CISCO Systems, Norway. It contains 132 test cases and 74 machines. The duration of test cases varies from 13 s to 4 h, where the vast majority has a duration between 100 s and 800 s. The IPS test suite aims at testing a distributed paint control system for complex industrial robots, developed at ABB Robotics, Norway. It contains 33 test cases, with duration ranging from 1 s to 780 s, and 16 distinct machines. There are two global resources for this test suite, an airflow meter and a simulator for an optical encoder. *TV-everywhere* is a mobile application that allows users to watch TV on tablets, smart phones, and laptops. Its test suite only contains manual test cases, but, in our benchmark, it serves as a useful example of a test suite with a large number of constraints limiting the number of possible machines for each test case.

Based on data from the three industrial test suites, we composed 14 groups of test suites, denoted TS1-TS14, with randomized assignments of test cases to machines and exclusive usages of global resources. Let $|T|$ be the number of test cases, and $|M|$ be the number of machines, and $|R| = \{3, 5, 10\}$ be the number of resources. Table 2 gives an overview of the groups of test suites. For test suite TSx, we write TSxR3, TSxR5, or TSxR10 to indicate the number of resources.

For each of the $14 \cdot 3$ variants, we generated 20 random test suites. The duration of each test case was chosen randomly between 1 s and 800 s, and each test case had a 30% chance of using a global resource. The number of resources was chosen randomly between 1 and $|R|$. A total of 80% of the tests were considered to be executable on all machines, while the remaining 20% were executable on a smaller subset of machines. For these tests, the number of machines on which each test case could be executed was selected randomly between 1% and 40% of the number of available machines. This means that a test case was executable either on all machines (part of the 80% group) or only on at most 40% of the machines. In total, we generated $14 \cdot 3 \cdot 20 = 840$ different test suites.

---

[6] All generated instances are available in CSPLib, a library of test problems for constraint solvers [25].

6.2 RQ1: How Does TC-Sched Compare with Simpler Scheduling?

To compare our TC-Sched method with the *greedy* and *random* methods, we recorded the first solution, C_f^*, found by TC-Sched. We also recorded the last solution, C_l^*. This is either a proved optimal solution, or the best solution found after 5 min of solving time. For each of the 840 test suites, we computed the differences between the *random* and *greedy*, C_f^* and *greedy*, and C_l^* and *greedy*, where *greedy* is the baseline of 100%. The results show that *random* is 30%–60% worse than *greedy*, which means that *random* can clearly be discarded from further analysis. Our findings are summarized in Fig. 4, showing the difference between TC-Sched and *greedy*. For all test suites but the hardest subset of TS1 and some instances of TS2, C_f^* is better than *greedy*. We also observe that for larger test suites, i.e., TS11–TS14, there is only a marginal difference between C_f^* and C_l^*. Hence, running the solver for a longer time has only little benefit.

Furthermore, to evaluate the effectiveness of the test case duration splitting search strategy, we compared it to standard strategies available in SICStus Prolog's `clpfd` with the same constraint model on the test suites TS1 and TS14. The search first enumerates on the machine assignments increasingly, i.e. without load-balancing, and afterwards assigns end times via domain splitting by bisecting the domain, starting from the earliest end times. As variable selection

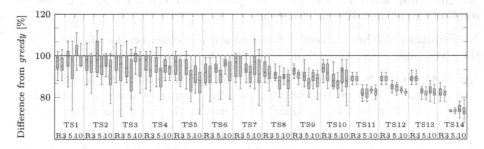

Fig. 4. The differences in schedule execution times produced by the different methods for test suites TS1–TS14, with *greedy* as the baseline of 100%. The blue is the difference between the first solution C_f^* and *greedy* and the red shows the difference between the final solution C_l^* and *greedy*. (Color figure online)

Table 2. Randomly generated test suites.

# machines	# of tests					
	20	30	40	50	100	500
100	-	-	-	-	-	TS11
50	-	-	-	-	TS8	TS12
20	-	TS2	TS4	TS6	TS9	TS13
10	TS1	TS3	TS5	TS7	TS10	TS14

strategies, we tested both the default setting, selecting the leftmost variable, and a first-fail strategy, selecting the variable with the smallest domain. Additionally, we tried sorting the variables by decreasing resource usage.

All variants of the standard searches performed substantially worse than test case duration splitting, with first-fail search on sorted variables being the best. After finding an initial solution, further improvements are rare and the makespan of the final solution is in average 4 times larger compared to using test case duration splitting with the same time contract of 5 min.

6.3   RQ2: Will Longer Solving Time Reduce the Total Execution Time?

RQ2 aims at finding an appropriate trade-off between the time spent in solving the constraint model, T_s, and the time spent in executing the schedule, C^*. As mentioned in Sect. 1, the round-trip time is critical in CI and has to be kept low. It is therefore crucial to determine the most appropriate timeout for the constraint optimizer. The ultimate goal being to generate a schedule which is quasi-optimal w.r.t. total execution time, $T_t = T_s + C^*$.

As mentioned above, TC-Sched can be given a time-contract for finding a quasi-optimal solution when minimizing the execution time of the schedule. More precisely, with this time-constrained process four outcomes are possible.

No Solution with Proof: TC-Sched proves that the OTS problem has no solution due to unsatisfiable constraints.

No Solution Without Proof: TC-Sched was not able to find a solution within the given time. Thus, there could be a solution, but it has not been found.

Quasi-Optimal Solution: At the end of the time-contract, a solution is returned, but TC-Sched was interrupted while trying to prove its optimality. Such a best-effort solution is usually sufficient in the examined industrial settings.

Optimal Solution: Before the end of the time-contract, TC-Sched returns an optimal solution along with its proof. This is obviously the most desired result.

Each solution i generated by TC-Sched can be represented by a tuple $(C_i^*, T_{s,i})$ where C_i^* is the makespan of solution i and $T_{s,i}$ is the time the solver spent finding solution i. The goal of RQ2 is to find the value of $T_{s,i}$ that minimizes $(C_i^* + T_{s,i}), \forall i$ and use this value as the time-contract.

To answer RQ2, we executed TC-Sched on all 840 test suites, with a time-contract of 5 min. During this process, we recorded all intermediate search results to calculate the optimal value of T_s for each test suite.

Figure 5 shows the distribution in solving time for the first solution found by TC-Sched, the last solution and also how the optimal value of T_s is distributed. For the group of 600 test suites containing up to 100 test cases (TS1–TS10), the results show that a solution that minimizes the total execution time, noted T_t, is found in $T_s < 5\,\mathrm{s}$ for 96.8% of the test suites. If we extend the search time to $T_s < 10\,\mathrm{s}$, the number grows to 98% of the test suites. For this group, the worst case optimal solving time was $T_s = 122.3\,\text{s}$. We see that a solution is always found in less than 0.1 s. For the group of 240 test suites containing 500 test cases (TS11–TS14), the results show that a solution that minimizes T_t is found in $T_s < 120\,\text{s}$ for 97.5% of the test suites. A solution minimizing T_t is found in less than 240 s for all test suites, except one instance with $T_t = 264\,\text{s}$.

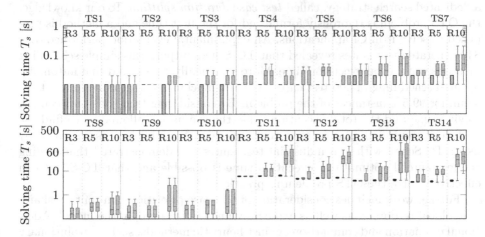

Fig. 5. The black boxes show the distribution in solving time, T_s, for the first solution found by TC-Sched. The blue boxes show the distribution in T_s where the total execution time, T_t, is optimal. Finally, the red boxes show the distribution in T_s for the last solution found by TC-Sched, which can be the optimal value or the last value found before timeout. The timeout was set to 5 min. (Color figure online)

An increased investment in the solving part does not seem to necessarily pay off if one considers the total execution time. The reported experiments give hints to evaluate and select the optimal test contract for the solving part.

6.4 RQ3: Can TC-Sched Efficiently Solve Industrial OTS Problems?

To answer RQ3, we consider two of the three industrial case studies, namely, IPS and VCS. These case studies are composed of automated test scripts, which makes the application of the TC-Sched method especially pertinent.

In both case studies, the guaranteed optimal solution is already found as the first solution in less than 200 ms. This avoids the necessity to compromise between C^* and T_s for these industrial applications.

When applying TC-Sched to the IPS test suite, we find the optimal solution, $C^* = 780\,\text{s}$, at $T_s = 10\,\text{ms}$. For the VCS test suite, the optimal solution, $C^* = 14637\,\text{s}$ is found at $T_s = 160\,\text{ms}$.

In summary, TC-Sched can easily be applied to both VCS and IPS, and in both cases, the best result is achieved when C^* is minimized and T_s is neglected.

7   Conclusion

This paper introduced TC-Sched, a time-aware method for solving the optimal test suite scheduling (OTS) problem, where test cases can be executed on multiple execution machines with non-shareable global resources. TC-Sched exploits the CUMULATIVES global constraint and a time-aware minimization process, and a dedicated search strategy, called *test case duration splitting*. To our knowledge, the OTS problem is rigorously formalized for the first time and a method is proposed to solve it in CI applications. An experimental evaluation performed over 840 generated test suites revealed that TC-Sched outperforms simple scheduling methods w.r.t. total execution time. More specifically, we showed that automatic optimal scheduling of 500 test cases over 100 machines is reachable in less than 4 min for 99.5% instances of the problem. By considering trade-offs between the solving time and the total execution time, the evaluation allowed us to find the best compromise to allocate time-contracts to the solving process. Finally, by using TC-Sched with two industrial test suites, we demonstrated that finding the guaranteed optimal test execution time is possible and that TC-Sched can effectively solve the OTS problem in practice.

Further work includes consideration of test case priorities, non-unitary shareable global resources, as well as explicit symmetry breaking in the model. Additional evaluation and comparison against heuristic methods, such as evolutionary algorithms, or Mixed-Integer Linear Programming could extend the presented work and support the integration of TC-Sched in practical CI processes.

References

1. Aggoun, A., Beldiceanu, N.: Extending CHIP in order to solve complex scheduling and placement problems. Math. Comput. Modell. **17**(7), 57–73 (1993)
2. Baptiste, P., Le Pape, C., Nuijten, W.: Constraint-Based Scheduling: Applying Constraint Programming to Scheduling Problems, vol. 39. Springer Science & Business Media, Berlin (2001)
3. Beck, J.C., Feng, T.K., Watson, J.P.: Combining constraint programming and local search for job-shop scheduling. INFORMS J. Comput. **23**(1), 1–14 (2011)
4. Behnke, D., Geiger, M.J.: Test instances for the flexible job shop scheduling problem with work centers. Technical report RR-12-01-01, Helmut-Schmidt University, Hamburg, Germany (2012)
5. Beldiceanu, N., Carlsson, M.: A new multi-resource *cumulatives* constraint with negative heights. In: Hentenryck, P. (ed.) CP 2002. LNCS, vol. 2470, pp. 63–79. Springer, Heidelberg (2002). doi:10.1007/3-540-46135-3_5
6. Brandimarte, P.: Routing and scheduling in a flexible job shop by tabu search. Ann. Oper. Res. **41**(3), 157–183 (1993)
7. Brucker, P., Knust, S.: Complex Scheduling (GOR-Publications). Springer-Verlag New York Inc., Secaucus (2006)
8. Brucker, P., Drexl, A., Möhring, R., Neumann, K., Pesch, E.: Resource-constrained project scheduling: notation, classification, models, and methods. Eur. J. Oper. Res. **112**(1), 3–41 (1999)

9. de Campos, J., Arcuri, A., Fraser, G., de Abreu, R.: Continuous test generation: enhancing continuous integration with automated test generation. In: ASE 2014, Västerås, Sweden, pp. 55–66 (2014)

10. Carlsson, M., Ottosson, G., Carlson, B.: An open-ended finite domain constraint solver. In: Glaser, H., Hartel, P., Kuchen, H. (eds.) PLILP 1997. LNCS, vol. 1292, pp. 191–206. Springer, Heidelberg (1997). doi:10.1007/BFb0033845

11. Carlsson, M., et al.: SICStus Prolog user's manual, release 4. Technical Report, SICS - Swedish Institute of Computer Science (2007)

12. Do, H., Mirarab, S., Tahvildari, L., Rothermel, G.: The effects of time constraints on test case prioritization: a series of controlled experiments. IEEE Trans. Soft. Eng. **36**(5), 593–617 (2010)

13. Duvall, P.M., Matyas, S., Glover, A.: Continuous Integration: Improving Software Quality and Reducing Risk. Pearson Education, London (2007)

14. Elbaum, S., Rothermel, G., Penix, J.: Techniques for improving regression testing in continuous integration development environments. In: FSE 2014 (2014)

15. Fowler, M., Foemmel, M.: Continuous integration (2006). http://martinfowler.com/articles/continuousIntegration.html

16. Gotlieb, A., Marijan, D.: Flower: optimal test suite reduction as a network maximum flow. In: ISSTA 2014, San José, CA, USA, pp. 171–180 (2014)

17. Hao, D., Zhang, L., Wu, X., Mei, H., Rothermel, G.: On-demand test suite reduction. In: ICSE 2012, pp. 738–748 (2012)

18. Hartmann, S., Briskorn, D.: A survey of variants and extensions of the resource-constrained project scheduling problem. Eur. J. Oper. Res. **207**(1), 1–14 (2010)

19. Hartmann, S., Kolisch, R.: Experimental evaluation of state-of-the-art heuristics for the resource-constrained project scheduling problem. Eur. J. Oper. Res. **127**(2), 394–407 (2000)

20. Herroelen, W., De Reyck, B., Demeulemeester, E.: Resource-constrained project scheduling: a survey of recent developments. Comput. Oper. Res. **25**(4), 279–302 (1998)

21. Kolisch, R., Hartmann, S.: Experimental investigation of heuristics for resource-constrained project scheduling: an update. Eur. J. Oper. Res. **174**(1), 23–37 (2006)

22. Kreter, S., Schutt, A., Stuckey, P.J.: Modeling and solving project scheduling with calendars. In: Pesant, G. (ed.) CP 2015. LNCS, vol. 9255, pp. 262–278. Springer, Cham (2015). doi:10.1007/978-3-319-23219-5_19

23. Lin, C., Tang, K., Kapfhammer, G.: Test suite reduction methods that decrease regression testing costs by identifying irreplaceable tests. Inf. Softw. Technol. **56**, 1322–1344 (2014)

24. Marijan, D., Gotlieb, A., Sen, S.: Test case prioritization for continuous regression testing: an industrial case study. In: ICSM 2013, Eindhoven, The Netherlands (2013)

25. Mossige, M.: CSPLib problem 073: test scheduling problem. http://www.csplib.org/Problems/prob073

26. Mossige, M., Gotlieb, A., Meling, H.: Using CP in automatic test generation for ABB robotics' paint control system. In: O'Sullivan, B. (ed.) CP 2014. LNCS, vol. 8656, pp. 25–41. Springer, Cham (2014). doi:10.1007/978-3-319-10428-7_6

27. Orso, A., Rothermel, G.: Software testing: a research travelogue (2000–2014). In: FOSE 2014, Hyderabad, India, pp. 117–132 (2014)

28. Orso, A., Shi, N., Harrold, M.J.: Scaling regression testing to large software systems. In: FSE 2014, pp. 241–251. ACM Press, Newport Beach (2004)

29. Schutt, A., Feydy, T., Stuckey, P.J., Wallace, M.G.: Why cumulative decomposition is not as bad as it sounds. In: Gent, I.P. (ed.) CP 2009. LNCS, vol. 5732, pp. 746–761. Springer, Heidelberg (2009). doi:10.1007/978-3-642-04244-7_58

30. Schutt, A., Chu, G., Stuckey, P.J., Wallace, M.G.: Maximising the net present value for resource-constrained project scheduling. In: Beldiceanu, N., Jussien, N., Pinson, É. (eds.) CPAIOR 2012. LNCS, vol. 7298, pp. 362–378. Springer, Heidelberg (2012). doi:10.1007/978-3-642-29828-8_24

31. Schutt, A., Feydy, T., Stuckey, P.J.: Scheduling optional tasks with explanation. In: Schulte, C. (ed.) CP 2013. LNCS, vol. 8124, pp. 628–644. Springer, Heidelberg (2013). doi:10.1007/978-3-642-40627-0_47

32. Schutt, A., Feydy, T., Stuckey, P.J., Wallace, M.G.: Solving RCPSP/max by lazy clause generation. J. Sched. **16**(3), 273–289 (2013)

33. Siala, M., Artigues, C., Hebrard, E.: Two clause learning approaches for disjunctive scheduling. In: Pesant, G. (ed.) CP 2015. LNCS, vol. 9255, pp. 393–402. Springer, Cham (2015). doi:10.1007/978-3-319-23219-5_28

34. Simonis, H., O'Sullivan, B.: Search Strategies for rectangle packing. In: Stuckey, P.J. (ed.) CP 2008. LNCS, vol. 5202, pp. 52–66. Springer, Heidelberg (2008). doi:10.1007/978-3-540-85958-1_4

35. Stolberg, S.: Enabling agile testing through continuous integration. In: AGILE 2009, pp. 369–374. IEEE (2009)

36. Szeredi, R., Schutt, A.: Modelling and solving multi-mode resource-constrained project scheduling. In: Rueher, M. (ed.) CP 2016. LNCS, vol. 9892, pp. 483–492. Springer, Cham (2016). doi:10.1007/978-3-319-44953-1_31

37. Taillard, E.: Benchmarks for basic scheduling problems. Eur. J. Oper. Res. **64**(2), 278–285 (1993)

38. Walcott, K.R., Soffa, M.L., Kapfhammer, G.M., Roos, R.S.: Time-aware test suite prioritization. In: ISSTA 2006, Portland, Maine, USA, pp. 1–12 (2006)

39. Zhang, L., Hou, S., Guo, C., Xie, T., Mei, H.: Time-aware test-case prioritization using integer linear programming. In: ISSTA 2009, Chicago, IL, USA, pp. 213–224 (2009)

Integrating ILP and SMT for Shortwave Radio Broadcast Resource Allocation and Frequency Assignment

Linjie Pan[1,4], Jiwei Jin[6], Xin Gao[2], Wei Sun[5], Feifei Ma[1,3,4]([✉]),
Minghao Yin[2]([✉]), and Jian Zhang[1,4]

[1] State Key Laboratory of Computer Science,
Institute of Software Chinese Academy of Sciences, Beijing, China
maff@ios.ac.cn
[2] College of Computer Science, Northeast Normal University, Changchun, China
ymh@nenu.edu.cn
[3] Laboratory of Parallel Software and Computational Science,
Institute of Software Chinese Academy of Sciences, Beijing, China
[4] University of Chinese Academy of Sciences, Beijing, China
[5] Administration Bureau of Radio Stations, State Administration of Press,
Publication, Radio, Film and Television of the People's Republic of China,
Beijing, China
[6] Shan Dong Jiaotong University, Jinan, China

Abstract. Shortwave radio broadcasting is the principal way for broadcasting of voice in many countries. The broadcasting quality of a radio program is determined not only by the parameters of the transmission device, but also by the radio frequency. In order to optimize the overall broadcasting quality, it is desirable to designate both devices and frequencies to radio programs, subject to various constraints including the non-interference of radio programs. In this paper, we propose a two-phase approach to this constrained optimization problem. It integrates ILP and SMT solving, as well as a local search algorithm. These methods are evaluated using real data, and the results are promising.

Keywords: Integer Linear Programming · Satisfiability modulo theories · Shortwave radio broadcast

1 Introduction

Shortwave radio is a significant medium in long distance broadcasting transmission, which uses shortwave frequencies ranging from 2 to 30 megahertz(MHz). [5] introduced the history of shortwave radio broadcasting. Nowadays, it remains the principal way for broadcasting of voice in many countries. There are various factors affecting the broadcasting quality of shortwave radio programs. How to arrange these factors properly is critical to shortwave broadcasting.

© Springer International Publishing AG 2017
J.C. Beck (Ed.): CP 2017, LNCS 10416, pp. 405–413, 2017.
DOI: 10.1007/978-3-319-66158-2_26

In the past, the staff with the Division of Radio Frequency Assignment of State Administration of Press,Publication,Radio,Film and Television (SAP-PRFT) of the People's Republic of China have been managing the allocation of broadcast resource manually. In [10], Ma et al. studied the shortwave radio broadcast resource allocation problem (SRBRA), which concerns how to allocate proper transmission devices to radio programs so that the overall broadcasting quality is optimized. They proved the NP-hardness of the SRBRA problem, and proposed a Pseudo-Boolean formulation and a local search algorithm.

In the SRBRA problem [10], the frequencies for the programs were assigned in advance. In real applications, it may be necessary to find a suitable frequency for each program, without introducing any interference among the programs. The frequency assignment problem (FAP) is another important problem in broadcasting transmission [1]. In the literature, FAP has been solved via several kinds of techniques, such as CSP and Local Search [8,9,11]. But in our application, FAP interleaves with broadcast resource allocation, hence cannot be solved separately. Thus we extend the SRBRA problem further to embody frequency assignment. In [10], only 87 programs were used in the empirical evaluation. (There are 87 programs in a single region.) But in total, there are 948 programs, if all regions are considered. This paper tries to deal with such challenges and investigates new approaches to the extended SRBRA problem.

The contributions of this paper include (1) extending the SRBRA problem to the Shortwave Radio Broadcast Resource Allocation and Frequency Assignment Problem (SRBRAFA), which involves both device allocation and frequency assignment, and (2) developing a two-phase approach to the SRBRAFA problem which integrates Integer Linear Programming (or local search) with Satisfiability Modulo Theories (SMT) [4]. Our methods are evaluated using real data from SAPPRFT, and the results are promising.

2   Problem Description

Given a set of programs $\mathcal{P} = \{P_1, P_2, ..., P_n\}$, a set of transmission devices $\mathcal{D} = \{D_1, D_2, ..., D_m\}$, and a set of frequencies $\mathcal{F} = \{F_1, F_2, ..., F_l\}$, the SRBRAFA problem involves allocating devices and assigning frequencies to programs, and maximizing the broadcasting quality. It is an extension of the SRBRA problem which only allocates devices to programs [10]. We use $A_i = <P_i, D_j, F_k>$ to represent the allocation of device D_j and frequency F_k to program P_i. The allocations should not conflict with each other or interfere with each other.

Conflicting Allocations. A program, which has predetermined target area and time span, can only be transmitted with one device and one frequency. If two programs P_i and P_j overlap by the broadcasting time span, then they are called overlapping programs, denoted by $overlap(P_i, P_j)$. A transmission device is assembled by a transmitter and an antenna. If two devices D_i and D_j share the same transmitter or antenna, then they are called conflicting devices, denoted by $conflict(D_i, D_j)$. For any two allocations $A_i = <P_i, D_j, F_k>$ and

$A_{i'} = <P_{i'}, D_{j'}, F_{k'}>$, if $overlap(P_i, P_{i'})$ and $conflict(D_j, D_{j'})$ hold, then they are called conflicting allocations.

Interfering Allocations. The broadcasting quality of an allocation at a monitoring site is measured by field strength and circuit reliability. Their values can be calculated through dedicated programs such as REC533 [2] or VOACAP [3]. The broadcasting quality at a monitoring site is considered to be acceptable by the SAPPRFT if the field strength is above 38 dB. The site is qualified if the field strength is above 55 dB and the circuit reliability is above 70%. Suppose that there are two allocations A_i and $A_{i'}$, they have a monitoring site in common and the field strengths at the site are both acceptable. If the absolute difference between the field strengths of these two allocations is less than 18 dB, and the absolute difference of their frequencies is no larger than 5 kHz, then the two allocations will interfere each other and weaken the broadcasting quality.

Bands and Frequencies. A band is a frequency interval, denoted by B with subscript. According to the requirement of SAPPRFT, frequencies in the same band have equivalent quality in broadcasting.

Diplomatic Programs. Besides the domestic program broadcasted by the SAPPRFT of China, there are also diplomatic programs broadcasted by other countries and regions with the fixed devices and frequencies. Unless otherwise specified, the term program is referred to as domestic program in this paper.

Optimization Goal. For an allocation $<P_i, D_j, F_k>$, if at least 60% of the sites in the target area of P_i (R_i) are acceptable, then the allocation is admissible. We use $N_{<i,j,k>}$ to represent the number of qualified sites in R_i. The optimization goal of the SRBRAFA problem is to maximize the total coverage rate $(\sum_{P_i \in \mathcal{P}} N_{<i,j,k>}/|R_i|)$ in the target areas of all programs. In [10], Ma et al. use the total number of qualified sites as the optimization goal. One drawback of this objective function is that the programs with large target areas will dominate those with small target areas. However, according to the Division of Radio Frequency Assignment of SAPPRFT, all programs are equally important. So we use the total coverage rate as the optimization goal in this paper.

In summary, the SRBRAFA problem can be defined in the following way.

Definition 1. *(The Shortwave Radio Broadcast Resource Allocation and Frequency Assignment Problem (SRBRAFA)). Given n radio programs, m transmission devices and l frequencies, for each program P_i select a device D_j and a frequency f_k such that:*

- *The allocation $<P_i, D_j, F_k>$ is admissible.*
- *For any two allocations A_i and A_j, A_i and A_j don't conflict with each other.*
- *For any two allocations A_i and A_j, A_i and A_j don't interfere with each other.*
- *The total coverage rate $(\sum_{P_i \in \mathcal{P}} N_{<i,j,k>}/|R_i|)$ is maximized.*

Since the decision version of the SRBRA problem, which is proved to be NP-complete [10], is a special case of the decision version of the SRBRAFA problem, the SRBRAFA problem is NP-hard.

3   The Two-Phase Approach

3.1   The Framework

In the previous section, we know that the broadcasting quality, i.e. field strength and circuit reliability is determined by device and band, not by frequency. This observation motivates us to use band instead of frequency for assignment in the first phase. After the allocations of devices and the assignments of bands, we will assign frequencies with consideration for interference. There are two advantages to the two-phase approach. Firstly, since it uses band instead of frequency in the first phase, the number of variables and constraints in the model is reduced. Secondly, it only takes consideration of interference in the second phase which means the number of allocation pairs which are potentially interfering is reduced.

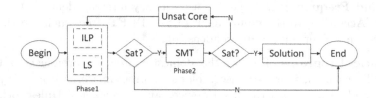

Fig. 1. The framework of the two-phase approach

The framework of the two-phase approach is shown in Fig. 1. In phase 1, an ILP-based method and a local search method are designed to solve problems under different scopes. They allocate one device and assign one band to each program, assuring no conflicting allocation exists and the total coverage rate is maximized. In phase 2, the algorithm constructs an SMT model based on the results of phase 1 to assign frequencies to programs with the condition that interference is not admitted. If a solution is found, it will be returned as the final solution. Otherwise, a new constraint representing that the previous results is not allowed will be added to phase 1 to avoid being stuck on these false allocations. The process repeats until the algorithm gets a final solution or no solution if the model of phase 1 is unsatisfiable.

3.2   Phase 1: Device Allocation and Band Assignment

The ILP Model. We first introduce two sets of 0–1 integer variables $\{Y_{i,j}\}$ and $\{Z_{i,k}\}$ to indicate whether device D_j and band B_k is allocated to program P_i respectively. For clarity, we also introduce two sets $\mathcal{QY}_i$ and $\mathcal{QZ}_i$ to represent available devices and bands for program P_i respectively:

$$\mathcal{QY}_i = \{j | \exists k, allocation <P_i, D_j, B_k> is\ admissible\}$$
$$\mathcal{QZ}_i = \{k | \exists j, allocation <P_i, D_j, B_k> is\ admissible\}$$

Recall that $N_{<i,j,k>}$ is the number of qualified sites, the objective function is as follows (**quadratic**):

$$Maximize \sum_{P_i \in \mathcal{P}} \sum_{j \in \mathcal{QY}_i} \sum_{k \in \mathcal{QZ}_i} (N_{<i,j,k>}/|R_i| \times Y_{i,j} \times Z_{i,k}) \tag{1}$$

There are two kinds of linear integer constraints. One represents that one program is allocated only one device and is assigned only one band:

$$\sum_{j \in \mathcal{QY}_i} Y_{i,j} = 1, \ \forall P_i \in \mathcal{P} \tag{2}$$

$$\sum_{k \in \mathcal{QZ}_i} Z_{i,k} = 1, \ \forall P_i \in \mathcal{P} \tag{3}$$

The other represents that no conflicting allocation is allowed.

$$Y_{i,u} + Y_{j,v} \leq 1, \ \forall overlap(P_i, P_j), \ conflict(D_u, D_v), u \in \mathcal{QY}_i, v \in \mathcal{QY}_j \tag{4}$$

The objective function (1) is a quadratic objective. In order to improve the performance of the approach, we rewrite it to a linear objective by introducing a set of variables $\{X_{i,j,k}\}$ to indicate whether device D_j and frequency F_k are allocated to P_i. We introduce $\mathcal{Q}_i$ to represent all pairs $<j,k>$ which make allocation $<P_i, D_j, B_k>$ admissible for P_i:

$$\mathcal{Q}_i = \{<j,k>|allocation \ <P_i, D_j, B_k> \ is \ admissible\}$$

The linear objective is as follows (**linear**):

$$Maximize \sum_{P_i \in \mathcal{P}} \sum_{<j,k> \in \mathcal{Q}_i} (N_{<i,j,k>}/|R_i| \times X_{i,j,k}) \tag{5}$$

In order to build connection between the two groups of variables, we need the following constraints to represent $X_{i,j,k} \leftrightarrow Y_{i,j} \wedge Z_{i,k}$:

$$Y_{i,j} + Z_{i,k} - X_{i,j,k} \leq 1, \ \forall <j,k> \in \mathcal{Q}_i \tag{6}$$

$$Z_{i,k} - X_{i,j,k} \geq 0, \ \forall <j,k> \in \mathcal{Q}_i \tag{7}$$

$$Y_{i,j} - X_{i,j,k} \geq 0, \ \forall <j,k> \in \mathcal{Q}_i \tag{8}$$

The Local Search Method. The local search method is an extension to the one in [10], which only allocates devices to programs. We modify the search procedure for device allocation and band assignment.

The local search method introduced in [10] consists of three steps, i.e. Construct, Swap and Substitute. Band assignment is completed in the process of Construct. For each unassigned program P_i, if $<P_i, D_j, B_k>$ is admissible, then we assign the band B_k and D_j to P_i so that $N_{<i,j,k>}$ is maximized and $<P_i, D_j, B_k>$ is consistent with the solution $\mathcal{S}$. That is to say, we apply a greedy strategy in band assignment. Similarly, in the process of Swap and Substitute, if we allocate a new device to a program, then we choose the band which can maximize the coverage rate of the program.

3.3 Phase 2: Frequency Assignment

In Phase 2, we determine the frequencies of the programs on the basis of the band assignment in Phase 1. The potential interfering program pairs (denoted by $IP(P_i, P_j)$) can be derived from the result of Phase 1. Suppose P_i and P_j are a pair of such programs, whose frequencies are denoted by F_{P_i} and F_{P_j} respectively. Generally, the domain of frequency is limited to the multiples of five, such as 6015 kHz and 7200 kHz. In order to avoid interference between P_i and P_j, the difference between F_{P_i} and F_{P_j} should be larger than 5, or formally:

$$F_{P_i} - F_{P_j} > 1 \vee F_{P_j} - F_{P_i} > 1, \ \forall IP(P_i, P_j), \ P_i \in \mathcal{P}, \ P_j \in \mathcal{P} \tag{9}$$

Note that the difference of frequency between interfering programs is larger than 1 instead of 5 since we divide the value of frequency by 5 in calculation.

Recall that there are diplomatic programs with fixed devices and frequencies. In order to deal with the interference of such programs, we divide the band of a program into several domains. Suppose that the band assigned to program P_i is $[7000, 7040]$. In order to prevent interference from a diplomatic program with frequency 7020 kHz, F_{P_i} should be greater than 7025 kHz or less than 7015 kHz. As a result, the domain of F_{P_i} is divided into two intervals, $[7000, 7010]$ and $[7030, 7040]$. We denote the collection of the domains for F_{P_i} by $\mathcal{C}_i$. Each $C \in \mathcal{C}_i$ is an interval $[f, f']$ of frequencies. The following constraint ensures that F_{P_i} should fall in one of these intervals.

$$\bigvee_{C \in \mathcal{C}_i} f \leq F_{p_i} \leq f', \ \forall P_i \in \mathcal{P} \tag{10}$$

The above constraints naturally form an SMT formula on difference logic (SMT(DL)). If the SMT formula is unsatisfiable, it suggests that the allocations in Phase 1 would inevitably lead to interference. By extracting the unsatisfiable core of the SMT formula, we can identify the allocations responsible for this inconsistency. Suppose $\mathcal{F}^{UC}$ is the set of frequency variables involved in the unsatisfiable core, then $\mathcal{X}^{IC} = \{X_{i,j,k} | F_{P_i} \in \mathcal{F}^{UC}, X_{i,j,k} = 1\}$ is the set of allocations with interference. In order to avoid the same inconsistency, we add the following constraint to the ILP model:

$$\sum_{X_{i,j,k} \in \mathcal{X}^{IC}} X_{i,j,k} \leq |\mathcal{X}^{IC}| - 1 \tag{11}$$

This trick also applies to the local search procedure by adding the unsat core to a taboo list.

4 Experimental Results and Analysis

This section evaluates the proposed approach on the entire data set of the Division of Radio Frequency Assignment of SAPPRFT. There are 948 programs in total. The total number of transmission devices is 7061, on the premise that the transmitters and antennas located in the same shortwave radio station can be fully connected. However, due to the limitation of the current circuits, only 873 devices are available in practice. So our experiments were conducted on two sets of devices: the practical devices, and the fully connected devices. We employ CPLEX [6] as the ILP solver, and Z3 [7] for SMT solving. All instances are available on the website[1]. The experiments were performed in windows 7 on 2.8 GHz Intel processor with 16 GB RAM.

Tables 1 and 2 show the comparison of ILP against local search (LS) in Phase 1, on the instances with practical devices, and the instances with fully connected

Table 1. Experimental results on the practical devices

$|\mathcal{P}|$	$|\mathcal{D}|$	ILP		Local search	
		Obj	Time(s)	Max(avg)	Time(s)
100	100	0.706	1.201	0.651(0.632)	0.06
100	200	0.763	1.716	0.699(0.682)	0.01
200	100	-	2.59	-	-
200	200	0.716	7.769	-	-
200	300	0.802	7.191	0.707(0.697)	0.01
300	200	-	9.797	-	-
300	300	0.795	21.403	-	-
300	400	0.855	24.944	0.741(0.730)	0.03
400	300	0.753	39.281	-	-
400	400	0.836	57.346	-	-
400	500	0.851	50.576	0.711(0.702)	0.05
500	400	0.810	111.478	-	-
500	500	0.835	88.499	0.692(0.676)	5.84
500	600	0.850	82.977	0.715(0.702)	0.12
600	500	0.823	153.661	-	-
600	600	0.844	135.736	-	-
600	700	0.856	126.439	0.713(0.70)	0.2
700	600	0.841	240.039	-	-
700	700	0.855	186.952	0.707(0.701)	0.19
700	800	0.870	250.569	0.717(0.711)	0.28
800	700	0.849	248.26	0.708(0.693)	0.24
800	800	0.863	333.359	0.719(0.709)	0.27
800	873	0.869	385.182	0.726(0.715)	0.36
948	800	0.847	447.145	0.703(0.690)	0.7
948	873	0.854	599.059	0.707(0.696)	0.86

---

[1] http://lcs.ios.ac.cn/~maff/.

Table 2. Experimental results on the fully connected devices

$|\mathcal{P}|$	$|\mathcal{D}|$	ILP		Local search	
		Obj	Time(s)	Max(avg)	Time(s)
100	1000	0.663	63.679	-	-
200	2000	0.790	387.148	0.624(0.611)	0.87
300	3000	0.788	1567.12	0.606(0.590)	0.43
400	4000	OM	OM	0.596(0.587)	0.71
500	5000	OM	OM	0.702(0.693)	1.19
600	6000	OM	OM	0.701(0.693)	2.07
700	7000	OM	OM	0.713(0.703)	3.23
948	7061	OM	OM	0.694(0.684)	4.6

devices respectively. Since all instances are solved in a single iteration of the two phases, and Phase 2 only took less than one second, we only perform the comparison for Phase 1. The instances in Table 1 are randomly taken from 948 programs and 873 practical devices. In Table 2, the instances are randomly taken from 948 programs and 7061 fully connected devices. For clarity, the obj shown in the table is the average of coverage rate rather than the sum of coverage rate. The time limit for LS is 10 s and for CPLEX is 3600 s. The LS method is executed 10 times for each instance, and both the maximum and average rates are listed. The average time for LS to reach a locally optimal solution is also listed. The symbol – indicates the instance has no solution, or LS failed to find a solution, and OM indicates that CPLEX ran out of memory.

We can observe from Table 1 that CPLEX can solve all these instances within 10 min. By constrast, LS failed on nearly half of the instances. For the rest, the coverage rates provided by LS is less optimal than CPLEX, but the time for LS to find the locally optimal solution within the time limit is always less than 1 second. The reason for the unsatisfactory performance of LS on these instances is that the ratios of the numbers of programs to the numbers of devices are much larger than those in [10], making it very hard for a stochastic algorithm to find a legal solution. In Table 2, there are much more devices in each instance. CPLEX ran out of memory for the larger instances, while LS can always provide a solution very quickly for most of the instances besides the first one.

5   Conclusions

In this paper, we studied the SRBRAFA problem. We proposed a two-phase approach integrating ILP (or local search) with SMT to solve the problem. The approach is evaluated using real data from the Division of Radio Frequency Assignment of SAPPRFT, and the results are promising. In the current artificial plan, the average coverage rate is only 0.534, and there are as many as 40 pairs of interfering programs. By contrast, we can achieve the optimal coverage rate 0.854 with CPLEX, and 0.696 with local search. Moreover, no interference exists.

Overall, the ILP method can solve the current problem of SAPPRFT completely. It achieves optimal solution for real-world instances with practical devices (up to 873 devices). But on instances with fully connected devices (up to 7061 devices), ILP doesn't scale and we use local search as an alternative. In the future, our aim is to solve larger-scale instances with better approach.

Acknowledgements. The authors are supported in part by the CAS/SAFEA International Partnership Program for Creative Research Teams and the National Science Foundation of China under Grant 61370156. Besides, we are grateful to the anonymous reviewers for their helpful comments.

References

1. Frequency Assignment Problem. http://fap.zib.de/
2. General information on the REC533 propagation prediction model. http://www.voacap.com/itshfbc-help/rec533-general.html
3. VOACAP Quick Guide. http://www.voacap.com
4. Barrett, C.W., Sebastiani, R., Seshia, S.A., Tinelli, C.: Satisfiability modulo theories. Handb. Satisf. **185**, 825–885 (2009)
5. Conrad, F.: Short-wave radio broadcasting. Proc. Inst. Radio Eng. **12**(6), 723–738 (1924)
6. ILOG CPLEX 11.0 users manual. ILOG S.A., Gentilly, France, p. 32 (2007)
7. Moura, L., Bjørner, N.: Z3: an efficient SMT solver. In: Ramakrishnan, C.R., Rehof, J. (eds.) TACAS 2008. LNCS, vol. 4963, pp. 337–340. Springer, Heidelberg (2008). doi:10.1007/978-3-540-78800-3_24
8. Hao, J.-K., Dorne, R., Galinier, P.: Tabu search for frequency assignment in mobile radio networks. J. Heuristics **4**(1), 47–62 (1998)
9. Idoumghar, L., Debreux, P.: New modeling approach to the frequency assignment problem in broadcasting. IEEE Trans. Broadcast. **48**(4), 293–298 (2002)
10. Ma, F., Gao, X., Yin, M., Pan, L., Jin, J., Liu, H., Zhang, J.: Optimizing shortwave radio broadcast resource allocation via pseudo-boolean constraint solving and local search. In: Rueher, M. (ed.) CP 2016. LNCS, vol. 9892, pp. 650–665. Springer, Cham (2016). doi:10.1007/978-3-319-44953-1_41
11. Yokoo, M., Hirayama, K.: Frequency assignment for cellular mobile systems using constraint satisfaction techniques. In: Proceedings of the IEEE Vehicular Technology Conference, vol. 2, pp. 888–894 (2000)

Constraint-Based Fleet Design Optimisation for Multi-compartment Split-Delivery Rich Vehicle Routing

Tommaso Urli$^{(\boxtimes)}$ and Philip Kilby

CSIRO Data61, Australian National University (ANU),
Tower A, Level 3, 7 London Circuit, Canberra, ACT 2601, Australia
{tommaso.urli,philip.kilby}@data61.csiro.au

Abstract. We describe a large neighbourhood search (LNS) solver based on a constraint programming (CP) model for a real-world rich vehicle routing problem with compartments arising in the context of fuel delivery. Our solver supports both single-day and multi-day scenarios and a variety of real-world aspects including time window constraints, compatibility constraints, and split deliveries. It can be used both to plan the daily delivery operations, and to inform decisions on the long-term fleet composition. We show experimentally the viability of our approach.

1 Introduction

The vehicle routing problem (VRP) is considered "one of the biggest success stories in operations research" [19]. One of the reasons behind this success is its direct applicability in industrial contexts, which motivated researchers and practitioners to extend the original formulation to include aspects arising in real-world scenarios (see [14]). Examples of such extensions are, for instance, time windows, multiple commodities, driver breaks, and heterogeneous fleets, for which dedicated formulations and techniques have been proposed. In a typical industrial application, several of such extensions coexist, along with client-specific constraints. The space of possible combinations is enormous. The effort to model and solve increasingly complex VRP variants generated a family of problems known as "rich" VRPs (see [7] for a survey).

In this paper we present an industrial application in the context of fuel-distribution. Our work stands at the intersection of three important VRP extensions: multi-compartment VRP (MCVRP, see [12]), split delivery VRP (SDVRP, see [1]), and fleet size and mix VRP (FSMVRP, see [18]). The defining feature of MCVRPs is that the vehicles have a number of isolated compartments, allowing them to carry different goods without mixing them. Such problems arise in the contexts of food and fuel delivery. SDVRP is a generalisation of the classic VRP in which the demand of one customer can be cumulatively satisfied by multiple vehicles. It has been shown [16] that SDVRPs can yield substantial savings compared to classic VRPs. Finally, the FSMVRP is a generalisation of the standard VRP in which, in addition to the routing aspects, one also needs to

© Springer International Publishing AG 2017
J.C. Beck (Ed.): CP 2017, LNCS 10416, pp. 414–430, 2017.
DOI: 10.1007/978-3-319-66158-2_27

decide the fleet composition considering both fixed costs, e.g., acquisition, and operational costs, e.g., fuel and salaries. The main contributions of this work are the following. First, we introduce a constraint programming (CP) formulation of a multi-day multi-compartment fleet size and mix rich vehicle routing problem with split deliveries. Our formulation supports several real-world aspects, such as an heterogeneous fleet, time window constraints, compatibility constraints, and route length limits. Second, we describe a constraint-based large neighbourhood search (LNS) procedure tailored around the aforementioned model. Third, we propose two different fleet design techniques that rely on the developed solver. We compare the proposed approaches experimentally on real-world data, and draw some important conclusions the applicability of the approaches and on the most promising future directions of this work.

2   Related Work

In this section we survey most significant literature on MCVRP, restricting ourselves to fuel distribution applications. We also briefly review the most important works on SDVRP and FSMVRP.

2.1   Multi-compartment Vehicle Routing

[5] describe the modelling of a US-based fuel distribution infrastructure. They model the dispatch process as an augmented assignment problem in which round trips must be assigned to trucks. Routing aspects are not considered, nor are split deliveries. The problem is initially solved using mixed integer programming (MIP) however, a heuristic approach is proposed to solve problems of practical size. In [4], the authors extend [5] by introducing features such as less-than-truckload orders (multi-stop routes), routing of vehicles, and assignment of orders to bulk terminals before the dispatch.

Later works follow the rich VRP tradition and consider a variety of side constraints. [6] describe an oil distribution supply chain in the Netherlands. The authors take into account several strategic and tactical level decisions, i.e., depot location, customer selection, and fleet design. [2] present a branch-and-price algorithm for the solution of a fuel distribution problem for a small company. The company controls a central depot and a fleet of heterogeneous vehicles with compartments. The problem is subject to several limitations, for instance each compartment must be either completely full or completely empty at all times. [8] formalise the petrol station replenishing problem (PSRP), and present an exact method to solve it. They consider an unlimited fleet of vehicles with different compartment configurations, thus providing insights on what is the best fleet on each day. In [9], the authors describe a heuristic approach to optimise the replenishment of a number of petrol stations as soon as the demand of any of the fuel types exceeds the available stock. The work is similar to [8], but cast in a multi-period context. As for earlier works, the model assumes that compartments must

be emptied completely at the delivery location[1]. [25] study a situation very close
to the one described above. They look at a multi-compartment vehicle routing
problem with incompatible product types and uncertain demands. They formulate the problem as a two-stage stochastic optimisation problem, with a recourse
action of returning to depot to unload or restock. They describe a memetic
algorithm to find feasible solutions for problems with up to 484 customers. [10]
describe a multi-depot fuel delivery problem. The solution method is based on
the concept of *trips*, i.e., routes from a depot through one or more service stations and back to the depot. A trip selection heuristic is used to simultaneously
choose trips and to concatenate trips into a workday for a given driver.

The literature includes also some more general formulations of MCVRP. [12]
generalise the type of problems seen in fuel and food delivery, and define a so-
called VRP with compartments (VRPC). Each compartment can have capacity
constraints, compatibility constraints with given products, and compatibility
constraints between different products in the same compartment. The authors
propose a heuristic based on Adaptive Large Neighbourhood Search (ALNS, see
[28]). [21] describe a VRP with compartments where the compartment partitions
can move. Hence, the number of compartments can vary, and the capacity of each
can be changed subject to a constraint on the sum. A Variable Neighbourhood
Search (VNS) procedure is proposed to solve the resulting problem. Finally, [22]
describe a VRP with multiple compartments, multiple product types, multiple
trips for each vehicle, and multiple time windows. A MIP formulation is solved
using a commercial solver for up to 15 customers.

2.2 Split Delivery Vehicle Routing

Split delivery vehicle routing problems (SDVRP) have received a notable amount
of attention in the literature. We mention here only the seminal works and some
more recent surveys and results. The first MIP formulation of SDVRP was initially proposed in [16,17]. In [15] the formulation was extended with several
additional valid cuts and sub-tour elimination constraints that are shown to be
effective in reducing the integrality gap at the root node. [1] provide an extensive survey of split delivery vehicle routing, along with a classic mixed integer
programming formulation and a thorough discussion of complexity aspects and
lower bounds. [3] describe a *scatter search* approach to solve the fleet mix problem for the daily split delivery vehicle routing problem. The proposed approach
includes some fleet design aspects, however these are limited to single-day and
single-commodity scenarios.

2.3 Fleet Size and Mix Vehicle Routing

[11] present a heuristic solution to FSMVRP for the capacitated VRP with time
windows. The approach is based on a linear programming formulation of the

---

[1] Such constraints are rather common in the literature, and reflect the presence or
absence of *flow* (or *debit*) *meters* at the replenishing plant or on the vehicles.

problem. [32] propose a component-based framework to solve a variety real-world VRP variants, including FSMVRP. The approach doesn't handle split deliveries or multiple commodities, however it appears to be very well engineered to solve a range of problems of industrial interest. While the literature on FSMVRP is ample, most of the existing approaches focus on single-day and single-commodity delivery problems, and only a small amount of works mention the tactical aspects. The main reason for this is that many FSMVRP approaches target rich VRP formulations, in which even single-day scenarios are complex to solve.

One of the main contributions of our work is to bring together three complex extensions of the VRP problem. To the best of our knowledge, although there is a lot of existing literature on each independent extension, no prior work exists at their intersection.

3 Problem Definition

In this section we describe the problem as presented to us by our industrial partner. The problem is initially cast in a single-day scenario and then extended to multiple days.

Our client serves a number of petrol stations $c \in C$ across Australia, and is responsible for replenishing them with different fuel types $k \in K$. In the following, we will make use of the term *customer* to refer to a petrol station. Each customer c corresponds to a location $l_c \in L$, we will therefore often refer to customer as locations and vice versa. The full set of locations L comprises the customers C and the depot p, i.e., $L = C \cup \{p\}$.

On a given day each customer C can issue a request for one or more fuels. Each request is defined by

- a time window $[\underline{tw}_c, \overline{tw}_c]$ for the start of the activity,
- a demand $q_{c,k} \geq 0, \forall k \in K$, and
- a set B_c of compatible *route types* onto which the visit can be scheduled.

A route type $b \in B$ combines a vehicle type $t_b \in T$ and a time window defined by a earliest departure time $\underline{tw}_b$ and a final arrival time $\overline{tw}_b$, representing a working shift, e.g., morning or afternoon. The existence of a route of some type b with an associated vehicle $v \in V$ of type t_b means that that vehicle is available to carry out deliveries in the specified shift. A maximum duration $\overline{m}_b$ for the route is provided to enforce driver break policies. With a slight abuse of notation, all the properties of a vehicle type $t \in T$, e.g., number of compartments, capacity, etc., are inherited by the routes $B_t \in \{b \in B \mid t_b = t\}$. Note that a vehicle $v \in V$ can be associated with multiple routes as long as they don't overlap, e.g., the same vehicle can have a morning route and an afternoon route.

The requests are fulfilled by dispatching a fleet of vehicles from the depot to the customers. Because of the limited capacity of the compartments, each vehicle might do several round trips per day. The fleet of vehicles is heterogeneous, each vehicle being of a given type $t \in T$ with different characteristics

- the number of compartments u_t,
- the total capacity $g_{t,k}$ for each fuel $k \in K$,
- the time needed for loading the vehicle $\bar{o}_t$,
- the time needed for unloading the vehicle $\underline{o}_t$,
- the dollar cost per metric (see below),
- the (amortised) daily fixed dollar cost f_t for owning the vehicle, and
- the maximum number $\bar{a}_t$ of vehicles of type t allowed in the fleet[2].

Informally, the rules for the deliveries are the following: (i) different fuels must be carried in different compartments, (ii) the assignment of fuels to the compartments of a vehicle can be changed, but only when the vehicle stops at the depot as the compartments must be washed, (iii) the demand of each customer must be completely fulfilled, but more than one vehicle can be used to do so, (iv) each activity at a given customer c must begin within the specified time window, and (v) each visit to a customer c must be scheduled on a compatible route $r \in B_c \subset B$.

The set of locations L represent the nodes of a connected digraph. For every edge $e = (l_1, l_2), \forall l_1, l_2 \in L$ a set of weights is provided. Each weight corresponds to a distance separating the two locations expressed according to some *metric*, e.g., seconds or meters. One among all the metrics is marked as the *time metric* and needs special handling. In addition to these, our formulation supports any number of *vehicle-dependent metrics* defined as linear combinations of base metrics whose weights depend on the type of the vehicle. One such metric represents the overall cost, in Australian dollars (AUD).

The objective of the problem is to minimise the total amortised fixed costs for owning the selected fleet of vehicles, and the operation costs measured as the dollar metric.

Multi-day Extension. In the context of fuel distribution, the problem described above arises daily. When considering a multi-day scenario, all of the above definitions and rules still apply, the only difference being how the fleet-specific costs are handled. In a single-day scenario, the fleet is composed by the vehicles needed to serve the demand of the given day. In a multi-day scenario, the fleet is composed by all vehicles used *at least once* on *any* day. Since the amortised fixed costs are daily, we need to multiply the fixed cost of the fleet by the number of days in the horizon. In the next section, we will see that this aspect of the problem can be represented very naturally in our CP model.

4   Model

We model our problem with a step-based formulation similar in spirit to the one used in [13]. Unlike classic CP successor-based VRP models, where for each customer $c \in C$ a *successor* variable encodes the next visit in a route, in a step-based formulation the routes are sequences of visits of fixed length (see Fig. 1).

---

[2] This limit allows to optimise the routing using an existing fleet.

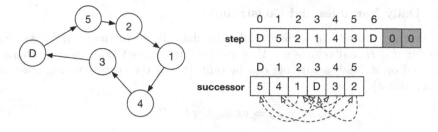

Fig. 1. Step-based model vs. successor-based model.

The main advantages of the step-based formulation is that we can trivially model split deliveries. In contrast, with a successor-based formulation we would have to decide in advance how many times a certain location can be visited. This aspect is important for fuel distribution applications, where the length of the round trips to the depot is usually small (typically between 2 and 4 stops), and several round trips can be performed in a day. A disadvantage of this modelling is that we need to explicitly provide bounds on the maximum length of a route, and introduce a virtual "null" location to model routes that are shorter than the bound (see Fig. 1). Another difference with the standard VRP models is that we make a distinction between routes and vehicles. This allows, for instance, to use the same vehicle in different shifts, which makes handling the fleet design aspect easier.

Our model needs therefore two parameters, $\bar{r}$ and $\bar{s}$ which represent, respectively, the maximum number of routes that we are allowed to use (coming from the FSMVRP formulation) and the maximum length of such routes (coming from the step-based formulation). If such parameters are not provided, the solver tries to guess them by heuristically constructing a feasible solution, and then setting $\bar{r}$ and $\bar{s}$ as a function of the identified number and length of routes.

4.1   Horizon-Wide Variables and Constraints

In our problem, the fleet is represented by an array v_type of integer variables of length $\bar{v}$. Each variable in the array encodes a vehicle type $t \in T \cup \{ \maltese \}$, where $\maltese$ represents a zero-cost and zero-capacity *null vehicle* that unused routes can be assigned to. Additionally, an auxiliary binary variable array used models whether a vehicle is part of the overall fleet. A small set of fleet constraints keep these variables consistent. Constraint 1 says that if a vehicle is used then it has a proper type, and Constraint 2 limits the number of vehicles of a given type according to the input data.

$$\mathsf{used}_v \Leftrightarrow \mathsf{v\_type}_v \neq \maltese, \quad \forall v \in V \tag{1}$$

$$| \{v \in V \mid \mathsf{v\_type}_v = t\} | \leq \bar{a}_t, \quad \forall t \in T \tag{2}$$

4.2   Daily Variables and Constraints

The first set of daily constraints links the daily fleets to the overall fleet. For each day $d \in H$ and vehicle $v \in V$, a variable array $\texttt{used\_on}_{d,v}$ models whether v is used on d. Such a variable can be only true if the corresponding $\texttt{used}_v$ is (Constraint 3).

$$\texttt{used\_on}_{d,v} \Rightarrow \texttt{used}_v, \quad \forall d \in H, v \in V \tag{3}$$

For each route $r \in R_d$, $d \in H$ we have an integer variable $\texttt{vehicle}_{d,r}$ mapping it to a vehicle in the fleet, and an integer variable $\texttt{r\_type}_{d,r}$ with domain $B \cup \{\textonehalf\}$ representing its type, where $\textonehalf$ represents a *null route*, i.e., a route which is not used. Constraint 4 makes sure that a route is always assigned to a vehicle of the right type (the only compatible vehicle for $\textonehalf$ being $\mathbf{\Psi}$).

$$\texttt{v\_type}_{\texttt{vehicle}_{d,r}} = t_{\texttt{r\_type}_{d,r}} \tag{4}$$

Constraint 5 ensures that all vehicles marked as used on some day $d \in H$ execute at least one route on that day. Additionally, we use a scheduling constraint for unary resources (see [33]) to guarantee that a vehicle is used by at most one route at a time.

$$\texttt{used\_on}_{d,v} \Leftrightarrow | \{r \in R_d \mid \texttt{vehicle}_{d,r} = v\} | > 0, \quad \forall v \in V, d \in H. \tag{5}$$

The next set of variables and constraints concern the structure of the routes. First, an array of integer variables $\texttt{visit}_{d,r}$ of length $\overline{s}_d$ models the list of successive locations visited by route r on day d. To represent routes shorter than $\overline{s}_d$ we include a *null location* $\mathbf{\natural}$, where nothing can happen, in the domains of the $\texttt{visit}_{d,r}$ variables. Since $\texttt{visit}_{d,r}$ is in fact a *string*, we can use regular language membership constraints (see [26]) to make sure that the routes have the correct structure. Constraint 6 imposes that a route be either completely empty ($\overline{s}_d$ visits to $\mathbf{\natural}$) or have the structure of a feasible route, i.e., a visit to the depot, followed by a visit to a customer, followed by any number of visits to the depot (to refill) or customers, ended by a visit to the depot followed by zero or more visits to $\mathbf{\natural}$.

$$\texttt{visit}_{d,r} \in \textbf{reg}\left((pC_d(C_d \mid pC_d) * p\natural*) \mid \natural\{\overline{s}_d\} \right), \quad \forall d \in H, r \in R_d \tag{6}$$

Constraint 7 guarantees that, if the first visit is to $\mathbf{\Psi}$, then the route is of type $\textonehalf$ (and as such must be assigned to a vehicle of type $\mathbf{\Psi}$).

$$\texttt{visit}_{d,r,0} = \natural \Leftrightarrow \texttt{r\_type}_r = \textonehalf \tag{7}$$

On top of the above constraints, we define a number of additional redundant constraints whose aim is to reduce the search space. For instance, no customer can be visited in two consecutive steps. Since these constraints are rather obvious, we exclude them from the present discussion.

The above constraints and variables deal with the routing aspects of the problem. We also need to model the activity performed at each step of the route. For each $r \in R_d$, for each fuel $k \in K$, and for each step $s \in S_d$ we therefore introduce two integer variable arrays, $\mathtt{load}_{d,r,k,s}$ and $\mathtt{activity}_{d,r,k,s}$, respectively modelling the amount of fuel k that is *left* on the truck executing route r after the visit at step s, and the amount of fuel k that is *transferred* from the truck executing route r during the visit at step s. The two sets of variables are connected in the obvious way, i.e., the activity at a given step is the difference in load between the current and the previous step. We require the following activity constraints. First, the activity at 1 and final depots (visits to depots followed by visits 1, or whose index is the last in the route) must always be zero. The activity for any commodity at the starting and intermediate visits to the depot must be *greater or equal* to zero, and the total sum of activity must be *strictly greater* than zero. Similarly, at the petrol stations the activity for each commodity has to be *less than or equal* to zero, but the sum of the activity must be *strictly less* than zero. The trucks must always be completely emptied before any visit to the depot.

Since a vehicle is organised in compartments that can only transport one type of fuel at a time, the capacity constraints are slightly more complex than in a traditional VRP. For each $r \in R_d$, for each step $s \in S_d$, and for each fuel type $k \in K$ an array of integer variables $\mathtt{compartments}_{d,r,k,s}$ model how many compartments are dedicated, at each step, to each fuel. Constraint 8 enforces that the sum of the compartments dedicated to each fuel to be less than or equal to the number of compartments on each vehicle. Constraint 9 states that the total load of a given fuel must be less than or equal to the capacity of the compartments for that fuel multiplied by the number of compartments dedicated to it.

$$\sum_{k \in K} \mathtt{compartments}_{d,r,k,s} \leq u_{\mathtt{r\_type}_r}$$

$$\forall d \in H, r \in R_d, s \in S_d \tag{8}$$

$$\mathtt{load}_{d,r,k,s} \leq \mathtt{compartments}_{d,r,k,s} \cdot \left(g_{\mathtt{r\_type}_r,k} / u_{\mathtt{r\_type}_r} \right)$$

$$\forall d \in H, r \in R_d, s \in S_d, k \in K \tag{9}$$

Our model allows to change the allocation of compartments by means of the above variables; Constraint 10 restricts such changes in the visits to the depot.

$$\mathtt{compartments}_{d,r,k,s} \neq \mathtt{compartments}_{d,r,k,s-1} \Rightarrow \mathtt{visit}_{d,r} = p$$

$$\forall d \in H, r \in R_d, s \in S_d, s \neq 0, k \in K \tag{10}$$

Another set of constraints maintains the values of the cumulative metrics along the routes in the obvious way based on the distance matrix provided as input. All the metrics are initialised to zero at step $s = 0$, except for the time metric which is initialised to the earliest start time $\underline{tw}_{\mathtt{r\_type}_r}$. As for the time metric, at each visit we also consider the loading time $\overline{o}$ or the unloading $\underline{o}$ time, according to whether the activity is positive or negative. Time windows constraints are enforced at each step by bounding the time variable according to the time window of the location being served at that step. Arbitrary waiting times are implemented by constraining the time metric at a step s to be *greater or equal* to its value at $s-1$ plus the travel time between the two locations. Finally, for each route and each vehicle-dependent metric, we keep a variable whose value is computed from the value of the base metrics at the last step of the route, and the vehicle-dependent coefficients. Note that one of such metrics is used to represent the routing costs of the solution. A final set of constraints ensures that all the demand is satisfied. Since our model supports split deliveries, these constraints must make sure that the sum of the activity carried out at each customer $c \in C$ is exactly the negation of the demand of c. These constraints are rather trivial and we won't describe them in detail here.

The objective of the model is the sum of the variables for each day $d \in H$ and $r \in R_d$ representing the routing costs, and the v_type variables weighted by the coefficients f_t with $t \in T$ and multiplied by h (the length of the horizon).

5   Search Strategy

Our search strategy is a large neighbourhood search (LNS) scheme based on the above model and on two custom branching strategies. Each strategy is composed of a variable selection heuristic and a set of value selection heuristics. Both branching strategies terminate when all the demand on a given day is satisfied. The remaining free variables are assigned the minimum value in their domain.

It should be noted that our branching strategies are incomplete: decision variables are sometimes assigned heuristically in the hope of obtaining feasibility earlier, but some of the values in their domains are never tried. This is somewhat similar to the concept of "streamlined constraint reasoning" [20], a technique to prioritise promising areas of the search space. Streamlining uses additional artificial constraints to enforce properties that are satisfied in solutions of smaller problem instances, and has obtained good scaling properties on some domains. Both approaches have their merits. Streamlining can, in principle, be complete since it prioritises the search by only postponing the exploration of non-promising parts of the search space. On the other hand, incomplete branching heuristics are easier to design, and they don't necessarily require studying the properties of solutions to smaller instances of the problem.

Route-First Branching. The first strategy aims at fully constructing a route before moving on to the next one. The variable selection heuristic first iterates over the existing routes, trying to identify one in which the visit at the last step has not yet been fixed. If no such route exists, a new route must be created and the variable to be branched upon is r_type$_r$ where r is the index of the first unused route. Conversely, if such a route r exists, the heuristic tries to identify the first step s which is *incomplete*, i.e., for which one among visit$_{r,s}$, activity$_{r,s,k}$ where $k \in K$, and metric$_{\text{time},r,s}$ (in this order) has not been assigned. The heuristic branches on the corresponding variable. Once a variable x has been selected for branching, a value selection heuristic sorts the values in its domain depending on the type if x.

- The heuristic for $\texttt{r\_type}_r$ sorts the route types based on how many unsatisfied customers are compatible with each route type. Route types with a larger number of unsatisfied compatible customers are tried first.
- The heuristic for $\texttt{visit}_{r,s}$ sorts the remaining unsatisfied visits lexicographically based on (i) the earliest start time for the activities at the customer, and (ii) the maximum amount of activity that can be carried out at the customer. If there is no residual capacity on the vehicle executing the current route, the next visit is (by propagation) the depot.
- The heuristic for $\texttt{activity}_{r,s,k}$ sorts the commodities based on the amount that can be unloaded at the current customer, and then *sets* the unloaded amount to the maximum value in the domain (this heuristic, therefore, introduces incompleteness since intermediate quantities are never tried).
- The heuristic for $\texttt{metric}_{\texttt{time},r,s}$ assigns the smallest value in the domain of the variable (this is always a safe assumption).

Customer-First Branching. The second strategy tries to completely satisfy a customer (the *current customer*) before moving on to the next. Because a customer's demand may be fulfilled collectively by more than a route, an unknown number of branching steps may be needed to completely satisfy it. For this reason, we use a temporary *cache* (*curr*) to keep track of the customer being currently handled, and we keep its value fixed throughout the branching steps until the work on customer is completed. If the current customer has been selected but not completely satisfied, the heuristic looks for an existing route in which the $\texttt{activity}_{r,s,k}$ (for some $k \in K$) at the last visit has not been assigned. Since, by design, the previously selected customer must have been completely satisfied, such a visit must be to the current customer. If such a route exists, the heuristic chooses $\texttt{activity}_{r,s,k}$ as the variable to branch upon. A similar step is performed to set the $\texttt{metric}_{\texttt{time},r,s}$ variable. If all the last visits of all the existing routes have been finalised, the heuristic tries to find a compatible existing route where to insert a new visit to the current customer. During this search if the heuristic finds a route compatible with the visit, but without capacity left, it inserts a visit to the depot (a *refill*) so as to restore the capacity. If, at the end of this search, there is at least one route where a visit to the current customer can be inserted, the heuristic branches on the insertion. This branching is somewhat reversed with respect to a classic CP branching, in that the value of the $\texttt{visit}_{r,s}$ variable is known in advance, however the route r and the step s indices have to be chosen among the compatible ones. If there are no existing compatible routes where to insert a visit to the current customer, the branching strategy branches on $\texttt{r\_type}_r$ where r is the first unused route, thus initialising a new route of type compatible with the current customer. The following value selection heuristics are used to decide how to finalise the variable assignments and choose the next *current customer*.

- The heuristic for *curr* sorts the unsatisfied customers based on their earliest time window start. In this sense, we are trying to pack the routes so that there is the smallest amount of slack time between the visits.

– The branching on r tries each compatible open route in turn, and appends a visit to the current customer as the last step of the selected route.
– Finally, the heuristic for $\texttt{r\_type}_r$ sorts the route types according to the earliest time in which it is possible to start the activity at the current customer.

The value selection heuristics for $\texttt{activity}_{r,s,k}$ and $\texttt{metric}_{time,r,s}$ are identical to the ones in the route-first branching strategy.

None of the above branching strategies ever branches on the $\texttt{compartments}_{d,r,k,s}$ variables. When the activity for some commodity k at a given step is assigned, the lower bound on the number of compartments dedicated to k is updated by the propagation engine.

5.1 Large Neighbourhood Search

Large neighbourhood search (LNS, see [27,30]) is a local search meta-heuristic based on the principle that exploring large neighbourhoods of a solution helps avoiding local optima. To reduce the time needed to explore the neighbourhood, filtering techniques, e.g., constraint propagation [23], are often used. In our approach, we use the model presented in Sect. 4 to prune the set of explored neighbours of a solution during the search. At each iteration, a destroy step is performed to unassign part of the incumbent solution, and a repair step is performed to re-optimise it. To obtain the initial solution, we start from an empty solution and we apply the route-first branching strategy described in Sect. 5 until termination.

Destroy Step. We employ four different destroy strategies. Three of these are used both in single-day and multi-day scenarios, one is available in multi-day scenarios only. A parameter dr (the *destruction rate*) controls, albeit indirectly[3], how many variables are unassigned at each destroy step. In our approach dr is initialised to a value $\underline{dr}$, a parameter of the solver, and can increase during search up to $\overline{dr}$, also a parameter of the solver.

The destruction strategies are the following

– **Destroy dr routes.** We choose dr routes uniformly at random, and reset all the relative variables to their original domains. The first unused route, if available, is also destroyed.
– **Destroy dr vehicles.** We choose dr vehicles uniformly at random, and reset all the variables relative to the routes assigned to these vehicles to their original domains.
– **Destroy dr days.** The solver chooses dr days uniformly at random, and resets all of the variables to their original domains. Only in multi-day mode.

---

[3] Depending on the destruction strategy, the same value of dr can cause different numbers of variables to be relaxed. The per-variable timeout used in the repair step mitigates this disparity. Moreover, relaxing the problem "semantically" rather than randomly allows us to preserves the structure of the solution.

– **Repack fleet.** We reset all the fleet variables, i.e., vehicle types and assignment of routes to vehicles, to achieve a better reassignment of routes to compatible vehicles.

The relative probabilities of choosing one strategy over the others are parameters to the solver.

Repair Step. Once a solution has been partially destroyed it is re-optimised applying the two branching strategies presented in Sect. 5. These are used in a branch & bound scheme to reassign the variables unassigned in the destroy step. The branching heuristics are chosen at random according to probabilities that are parameters to the solver. The branch & bound procedure is run to a time limit $t_{max} = t_{var} \cdot n_{free}$ where t_{var} is a parameter of the solver, and n_{free} is the number of variables that have been relaxed in the destroy step. At each repair step we constrain the cost of the next solution to be lower than the cost of the incumbent. Of course it is not guaranteed that a new solution is found during the repair step. If the repair step fails, a counter ii (idle iterations) is increased. When the number of idle iterations exceeds a threshold ii_{max}, the destroy rate dr is increased by one. When an improving solution is found, or when dr reaches $\overline{dr}$, the dr is reset to $\underline{dr}$[4].

6   Fleet Design

We present two different approaches to address the fleet design problem. Both are based on the CP model and LNS strategy presented above, but differ in terms of computational requirements.

Union Fleet. The first method is composed of two phases. In the first phase, we run the solver on each day of the horizon independently, so as to generate a specialised fleet for each day. We encode such fleets as multi-sets, e.g., if on a given day i the solver identified an i-specific fleet with 2 vehicles of type A and 1 vehicle of type B, we represent such fleet with the multi-set $F_i = \{A : 2, B : 1\}$. In the second phase, we compute the multi-set union of the fleets, and obtain an *union fleet* that is guaranteed to be feasible across the whole horizon. Building on the previous example, suppose that on another day j the solver identified the j-specific fleet $F_j = \{A : 1, B : 3\}$, then the union fleet is $F_o = F_i \cup F_j = \{A : 2, B : 3\}$.

Multi-day Fleet. The second method is a generalisation of [24], where a multi-day fleet size and mix problem is solved to generate a *multi-day fleet* that is guaranteed to work on the whole horizon. The problem considered here is much

---

[4] Because only a limited number of attempts is made at each dr level, restarting the search with $dr = \underline{dr}$ allows us to try (again) non-expensive relaxations in case we previously missed a possibility for improvement.

richer than the one in [24] but the multi-day fleet design approach is essentially the same. Note that the Pareto-based pre-processing step used in [24] to reduce the amount of days that need to be considered cannot be used in the present scenario.

7   Experimental Analysis

In this section we present experimental results obtained on a set of benchmark instances based on real-world data.

The CP model and the branching heuristics were implemented in C++ and modelled in GECODE 5.0.0 [29]. The LNS solver was implemented through the GECODE-LNS search engine[5]. All the experiments were run on an UBUNTU 16.04.1 machine equipped with an Intel® Core™ i7-4770 at 3.40 GHz and 16 GB of RAM. The results presented in this section are based on benchmark instances based on real-world data provided to us by our industrial partner. The data represents the demand requested to a fuel distribution centre in Queensland, Australia, in the span of one month. Since some of the days had zero demand, the set of instances contains only 25 days. On each day, the set of customers may change, and the number of visits varies between 1 and 30.

7.1   Results on Individual Days

Table 1 reports the results obtained by running our solver 10 times on each single-day instance using a timeout of 10 min. The instances are sorted by number of customers. We report both the mean overall cost ($\overline{\text{cost}}$) and its standard deviation (σ_{cost}), as well as a breakdown of the costs ($\overline{\text{cost}}_{\text{routing}}$ and $\overline{\text{cost}}_{\text{fleet}}$) and the median number of routes and vehicles required by the solutions. The last column is a sparkline [31] showing the convergence of our search algorithm on a random run of each instance.

Such experiments form the basis upon which we built our union fleet. Note that the total fleet cost reported in the table is the sum of the fixed costs of the daily fleets, and thus represents a lower bound to the fixed costs, which is in general not obtainable by a single fleet.

7.2   Union Fleet

The union fleet generated is reported in Table 2. Such a fleet has a fixed cost of 59, 177 AUD (amortised over 25 days) and a mean routing cost of 50, 619 AUD. The total routing cost has been obtained by running the single-day instances using only vehicles from the union fleet with fixed costs equal to zero. Note that the routing cost found is smaller than the one reported in Table 1, because (1) small-demand days now can access larger vehicles for free, and (2) the search space is much smaller, because we have fewer daily fleets to choose amongst. Overall, the mean total cost of owning and operating the fleet on the whole horizon is therefore 109, 796 AUD.

---

[5] Available at https://github.com/tunnuz/gecode-lns.

Table 1. Aggregated results over 10 runs (with different random seeds) of 10 min on every single-day instance in fleet-size and mix mode.

Instance	Stat.	$\overline{\text{cost}}$	σ_{cost}	$\overline{\text{cost}}_{\text{routing}}$	$\overline{\text{cost}}_{\text{fleet}}$	M_{routes}	M_{vehicles}	Convergence
INST-13	1	750.93	0.00	378.33	372.60	1	1	
INST-6	3	580.12	0.00	207.52	372.60	1	1	
INST-20	3	957.26	0.00	584.66	372.60	2	1	
INST-28	4	1060.21	0.00	687.61	372.60	1	1	
INST-21	6	1582.72	44.29	1210.12	372.60	2	1	
INST-14	8	3379.90	56.76	2538.54	841.36	4	2	
INST-22	8	4649.36	41.59	3472.66	1176.70	5	3	
INST-24	8	1830.22	27.68	988.86	841.36	3	2	
INST-8	9	2469.93	90.53	1724.73	745.20	3	2	
INST-25	9	2408.67	115.14	1652.87	755.80	3	2	
INST-17	10	1814.35	26.62	1059.54	754.82	3	2	
INST-16	11	3531.78	183.43	2637.54	894.24	4	2	
INST-5	12	2818.32	117.05	2066.06	752.27	4	2	
INST-4	12	5306.76	132.70	4092.80	1213.96	6	3	
INST-11	12	3921.77	137.67	2782.33	1139.44	5	3	
INST-3	13	2549.90	95.39	1795.08	754.82	4	2	
INST-18	13	5697.15	453.40	4332.95	1364.20	6	3	
INST-12	13	2738.46	159.49	1941.86	796.60	4	2	
INST-10	13	3147.88	32.50	2373.83	774.05	4	2	
INST-9	14	3876.72	133.30	2734.95	1141.77	4	3	
INST-15	14	2538.77	161.92	1793.57	745.20	4	2	
INST-23	15	3735.99	173.88	2618.19	1117.80	5	3	
INST-26	16	3321.97	148.53	2258.25	1063.72	4.5	3	
INST-19	17	3118.33	144.19	1970.48	1147.85	5	3	
INST-27	30	9384.37	198.81	6972.15	2412.22	11	6	
Total	–	77171.9	–	54875.5	22296.4	-	-	-

7.3   Multi-day Fleet

We also run the solver on a multi-day instance generated by merging the 25 days in the considered benchmark. In order to carry out a fair comparison with the union fleet approach, we use a timeout of 250 min (corresponding to the total time used to solve the 25 independent runs used to generate the union fleet). The best multi-day fleet generated from 10 repetitions of such experiment is reported in Table 3. Notably, the obtained fleet has the same number of vehicles as the union fleet, but a different composition. This reflects in the fixed cost (58, 294 AUD) which is ~1.5% smaller than the one of the union fleet, and also in the routing cost (50, 805 AUD) which is, as expected, slightly higher. The mean total cost of owning and operating the fleet on the whole horizon is therefore 109, 099 AUD, only a small improvement over the union fleet.

Table 2. Union fleet specification.

Vehicle type	Amount
VEH-4	2
VEH-20	2
VEH-26	1
VEH-60	1

Table 3. Multi-day fleet specification.

Vehicle type	Amount
VEH-4	2
VEH-20	3
VEH-26	0
VEH-60	1

7.4   Discussion

We need to make some observation about the computational aspects of our fleet design approaches. In the union fleet, each day is solved independently. This allows to leverage multi-core or cluster architectures to handle large planning horizons, e.g., years. Conversely, distributing the computational load within a single multi-day run is much harder and depends on the capabilities of the underlying CP solver. Moreover, the amount of memory required to solve a multi-day problem of n days is usually much higher than the one required to solve n single-day problems. This can slow down the convergence of the multi-day fleet approach towards a good quality solution. In terms of convergence, memory is not the only issue. In order to reduce the size of a fleet, a multi-day solver must first get all the days to agree on using a smaller fleet. The probability of this happening is rather small. In fact the union fleet approach obtained smaller fleets more often than the multi-day fleet approach.

The tractability of the union fleet approach suggests that decomposing the overall fleet design problem is a promising strategy as the horizon grows longer. However, it is clear that the independence of the runs prevents the solver from optimising globally. This suggests that a decomposition approach in which the single-day runs can share knowledge about the overall fleet could obtain better results than any of the proposed approaches.

8   Conclusions

In this paper we proposed a rich VRP model based on a step-based formulation to solve a fuel delivery problem at the request of a fuel distribution company. Along with the operational level aspects, such as the loading and routing of vehicles, we also considered fleet design decisions that belong to the tactical level problem, and that have a long term impact on the company operations.

We have implemented our model as a CP-based large neighbourhood search (LNS) solver, and proposed two different methods for automatically obtaining feasible fleet designs to support the operations across the planning horizon. We compared these approaches on a set of real-world problem instances provided to us by our industrial partner. Our results suggest that the most promising direction to inform fleet design decisions might be problem decomposition methods, which can retain both scalability and quality of solutions.

References

1. Archetti, C., Speranza, M.G.: The split delivery vehicle routing problem: a survey. In: Golden, B., Raghavan, S., Wasil, E. (eds.) The Vehicle Routing Problem: Latest Advances and New Challenges, Operations Research/Computer Science Interfaces, vol. 43, pp. 103–122. Springer, Boston (2008). doi:10.1007/978-0-387-77778-8_5
2. Avella, P., Boccia, M., Sforza, A.: Solving a fuel delivery problem by heuristic and exact approaches. Eur. J. Oper. Res. **152**(1), 170–179 (2004)
3. Belfiore, P., Yoshizaki, H.T.: Heuristic methods for the fleet size and mix vehicle routing problem with time windows and split deliveries. Comput. Ind. Eng. **64**(2), 589–601 (2013)
4. Brown, G.G., Ellis, C.J., Graves, G.W., Ronen, D.: Real-time, wide area dispatch of mobil tank trucks. Interfaces **17**(1), 107–120 (1987)
5. Brown, G.G., Graves, G.W.: Real-time dispatch of petroleum tank trucks. Manag. Sci. **27**(1), 19–32 (1981)
6. van der Bruggen, L., Gruson, R., Salomon, M.: Reconsidering the distribution structure of gasoline products for a large oil company. Eur. J. Oper. Res. **81**(3), 460–473 (1995)
7. Caceres-Cruz, J., Arias, P., Guimarans, D., Riera, D., Juan, A.A.: Rich vehicle routing problem: survey. ACM Comput. Surv. (CSUR) **47**(2), 32 (2015)
8. Cornillier, F., Boctor, F.F., Laporte, G., Renaud, J.: An exact algorithm for the petrol station replenishment problem. J. Oper. Res. Soc. **59**(5), 607–615 (2008)
9. Cornillier, F., Boctor, F.F., Laporte, G., Renaud, J.: A heuristic for the multi-period petrol station replenishment problem. Eur. J. Oper. Res. **191**(2), 295–305 (2008)
10. Cornillier, F., Boctor, F.F., Renaud, J.: Heuristics for the multi-depot petrol station replenishment problem with time windows. Eur. J. Oper. Res. **220**, 361–369 (2012)
11. Dell'Amico, M., Monaci, M., Pagani, C., Vigo, D.: Heuristic approaches for the fleet size and mix vehicle routing problem with time windows. Transp. Sci. **41**(4), 516–526 (2007)
12. Derigs, U., Gottlieb, J., Kalkoff, J., Piesche, M., Rothlauf, F., Vogel, U.: Vehicle routing with compartments: applications, modelling and heuristics. OR Spectr. **33**(4), 885–914 (2011)
13. Di Gaspero, L., Rendl, A., Urli, T.: Balancing bike sharing systems with constraint programming. Constraints **21**(2), 318–348 (2016)
14. Drexl, M.: Rich vehicle routing in theory and practice. Logist. Res. **5**(1–2), 47–63 (2012)
15. Dror, M., Laporte, G., Trudeau, P.: Vehicle routing with split deliveries. Discret. Appl. Math. **50**(3), 239–254 (1994)
16. Dror, M., Trudeau, P.: Savings by split delivery routing. Transp. Sci. **23**(2), 141–145 (1989)
17. Dror, M., Trudeau, P.: Split delivery routing. Nav. Res. Logist. (NRL) **37**(3), 383–402 (1990)

18. Golden, B., Assad, A., Levy, L., Gheysens, F.: The fleet size and mix vehicle routing problem. Comput. Oper. Res. **11**(1), 49–66 (1984)
19. Golden, B., Raghavan, S., Wasil, E. (eds.): Preface. Operations Research/ Computer Science Interfaces, vol. 43. Springer, US (2008)
20. Gomes, C., Sellmann, M.: Streamlined constraint reasoning. In: Wallace, M. (ed.) CP 2004. LNCS, vol. 3258, pp. 274–289. Springer, Heidelberg (2004). doi:10.1007/ 978-3-540-30201-8_22
21. Henke, T., Speranza, M.G., Wäscher, G.: The multi-compartment vehicle routing problem with flexible compartment sizes. Eur. J. Oper. Res. **246**(3), 730–743 (2015)
22. Kabcome, P., Mouktonglang, T.: Vehicle routing problem for multiple product types, compartments, and trips with soft time windows. Int. J. Math. Math. Sci. **2015** (2015)
23. Kilby, P., Shaw, P.: Vehicle routing. In: Handbook of Constraint Programming, pp. 799–834 (2006)
24. Kilby, P., Urli, T.: Fleet design optimisation from historical data using constraint programming and large neighbourhood search. Constraints **21**(1), 2–21 (2016)
25. Mendoza, J.E., Castanier, B., Guéret, C., Medaglia, A.L., Velasco, N.: A memetic algorithm for the multi-compartment vehicle routing problem with stochastic demands. Comput. Oper. Res. **37**(11), 1886–1898 (2010)
26. Pesant, G.: A regular language membership constraint for finite sequences of variables. In: Wallace, M. (ed.) CP 2004. LNCS, vol. 3258, pp. 482–495. Springer, Heidelberg (2004). doi:10.1007/978-3-540-30201-8_36
27. Pisinger, D., Ropke, S.: Large neighborhood search. In: Gendreau, M., Potvin, J.Y. (eds.) Handbook of Metaheuristics. ISOR, vol. 146, pp. 399–419. Springer, Boston (2010). doi:10.1007/978-1-4419-1665-5_13
28. Ropke, S., Pisinger, D.: An adaptive large neighborhood search heuristic for the pickup and delivery problem with time windows. Transp. Sci. **40**(4), 455–472 (2006)
29. Schulte, C., Tack, G., Lagerkvist, M.Z.: Modeling and programming with gecode. In: Schulte, C., Tack, G., Lagerkvist, M.Z. (eds.) Modeling and Programming with Gecode. Self-published (2015). Corresponds to Gecode 4.4.0
30. Shaw, P.: Using constraint programming and local search methods to solve vehicle routing problems. In: Maher, M., Puget, J.-F. (eds.) CP 1998. LNCS, vol. 1520, pp. 417–431. Springer, Heidelberg (1998). doi:10.1007/3-540-49481-2_30
31. Tufte, E.R.: Beautiful Evidence. Graphics Press LLC, Cheshire (2006)
32. Vidal, T., Crainic, T.G., Gendreau, M., Prins, C.: A unified solution framework for multi-attribute vehicle routing problems. Eur. J. Oper. Res. **234**(3), 658–673 (2014)
33. Vilím, P.: Global constraints in scheduling. Ph.D. thesis, Charles University in Prague, Faculty of Mathematics and Physics, Department of Theoretical Computer Science and Mathematical Logic, KTIML MFF, Universita Karlova, Praha 1, Czech Republic (2007)

Integer and Constraint Programming for Batch Annealing Process Planning

Willem-Jan van Hoeve$^{(\boxtimes)}$ and Sridhar Tayur

Tepper School of Business, Carnegie Mellon University, 5000 Forbes Avenue,
Pittsburgh, PA 15213, USA
{vanhoeve,stayur}@andrew.cmu.edu

Abstract. We describe an optimization application in the context of
steel manufacturing, to design and schedule batches for annealing furnaces. Our solution approach uses a two-phase decomposition. The first
phase groups together orders into batches using a mixed-integer linear
programming model. The second phase assigns the batches to furnaces
and schedules them over time, using constraint programming. Our solution has been developed for operational use in two plants of a steel manufacturer in North America.

1 Introduction

We present an application of optimization technology for a steel manufacturer in
North America, that operates two plants with annealing capability. Annealing
is used in the steel industry as a heat treatment to modify the structure of the
metals. For example, it may remove stresses, soften the steel, or refine the grain
structure. In our case, the annealing process is performed in box furnaces, which
can hold a specific number of steel coils. The furnaces are a primary (bottleneck)
resource of the plants, which means that related operations are scheduled subject
to the furnace annealing schedules. Since the existing approach for creating and
scheduling batches for the box furnaces requires substantial manual interaction,
the purpose of our project was to automate and improve this process.

Optimization models for batch design and scheduling in steel plants have
been proposed before. For example, in [6] a mixed-integer programming (MIP)
model is proposed for annealing batch scheduling in a general industrial setting,
with fixed batches. Their allocation and scheduling of batches to furnaces is
similar to our setting, but the authors consider different resources such as crane
movements as well. In [4] a genetic algorithm to this problem is proposed. An
excellent survey of batch scheduling is presented in [5].

In [3], a decomposition approach is proposed for a more general steel production problem (not just annealing). Conceptually, we follow a similar approach. A
main difference is that we utilize constraint programming (CP) for the scheduling of batches, whereas [3] uses MIP. Furthermore, because we work with a more
specific application, we can streamline our batch design MIP model using additional constraints. Lastly, [2] combines the batch composition and scheduling,

© Springer International Publishing AG 2017
J.C. Beck (Ed.): CP 2017, LNCS 10416, pp. 431–439, 2017.
DOI: 10.1007/978-3-319-66158-2_28

Table 1. Main sets for the problem.

O^L	Set of locked orders
O^W	Set of work orders
O^P	Set of planned orders
O	Set of all orders $O^L \cup O^W \cup O^P$
A	Set of anneal cycles
F	Set of furnaces
H	Integer time horizon (in minutes)

Table 2. Characteristics of the orders.

$c_i \in A$	Anneal cycle of $i \in O$
$w_i \in \mathbb{R}^+$	The weight of one coil in $i \in O$
$n_i \in [K]$	The number of coils in $i \in O$
$r_i \in H$	Release date of $i \in O$
$d_i \in H$	Due date of $i \in O$
$g_i \subseteq F$	Furnace group of $i \in O^L$
$\Pi \subset O \times O$	Set of precedences (i, j)

as we do in this work. Their batch design is restricted to the size of the batch, however, and does not group together different orders for example.

In summary, the main novelty of our approach is the combination of a rich batch design problem (solved with MIP) with a batch scheduling problem (solved with CP). Decomposing the problem into two parts (MIP for batch design and CP for scheduling) was crucial to make the approach scalable: Our two-phase approach scales to problems with (at least) 22 furnaces and 600 orders, creating a detailed schedule for about 7 days, within 15 to 30 min of computing time.

2   Problem Description

The input to our problem is a set of annealing orders O. Each order consists of a number of steel coils that need to be annealed using a specific recipe (anneal cycle) in box furnaces. The set of all anneal cycles is denoted by A. The set of all furnaces is denoted by F. The orders are grouped together in batches of a fixed maximum size, depending on the furnace capacity. For our application, the furnace capacity differs per plant, but the furnaces have uniform capacity K (total number of coils that can be loaded per batch) for a given plant.

After the batches have been created, they need to be allocated and scheduled on the available furnaces, given a discrete time horizon H. Table 1 summarizes the main sets of our problem, including a partition of the orders in three types:

- Locked orders (denoted by the set O^L) have been committed in the previous planning phase. They have been grouped together in a batch, and assigned to a furnace group for execution. These batches cannot be changed and must be scheduled as soon as possible.
- Work orders (denoted by the set O^W) are partially committed in the previous planning phase. They have a fixed number of coils that cannot be changed, but their batches have not yet been decided.
- Planned orders (denoted by the set O^P) are not yet committed, but are available to be scheduled. The given number of coils of a planned order may be reduced to complete the size of a batch, but not split into separate orders.

Each order has a number of characteristics, as presented in Table 2.[1] For each order, the number of coils (n_i) is at most K. Also, for some orders the due date d_i may come before the release date r_i. These orders are identified as late, and are given high preference to be scheduled as early as possible. Only orders for which the due date is at most a given date D (for example, day two) are required to be scheduled. Lastly, there exist pairwise precedence relations (i, j) for $i, j \in O$: order j must be scheduled at least three days after order i finishes. In fact, i and j represent two annealing operations for the same set of coils.

Table 3. Characteristics of the anneal cycles.

$p_{a,f} \in \mathbb{N}$	Processing time of anneal cycle $a \in A$ on furnace $f \in F$ (in minutes)
$t_{a,a',f} \in \mathbb{N}$	Sequence-dependent switchover time from anneal cycle a to a' on furnace f (in minutes)
$C_{a,a'} \in \{0,1\}$	Whether $a \in A$ and $a' \in A$ are compatible and can be combined
$\Gamma_{a,a'} \in \{a, a'\}$	Anneal cycle that determines the processing time of the combined cycle, for $a, a' \in A$ such that $C_{a,a'} = 1$

The characteristics of the anneal cycles are given in Table 3. Some orders with different anneal cycles can be combined in one batch, if they are compatible. In that case, one of the cycles will determine the anneal recipe for the compatible batch. At most one other compatible cycle can be added to a batch. An order is not allowed to be both reduced and added as a compatible order to a batch.

Since not all anneal recipes can be performed on each furnace, we introduce a Boolean parameter $T_{f,a}$ to indicate whether furnace $f \in F$ can perform anneal cycle $a \in A$. Furthermore, each furnace can only handle a maximum of M 'heavy' coils that have a weight of at least W, within one batch. The maximum coil weight that can be processed on f is denoted by $w_f^{\max}$.

The problem is to (1) group together orders into batches (the batch design problem), and (2) allocate the batches to furnaces and sequence them over time (the batch scheduling problem). The purpose of the solution is to create a near-term schedule (about six days), in which we will operationally commit batches that are scheduled in the first two days. Therefore, the qualitative goals are:

- Minimize furnace idle time, especially in the first two to three days;
- Minimize unfilled furnace capacity (i.e., maximize the batch sizes), especially in the first two to three days;
- Minimize the number of late coils.

---

[1] We use the common shorthand $[n]$ to denote the set $\{1, \dots, n\}$ for an integer n.

Table 4. Variables used in the mixed-integer programming model.

$x_{i,a,k} \in \{0,1\}$	Allocate order i to $b_{a,k}$, for $i \in O^W \cup O^P, a \in A$ such that $c_i = a$ and $k \in [N_a]$
$y_{i,a,k} \in \{0,1\}$	Allocate i as *reduced* order to $b_{a,k}$, for $i \in O^P, a \in A$ such that $c_i = a$ and $k \in [N_a]$
$z_{i,a,k} \in \{0,1\}$	Allocate i as *compatible* order to $b_{a,k}$, for $i \in O^W \cup O^P, a \in A$ such that $C_{c_i,a} = 1$, and $k \in [N_a]$
$y'_{i,a,k} \in [0, \frac{(K-1)}{K}]$	Fraction of coils from (reduced) order i that will be used in $b_{a,k}$, for $i \in O^P$, $a \in A$ such that $c_i = a$ and $k \in [N_a]$
$u_{a,k} \in \{0,1\}$	Whether $b_{a,k}$ is in use, for $a \in A$, $k \in [N_a]$
$z'_{a,k,a'} \in \{0,1\}$	Whether $b_{a,k}$ is in use as compatible batch for a', for $a \in A$, $a' \in A, a \neq a', C_{a,a'} = 1, k \in [N_a]$
$e_{a,k} \in \{0, \ldots, K-1\}$	Number of empty positions in $b_{a,k}$, for $a \in A$, $k \in [N_a]$
$l_{a,k} \in \{0,1\}$	Whether $b_{a,k}$ contains a late order, for $a \in A$, $k \in [N_a]$
$m_{a,k} \geq 0$	Maximum release date of orders in $b_{a,k}$, for $a \in A, k \in [N_a]$
$m'_{a,k} \geq 0$	Release date violation of first-day batches $b_{a,k}$, for $a \in A$, $k \in [N_a], k \leq k_a^I$

3   Phase 1: Batch Design

We next describe our MIP model for the batch design problem. We recall that the locked orders in O^L are already grouped in batches, so we consider here the work orders O^W and planned orders O^P. The first step is to define the possible batches that we can assign the orders to. Since orders will be grouped by anneal cycle, we create a set $\{b_{a,1}, \ldots, b_{a,N_a}\}$ for $a \in A$, representing the possible batches for anneal cycle a. The size N_a of this set depends on the total number of coils that can be assigned to cycle a, the furnace capacity K, and the precedence constraints between orders, and is computed in advance.

Our MIP model will not keep track of time in full detail, as this will be the responsibility of the batch scheduling model. However, we do need to take timing considerations into account, for example by aiming to group together orders with the same release date. We will therefore associate an earliest release date with each batch, and our model is designed to create batches such that the release date of $b_{a,k}$ is at most the release date of $b_{a,k+1}$, if both are used.

In addition, we wish to avoid grouping first-day orders with orders that have a later release date, to create enough batches to schedule on the first day of the horizon. To that end, we identify all orders (for a given anneal cycle $a \in A$) that can start on the first day of the horizon, and determine (approximately) the number of batches that can start on the first day as $k_a^I = \lfloor (\sum_{i \in O^W \cup O^P : c_i = a, r_i \leq 1} n_i)/K \rfloor$. We will use this information to group together first-day orders (only) in as many batches as possible.

The MIP model is as follows (the variables are presented in Table 4):

$$\min \quad \sum_{a\in A, k\in[N_a]} (\gamma^l l_{a,k} + \gamma^e_{a,k} e_{a,k}) + \sum_{\substack{i\in O^P, a\in A, k\in[N_a]:\\ c_i=a}} \gamma^y y_{i,a,k} +$$

$$\sum_{\substack{i\in O^W\cup O^P,\\ a\in A, k\in[N_a]: C_{c_i,a}=1}} \gamma^z z_{i,a,k} + \sum_{\substack{a\in A, k\in[N_a]:\\ k\le k^I_a}} \gamma^m m'_{a,k} \tag{1}$$

$$\text{s.t.} \quad \sum_{a\in A, k\in[N_a]:c_i=a} x_{i,a,k} + \sum_{a\in A, k\in[N_a]:C_{c_i,a}} z_{i,a,k} \le 1 \quad \forall i\in O^W \tag{2}$$

$$\sum_{a\in A, k\in[N_a]:c_i=a} (x_{i,a,k}+y_{i,a,k}) + \sum_{a\in A, k\in[N_a]:C_{c_i,a}} z_{i,a,k} \le 1 \quad \forall i\in O^P \tag{3}$$

$$\sum_{\substack{i\in O^P\cup O^W:\\ c_i=a, d_i<r_i}} (x_{i,a,k}+y_{i,a,k}) + \sum_{\substack{i\in O^P\cup O^W:\\ C_{c_i,a}=1, d_i<r_i}} z_{i,a,k} \le Kl_{a,k} \quad \forall a\in A, k\in[N_a] \tag{4}$$

$$\sum_{i\in O:c_i=a} (n_i x_{i,a,k} + n_i y'_{i,a,k}) + \sum_{i\in O:C_{c_i,a}} n_i z_{i,a,k} = Ku_{a,k} - e_{a,k} \quad \forall a\in A, k\in[N_a] \tag{5}$$

$$y'_{i,a,k} \le y_{i,a,k}(n_i-1)/n_i \quad \forall i\in O^P, a\in A, c_i=a, k\in[N_a] \tag{6}$$

$$y'_{i,a,k} \ge y_{i,a,k}/n_i \quad \forall i\in O^P, a\in A, c_i=a, k\in[N_a] \tag{7}$$

$$\sum_{i\in O^P:c_i=a} y_{i,a,k} \le 1 \quad \forall a\in A, k\in[N_a] \tag{8}$$

$$\sum_{i\in O:c_i=a'} n_i z_{i,a,k} \le z'_{a,k,a'}(K-1) \quad \forall a\in A, k\in[N_a], a'\in A: C_{a,a'}=1 \tag{9}$$

$$\sum_{a'\in A:C_{a,a'}=1} z'_{a,k,a'} \le 1 \quad \forall a\in A, k\in[N_a] \tag{10}$$

$$\sum_{\substack{a\in A, k\in[N_a]:\\ c_i=a, w_i\ge W}} n_i(x_{i,a,k}+y'_{i,a,k}) + \sum_{\substack{a\in A, k\in[N_a]:\\ C_{c_i,a}, w_i\ge W}} n_i z_{i,a,k} \le M \quad \forall a\in A, k\in[N_a] \tag{11}$$

$$\sum_{\substack{a\in A,\\ k\in[N_a]:\\ c_i=a}} k(x_{i,a,k}+y_{i,a,k}) + \sum_{\substack{a\in A,\\ k\in[N_a]:\\ C_{c_i,a}=1}} kz_{i,a,k} \le \sum_{\substack{a\in A,\\ k\in[N_a]:\\ c_j=a}} k(x_{j,a,k}+y_{j,a,k}) + \sum_{\substack{a\in A,\\ k\in[N_a]:\\ C_{c_j,a}=1}} kz_{j,a,k}$$

$$-1 + (1+N_{c_i})\left(1 - \sum_{\substack{a\in A,\\ k\in[N_a]:\\ c_j=a}} (x_{j,a,k}+y_{j,a,k}) + \sum_{\substack{a\in A,\\ k\in[N_a]:\\ C_{c_j,a}=1}} z_{j,a,k}\right) \quad \substack{\forall(i,j)\in\Pi:\\ c_i=c_j\vee C_{c_i,c_j}=1}$$

$$\tag{12}$$

$$\sum_{\substack{a\in A,\\k\in[N_a]:\\c_j=a}} (x_{j,a,k}+y_{j,a,k})+ \sum_{\substack{a\in A,\\k\in[N_a]:\\C_{c_j,a}=1}} z_{j,a,k} \leq \sum_{\substack{a\in A,\\k\in[N_a]:\\c_i=a}} (x_{i,a,k}+y_{i,a,k})+ \sum_{\substack{a\in A,\\k\in[N_a]:\\C_{c_i,a}=1}} z_{i,a,k} \quad \forall (i,j)\in \Pi$$

$$(13)$$

$$\sum_{\substack{a\in A,\\k\in[N_a]:\\c_i=a}} x_{i,a,k} + \sum_{\substack{a\in A,\\k\in[N_a]:\\C_{c_i,a}}} z_{i,a,k} \geq \sum_{\substack{a\in A,\\k\in[N_a]:\\c_j=a}} x_{j,a,k} \sum_{\substack{a\in A,\\k\in[N_a]:\\C_{c_j,a}}} z_{j,a,k} \quad \begin{array}{c} \forall i,j\in O^W\cup O^P:\\ (c_i=c_j\vee C_{c_i,c_j}=1)\\ d_i<d_j \end{array} \quad (14)$$

$$\sum_{\substack{a\in A,\\k\in[N_a]:\\c_i=a}} y_{i,a,k} \geq \sum_{\substack{a\in A,\\k\in[N_a]:\\c_j=a}} y_{j,a,k} \quad \begin{array}{c} \forall i,j\in O^P:\\ (c_i=c_j\vee C_{c_i,c_j}=1)\\ d_i<d_j \end{array} \quad (15)$$

$$K\left(\sum_{\substack{i\in O^W\cup O^P:\\c_i=a}} x_{i,a,k} + \sum_{\substack{i\in O^W\cup O^P:\\C_{c_i=a}=1}} z_{i,a,k} \right) \geq \sum_{\substack{i\in O^W\cup O^P:\\c_i=a}} x_{i,a,k+1} + \qquad (16)$$

$$\sum_{\substack{i\in O^W\cup O^P:\\C_{c_i=a}=1}} z_{i,a,k+1} \quad \forall a\in A, k\in[N_a-1]$$

$$\sum_{\substack{a\in A,k\in[N_a]:\\c_i=a}} x_{i,a,k} + \sum_{\substack{a\in A,k\in[N_a]:\\C_{c_i,a}}} z_{i,a,k} = 1 \quad \forall i\in O^W\cup O^W: d_i\leq D \quad (17)$$

$$m_{a,k} \geq r_i(x_{i,a,k}+y_{i,a,k}) \quad \forall a\in A, k\in[N_a], i\in O^W\cup O^P: c_i=a \quad (18)$$

$$m_{a,k} \geq r_i z_{i,a,k} \quad \forall a\in A, k\in[N_a], i\in O^W\cup O^P: C_{c_i,a}=1 \quad (19)$$

$$m_{a,k} \leq m_{a,k+1} \quad \forall a\in A, k\in[N_a-1] \quad (20)$$

$$m_{a,k} \leq 1+m'_{a,k} \quad \forall a\in A, k\in[N_a]: k\leq k_a^I \quad (21)$$

The objective (1) is a weighted sum of penalty terms representing the number of batches with late orders, the number of empty slots, the number of reduced orders, the number of compatible orders, and the tardiness of first-day orders. Each term has its associated penalty parameter, denoted by $\gamma^l, \gamma_{a,k}^e, \gamma^y, \gamma^z$, and γ^m, respectively (listed here by decreasing emphasis). Parameter $\gamma_{a,k}^e$ is defined as $\gamma_{a,k}^e = 1+0.1(N_a-k)/N_a$. It decreases for larger k, giving higher priority to filling earlier batches. The other penalty parameters are taken in $[0.01, 1.0]$.

Constraints (2) and (3) allocate the work orders and planned orders, respectively. Constraints (4) define a late batch with respect to the late orders. Constraints (5) define the capacity of the furnaces and the number of empty slots. Constraints (6) and (7) link the fraction of coils for reduced orders to the associated binary variable, while constraints (8) ensure that each planned order can be used at most once as reduced order. Constraints (9) define compatible batches with respect to compatible orders, while constraints (10) ensure that each batch can be compatible with at most one other anneal cycle. Constraints (11) limit the number of heavy coils that can be allocated to a batch. Constraints (12) and (13) represent the precedence constraints. The model ensures feasibility by allocating orders i and j, with $(i, j) \in \Pi$, to different batches. If they are allocated to different anneal cycles, this is already guaranteed. Otherwise, if $c_i = c_j$ or $C_{c_i, c_j} = 1$, constraints (12) ensure that order i is placed in a batch with a lower index k than j. Constraints (13) ensure that j is not allocated if i is not. Constraints (14) and (15) state that if two orders are equivalent in terms of anneal cycle, then the one with earlier due date is allocated first, for work orders and planned orders, respectively. Constraints (16) impose that batch $b_{a,k+1}$ can only be used if $b_{a,k}$ is. Constraints (17) ensure that all near-term orders (with $d_i \leq D$) are allocated. Constraints (18) and (19) define the release dates of the batches. Constraints (20) ensure that batches are ordered by non-decreasing release date. Constraints (21) define the release violation ($m'_{a,k}$) of first-day batches.

4   Phase 2: Batch Scheduling

For the batch scheduling problem we employ a constraint programming model, following the constraint-based scheduling concepts of activities and resources [1]. In the model syntax below, we follow the definitions of AIMMS [7]. An *activity* represents a task to be scheduled over time, by means of four implied variables: begin, length, end, and presence. When an activity is *optional*, its presence can be either 0 (absent) or 1 (present). When an activity is present, it respects the relation 'begin+length=end'. *Resources*, for example machines, represent constraints on activities. In this paper, we only consider sequential (or disjunctive) resources that limit the execution to at most one activity at a time. In addition, the sequential resource can enforce sequence-dependent setup (or switchover) times between two consecutive activities.

Table 5. Characteristics of the batches and their definition.

$\bar{w}_b \in \mathbb{R}^+$	$\max_{i \in \mathcal{O}(b)} w_i$	$\bar{c}_b \in A$	if $z'_{a,k,a'} = 1$ then $\Gamma_{a,a'}$ else a, for $b = b_{a,k}$
$\bar{n}_b \in [K]$	$\sum_{i \in \mathcal{O}(b)} n_i$	$\bar{g}_b \subseteq F$	g_i for any $i \in \mathcal{O}(b)$ if $b \in B^L$
$\bar{r}_b \in H$	$\max_{i \in \mathcal{O}(b)} r_i$	$\bar{\Pi} \subset B \times B$	(b, b') if $\exists i \in \mathcal{O}(b), i' \in \mathcal{O}(b') : (i, i') \in \Pi$
$\bar{d}_b \in H$	$\min_{i \in \mathcal{O}(b)} d_i$		

Our model is based on the problem definition from Sect. 2 and the MIP solution from Sect. 3. We first define the set of batches B as the union of the locked batches (denoted by B^L) and the batches $b_{a,k}$ that are used in the MIP model, i.e., for which $u_{a,k} = 1$ (for $a \in A, k \in [N_a]$). We use $\mathcal{O}(b) \subseteq O$ to denote the orders that are assigned to batch b. We define the associated parameters of a batch in Table 5. We introduce the following activities:

L_b: the execution of batch b, for $b \in B$

$L_{b,f}^{opt}$: the execution of batch b on furnace f, for $b \in B, f \in F : \bar{w}_b \leq w_f^{max}$, $T_{f,\bar{c}_b} = 1$, and $f \in \bar{g}_b$ if $b \in B^L$

We define the schedule domain for both types of activities as $[\bar{r}_b, H]$, for $b \in B$. Activities $L_{b,f}^{opt}$ are optional, while activities L_b must always be present. The processing time for both L_b and $L_{b,f}^{opt}$ is given by $p_{\bar{c}_b,f}$.

We introduce a sequential resource R_f for each furnace $f \in F$. As arguments it receives the activities $L_{b,f}^{opt}$, as well as the sequence-dependent switchover times $t_{a,a',f}$. These resources will ensure that all activities $L_{b,f}$ that are present do not overlap in time, and respect the switchover times.

The additional constraints for our CP model are as follows:

$$\texttt{Alternative}(L_b, \{L_{b,f}^{opt}\}_{f \in F}, 1) \quad \forall b \in B \tag{22}$$

$$\texttt{EndBeforeBegin}(L_b, L_{b'}, 4320) \quad \forall (b, b') \in \bar{\Pi} \tag{23}$$

$$\texttt{EndBeforeBegin}(L_{b,f}^{opt}, L_{b',f}^{opt}) \quad \forall b, b' \in B, f \in F : b \neq b', \bar{r}_b = \bar{r}_{b'}, \tag{24}$$
$$\bar{d}_b < \bar{d}_{b'}, \bar{n}_b = K, (b', b) \notin \bar{\Pi}$$

$$(L_b).\texttt{end} \leq \bar{d}_i + v_b \quad \forall b \in B \tag{25}$$

The $\texttt{Alternative}$ constraints (22) ensure that exactly one optional activity $L_{b,f}^{opt}$ is present for each $b \in B$, and it coincides with the execution of activity L_b. The precedence constraints are defined by constraints (23).[2] Constraints (24) are added to streamline the solutions, by ordering pairs of activities on a furnace by due date. Lastly, we introduce an integer variable v_b, for $b \in B$, that represents its due date violation (or tardiness), as defined in constraint (25). In this constraint (and in the objective function below), .$\texttt{end}$ is used to retrieve the end variable of an activity.

We conclude with the objective function, which is a weighted sum of completion times and tardiness over different groups of activities:

$$\min \quad \bar{\gamma}^L \sum_{b \in B^L} (L_b).\texttt{end} + \bar{\gamma}^I \sum_{b \in B^I} \bar{n}_b (L_b).\texttt{end} + \bar{\gamma}^V \sum_{b \in B} \bar{n}_b v_b \tag{26}$$

Here, B^I refers to the batches that contain first-day orders (these refer to the batches $b_{a,k}$ with $k \leq k_a^I$). The weights $\bar{\gamma}^L$, $\bar{\gamma}^I$, and $\bar{\gamma}^V$ have decreasing value in the range $(0, 100]$, placing most emphasis on locked orders and first-day orders.

5   Implementation and Results

We implemented our MIP and CP models in the optimization modeling system AIMMS, using IBM ILOG CPLEX 12.4 for solving the MIP and CP models. We also used AIMMS to build an end-user interface for operational use.

---

[2] Recall that b' must be scheduled at least three days (4320 min) after b finishes.

The purpose of our tool is to plan, or revisit, the design and scheduling of annealing batches on a daily basis. A typical run may receive about 600 orders (with 100 locked orders, in 60 batches). The batch design MIP model for the remaining 500 orders has about 20,000 variables (18,000 integer) and 30,000 constraints. A feasible integer solution (with optimality gap 30%) is found within a couple seconds, while a near-optimal solution (1% optimality gap) is found within 15 min. The model creates about 190 batches, which together with the locked batches makes around 250 batches to be scheduled. For about 15 available furnaces, the resulting CP model has about 8,000 variables and 25,000 constraints. It finds a feasible solution instantly, and typically returns solutions in about 2 min that are not necessarily optimal, but are considered of good quality by the client with respect to the goals listed in Sect. 2.

6   Conclusion

We introduced a two-phase optimization approach to design and schedule batches for box annealing furnaces. We first allocate a given set of orders to batches using a mixed-integer programming model. We then solve the batch scheduling problem using a constraint programming model. Using this decomposition, our approach is able to compute operational schedules for a one-week planning horizon within 15 to 30 min of computation time.

References

1. Baptiste, P., Le Pape, C., Nuijten, W., Scheduling, C.-B.: Applying Constraint Programming to Scheduling Problems. Kluwer Academic Publishers, Dordrecht (2001)
2. Castro, P.M., Erdirik-Dogan, M., Grossmann, I.E.: Simultaneous batching and scheduling of single stage batch plants with parallel units. AIChE J. **54**(1), 183–193 (2008)
3. Harjunkoski, I., Grossmann, I.E.: A decomposition approach for the scheduling of a steel plant production. Comput. Chem. Eng. **25**(11–12), 1647–1660 (2001)
4. Liu, Q., Wanga, W., Zhanb, H., Wanga, Z., Liua, R.: Optimal scheduling method for a bell-type batch annealing shop and its application. Control Eng. Pract. **13**(10), 1315–1325 (2005)
5. Méndez, C.A., Cerdá, J., Grossmann, I.E., Harjunkoski, I., Fahl, M.: State-of-the-art review of optimization methods for short-term scheduling of batch processes. Comput. Chem. Eng. **30**, 913–946 (2006)
6. Moon, S., Hrymak, A.N.: Scheduling of the batch annealing process - deterministic case. Comput. Chem. Eng. **23**(9), 1193–1208 (1999)
7. Roelofs, M., Bisschop, J.J.: AIMMS 4.2: The Language Reference. AIMMS B.V. (2014)

Machine Learning and CP Track

Minimum-Width Confidence Bands via Constraint Optimization

Jeremias Berg[1]([✉]), Emilia Oikarinen[2], Matti Järvisalo[1], and Kai Puolamäki[2]

[1] HIIT, Department of Computer Science, University of Helsinki, Helsinki, Finland
{jeremias.berg,matti.jarvisalo}@helsinki.fi
[2] Finnish Institute of Occupational Health, Helsinki, Finland

Abstract. The use of constraint optimization has recently proven to be a successful approach to providing solutions to various NP-hard search and optimization problems in data analysis. In this work we extend the use of constraint optimization systems further within data analysis to a central problem arising from the analysis of multivariate data, namely, determining minimum-width multivariate confidence intervals, i.e., the minimum-width confidence band problem (MWCB). Pointing out drawbacks in recently proposed formalizations of variants of MWCB, we propose a new problem formalization which generalizes the earlier formulations and allows for circumvention of their drawbacks. We present two constraint models for the new problem in terms of mixed integer programming and maximum satisfiability, as well as a greedy approach. Furthermore, we empirically evaluate the scalability of the constraint optimization approaches and solution quality compared to the greedy approach on real-world datasets.

1 Introduction

The use of constraint programming systems has recently proven to be a successful approach to providing solutions to various NP-hard search and optimization problems in data mining and machine learning, using a variety of constraint optimization paradigms such as constraint programming (CP), mixed integer programming (MIP), Boolean satisfiability (SAT), maximum satisfiability (MaxSAT), and answer set programming (ASP). Compared to the more typical in-exact, problem-specific local search style algorithms, the benefits of constraint reasoning and optimization lie on one hand in the ability to provide *provably optimal solutions*, translating into more accurate solutions to the data analysis task at hand, and on the other hand by generality of algorithmic solutions resulting from the declarative approach, allowing for capturing different problem variants simply by enforcing additional or slightly modified constraints.

This work was financially supported by Academy of Finland (grants 251170 COIN, 276412, 284591, 288814); Tekes (Revolution of Knowledge Work); and DoCS Doctoral School in Computer Science and Research Funds of the University of Helsinki.

© Springer International Publishing AG 2017
J.C. Beck (Ed.): CP 2017, LNCS 10416, pp. 443–459, 2017.
DOI: 10.1007/978-3-319-66158-2_29

In this work we extend the use of constraint optimization systems further within data analysis to a central problem arising from the analysis of multivariate data, namely, determining minimum-width multivariate confidence intervals. Confidence intervals are commonly used to summarize distributions over reals, e.g., to denote ranges of data, to specify accuracies of estimates of parameters, or in Bayesian settings to describe the posterior distribution. Represented with an upper and a lower bound, confidence intervals are also easy to interpret together with the data. In contrast to p-values, which only convey information about statistical significance—a problem which has been long and acutely recognized in many disciplines [4, 20, 23, 25]—confidence intervals give information on both the statistical significance of the result as well as the effect size. Indeed, statistically significant results can be meaningless in practice due to the small effect size [23, 25]. The proposed solution is not to report p-values at all, but use confidence intervals instead [20]. Optimizing the width of multivariate confidence intervals is NP-hard, and furthermore, expectedly even hard to approximate [11], which motivates the use of constraint optimization systems for the task.

The problem of estimating the confidence interval of a distribution based on a finite-sized sample from the distribution has been extensively studied, see e.g. [7]. However, most effort has focused on describing a single univariate distribution over real numbers, and there are surprisingly few approaches to multivariate confidence intervals. In the time series domain, multivariate confidence intervals [9, 11], namely *confidence bands*, have been defined in terms of the *minimum-width envelope* (MWE). The only exact approach to MWE (in this paper denoted by MWCB(k) as motivated later on) that we are aware of is the very recent integer programming model of [21]. However, as explained in [10], solving MWE can result in very conservative confidence bands when there are local deviations from what constitutes as normal behaviour in the data at hand. To overcome this, an alternative definition as what we will refer to as MWCB(k, s) was recently introduced in [10], and a greedy approach to solving this variant was provided; to the best of our understanding no exact algorithms for MWCB(k, s) have been proposed. However, as we will shortly explain, MWCB(k, s) can result in optimal solutions exhibiting extremely narrow parts in the confidence band even when there is no clear explanation for this behaviour in the data.

In this paper, we focus on the combinatorial variant of the minimum-width confidence band problem, and specifically, on constraint optimization approaches to obtaining optimal solutions to a new variant MWCB(k, s, t) of the multivariate minimum-width confidence interval problem. In more detail, *our contributions* are the following. (i) We demonstrate that minimum-width (k, s)-confidence bands as defined in [10] tend to have very narrow parts without a clear intuitive meaning. (ii) We propose an alternative definition to overcome this undesirable property, denoted as the MWCB(k, s, t) problem. (iii) As a novel application domain of declarative constraint optimization, we provide two constraint models for MWCB(k, s, t) in terms of MIP and MaxSAT as the constraint languages of choice at this time. (iv) We also provide a greedy algorithm for MWCB(k, s, t), which provides more scalability, but also allows for analyzing the benefits of exact constraint optimization for the problem in terms of the quality of obtained solutions. (v) To this end, we present an overview of an empirical evaluation on the scalability of the constraint optimization approaches and solution quality.

The rest of this paper is organized as follows. We start with the previously proposed variants MWCB(k) and MWCB(k, s) of the multivariate confidence interval problem, pointing out their drawbacks, and motivated by these we propose the focus of this work, the MWCB(k, s, t) problem (Sect. 2). We then introduce constraint optimization models for MWCB(k, s, t) in terms of MIP and MaxSAT (Sect. 3), as well as a first greedy approach to MWCB(k, s, t) for comparing with the constraint models (Sect. 4). Overview of an empirical evaluation of the constraint models is presented in Sect. 5 using real-world time series datasets, and related work discussed further in Sect. 6.

2   The Minimum-Width Confidence Band Problem

We consider a set of n data vectors x_i, each of length m, represented by a matrix $X \in \mathbb{R}^{n \times m}$. Let $x_{ij} \in X$ denote the j-th element of x_i. The data X can, for instance, represent time series data, i.e., each x_i is a sequence of values taken at successive points in time. (In the general setting, preprocessing may be necessary as one needs to make sure that the variables or at least their scales are comparable.)

A *confidence band* is defined as a pair (l, u) of vectors, where $l, u \in \mathbb{R}^m$ and $l_j \leq u_j$ for all j. The size of the confidence band $CB = (l, u)$ is SIZE(CB) = $\sum_{j=1}^{m}(u_j - l_j)$. In order to capture the relationship between a confidence band and a dataset, we use the concept of an *error* of a confidence band w.r.t. the data at hand. An indicator function (unity if the condition $\square$ is satisfied and zero otherwise) is denoted by $I[\square]$.

Definition 1. *Given a data vector x_i and a confidence band $CB = (l, u)$, the error of x_i w.r.t. CB is the number of points in x_i that lie outside of CB, i.e., ERROR(x_i, CB) = $\sum_{j=1}^{m} I[x_{ij} < l_j \vee u_j < x_{ij}]$.*

There are several possible ways to control the error. In [9,11] a data vector is considered an *outlier* or *extreme* if it is outside the confidence band in at least one dimension. In the minimum-width confidence band (MWCB(k)) problem the number of extreme data vectors is controlled by a parameter k.

Definition 2 (MWCB(k)). [9,11] *Given a dataset $X \in \mathbb{R}^{n \times m}$, any confidence band $CB^* \in$ arg min SIZE(CB) over those CB for which $\sum_{i=1}^{n} I[$ERROR(x_i, CB) > 0] $\leq k$ is a solution to the MWCB(k) problem.*

The above definition results in well-defined confidence bands which gives the user control over the error in analogy to the family-wise error rate. However, as argued in [10], the resulting confidence bands can be too conservative in cases in which there are local deviations from what constitutes as normal behaviour in the data at hand. To overcome this feature, a relaxed variant of the MWCB(k) problem allowing local deviations from the confidence band was recently proposed [10,24].

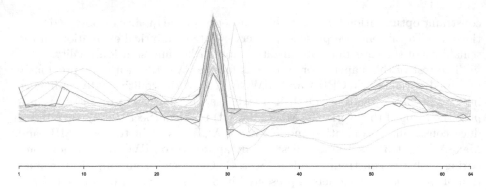

Fig. 1. An example of time series data with $n = 50$ time series of length $m = 64$ (represented with purple lines) for which an MWCB(k, s) confidence band (green lines) has very narrow parts, while the respective MWCB(k, s, t) confidence band (orange lines) does not. Here we have used $k = t = 5$ and $s = 6$, each value representing approximately 10% of the respective dimension. (Color figure online)

Definition 3 (MWCB(k, s)). [10] *Given a dataset $X \in \mathbb{R}^{n \times m}$ and two integers k and s, any confidence band $CB^* \in \arg\min \text{SIZE}(CB)$ over those CB for which $\sum_{i=1}^{n} I[\text{ERROR}(x_i, CB) > s] \leq k$ is a solution to the MWCB(k, s) problem.*

Now, one may observe that a solution to the MWCB(k, s) problem can be very narrow at places. This was in fact pointed out in [10], where it was further argued that in real datasets with non-trivial marginal distributions and correlation structure this was unlikely to happen, and the confidence band would be approximately of similar width across columns. However, we note that optimal solutions to the MWCB(k, s) problem on real data are likely to contain narrow intervals with no clear explanation. For example, consider Fig. 1. The data consists of 50 time series of length 64 sampled from the MITDB data (see Sect. 5 for details). The green lines represent a confidence band that is a solution to MWCB(5, 6) problem. We can observe that the confidence band is very narrow at the peak, i.e., around the time interval [25, 30]. One should notice that we use a real data set here to demonstrate the unwanted behaviour, and obviously it is not difficult to craft synthetic instances for which an optimal solution to the MWCB(k, s) problem has extremely narrow parts in the confidence band.

These observations suggest that there should be a mechanism to control the amount of column-wise error in addition to the row-wise constraints, and to this end we propose the concept of a *minimum-width (k, s, t)-confidence band* in terms of the MWCB(k, s, t) problem as follows.

Definition 4 (MWCB(k, s, t)). *Given a dataset $X \in \mathbb{R}^{n \times m}$ and integers k, s, and t, any confidence band $CB^* \in \arg\min \text{SIZE}(CB)$ over those $CB = (l, u)$ for which $\sum_{i=1}^{n} I[\text{ERROR}(x_i, CB) > s] \leq k$ and $\sum_{i=1}^{n} I[x_{ij} < l_j \vee x_{ij} > u_j] \leq t$ for all $1 \leq j \leq m$, is a solution to the MWCB(k, s, t) problem.*

As straightforward connection between the MWCB(k, s, t) and MWCB(k, s) problems is the following.

Proposition 1. *A confidence band CB for a dataset $X \in \mathbb{R}^{n \times m}$ is a solution to the MWCB(k, s) problem iff it is a solution to the MWCB(k, s, n) problem.*

The additional parameter t gives the user control over the amount of outliers allowed column-wise. If local deviations are likely to not to happen too often, setting the value of t equal to, or slightly larger than, k is a reasonable choice. For example, in Fig. 1, the orange lines represent a confidence band that is a solution to MWCB(5, 6, 5). One can observe that, indeed, for each time point a majority (90%, i.e., 45 out of 50 to be exact) of the time series are inside the confidence band. Furthermore, at most 10% of the time series deviate from the confidence band in more than 6 time points. Based on experimentation, it seems that for the real datasets we consider in this work, the size of the confidence band is approximately the same for $t \in \{k, k+1, \ldots, 2k\}$ (assuming $k \ll n$), and thus a conservative choice, e.g., $t = (1 + \epsilon)k$ for $\epsilon < 1$, seems to be a reasonable one.

3   Constraint Optimization Models for MWCB(k, s, t)

Recently, a MIP model for the MWCB(k) problem was proposed in [21]. However, to the best of our knowledge, no efficient exact algorithms for solving the MWCB(k, s) problem (nor the more general MWCB(k, s, t) problem) exist. Two heuristic algorithms are provided in [10], with no guarantee of solution quality. Korpela et al. [10] do provide a MIP model for the special case of one-sided confidence bands. However, this model is only used to show an approximability result and does not yield a practically efficient method, even for the special case.

In the following we present two constraint optimization models for MWCB(k, s, t), one using mixed integer programming and the other using maximum satisfiability. For notation, let $\mathcal{N} = \{1, \ldots, n\}$ and $\mathcal{M} = \{1, \ldots, m\}$. Both of our constraint models use a column ordering for the data X. Thus, we assume that we have an ordering for each of the columns using *dense-rank*[1] (as provided in R) and denote by r_j^{max} the maximum rank in column j. In the following, for a given $r \in \{1, \ldots, r_j^{max}\}$, we use (r) to denote the index i such that x_{ij} has rank r in column j.

3.1   Mixed Integer Programming Model

For our MIP model, we use the *band-wise reduction procedure* suggested in [22], similarly to [21]. However, in our model, instead of looking for whole data vector to exclude from the confidence band, we need to allow the exclusion of individual data points while maintaining both the column-wise and the row-wise constraints.

Our MIP model is presented in detail in Fig. 2. We introduce variables $l_j, u_j \in \mathbb{R}$ for each $j \in \mathcal{M}$ for the confidence band, and the objective (1) is to minimize

---

[1] In case of ties, both elements get the same rank r and the next greatest element gets rank $r + 1$.

MINIMIZE

$$\sum_{j=1}^{m}(u_j - l_j) \tag{1}$$

SUBJECT TO

$$l_j \leq x_{(t+1)j} \qquad \forall j \in \mathcal{M} \tag{2}$$

$$u_j \geq x_{(r_j^{max}-t)j} \quad \forall j \in \mathcal{M} \tag{3}$$

$$l_j - M_j^l d_{ij} \leq x_{ij} \qquad \forall i \in \mathcal{N}, j \in \mathcal{M} \tag{4}$$

$$u_j + M_j^u d_{ij} \geq x_{ij} \qquad \forall i \in \mathcal{N}, j \in \mathcal{M} \tag{5}$$

$$\sum_{i=1}^{n} d_{ij} \leq t \qquad \forall j \in \mathcal{M} \tag{6}$$

$$\sum_{j=1}^{m} d_{ij} - (m-s)y_i \leq s \qquad \forall i \in \mathcal{N} \tag{7}$$

$$\sum_{i=1}^{n} y_i \leq k \tag{8}$$

$$y_i, d_{ij} \in \{0,1\} \qquad \forall i \in \mathcal{N}, j \in \mathcal{M}$$

$$l_j, u_j \in \mathbb{R} \qquad \forall j \in \mathcal{M}$$

Fig. 2. Mixed integer programming model for MWCB(k, s, t).

the size of the confidence band, i.e., the sum of $(u_j - l_j)$'s over all columns j. We introduce $n \times m$ binary variables d_{ij} with the interpretation $d_{ij} = 1$ iff the jth element of x_i is outside the confidence band, i.e., $x_{ij} < l_j$ or $x_{ij} > u_j$. Furthermore, we use n binary variables y_i with the interpretation $y_i = 1$ iff x_i is outside the confidence band in at least s positions, i.e., $\sum_{j=1}^{m} I[x_{ij} < l_j \vee x_{ij} > u_j] > s$.

For the band-wise reduction procedure [22], we can make use of the following observation: since we have the column-wise constraint that at most t data points can be outside the confidence band at each column, we know that the value with rank $t + 1$ has to be inside the lower band. Thus, we include the constraints (2). Respective constraints for the upper band are provided in (3). Next, the constraints (4) (resp. (5)) encode the choice that either a value x_{ij} is inside the lower (resp. upper) band or it is outside. Here we use constant vectors $M_l = (M_1^l, \ldots, M_m^l)$ and $M^u = (M_1^u, \ldots, M_m^u)$ defined as

$$M_j^l = x_{(t+1)j} - \min(x_{1j}, \ldots, x_{nj}) \text{ and}$$
$$M_j^u = \max(x_{1j}, \ldots, x_{nj}) - x_{(r_j^{max}-t)j}.$$

Now, if $d_{ij} = 1$, the constraints are de-activated, and if $d_{ij} = 0$ the constraints (4) and (5) together ensure that $l_j \leq x_{ij} \leq u_j$. Here we once more use the property that at most t values can be outside the confidence band. The constraints (6) enforce this. We use the constraints (7) to represent the relationship between d_{ij}'s and y_i. If $y_i = 0$, then at most s variables d_{ij} for each $j \in \mathcal{M}$ can have value 1. On the other hand, if $y_i = 1$, then each constraint (7) reduces to $\sum_{j=1}^{m} d_{ij} \leq m$ which is always satisfied. Finally, the constraint (8) makes sure that at most k data vectors have more that s elements outside the confidence band.

HARD CLAUSES

$$\text{CNF}(\textstyle\sum_{i=1}^{n} y_i \leq k) \tag{9}$$

$$\text{CNF}((\textstyle\sum_{j=1}^{m} d_{ij} > s) \rightarrow y_i) \quad \forall i \in \mathcal{N} \tag{10}$$

$$\text{CNF}(\textstyle\sum_{i=1}^{n} d_{ij} \leq t) \quad \forall j \in \mathcal{M} \tag{11}$$

$$\text{CNF}(\textstyle\sum_{r=1}^{t+1} l_j^r = 1) \quad \forall j \in \mathcal{M} \tag{12}$$

$$\text{CNF}(\textstyle\sum_{r=r_j^{max}-t}^{r_j^{max}} u_j^r = 1) \quad \forall j \in \mathcal{M} \tag{13}$$

$$\neg d_{(r)j} \rightarrow \textstyle\bigwedge_{h=(r+1)}^{t+1} \neg l_j^h \quad \forall r \in \mathcal{R}_m, j \in \mathcal{M} \tag{14}$$

$$\neg d_{(r)j} \rightarrow \textstyle\bigwedge_{h=r_j^{max}-t}^{(r-1)} \neg u_j^h \quad \forall r \in \mathcal{R}_M^j, j \in \mathcal{M} \tag{15}$$

SOFT CLAUSES

$$(\neg l_j^r \vee \neg u_j^h) \quad \forall j \in \mathcal{M}, \forall r \in \mathcal{R}_m \tag{16}$$

$$\forall h \in \mathcal{R}_M^j$$

WEIGHTS

$$w((\neg l_j^r \vee \neg u_j^h)) = x_{(h)j} - x_{(r)j} \tag{17}$$

Fig. 3. The base clauses in our MaxSAT encoding for MWCB(k, s, t).

3.2 Maximum Satisfiability

Before presenting our second constraint optimization model for MWCB(k, s, t), we give a brief background on maximum satisfiability. For a more extensive review we direct the reader to [2].

For a Boolean variable x there are two literals, the positive literal x and the negative literal $\neg x$. A clause is a disjunction ($\vee$) of literals and a conjunctive normal form (CNF) formula is a conjunction ($\wedge$) of clauses. Equivalently, a clause is a set of literals and a CNF formula a set of clauses. A truth assignment τ is a function from Boolean variables to $\{0, 1\}$. A truth assignment τ satisfies a clause C ($\tau(C) = 1$) if it assigns a positive literal $x \in C$ to 1 or a negative literal $\neg x \in C$ to 0, and else τ falsifies the clause ($\tau(C) = 0$). Assignment τ satisfies a CNF formula F if it satisfies all clauses in F. An instance of the (weighted partial) maximum satisfiability (MaxSAT) problem (F_h, F_s, w) consists of two CNF formulas, the set of hard clauses F_h and the set of soft clauses F_s, together with a function $w : F_s \rightarrow \mathbb{N}$ assigning a positive weight to each soft clause. Any truth assignment τ that satisfies all hard clauses is a solution to the MaxSAT problem. A solution τ is optimal if it minimizes the sum of the weights of the soft clauses it falsifies, i.e., if $\sum_{C \in F_s} (1 - \tau(C))w(C) \leq \sum_{C \in F_s} (1 - \tau'(C))w(C)$ for all solutions τ'.

Our MaxSAT model makes extensive use of cardinality networks [1]. For our purposes, given a set of literals L, a literal l_B and an integer bound K, a cardinality network produces a set of clauses

$$\text{CNF}(\sum_{l \in L} l > K \rightarrow l_B)$$

that encodes the property that whenever more than K literals from the set L are assigned to 1, then so is the literal l_B. We use $\mathrm{CNF}(\sum_{l \in L} l \leq K)$ as shorthand for the CNF formula $\mathrm{CNF}(\sum_{l \in L} l > K \rightarrow l_B) \wedge (\neg l_B)$. Notice that the clauses in $\mathrm{CNF}(\sum_{l \in L} l \leq K)$ together essentially enforce that at most K literals of the set L can be assigned to 1. As an important special case we also use $\mathrm{CNF}(\sum_{l \in L} l = 1)$ as shorthand for $\mathrm{CNF}(\sum_{l \in L} l \leq K) \wedge (\bigvee_{l \in L} l)$, enforcing that exactly one of the literals in L has to be assigned to 1.

Figure 3 gives the clauses in our MaxSAT encoding. We start by describing the intuition behind the Boolean variables used. Note that for every solution $CB = (l, u)$ to the MWCB(k, s, t) problem and every $j \in \mathcal{M}$, there exists a $r \in \{1, \ldots, t+1\}$ (resp. $r \in \{r_j^{max} - t, \ldots, r_j^{max}\}$) such that $l_j = x_{(r)j}$ (resp. $u_j = x_{(r)j}$). For each column j, we use $\mathcal{R}_m^j$ to denote the set of possible r for which $l_j = x_{(r)j}$ can hold. Since at most t points can lie outside the lower band for any $j \in \mathcal{M}$, we have $\mathcal{R}_m^j = \mathcal{R}_m = \{1, \ldots, t+1\}$. Similarly we use $\mathcal{R}_M^j = \{r_j^{max} - t, \ldots, r_j^{max}\}$ to denote the set of possible indices r for which $u_j = x_{(r)j}$ can hold. We introduce variables l_j^r and u_j^h for $j \in \mathcal{M}$, $r \in \mathcal{R}_m$ and $h \in \mathcal{R}_M^j$ with the interpretation $l_j^r = 1$ (resp. $u_j^h = 1$) iff $l_j = x_{(r)j}$ (resp. $u_j = x_{(h)j}$). In addition, we use the variables d_{ij} and y_i with the same semantics as in the MIP model.

Next we describe the hard clauses enforcing these semantics. The constraints (9) enforce that at most k data vectors are outside the confidence band in more than s elements. The constraints (10) enforce the correct semantics for the y_i variables, i.e., whenever x_i lies outside the confidence band in more than s elements, the variable y_i is also set to true. Next, the constraints (11) enforce that at most t data points lie outside the confidence band in each column. The constraints (12) and (13) enforce that the value of l_j and the value of u_j is uniquely defined in each column j, i.e., exactly one of the l_j^r and u_j^h variables are true for each j. The constraints (14) enforce the correct semantics for the l_j^r variables: whenever a data point $x_{(r)j}$ is inside the lower confidence band l_j, i.e., $d_{(r)j} = 0$, then the value of l_j is at most the value of $x_{(r)j}$. In order to get shorter clauses in the final MaxSAT instance, we use instead an equivalent condition stating that whenever $d_{(r)j} = 0$, the value of l_j is not equal to $x_{(r')j}$ for any $r' \in \{r+1, \ldots, t+1\}$. The constraints (15) enforce a similar condition for the u_j^h variables. The soft clauses (16) enforce that the confidence band defined by the l_j^h and u_j^h variables is of minimum size. For a fixed column j, the clause $(\neg l_j^r \vee \neg u_j^h)$ is falsified if both l_j^r and u_j^h are true, corresponding to $l_j = x_{(r)j}$ and $u_j = x_{(h)j}$. The cost of the clause is set to be $x_{(h)j} - x_{(r)j} = u_j - l_j$, i.e., the contribution of that column to the size of the final confidence band. Notice that due to the hard clauses in the encoding, exactly one soft clause per column will be falsified.

Redundant Constraints. The clauses just described are enough to guarantee soundness. However, the encoding also includes redundant clauses meant to improve performance of the MaxSAT algorithms. These are based on the fact that at most t data points can lie outside the confidence band in each column.

For a fixed column j this implies that there are certain pairs of indices $r \in \mathcal{R}_m$, $h \in \mathcal{R}_M^j$ for which the variables u_j^h and l_j^r cannot both be set to true.

As an example, the variables u_j^r and l_j^{t+1} for $r = r_j^{max} - t$ cannot be set to true simultaneously, since this would require $2t$ data points, namely $x_{(1)j}, \ldots, x_{(t)j}$ and $x_{(r_j^{max})j}, \ldots, x_{(r_j^{max} - (t-1))j}$ to be outside of the confidence bands in column j. Hence the clause $(\neg u_j^r \vee \neg l_j^{t+1})$ for $r = r_j^{max} - t$ is always satisfied, making it redundant as a soft clause in our encoding. However, we can instead introduce it as a hard clause to improve propagation during search. Generalizing the above observation, for a fixed variable l_j^r we introduce the clause $(\neg l_j^r \vee \neg u_j^h)$ as hard clause instead of a soft one for all $h \in \{r_j^{max} - t, \ldots, r_j^{max} - (t - (r-1))\}$.

4   A Greedy Approach to MWCB(k, s, t)

In this section we present a greedy algorithm for finding (typically non-optimal) solutions for the MWCB(k, s, t) problem. The overall idea is to exclude individual data points greedily as long as the row-wise and column-wise constraints remain satisfied. The general idea is similar to the greedy algorithm proposed in [11], but instead of excluding a data vector fully, we consider excluding a single data point at a time.

The greedy algorithm is presented in pseudocode as Algorithm 1. We use an ordering structure R (line 2) that allows us $O(1)$ time access to the largest and the second to largest (resp. the smallest and the second to smallest) element in each column. The ordering structure consists of a doubly-linked list R_j for each column j, and can be initialized in $O(mn \log n)$ time. We use vectors rmd_C and rmd_R to keep track of the number of excluded values for each column and row, respectively, as well as a counter rmd_{cnt} to keep track of the number of rows for which more than s elements are excluded. Let gains(R) on line 4 be a method returning the possible gains for each of the columns, i.e., $x_{(2)j} - x_{(1)j}$ for the lower band and $x_{(r_j^{max})j} - x_{(r_j^{max}-1)j}$ for the upper band in $O(m)$ time. The values are stored in a priority queue G in $O(m)$ time.

The main part of the algorithm is the while-loop on lines 5–15. The loop is repeated at most $O(mn)$ times and each iteration takes $O(\log m)$ time. We use a Boolean $b \in \{0, 1\}$ to keep track of whether the current gain is obtained from the lower band ($b = 0$) or the upper band ($b = 1$). At each iteration the element with largest gain is used as a candidate for removal. On line 7, $\mathrm{idx}(R_j, b)$ returns the index of the currently highest/lowest ranked value in R_j. On line 8, we have the condition under which it is possible to exclude a value from the confidence band (realized by removing the respective element from the ordering structure R). In every case we have to maintain the column-wise constraint, i.e., check that less than t values have been excluded from the column in previous steps. Furthermore, the row-wise constraints can be satisfied in two ways. The first condition $\mathrm{rmd}_R(i) \neq s$ summarizes two cases: if $\mathrm{rmd}_R(i) < s$, it is always safe to exclude the candidate. On the other hand, if $\mathrm{rmd}_R(i) > s$, then the data vector with index i is already among the k possible outliers with more than s elements excluded, and thus the current value can be excluded as well. The remaining

```
input  : dataset X ∈ ℝⁿˣᵐ, integers k,s,t
output: CB ∈ ℝᵐˣ²
1 begin
2 |   R ←ordering structure for observations in X
3 |   rmd_C ←zeros(m); rmd_R ←zeros(n); rmd_cnt ← 0
4 |   G ←priorityQueue(gains(R))
5 |   while G ≠ ∅ do
6 |   |   (val, j, b) ←getMaximumElement(G)
7 |   |   i ←idx(R_j, 1, b)
8 |   |   if rmd_C(j) < t and (rmd_R(i) ≠ s or rmd_cnt < k) then
9 |   |   |   R ←remove(R_j, b)
10|   |   |   if rmd_R(i) == s then
11|   |   |   |   rmd_cnt++
12|   |   |   rmd_R(i)++
13|   |   |   rmd_C(j)++
14|   |   |   val ← value(R_j, 1, b)− value(R_j, 2, b)
15|   |   |   G.add(key=val, col=j, bit=b)
16|   for j ∈ 1 : m do
17|   |   CB(j, :) ← [value(R_j, 1, 0), value(R_j, 1, 1)]
18|   return CB
```

Algorithm 1. Greedy algorithm for $MWCB(k, s, t)$.

case $rmd_R(i) == s$ requires that no more than $k - 1$ data vectors have more than s elements excluded.

The respective counters are then updated (lines 10–13). On lines 14 and 15, a new candidate gain is computed and pushed to the priority queue. Here $val(R_j, r, b)$ is the value of the lowest ($b = 0$) or the highest ($b = 1$) ranked element still present in R_j for $r = 1$, resp. the value of the second lowest/highest ranked element for $r = 2$. Finally, the confidence band to be returned is directly obtained from the ordering structure. The overall time complexity for the greedy algorithm is $O(mn \log mn)$ and memory complexity $O(mn)$.

5 Experiments

We present an overview of an empirical evaluation on the scalability of the MIP and MaxSAT models using state-of-the-art solvers on $MWCB(k, s, t)$ instances constructed from real-world time series datasets, as well as on the relative quality of solutions provided by exact constraint optimization and the greedy approach.

For the experimental evaluation, we used the state-of-the-art commercial mixed integer programming system CPLEX version 12.7.1 from IBM [8], and the MaxSAT solvers QMaxSAT [12], MSCG [19], and MaxHS [3]. The MaxSAT solvers are representatives of state-of-the-art solvers based on different types of algorithms: QMaxSAT is a so-called model-guided SAT-based solver (using a SAT solver to search for increasingly good solutions until no better solutions can be found), MSCG is a core-guided SAT-based solver (using a SAT solver to extract and rule out unsatisfiable cores of a MaxSAT instance until a satisfying assignment is found), and MaxHS is a hybrid SAT-IP solver for MaxSAT, implementing a so-called implicit hitting set approach. The experiments were run on 2.83-GHz Intel Xeon E5440 quad-core machines with 32-GB RAM and Debian GNU/Linux 8 using a per-instance timeout of 3600 s.

We implemented the greedy procedure (recall Sect. 4) in R. As the greedy procedure has much better running time scalability than the constraint solvers on finding provably-optimal solutions, here we focus on comparing the improvements in solution costs provided by the exact approaches to those provided by the greedy procedure.

5.1 Datasets

For the experiments, we obtained benchmark instances based on the following real-world time series datasets.

Milan temperature data (MILAN). We use the `max-temp-milan` dataset from [11]. The raw data is obtained from Global Historical Climatology Network (GHCN) daily dataset [16,17] from US National Oceanic and Atmospheric Administration's National Climatic Data Center (NOAA NCDC)[2]. The preprocessed data contains average monthly maximum temperatures for a station located in Milan for the years 1763–2007, resulting in $n = 245$ time series with length $m = 12$.

UCI-Power data (POWER). The UCI-Power dataset is the individual household electric power consumption data[3] from the UCI machine learning repository [13]. It consists of hourly averages of the variable `active.power`, resulting in a dataset with $n = 1417$ time series with $m = 24$ time points.

Heartbeat data (MITDB). We use the preprocessed datasets `heartbeat-normal` and `heartbeat-pvc` from [11]. These datasets are obtained from the MIT-BIH arrhythmia database available at Physionet [5]. The data contains annotated 30-min records of normal and abnormal heartbeats [18]. There are 1507 observations in `heartbeat-normal` and 520 observations in `heartbeat-pvc` both with $m = 253$ time points.

As a preprocessing step, we shifted the data so that all values are non-negative. To assess the scalability of our constraint models, we used the datasets to produce instances with varying dimensions. To obtain an instance with n' time series, we randomly sample n' time series from the respective dataset. For the heartbeat data, we create instances with n' observations by sampling at random $0.9n'$ time series from `heartbeat-normal` and $0.1n'$ time series from `heartbeat-pvc`. Furthermore, in order to obtain instances with $m' < 253$ while maintaining the autocorrelation structure, we take every jth time point for $j \in \{2, 4, 6, 8, 10\}$. Table 1 summarizes the datasets and the parameters of the instances sampled from the datasets.

---

[2] http://www.ncdc.noaa.gov/.

[3] http://archive.ics.uci.edu/ml/datasets/Individual+household+electric+power+consumption.

Table 1. Summary of datasets and the instances generated.

Dataset	Sample sizes for n	Sample sizes for m
MILAN ($n = 245$, $m = 12$)	50, 100, 150, 200, 245	12
POWER ($n = 1417$, $m = 24$)	200, 400, 600, 800, 1000	24
MITDB ($n = 2027$, $m = 253$)	100, 150, 200, 250, 300	26, 32, 43, 64, 127, 253

5.2   Results

Due to the large parameter value space, we will present selected views on the results which provide interesting insights into the performance of the MIP and MaxSAT approaches for the problem, as well as quality of solutions obtained.

Scalability of the Exact Approaches. We start the overview of the empirical results with the scalability of the exact approaches, i.e., CPLEX on the proposed MIP model and the three considered state-of-the-art MaxSAT solvers on the MaxSAT encoding.

Results from this comparison are provided in Figs. 4 and 5. Figure 4 (left) shows the number of instances solved by each of the four solvers under different per-instance time limits on instances based on the MILAN dataset using the parameter values $k \in \{0.01n, 0.02n, 0.05n\}$, $s \in \{1,2\}$, and $t \in \{k, k+2\}$, rounding values below 1 to 1, and the rest to the nearest integer. To increase the number of instances, we used $k \in \{1, 2, 3\}$ for the smallest value $n = 50$. This results in 60 instances in total. Out of the three MaxSAT solvers, the model-guided QMaxSAT performs the best. However, CPLEX on the MIP model solves each of the instances very fast, surpassing in performance the MaxSAT solvers on the MaxSAT encoding.

Based on this, we take a further look at the performance of CPLEX and QMaxSAT as the two most promising out of the considered solvers. Figure 4 (right) shows the number of solved instances generated from all of the three benchmark datasets (MILAN, POWER, and MITDB) using parameter values

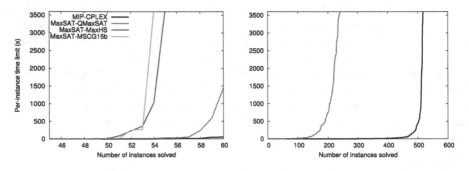

Fig. 4. Comparison of solver scalability. Left: MILAN dataset, right: all datasets.

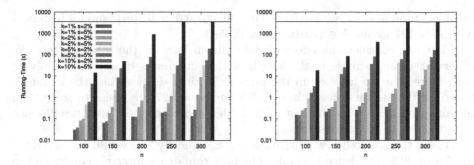

Fig. 5. The solving times for the instances sampled from the MITDB data with $m = 43$ using the MIP model. Left: $t = k$, right: $t = k + 2$.

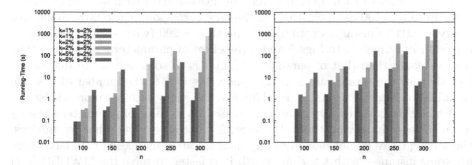

Fig. 6. The solving times for the instances sampled from the MITDB data with $m = 127$ using the MIP model. Left: $t = k$, right: $t = k + 2$.

$k \in \{0.01n, 0.02n, 0.05n\}$, $t \in \{k, k + 2\}$ and $s = \{0.01m, 0.02m, 0.05m\}$ (720 instances in total), and gives further support for the fact that CPLEX dominates in performance the MaxSAT approach.

One should note that the value of t has a direct impact on the size of the search space. Values $t < k$ are allowed by the definition, but can result in unintuitive solutions. Thus, we consider values $t \geq k$. As an increase in the value of t can intuitively drastically increase the hardness of an instance (in the worst case all nm values need to be considered for removal), we assessed the effect of t on the solution quality, i.e., the size of the optimal confidence bands. Experimentation with instances from the MITDB and POWER dataset with 200–400 time series showed that increasing the value of t from k to $2k$ for $k = 0.05n$ decreased the size of an optimal solution by less that 1%. Thus, to assess the scalability of our approach, we chose to use the conservative values $t \in \{k, k + 2\}$. One should note, however, that the best value for t depends on the dataset at hand and the expected distribution of local and global outliers.

Figures 5 and 6 give further insights into the scalability of the MIP approach, using instances based on the MITDB dataset with $m = 43$ and $m = 127$. As parameters values we consider $s = \{0.02m, 0.05m\}$ and $t \in \{k, k + 2\}$. For instances with $m = 43$, we use $k \in \{0.01n, 0.02n, 0.05n, 0.10n\}$, and for instances with $m = 127$ we use $k \in \{0.01n, 0.02n, 0.05n\}$.

First, we consider the effect of data dimensions on the solving time. We observe that an increase in the length of the time series m affects the solving time more than an increase in the number of time series n, i.e., the instances based on the MITDB data with $n = 250$ and $m = 127$ are easier to solve than instances with $n = 100$ and $m = 253$ (detailed results for MITDB-$m253$ not reported due to space constraints).

As for the scalability w.r.t. parameter k, typically one would be interested in 95% or 90% confidence intervals. The 95% confidence intervals correspond to setting $k = \lfloor 0.05 \rfloor$, and our MIP model can handle $k = \lfloor 0.05n \rfloor$ with instances up to $n = 300$ and $m = 127$. For the instances based on MITDB data with $m = 253$, instances with up to $n = 100$ and POWER data with $m = 24$ instances with up to $n = 600$ can be solved. In the case corresponding to 90% confidence intervals, MITDB instances with $m = 43$ up to $n = 200$ (with $s = 0.05\,m$) can be solved before timeout. In Figs. 5 and 6 the effect of parameter s on solving time seems smaller than that of parameter k. This is to some extend to due to the fact that for the actual values used, namely $1 \leq s \leq 6$, the number of possible combinations stays reasonable. For larger s, the effect becomes more visible.

Finally, as expected, typically an increase on the value of parameter t results in an increase in solving times. However, in contrast to the other parameters, the solving times do not monotonically increase upon increasing t. In fact, there are some instances with $t > t'$ for which it is faster to solve the $\mathrm{MWCB}(k, s, t)$ problem than the $\mathrm{MWCB}(k, s, t')$ problem, e.g., the MITDB-$m43$ instance with $n = 200$.

Overall, based on the empirical results, CPLEX on the proposed MIP model for $\mathrm{MWCB}(k, s, t)$ scales reasonably well on the real-world datasets under various parameter value combinations.

Solution Quality: Exact vs Greedy. Finally, we look at the relative quality of solutions obtained on one hand using the exact MIP approach and, on the other hand, using the greedy algorithm presented in Sect. 4. Here we focus on the question of whether the higher computational cost of exact optimization pays off by offering in cases better solutions than the greedy approach.

As witnessed by the results presented in Fig. 7, the optimal solutions are in cases non-negligibly better than those provided by the greedy approach. In more detail, the histograms in Fig. 7 show the counts of the relative costs of greedy and optimal solutions, defined as $\mathrm{SIZE}(CB_{gr})/\mathrm{SIZE}(CB_{opt})$, for instances based on the MITDB ($m = 43$) dataset with $t = k$ (left) and $t = k + 2$ (right).

We observe that there are greedy solutions that have a cost of up to approximately 127.5% of that of the optimal solution, while on average the cost of the greedy solution is 108% of the optimum for the MITDB-$m43$ instances.

Furthermore, we observed that the MIP approach can provide solutions with a low cost (without proving them optimal) often much faster than what it takes for CPLEX to prove the solutions found optimal. In detail, for 91 out of the 120

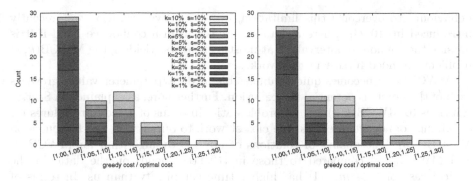

Fig. 7. Comparison of the relative cost of greedy and optimal solutions of solutions for the MITDB instances with $m = 43$ and $n \in \{100, 150, 200, 250, 300\}$. Left: $t = k$, right: $t = k + 2$. The other parameter values used were $k \in \{0.01n, 0.02n, 0.05n, 0.1n\}$ and $s \in \{0.02m, 0.05m, 0.1m\}$. The relative cost is provided for the 110 (out of 120) instances for which a provably optimal solution is found in 3600 s.

instances considered in Fig. 7, CPLEX provided a provably optimal solution in less than one minute on our MIP model. For 28 out of the remaining 29 instances, CPLEX provided within 60 s solutions with 7% lower cost on average compared to the solutions provided by the greedy algorithm. Thus we observed that even in cases in which an optimal solution cannot be found fast, our MIP model can be typically used to obtain better than greedy solutions relatively fast.

These observations motivate the exact approach presented in this work, as well as future work on ways of further improving the scalability of exact approaches for the MWCB(k, s, t) problem. On the other hand, if solutions to very large instances of MWCB(k, s, t) are needed very fast, our greedy algorithm is also a viable option.

6   Related Work

The univariate confidence interval of a distribution based on a finite-sized sample from the distribution has been extensively studied (see, e.g., [7]). However, there are surprisingly few approaches to multivariate confidence intervals and most of the effort has been focused on describing univariate distributions. Another alternative are the *confidence regions* (see, e.g., [6]), which however require making assumptions about the underlying distributions or which cannot be described simply by upper and lower bounds; e.g., confidence regions for multivariate Gaussian data are ellipsoids.

In the time series domain, multivariate confidence intervals [9,11], namely *confidence bands* have been defined in terms of the *minimum-width envelope* (MWE) problem: a time series is within a confidence band if it is within the confidence interval of *every time point*, also see [14,15,21] for similar approaches. While this definition has desirable properties, it can result in very conservative confidence bands if there are local deviations from what constitutes as normal behaviour. To overcome this limitation, an alternative definition was recently introduced in [10,24], where a data vector is within a confidence band if it is outside the confidence intervals of at most s elements, yielding the MWCB(k, s) problem extended further in this work.

MWCB(k, s) becomes quickly unfeasible as data/parameter values grow, as each of the points is potential for exclusion. Furthermore, as explained in Sect. 2, solutions to MWCB(k, s) can be problematic. In terms of greedy procedures for obtaining confidence intervals, the closest work to ours is [10] which focuses on MWCB(k, s). The quality of solutions of our greedy algorithm for MWCB(k) and MWCB(k, s) compared to those in [10,11] depends on the data. In the typical case of $n > m$, [11] has higher time complexity than us. In terms of using exact constraint optimization to determining confidence bands, the only and closest work to our is [21] where a MIP model is provided for MWCB(k); we generalize here to MWCB(k, s, t). Our approach applies also to the special case of MWCB(k), although for capturing at the same time the more general setting considered here we use $n \times m$ binary variables (as compared to n binary variables in [21]).

7   Conclusions

We focused on the combinatorial optimization problem of determining tight (minimum-width) multivariate confidence bands as a central yet NP-hard optimization problem in data analysis. Pointing out drawbacks in earlier characterizations of the problem, we proposed a generalization MWCB(k, s, t) circumventing some of the earlier drawbacks. We proposed two constraint models allowing for exactly solving instances of MWCB(k, s, t), as well as a greedy algorithm for the problem. We studied the scalability of mixed integer programming and maximum satisfiability solvers on the respective constraint models, and observed that mixed integer programming especially provides good scalability on MWCB(k, s, t) instances based on real-world data. The greedy algorithm, on the other hand, can provide relatively good solutions very fast. However, we also showed empirically that the optimal solutions provided by the exact constraint-based approach can at times provide noticeably better solutions than the greedy approach, and can also provide relatively fast better quality solutions (without proving optimality). The study of potential alternative characterizations (e.g., objective functions) of the minimum-width confidence band problem which would still have the same benefits as MWCB(k, s, t) compared to MWCB(k) and MWCB(k, s) is one interest aspect for further work.

References

1. Asín, R., Nieuwenhuis, R., Oliveras, A., Rodríguez-Carbonell, E.: Cardinality networks: a theoretical and empirical study. Constraints **16**(2), 195–221 (2011)
2. Biere, A., Heule, M., van Maaren, H., Walsh, T.: Handbook of Satisfiability. Frontiers in Artificial Intelligence and Applications, vol. 185. IOS Press, Amsterdam (2009)

3. Davies, J., Bacchus, F.: Exploiting the power of MIP solvers in MAXSAT. In: Järvisalo, M., Van Gelder, A. (eds.) SAT 2013. LNCS, vol. 7962, pp. 166–181. Springer, Heidelberg (2013). doi:10.1007/978-3-642-39071-5_13
4. Gardner, M.J., Altman, D.G.: Confidence intervals rather than P values: estimation rather than hypothesis testing. Br. Med. J. (Clin. Res. Ed.) **292**(6522), 746–750 (1986)
5. Goldberger, A.L., Amaral, L.A.N., Glass, L., Hausdorff, J.M., Ivanov, P.C., Mark, R.G., Mietus, J.E., Moody, G.B., Peng, C.K., Stanley, H.E.: PhysioBank, PhysioToolkit, and PhysioNet: components of a new research resource for complex physiologic signals. Circulation **101**(23), e215–e220 (2000)
6. Guilbaud, O.: Simultaneous confidence regions corresponding to Holm's step-down procedure and other closed-testing procedures. Biom. J. **50**(5), 678 (2008)
7. Hyndman, R.J., Fan, Y.: Sample quantiles in statistical packages. Am. Stat. **50**(4), 361–365 (1996)
8. IBM ILOG: CPLEX optimizer (2017). http://www-01.ibm.com/software/commerce/optimization/cplex-optimizer/
9. Kolsrud, D.: Time-simultaneous prediction band for a time series. J. Forecast. **26**(3), 171–188 (2007)
10. Korpela, J., Oikarinen, E., Puolamäki, K., Ukkonen, A.: Multivariate confidence intervals. In: Proceedings of SDM, pp. 696–704. SIAM (2017)
11. Korpela, J., Puolamäki, K., Gionis, A.: Confidence bands for time series data. Data Min. Knowl. Discov. **28**(5–6), 1530–1553 (2014)
12. Koshimura, M., Zhang, T., Fujita, H., Hasegawa, R.: QMaxSAT: a partial Max-SAT solver. J. Satisf. Boolean Model. Comput. **8**(1/2), 95–100 (2012)
13. Lichman, M.: UCI machine learning repository (2013). http://archive.ics.uci.edu/ml
14. Liu, W., Jamshidian, M., Zhang, Y., Bretz, F., Han, X.: Some new methods for the comparison of two linear regression models. J. Stat. Plan. Inference **137**(1), 57–67 (2007)
15. Mandel, M., Betensky, R.A.: Simultaneous confidence intervals based on the percentile bootstrap approach. Comput. Stat. Data Anal. **52**(4), 2158–2165 (2008)
16. Menne, M., Durre, I., Korzeniewski, B., McNeal, S., Thomas, K., Yin, X., Anthony, S., Ray, R., Vose, R., Gleason, B., Houston, T.: Global Historical Climatology Network – Daily (GHCN-Daily), version 3.11 (2012)
17. Menne, M., Durre, I., Vose, R., Gleason, B., Houston, T.: An overview of the global historical climatology network-daily database. J. Atmos. Ocean. Technol. **29**, 897–910 (2012)
18. Moody, G.B., Mark, R.G.: The impact of the MIT-BIH arrhythmia database. IEEE Eng. Med. Biol. Mag. **20**(3), 45–50 (2001)
19. Morgado, A., Dodaro, C., Marques-Silva, J.: Core-guided MaxSAT with soft cardinality constraints. In: O'Sullivan, B. (ed.) CP 2014. LNCS, vol. 8656, pp. 564–573. Springer, Cham (2014). doi:10.1007/978-3-319-10428-7_41
20. Nuzzo, R.: Scientific method: statistical errors. Nature **506**, 150–152 (2014)
21. Schüssler, R., Trede, M.: Constructing minimum-width confidence bands. Econ. Lett. **145**, 182–185 (2016)
22. Staszewska-Bystrova, A., Winker, P.: Constructing narrowest pathwise bootstrap prediction bands using threshold accepting. Int. J. Forecast. **29**(2), 221–233 (2013)
23. Trafimow, D., Marks, M.: Editorial. Basic Appl. Soc. Psychol. **37**(1), 1–2 (2015)
24. Wolf, M., Wunderli, D.: Bootstrap joint prediction regions. J. Time Ser. Anal. **36**(3), 352–376 (2015)
25. Woolston, C.: Psychology journal bans P values. Nature **519**, 9 (2015)

Constraint Programming for Multi-criteria Conceptual Clustering

Maxime Chabert[1,2]($\boxtimes$) and Christine Solnon[1]

[1] Université de Lyon, INSA Lyon, LIRIS, 69622 Villeurbanne, France
maxime.chabert@liris.cnrs.fr
[2] Infologic, Bourg-lès-Valence, France

Abstract. A conceptual clustering is a set of formal concepts (*i.e.*, closed itemsets) that defines a partition of a set of transactions. Finding a conceptual clustering is an $\mathcal{NP}$-complete problem for which Constraint Programming (CP) and Integer Linear Programming (ILP) approaches have been recently proposed. We introduce new CP models to solve this problem: a pure CP model that uses set constraints, and an hybrid model that uses a data mining tool to extract formal concepts in a preprocessing step and then uses CP to select a subset of formal concepts that defines a partition. We compare our new models with recent CP and ILP approaches on classical machine learning instances. We also introduce a new set of instances coming from a real application case, which aims at extracting setting concepts from an Enterprise Resource Planning (ERP) software. We consider two classic criteria to optimize, *i.e.*, the frequency and the size. We show that these criteria lead to extreme solutions with either very few small formal concepts or many large formal concepts, and that compromise clusterings may be obtained by computing the Pareto front of non dominated clusterings.

1 Introduction

Clustering is a non-supervised classification approach which aims at partitioning a set of objects into homogeneous clusters. Conceptual clustering provides, in addition to clusters, a description of clusters by means of formal concepts [5,15]. In this paper, we introduce new Constraint Programming (CP) models to solve this problem, and we evaluate these models on classical academic instances, but also on a new set of instances that comes from a real application case.

Presentation of the Applicative Context. Enterprise Resource Planning (ERP) softwares are generic softwares for managing companies. They address many functional goals ranging from commercial management to production or stock management [11]. While the same ERP can be used by many companies, it has to be customized specifically to fit each company needs. This is done thanks to parameters which are used to customize the ERP functionalities to each company depending on his structural and organizational needs. However, the large range

© Springer International Publishing AG 2017
J.C. Beck (Ed.): CP 2017, LNCS 10416, pp. 460–476, 2017.
DOI: 10.1007/978-3-319-66158-2_30

of functional goals makes the customization process complex and time consuming. We have studied the customization process of the *Copilote* ERP, developed by *Infologic* and specialized in food industry management. As pointed out in [21], it appears that most of the time of the customization process is dedicated to the parameterization step: this step basically involves assigning values to parameters in such a way that the ERP fulfills the client needs. The complexity of this step comes from the fact that the ERP has a large number of parameters with strong implicit interactions. Moreover, several studies have shown that this time-consuming parameterization step is less important than human factors (user training, personalized support, for example) [1,16]. Therefore, an important challenge is to reduce the time needed to parameterize an ERP in order to spend more time on human factors.

To reduce the time needed to parameterize an ERP, our goal is to (partially) automate this step. To this aim, we have collected a database of existing parameter settings, corresponding to recent installations of the *Copilote* ERP for 400 clients. We propose to identify relevant groups of parameter settings, and to associate them with functional needs. As many functional needs are common to several clients, these parameter setting groups will be reused when customizing the ERP for a new client with similar needs. To identify relevant parameter setting groups, we propose to partition the database of parameter settings into clusters. As we do not have a relevant measure to evaluate the similarity of different parameter settings, and as most parameters are symbolic ones, we propose to use conceptual clustering to achieve this task: this approach does not assume that there exists a similarity measure, and allows us to describe each cluster by a set of parameter values which are shared by all parameter settings in the cluster.

Contributions and Organization of the Paper. We introduce the background on conceptual clustering in Sect. 2. In particular, we describe two declarative approaches which have been recently proposed: [4], that uses Constraint Programming (CP), and [17], that uses Integer Linear Programming (ILP). These declarative approaches are very relevant in our applicative context because they allow us to easily model new constraints or objective functions: a main issue is to find relevant objective functions and constraints to extract setting concepts which make sense for our ERP experts.

In Sect. 3, we introduce two new CP models to compute optimal conceptual clusterings: the first one is a full CP model; the second one is an hybrid model that uses a dedicated tool to extract formal concepts and uses CP to select the subset of formal concepts that defines an optimal partition.

We experimentally evaluate these models in Sect. 4. This evaluation is done on some classical academic instances. We also introduce a new benchmark composed of seven instances that have been extracted from our database of parameter settings. In this first evaluation, we mainly consider two objective functions to optimize: the size of the concepts (corresponding to the number of parameters that are common to several clients), and the frequency of the concepts (corresponding to the number of clients that share a common setting). We evaluate scale-up properties of the different approaches when the number of clusters is fixed and when it is not fixed, for each of these objectives separately.

The two objective functions that we consider are complementary and are related to the number of clusters: when maximizing concept sizes, optimal clusterings have many clusters of small frequencies; when maximizing concept frequencies, optimal clusterings have very few clusters of small sizes. In Sect. 5, we propose to compute the Pareto front of all non-dominated solutions, and we introduce and compare different ways for achieving this with our CP models.

2   Background on Conceptual Clustering

Formal Concepts. Formal Concept Analysis is a way of grouping together objects sharing a same set of attribute values [5]. In this paper, we use the transactional database terminology: objects are called *transactions*, and attribute values are called *items*. More formally, let $\mathcal{T}$ be a set of m transactions (numbered from 1 to m), $\mathcal{I}$ a set of n items (numbered from 1 to n), and $\mathcal{R} \subseteq \mathcal{T} \times \mathcal{I}$ a binary relation that relates transactions to items: $(t, i) \in \mathcal{R}$ denotes the fact that transaction t has item i. We note $itemset(t)$ the set of items associated with t, *i.e.*, $\forall t \in \mathcal{T}, itemset(t) = \{i \in \mathcal{I} : (t, i) \in \mathcal{R}\}$. Given a set E, we note $\mathcal{P}(E)$ the set of all its subsets, and $\#E$ its cardinality.

The *intent* of a subset $T \subseteq \mathcal{T}$ of transactions is the intersection of their itemsets, *i.e.*, $intent(T) = \cap_{t \in T} itemset(t)$. The *extent* of a set $I \subseteq \mathcal{I}$ of items is the set of transactions whose itemsets contain I, *i.e.*, $extent(I) = \{t \in \mathcal{T} : I \subseteq itemset(t)\}$. These two operators induce a Galois connection between $\mathcal{P}(\mathcal{T})$ and $\mathcal{P}(\mathcal{I})$, *i.e.*, $\forall T \subseteq \mathcal{T}, \forall I \subseteq \mathcal{I}, T \subseteq extent(I) \Leftrightarrow I \subseteq intent(T)$. A *formal concept* is a couple $(T, I) \in \mathcal{P}(\mathcal{T}) \times \mathcal{P}(\mathcal{I})$ such that $T = extent(I)$ and $I = intent(T)$. We note $\mathcal{F}$ the set of all formal concepts. The *frequency* of a formal concept (T, I) is the number of transactions, *i.e.*, $freq(T, I) = \#T$, and its *size* is the number of items, *i.e.*, $size(T, I) = \#I$.

For example, we display in Table 1 a transactional dataset and its associated set $\mathcal{F}$ of formal concepts. The couple $(\{t_1, t_2\}, \{i_1\})$ is not a formal concept because $intent(\{t_1, t_2\}) = \{i_1, i_4\} \neq \{i_1\}$ and $extent(\{i_1\}) = \{t_1, t_2, t_5\} \neq \{t_1, t_2\}$.

Table 1. Left: example of transactional dataset with $m = 5$ transactions and $n = 4$ items. Right: set $\mathcal{F}$ of formal concepts for this dataset.

	i_1	i_2	i_3	i_4
t_1	1	0	0	1
t_2	1	0	1	1
t_3	0	1	0	1
t_4	0	1	1	0
t_5	1	0	1	0

C	intent	extent	freq.	size
c_1	$\{i_1\}$	$\{t_1, t_2, t_5\}$	3	1
c_2	$\{i_2\}$	$\{t_3, t_4\}$	2	1
c_3	$\{i_3\}$	$\{t_2, t_4, t_5\}$	3	1
c_4	$\{i_4\}$	$\{t_1, t_2, t_3\}$	3	1
c_5	$\{i_1, i_3\}$	$\{t_2, t_5\}$	2	2

C	intent	extent	freq.	size
c_6	$\{i_1, i_4\}$	$\{t_1, t_2\}$	2	2
c_7	$\{i_2, i_3\}$	$\{t_4\}$	1	2
c_8	$\{i_2, i_4\}$	$\{t_3\}$	1	2
c_9	$\{i_1, i_3, i_4\}$	$\{t_2\}$	1	3

Formal Concepts and Closed Itemset Mining. Formal concepts correspond to *closed itemsets* [18] and the set $\mathcal{F}$ may be computed by using algorithms dedicated to the enumeration of frequent closed itemsets, provided that the frequency threshold is set to 1. In particular, LCM [24] is able to extract all formal concepts of $\mathcal{F}$ in linear time with respect to $\#\mathcal{F}$. As there is usually a huge number of closed itemsets, we may add constraints or optimization criteria to identify relevant concepts. For example, we may search for closed itemsets whose frequency is greater than some threshold and whose size is maximal. We may also combine several criteria, and search for the Pareto front of non dominated formal concepts (where a concept c_1 is dominated by another concept c_2 if c_2 is at least as good as c_1 on all criteria but one, and strictly better on this last criterion). This Pareto front is also called the *skyline* [2].

CP for Itemset Mining. Using CP to model and solve itemset search problems is a topic which has been widely explored during the last ten years [7,8,12,20]. Indeed, CP allows one to easily model various constraints on the searched itemsets, corresponding to application-dependent constraints for example. These constraints are used to filter the search space during the mining process, and allow CP to be competitive with dedicated mining tools such as LCM. Most recently, [14] introduced a global constraint for extracting frequent closed itemsets. This global constraint enforces domain consistency in polynomial time, and it is quite competitive with LCM: if it is an order slower on basic queries, it is more efficient for complex queries where extra constraints are added. Also, [23] proposed to use CP to extract skyline patterns, *i.e.*, non-dominated patterns according to several criteria: they use a dynamic approach, where constraints are added each time a new solution is found in order to forbid solutions dominated by it.

Conceptual Clustering. Clustering is an unsupervised classification approach the goal of which is to group objects into homogeneous clusters. Conceptual clustering provides, in addition to clusters, a description of clusters by means of formal concepts: each cluster corresponds to a formal concept. More precisely, a conceptual clustering is a set of k formal concepts $\mathcal{C} = \{(T_1, I_1), \ldots, (T_k, I_k)\}$ such that $\{T_1, \ldots, T_k\}$ is a partition of the set $\mathcal{T}$ of transactions.

Different optimization criteria may be considered. In this article, we consider two classical criteria: *minFreq*, to maximize the minimal frequency of a cluster; and *minSize*, to maximize the minimal size of a cluster. For example, let us consider the set $\mathcal{F}$ of formal concepts displayed in Table 1. Two examples of clusterings are $P_1 = \{c_1, c_2\}$ and $P_2 = \{c_1, c_7, c_8\}$. According to *minFreq*, the best clustering is P_1 (as the minimal frequency is 2 for P_1 and 1 for P_2). According to *minSize*, both clusterings are equivalent as their minimal size is 1.

The number k of clusters is an important parameter which has a great influence on the size and the frequency of the clusters: small values for k favor clusters with larger frequencies and smaller sizes, whereas large values favor clusters with smaller frequencies and larger sizes.

CP for Conceptual Clustering. Conceptual clustering is related to closed item-set mining, as each cluster corresponds to a closed itemset. However, the goal is no longer to find closed itemsets that satisfy some constraints or optimize some criteria, but to find a subset of closed itemsets which partitions the set of trans-actions and optimizes some criteria. Conceptual clustering is a special case of k-pattern set mining, as introduced in [9]: the conceptual clustering problem is defined by combining a cover and a non-overlapping constraint, and a CP model based on boolean variables is proposed to solve this problem.

[3] describes a CP model for clustering problems where a dissimilarity mea-sure between objects is provided. In this case, the goal is to find a partition of the objects which satisfies some constraints and optimizes an objective func-tion defined by means of this dissimilarity measure. This CP model has been extended to conceptual clustering in [4]. Experimental results reported in [4] show that this model outperforms the binary model of [7]. This model assumes that the number of clusters is defined by a constant k. There is an integer vari-able G_t for each transaction $t \in \mathcal{T}$: G_t represents the cluster of t and its domain is $D(G_t) = [1, k]$. Symmetries (due to the fact that cluster numbers may be swapped) are broken by adding a $precede(G, [1, k])$ constraint [13]. Each cluster is enforced to have at least one transaction by the constraint: $atLeast(1, G, k)$. For each cluster $c \in [1, k]$, a set variable $Intent_c$ represents the intent of the set of transactions in c, *i.e.*, the intersection of their itemsets. Its domain is the set of all possible itemsets, *i.e.*, $D(Intent_c) = \mathcal{P}(\mathcal{I})$. The extent constraint is expressed by: $\forall c \in [1, k], \forall t \in \mathcal{T}, G_t = c \Leftrightarrow Intent_c \subseteq itemset(t)$. It is implemented thanks to $k \times m$ reified constraints (with $m = \#\mathcal{T}$). The intent constraint is expressed by: $\forall c \in [1, k], Intent_c = \cap_{t \in \mathcal{T}, G_t = c} itemset(t)$. It is implemented thanks to k constraints, and each of these k constraints needs n reified domain constraints to build the set of all transactions in cluster c, and a set element global con-straint to select the corresponding itemsets and intersect them. An objective variable is introduced. Depending on the optimization criterion, this variable is constrained to be equal either to the minimal cardinality of all $Intent_c$ variables (*minSize*), or the minimal number of G variables assigned to a same cluster thanks to $atLeast(obj, G, c)$ constraints (*minFreq*).

The model proposed in [4] assumes that the number of clusters is fixed to a constant value k. It may easily be extended to the case where this number is not known, by introducing an integer variable k, whose domain is bounded between 2 and $m - 1$. However, performance is degraded when k is not fixed.

ILP for Conceptual Clustering. [17] proposes to compute conceptual clusterings by combining two exact techniques: in a first step, a dedicated closed itemset mining tool (*i.e.*, LCM [24]) is used to compute the set $\mathcal{F}$ of all formal concepts and, in a second step, ILP is used to select a subset of $\mathcal{F}$ that is a partition of the set $\mathcal{T}$ of transactions and that optimizes some given criterion. More precisely, for each formal concept $f \in \mathcal{F}$, there is a binary variable x_f such that $x_f = 1$ iff f is selected. The subset of selected formal concepts is constrained to define a partition of $\mathcal{T}$ by posting the constraint: $\forall t \in \mathcal{T}, \sum_{f \in \mathcal{F}} a_{tf} x_f = 1$, where $a_{tf} = 1$ if transaction t belongs to the extension of concept f, and 0 otherwise. Contrary to the CP approaches of [3,7], the number of clusters is not fixed and it is a variable k which is constrained to be equal to the number of selected concepts by posting the constraint: $k = \sum_{f \in \mathcal{F}} x_f$. In [17], the goal is to maximize the sum of the sizes of the selected concepts. Therefore, the objective function to maximize is: $\sum_{f \in \mathcal{F}} v_f x_f$ where v_f is the size of the concept f. If the case is not explicitly discussed in [17], we may easily extend this ILP model to maximize the minimal size (resp. frequency) of the selected concepts: we introduce a variable v_{min} and enforce this variable to be smaller than or equal to the size (resp. the frequency) of the selected concepts by adding the constraint $\forall t \in T, v_{min} \leq v_f x_f + M(1 - x_f)$, where M is a positive constant greater than the largest possible size (resp. frequency), and v_f is the size (resp. frequency) of the concept f.

3 New CP Models

In this section, we introduce two new CP models for computing optimal conceptual clusterings. The first model (described in Sect. 3.1) may be seen as an improvement of the CP model of [3]. The second model (described in Sect. 3.2) follows the two step approach of [17]: the first step is exactly the same; the second step uses CP to select formal concepts. These models are experimentally evaluated and compared with the approaches of [3,17] in Sect. 4.

For both models, we do not assume that the number of clusters is fixed: k is a variable whose domain is $[k_{min}, k_{max}]$, where k_{min} and k_{max} are two given bounds such that $2 \leq k_{min} \leq k_{max} < m$.

3.1 New Full CP Model

Like the CP model of [3], we use G_c integer variables to model clusters. However, we associate the *Intent* set variables to transactions instead of associating them to clusters. This simplifies the propagation of the intent constraint. Another reason for associating *Intent* set variables to transactions instead of clusters is that, when k is strictly lower than k_{max}, each $Intent_c$ set variable such that $c > k$ should be empty. This would imply to use reification to compute the minimal intent size, as we must not consider $Intent_c$ variables such that $c > k$ (otherwise, the minimal size would be equal to 0 whenever $k < k_{max}$). Also, we introduce new set variables to explicitly model extents and these set variables are associated with transactions to ease the computation of the minimal frequency. Finally, we introduce redundant set variables which model concept extents and are associated with clusters: these variables are used to add a redundant partition constraint which improves the solution process.

More formally, we use the following variables:

- an integer variable k (with $D(k) = [k_{min}, k_{max}]$), which represents the number of clusters;
- for each transaction $t \in T$:
  - an integer variable G_t (with $D(G_t) = [1, kMax]$), which represents the cluster of t;

- a set variable $Intent_t$ (with $D(Intent_t) = \mathcal{P}(itemset(t)))$, which represents the set of items in the intent of the cluster of t;
- a set variable $Extent_t$ (with $D(Extent_t) = \mathcal{P}(\mathcal{T}))$, which represents the set of transactions in the extent of the cluster of t;
- for each cluster $c \in [1, k_{max}]$, a set variable $ExtentCluster_c$ (with $D(ExtentCluster_c) = \mathcal{P}(\mathcal{T}))$, which represents the set of transactions in c;
- two integer variables $minFreq$ (with $D(minFreq) = [1, m-1]$) and $minSize$ (with $D(minSize) = [1, n-1]$), which represent the minimal frequency and size, respectively.

As proposed in [3,4], we break symmetries (due to the fact that clusters may be swapped) by posting the constraint: $precede(G, [1, k_{max}])$.

We relate $extent_t$ and $extentCluster_c$ variables by posting the constraint: $\forall t \in \mathcal{T}, Extent[t] = ExtentCluster[G_t]$, and we relate $ExtentCluster_c$ and G_t variables by posting the constraint

$$\forall t \in \mathcal{T}, \forall c \in [1, k_{max}], t \in ExtentCluster_c \Leftrightarrow G_t = c$$

We add a redundant partition constraint that enforces $extent$ to be a partition of $\mathcal{T}$: $partition(ExtentCluster, \mathcal{T})$. This constraint is redundant because each transaction is already enforced to belong to exactly one cluster by G variables. However, its propagation both reduces the search space and the CPU time.

We reify $m(m-1)/2$ equality constraints between G variables to ensure that two transactions are in a same cluster iff they have the same intent, and this intent is included in their itemsets: $\forall \{t_1, t_2\} \subseteq \mathcal{T}$

$$(G_{t_1} = G_{t_2}) \Leftrightarrow (Intent_{t_1} = Intent_{t_2}) \Leftrightarrow (Intent_{t_1} \subseteq itemSet(t_2))$$

This constraint ensures the extent property as any transaction t_1 such that $itemset(t1) \supseteq Intent_{t2}$ is constrained to be in the same cluster as t_2. However, this constraint only partially ensures the intent property: for each transaction t, it ensures $Intent_t \subseteq \cap_{t' \in \mathcal{T}, G_t = G_{t'}} itemset(t')$ whereas the intent property requires that $Intent_t$ is equal to the itemset intersection. However, given any solution that satisfies the constraint $Intent_t \subseteq \cap_{t' \in \mathcal{T}, G_t = G_{t'}} itemset(t')$, we can easily transform it to ensure that it fully satisfies the intent property by adding to $Intent_t$ every item $i \in (\cap_{t' \in \mathcal{T}, G_t = G_{t'}} itemset(t')) \setminus Intent_t$. Hence, each time a solution is found, for each cluster c, we compute its actual intent by intersecting the intersection of all its transaction itemsets. This ensures that each cluster actually is a formal concept, and therefore this ensures correction. Completeness is ensured by the fact that our constraint is a relaxation of the initial constraint.

Finally, we compute the minimal size and frequency by posting the constraints: $minSize = \min_{t \in \mathcal{T}} \#Intent_t$ and $minFreq = \min_{t \in \mathcal{T}} \#Extent_t$.

The search strategy depends on the objective function: if the goal is to maximize $minFreq$, then we first assign k to its lower values (as a smaller number of clusters usually leads to concepts with larger frequencies), whereas if the goal is to maximize $minSize$, then we first assign k to its higher values (as a larger number of clusters usually leads to concepts with larger sizes).

3.2 New Hybrid Model

This model solves the problem in two steps as in [17]: in a first step, we extract the set $\mathcal{F}$ of all formal concepts with a dedicated tool (LCM), and in a second step we use CP to select the subset of $\mathcal{F}$ forming an optimal clustering.

We have designed and compared several CP models for this second step. In particular, we have designed a model that associates a set variable $Extent_c$ with each cluster c (such that $Extent_c$ contains all transactions in the extent of c), and then post a *partition* global constraint on these variables to ensure that they form a partition of $\mathcal{T}$. This model is always outperformed by the model described below.

Our CP model for the second step uses the following variables:

- an integer variable k (with $D(k) = [k_{min}, k_{max}]$), which represents the number of clusters (*i.e.*, the number of selected concepts);
- a set variable P (with $D(P) = \mathcal{P}(\mathcal{F})$), which represents the set of selected formal concepts that define an optimal clustering;
- for each transaction $t \in \mathcal{T}$, an integer variable $Concept_t$ (with $D(Concept_t) = \{f \in \mathcal{F} \mid t \in extent(f)\}$), which represents the concept that contains t in its extent (each transaction must belong to exactly one selected concept);
- two integer variables $minFreq$ (with $D(minFreq) = [1, m-1]$) and $minSize$ (with $D(minSize) = [1, n-1]$), which represent the minimal frequency and size, respectively.

To ensure that $Concept_t$ belongs to P, for each transaction $t \in \mathcal{T}$, we post the constraint: $\forall t \in \mathcal{T}, member(Concept_t, P)$.

To ensure that the selected concepts define a partition of $\mathcal{T}$, we ensure that each transaction t is contained in the extent of exactly one selected formal concept. To this aim, we compute, for each transaction t, the set $CF(t)$ of all the concepts of $\mathcal{F}$ that contain t in their extent, *i.e.*,

$$\forall t \in \mathcal{T}, CF(t) = \{f \in \mathcal{F} : t \in extent(f)\}$$

and we ensure that the set variable P contains exactly one element of $CF(t)$ by posting the constraint: $\forall t \in \mathcal{T}, \#(CF(t)) \cap P) = 1$.

Finally, we compute the minimal size and frequency by posting the constraints: $minSize = \min_{t \in \mathcal{T}} \#intent(C_t)$ and $minFreq = \min_{t \in \mathcal{T}} \#extent(C_t)$.

The number of clusters of the solution is constrained in two different ways according to the objective function:

- If the goal is to maximize $minFreq$, optimal solutions often have a small number of clusters and, in this case, we ensure that k is equal to the number of distinct values contained in C by posting the constraint: $nValue(C, k)$.
- If the goal is to maximize $minSize$, optimal solutions often have a large number of clusters and, in this case, we ensure that k is equal to the cardinality of P by posting the constraint: $\#P = k$.

The search strategy also depends on the objective function. The idea is to first select concepts with large sizes (resp. frequency) when the goal is to maximize $minSize$ (resp. $minFreq$). To this aim, we sort formal concepts by decreasing size (resp. frequency), and use this order as value ordering heuristic for C_t. We use a *First fail* strategy to select the variable with the smallest domain as next variable.

Furthermore, when the goal is to maximize $minFreq$, we use the *ObjectiveStrategy* proposed by Choco [19], which performs a dichotomous branching over the domain of $minFreq$.

4   Experimental Comparison for Single Objective Problems

We compare our new models with the CP model of [3] and the ILP model of [17] for computing conceptual clusterings that optimize a single objective.

Experimental Protocol. All experiments were conducted on Intel(R) Core(TM) i7-6700 with 3.40 GHz of CPU and 65 GB of RAM. The approach of [4] (called *FullCP1*) is implemented in Gecode v4.3. The approach of [17] (called *ILP*) uses LCM to extract formal concepts and Cplex v12.7 to solve the selection problem. Our CP model described in Sect. 3.1 (called *FullCP2*) is implemented in Choco v.4.0.3 [19]. Our hybrid approach described in Sect. 3.2 (called *HybridCP*) uses LCM v5.3 to extract formal concepts and Choco v.4.0.3 to solve the selection problem. We have limited the CPU time of each run to 1000 s.

Test Instances. We consider four classical machine learning instances, coming from the UCI database: zoo, vote, tic-tac-toe, and soybean. We also introduce seven new instances (called ERPi, with $i \in [1, 7]$) that have been extracted from our ERP database. Our ERP database contains 400 parameter settings: each setting corresponds to the customization of the ERP for a different client, and specifies the values of almost 450 different parameters. Each parameter can only take a finite number of values, and most of them are symbolic attributes. For each parameter/value couple, we have created a boolean item (set to 1 iff the parameter is assigned to the value in the setting). We have split this database into smaller ones by focusing on different groups of parameters, thus obtaining seven instances of various sizes[1]. All instances are described in Table 2.

Computation of $\mathcal{F}$ with LCM. Table 2 displays the time spent by LCM to extract all formal concepts, for each instance. This time is proportional with the size of $\mathcal{F}$, as the complexity of LCM is linear with respect to $\#\mathcal{F}$. CPU time is smaller than one second for all instances but two, and it is smaller than seven seconds for the instance that has the largest number of formal concepts (soybean).

---

[1] These instances are available on http://liris.cnrs.fr/csolnon/ERP.html.

Table 2. Test instances: each row gives the number of transactions ($\#\mathcal{T}$), the number of items ($\#\mathcal{I}$), the density (d), the number of formal concepts ($\#\mathcal{F}$), and the CPU time (in seconds) spent by LCM to extract all formal concepts.

Instance	$\#\mathcal{T}$	$\#\mathcal{I}$	$d(\%)$	$\#\mathcal{F}$	Time
ERP1	50	27	48	1 580	0.01
ERP2	47	47	58	8 1337	0.03
ERP3	75	36	51	10 835	0.03
ERP4	84	42	45	14 305	0.05
ERP5	94	53	51	63 633	0.28
ERP6	95	61	48	71 918	0.45
ERP7	160	66	45	728 537	5.31

Instance	$\#\mathcal{T}$	$\#\mathcal{I}$	$d(\%)$	$\#\mathcal{F}$	Time
zoo	59	36	44	4 567	0.01
vote	341	48	34	227 031	0.54
tic-tac-toe	958	27	33	42 711	0.05
soybean	303	116	29	817 534	6.7

Comparison of Scale-Up Properties of the Different Approaches. Table 3 displays the times for computing optimal clusterings when the goal is to maximize *minFreq* (upper part) or *minSize* (lower part). We report times when k is fixed to 2, 3, and 4, respectively, and then when k is not fixed (with $k_{min} = 2$ and $k_{max} = m - 1$). Times displayed for ILP and HybridCP both include the time spent by LCM to extract all formal concepts.

Table 3. Times when the goal is to optimize *minFreq* (upper part) and *minSize* (lower part): each line gives the time of the four approaches when k is fixed to 2, 3, and 4, respectively, and when k is not fixed (N). '-' is reported when time exceeds 1000 s.

	ILP				FullCP1				FullCP2				HybridCP			
	$k=2$	$k=3$	$k=4$	N	$k=2$	$k=3$	$k=4$	N	$k=2$	$k=3$	$k=4$	N	$k=2$	$k=3$	$k=4$	N
Objective = Maximize *minFreq*																
ERP1	0.2	0.9	1.0	0.8	0.0	0.0	0.3	0.2	0.2	0.7	4.3	0.3	0.2	0.9	1.4	0.3
ERP2	1.5	2.7	**2.3**	1.0	0.0	0.4	4.8	0.5	0.1	**0.2**	2.8	**0.1**	4.4	1.5	4.6	0.3
ERP3	1.5	2.5	3.2	1.7	0.0	0.3	20.0	2.4	0.2	1.5	**1.6**	0.3	9.2	24.7	2.4	0.6
ERP4	7.5	15.0	**20.9**	13.6	0.0	0.3	36.6	1.2	0.3	2.8	37.9	**0.4**	1.4	100.6	153.1	0.8
ERP5	12.5	18.3	**83.7**	18.3	0.0	1.4	773.6	125.3	0.5	5.0	91.9	**1.5**	172.2	634.4	-	10.6
ERP6	52.6	145.8	339.6	143.3	0.0	10.3	302.7	51.7	0.5	**2.7**	**101.1**	**1.1**	8.6	-	-	8.0
ERP7	-	-	-	-	0.0	82.9	-	973.4	2.8	**26.8**	**742.9**	**5.0**	-	-	-	-
Zoo	1.0	2.2	3.0	1.5	0.0	0.0	0.8	0.1	0.2	0.2	4.5	0.3	0.5	0.6	1.0	0.2
Vote	40.6	-	-	55.2	0.0	2.0	292.6	-	1.6	19.2	370.5	33.1	17.8	150.0	**95.4**	**20.8**
Tic-tac-toe	61.3	80.6	-	718.6	0.2	0.3	106.0	-	32.5	75.9	-	179.7	10.9	25.2	-	**33.3**
Soybean	-	-	-	0.1	160.1	-	-	1.4	**7.9**	**166.0**	-	63.7	980.2	-	-
Objective = Maximize *minSize*																
ERP1	0.2	0.3	**0.3**	0.4	0.0	0.1	1.0	0.2	0.3	0.7	2.5	0.2	0.2	0.4	1.6	**0.1**
ERP2	1.7	1.6	1.6	0.8	0.0	0.5	19.9	-	0.1	**0.2**	**0.8**	**0.0**	4.6	17.7	7.2	0.1
ERP3	1.6	1.6	**1.7**	1.2	0.0	0.6	252.9	-	0.3	2.0	7.0	**0.1**	9.6	42.5	61.6	0.2
ERP4	7.5	8.3	**7.2**	18.3	0.0	0.8	184.8	2.1	0.5	4.6	34.4	0.5	22.0	103.4	329.5	**0.3**
ERP5	13.1	21.1	**40.6**	12.5	0.0	2.2	-	-	0.6	6.1	58.4	**0.3**	-	-	-	1.5
ERP6	63.4	93.3	648.3	-	0.0	14.2	**9.6**	7.2	0.8	**7.5**	54.2	**0.5**	645.0	-	-	1.9
ERP7	-	-	-	-	0.0	191.1	-	47.2	4.4	69.2	**682.5**	**2.3**	-	-	-	39.5
Zoo	1.1	0.9	**1.2**	2.0	0.0	0.0	1.4	0.5	0.2	1.7	7.7	**0.1**	0.7	6.8	9.4	**0.1**
Vote	40.8	243.5	249.7	-	0.0	3.9	969.4	-	3.3	12.2	**191.8**	20.4	16.2	69.0	-	**17.2**
Tic-tac-toe	60.7	80.4	-	254.5	0.4	0.3	**105.6**	-	33.2	54.1	-	-	10.9	25.9	-	**18.7**
Soybean	-	-	-	-	0.0	145.7	-	-	2.5	**7.1**	**93.3**	22.2	93.4	460.6	-	342.4

For all approaches and all instances, time increases when increasing k from 2 to 4. FullCP1 and FullCP2 are more efficient when $k = 2$ than when k is not fixed, and they are more efficient when k is not fixed than when $k = 4$. HybridCP approach needs more time to solve the problem when $k = 2$ than when k is not fixed for all instances but three (vote, tic-tac-toe and soybean) whereas ILP approach needs more time only for the smallest ERP instances.

When $k = 2$, the best approach is FullCP1, which is able to solve all instances in less than 0.1 s (except tic-tac-toe). However, when increasing k from 2 to 3, times of FullCP1 are strongly increased (up to 191 s for ERP7 with $minSize$) and FullCP1 is outperformed by FullCP2 for 8 instances. When further increasing k to 4, FullCP2 becomes the only approach able to solve all instances but tic-tac-toe, though ILP is able to solve 8 instances quicker.

When k is not fixed, the best performing approaches are fullCP2 (which is able to solve all instances but soybean for $minFreq$) and HybridCP (which is able to solve all instances for $minSize$).

Maximization of the Sum of Sizes. In [17], the objective function to maximize is the sum of the sizes of the selected concepts, and the proposed ILP model scales well for this objective: when k is not fixed, the time needed to find the optimal solution is 0.1, 0.4, 0.7, 1.7, 5.9, 14.0, and 183.2 for ERP1 to ERP7, respectively, and it is 0.3, 51.9, 32.0, and 120.0 for zoo, vote, tic-tac-toe, and soybean, respectively. Hence, ILP is more efficient for maximizing the sum of the sizes than for maximizing $minSize$. None of the CP models considered here scales well when the goal is to maximize the sum of the sizes, and they are far slower than ILP in this case: they usually very quickly find the optimal solution, but they are not efficient to prove optimality. However, we noticed that the optimal solutions found with the two criteria sum of sizes and $minSize$ are very similar (and often equal). Indeed, when maximizing the minimal size, we also tend to maximize the size of all concepts.

Comparison of Frequencies, Sizes, and Number of Clusters of Optimal Solutions. Table 4 displays the values of $minFreq$, $minSize$, and k for the optimal solutions (tic-tac-toe does not have clusterings when $k \in \{2, 4\}$, and soybean does not have clusterings when $k = 2$). It shows us that when k is fixed, the optimal values of $minFreq$ and $minSize$ often greatly vary when modifying the value of k. For example, let us consider instance ERP4: $minFreq$ decreases from 42 to 27 and 18 (resp. $minSize$ increases from 3 to 6 and 8) when k is increased from 2 to 3 and 4. From an applicative point of view, finding the relevant value for k is not straightforward. When k is not fixed, we obtain extreme solutions: when the goal is to maximize $minFreq$, there are only 2 clusters (except for tic-tac-toe, as there is no solution with $k = 2$), and when the goal is to maximize $minSize$, there are $m - 1$ clusters for all instances but 4 (ERP2, ERP3, ERP5 and vote), whereas for the remaining instances the value of k is rather high.

Table 4 also displays the values of $minFreq$, $minSize$, and k when optimizing the two criteria in a lexicographic order and not fixing k. Let us call Freq+Size the solution that maximizes $MinFreq$ while breaking ties with $MinSize$, and

Table 4. Experimental comparison of frequencies, sizes, and number of clusters: each line displays the optimal values of *minFreq* and *minSize* when k is set to 2, 3, and 4, and when k is not fixed (N), followed by the value of k in the optimal solution (in brackets). Finally, it displays the optimal values of *minFreq* and *minSize* when maximizing *minFreq* and breaking ties with *minSize* (Freq+Size), and when maximizing *minSize* and breaking ties with *minFreq* (Size+Freq). In this case, k is not fixed and its value in the optimal solution is displayed in brackets.

k	Maximize *minFreq*					Maximize *minSize*					Freq+Size			Size+Freq		
	2	3	4	N	(k)	2	3	4	N	(k)	Freq	Size	(k)	Freq	Size	(k)
ERP1	21	14	11	21	(2)	4	5	6	12	(49)	21	4	(2)	1	12	(49)
ERP2	21	15	11	21	(2)	6	11	13	16	(42)	21	6	(2)	2	16	(8)
ERP3	31	22	18	31	(2)	3	4	6	12	(59)	31	2	(2)	1	12	(59)
ERP4	42	27	18	42	(2)	3	6	8	18	(83)	42	3	(2)	1	18	(83)
ERP5	41	30	22	41	(2)	3	6	7	16	(79)	41	2	(2)	1	16	(79)
ERP6	42	31	22	42	(2)	8	11	9	28	(94)	42	8	(2)	1	28	(94)
ERP7	70	50	35	70	(2)	5	8	11	29	(159)	70	5	(2)	1	29	(159)
Zoo	28	19	14	28	(2)	2	3	5	15	(58)	28	2	(2)	1	15	(58)
Vote	102	34	34	102	(2)	1	1	1	15	(317)	102	1	(2)	1	5	(317)
Tic-tac-toe		250		250	(3)	1			7	(957)	250	1	(3)	-	—	—
Soybean		2	5	—	-		1	1	6	(302)	—	-	-	1	6	(302)

Size+Freq the solution that maximizes *MinSize* while breaking ties with *Min-Freq*. These solutions correspond to very different situations: for Freq+Size, k is always equal to 2 (except for tic-tac-toe) and *minSize* is rather low (ranging between 1 and 8); for Size+Freq, k is equal to $m - 1$ for 6 instances, and rather large for the other instances, while *minFreq* is always equal to 1, except for ERP2. From an applicative point of view, these solutions are not very interesting, and we need to find better compromises between size and frequency.

5   Multi-criteria Optimization

As the two optimization criteria tend to produce extreme solutions which are not very meaningful for our application, we propose to compute the Pareto front of non dominated solutions. A clustering C_1 is dominated by another clustering C_2 if the size and the frequency of C_1 are smaller than or equal to the size and the frequency of C_2. Non dominated solutions correspond to different compromises between the two criteria. The two extrema of the Pareto front are the solutions called Size+Freq and Freq+Size in the previous Section. In Sect. 5.1, we experimentally evaluate the efficiency of our two CP models for computing these extrema. Then, in Sect. 5.2, we propose and compare different approaches for computing the whole Pareto front.

5.1   Computation of Extrema Solutions

Table 5 displays the time spent by FullCP2 and HybridCP to compute the two extrema solutions of the Pareto front: Size+Freq is computed by first maximizing *MinSize*, fixing *MinSize* to its optimal value, and then maximizing *MinFreq*; Freq+Size is obtained by first maximizing *MinFreq*, fixing *MinFreq* to its optimal value, and then maximizing *MinSize*.

Table 5. Times needed to compute Size+Freq (left part) and Freq+Size (rightpart) for FullCP2 and HybridCP. For each solution, we first give the time needed to optimize the first criterion (1st), then the time needed to optimize the second criterion (2nd), and finally the total time ('-' if total time exceeds 1000 s).

Instance	Size+Freq						Freq+Size					
	FullCP2			HybridCP			FullCP2			HybridCP		
	1st	2nd	Total	1st	2nd	Total	1st	2nd	Total	1st	2nd	Total
ERP1	0.1	0.2	**0.3**	0.3	0.3	0.6	0.4	0.2	0.7	0.2	0.1	**0.4**
ERP2	-	-	-	0.1	0.3	**0.4**	0.2	0.1	**0.3**	0.3	0.3	0.6
ERP3	-	-	-	0.2	0.4	**0.6**	0.4	0.3	**0.7**	0.7	0.3	1.0
ERP4	0.4	6.9	7.3	0.6	1.5	**2.1**	0.5	0.5	1.0	0.8	0.1	**0.9**
ERP5	-	-	-	1.6	34.7	**36.3**	1.1	0.6	**1.6**	10.6	13.0	23.5
ERP6	0.5	1.4	**1.9**	4.0	94.2	98.2	1.6	0.7	**2.3**	8.0	59.8	67.9
ERP7	3.7	975.1	**978.8**	-	-	-	6.3	2.9	**9.2**	—	—	-
Zoo	0.1	1.0	1.1	0.2	0.2	**0.4**	0.4	0.3	0.7	0.3	0.0	**0.3**
Vote	20.0	6.2	**26.2**	18.3	648.7	667.0	26.5	13.5	**40.0**	—	—	-
Tic-tac-toe	-	-	-	-	-	-	230.9	152.1	383.0	31.6	19.1	**50.7**
Soybean	-	-	-	325.0	342.1	**667.1**	—	—	-	—	—	-

For Freq+Size, FullCP2 outperforms HybridCP on 6 instances and it is able to solve all instances except soybean while HybridCP is not able to solve ERP7, vote and soybean. However, for Size+Freq, FullCP2 is not able to solve 5 instances, while HybridCP is able to solve all instances but 2.

5.2   Computation of the Pareto Front

[3] describes a CP approach to compute non dominated bi-criteria clusterings by iteratively solving single criterion optimization problems while alternating between the two criteria. [6] describes a more dynamic CP approach to compute Pareto front: the idea is to search for all solutions, and dynamically add a constraint each time a new solution is found to prevent the search from computing solutions that are dominated by it. This idea has been improved in [10, 22]. We have experimentally compared these two approaches, and found that the dynamic approach of [6] is more efficient than the static approach of [3] for our problem. Hence, we only consider this approach in this section. It proceeds as

follows: we build an initial model as described in Sect. 3, and ask the solver to search for all solutions. Each time a solution *sol* is found, we dynamically post the constraint $(minFreq > f) \lor (minSize > s)$ where f and s are the values of *minFreq* and *minSize* in *sol*, and go on the search for all solutions. The search stops when there is no more non-dominated solutions.

We have evaluated this dynamic approach with FullCP2 and HybridCP. FullCP2 is able to solve ERP1 in ten minutes, but fails to solve all other instances within a time limit of two hours. HybridCP is much more efficient, and is able to solve 6 instances within this time limit. Hence, we only consider HybridCP in our experiments.

We have compared different variants of this dynamic approach. For all variants, we use the two extrema solutions to reduce the search space in an *a priori* way. More precisely, let $f_{\text{Freq+Size}}$ and $s_{\text{Freq+Size}}$ (resp. $f_{\text{Size+Freq}}$ and $s_{\text{Size+Freq}}$) be the value of *minFreq* and *minSize* in the solution Size+Freq (resp. Freq+Size). We set the domain of *minFreq* to $[f_{\text{Size+Freq}} + 1, f_{\text{Freq+Size}} - 1]$ and the domain of *minSize* to $[s_{\text{Freq+Size}} + 1, s_{\text{Size+Freq}} - 1]$.

The first two variants correspond to the dynamic approach described below, with different search heuristics: $freq_{seq}$ (resp. $size_{seq}$) uses the search heuristics dedicated to the *minFreq* (resp. *minSize*) objective as described in Sect. 3.2. These two variants find complementary solutions at the beginning of the search process: $freq_{seq}$ first finds clusterings with large frequencies, whereas $size_{seq}$ first finds clusterings with large sizes. Hence, the variant $freqSize_{par}$ takes advantage of this complementarity and launches the two variants in two parallel threads which communicate their solutions to update the non-dominated area: each time a solution is found by one thread, it dynamically adds constraints to filter the solutions dominated by this solution, and it also checks whether the other thread has found new solutions and dynamically adds constraints if ever.

The variant $freq+D_{seq}$ (resp. $size+D_{seq}$) decomposes the problem into two subproblems by separating the domain of *minFreq* (resp. *minSize*) in two equal parts. The two subproblems are solved sequentially. We first solve the subproblem corresponding to the upper part of the domain, as no solution of this problem can be dominated by a solution of the other subproblem. Then, we solve the second subproblem while preventing it from computing solutions that are dominated by the solutions of the first subproblem by adding constraints. The variants $freq+D_{par}$ and $size+D_{par}$ are similar to $freq+D_{seq}$ and $size+D_{seq}$: the only difference is that they solve the two subproblems in two parallel threads. In this case, only one subproblem (the one with the upper part of the domain) communicates its solutions to the other thread.

Table 6 compares times of these different variants on 6 instances with a time limit of 2 h (all other instances cannot be solved in less than 2 h). For all instances, $size_{seq}$ is much more efficient than $freq_{seq}$. Launching these two approaches in two parallel threads does not pay off: $freqSize_{par}$ is faster than $size_{seq}$ for only two instances. This may come from the fact that $freq_{seq}$ is really not efficient compared to $size_{seq}$. Decomposing the problem into two subproblems appears to be a better idea, even when solving the two subproblems sequentially, for the *freq*-based variants: $freq{+}D_{seq}$ is always much faster than $freq_{seq}$. However, $size{+}D_{seq}$ is faster than $size_{seq}$ for two instances only. Finally, solving the two subproblems on two parallel threads always pays-off for the *freq*-based variants, whereas it degrades the solving time for 4 instances for the *size*-based variants. Figure 1 displays the Pareto fronts for ERP4 and ERP5.

Table 6. Times to find all non-dominated solutions with different variants ('-' if time exceeds 2 h). $\frac{seq}{par}$ gives the speed-up between sequential and parallel variants (for $freqSize_{par}$, we consider the best sequential time).

	ERP1		Zoo		ERP4		ERP3		ERP2		ERP5	
	Time	$\frac{seq}{par}$	Time	$\frac{seq}{par}$	Time	$\frac{seq}{par}$	Time	$\frac{seq}{par}$	Time	$\frac{seq}{par}$	time	$\frac{seq}{par}$
$freq_{seq}$	2.3		24.4		6048.7		145.7		41.7		5542.5	
$size_{seq}$	1.4		8.6		211.8		42.3		**11.1**		**873.9**	
$freqSize_{par}$	1.4	1.0	7.0	1.2	216.5	1.0	36.0	1.2	12.7	0.9	1029.9	0.8
$freq{+}D_{seq}$	**0.8**		4.7		236.8		15.0		12.6		1579.6	
$freq{+}D_{par}$	**0.8**	1.0	**4.5**	1.0	210.0	1.1	**10.2**	1.5	11.8	1.1	1006.9	1.6
$size{+}D_{seq}$	**0.8**		10.6		264.3		24.8		12.4		2912.9	
$size{+}D_{par}$	1.1	0.7	15.0	0.7	**166.0**	1.6	59.1	0.4	24.4	0.5	2131.6	1.4

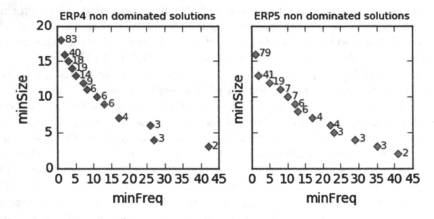

Fig. 1. Pareto front of ERP4 and ERP5: each point (x, y) corresponds to a non-dominated solution with $x = minFreq$ and $y = minSize$. The number k of clusters of the solution is displayed close to the point.

6   Conclusion

We have introduced new CP models for computing optimal conceptual clusterings. These models are able to quickly find solutions that maximize either the minimal size or the minimal frequency, even when the number of clusters is not fixed. Computing the Pareto front for these two criteria is a more challenging problem, and our CP models are able to solve this problem in less than two hours for six instances only. Further work will mainly aim at improving this. In particular, we plan to combine different decompositions to obtain more than two subproblems, *e.g.*, both decompose the domains of *minFreq* and *minSize* to obtain four subproblems that may be solved in four parallel threads. We also plan to evaluate scale-up properties of ILP for this problem, and combine ILP with CP if we observe complementary performance. Finally, we plan to evaluate the interest of combining our CP model with the propagation algorithm of [14].

Acknowledgments. We thank Jean-Guillaume Fages and Charles Prud'homme for enriching discussions on Choco, and authors of [4] for sending us their Gecode code.

References

1. Ahmad, M.M., Cuenca, R.P.: Critical success factors for ERP implementation in SMEs. Robot. Comput.-Integr. Manuf. **29**(3), 104–111 (2013)
2. Börzsönyi, S., Kossmann, D., Stocker, K.: The skyline operator. In: Proceedings of the 17th IEEE International Conference on Data Engineering, pp. 421–430 (2001)
3. Dao, T.B.H., Duong, K.C., Vrain, C.: Constrained clustering by constraint programming. Artif. Intell. **244**, 70–94 (2015)
4. Dao, T.B.H., Lesaint, W., Vrain, C.: Clustering conceptuel et relationnel en programmation par contraintes. In: JFPC 2015, Bordeaux, France, June 2015
5. Ganter, B., Wille, R.: Formal Concept Analysis: Mathematical Foundations. Springer, Heidelberg (1997). doi:10.1007/978-3-642-59830-2
6. Gavanelli, M.: An algorithm for multi-criteria optimization in CSPs. In: Proceedings of the 15th European Conference on Artificial Intelligence, ECAI 2002, Amsterdam, The Netherlands, pp. 136–140. IOS Press (2002)
7. Guns, T.: Declarative pattern mining using constraint programming. Constraints **20**(4), 492–493 (2015)
8. Guns, T., Nijssen, S., De Raedt, L.: Itemset mining: a constraint programming perspective. Artif. Intell. **175**(12–13), 1951–1983 (2011)
9. Guns, T., Nijssen, S., De Raedt, L.: k-Pattern set mining under constraints. IEEE Trans. Knowl. Data Eng. **25**(2), 402–418 (2013)
10. Hartert, R., Schaus, P.: A support-based algorithm for the bi-objective pareto constraint. In: Proceedings of the Twenty-Eighth AAAI Conference on Artificial Intelligence, AAAI 2014, pp. 2674–2679. AAAI Press (2014)
11. Hossain, L.: Enterprise Resource Planning: Global Opportunities and Challenges. IRM Press, Hershey (2001)
12. Khiari, M., Boizumault, P., Crémilleux, B.: Constraint programming for mining n-ary patterns. In: Cohen, D. (ed.) CP 2010. LNCS, vol. 6308, pp. 552–567. Springer, Heidelberg (2010). doi:10.1007/978-3-642-15396-9_44
13. Law, Y.C., Lee, J.H.M.: Global constraints for integer and set value precedence. In: Wallace, M. (ed.) CP 2004. LNCS, vol. 3258, pp. 362–376. Springer, Heidelberg (2004). doi:10.1007/978-3-540-30201-8_28

14. Lazaar, N., Lebbah, Y., Loudni, S., Maamar, M., Lemière, V., Bessiere, C., Boizumault, P.: A global constraint for closed frequent pattern mining. In: Rueher, M. (ed.) CP 2016. LNCS, vol. 9892, pp. 333–349. Springer, Cham (2016). doi:10.1007/978-3-319-44953-1_22

15. Michalski, R.S.: Knowledge acquisition through conceptual clustering: a theoretical framework and an algorithm for partitioning data into conjunctive concepts. Report, Department of Computer Science, University of Illinois at Urbana-Champaign (1980)

16. Motwani, J., Subramanian, R., Gopalakrishna, P.: Critical factors for successful ERP implementation: exploratory findings from four case studies. Comput. Ind. **56**(6), 529–544 (2005)

17. Ouali, A., Loudni, S., Lebbah, Y., Boizumault, P., Zimmermann, A., Loukil, L.: Efficiently finding conceptual clustering models with integer linear programming. In: Proceedings of the Twenty-Fifth International Joint Conference on Artificial Intelligence, IJCAI 2016, New York, NY, USA, 9–15 July 2016, pp. 647–654 (2016)

18. Pasquier, N., Bastide, Y., Taouil, R., Lakhal, L.: Discovering frequent closed itemsets for association rules. In: Beeri, C., Buneman, P. (eds.) ICDT 1999. LNCS, vol. 1540, pp. 398–416. Springer, Heidelberg (1999). doi:10.1007/3-540-49257-7_25

19. Prud'homme, C., Fages, J.-G., Lorca, X.: Choco Documentation. TASC, INRIA Rennes, LINA CNRS UMR 6241, COSLING S.A.S. (2016)

20. De Raedt, L., Guns, T., Nijssen, S.: Constraint programming for itemset mining. In: Proceedings of the 14th ACM SIGKDD International Conference on Knowledge Discovery and Data Mining, Las Vegas, Nevada, USA, 24–27 August 2008, pp. 204–212 (2008)

21. Robert, L., Davis, A.R., McLeod, A.: ERP configuration: does situation awareness impact team performance? In: 2011 44th Hawaii International Conference on System Sciences (HICSS 2011), pp. 1–8 (2011)

22. Schaus, P., Hartert, R.: Multi-objective large neighborhood search. In: Schulte, C. (ed.) CP 2013. LNCS, vol. 8124, pp. 611–627. Springer, Heidelberg (2013). doi:10.1007/978-3-642-40627-0_46

23. Ugarte, W., Boizumault, P., Crémilleux, B., Lepailleur, A., Loudni, S., Plantevit, M., Raïssi, C., Soulet, A.: Skypattern mining: from pattern condensed representations to dynamic constraint satisfaction problems. Artif. Intell. **244**, 48–69 (2017)

24. Uno, T., Asai, T., Uchida, Y., Arimura, H.: An efficient algorithm for enumerating closed patterns in transaction databases. In: Suzuki, E., Arikawa, S. (eds.) DS 2004. LNCS, vol. 3245, pp. 16–31. Springer, Heidelberg (2004). doi:10.1007/978-3-540-30214-8_2

A Declarative Approach to Constrained Community Detection

Mohadeseh Ganji[1,2($\boxtimes$)], James Bailey[1], and Peter J. Stuckey[1,2]

[1] Department of Computing and Information Systems,
University of Melbourne, Melbourne, Australia
sghasempour@student.unimelb.edu.au, {baileyj,pstuckey}@unimelb.edu.au
[2] Data61, CSIRO, Canberra, Australia

Abstract. Community detection in the presence of prior information or preferences on solution properties is called semi-supervised or constrained community detection. The task of embedding such existing kinds of knowledge effectively within a community discovery algorithm is challenging. Indeed existing approaches are not flexible enough to incorporate a variety of background information types. This paper provides a framework for semi-supervised community detection based on constraint programming modelling technology for simultaneously modelling different objective functions such as modularity and a comprehensive range of constraint types including community level, instance level, definition based and complex logic constraints. An advantage of the proposed framework is that, using appropriate solvers, optimality can be established for the solutions found. Experiments on real and benchmark data sets show strong performance and flexibility for our proposed framework.

1 Introduction

Community detection is the task of identifying densely connected sub-graphs in networks. Although most research on community detection has focused on unsupervised learning, which only relies on network topology, there is emerging interest in semi-supervised or constrained community detection which benefits from existing side information as well. This can result in more efficient and actionable solutions.

More generally, the increasing flexibility of data mining techniques to deal with complex constraints has attracted increasing attention. The topic of constrained-based mining aims to develop data mining techniques that can handle complex and domain-specified constraints. This has been shown to be possible for some data mining tasks such as pattern and sequence mining, item set mining and constrained clustering using constraint solving technology (e.g. [14,18,27]).

There are two main motivations for constrained (semi-supervised) community detection:

Quality Solutions: The community detection process can benefit from prior information to improve the quality of solutions. For example, the supervision

© Springer International Publishing AG 2017
J.C. Beck (Ed.): CP 2017, LNCS 10416, pp. 477–494, 2017.
DOI: 10.1007/978-3-319-66158-2_31

effect has been studied in the presence of noisy links in the network and it has been shown that semi-supervised community detection approaches are usually more robust to noise than topology-based approaches [15].

Figure 1 illustrates the effect of supervision on the quality of solutions of two different community detection problems. Figure 1a is a network of a clique of size 50 connected to two small cliques of size 5. The left figure shows how pure modularity maximization merges the two small cliques to one community while adding a supervision constraint to force the number of communities to be 3 leads to more meaningful communities (right figure). Figure 1b shows a circle structure of 30 cliques of size 5. The communities found using the modularity criterion (left figure) are unconvincing, since adjacent cliques are grouped into one community. Imposing a size constraint of between 5 to 9 on this community detection problem reveals the intuitively correct communities (right figure).

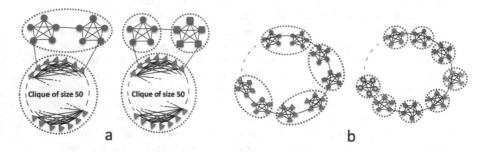

Fig. 1. The effect of adding (a) a bound on the number of communities and (b) bound on size of communities. Each figure shows partitions before (left) and after (right) applying the constraints.

Complex Problem Solving: Constrained community detection is the only way to tackle some challenging problems, in which different types of constraints must be satisfied at the same time. Constraints arise by the imposition of user preferences on community properties, or natural complexity of some problems with a variety of requirements to be satisfied simultaneously. As an example, consider finding groupings for a class of students engaged on a project. One may desire to balance the number of female and male students in each group. To make teamwork smoother, we may also require that everyone has several of his Facebook friends or classmates in the group. These are examples of user and ad hoc constraints to help detect communities with desired properties.

Some attempts have been made to adapt existing community detection algorithms to incorporate background knowledge [2,15]. However, they are limited in the types of prior information that can be used. This lack of flexibility narrows the scope of problems they can tackle. For example, some algorithms can only incorporate pairwise instance level supervision [2,15] while others are only capable of finding size-constrained communities [11]. When more than one type of supervision exists for a community detection task, *it is not clear how to integrate the results of such different algorithms.* In other words, none of the existing approaches and algorithms has the flexibility to be able to solve complex problems by incorporating both classic and ad-hoc types of supervision and user-defined constraints at the same time. In this paper, *we propose a generic constraint programming framework for the constrained community detection task with the flexibility to be able to capture a wide variety of possible supervision types.*

Constraint programming (CP) [35] is a paradigm for modelling solving combinatorial optimization problems where relations between variables are represented by constraints. One of the strengths of the CP approach is that the constraints can be arbitrary. CP provides state of the art solutions to many industrial scheduling and routing problems, and has been successfully used for constrained data mining problems [18]. Constraint programming modelling technology has the power to express different sorts of logical and mathematical constraints, which provides flexibility in modelling a variety of classic and ad-hoc user defined constraints for constrained community detection. We can use constraint programming modelling technology without committing to a particular solving methodology. If we use complete solving methods, it has the advantage of being able to prove optimality of a solution. But we can also use incomplete solving methods.

Many community detection problems are of large size and cannot currently be solved to optimality using complete methods in reasonable time. However, finding optimal solutions to small constrained community detection problems can give us better insight about the task itself and the characteristics of optimal communities. In addition, there are some small data problems which are naturally very complex and sufficiently important to justify the resources required to find an optimal solution. When dealing with bigger problems, existing CP systems can find feasible solutions to complex community detection problems even without hope of proving optimality. Complete solver technology can often find good feasible solutions in a reasonable time, and continue to improve them given more time. Using incomplete solver technology can often generate better solutions in less time, although we give up the possibility of proving optimality. But in any case there is no competing approach we are aware of for tackling complex constrained community detection problems.

A summary of our contributions in this paper are:

- We examine new types of community level constraints based on community definitions and complex logic constraints to dynamically capture properties of interest during the community detection process.
- We show how to model constrained community detection problems with modularity maximization or other objective functions using constraint programming models. This allows the expression of instance level, community level, definition based and other ad hoc and complex logic constraints.
- We demonstrate via experiments that the framework, using modern complete solving technology, can effectively solve smaller scale, complex problems to

optimality. And it is the only approach to solve very complex real world constrained community detection problems.
- We demonstrate we can use the same models, using incomplete solving technology to scale to larger problems, although we may give up the possibility of satisfiying all constraints.

2   Background

Given a graph $G = (V, E)$ of vertices V and edges E, many community detection algorithms optimize a criterion such as *modularity* [30] to find the best communities. Two vertices v_1 and v_2 are in the same community if $x(v_1) = x(v_2)$.

The modularity value of a partition is given by Eq. (1) where W is the modularity matrix which quantifies the deviation of the network from randomness: $W_{ij} = (A_{ij} - \frac{d_i \times d_j}{2|E|})$ where A_{ij} is 1 if $(i, j) \in E$ and 0 otherwise and d_i is the degree of node $i \in V$. The modularity of a partition is the summation of the modularities between pairs of the same community.

$$Q = \frac{1}{2|E|} \sum_{i,j} W_{ij}(x(i) = x(j))$$ (1)

It has been shown that finding a partition with maximum modularity is an NP-hard problem [7]. One of the main heuristics for modularity maximization was proposed by Blondel *et al.* [6]. It is a greedy hierarchical algorithm which merges communities in each phase to improve the partition's modularity value and continues till no more improvement is possible. Aloise *et al.* [3] introduced a column generation based exact method for finding communities by modularity maximization which can solve small size problems to optimality.

Rather than optimizing a criterion, in other schemes, any partition satisfying some conditions is a solution to the community detection task. Two of such conditional definitions were proposed by Radicchi *et al.* [34], termed "communities in the strong and weak sense." Let $nbh(v) = \{v' \mid (v, v') \in E\}$ be the neighbours of v in G. Given a community mapping x, the *in-degree of a vertex* v, $in(v)$ is the number of neighbours in the same community, i.e. $in(v) = |nbh(v) \cap \{v' \in V | x(v) = x(v')\}|$. Similarly the *out-degree of a vertex* v, $out(v)$ is the number of neighbours in a different community, i.e. $out(v) = |nbh(v) \cap \{v' \in V | x(v) \neq x(v')\}|$. A sub-graph S is a *strong community* if and only if each vertex in S has more in-degree than out-degree:

$$\forall v \in S \quad in(v) > out(v)$$ (2)

A sub-graph S is a *community in the weak sense* if and only if the sum of the internal degrees of the community is larger than the sum of its external degrees:

$$\sum_{v \in S} in(v) > \sum_{v \in S} out(v)$$ (3)

Later on, Hu *et al.* [19] introduced a comparative definition of community or semi-strong community. Sub-graph S is a *community in the semi-strong sense* if and only if all its vertices have more neighbours within the community than the maximum number of neighbours in any other community, where m is the number of communities.

$$\forall v \in S \quad in(v) > \max_{\substack{t=1,\ldots,m \\ t \neq x(v)}} |nbh(v) \cap \{v' \in V \mid x(v') = t\}| \tag{4}$$

Similar to the above mentioned criteria, Cafieri *et al.* [8] defined another relaxed version of strong community called communities in the almost strong sense and they designed a heuristic algorithm based on a set of rules to find such communities in networks. Sub-graph S is a *community in the almost strong sense* if and only if each of its vertices of degree other than two shares more edges within the sub-graph S than with the rest of the network.

$$\begin{aligned} \forall v \in S | |nbh(v)| \neq 2 \quad & in(v) > out(v) \\ \forall v \in S | |nbh(v)| = 2 \quad & in(v) > 0 \end{aligned} \tag{5}$$

Among different supervision types for constrained community detection, size-constrained community detection has been studied by Ciglan and Nørvåg [11]. They proposed a greedy algorithm based on label propagation to find sized constrained communities based on the semi-strong community definition.

Background knowledge can also be represented as known labels and pairwise constraints which model whether a pair of vertices must lie within the same community (*must-link or ML*) or lie within different communities (*cannot-link or CL*). Allahverdyan *et al.* [2] studied the problem of community detection in networks where community assignments for a fraction of vertices are known in advance. They designed a so called *planted bisection graph model* and investigated the effect of such supervision scheme on detectability threshold of communities.

Eaton and Mansbach [15] proposed a spin-glass model for incorporating pairwise constraints in a modularity maximization scheme. Their model penalizes partitions violating the guidance by adding/subtracting a fixed term to modularity value of pairs involved in must-link/cannot-link constraints.

Cafieri *et al.* [9] extended the column generation model in [3] for modularity maximization to incorporate cohesion constraints as in general, it is recognized that communities found by modularity maximization do not necessarily satisfy variations of strong community conditions (cohesions). However, their column generation algorithm doesn't incorporate other constraint types.

Although there is a lack of flexibility in encoding different supervision types in constrained community detection approaches, there are some studies in constrained clustering schemes to address the flexibility and exact solving technology. Babaki *et al.* [4] incorporated pairwise constraints in an exact column generation scheme for minimum-sum-of-square constrained clustering with pairwise constraints. Berg and Järvisalo [5] proposed a MAXSAT approach for constrained correlation clustering. Davidson *et al.* [12] proposed a SAT based approach and Duong *et al.* [13] used constraint programming to encode several instance and cluster level constraints in clustering problems.

However, in spite of the high level similarity of clustering and community detection tasks, their optimization criteria are often different i.e. modularity vs sum-of-squared-distances. In addition, clustering methods rely on measures of distance between two points, rather than measures based on the network configuration. Hence community detection typically involves very different constraints which are not applicable in clustering schemes, for example, community definition based constraints, such as strong/weak community constraints. Minimal distance and max diameter constraints which are important for constrained clustering do not usually make sense for community detection. Much of the other focus of work on constrained clustering, e.g. within cluster sum of dissimilarities, is often not applicable to communities since there is no standard notion of dissimilarity. In addition, capturing complex community level constraints requires a dual viewpoint to the partitioning problem which makes the modelling of constrained community detection different to constrained clustering.

This paper addresses the existing gap in the literature for constrained community detection to propose a flexible and generic CP based framework to handle a variety of constraint types at the same time.

3   Preliminaries

Constraint programming [35] is an effective and generic paradigm to address and solve constraint satisfaction problems (CSP), or constraint optimisation problems (COP). A CSP $P = (X, D, C)$ consists of a set of variables X, a finite *domain* D for each variable $x \in X$ that defines the possible values that it can take, and a set of *constraints* C. A COP is a CSP together with an objective function f which maps a solution θ on X to an objective value $f(\theta)$. The aim is then to find a solution that maximizes (or w.l.o.g. minimizes) the objective function.

The strength of constraint programming arises from the ability to combine arbitrary different constraints in the same model. This naturally gives rise to very expressive modelling. Traditional complete CP solvers are able to tackle this heterogeneous constraint solving problem by using propagators to infer information from individual constraints, communicating through shared variable domains.

Because of this approach the community has developed many *global constraints* which define important combinatorial substructures that reoccur in many problems, and algorithms to propagate them. An example is the constraint `alldifferent`$([x_1, ..., x_n])$ which requires all the variables $x_1, ..., x_n$ to be pairwise distinct. Global constraints have a custom propagator able to exploit the semantics of constraints. This leads to more efficient solving than if one would decompose that constraint as the conjunction of several simple logical or mathematical constraints.

The existence of global constraints further enriches the modelling capabilities for CP, as we try to understand discrete optimization problems as a combination of combinatorial substructures. It is this rich modelling approach that will allow us to express constrained community detection problems succinctly.

We will make use of a few global constraints in capturing community detection problems.

The `global_cardinality_low_up` global constraints is a generalization of `alldifferent` constraint. The `global_cardinality_low_up`($[x_1, \ldots, x_n], [d_1, \ldots, d_m], [l_1, \ldots, l_m], [u_1, \ldots, u_m]$) requires each value d_j is assigned to at least l_j and at most u_j of the variables $x_1, \ldots x_n$ for each $1 \leq j \leq m$.

The `value_precede_chain`($[d_1, \ldots, d_m], [x_1, \ldots, x_n]$) requires d_i precedes d_{i+1} in $[x_1, \ldots, x_n]$ for each $1 \leq i \leq m - 1$. This global constraint is very useful in value symmetry breaking of CP models and avoiding multiple symmetric representations of the same solution.

4   Constraint Based Community Detection

We now show how we can model constrained community detection problems using constraint programming. We study four main categories of constraints including instance level, community level, definition based and complex logic constraints which have not been simultaneously applied for constrained community detection before.

The input to our CP model is the network's number of vertices (n) and, if needed, the modularity matrix (or some variations, e.g. the generalized modularity matrix [16]) denoted by W which is used for building the objective function. We also assume a maximum number of communities parameter m, which by default can be n, and a description of the adjacency relation either as an adjacency matrix, A, or the neighbourhood function nbh where $nbh(v)$ are the vertices adjacent to v. Without loss of generality, we assume vertices are indexed from 1 to n and we refer to them by their index.

4.1   Decision Variables

The critical decisions of the problem are for each vertex which community it belongs to. A one dimensional array x represents the communities to which each (index) vertex belongs. The length of x is equal to the number of vertices n. The domains of x are 1..m where m is the maximum number of communities possible (in the worst case it can be n). While using $|x|$ variables is the most natural way to model the problem (it directly encodes the community mapping) and enables us to use efficient global constraints, it has limitations in modelling some of the community level constraints.

There is a dual viewpoint for the problem. For each community, describe which vertices are contained in that community. We denote this by an array of sets of vertices S indexed from 1 to m. Since we have multiple viewpoints concerning the same decisions, we need to connect them via channeling constraints [10], as follows:

$$\forall j \in 1..m, \quad \forall i \in V \quad i \in S[j] \Leftrightarrow x[i] = j \tag{6}$$

Note that if the dual viewpoint is not needed to express any constraints, then the S variables and the channelling constraints (6) can be omitted.

4.2 Constraints on Representation

In the solution represented by an array x, more than one representation exists for a unique solution. e.g. $x = [1, 1, 2, 2, 2]$ and $x = [2, 2, 1, 1, 1]$. This is called *value symmetry* [33] which can dramatically effect the ability of complete solvers to find solutions, and in particular to prove optimality of solutions. For any solution with k communities there are $k!$ symmetric solutions by permuting the community numbers. To avoid these situations we use the global constraint value_precede_chain($[i | i \in 1..m], x$) which ensures that no community i can have a vertex j unless all communities $1...i - 1$ have at least one lower numbered vertex (less than j) as a member. This constraint enforces a unique community numbering for any particular partition. It can be viewed as a lexicographic ordering constraint on the assignment of vertices to communities. This value symmetry removal is essential for efficiency of the complete solution methods. Note that the addition of the symmetry breaking constraint is typically counter productive for incomplete solving methods, since it creates an artificial constraint that they must satisfy. It is omitted (rewritten away by preprocessing) when we use incomplete solving methods.

4.3 Objective

The modularity objective is defined as follows:

$$OBJ = \sum_{i,j \in V} (x[i] = x[j]) * W[i, j]$$

One of the advantages of the CP system is the ability to encode logical expressions. The expression $(x[i] = x[j])$ will evaluate to 1 if the expression is true or 0 otherwise.

Although the primary objective function studied in this paper is modularity maximization, the CP framework can encode a variety of other complex arbitrary objective functions. For example, minimizing the differences in sizes of the communities is encoded as minimizing the variable OBJ:

$$OBJ = \max_{t \in 1..m} |S[t]| - \min_{t \in 1..m} |S[t]|$$

4.4 Modelling Instance Level Supervision

Instance level supervision is usually represented by pairwise must-link and cannot-link constraints which is given to the model by a set of pairs of indices denoted by ML (CL). Similar to [14], the pairwise constraints for constrained community detection then can be encoded as follows:

$$\forall (m_1, m_2) \in ML \qquad x[m_1] = x[m_2]$$
$$\forall (c_1, c_2) \in CL \qquad x[c_1] \neq x[c_2]$$

4.5 Modelling Community Level Supervision

The CP framework enables incorporating a vast range of community level constraint types. For instance:

– Maximum number of communities: This is implicit in the representation given by the integer m.
– Minimum number of communities: We can enforce that the first l communities are non-empty by simply

$$\forall i \in 1..l \qquad |S[i]| > 1$$

We can do the same without using the dual viewpoint using the global cardinality constraint

$$\texttt{global\_cardinality\_low\_up}(x, [i|i \in 1..l], [1|i \in 1..l], [n|i \in 1..l])$$

– Minimum and maximum community size:

$$\forall i \in 1..m \quad (|S[i]| \geq minsize) \wedge (|S[i]| \leq maxsize)$$

Again we can avoid the dual viewpoint using global cardinality constraints. We can also set just a minimum or maximum by using a trivial bound for the other end of the range (0, or n).

– Minimum community size and unknown number of communities: For this combination of constraints we need to make use of the dual viewpoint.

$$\forall i \in 1..m \quad |S[i]| \geq minsize \vee |S[i]| = 0$$

– Minimum separation between communities: this is defined based on the maximum number of edges between communities which can be set to be less than a predefined threshold T as follows.

$$\forall l, l' \in 1..m \text{ where } l < l' \qquad \sum_{i \in S[l], j \in S[l']} A[i,j] < T$$

– Distribution of different tags in communities. This is an example of user preferences in networks with known tags. For example, consider a scientific collaboration network in which each vertex has a tag: student (St), faculty (F) or research staff (R). The university is interested in the collaboration communities where the ratio of students to faculty is less than p in each group.

$$\forall l \in 1..m \sum_{i \in S[l]} (i \in St) < p \times \sum_{i \in S[l]} (i \in F)$$

4.6   Modelling Definition Based Constraints

We can encode various community definitions as constraints to the CP model. In this case, the CP model will find the partitions satisfying the local community definitions with the maximum possible modularity (objective) value. This flexibility enables us to benefit from both categories of definitions. Below we model some community definition constraints for our CP framework (recall the definitions from Sect. 2).

- Communities in the strong sense (Eq. 2):

$$\forall i \in V \sum_{j \in nbh[i]} (x[i] = x[j]) > |nbh[i]|/2$$

- Communities in the weak sense (Eq. 3):

$$\forall t \in 1..m \sum_{i,j \in S[t]} A[i,j] > \sum_{i \in S[t], j \in nbh[i]} 1 - (j \in S[t])$$

- Communities in the semi-strong sense (Eq. 4):

$$\forall i \in V, \quad \forall t \in 1..m \quad t \neq x[i] \longrightarrow$$
$$\sum_{j \in nbh[i]} (x[i] = x[j]) > \sum_{j \in nbh[i]} (j \in S[t])$$

- Communities in the almost-strong sense (Eq. 5):

$$\forall i \in V \ |nbh[i]| > 2 \quad \rightarrow \quad \sum_{j \in nbh[i]} (x[i] = x[j]) > \sum_{j \in nbh[i]} (x[i] \neq x[j])$$
$$\forall i \in V \ |nbh[i]| = 2 \quad \rightarrow \quad \sum_{j \in nbh[i]} (x[i] = x[j]) > 0$$

The CP framework can provide further flexibility. Using implication one can require a proportion of the network to follow a community definition based on other constraints.

4.7   Modelling Complex Logic Constraints

Unlike any other existing approaches for semi-supervised community detection, CP modelling technology can enable encoding complex logic supervision such as conjunction, disjunctions, negation and implication of any instance level, community level and definition based constraints. For example, the constraint "when instance i belongs to a community, the size of that community should be bounded by α and β" can be modelled as follows.

$$\forall t \in 1..m \quad (i \in S[t]) \rightarrow (|S[t]| < \beta) \wedge (|S[t]| > \alpha)$$

The above constraint is just an illustration that any logic constraint can be captured by a CP modelling framework. In practice, complex logic constraints may arise in different applications in real world problems. For example, in power grid networks, complex requirements may have to be imposed to implement strategies for network reliability and improving the network behaviour in cascading events [32].

5   Experimental Results

In this section, we present experiments on real and benchmark data sets to evaluate the performance and flexibility of the proposed framework. The constrained community detection problems are written in Minizinc 2.0.12 [28]. All the experiments are performed using the Gecode [37] or OSCAR CBLS [31] CP solver with a timeout of one hour on a Macbook with 8 GB RAM and 2.7 GHz Intel Core i5.

The main questions we address in this section are:

Q1: How the solutions found by the CP framework compare with other constrained community detection methods?
Q2: Can the quality of the solutions be improved by adding community level supervision to instance level constraints?
Q3: Can the modelling framework enable us to simultaneously encode different constraint types on a complex real problem?
Q4: How scalable can approaches based on constraint programming modelling be?

5.1   Comparison to Other Methods

The proposed modelling framework can encode a variety of objectives as well as classic and arbitrarily complex instance and community level constraints at the same time while there is no other approach in the literature with such ability for semi-supervised community detection. However, to address the question Q1 and compare the proposed modelling framework to an state-of-the-art algorithm in constrained community detection, we limit the supervision type to only ML and CL pairwise constraints and set the objective to modularity maximization. There exist some approximate approaches for incorporating pairwise constraints in modularity optimization scheme [15,24,39]. Here we compare the proposed framework with the spin-glass model [15], discussed in Sect. 2, because it is based on modularity and it is shown to perform better than some other approaches [15]. To implement this approach, we set the parameters according to [15] and used GenLouvain algorithm [20] for optimizing the spin-glass model.

Since we have the ground truth of the data sets, for evaluating quality of the solutions, we use the Normalized Mutual Information (NMI) measure (Eq. 7) proposed by Danon $et\ al.$ [22].

$$I_{norm}(A, B) = \frac{-2 \sum_{i=1}^{CA} \sum_{j=1}^{CB} N_{ij} \log(N_{ij} N / N_{i.} N_{.j})}{\sum_{i=1}^{CA} N_{i.} \log(N_{i.}/N) + \sum_{j=1}^{CB} N_{.j} \log(N_{.j}/N)} \qquad (7)$$

In Eq. (7), A represents the real communities and B represents the detected communities while CA and CB are the number of communities in A and B respectively. In this formula, N is the confusion matrix with rows representing the original communities and columns representing the detected communities. The value of N_{ij} is the number of common vertices that are in the original community i but found in community j. The sum over the ith row is denoted by $N_{i.}$ and the sum over the jth column is denoted by $N_{.j}$.

For each data set listed in Table 1, in the second column section we give the size, number of constraints, the ground truth number of communities k and maximum number of communities use in the CP model m. For each data set, we generated 5 different sets of random constraints (equally divided to ML and CL) based on the ground truth. To have more rigorous comparison, we executed the spin-glass model 50 times on each data set and reported the NMI correspond to the best solution (the highest number of constraints satisfied and modularity score) in the third column section of Table 1. In addition, we also report the run time and number of constraint violations for the spin-glass method. The average NMI and runtime of the complete solver Gecode on the model are reported in the fourth column of Table 1. In the model, the maximum number of communities m are set according to the solution of the corresponding unconstrained modularity maximization problem. Note that the solutions found by the CP framework satisfy all of the constraints, hence the number of constraint violations is zero. The P-value corresponding to Friedman statistical test is reported in the last line of Table 1. The null hypothesis of this test is two algorithms have no significant difference in their performance. This hypothesis is rejected based on the very small p-values, indicating that Gecode applied to the model statistically significantly outperforms the spin-glass method in solution quality.

Using Gecode on the CP model finds high quality solutions while it is often very fast as well. In addition, Gecode can prove optimality in reasonable time for solutions of smaller problems. For bigger problems sizes such as Political blogs, Gecode could not prove optimality within one hour. Still Gecode could find a better solution (NMI = 0.88) than the spin-glass solution (NMI = 0.59) in 17 s and it kept searching for better solutions within the timeout. This shows the

Table 1. Comparison to other methods and effect of community level constraints

Data	n	#const	k	m	Spin-glass (pairwise)			CP (pairwise)		CP (pairwise + community-level)		
					NMI	Time	#viol	NMI	Time	NMI	Time	Supervision type
Sampson [36]	25	24	2	4	0.82	<1	0.6	0.88	<1	0.96	<1	Weak
Strike [26]	24	24	3	5	0.72	<1	1.5	0.78	<1	0.81	<1	# of community
Zachary [38]	34	34	2	4	0.85	<1	0.19	0.87	<1	0.9	<1	Weak
Mexican [17]	35	34	2	4	0.32	<1	0.44	0.45	60	0.53	35	Weak
Dolphin [23]	62	124	2	5	0.81	<1	4.76	0.95	<1	0.98	<1	Almost strong
Adjacent words [29]	112	224	2	4	0.05	<1	51.92	0.52	73	0.8	1.7	Cardinality
Political books [21]	105	210	3	5	0.67	<1	18.7	0.94	15	0.95	2.5	Cardinality
Political blogs [1]	1490	2980	2	4	0.59	1.2	145.6	0.88	3600(17)	0.97	142(14)	# of community
p-value					0.0047			Baseline		0.0047		

ability of the CP modelling framework to produce promising results even in big problem instances in reasonable time where proving optimality is not possible.

5.2   Effect of Community Level Supervision

To address the second question, we add community level constraints to pairwise ML and CL constraints for the data sets of Table 1 agreeing with their ground truth. The results of adding community level constraints including communities in weak and almost-strong sense (Eqs. 3 and 5), number of communities and size (cardinality) constraints are shown in the last column section of Table 1. For the cardinality constraint, we generated 10 sets of random samples from the ground truth and set the minimum and maximum size of the communities according to the observed number of samples from each community.

Comparing the last two column sections of Table 1 shows that adding community level supervision can lead the community detection process to more accurate solutions, while often enhancing the runtime of the optimization algorithm. For instance, in the Political Blogs dataset, adding the number of community constraints leads to a significant improvement in runtime and solution quality comparing to considering just pairwise constraints. The significancy of the results are verified using Friedman statistical test and the very low p-value.

5.3   Case Study

To address question Q3, we consider contacts and friendship relations between students in a high school in Marseilles, France, in December 2013. Students were asked to record their contacts with other students in a diary and also list their friends at school. The Facebook network is available for a subset of these students. Gender, student ID and the teaching class of each student is also reported. We consider the network of 81 students whose Facebook and friendship information is known [25].

We consider a hypothetical scenario in which the School Principal wants to group the students based on their Facebook communications network, while at the same time, desiring certain properties to hold based on students' friendships and contacts information. For example, students who had more than 1 h contact during the data collection period, must be assigned to the same community and students who declared friendship and had contact during the data collection period also must be in the same community. These kinds of properties can be captured by a set of must-link constraints. 50 must-link constraint were extracted based on the above mentioned requirements on individual student assignments. For balancing group populations, suppose the School Principal wants each community to have 15–25 students and having 4 groups of students is desired. The School Principal also wants to balance the gender distribution in groups by requiring the difference in number of male and female students to be less than or equal to 5. To make students feel more comfortable, it is also required that each student has at least two other students from their original class for the new group they are assigned into.

Table 2. Community profiling of the case study

Methods	Size	Gender	Class distribution
Spin glass	39	F = 29, M = 10	4, 7, 25, 2, 1
	38	F = 15, M = 23	3, 1, 20, 5, 9
	4	F = 3, M = 1	4
CP just ML	40	F = 23, M = 17	3, 2, 12, 18, 5
	30	F = 20, M = 10	1, 5, 13, 3, 6, 1, 1
	11	F = 4, M = 7	2, 1, 8
CP final	25	F = 12, M = 13	3, 17, 5
	25	F = 15, M = 10	4, 4, 14, 3
	15	F = 10, M = 5	3, 3, 9
	16	F = 10, M = 6	3, 5, 5, 3

There is no existing community detection algorithm capable of automatically incorporating these complex instance and community level constraints at the same time. For example, the spin-glass model [15] can only incorporate must-link and cannot-link constraints. The best solution found by the spin-glass model from 1000 executions is shown in first row section of Table 2 satisfying all the must-link constraints. The size column shows the cardinality of each community and the next column shows the number of females and males in each community respectively. The class distribution column shows the number of classmates based on original class labels for each of the new communities. Based on this information, it is clear that the solution found by the spin-glass model violates all the other constraints.

Real world problems such as this example often include a large number of varying requirements on communities which often cannot be satisfied by existing approaches. However, our modelling framework can easily deal with various complex constraints at the same time. Within the timeout of one hour, a solution of the model just considering must-link constraints (to compare with the spin-glass solution), using the complete solver Gecode, and a full solution also incorporating size, number of community, gender and class distribution constraints are shown in the second and third rows of Table 2. As shown in Table 2, the solution found using Gecode satisfies all the constraints required by the School Principal.

5.4   Scalability

One of the advantages of using a solver-independent modelling framework is that we do not commit to a particular solving technology. While complete solving methods are effective on constrained supervision problems which are not too large, by their nature they do not scale as well as incomplete methods. We can use an incomplete solver to tackle the same model. Indeed since we use

MiniZinc [28] we can send the same model to both solvers.[1] Incomplete solvers are typically more scalable, but may struggle to satisfy all the constraints of the problems.

We consider solving the problem using the Oscar CBLS solver [31]. On the smaller examples of Table 1 Gecode is uniformly better than Oscar, in both time to solve and NMI of resulting solution, but as the size of the graphs grow Oscar becomes quicker to find solutions.

Figure 2 shows the best modularity value of the solution over time using Gecode, Oscar and the spin-glass on the Political Blogs example. The spin-glass method repeatedly run keeping either the lexicographic best solution (modularity, number of constraints satisfied) or the lexicographic best solution in the other order. Clearly the spin-glass concentrates on modularity maximization, and never satisfies all constraints (never finds a feasible solution), and reaches much higher modularity values.

Figure 3 shows the plot of the same solutions, but here ranked on NMI value. Clearly the spin glass method never achieves a good NMI, but tracking solutions that satisfy more constraints is preferable for NMI. Oscar quickly gets a good solution to the problem with high NMI and gradually improves. Interesting Gecode actually finds the best solution in terms of NMI as its first solution, then gradually degrades as it optimizes the modularity objective. The first solution found by Gecode requires 24 s, while the first solution found by Oscar requires only 3.5 s.

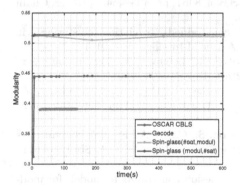

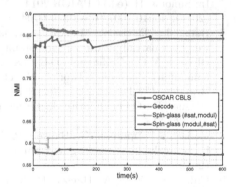

Fig. 2. Modularity over time for Political Blogs data set using Gecode, Oscar and spin-glass methods

Fig. 3. NMI over time for Political Blogs data set using Gecode, Oscar and spinglass methods

---

[1] Note that we also tried running MIP solvers on the models, but they were non-competitive, which is unsurprising since the linear relaxation of these problems is very weak.

6 Conclusion

The challenging problem of constrained community detection with a variety of constraint types and objective functions has been explored in this paper. We proposed a generic framework based on constraint programming modelling approach, which enables including a variety of instance and community level, definition based and complex logic supervision types as constraints. Our models are able to prove optimality of the solutions, when using complete solving methods, and in our experiments we have shown it can work with real networks and complex problems. An obvious direction for future work is to consider specialized propagators for community definitions, for example a stronger propagator for globally strong communities seems clearly possible.

References

1. Adamic, L.A., Glance, N.: The political blogosphere and the 2004 US election: divided they blog. In: Proceedings of Link Discovery, pp. 36–43. ACM (2005)
2. Allahverdyan, A.E., Ver Steeg, G., Galstyan, A.: Community detection with and without prior information. Europhys. Lett. **90**(1), 18002 (2010)
3. Aloise, D., Cafieri, S., Caporossi, G., Hansen, P., Perron, S., Liberti, L.: Column generation algorithms for exact modularity maximization in networks. Phys. Rev. E **82**(4), 046112 (2010)
4. Babaki, B., Guns, T., Nijssen, S.: Constrained clustering using column generation. In: Simonis, H. (ed.) CPAIOR 2014. LNCS, vol. 8451, pp. 438–454. Springer, Cham (2014). doi:10.1007/978-3-319-07046-9_31
5. Berg, J., Järvisalo, M.: Cost-optimal constrained correlation clustering via weighted partial maximum satisfiability. Artificial Intelligence (2015)
6. Blondel, V.D., Guillaume, J.-L., Lambiotte, R., Lefebvre, E.: Fast unfolding of communities in large networks. JSTAT **2008**(10), P10008 (2008)
7. Brandes, U., Delling, D., Gaertler, M., Gorke, R., Hoefer, M., Nikoloski, Z., Wagner, D.: On modularity clustering. IEEE Trans. Knowl. Data Eng. **20**(2), 172–188 (2008)
8. Cafieri, S., Caporossi, G., Hansen, P., Perron, S., Costa, A.: Finding communities in networks in the strong and almost-strong sense. Phys. Rev. E **85**(4), 046113 (2012)
9. Cafieri, S., Costa, A., Hansen, P.: Adding cohesion constraints to models for modularity maximization in networks. J. Complex Netw. **3**(3), 388–410 (2015)
10. Choi, C.W., Lee, J.H.M., Stuckey, P.J.: Removing propagation redundant constraints in redundant modeling. ACM Trans. Comput. Log. **8**(4), 23 (2007)
11. Ciglan, M., Nørvåg, K.: Fast detection of size-constrained communities in large networks. In: Chen, L., Triantafillou, P., Suel, T. (eds.) WISE 2010. LNCS, vol. 6488, pp. 91–104. Springer, Heidelberg (2010). doi:10.1007/978-3-642-17616-6_10
12. Davidson, I., Ravi, S.S., Shamis, L.: A SAT-based framework for efficient constrained clustering. In: SIAM Data Mining, pp. 94–105 (2010)
13. Dao, T.-B.-H., Duong, K.-C., Vrain, C.: A declarative framework for constrained clustering. In: Blockeel, H., Kersting, K., Nijssen, S., Železný, F. (eds.) ECML PKDD 2013. LNCS, vol. 8190, pp. 419–434. Springer, Heidelberg (2013). doi:10.1007/978-3-642-40994-3_27

14. Duong, K.-C., Vrain, C., et al.: Constrained clustering by constraint programming. Artificial Intelligence (2015)
15. Eaton, E., Mansbach, R.: A spin-glass model for semi-supervised community detection. In: AAAI, Citeseer (2012)
16. Ganji, M., Seifi, A., Alizadeh, H., Bailey, J., Stuckey, P.J.: Generalized modularity for community detection. In: Appice, A., Rodrigues, P.P., Santos Costa, V., Gama, J., Jorge, A., Soares, C. (eds.) ECML PKDD 2015. LNCS, vol. 9285, pp. 655–670. Springer, Cham (2015). doi:10.1007/978-3-319-23525-7_40
17. Gil-Mendieta, J., Schmidt, S.: The political network in Mexico. Soc. Netw. **18**(4), 355–381 (1996)
18. Guns, T., Dries, A., Nijssen, S., Tack, G., De Raedt, L.: MiningZinc: a declarative framework for constraint-based mining. Artif. Intell. **244**, 6–29 (2017)
19. Hu, Y., Chen, H., Zhang, P., Li, M., Di, Z., Fan, Y.: Comparative definition of community and corresponding identifying algorithm. Phys. Rev. E **78**(2), 026121 (2008)
20. Jutla, I.S., Jeub, L.G.S., Much, P.J.: A generalized Louvain method for community detection implemented in MATLAB (2012). http://netwiki.amath.unc.edu/GenLouvain
21. Krebs, V.: www.orgnet.com/
22. Daz-Guilera, A., Danon, L., Arenas, A.: The effect of size heterogeneity on community identification in complex networks. JSTAT **2006**, P11010 (2006)
23. Lusseau, D., Schneider, K., Boisseau, O.J., Haase, P., Slooten, E., Dawson, S.M.: The bottlenose dolphin community of doubtful sound features a large proportion of long-lasting associations. Behav. Ecol. Sociobiol. **54**(4), 396–405 (2003)
24. Ma, X., Gao, L., Yong, X., Lidong, F.: Semi-supervised clustering algorithm for community structure detection in complex networks. Phys. A: Stat. Mech. Appl. **389**(1), 187–197 (2010)
25. Mastrandrea, R., Fournet, J., Barrat, A.: Contact patterns in a high school: a comparison between data collected using wearable sensors, contact diaries and friendship surveys. PloS ONE **10**(9), e0136497 (2015)
26. Michael, J.H.: Labor dispute reconciliation in a forest products manufacturing facility. Forest Prod. J. **47**(11/12), 41 (1997)
27. Negrevergne, B., Guns, T.: Constraint-based sequence mining using constraint programming. In: Michel, L. (ed.) CPAIOR 2015. LNCS, vol. 9075, pp. 288–305. Springer, Cham (2015). doi:10.1007/978-3-319-18008-3_20
28. Nethercote, N., Stuckey, P.J., Becket, R., Brand, S., Duck, G.J., Tack, G.: MiniZinc: towards a standard CP modelling language. In: Bessière, C. (ed.) CP 2007. LNCS, vol. 4741, pp. 529–543. Springer, Heidelberg (2007). doi:10.1007/978-3-540-74970-7_38
29. Newman, M.E.J.: Finding community structure in networks using the eigenvectors of matrices. Phys. Rev. E **74**(3), 036104 (2006)
30. Newman, M.E.J., Girvan, M.: Finding and evaluating community structure in networks. Phys. Rev. E **69**(2), 026113 (2004)
31. OscaR Team: OscaR: Scala in OR (2012). https://bitbucket.org/oscarlib/oscar
32. Pahwa, S., Hodges, A., Scoglio, C., Wood, S.: Topological analysis of the power grid and mitigation strategies against cascading failures. In: 2010 4th Annual IEEE on Systems Conference, pp. 272–276. IEEE (2010)
33. Puget, J.-F.: Symmetry breaking revisited. In: Van Hentenryck, P. (ed.) CP 2002. LNCS, vol. 2470, pp. 446–461. Springer, Heidelberg (2002). doi:10.1007/3-540-46135-3_30

34. Radicchi, F., Castellano, C., Cecconi, F., Loreto, V., Parisi, D.: Defining and identifying communities in networks. Proc. Nat. Acad. Sci. **101**(9), 2658–2663 (2004)
35. Rossi, F., van Beek, P., Walsh, T.: Handbook of CP. Elsevier, Amsterdam (2006)
36. Sampson, S.F.: A novitiate in a period of change: an experimental and case study of social relationships. Cornell University (1968)
37. Schulte, C., et al.: Gecode (2016). http://www.gecode.org/
38. Zachary, W.W.: An information flow model for conflict and fission in small groups. J. Anthropol. Res. **33**, 452–473 (1977)
39. Zhang, Z.-Y.: Community structure detection in complex networks with partial background information. EPL (Europhys. Lett.) **101**(4), 48005 (2013)

Combining Stochastic Constraint Optimization and Probabilistic Programming
From Knowledge Compilation to Constraint Solving

Anna L.D. Latour[1]([✉]), Behrouz Babaki[2], Anton Dries[2], Angelika Kimmig[2], Guy Van den Broeck[3], and Siegfried Nijssen[4]([✉])

[1] LIACS, Leiden University, Leiden, The Netherlands
a.l.d.latour@liacs.leidenuniv.nl
[2] Department of Computer Science, KU Leuven, Leuven, Belgium
[3] Computer Science Department, UCLA, Los Angeles, USA
[4] ICTEAM, Université Catholique de Louvain, Louvain-la-Neuve, Belgium
siegfried.nijssen@uclouvain.be

Abstract. We show that a number of problems in Artificial Intelligence can be seen as Stochastic Constraint Optimization Problems (SCOPs): problems that have both a stochastic and a constraint optimization component. We argue that these problems can be modeled in a new language, SC-ProbLog, that combines a generic Probabilistic Logic Programming (PLP) language, ProbLog, with stochastic constraint optimization. We propose a toolchain for effectively solving these SC-ProbLog programs, which consists of two stages. In the first stage, decision diagrams are compiled for the underlying distributions. These diagrams are converted into models that are solved using Mixed Integer Programming or Constraint Programming solvers in the second stage. We show that, to yield linear constraints, decision diagrams need to be compiled in a specific form. We introduce a new method for compiling small Sentential Decision Diagrams in this form. We evaluate the effectiveness of several variations of this toolchain on test cases in viral marketing and bioinformatics.

1 Introduction

Two important areas in Artificial Intelligence are those of *probabilistic reasoning* and *constraint optimization*. *Constraint optimization* problems involve finding the best assignment to given variables satisfying constraints on these variables. The best-known *probabilistic inference* problems are arguably those that involve calculating the marginal conditional probability $P(X \mid Y)$ for given sets of variable assignments X and Y in a probability distribution.

In recent years it has become increasingly clear that these areas are closely related to each other. For example: calculating $P(X \mid Y)$ can be understood as

A.L.D. Latour wishes to thank KU Leuven, since the inspiration for this work came during a research visit to its Computer Science Department. She also wishes to thank Université catholique de Louvain, since the work itself was done during a research visit to its ICTEAM institute.

weighted model counting, i.e., calculating a weighted sum over all assignments to variables that satisfy constraints [9]. Similarly, maximum a posteriori (MAP) inference, the problem of computing the most likely assignment to given variables in a distribution, can be seen as a constraint optimization task [23]. Optimization problems over distributions are closely linked to constraint optimization problems under soft constraints [4]. *Mixed networks* essentially combine probabilistic graphical models and constraint networks [18].

One combination of constraint programming (CP) and probabilistic inference is the focus of this paper: *stochastic constraint programming* (SCP) [27], which is closely related to *chance constraint programming* [8] and *probabilistic constraint programming* [25]. The key idea in SCP is to introduce *stochastic constraints* and *stochastic optimization criteria* in CP. An example of a stochastic constraint is that the probability of the occurrence of an event should not exceed a threshold.

Three key limitations of the state of the art of SCP are the basis for this work. First: most publications on SCP are focused on specific types of problems: scheduling and planning problems, typically (see [1,17] for some recent examples). Second: there is no generic language for modeling Stochastic Constraint Optimization Problems (SCOPs). Third: there is no automatic toolchain for solving SCOPs written in such a modeling language. The aim of this work is to advance the state of the art in SCP on these dimensions.

We will use two motivating examples to illustrate that SCP is not only useful in planning and scheduling, but also in data mining and bioinformatics:

Viral marketing [16]. We are given a social network of individuals whose trust relationships are probabilistic: the behaviour of one person inspires each of their friends to do the same with a certain probability. We have budget to distribute marketing material to k nodes in this network. Which people do we target for marketing to such that we (indirectly) influence the largest expected number of people?

Signaling-regulatory pathway inference [21]. We are given a network of genes, proteins and their interactions, where the interactions are probabilistic. Furthermore, we are given knock-out pairs: pairs of nodes for which positive or negative change in the expression level of one node is observed when the other node is knocked out. Paths of interactions can explain the positive or negative effect of one node on another. In order to better understand these interactions, we want to extract the part of the network that best explains the positive effect (theory compression [13]). We ask: which interactions should we select such that in the resulting extracted network the expected number of positive effects is maximized, but the expected number of negative effects is limited by a constant?

Clearly, these problems also involve a combination of constraint optimization and probabilistic reasoning. They can be considered instances of SCP, as they involve finding an assignment to discrete variables, such that a probabilistic optimization criterion is maximized and a probabilistic constraint is satisfied.

A specific property of these problems is however that the decision problem is specified over a very different type of distribution than common in existing

SCP systems: *probabilistic networks*, i.e., networks in which edges exist with a certain probability. To the best of our knowledge, no tools currently exist that are sufficiently general that they allow for modeling and solving these SCOPs. The second aim of this paper is to introduce a system that can be used to model and solve these SCOPs, and potentially many other SCOPs. As common in CP, our system consists of two components: a modeling and a solving component.

For the modeling component we propose to exploit the fact that in recent years, significant progress has been made in the development of *probabilistic programming languages*[1]. These languages allow programmers to program distributions. Until now, however, they have rarely been linked to constraint programming. In this paper, we expand a probabilistic programming language, ProbLog [14], which is especially suited for programming distributions over probabilistic networks, such that it can be used to formalize SCOPs as well; we call the resulting language SC-ProbLog (Stochastic Constraint Probabilistic Logic Programming). This extension of ProbLog builds on an earlier version of ProbLog for solving *decision-theoretic* problems (DT-ProbLog) [26]; compared to DT-ProbLog, SC-ProbLog adds support for hard constraints.

For the solving component we propose to build a toolchain on technology that is taken both from the probabilistic reasoning and constraint programming literature. For the probabilistic reasoning component, we focus on the compilation of *Sentential Decision Diagrams* (SDDs) [12], as they are known to lead to smaller representations of distributions than for instance *Ordered Binary Decision Diagrams* (OBDDs) [7]. We use these SDDs to generate arithmetic circuits (ACs) and formalize deterministic constraints based on these ACs. For constraint solving we use both CP solvers and Mixed Integer Programming (MIP) solvers. A key technical contribution of this paper is that we show that SDDs need to satisfy strict criteria in order for them to yield linear representations of probabilistic constraints. We introduce a new algorithm for minimizing SDDs within this normal form. This allows us to reduce the size of the resulting ACs.

This paper is organized as follows. First, we introduce the range of SCOPs that are the focus of this work, showing by example how problems can be modeled in the proposed SC-ProbLog language. In Sect. 3 we provide background on how probabilities are defined and calculated in ProbLog, which is necessary to understand the first stage of our proposed method. In Sect. 4 we describe our method: we introduce the aforementioned normal form and our new SDD minimization algorithm. Experiments are presented in Sect. 5.

2   Modeling Problems in SC-ProbLog: An Example

As common in (one-stage) SCP [27], we assume given two types of variables: *decision variables* (denoted as d_i) and mutually independent *stochastic variables* (denoted as t_i). The aim is to find an assignment to the decision variables, such that stochastic constraints and optimization criteria are satisfied.

---

[1] See http://probabilistic-programming.org/ for a recent list of systems.

Constraints and optimization criteria are considered to be stochastic if their definition involves the use of stochastic variables.

We consider a limited choice of constraints and variables in this work. First, we restrict our attention to problems in which all variables take Boolean values. As a consequence, each stochastic variable is independently true or false with a given a probability. Second, we only consider constraints of the following kind:

$$\sum_i r_i v_i \leq \theta \text{ and/or } \sum_i r_i v_i \geq \theta, \tag{1}$$

where v_i represents either a decision variable d_i or the conditional probability $P_i(\varphi_i \mid \sigma_i)$ that a stochastic Boolean formula φ_i evaluates to true given an assignment to decision variables σ_i. We let $r_i \in \mathbb{R}$ be a reward for decision variable d_i or formula φ_i evaluating to true, and let θ be a constant threshold. This constraint can be thought of as expressing a bound on *expected utilities*: we sum rewards for events, each of which could happen with a certain probability, given an assignment to the decision variables. Whether an event happens in a certain situation, is expressed using a Boolean logical formula φ_i that includes the stochastic variables; hence the formula φ_i is only true with a certain probability.

For reasons of simplicity, we limit ourselves in this paper to the case that $r_i = 1$, although it is trivial to extend our approach to settings in which $r_i \neq 1$. Optimization criteria are of a similar linear form.

The viral marketing problem [16] is an example of a SCOP in this class of SCOPs. We illustrate this on the network of Fig. 1. The nodes represent people; they are either targeted directly in a marketing campaign or not (the decisions). The (undirected) edges represent probabilities that one person trusts another, and vice versa. These probabilities are indicated by variables such as p_{ab} on the edges of the graph. We formalize this problem as a SCOP as follows:

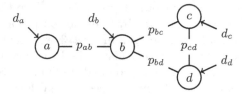

Fig. 1. A social network with a viral marketing problem superimposed on it. Nodes are people, undirected edges indicate trust relationships, where the probability that person i and person j trust each other is p_{ij}. The decision whether or not to target person i directly is indicated by variable d_i.

- for each node i in the graph we create a decision variable d_i;
- for each edge (i, j) in the graph we create a stochastic variable t_{ij}; the probability that the variable t_{ij} is true is equal to that of the edge, p_{ij};
- as constraint we impose the requirement that $\sum_i d_i \leq k$;

- as optimization criterion we use the function $\sum_i P(\varphi_i \mid d_1, \ldots, d_n)$; intuitively, the aim is that $P(\varphi_i \mid d_1, \ldots, d_n)$ represents the conditional probability that node i is reached if an advertisement is sent to exactly those people indicated by the variables $d_1, \ldots, d_n$. By summing these probabilities, we obtain an expected number of persons that is reached.

An important idea is hence to formalize the probability that a person is reached as the probability that some given logical formula φ_i evaluates to true given an assignment to decision variables.

We propose the development of a language, SC-ProbLog, for writing down these constraints and the distributions $P(\varphi_i \mid d_1, \ldots, d_n)$ in a systematic manner. This language extends the ProbLog language [14,15]. An example of a program in SC-ProbLog is given below. Lines 1–9 are written in ProbLog; lines 10–14 are specific to SC-ProbLog. As this example demonstrates, ProbLog's notation is similar to that of Prolog; its main extension is the ability to add probabilities to facts (lines 5 and 6). These facts become stochastic variables.

```
1. % Background knowledge
2. person(a).                      person(c).
3. person(b).                      person(d).

4. % Probabilistic facts
5. 0.7::directed(a,b).             0.4::directed(b,d).
6. 0.2::directed(b,c).             0.6::directed(c,d).

7. % Relations
8. trusts(X,Y) :- directed(X,Y).   buys(X) :- marketed(X).
9. trusts(X,Y) :- directed(Y,X).   buys(X) :- trusts(X,Y), buys(Y).

10. % Decision variables
11. ?::marketed(P) :- person(P).

12. % Constraints and optimization criteria
13. { marketed(P) => 1 :- person(P). } 8.
14. #maximize { buys(P) => 1 :- person(P). }.
```

The example program reflects several assumptions in lines 8–9. First, the trust relationship is bidirectional. Second, if a person is targeted directly, they will certainly buy the product. Third, if a person i trusts another person j and j buys the product, then i buys the product.

Traditional ProbLog would allow for the calculation of a success probability for a given *query*, such as :- buys(a)., based on lines 1–9, for a given set of facts marketed(X).

In the syntax of lines 10–14, we draw inspiration from DT-ProbLog, a version of ProbLog with support for optimization, but not constraints [26], and Answer Set Programming, to formalize constraints. Line 11 defines a decision variable for each person; it defines a search space of facts that can be added to the ProbLog program. Subsequently, we specify optimization criteria and constraints. Line 13 defines a reward (a weight r_i) of 1 for each person that marketing materials are sent to, and we bound the number of targeted persons to 8. Line 14 adds a probabilistic query buys(P). for each person P to the optimization criterion. Here we effectively maximize the expected number of people that buy the product.

3   Background

To understand the model in the previous setting, and to understand our newly proposed method, it is important to understand in more detail how the calculation of a conditional probability in ProbLog can be formalized as calculating the probability that a formula over decision variables and stochastic variables evaluates to true. We will use our earlier example to illustrate this. For a full introduction, the reader is referred to the literature [14].

As an example we consider calculating the probability that person a in our network buys a product, given decision variables for each person. The key insight is that for the query buys(a), the following *grounded* formula in Disjunctive Normal Form (DNF) can be constructed:

$$\varphi_a = d_a \vee (t_{ab} \wedge d_b) \vee (t_{ab} \wedge t_{bc} \wedge d_c) \vee (t_{ab} \wedge t_{bd} \wedge d_d) \\ \vee (t_{ab} \wedge t_{bd} \wedge t_{dc} \wedge d_c) \vee (t_{ab} \wedge t_{bc} \wedge t_{cd} \wedge d_d), \tag{2}$$

This formula can be derived using *Selective Linear Definite clause resolution*, or SLD-resolution [3,14], from the original ProbLog program. For example: the clause $(t_{ab} \wedge d_b)$ reflects the possibility that person a buys the product if it is marketed to b and the edge between nodes a and b is present. As earlier, d_b is a decision variable; t_{ab} is a stochastic variable with probability p_{ab} of being true.

Assume that the product is only marketed to person d. In this case, the formula reduces to $\varphi_a = (t_{ab} \wedge t_{bd}) \vee (t_{ab} \wedge t_{bc} \wedge t_{cd})$. What is now the probability that person a will buy the product? The key idea that underlies both SCP and ProbLog is that the stochastic variables are considered to be true with a probability that is independent from the other stochastic variables. One possible *model* for formula φ_a is: $t_{ab} = t_{bd} = \top$, $t_{bc} = t_{cd} = \bot$. The probability for this model (its *weight*) is $p_{ab} \times p_{bd} \times (1 - p_{bc}) \times (1 - p_{cd})$. The *probability* of the query φ_a is defined to be sum of the weights of all the models of the above formula. Hence, this problem is a *weighted model counting* (WMC) problem [9].

Calculating the WMC by enumerating all models is usually not efficient. A more efficient calculation is the following: $p_{ab} \times p_{bd} + p_{ab} \times (1 - p_{bd}) \times p_{bc} \times p_{cd}$. The first product corresponds to the first possible path, the second product to the second path. Note that this formula includes a term $(1 - p_{bd})$. This term is necessary as we would otherwise count the model $t_{ab} = t_{bd} = t_{bc} = t_{cd} = \top$ twice. This problem is known as the *disjoint sum* problem.

As the previous example makes clear, computing the probability of a DNF formula is hard due to the disjoint sum problem; in general, it is known to be #P-complete [24]. This makes solving this type of SCP particularly hard. However, several practical approaches have been proposed to make WMC feasible in practice. One such approach is based on *compiling* the logical formula into a *decision diagram*, and constructing an AC from this diagram [11]. Two well-studied types of decision diagrams are *Ordered Binary Decision Diagrams* (OBDDs) [7] and *Sentential Decision Diagrams* (SDDs) [12]. The latter type of decision diagrams has recently been shown to generalize OBDDs, and can be exponentially more compact [6]. For this reason, we focus on SDDs.

An SDD consists of decompositions, disjunctions and terminals (see Fig. 2 for an example SDD for a formula f that is similar to the formula considered earlier, but that illustrates the concept of SDDs better). A decomposition consists of a *prime* p and a *sub* s, and one decomposition represents the logical formula $(p \wedge s)$. Disjunction nodes represent the disjunction of two or more decompositions. The shape of the SDD is completely determined by a tree structure over the variables present in it. This tree structure is called a *vtree* [22]. Two examples of vtrees are given in Fig. 2. A vtree induces a *total variable order* for an SDD when traversed from left to right. We now discuss how vtrees relate to SDDs.

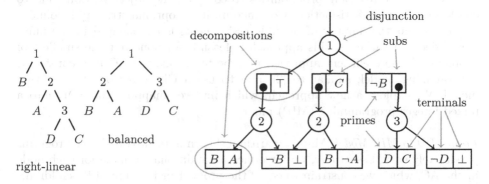

Fig. 2. Two examples of vtrees (left, center), each for variable order $B < A < D < C$. An SDD (right) for logic formula $f = (A \wedge B) \vee (B \wedge C) \vee (C \wedge D)$, which respects the balanced vtree. Example from Darwiche [12].

All disjunctions are required to *respect* specific nodes in the corresponding vtree. A disjunction respects a vtree node i if for all its child decompositions, each variable occurring in the sub-SDD rooted at the prime (sub) of the decomposition occurs in the sub-vtree rooted at the left (right) child of i. Thus, the disjunctions labeled '2' in Fig. 2 each respect vtree node 2 in the balanced vtree shown in the same figure. An SDD that respects a right-linear vtree is essentially an OBDD [12]; hence, SDDs generalize OBDDs. As with OBDDs, the size of an SDD is influenced by the total variable order that is induced by the vtree it respects. The shape of the vtree also influences the size of that SDD.

Once the SDD is compiled, WMC can be performed in time linear in the size of the SDD. In a bottom-up fashion the SDD is first turned into an *arithmetic circuit* (AC). In this AC, we assign the appropriate probabilities and decision values to the leafs of the circuit. The transformation of the SDD into an AC is simple: each decomposition node is replaced by a product node between its prime and its sub; each disjunction node is replaced by a summation node over the child nodes.[2] The properties of an SDD ensure that the disjoint sum problem is taken care of in the resulting circuit.

---

[2] This method was used for counting models of a Boolean formula in *decomposable Deterministic Negation Normal Form (d-DNNF)* [11], and can be applied to SDDs because SDDs are a proper subset of d-DNNFs [12].

4   Approach

We first make some observations, then aggregate them in a proposed algorithm.

SCOP Solving with MIP Solvers. Given an SC-ProbLog program that models a certain SCOP instance, the naive way of solving this SCOP is the following. Compile each of the queries present in the program into an AC containing decision variables and stochastic variables. For each possible assignment to the decision variables, fill in their values in the AC. Calculate the probabilities using the AC. Use the resulting probabilities to compute the objective value and to check for constraint satisfaction. Continue until the optimal strategy is found.

Given that the number of possible assignments is exponential in the number of decision variables, this approach is feasible for none but the smallest of problems. A more efficient approach may be to encode the AC in a constraint programming model, similar to [2], and to use a CP solver on the resulting model. We explore a new approach, which involves mapping the SDD into a mixed integer programming (MIP) model.

From SDD to MIP Model. Mapping arithmethic circuits into quadratic programs is relatively easy. Essentially, we introduce an additional variable for each node in the AC, which we constrain to equal the product or the sum of its children.

For MIP solvers the quadratic constraints in this naïve model can however be problematic. As the constraints can be shown to be nonpositive semidefinite, we cannot apply QCQP solvers either. It is important that we are able to *linearize* the products in our model, i.e., that we can transform the model in a set of equivalent linear constraints. As a short reminder, a constraint of the form $a = b \times c$ can be linearized in these cases[3]: (1) at least one of the two variables in $\{b, c\}$ is a constant; (2) at least one of the two variables in $\{b, c\}$ is a Boolean variable. Therefore, we need to ensure that in a decomposition node of the SDD, variables representing the two children satisfy these requirements.

Special vtrees. Next, we show that it suffices to constrain the vtrees to ensure that SDDs can be linearized. Recall that for each SDD decomposition node, the respected vtree determines the variables that can occur in the prime and in the sub. We observe the following: if all left-hand (right-hand) descendants of an internal vtree node n are stochastic variables, then for each SDD decomposition node m whose parent respects n, it holds that all variables occurring in m's prime (sub) are stochastic as well. A similar property holds for decision variables.

If a prime contains only probabilities, which can be considered as constants for the model, we can precompute the corresponding value for the prime, effectively eliminating the MIP model variable associated with that prime. Similarly: since we can linearize all operations on Boolean variables [19], any prime containing only decision variables can be expressed by a Boolean variable with linear

---

[3] Using the big M-approach [19] with $M \leq 1$, as all real values are probabilities.

relations to other variables. Thus, in each of these two cases, the expression represented by the prime can be linearized and hence the product represented by the SDD decomposition node as well. The same holds for subs.

This leads us to define the concept of *mixed* and *pure* nodes in a vtree. A *pure* node is an internal node whose leaf descendants all are variables of the same type (either stochastic or decision), while a *mixed* node is an internal node that has leaf descendants of both types. We state that an SDD can be linearized into a MIP model if the vtree that it respects has the *single mixed path* property.

Definition 1. *Given a vtree on variables of two distinct classes (e.g. decision and stochastic). This vtree has the* single mixed path (SMP) *property (and is called an SMP vtree) if, for each of its internal nodes n, the following holds: either both children of n are pure nodes, or one child of n is pure and the other child is mixed. As a consequence, if an SMP vtree has mixed nodes, all mixed nodes occur on the same path from the root of the vtree to the lowest mixed node.*

Minimizing SDDs. Recall that SDDs that respect right-linear vtrees are essentially OBDDs. One can easily verify that a right-linear vtree has the SMP property: if it has an SMP, it is on the right spine of the vtree. From this follows that OBDDs can be linearized. However: right-linear vtrees generally do not yield the smallest SDDs. Since the size of the SDD determines the size of the resulting MIP model, and thus the solving time, small SDDs are preferable as input for the MIP model builder.

Choi and Darwiche have proposed a local search algorithm for SDD minimization [10]. This algorithm considers three operations on the vtree: *right-rotate*, *left-rotate* (each well-known operations on binary trees) and *swap*. When a swap operation is applied to an internal node, the sub vtrees rooted at its children are swapped. Given a (sub) vtree, the greedy local search algorithm of Choi and Darwiche loops through its neighbourhood of different vtrees by applying consecutive rotate and swap operations, trying to find a vtree that yields a smaller SDD. Since OBDD minimization is NP-hard [5], we expect SDD minimization to also be NP-hard, but we are not aware of any published proof of this.

Generally, this minimization produces vtrees that do *not* have the SMP property, even if the initial vtree did; the rotate moves may remove this property.

A desirable property of Choi and Darwiche's algorithm is the following: the three local moves considered are sufficient to turn *any* vtree on a certain set of variables into *any* other vtree on the same set of variables. Consequently, the local moves in principle allow complete traversal of the search space of vtrees.

Here, we propose a simple modification of Choi and Darwiche's algorithm: we use the same local moves as their algorithm does, but any move that leads to a vtree that violates the SMP property is immediately rejected.

While this modification is conceptually easy, a relevant fundamental question is whether under this modification it is still possible to traverse the space of SMP vtrees on a fixed set of variables completely. We show that this is indeed the case.

In the following we refer to the leaf node that represents the variable that is lowest in the order associated with a vtree as LL (lowest leaf).

Lemma 1. *Let v be the parent and x the grandparent of the LL in an SMP vtree. Right rotate on x maintains the SMP property for the vtree rooted at v.*

Proof. Consider the left SMP vtree in Fig. 3. Given that this vtree satisfies the SMP property by assumption, sub vtrees b and c cannot both be mixed, but one of them can be. Now consider the following cases:

Both b and c are pure and of the same class as LL: Lemma 1 holds trivially.
Both b and c are pure, not each of the same class as LL: Any class
   assignment to b and c will preserve the SMP property.
Node b is pure, node c is mixed: Since b is of the same class as LL (by
   assumption), node v is pure and node x is mixed. After applying right-rotate
   on node v, both v and x are mixed, and the SMP property is preserved.
Node b is mixed, node c is pure: Node c can belong to any class, since both
   node v and node x are mixed before as well as after applying right-rotate to
   v, preserving the SMP property under rotation.

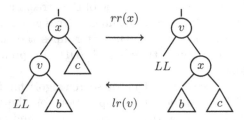

Fig. 3. Rotate operations on an SMP vtree. Node LL is the lowest variable in the variable induced by these vtrees. Nodes v and x are internal; b and c are sub vtrees.

Note that the SMP vtree described above may be a sub vtree of a larger vtree. The fact that the right-rotate operation does not change the nature (mix or pure) of the root of this sub vtree, leads to the following corollary:

Corollary 1. *A right-rotate operation on the grandparent of the LL node does not change the SMP status of the full vtree.*

Lemma 2. *Given an SMP vtree with node LL in order $\mathcal{O}$. We can always obtain an SMP vtree on the same order $\mathcal{O}$ in which the LL is the left child of the root, through a series of right-rotate operations, without ever in the process transforming it into a vtree that violates the SMP property.*

Proof. A right-rotate operation on an internal vtree node decreases its left child's distance to the root of the vtree by one. Repeated applications of right-rotate on LL's grandparent ultimately makes LL's parent the vtree's root. By Lemma 1 and Corollary 1, the SMP status of the vtree never changes in this process.

Lemma 3. *Given an SMP vtree on order $\mathcal{O}$, we can always obtain a right-linear vtree on the same order, through a series of right-rotate operations, without ever in the process transforming it into a vtree that violates the SMP property.*

Proof. By Lemma 2 we can turn any SMP vtree in one for which the LL is the left child of the root. This vtree can be made right-linear by recursively applying this method to the root's right child.

Lemma 4. *A right-linear SMP vtree with variable order $\mathcal{O}$ can be transformed in any SMP vtree on the same variable order by a series of left-rotate operations without ever in the process transforming into a vtree without the SMP property.*

Proof. Since left-rotate is the dual operation of right-rotate, a sequence of right-rotate moves transforming any vtree to a right-linear one through right-rotate operations, can simply be reversed through left-rotate operations to turn a right-linear vtree in any other (on the same variable order).

Note that rotate operations preserve the variable order in the vtree, only changing its shape. However, the space of possible vtrees on a fixed set of variables is larger, since different variable orders exist. The order of variables is changed by the application of swap operations.

Lemma 5. *Any right-linear vtree on variable order $\mathcal{O}$ can be transformed into a right-linear vtree on any other variable order $\mathcal{O}'$ through a series of rotate and swap operations without ever in the process transforming into a vtree that violates the SMP property.*

Proof. Observe that any right-linear vtree satisfies the SMP property. Observe that if we can reverse the mutual order of two adjacent variables (e.g. $A < B < C < D$ becomes $A < C < B < D$), we can create any variable order by repeatedly reversing the orders of adjacent variables.

This order reversal is simple. Suppose that node b in the right vtree of Fig. 3 is a single variable, as is LL. We can make LL and b swap places by applying a left-rotate on v, resulting in the left vtree of Fig. 3, and then applying a swap operation on v, followed by a right-rotate operation on x.

Theorem 1. *Any SMP vtree can be transformed into any other SMP vtree on the same variable through a series of rotation and swap moves, without ever in the process transforming into a vtree that does not have the SMP property.*

We conclude that an SMP-preserving minimization algorithm that applies only swap and rotate operations can in principle convert any SMP vtree into any other SMP vtree on the same variables.

Summary. These observations spark the following algorithm for solving SCOPs:

1. ground formulas for the queries present in the SCOP;
2. compile SMP vtree respecting SDDs for all these queries (ProbLog's default mechanism uses right-linear vtrees, so this is automatically satisfied);

3. apply the SMP-preserving local search algorithm to minimize these SDDs;
4. convert the SDDs into arithmetic circuits and then into sets of constraints;
5. add the optimization criterion and linear constraints of the SCOP to the MIP model, ensuring e.g. that for an upper-bounded stochastic constraint the model variables representing the root of each relevant query are added using a linear model constraint of the form $\sum_i r_i v_i \leq \theta$;
6. apply a MIP solver or a CP solver to find a solution.

For CP solvers, the unconstrained minimization algorithm can be used to obtain smaller SDDs. ProbLog's compilation strategy yields SDDs respecting right-linear vtrees. Thus, without minimization, the SDDs are essentially OBDDs.

5 Experiments

We state some questions that we wish to answer for the approach described in the previous section. Then we describe the experiments we use to answer these questions.

Questions. Recall that the size of a MIP or CP model is linear in the size of the SDDs it is built on. We expect smaller models to be faster to solve. However: minimizing an SDD takes time. Furthermore, when quadratic constraints are allowed, we expect to obtain smaller SDDs; however, solving quadratic problems using CP may take longer than solving MIPs. We pose the following questions:

(Q2) How do SDD sizes depend on the choice of minimization algorithm?
(Q3) How do the calculation times for the full toolchain compare for CP and MIP solvers, with and without appropriate minimization?
(Q4) How do the computation times for different phases of the algorithm compare to each other?

To answer these questions, and to demonstrate that SC-ProbLog programs can be solved in practice, we apply our algorithms to different SCOPs. Of course, the constraints determine problem hardness, which begs the question:

(Q1) Which threshold settings are useful for an evaluation of the solving times?

Description of Test Data. Our experiments focus on two types of real data sets: a social network and a gene-protein interaction network. As **social network** we use the *High-energy theory collaborations network* [20], which was also used in earlier publications on viral marketing [16]. This collaboration network of 7610 authors (nodes) has 15751 undirected weighted edges, which we turn into probabilities following Kempe's approach [16]. Initial experiments showed that the full network is too large to ground the problem's programs. We use *Gephi*'s[4] implementation of the Louvain algorithm for weighted community detection to

---

[4] Available at https://gephi.org/.

extract communities. We consider two specific communities, referred to as **hep-th47** and **hep-th5**. Compared to our earlier viral marketing ProbLog program, in our experiments we include additional stochastic variables such that a person does not automatically buy a product if it is marketed to them.

As **DNA-protein and protein-protein interaction network** we use the *Signaling-regulatory Pathway INference* [21] (or *SPINE*) network, with 4696 nodes representing genes and proteins. It contains 15147 undirected protein-protein edges, and 5568 directed protein-gene edges. The set provides probabilities for both the undirected protein-protein edges, and the directed protein-gene edges. We again use *Gephi*'s community detection, where we take care to ensure that both negative and positive knockout pairs are contained in our samples. We consider models referred to as **spine16** and **spine27** in our experiments. We use a specific path definition that requires paths to end in a protein-DNA edge.

Optimization and Constraint Settings. We consider several combinations of optimization and constraint settings on the programs described above. We use the following abbreviations. **maxSumProb** denotes a maximization over stochastic variables, while **maxTheory** denotes a maximization over the sum of decision variables set to true (theory size). For constraints we use these abbreviations: **ubSumProb** denotes a constraint in which we impose an upper bound on an expectation; **ubTheory** denotes a constraint in which we impose an upper bound on the theory size. We also define minimization and lower bound counterparts of these settings. Table 1 lists the four datasets that we use, along with the tasks we evaluate on each dataset. For instance, the combination (**maxSumProb**, **ubTheory**) is the viral marketing setting we considered earlier in this paper.

Software and Hardware. We use `Gurobi` 6.52 as MIP solver and `Gecode` 5.0.0 as CP solver[5]. For each phase of the toolchain (grounding of the program, SDD compilation, building of the constraint model and solving it) we use a timeout on our experiments of 3600 s. They were implemented in `Python 3.4`, using `ProbLog 2.1`[6] for the grounding of programs. `ProbLog 2.1` uses version 1.1.1 of UCLA's `sdd` library[7], which is implemented in `C`, for SDD compilation. They were run on a machine with an Intel Xeon E5-2630 processor and 512 GB RAM, under Red Hat 4.8.3-9.

Results. To answer **(Q1)**, Fig. 4 shows solving times for the **hep-th47** problem in the (**maxSumProb**, **ubTheory**) setting, for different thresholds. As expected, we find that thresholds that are not very strict or loose, require the longest solving times. We performed similar experiments for the other problem settings to systematically identify the threshold for which each problem was the hardest, which we then chose as test cases for the SCOP solving method comparison.

To answer **(Q2)**, Fig. 5 shows a comparison of the size reductions obtained by the SMP-minimization algorithm and the default minimization algorithm

---

[5] Available at www.gurobi.com and www.gecode.org.
[6] Availabe at https://dtai.cs.kuleuven.be/problog/.
[7] Available at http://reasoning.cs.ucla.edu/sdd/.

Table 1. Performance in seconds of the different methods on the hardest instances of the testcases for the full toolchain. We give the problem set, optimization and constraint setting, number of decision variables n_d, number of ProbLog queries n_q that comprise the objective function and/or constraint, threshold θ and objective value v_{obj} (N/A denotes a problem that has no solution for that threshold). We show the solving times for the default SDD with no minimization (t_{none}), SMP minimization (t_{smp}) and default minimization ($t_{default}$) for Gurobi and Gecode. We indicate a timeout with t/o.

Instance			Characteristics				Gurobi		Gecode	
Problem	opt.	cst.	n_d	n_q	θ	v_{obj}	t_{none}	t_{smp}	t_{none}	$t_{default}$
spine16	maxSumPr.	ubTh.	36	23	15	14.40	3.9	3.4	1389.5	591.4
spine16	minTh.	lbSumPr.	36	23	6.9	8	4.1	3.9	70.9	31.4
spine27	maxSumPr.	ubSumPr.	86	26	1.3	9.51	443.2	471.3	t/o	t/o
spine27	maxSumPr.	ubTh.	76	13	25	10.18	5.9	5.6	t/o	t/o
spine27	maxTh.	ubSumPr.	71	13	6.5	52	23.3	21.9	222.9	8.6
spine27	minTh.	lbSumPr.	76	13	6.5	8	4.7	5.7	t/o	1878.2
hep-th47	maxSumPr.	ubTheory	20	20	10	3.21	545.83	412.7	t/o	130.9
hep-th47	minTh.	lbSumPr.	20	20	2	6	188.61	163.8	2859.9	6.9
hep-th5	maxSumPr.	ubTh.	33	10	20	2.81	2076.83	1185.7	t/o	t/o
hep-th5	minTh.	lbSumPr.	33	10	5	N/A	364.62	346.4	t/o	t/o

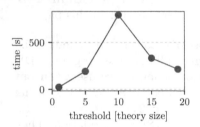

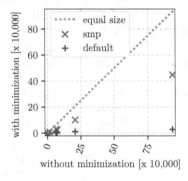

Fig. 4. Example of performance of Gurobi with non-minimized SDD on different thresholds, for problem **hep-th47** with **maxSumProb, ubTheory**.

Fig. 5. Comparison of size reduction by SDD minimization algorithms.

provided by the sdd library. We find that the SMP minimization algorithm typically halves the size of the initial SDD. The default minimization typically reduces the size of the SDD by one or two orders of magnitude.

To answer (**Q3**), we summarize the performance of the four methods on our test cases in Table 1. For the **hep-th5** problem we selected the ten highest-degree nodes for the queries, since the program could not be grounded within one hour if we selected all 33 nodes in the problem for querying. This reduced

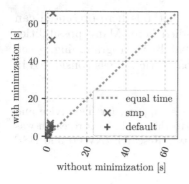

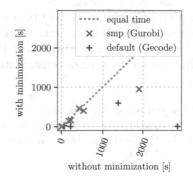

Fig. 6. Comparison of SDD compilation times.

Fig. 7. Comparison of full toolchain solving times for the two solvers.

the grounding time to about 120 s. For the other test cases we have selected all queries in the problem, with grounding times in the range of 1–5 s.

We observe that without any minimization of the SDD, `Gurobi` consistently outperforms `Gecode`. Furthermore, we observe that the difference made by SDD minimization is larger for the `Gecode` methods than for the `Gurobi` methods. This can largely be explained by the results in Fig. 5, and by those in Fig. 6, which answer question **(Q4)**. The latter show that generally, compiling SDDs is a matter of seconds, whether they are being minimized or not. The exception is the **hep-th5** problem, which takes tens of seconds to compile into an SDD when using SMP minimization. Observe from the table that minimization is still useful here, as it reduces solving time enough to make up for the extra minimization time. We note that the minimization algorithms are based on heuristics, and minimization speed-up may lie in the improvement of these heuristics.

Finally, Fig. 7 shows that the time that is gained during the optimization part of the entire solving chain, can be orders of magnitude larger than the time lost by minimizing the SDD. We do note that, since compiling the SDD can be done in seconds, this effect is less noticable for the smaller problems.

6   Conclusions

We introduced a specific class of SCOPs, in which we can impose constraints and optimization criteria based on expected utilities over probabilistic programs. We demonstrated that a viral marketing problem and a problem in bioinformatics can be considered instances of such SCOPs. We showed how generic probabilistic programming technology can be combined with constraint optimization solvers to solve these problems, and introduced an SDD minimization algorithm that preserves properties that ensure linearizability of the SDD to a MIP model, while reducing the size of the SDD. While the results are encouraging, an important remaining challenge is scalability; local search and sampling algorithms could be of interest here for the probability calculation, the optimization, and the minimization of circuit sizes. We believe that the methods here presented can also be applied in other contexts than those studied here. Many possibilities remain for the further integration of CP and probabilistic programming, given the limitations on the type of constraints and probabilistic models considered in this work.

Acknowledgements. We thank Luc De Raedt for his support, for his advice and for the numerous other ways in which he contributed to this work. This research was supported by the Netherlands Organisation for Scientific Research (NWO) and NSF grant #IIS-1657613.

References

1. Babaki, B., Guns, T., De Raedt, L.: Stochastic constraint programming with and-or branch-and-bound. In: Proceedings of the Twenty-Sixth International Joint Conference on Artificial Intelligence (2017, to appear). doi:10.24963/ijcai.2017/76
2. Babaki, B., Guns, T., Nijssen, S., De Raedt, L.: Constraint-based querying for Bayesian network exploration. In: Fromont, E., De Bie, T., van Leeuwen, M. (eds.) IDA 2015. LNCS, vol. 9385, pp. 13–24. Springer, Cham (2015). doi:10.1007/978-3-319-24465-5_2
3. Ben-Ari, M.: Mathematical Logic for Computer Science, 3rd edn. Springer Publishing Company, Incorporated, Heidelberg (2012). doi:10.1007/978-1-4471-4129-7
4. Bistarelli, S., Rossi, F.: Semiring-based soft constraints. In: Degano, P., De Nicola, R., Meseguer, J. (eds.) Concurrency, Graphs and Models. LNCS, vol. 5065, pp. 155–173. Springer, Heidelberg (2008). doi:10.1007/978-3-540-68679-8_11
5. Bollig, B., Wegener, I.: Improving the variable ordering of OBDDs is NP-complete. IEEE Trans. Comput. **45**(9), 993–1002 (1996). doi:10.1109/12.537122
6. Bova, S.: SDDs are exponentially more succinct than OBDDs. In: Proceedings of the Thirtieth AAAI Conference on Artificial Intelligence, AAAI 2016, pp. 929–935. AAAI Press (2016)
7. Bryant, R.E.: Graph-based algorithms for Boolean function manipulation. IEEE Trans. Comput. **35**(8), 677–691 (1986). doi:10.1109/TC.1986.1676819
8. Charnes, A., Cooper, W.W.: Chance-constrained programming. Manag. Sci. **6**, 73–79 (1959)
9. Chavira, M., Darwiche, A.: On probabilistic inference by weighted model counting. Artif. Intell. **172**(6–7), 772–799 (2008). doi:10.1016/j.artint.2007.11.002
10. Choi, A., Darwiche, A.: Dynamic minimization of sentential decision diagrams. In: Proceedings of the Twenty-Seventh AAAI Conference on Artificial Intelligence, AAAI 2013, pp. 187–194. AAAI Press (2013)
11. Darwiche, A.: On the tractable counting of theory models and its application to truth maintenance and belief revision. J. Appl. Non-Class. Log. **11**(1–2), 11–34 (2001). doi:10.3166/jancl.11.11-34
12. Darwiche, A.: SDD: a new canonical representation of propositional knowledge bases. In: Proceedings of the Twenty-Second International Joint Conference on Artificial Intelligence, IJCAI 2011, vol. 2, pp. 819–826. AAAI Press (2011). doi:10.5591/978-1-57735-516-8/IJCAI11-143
13. De Raedt, L., Kersting, K., Kimmig, A., Revoredo, K., Toivonen, H.: Compressing probabilistic Prolog programs. Mach. Learn. **70**(2), 151–168 (2008). doi:10.1007/s10994-007-5030-x

14. De Raedt, L., Kimmig, A., Toivonen, H.: ProbLog: a probabilistic Prolog and its application in link discovery. In: Proceedings of the 20th International Joint Conference on Artifical Intelligence, IJCAI 2007, pp. 2468–2473. Morgan Kaufmann Publishers Inc., San Francisco (2007)
15. Fierens, D., Van den Broeck, G., Renkens, J., Shterionov, D., Gutmann, B., Thon, I., Janssens, G., De Raedt, L.: Inference and learning in probabilistic logic programs using weighted Boolean formulas. Theory Pract. Log. Program. **15**(03), 358–401 (2015). doi:10.1017/S1471068414000076
16. Kempe, D., Kleinberg, J., Tardos, É.: Maximizing the spread of influence through a social network. In: Proceedings of the Ninth ACM SIGKDD International Conference on Knowledge Discovery and Data Mining, KDD 2003, pp. 137–146. ACM, New York (2003). doi:10.1145/956750.956769
17. Lombardi, M., Milano, M.: Allocation and scheduling of conditional task graphs. Artif. Intell. **174**(7–8), 500–529 (2010). doi:10.1016/j.artint.2010.02.004
18. Mateescu, R., Dechter, R.: Mixed deterministic and probabilistic networks. Ann. Math. Artif. Intell. **54**(1–3), 3–51 (2008). doi:10.1007/s10472-009-9132-y
19. McKinnon, K.I.M., Williams, H.P.: Constructing integer programming models by the predicate calculus. Ann. Oper. Res. **21**(1), 227–245 (1989). doi:10.1007/BF02022101
20. Newman, M.E.J.: The structure of scientific collaboration networks. Proc. Natl. Acad. Sci. **98**(2), 404–409 (2001). doi:10.1073/pnas.021544898
21. Ourfali, O., Shlomi, T., Ideker, T., Ruppin, E., Sharan, R.: SPINE: a framework for signaling-regulatory pathway inference from cause-effect experiments. Bioinformatics **23**(13), i359–i366 (2007). doi:10.1093/bioinformatics/btm170
22. Pipatsrisawat, K., Darwiche, A.: New compilation languages based on structured decomposability. In: Proceedings of the 23rd National Conference on Artificial Intelligence, AAAI 2008, vol. 1, pp. 517–522. AAAI Press (2008)
23. Riedel, S.: Improving the accuracy and efficiency of MAP inference for Markov logic. In: Proceedings of the 24th Conference in Uncertainty in Artificial Intelligence, UAI 2008, Helsinki, Finland, 9–12 July 2008, pp. 468–475 (2008)
24. Roth, D.: On the hardness of approximate reasoning. Artif. Intell. **82**(1–2), 273–302 (1996). doi:10.1016/0004-3702(94)00092-1
25. Tarim, S.A., Hnich, B., Prestwich, S.D., Rossi, R.: Finding reliable solutions: event-driven probabilistic constraint programming. Ann. OR **171**(1), 77–99 (2009). doi:10.1007/s10479-008-0382-6
26. Van den Broeck, G., Thon, I., van Otterlo, M., De Raedt, L.: DTPROBLOG: a decision-theoretic probabilistic Prolog. In: Proceedings of the Twenty-Fourth AAAI Conference on Artificial Intelligence, AAAI 2010, pp. 1217–1222. AAAI Press (2010)
27. Walsh, T.: Stochastic constraint programming. In: Proceedings of the 15th European Conference on Artificial Intelligence, ECAI 2002, Lyon, France, July 2002, pp. 111–115 (2002)

Learning the Parameters of Global Constraints Using Branch-and-Bound

Émilie Picard-Cantin[1]([⊠]), Mathieu Bouchard[2], Claude-Guy Quimper[1], and Jason Sweeney[2]

[1] Université Laval, Quebec City, Canada
emilie.picard-cantin.1@ulaval.ca, claude-guy.quimper@ift.ulaval.ca
[2] PetalMD, Quebec City, Canada
mathbouchard@gmail.com, jason.pierre.sweeney@gmail.com

Abstract. Precise constraint satisfaction modeling requires specific knowledge acquired from multiple past cases. We address this issue with a general branch-and-bound algorithm that learns the parameters of a given global constraint from a small set of positive solutions. The idea is to cleverly explore the possible combinations taken by the constraint's parameters without explicitly enumerating all combinations. We apply our method to learn parameters of global constraints used in timetabling problems such as SEQUENCE and SUBSETFOCUS. The later constraint is our adaptation of the constraint FOCUS to timetabling problems.

Keywords: Constraint acquisition · Timetabling · Machine learning · CSP · Global constraints · Brand-and-Bound

1 Introduction

Modeling a constraint satisfaction problems requires specific knowledge acquired from multiple past cases, each model being different from the last. For example, CSPs of nurse timetabling problems for two different hospitals most likely use similar constraints but with different parameters. A hospital might require to work in the emergency ward no more than 3 days out of 7 while another hospital might set the limit to no more than 4 days out of 9. A system able to determine the parameters that created a set of existing solutions would greatly speedup the modeling process. This explains the popularity of modeling automation in the recent years.

A global constraint has variables, encoding solutions, and known parameters that define the relation between the variables. In our context, we are given a global constraint and example solutions, so the variables are known but the parameters are unknown. We want to learn the parameters that generated the examples.

The main contribution of this paper is an algorithm that learns the parameters of given global constraints from a small pool of examples. This algorithm can be applied to constraints such as AMONG and SEQUENCE, commonly used in

© Springer International Publishing AG 2017
J.C. Beck (Ed.): CP 2017, LNCS 10416, pp. 512–528, 2017.
DOI: 10.1007/978-3-319-66158-2_33

timetabling. We use a branch-and-bound to quickly and cleverly travel through all the combinations of parameters of the constraint and to determine which combination best describes the given examples.

We also introduce SUBSETFOCUS, a constraint useful in the modelization of timetabling problems, and we show how to learn its parameters from examples.

2   Background

We present the global constraints that are studied in this paper. We then report important past contributions related to the automation of constraint modeling and to the learning of parameters for global constraints.

2.1   Global Constraints

Let $C(X, a)$ be a global constraint where $X = [X_1, \ldots, X_n]$ are integer variables denoted in upper case and $a = [a_1, \ldots, a_m]$ are parameters denoted in lower case. Let $S_C(a)$ be the solution set for C with parameters a and let $S_{C_i}(x, a) = S_C(a_1, \ldots, a_{i-1}, x, a_{i+1}, \ldots, a_m)$, where all a_j in a are fixed except for a_i.

We present global constraints that are studied in this paper. Let $s_{ij} = \{i, i+1, \ldots, j\}$ be a sequence. The constraint FOCUS(X, Y, l, k) [28] is satisfied if and only if X and Y are such that there exists a set S of disjoint sequences s_{ij} such that

1. $|S| \leq Y$
2. $X_b > k \iff \exists s_{ij} \in S$ s.t. $b \in s_{ij}, \quad \forall b \in [1, \ldots, n]$
3. $j - i + 1 \leq l, \quad \forall s_{ij} \in S$

FOCUS controls the number and the length of subsequences of variables greater than k.

Example 1. Suppose we have FOCUS($[X_1, \ldots, X_7], Y, l = 2, k = 1$) with $\text{dom}(X_i) = \{1, 2, 3, 4\}$ and $\text{dom}(Y) = \{2, 3\}$. Then, $[2, 1, 3, 3, 1, 1, 1]$ and $[2, 3, 4, 1, 1, 2, 1]$ are solutions because $S = \{s_{11}, s_{34}\}$ and $S = \{s_{12}, s_{33}, s_{66}\}$ are valid sets. $[2, 1, 2, 1, 2, 1, 2]$ is not a solution because the cardinality of the only possible set $S = \{s_{11}, s_{33}, s_{55}, s_{77}\}$ is greater than $\max(\text{dom}(Y))$.

Some global constraints are applied to a subset of values in their original definition. We choose to encode this set with a vector z. $z_v = 1$ if the value v is in the set and $z_v = 0$ otherwise. In other words, the vector z is the bitset encoding of a set. By abuse of notation, we will consider z sometimes as a binary vector, sometimes as a set of values.

We propose a new constraint SUBSETFOCUS(X, l, m, z) (SF), a generalization of FOCUS, that controls the number and the length of subsequences of variables that belong to a set of values. SUBSETFOCUS(X, l, m, z) is satisfied if X is such that there exists S, a set of disjoint sequences s_{ij}, such that:

1. $|\mathcal{S}| \leq m$
2. $X_b = v \wedge z_v = 1 \iff \exists s_{ij} \in \mathcal{S}$ s.t. $b \in s_{ij}, \quad \forall b \in [1, \ldots, n]$
3. $j - i + 1 \leq l, \quad \forall s_{ij} \in \mathcal{S}$
4. $s_{ij} \in \mathcal{S} \Rightarrow s_{j+1,j'} \notin \mathcal{S}, \quad \forall j' \geq j + 1.$

In other words, SUBSETFOCUS($\boldsymbol{X}, l, m, \boldsymbol{z}$) is satisfied if the assignment $\boldsymbol{X}$ has fewer than m stretches of maximum length l with values in the set defined by $\boldsymbol{z}$. SUBSETFOCUS can be used to limit the number of stretches of night shifts to m while limiting the length of the stretches to l in a medical timetabling problem.

We introduce SUBSETFOCUS to fulfill a request from PetalMD, the company financing this research, to create medical schedules with clusters of night shifts. The set $\mathcal{S}$ represents sequences of consecutive night shifts. In our context, two consecutive night shifts should not be considered as two separate stretches of work as it could be with Focus. Hence the fourth condition of SubsetFocus.

Example 2. Suppose we have SUBSETFOCUS($[X_1, \ldots, X_7], l = 2, m = 2, \boldsymbol{z} = [0, 1, 1, 1]$) with dom($X_i$) = $\{1, 2, 3, 4\}$. Then, $[2, 1, 3, 3, 1, 1, 1]$ is a solution because $\mathcal{S} = \{s_{11}, s_{34}\}$ is a valid set. $[2, 3, 4, 1, 1, 1, 1]$ is not a solution because s_{13} is not a valid sequence according to condition 3 and all other combinations would violate condition 4.

AMONG($\boldsymbol{X}, l, u, \boldsymbol{z}$) [4] ensures that at most u and at least l variables take values in $\boldsymbol{z}$. SEQUENCE($\boldsymbol{X}, l, u, w, \boldsymbol{z}$) (SEQ) [4] ensures that for every subset of w consecutive variables in $[X_1, \ldots, X_n]$ AMONG($[X_i, \ldots, X_{i+w-1}], l, u, \boldsymbol{z}$) holds.

Let occ$_v$ = $|\{i : X_i = v\}|$ be the number of occurrences of value v in $\boldsymbol{X}$. GCC($\boldsymbol{X}, [l_1, \ldots, l_m], [u_1, \ldots, u_m]$) (or GLOBALCARDINALITY constraint) [30] ensures that occ$_v \in [l_v, u_v]$, for each value $v \in \{1, \ldots, m\}$. ATMOST-NVALUE($\boldsymbol{X}, k$) (ATLEASTNVALUE($\boldsymbol{X}, k$)) [26] ensures that the variables take at most (at least) k different values.

BALANCE($\boldsymbol{X}, b$) [9] ensures that the balance b is the difference between the most occurring value and the least occurring value, among assigned values only.

$$b = \max_{v \in \{X_i\}_{i=1}^n} \text{occ}_v - \min_{v \in \{X_i\}_{i=1}^n} \text{occ}_v .$$

The constraint ATMOSTBALANCE($\boldsymbol{X}, b$) [16] ensures that the balance is at most b. The variant ATMOSTBALANCE*($\boldsymbol{X}, b$) also takes into consideration non-occurring values, i.e.: $b \geq \max_v \text{occ}_v - \min_v \text{occ}_v$.

Pesant et al. [27] count the solutions for multiple global constraints. In particular, they propose a dynamic programming approach to count the solutions satisfying REGULAR($[X_1, \ldots, X_n], \mathcal{A}$), that ensures the word $[X_1, \ldots, X_n]$ belongs to the regular language described by the finite automaton $\mathcal{A}$. The idea is to encode REGULAR as a layered graph and then recursively count all paths, from the last layer to the first.

2.2   Constraint Acquisition

There exist multiple approaches to learn, from a set of solutions, which constraints form a model. The *model seeker* [5,8] learns a CSP from positive examples. It lists the global constraints satisfied by all examples using the *constraint seeker* [6,7], which uses multiple criteria to order the constraints according to pertinence. One criterion is the number of assignments that satisfy the constraint. If this number is small, that indicates that the solution come from a small subset of possible assignments. This is more likely to occur if the constraint was imposed.

Many propose interactive constraint acquisition systems learning a complete CSP from examples by asking queries to the user. Among the most recent publications: Bessiere et al. [10,14,15], Daoudi et al. [20], and Arcangioli and Lazaar [1].

Bessiere et al. [12,13], Barták et al. [2], and Charnley et al. [19] study the acquisition of implied constraints from a CSP in order to improve the resolution time. Bessiere et al. [11] use the examples to remove redundant constraints from the model during the constraint acquisition phase.

Kiziltan et al. [21], Little et al. [23], and Lopez and Lallouet [25] study the translation of a problem description written in natural language into a formal model.

Machine learning techniques can also be used to learn part of a model. Bonfietti et al. [17], Bartolini et al. [3], and Lombardi et al. [24] train neural networks or decision trees to recognize a solution to a problem and then embed the trained neural network or the trained decision trees into a global constraint.

Campigotto et al. [18] and Kolb [22] study the acquisition of the utility function of the optimization model.

Suraweera et al. [31] propose a system that learns parameters for template constraints previously defined from historical schedules. The quality of a parameter set is the distance to the real set of parameters that created the examples. They choose the parameters with the highest quality.

2.3   Learning Parameters of a Global Constraint

Picard-Cantin et al. [29] consider a global constraint $C(X, a)$ whose scope X is known and want to learn its parameters a from a small set of examples $E = \{e_1, \ldots, e_q\}$. Each parameter a_i must take a value from a predefined set of values called *domain* denoted $\mathrm{dom}(a_i)$. They list all combinations of parameters satisfied by the examples and choose the combination with the lowest probability of being satisfied, since it has the highest chance of being imposed by a mathematical model. Let $G_C(a)$ be the probability that a random assignment X satisfies $C(X, a)$. The goal is to solve the following optimization problem.

$$\min_{a} \quad G_C(a)$$
$$\text{s.t.} \quad \bigwedge_{e \in E} C(e, a)$$
$$a_i \in \mathrm{dom}(a_i)$$

The method assumes that the domains of $X_i \in \boldsymbol{X}$ are identical. Let $p_v = P[X_i = v]$, the probability of assigning the value v to any variable. Recall that $S_C(\boldsymbol{a})$ is the solution set for C with parameters $\boldsymbol{a}$. Therefore, $P[e] = \prod_{i=1}^{n} p_{e_i}$ and $G_C(\boldsymbol{a})$ is the sum of the probabilities of each solution.

$$G_C(\boldsymbol{a}) = \sum_{e \in S_C(\boldsymbol{a})} P[e] = \sum_{e \in S_C(\boldsymbol{a})} \prod_{i=1}^{n} p_{e_i} \tag{1}$$

Picard-Cantin et al. encode the constraints using an automaton that they transform into a Markov chain by adding on each transition the probability of reading the associated value. The Markov chain efficiently computes the probability $G_C(\boldsymbol{a})$. To solve the optimization problem, they iterate over all combinations of parameters $\boldsymbol{a}$ while avoiding *dominated* parameters.

With this method, Picard-Cantin et al. [29] learn the parameters of global constraints that can be encoded as REGULAR, such as SEQUENCE and AMONG. The drawback is that listing all combinations of parameters for a global constraint quickly becomes infeasible, specifically when a parameter is a set. For example, there are $\frac{n^2}{2} \sum_{i=1}^{k} \binom{k}{i}$ combinations of parameters for SUBSETFOCUS($[X_1, \ldots, X_n], l, m, [z_1, \ldots, z_k]$). In this case, having a clever way to explore those combinations becomes crucial.

3   Methodology

We present a more refined approach to explore the space of parameters. We also show how to compute $G_C(\boldsymbol{a})$ for the constraints described in Sect. 2.1.

We propose a general branch-and-bound to explore the values that can be given to the parameters. The objective of the algorithm is to find the parameter values minimizing the probability function G for a given global constraint. Therefore, the bounding algorithm needs to compute a lower bound on G.

The branch-and-bound is presented in Algorithm 1 and it requires two sub-algorithms specific to the global constraint for which we wish to learn the parameters. The first, FILTERPARAMETERS$_C([\text{dom}'(a_1), \ldots, \text{dom}'(a_m)], E)$, filters the domains of the parameters using the given examples. The second, COMPUTELOWERBOUND$_C([\text{dom}'(a_1), \ldots, \text{dom}'(a_m)])$, computes an optimist lower bound on the probability function G. Those two are called at each node of the search tree.

3.1   Monotonicity

We also identify a situation where a subset of parameters can be fixed to their extreme values for the computation of the lower bound on G, simplifying bound computation. These parameters are such that the probability function G is monotonic w.r.t. those parameters (see Sect. 3.1).

Let $F : D^n \to \mathbb{R}$ be a multivariate function. F is said to be *monotonic w.r.t.* a_i if and only if $F_i(x, \boldsymbol{a}) = F(a_1, \ldots, a_{i-1}, x, a_{i+1} \ldots, a_m)$ is monotonic.

Algorithm 1. BRANCHANDBOUND(C, $[\dom(a_1), \ldots, \dom(a_m)]$, E)

1 $S \leftarrow \{\langle 0, [\dom(a_1), \ldots, \dom(a_m)]\rangle\}$
2 $BestSolution \leftarrow null$
3 $BestBound \leftarrow \infty$
4 **while** $\min\{lb \mid \langle lb, [\dom(a_1), \ldots, \dom(a_m)]\rangle \in S\} < BestBound$ **do**
5      $\langle lb, [\dom(a_1), \ldots, \dom(a_m)]\rangle \leftarrow \underset{\langle lb, [\dom(a_1), \ldots, \dom(a_m)]\rangle \in S}{\arg\min} lb$
6   Remove $\langle lb, [\dom(a_1), \ldots, \dom(a_m)]\rangle$ from S
7   **if** $|\dom(a_i)| = 1 \, \forall i \in \{1, \ldots, m\}$ **then**
8         $BestSolution \leftarrow [\dom(a_1), \ldots, \dom(a_m)]$
9         $BestBound \leftarrow lb$
10   **else**
11   Choose a_i such that $|\dom(a_i)| > 1$
12   **for** $v \in \dom(a_i)$ **do**
13   **for** $j \in \{1, \ldots, m\} \setminus \{i\}$ **do**
14              $\dom'(a_j) = \dom(a_j)$
15           $\dom'(a_i) = \{v\}$
16   FILTERPARAMETERS$_C([\dom'(a_1), \ldots, \dom'(a_m)], E)$
17           $lb' \leftarrow$ COMPUTELOWERBOUND$_C([\dom'(a_1), \ldots, \dom'(a_m)])$
18           $S \leftarrow S \cup \{\langle lb', [\dom'(a_1), \ldots, \dom'(a_m)]\rangle\}$

19 **return** $BestSolution, BestBound$

Theorem 1. *A global constraint* $C(\boldsymbol{X}, \boldsymbol{a})$ *has a monotonic probability function* $G_C(\boldsymbol{a})$ *w.r.t.* a_i *if either*

$$x \leq y \Rightarrow S_{C_i}(x, \boldsymbol{a}) \subseteq S_{C_i}(y, \boldsymbol{a}), \quad \forall x, y \in \dom(a_i)$$

or

$$x \leq y \Rightarrow S_{C_i}(y, \boldsymbol{a}) \subseteq S_{C_i}(x, \boldsymbol{a}), \quad \forall x, y \in \dom(a_i).$$

Proof. Suppose that for all $x, y \in \dom(a_i)$, $x \leq y$ implies $S_{C_i}(x, \boldsymbol{a}) \subseteq S_{C_i}(y, \boldsymbol{a})$. Then, for all $x, y \in \dom(a_i)$ such that $x \leq y$.

$$G_C(a_1, \ldots, a_{i-1}, y, a_{i+1}, a_m)$$

$$= \sum_{e \in S_{C_i}(y, a)} P[e]$$

$$= \sum_{e \in S_{C_i}(x, a)} P[e] + \sum_{e \in S_{C_i}(y, a) \setminus S_{C_i}(x, a)} P[e]$$

$$\geq G_C(a_1, \ldots, a_{i-1}, x, a_{i+1}, a_m)$$

A similar argument holds when $x \leq y \Rightarrow S_{C_i}(y, \boldsymbol{a}) \subseteq S_{C_i}(x, \boldsymbol{a}), \forall x, y \in \dom(a_i)$. $\square$

If $G_C(\boldsymbol{a})$ is monotonic w.r.t a_i for all $i \in \{1 \ldots, m\}$, then we can easily compute the lower bound on $G_C(\boldsymbol{a})$ at each node of the branch-and-bound. We only need to fix the parameters to their extreme values. To ensure that the examples satisfy the constraint, we apply filtering techniques to prune the domains of the parameters. Note that the parameter n is always known in our case.

Corollary 1. $G_{\text{SUBSETFOCUS}}(n, l, m, \boldsymbol{z})$ *is monotonic w.r.t l and m, but not $\boldsymbol{z}$.*

Proof. For all $l_1 \leq l_2$, we have $j - i + 1 \leq l_1 \leq l_2$ for all $s_{ij} \in \mathcal{S}$, therefore $S_{\text{SUBSETFOCUS}}(n, \boldsymbol{X}, l_1, m, \boldsymbol{z}) \subseteq S_{\text{SUBSETFOCUS}}(n, \boldsymbol{X}, l_2, m, \boldsymbol{z})$. For all $m_1 \leq m_2$, we have $|\mathcal{S}| \leq m_1 \leq m_2$, so $S_{\text{SUBSETFOCUS}}(n, \boldsymbol{X}, l, m_1, \boldsymbol{z}) \subseteq S_{\text{SUBSETFOCUS}}(n, \boldsymbol{X}, l, m_2, \boldsymbol{z})$. By Theorem 1, $G_{\text{SF}}(n, l, m, \boldsymbol{z})$ is monotonic w.r.t parameters l and m.

Let $p_1 = 0.73$, $p_2 = 0.23$, and $n = 3$. We have $G_{\text{SUBSETFOCUS}}(3, 3, 1, [1, 1]) = 1.0$, $G_{\text{SUBSETFOCUS}}(3, 3, 1, [1, 0]) \approx 0.8561$, $G_{\text{SUBSETFOCUS}}(3, 3, 1, [0, 1]) \approx 0.9468$, and $G_{\text{SUBSETFOCUS}}(3, 3, 1, [0, 0]) = 1.0$. Therefore, $G_{\text{SUBSETFOCUS}}(n, l, m, \boldsymbol{z})$ is not monotonic w.r.t. any z_i.   □

Corollary 2. $G_{\text{AMONG}}(n, l, u, \boldsymbol{z})$ *is monotonic w.r.t. l and u, but not $\boldsymbol{z}$.*

Proof. For all l_1, l_2 such that $0 \leq l_1 \leq l_2 \leq u$, if $\sum_{v \in \boldsymbol{z}} \text{occ}_v \geq l_2$ then we have $\sum_{v \in \boldsymbol{z}} \text{occ}_v \geq l_1$. Therefore, $S_{\text{AMONG}}(n, \boldsymbol{X}, l_2, u, \boldsymbol{z}) \subseteq S_{\text{AMONG}}(n, \boldsymbol{X}, l_1, u, \boldsymbol{z})$. For all u_1, u_2 such that $u_1 \leq u_2 \leq n$, if $\sum_{v \in \boldsymbol{z}} \text{occ}_v \leq u_1$ then we have $\sum_{v \in \boldsymbol{z}} \text{occ}_v \leq u_2$. Meaning that $S_{\text{AMONG}}(n, \boldsymbol{X}, l, u_1, \boldsymbol{z}) \subseteq S_{\text{AMONG}}(n, \boldsymbol{X}, l, u_2, \boldsymbol{z})$. Therefore, by Theorem 1 $G_{\text{AMONG}}(n, l, u, \boldsymbol{z})$ is monotonic w.r.t. l and u.

Let $p_1 = 0.73$, $p_2 = 0.23$, and $n = 3$. We have $G_{\text{AMONG}}(3, 2, 2, [1, 0]) \approx 0.2878$, $G_{\text{AMONG}}(3, 2, 2, [1, 1]) = 0.0$, $G_{\text{AMONG}}(3, 3, 3, [1, 1]) = 1.0$ and $G_{\text{AMONG}}(3, 3, 3, [1, 0]) \approx 0.389$. Therefore, $G_{\text{AMONG}}(n, l, u, \boldsymbol{z})$ is not monotonic w.r.t. any z_i.   □

Corollary 3. $G_{\text{SEQ}}(n, l = 0, u, w, \boldsymbol{z})$ *is monotonic w.r.t. u and w, but not $\boldsymbol{z}$.*

Proof. For all $0 \leq u_1, u_2$ s.t. $u_1 \leq u_2 \leq n$, if $\sum_{v \in \boldsymbol{z}} \text{occ}_v \leq u_1$ then $\sum_{v \in \boldsymbol{z}} \text{occ}_v \leq u_2$ and $S_{\text{SEQ}}(n, \boldsymbol{X}, 0, u_1, w, \boldsymbol{z}) \subseteq S_{\text{SEQ}}(n, \boldsymbol{X}, 0, u_2, w, \boldsymbol{z})$. If $\sum_{v \in \boldsymbol{z}} \text{occ}_v \leq u$ for all subsequences of length w_2 and if $w_1 \leq w_2$, then $\sum_{v \in \boldsymbol{z}} \text{occ}_v \leq u$ for all subsequences of length w_1 and $S_{\text{SEQ}}(n, \boldsymbol{X}, 0, u, w_2, \boldsymbol{z}) \subseteq S_{\text{SEQ}}(n, \boldsymbol{X}, 0, u, w_1, \boldsymbol{z})$. Therefore, by Theorem 1 $G_{\text{SEQ}}(n, 0, u, w, \boldsymbol{z})$ is monotonic w.r.t. u and w. By Corollary 2, $G_{\text{SEQ}}(n, 0, u, w, \boldsymbol{z})$ is not monotonic w.r.t. $\boldsymbol{z}$ when $n = w = 3$.   □

Corollary 4. $G_{\text{SEQ}}(n, l, u = n, w, \boldsymbol{z})$ *is monotonic w.r.t. l and w, but not $\boldsymbol{z}$.*

Proof. For all l_1, l_2 such that $0 \leq l_1 \leq l_2 \leq n$, if $\sum_{v \in \boldsymbol{z}} \text{occ}_v \geq l_2$ for all subsequences of length w then $\sum_{v \in \boldsymbol{z}} \text{occ}_v \geq l_1$ and $S_{\text{SEQ}}(n, \boldsymbol{X}, l_2, u, w, \boldsymbol{z}) \subseteq S_{\text{SEQ}}(n, \boldsymbol{X}, l_1, u, w, \boldsymbol{z})$. If $w_1 \leq w_2$ and if $\sum_{v \in \boldsymbol{z}} \text{occ}_v \geq l$ for all subsequences of length w_1, then we have $\sum_{v \in \boldsymbol{z}} \text{occ}_v \geq l$ for all subsequences of length w_2 and $S_{\text{SEQ}}(n, \boldsymbol{X}, l, u, w_1, \boldsymbol{z}) \subseteq S_{\text{SEQ}}(n, \boldsymbol{X}, l, u, w_2, \boldsymbol{z})$. Therefore, $G_{\text{SEQ}}(n, l = 0, u, w, \boldsymbol{z})$ is monotonic w.r.t. u and w by Theorem 1. $G_{\text{SEQ}}(n, l, u = n, w, \boldsymbol{z})$ is not monotonic w.r.t. $\boldsymbol{z}$ by Corollary 2 using $n = w = 3$.   □

Corollary 5. $G_{\text{GCC}}(n, l, u)$ *is monotonic w.r.t. each* l_i *and each* u_i.

Proof. For any $v \in \{1, \ldots, m\}$, if $l_v \leq l'_v$ and $u_v \leq u'_v$, then $l_v \leq l'_v \leq \text{occ}_v \leq u_v \leq u'_{v1}$. Therefore, $S_{\text{GCC}}(n, l, u) \subseteq S_{\text{GCC}}(n, [l_1, \ldots, l_{v-1}, l'_v, l_{v+1}, \ldots, l_m], u)$ and $S_{\text{GCC}}(n, l, u) \subseteq S_{\text{GCC}}(n, l, [u_1, \ldots u_{v-1}, u'_v, u_{v+1}, \ldots u_m])$.   □

Corollary 6. *Both* $G_{\text{ATMOSTNVALUE}}(n, k)$ *and* $G_{\text{ATLEASTNVALUE}}(n, k)$ *are monotonic w.r.t.* k.

Proof. Solutions with at most k values form a subset of the solutions with at most $k + 1$ values. A symmetric argument holds for ATLEASTNVALUE.   □

Corollary 7. *Both* $G_{\text{ATMOSTBALANCE}}(n, b)$ *and* $G_{\text{ATMOSTBALANCE}^*}(n, b)$ *are monotonic w.r.t.* b.

Proof. Solutions with a balance no greater than b form a subset of solutions with a balance no greater than $b + 1$, regardless of how the balance is computed.   □

The previous theorem and corollaries are used in the computation of the bound on G in the branch-and-bound algorithm. These results show that we can temporarily fix the monotonic parameters to their extreme values to compute an optimist bound.

3.2   Bounding and Filtering Specific Constraints

We show how to implement the functions FILTERPARAMETERS$_C$ and COMPUTELOWERBOUND$_C$ for the constraints AMONG, SEQUENCE, SUBSETFOCUS, GCC, ATMOSTNVALUE, ATLEASTNVALUE, ATMOSTBALANCE, and ATMOSTBALANCE*.

SubsetFocus. We define the function COMPUTELOWERBOUND$_{\text{SUBSETFOCUS}}$ that computes a lower bound on the probability $G_{\text{SUBSETFOCUS}}(n, l, m, z)$. To take into consideration the other parameters, we consider the probability p that a variable X_i is assigned to a value v such that $z_v = 1$. During the branch-and-bound, the set z is only partially defined depending on which parameter z_v is assigned. However, this partial assignment allows bounding of the probability p as follows. The summations apply on instantiated parameters.

$$\alpha := \sum_{v : z_v = 1} p_v, \quad \beta := 1 - \sum_{v : z_v = 0} p_v, \quad \alpha \leq p \leq \beta$$

A solution e can be mapped to a binary vector y such that $y_i = 1 \iff e_i = v \land z_v = 1$. The satisfiability of the constraint can be tested simply by checking the number and the length of the stretches of ones in the vector y. Let $\gamma(e) = |\{i \mid e_i = v \land z_v = 1\}| = \sum_{i=1}^{n} y_i$ be the number of variables assigned to a value in the considered set for a given solution e. Let $A(n, l, m, k)$ be the number of vectors y of size n with exactly k components set to 1 that satisfies the constraint with parameters l and m. The probability $G_{\text{SUBSETFOCUS}}(n, l, m, \mathbf{z})$ can be computed as follows.

$$G_{\text{SF}}(n, l, m, \mathbf{z}) = \sum_{e \in S_{\text{SF}}(\mathbf{a})} P[e] = \sum_{e \in S_{\text{SF}}(\mathbf{a})} p^{\gamma(e)}(1-p)^{n-\gamma(e)} \tag{2}$$

$$= \sum_{k=0}^{n} A(n, l, m, k) p^k (1-p)^{n-k}. \tag{3}$$

Therefore, the probability $G_{\text{SUBSETFOCUS}}(n, l, m, \mathbf{z})$ can be bounded with

$$G_{\text{SF}}(n, l, m, \mathbf{z}) \geq \sum_{k=0}^{n} A(n, l, m, k) \min_{\alpha \leq p \leq \beta} p^k (1-p)^{n-k}.$$

Differentiating $x^k(1-x)^{n-k}$ yields the optima $k/n, 0, 1$. Since $0 \leq \alpha \leq \beta \leq 1$ and since k/n is a maximum, the bound used in the branching algorithm is

$$\sum_{k=0}^{n} A(n, l, m, k) \min_{p \in \{\alpha, \beta\}} p^k (1-p)^{n-k}. \tag{4}$$

We now have to show how to compute the function $A(n, l, m, k)$. Let $D(l, k, m)$ be the number of ways to split a sequence of length k into exactly m subsequences of length at most l. The function $D(l, k, m)$ can be recursively computed.

$$D(l, k, m) = \begin{cases} 1 & \text{if } k = m = 0 \\ 0 & \text{if } m = 0 \wedge k > 0 \\ 0 & \text{if } m > 0 \wedge k = 0 \\ \sum\limits_{j=1}^{\min(k,l)} D(l, k-j, m-1) & \text{otherwise} \end{cases}$$

To count the number of feasible binary vectors $\mathbf{y}$, we count how many ways we can segment a sequence of k 1s into exactly i sequences using the function $D(l, k, i)$. We insert $i-1$ zeros, one between each segment. There exist $\binom{n-k+1}{i}$ ways to insert the remaining $n - k - (i-1)$ zeros before or after the segments to obtain a vector of length n. Finally, we let the number of segments i vary between 1 and m.

$$A(n, l, m, k) = \sum_{i=1}^{m} D(l, k, i) \binom{n-k+1}{i}.$$

Since $G_{\text{SF}}(n, l, m, \mathbf{z})$ is monotonic w.r.t. m and l by Corollary 1, the function COMPUTELOWERBOUND$_{\text{SUBSETFOCUS}}$ returns the bound (4) by setting l and m to their smallest value in their domains.

During the branch-and-bound, we apply a minimal filtering to the domains of parameters l and m. Here is how we define FILTERPARAMETERS$_{\text{SUBSETFOCUS}}$. We suppose that all uninstantiated variables z_i could be instantiated to zero. In that situation, we compute the length of the largest stretch in an example $e \in E$

and set this length as a lower bound on $\text{dom}(l)$. To compute a lower bound on $\text{dom}(m)$, for each example $e \in E$, we create a binary vector $\boldsymbol{y}$ by setting $y_i = z_{e_i}$ if z_{e_i} is instantiated. When z_{e_i} is uninstantiated, we greedily assign a value to y_i to minimize the number of stretches in $\boldsymbol{y}$. We set $\min(\text{dom}(m))$ to the maximum number of stretches observed in one example. The parameters $\boldsymbol{z}$ are not filtered.

Lemma 1. *The bound on* $G_{\text{SUBSETFOCUS}}(n, l, m, \boldsymbol{z})$ *from* (4) *is tight when variables are instantiated.*

Proof. Since $\boldsymbol{z}$ are fixed, then $\sum_{v:z_v=1} p_v = \alpha = \beta = p$. Therefore, (4) is equal to (3) and the bound is tight. $\square$

Sequence. Consider the SEQUENCE constraint whose parameter l is known to be 0. We therefore need to learn parameters u, w, and $\boldsymbol{z}$. The probability $G_{\text{SEQ}}(n, u, w, \boldsymbol{z})$ can be bounded in a similar way as we did for SUBSETFOCUS. Let $B(n, u, w, k)$ be the number of binary vectors $\boldsymbol{y}$ of length n such that exactly k components are set to one and the sum of any subsequence of w components is at most u. Therefore, the function COMPUTELOWERBOUND$_{\text{SEQUENCE}}$ returns the following lower bound on $G_{\text{SEQ}}(n, u, w, \boldsymbol{z})$.

$$G_{\text{SEQ}}(n, u, w, \boldsymbol{z}) = \sum_{k=0}^{n} B(n, u, w, k) p^k (1-p)^{n-k} \tag{5}$$

$$\geq \sum_{k=0}^{n} B(n, u, w, k) \min_{p \in \{\alpha, \beta\}} p^k (1-p)^{n-k} \tag{6}$$

Since $G_{\text{SEQ}}(n, u, w, \boldsymbol{z})$ is monotonic w.r.t. u and w by Corollary 3, we compute $B(n, u, w, k)$ with u fixed to its minimal value and w fixed to its maximal value. We use dynamic programming to compute the function $B(n, u, w, k)$. Let $F(n, u, w, k, [s_1, \ldots, s_w])$ be the number of feasible binary vectors $\boldsymbol{y}$ of length n with exactly k components set to one and whose last w components are $[s_1, \ldots, s_w]$.

$$F(n, u, k, [s_1, \ldots, s_w]) = \begin{cases} 0 & \text{if } \sum_{i=1}^{n} s_i > \min(u, k) \\ 0 & \text{if } n - w + \sum_{i=1}^{n} s_i < k \\ 1 & \text{if } n = w \\ F(n-1, u, k - s_w, [0, s_1, \ldots, s_{w-1}]) \\ \quad + F(n-1, u, k - s_w, [1, s_1, \ldots, s_{w-1}]) & \text{otherwise} \end{cases}$$

$$B(n, u, w, k) = F(n + w, u, k, [\underbrace{0, \ldots, 0}_{w \text{ times}}])$$

The two first cases in function F occur when there are too many or two few ones in the sequence $[s_1, \ldots, s_w]$ to be extended to a feasible sequence of length n. The third case occurs when $n = w$ and $[s_1, \ldots, s_w]$ is a feasible solution. The last case computes the number of solutions with one fewer component. The function $B(n, u, w, k)$ gives the number of sequences of length $n + w$ whose last w components are null.

We define FILTERPARAMETERS$_\text{SEQUENCE}$. Once the parameter w is instantiated, the parameter u can be filtered. For every solution $e \in E$ and for every subsequence of length w in e, we count the number of variables that must belong to z. The largest encountered value gives a lower bound on u.

Lemma 2. *The bound on $G_\text{SEQ}(n, u, w, z)$ from (6) is tight when variables are instantiated.*

Proof. Since z are fixed, then $\sum_{v:z_v=1} p_v = \alpha = \beta = p$. Therefore, (6) is equal to (5) and the bound is tight. □

Among. The probability that a random assignment has exactly k variables assigned to a value v such that $z_v = 1$ follows a binomial distribution. The probability to satisfy the constraint AMONG with parameters l, u, and z is therefore

$$G_\text{AMONG}(n, l, u, z) = \sum_{k=l}^{u} \binom{n}{k} p^k (1-p)^{n-k} \tag{7}$$

As we did for SUBSETFOCUS, the function COMPUTELOWERBOUND$_\text{AMONG}$ returns the following lower bound.

$$G_\text{AMONG}(n, l, u, z) \geq \sum_{k=l}^{u} \binom{n}{k} \min_{p \in \{\alpha, \beta\}} p^k (1-p)^{n-k} \tag{8}$$

By Corollary 2, we can simply fix u to its minimal value and l to its maximal value for the bound computation.

During the search, one can filter the parameters l and u using the function FILTERPARAMETERS$_\text{AMONG}$. Let $M = \{v \mid z_v = 1\}$ be the set of values that are considered and $C = \{v \mid z_v = 1 \lor z_v$ is uninstantiated$\}$ be the set of values that might be considered. One can set the lower bound of dom(l) to $\min_{e \in E} |\{i \mid e_i \in M\}|$ and the upper bound of dom(u) to $\max_{e \in E} |\{i \mid e_i \in C\}|$.

Lemma 3. *The bound on $G_\text{AMONG}(n, l, u, z)$ from (8) is tight when variables are instantiated.*

Proof. Since z are fixed, then $\sum_{v:z_v=1} p_v = \alpha = \beta = p$. Therefore, 8 is equal to 7 and the bound is tight. □

Other Constraints. Unlike with SUBSETFOCUS, SEQUENCE, and AMONG, the parameters for GCC, ATMOSTNVALUE, ATLEASTNVALUE, ATMOSTBALANCE, and ATMOSTBALANCE* are all monotonic and independent. This means that the branch-and-bound finds the optimal solution for these last constraints without backtracking. For the GCC for example, the optimal choice is always $\min(\text{dom}(l_i))$ when branching on a l_i parameter and $\max(\text{dom}(u_i))$ when branching on a u_i parameter by Corollary 5. The algorithm finds the optimal solution on the first try with the help of the filtering algorithm for the parameters.

4   Experiments

We test our branch-and-bound algorithm on two different benchmarks, one for SUBSETFOCUS and one for SEQUENCE, both solved on an Intel Core i7 3.40 GHz with 4 Gb of RAM running Linux. All the code was written in Python, except for the brute force algorithm that learns the parameters of SEQUENCE, for which Picard-Cantin et al. [29] provided their code written in R.

Since AMONG is a specific case of SEQUENCE, we do not experiment on this constraint. Furthermore, we do not need to experiment on GCC, ATMOST-NVALUE, ATLEASTNVALUE, ATMOSTBALANCE or ATMOSTBALANCE* since the solution is found without backtracks.

Random timetables were generated for the benchmark. A sequence is the schedule of an employee where each component of the sequence corresponds to a task for a given time slot. The task 0 is a special task that corresponds to a day off. This value is known not to belong to the considered set and therefore $z_0 = 0$ for both experiments: the one with SUBSETFOCUS and the one with SEQUENCE.

Note that the generated example have no particular structure as they are chosen uniformly among solutions that satisfy either SUBSETFOCUS or SEQUENCE with the given parameters and therefore they have maximal entropy. The examples are not real, but they are possibly not tight, meaning that they satisfy constraints stricter than the one imposed.

4.1   SubsetFocus

The benchmark for SUBSETFOCUS is composed of 600 randomly generated instances. An instance is composed of a set of $d + 1$ values $\{0, \ldots, d\}$, a vector of assignment probability for the values $\boldsymbol{p} = [p_0, \ldots, p_d]$ ($p_v = P[X_i = v]$), a horizon n, and the set of parameters $(l, m, \boldsymbol{z})$. We generated instances with $d \in \{10, 30\}$, $l \in \{1, \ldots, 10\}$, $m \in \{1, \ldots, 10\}$, and $n \in \{100, 200, 300\}$. The vector $\boldsymbol{p}$ is generated such that $\sum_v p_v = 1$ and $\sum_{v:z_v=1} p_v \in \{0.2, 0.8\}$. Finally, for each instance defined by $(d, \boldsymbol{z}, \boldsymbol{p}, l, m, n)$, we generate ten (10) solutions that satisfy SUBSETFOCUS with parameters $l, m, \boldsymbol{z}$ and whose task occurrences are proportional to their probabilities $\boldsymbol{p}$. The learning algorithms are given $\{0, \ldots, d\}$, $\boldsymbol{p}$, n, and a subset of the ten solutions (according to the experiment). Their goal is to learn the parameters $(l, m, \boldsymbol{z})$ that generated the solutions.

We compare the branch-and-bound algorithm described earlier with a brute force algorithm that lists all possible sets $\boldsymbol{z}$ (with $z_0 = 0$), that computes lower bounds on l and m from examples, that uses the monotonicity to fix l and m to their minimums, and that finally computes the probability of each resulting combination of parameters for SUBSETFOCUS. This later technique is equivalent to the one proposed by Picard-Cantin et al. [29].

Figure 1 shows the resolution time to learn the parameters of SUBSETFOCUS from a single solution. The branch-and-bound dominates the brute force by solving every instance in fewer than 73 s while the brute force sometimes reaches the 6-min timeout. Figure 2 compares the number of times that the probability of the constraint for given parameters is computed using (3). That is once per node for the branch-and-bound and once per parameter combination for the brute force. Once again, the branch-and-bound dominates the brute force. Figure 3 shows how many times, in percentages, the algorithms correctly predict the parameters given the number of examples that are provided. Since both algorithms return the same parameters when they solve an instance under the time limit, we only consider the branch-and-bound algorithm for this analysis. The low success rates are not alarming since the initial constraints were overly permissive in many cases. Therefore, the examples produced tend to satisfy more restrictive constraints that the algorithms detect.

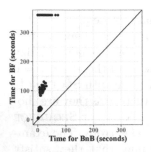 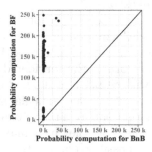 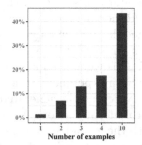

Fig. 1. Resolution time (log scale) for SUBSETFOCUS

Fig. 2. Number of probability computations for SUBSETFOCUS

Fig. 3. Percentage of correctly learned instances for SUBSETFOCUS

More details are given in Table 1 on the prediction quality of the learning algorithm according to the number of examples considered. For this comparison, we keep the three best predictions for each instance, meaning we keep the three sets of parameters for SUBSETFOCUS satisfying the examples and with the lowest probability $G_{\text{SUBSETFOCUS}}$. The rank of an instance is the place of the initial constraint in the prediction list. If the initial constraint was not in the best three sets of parameters returned by the algorithm, then we consider the rank to be ∞.

Table 1. Results for SUBSETFOCUS. Number of instances for which the initial constraint was ranked first, second, third or was not found.

Num. of examples	Rank of initial constraint				Num. of instances
	1	2	3	∞	
1	8	46	1	545	600
2	42	119	0	439	600
3	78	148	0	374	600
4	105	172	0	323	600
5	139	170	0	291	600
10	261	117	0	222	600

The results of the first column are represented in Fig. 3. From these results, we can say that if the learning algorithm has to give a short list of candidates from which the user would choose to add to the mathematical model, it would return the correct constraint in the first two choices 63% of the time, considering ten (10) examples. Moreover, the correct constraint seldom takes the third position.

4.2 Sequence

The benchmark for SEQUENCE is composed of 84 randomly generated instances. An instance is composed of a set of values $\{0, \ldots, d\}$, a vector of assignment probability for the values $\boldsymbol{p} = [p_0, \ldots, p_d]$ ($p_v = P[X_i = v]$), a horizon n, and the set of parameters $(u, w, \boldsymbol{z})$. We fix $l = 0$ for all instances.

We generated instances with $d \in \{10, 30\}$, $w \in \{5, 6, 7\}$, $u \in \{1 + 3k \mid 1 + 3k \leq w \wedge k \in \mathbb{N}\}$, and $n \in \{100, 200, 300\}$. The vectors $\boldsymbol{z}$ and $\boldsymbol{p}$ are generated such that $\sum_v p_v = 1$ and $\sum_{v : z_v = 1} p_v \in \{0.2, 0.8\}$. Finally, for each instance defined by $(d, \boldsymbol{z}, \boldsymbol{p}, u, w, n)$, we generate ten (10) solutions that satisfy SEQUENCE with parameters $u, w, \boldsymbol{z}$ and whose task occurrences are proportional to their probabilities $\boldsymbol{p}$.

Figure 4 shows the computation times. We can separate the instances into two subsets according to the number of values $d \in \{10, 30\}$. When the number of values is small, the brute force algorithm is faster because it does not have to keep track of partial problems like the branch-and-bound does. When the number of values is large, the brute force algorithm has too many combinations to test and the branch-and-bound is faster. Therefore, the branch-and-bound algorithm for SEQUENCE is more useful when the problem to solve is large. Figure 5 explains why the branch-and-bound is faster on larger instances as it shows that the number of probability computations is smaller for the branch-and-bound algorithm. Figure 6 shows how many times the algorithms correctly predict the parameters given the number of examples that are provided. As for SUBSETFOCUS, we give more information about the prediction quality of the algorithm in Table 2. For SEQUENCE, we observe that the correct constraint is returned by the learning algorithm among the first two choices 59.5% of the time when we consider ten examples.

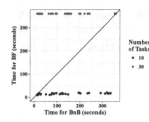

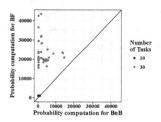

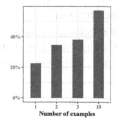

Fig. 4. Resolution time (log scale) for SEQUENCE

Fig. 5. Number of probability computations for SEQUENCE

Fig. 6. Percentage of correctly learned instances for SEQUENCE

Table 2. Results for Sequence. Number of instances for which the initial constraint was ranked first, second, third or was not found.

Num. of examples	Rank of initial constraint				Num. of instances
	1	2	3	∞	
1	19	7	0	58	84
2	29	7	0	48	84
3	32	8	0	44	84
10	48	2	0	34	84

4.3   Discussion

The proposed branch-and-bound is more efficient than the state-of-the-art algorithms and it can be applied to multiple global constraints. We explained how to compute a lower bound on G_C for eight global constraints.

A drawback of the technique is that it does not detect overly permissive constraints with either algorithm, the branch-and-bound or the brute force. This is due to the fact that a permissive constraint has a lot of solutions and those solutions might satisfy more restrictive constraints. To avoid overfitting and therefore learning constraints that are too restrictive, one needs to increase the number of examples. In our experiments, we obtained satisfying results with only ten examples. We didn't test overly permissive constraints as we are certain not to find them as the most probable constraints and probably not even in the top 3. Having prediction errors on permissive constraints is not as important as missing tight constraints since the former have a smaller impact on the solutions.

5   Conclusion

We showed how a branch-and-bound can be used to learn the parameters of global constraints from positive examples. We showed how the monotonicity can be exploited to obtain tight bounds on the probability that a random assignment satisfies a constraint. Some constraints have for parameter a set of values z. The set is used to limit the number of occurrences of its values. This parameter is not monotonic. Nevertheless, we presented a technique to bound the probability by counting the number of solutions that satisfy the constraint with a fixed number of values belonging to a set. Experiments show that the new technique is more time efficient than the state of the art algorithms, based on benchmarks inspired from timetabling problems.

References

1. Arcangioli, R., Lazaar, N.: Multiple constraint acquisition. In: Proceedings of the 2015 International Conference on Constraints and Preferences for Configuration and Recommendation and Intelligent Techniques for Web Personalization, CPCR+ITWP 2015, vol. 1440, pp. 16–20. CEUR-WS.org (2015)

2. Barták, R., Čepek, O., Surynek, P.: Discovering implied constraints in precedence graphs with alternatives. Ann. Oper. Res. **180**(1), 233–263 (2010)
3. Bartolini, A., Lombardi, M., Milano, M., Benini, L.: Neuron constraints to model complex real-world problems. In: Lee, J. (ed.) CP 2011. LNCS, vol. 6876, pp. 115–129. Springer, Heidelberg (2011). doi:10.1007/978-3-642-23786-7_11
4. Beldiceanu, N., Contejean, E.: Introducing global constraints in chip. Math. Comput. Model. **20**(12), 97–123 (1994)
5. Beldiceanu, N., Ifrim, G., Lenoir, A., Simonis, H.: Describing and generating solutions for the EDF unit commitment problem with the ModelSeeker. In: Schulte, C. (ed.) CP 2013. LNCS, vol. 8124, pp. 733–748. Springer, Heidelberg (2013). doi:10.1007/978-3-642-40627-0_54
6. Beldiceanu, N., Simonis, H.: A constraint seeker: finding and ranking global constraints from examples. In: Lee, J. (ed.) CP 2011. LNCS, vol. 6876, pp. 12–26. Springer, Heidelberg (2011). doi:10.1007/978-3-642-23786-7_4
7. Beldiceanu, N., Simonis, H.: Using the global constraint seeker for learning structured constraint models: a first attempt. In: The 10th International Workshop on Constraint Modelling and Reformulation (ModRef 2011), Perugia, Italy, pp. 20–34 (2011)
8. Beldiceanu, N., Simonis, H.: A model seeker: extracting global constraint models from positive examples. In: Milano, M. (ed.) CP 2012. LNCS, pp. 141–157. Springer, Heidelberg (2012). doi:10.1007/978-3-642-33558-7_13
9. Beldiceanu, N., Carlsson, M., Demassey, S., Petit, T.: Global constraint catalogue: past, present and future. Constraints **12**(1), 21–62 (2007)
10. Bessiere, C., Coletta, R., Daoudi, A., Lazaar, N., Mechqrane, Y., Bouyakhf, E.H.: Boosting constraint acquisition via generalization queries. In: ECAI, pp. 99–104 (2014)
11. Bessiere, C., Coletta, R., Freuder, E.C., O'Sullivan, B.: Leveraging the learning power of examples in automated constraint acquisition. In: Principles and Practice of Constraint Programming-CP 2004, pp. 123–137 (2004)
12. Bessiere, C., Coletta, R., Petit, T.: Acquiring parameters of implied global constraints. In: Principles and Practice of Constraint Programming-CP 2005, pp. 747–751 (2005)
13. Bessiere, C., Coletta, R., Petit, T.: Learning implied global constraints. In: IJCAI, pp. 44–49 (2007)
14. Bessiere, C., et al.: New approaches to constraint acquisition. In: Bessiere, C., De Raedt, L., Kotthoff, L., Nijssen, S., O'Sullivan, B., Pedreschi, D. (eds.) Data Mining and Constraint Programming. LNCS, vol. 10101, pp. 51–76. Springer, Cham (2016). doi:10.1007/978-3-319-50137-6_3
15. Bessiere, C., Koriche, F., Lazaar, N., O'Sullivan, B.: Constraint acquisition. Artif. Intell. (2015, in press)
16. Bessiere, C., Hebrard, E., Katsirelos, G., Kiziltan, Z., Picard-Cantin, É., Quimper, C.-G., Walsh, T.: The balance constraint family. In: O'Sullivan, B. (ed.) CP 2014. LNCS, vol. 8656, pp. 174–189. Springer, Cham (2014). doi:10.1007/978-3-319-10428-7_15
17. Bonfietti, A., Lombardi, M., Milano, M.: Embedding decision trees and random forests in constraint programming. In: Michel, L. (ed.) CPAIOR 2015. LNCS, vol. 9075, pp. 74–90. Springer, Cham (2015). doi:10.1007/978-3-319-18008-3_6
18. Campigotto, P., Passerini, A., Battiti, R.: Active learning of combinatorial features for interactive optimization. In: Coello, C.A.C. (ed.) LION 2011. LNCS, vol. 6683, pp. 336–350. Springer, Heidelberg (2011). doi:10.1007/978-3-642-25566-3_25

19. Charnley, J., Colton, S., Miguel, I.: Automatic generation of implied constraints. ECAI **141**, 73–77 (2006)
20. Daoudi, A., Lazaar, N., Mechqrane, Y., Bessiere, C., Bouyakhf, E.H.: Detecting types of variables for generalization in constraint acquisition. In: 2015 IEEE 27th International Conference on Tools with Artificial Intelligence (ICTAI), pp. 413–420. IEEE (2015)
21. Kiziltan, Z., Lippi, M., Torroni, P.: Constraint detection in natural language problem descriptions. In: Proceedings of the Twenty-fifth International Joint Conference on Artificial Intelligence, IJCAI 2016, New York, USA, 9–15 July 2016
22. Kolb, S.: Learning constraints and optimization criteria. In: Proceedings of the First Workshop on Declarative Learning Based Programming (2016)
23. Little, J., Gebruers, C., Bridge, D., Freuder, E.C.: Using case-based reasoning to write constraint programs. In: Rossi, F. (ed.) CP 2003. LNCS, vol. 2833, p. 983. Springer, Heidelberg (2003). doi:10.1007/978-3-540-45193-8_107
24. Lombardi, M., Milano, M., Bartolini, A.: Empirical decision model learning. Artif. Intell. **244**, 343–367 (2017). Elsevier
25. Lopez, M., Lallouet, A.: On learning CSP specifications. In: The 15th International Conference on Principles and Practice of Constraint Programming Doctoral Program Proceedings, p. 70 (2009)
26. Pachet, F., Roy, P.: Automatic generation of music programs. In: International Conference on Principles and Practice of Constraint Programming, pp. 331–345 (1999)
27. Pesant, G., Quimper, C.G., Zanarini, A.: Counting-based search: branching heuristics for constraint satisfaction problems. J. Artif. Intell. Res. **43**, 173–210 (2012)
28. Petit, T.: FOCUS: a constraint for concentrating high costs. In: Milano, M. (ed.) CP 2012. LNCS, pp. 577–592. Springer, Heidelberg (2012). doi:10.1007/978-3-642-33558-7_42
29. Picard-Cantin, É., Bouchard, M., Quimper, C.-G., Sweeney, J.: Learning parameters for the sequence constraint from solutions. In: Rueher, M. (ed.) CP 2016. LNCS, vol. 9892, pp. 405–420. Springer, Cham (2016). doi:10.1007/978-3-319-44953-1_26
30. Régin, J.C.: Generalized arc consistency for global cardinality constraint. In: Proceedings of the 13th National Conference on Artificial Intelligence, AAAI 1996, vol. 1, pp. 209–215. AAAI Press (1996)
31. Suraweera, P., Webb, G.I., Evans, I., Wallace, M.: Learning crew scheduling constraints from historical schedules. Transp. Res. Part C: Emerg. Technol. **26**, 214–232 (2013)

CoverSize: A Global Constraint
for Frequency-Based Itemset Mining

Pierre Schaus[1]([⊠]), John O.R. Aoga[1,2][iD], and Tias Guns[3,4]

[1] UCLouvain, ICTEAM, Louvain-la-Neuve, Belgium
{pierre.schaus,john.aoga}@uclouvain.be
[2] UAC, ED-SDI, Abomey-Calavi, Benin
[3] VUB Brussels, Brussels, Belgium
tias.guns@vub.be
[4] KU Leuven, Leuven, Belgium
tias.guns@cs.kuleuven.be

Abstract. Constraint Programming is becoming competitive for solving certain data-mining problems largely due to the development of global constraints. We introduce the CoverSize constraint for itemset mining problems, a global constraint for counting and constraining the number of transactions covered by the itemset decision variables. We show the relation of this constraint to the well-known table constraint, and our filtering algorithm internally uses the reversible sparse bitset data structure recently proposed for filtering table. Furthermore, we expose the size of the cover as a variable, which opens up new modelling perspectives compared to an existing global constraint for (closed) frequent itemset mining. For example, one can constrain minimum frequency or compare the frequency of an itemset in different datasets as is done in discriminative itemset mining. We demonstrate experimentally on the frequent, closed and discriminative itemset mining problems that the CoverSize constraint with reversible sparse bitsets allows to outperform other CP approaches.

1 Introduction

Frequent itemset mining (FIM) is one of the well-known and most studied data mining problems [8] and first introduced in [2]. Guns et al. [19] showed that FIM problems could be modelled and solved using Constraint Programming (CP) with the additional benefit that new constraints can easily be integrated into the models. Since then several CP (also SAT) approaches have been proposed for other data-mining problems such as frequent sequence mining [22,29], dominance-based pattern mining [28] and closed FIM [20,21,24].

The flexibility of adding constraints when using a generic CP solver typically comes at the cost of efficiency; a well-known tradeoff. We can hence look at itemset mining papers in terms of where they are on the efficiency versus generality scale. Most works in itemset mining focus primarily on efficiency [8,39], while typical constraint-based mining papers hard-code a select number of constraints

© Springer International Publishing AG 2017
J.C. Beck (Ed.): CP 2017, LNCS 10416, pp. 529–546, 2017.
DOI: 10.1007/978-3-319-66158-2_34

based on properties like (anti-)monotonicity [32]. Earlier papers on using CP for itemset mining focus mostly on generality and decompose the itemset mining constraints into many (reified) linear constraints [19] at the cost of efficiency. In line with recent works in CP for sequence mining [3,4,22], Lazaar et al. [24] have shown that a single global constraint for closed frequent itemset mining can outperform a decomposition approach. This comes at significant cost for generality though, because (1) by encapsulating all but the itemset variables, only syntactic constraints on the items can be added; (2) only *closed frequent* patterns can be found and adding syntactic constraints can have unwanted side-effects [6].

In this paper, we aim to maximize both generality and efficiency while employing a global constraint for itemset mining. We achieve this by introducing the *CoverSize* global constraint, which (1) computes a lower and upper-bound on the frequency and synchronizes this with a decision variable, meaning that the frequency can be used in other separate constraints; (2) internally, the filtering algorithm uses the reversible sparse bitset data structure which was introduced to efficiently filter table constraints [14]. This *CoverSize* constraint can be combined with a *CoverClosure* constraint to enforce the closed property [24,34] or with a discriminative optimization constraint (such as χ^2) to solve the correlated itemset mining problem as in [31].

In contrast to most constraints in CP, what is typical about global constraints for data mining is that they must be able to handle large amounts of data. A traditional global constraint that shares this property is the table constraint, which has a rich history in CP literature [5,12,25,35]. Its link with a global constraint for itemset mining (IM) is even stronger, as both can be seen as operating on a binary matrix; for IM the columns are *items* (Boolean variables) and for table the columns are *(variable, value) pairs*. The use of bitvectors and fast bitvector operations is common in itemset mining implementations, indeed, it was also used for the closed FIM constraint [24]. Related, a column-based bitvector representation for the table constraint was recently proposed [14], and the propagator was shown to outperform all other approaches. Inspired by this relation, we show that the *reversible sparse bitset* data structure that was devised for table can also be used to implement efficient itemset mining propagators. Our scaling experiments on large and sparse data indicate the benefit of this.

Furthermore, our proposed *CoverSize* global constraint propagates from the *item* variables to a variable representing the frequency and back. This means that the same constraint can be used to enforce a minimum and maximum frequency of a set in a table/database. We show that domain-consistent filtering is NP-hard, though good results can be obtained with weaker filtering. We showcase the added flexibility of such a choice by using it as a building block for modelling closed frequent itemset problem [24] and correlated itemset mining problem [31]. For *closed*, we argue for and propose a separate global constraint.

Our contributions are hence as follows: we show how advances in data structures for table constraints can benefit global constraints for itemset mining too; we propose that global constraints for itemset mining expose the frequency through a variable, and demonstrate how this allows, for example, to solve discriminative itemset mining too; and empirically our experiments with a generic CP solver show that this approach outperforms other CP approaches and is on par with a special-purpose CP solver, thereby decreasing the gap to the highly efficient specialized itemset miners.

2 Background

2.1 Itemset Mining

Frequent itemset mining is concerned with finding a set of *items* that appears frequently in a database of sets [2]. The database is often called a transaction database, and each entry in the database is called a transaction (such as a purchase of products). Itemset mining has applications in market basket analysis, web log mining, bio-informatics and more [1].

More formally, given a set $\mathcal{I}$ of n possible items and a transaction database of size m: $\mathcal{H} = \{(t, T) \mid t \in \{1, \dots, m\}, T \subseteq \mathcal{I}\}$. Figure 1a shows an example database with $\mathcal{I} = \{A, B, C, D\}$. The goal of the *frequent itemset mining problem* is to enumerate all sets $I \subseteq \mathcal{I}$ such that $|\{(t, T) \in \mathcal{H} \mid I \subseteq T\}| \geq \theta$ with θ a user-supplied threshold.

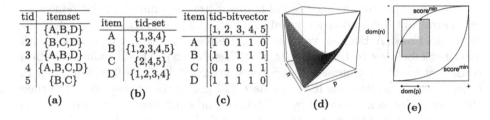

Fig. 1. (a–c) Three equivalent representations of itemset databases, (d) χ^2 ZDC function and (e) Filtering of the $ZDC(|\mathcal{D}^+|, |\mathcal{D}^-|, p, n, score)$ constraint. Note: $c+ = p$, $c- = n$, (a) Horizontal sparse - $\mathcal{H}$, (b) Vertical sparse - $\mathcal{V}$ (c) Vertical dense - $\mathcal{D}$

The set of transactions that contain the itemset $\{(t, T) \in \mathcal{H} \mid I \subseteq T\}$ is called the *cover*. Computing this set efficiently is a core aspect of itemset mining algorithms. Different algorithms have used different representations of the transaction database. Figure 1(a–c) shows three of them. In a vertical representation, the intersection is the key operation. Let $\mathcal{V}(i)$ be the set of transaction identifiers of item i, then $|\{(t, T) \in \mathcal{H} \mid I \subseteq T\}| = |\{\bigcap_{i \in I} \mathcal{V}(i)\}|$. In case of a vertical dense bitvector representation, efficient bitwise operations can be used for the intersection, which scales very well in practice.

Many other variations on frequent itemset mining have been investigated. For example, *closed* frequent itemset mining adds the additional restriction that an itemset must not have a superset with the same frequency: $\nexists I' \supset I : \{(t, T) \in \mathcal{H} \mid I' \subseteq T\} = \{(t, T) \in \mathcal{H} \mid I \subseteq T\}$; and *maximal* frequent itemset mining has additional restriction that an itemset must not have any frequent superset: $\nexists I' \supset I : |\{(t,T) \in \mathcal{H} \mid I' \subseteq T\}| \geq \theta$ [1]. Other constraints on items and transactions have been investigated as well [32].

In an optimization setting, one can search for the most or least frequent itemsets (typically under a number of other constraints) or find the most discriminating itemsets. Given two databases $\mathcal{H}^+$ and $\mathcal{H}^-$ (for example, from two consecutive months), the goal is to find the itemset(s) that best discriminate between the two; such as an itemset that is very frequent in one and barely frequent in the other. A range of *discriminative* measures, also called correlation measures, have been studied [27]. A property that we will exploit later is that these measures can be computed using just information on the frequency of the sets plus the total number of transactions of the databases. We can hence denote these measures by a function $f(|\mathcal{H}^+|, |\mathcal{H}^-|, p, n)$ where p, n represents the frequency of the itemset in the two databases.

Modeling Itemsets. Following [13], we use an array of Boolean decision variables $\mathcal{I} = [I_1, I_2, \ldots, I_n]$ to represent an itemset $X \subset \mathcal{I}$. Each I_i is a binary variable with domain $dom(I_i) = \{0, 1\}$ and an item $i \in X \iff I_i = 1$. We say that I_i is *unbound* if there is more than one value in $dom(I_i)$. I_i is *bound* to 1 (0) means the item i is part (not part) of the itemset. Hence, one assignment to $\mathcal{I}$ corresponds to one itemset.

The *decomposition* formulation of frequent itemset mining [19] introduces an extra array of Boolean decision variables $\mathcal{T} = [T_1, T_2, \ldots, T_m]$, one for each of the m transactions. A Boolean variable T_t indicates whether the transaction with identifier t belongs to the cover $\{(t, S) \in \mathcal{H} \mid I \subseteq S\}$. This is enforced with a constraint for every transaction as follows: $\forall (t, S) \in \mathcal{H} : T_t = 0 \iff \bigvee_{i \notin S} I_i$. In other words: if an item i is in the itemset and not in the transaction (t, S) then this transaction is not covered by the itemset and equivalently if a transaction is not covered none of the items i in the itemset do belong to (t, S). The size of the cover can then be constrained as follows: $\sum_t T_t \geq \theta$. This model is not domain consistent for the frequent itemset mining problem that aims to enumerate all frequent patterns for a certain θ. As suggested in [13], one can further add the redundant constraints $\forall i : I_i = 1 \implies (\sum_{(t,S) \in \mathcal{H}, i \in S} T_t) \geq \theta$ to achieve domain consistency for the frequent itemset problem: these constraints enforce that an item is only supported if adding it to the current itemset will not violate the frequency constraint.

2.2   Table Constraint and Reversible Sparse Bit-Sets

A table constraint enforces that an array of integer decision variables $[V_1, \ldots, V_n]$ corresponds to one of the provided tuples $\Gamma = \{(t, \tau)\} | t \in \{1, \ldots, m\}\}$, where t is the tuple identifier and each tuple $\tau = (v_1, \ldots, v_n)$ consists of n values corresponding to the n variables: $table([V_1, \ldots, V_n], \Gamma) \iff \exists (t, \tau) \in \Gamma : V_1 = \tau_1 \wedge \ldots \wedge V_n = \tau_n$. A key property to maintain is the set of tuples supported by the current domain: $currTable = \{(j, \tau) \in \Gamma \mid \tau_1 \in dom(V_1) \wedge \ldots \wedge \tau_n \in dom(V_n)\}$. In [14], a reversible sparse bitset was proposed to maintain the set of tuple indices during search. In the propagator, a dense vertical representation of Γ is used: for every variable/value combination $(V_i, v), v \in dom(V_i)$, a bitvector $support[V_i, v] = \{(j, \tau) \in \Gamma \mid \tau_i = v\}$ is precomputed that stores the tuple identifiers in which the pair (V_i, v) appears. The indices of $currTable$ and the consistency of each (V_i, v) is computed using bitwise operations, e.g. (V_i, v) is supported if $support[V_i, v] \cap currTable \neq \emptyset$.

We briefly recall the `RSparseBitSet` data structure [14] which we will use in our propagators. The pseudo-code of this data structure is given in Algorithm 1 and some illustrative methods are also shown. The Reversible Sparse BitSet represents a set as a bitset (array of 64-bit Long *words*) and is *"reversible"* means that it is able to restore itself on backtrack. The reversibility relies on a global trail mechanism well known in the folklore of constraint programming (see [23] for an introduction to trailing and time-stamping).

The originality of this structure is that it borrows the idea of reversible sparse-sets [37] to discard all-zero words. When a bitvector is sparse (contains many zero words), this can save unnecessary iterations and computations over those words.

The following class invariant is maintained to ignore zero words: the number of non-zero words is a reversible integer denoted `limit`; and the `limit` first entries of `index` are indexes to the non-zero words in the bitvector. All the words beyond that limit are the indexes of zero words.

For the intersect method, which is also crucial for itemset mining, one can see how this is maintained by exchanging a detected zero word with the last non-zero one before decreasing the `limit` (swapping).

Apart from skipping entire words, the bitvector representation allows using highly efficient operations over entire words such as *and* and *bitCount*.

3   Global Constraints for Frequency-Based Itemset Mining

There is a close relation between a table constraint that reasons over a binary representation of the table and itemset mining. Each variable/value pair (V_i, v) is a column and can be seen as an *item* (in the itemset mining problem), and internally a vertical dense representation of the table can be used. Because each tuple in table Γ is of size n, in a binary representation of the table there will be exactly n non-zero entries per row. Further knowing that there are exactly n variables that each must be assigned one value, one can see that checking whether the set representation of $V : \{(V_i, v) \mid V_i = v \in V\}$ is a *subset* of the set representation of a tuple $\tau : \{(V_i, \tau_i) \mid \tau_i \in \tau\}$ coincides with checking whether they can be *equal* as both sets will have equal length when V is fully assigned. The cover relation of itemset mining is hence equivalent to the table support relation in this case, and the table constraint can be seen as enforcing a minimum frequency constraint with $\theta = 1$.

Earlier work has proposed a single global constraint for minimum frequent closed itemset mining. For efficiency reasons, we propose to use the reversible sparse bitset to maintain the set of transactions that can still be covered. For generality reasons, we propose to separate the computation of the frequency from the minimum (or maximum) frequency restriction and to separate that from enforcing the closedness property.

3.1 Computing Frequency: The *CoverSize* Constraint

Given a set of boolean variables $\mathcal{I}$ representing the pattern (selected items), a vertical dense bitvector representation of the database $\mathcal{D}$ (see Fig. 1c for an example), and an integer variable c, the *CoverSize* global constraint enforces the relation

$$CoverSize([I_1,\ldots,I_n],\mathcal{D},c) \iff c = \left|\bigcap_{I_i=1} \mathcal{D}(I_i)\right|$$

such that c represents the number of bits set in the intersection of the vertical bitvectors ($\mathcal{D}$) of the selected items. Using bitwise operations it can be formulated as

$$CoverSize([I_1,\ldots,I_n],\mathcal{D},c) \iff c = \texttt{size}(\&_{I_i=1}\mathcal{D}(I_i)).$$

For example in Fig. 1c and for itemset $\{C,D\}$, $c = |\mathcal{D}(I_C)\,\&\,\mathcal{D}(I_D))| = 2$.

Lazaar et al. [24] have argued that a global constraint is preferred over a decomposition into a constraint per transaction because the many constraints that need to be handled create overhead for the solver. This was shown earlier in [30], which proposed a CP-inspired dedicated solver with a global constraint for (reified) matrix-wide operations over bitvector variables.

When not exporting the cover as individual Boolean variables, we can use an internal data structure to store the cover such as `RSparseBitSet`. Note that not exposing the cover also limits the generality of the approach: no constraints can be put on the cover so that constraints such as closedness, maximality, non-frequency-based quality measures, etc. either require changes to the global constraint, or a separate global constraint that recomputes the cover. However, there are a number of constraints that depend only on the size of the cover and hence for added flexibility we propose to hide the cover but expose the cover size.

Consistency of *CoverSize*. Theorem 1 is used to demonstrate that it is NP-hard to check the consistency for *CoverSize*.

Theorem 1. *Given a collection of sets $\{S_1,\ldots,S_n\}$, the problem of finding a subset of these such that their union is of fixed cardinality k is NP-hard.*

Proof. We build a reduction from the NP-hard exact cover by three sets (X3C) problem: given a collection $\{C_1,C_2,\ldots,C_n\}$ of 3-element subsets built from a universe X with $|X| = 3q$ (a multiple of 3), can we find exactly q subsets of C to cover X? We reduce this X3C problem into our problem such that $S_i = C_i \cup \{a_1^i,\ldots,a_{|X|+1}^i\}\ \forall i \in \{1,\ldots,n\}$. All the artificial a_j^i elements added are different and not in universe X. Each set S_i has thus a cardinality of $3+|X|+1 =$

$4 + 3q$. We are looking for a collection of sets such that their union is of size $k = q(4+3q)$. Only q sets can be selected; even when counting just the artificial elements ($|X| + 1 = 3q + 1$ per set), more than q sets is not possible because $k - ((q + 1) \cdot (3q + 1)) < 0$. Fewer then q sets is also not possible because $k - ((q - 1) \cdot (3q + 1)) > |X|$ and hence there would need to be more than $|X|$ unique elements in universe X to achieve cardinality k. For exactly q sets, one can verify that these q sets will cover $|X|$ after the removal of the $q(|X| + 1)$ added elements: $q(4 + 3q) - q(|X| + 1) = q(4 + 3q - 3q - 1) = 3q = |X|$. ∎

Corollary 1. *Given a collection of sets $\{S_1, \ldots, S_n\}$, the problem of finding a subset of these such that their intersection is of fixed cardinality k is NP-hard.*

Proof. We reduce the problem of Theorem 1. Let $X = \bigcup_i S_i$ be the universe and $\overline{S}_i = X \setminus S_i$ the complement set of S_i w.r.t. X. There exists a subset $\Omega \subseteq \{1, \ldots, n\}$ such that $\left|\bigcup_{i \in \Omega} S_i\right| = k$ if and only if $\left|\bigcap_{i \in \Omega} \overline{S}_i\right| = |X| - k$. ∎

Theorem 2. *Determining the satisfiability for CoverSize is NP-hard.*

Proof. The problem of Corollary 1 is reduced to finding a feasible solution for $CoverSize([I_1, \ldots, I_n], \mathcal{D}, c = k)$ with $\mathcal{D}(I_i)$ the bitvector representation for set S_i. ∎

Despite this hardness result, we can still propagate many conditions efficiently. The hardest part is to propagate from an upper bound on c to the item variables.

***CoverSize* Propagator.** We denote by $U = \{I_i \in \mathcal{I} \mid dom(I_i) = \{0,1\}\}$ the set of undecided items and by $P = \{I_i \in \mathcal{I} \mid dom(I_i) = \{1\}\}$ the set of included items. The filtering rules for *CoverSize* are:

1. (*Rule* 1) computes the maximum cover size (exact upper-bound) that corresponds to discarding all the undecided items: $\max(c) \leq \left|\bigcap_{I_j \in P} \mathcal{D}(I_j)\right|$.

2. (*Rule* 2) computes the minimum cover size (exact lower-bound) that corresponds to including all the undecided items $\min(c) \geq \left|\bigcap_{I_j \in (P \cup U)} \mathcal{D}(I_j)\right|$.

3. (*Rule* 3) discards item I_i if including it would result in a cover size that is below the minimum threshold. $\forall I_i \in U : \left|\bigcap_{I_j \in P} \mathcal{D}(I_j) \cap \mathcal{D}(I_i)\right| < \min(c) \implies I_i = 0$. This rule is also implemented in [24] and can be achieved in the decomposition with a redundant constraint for each item separately.

4. (*Rule* 4) detects mandatory items. If the lower-bound is equal to the maximum allowed cover size, then if the cover size lower-bound would increase while excluding an item I_i then this item I_i is mandatory. $\forall I_i \in U :$
$\left|\bigcap_{I_j \in (P \cup U)} \mathcal{D}(I_j)\right| = \max(c) \wedge \left|\bigcap_{I_j \in (P \cup U)} \mathcal{D}(I_j)\right| < \left|\bigcap_{I_j \in (P \cup U) \setminus \{I_i\}} \mathcal{D}(I_j)\right|$
$\implies I_i = 1$.

Algorithm 2 gives the filtering algorithm for the *CoverSize* constraint implementing the Rules 1–4. N denotes the newly bound item variables since the previous call to the `propagate` method. The algorithm is thus incremental.

The block at Line 5 updates the current cover to reflect the new items included in the itemset[1]. The second block at Line 8 filters out the items that if included would induce a cover size below the allowed threshold min(c). This corresponds to Rule 3.

Line 11 computes the upper-bound of the cover size according to Rule (1). The lower-bound (Line 12) is obtained by including all the unbound items according to Rule (2).

Line 13 is triggered when the smallest *possible* intersection size (*lb*) is the largest *allowed* size of the frequency variable (max(c)). In this case, all the items that are mandatory to reach the lower-bound can be forcefully included (this is not necessarily true when min(c) = max(c) as min(c) can be externally set resulting in min(c) > lb). An unbound item I is mandatory if it is the only item that does *not* contain a transaction that all the other unbound and included ones do; in that case lb would increase to lb' > max(c). In the algorithm, $m \leftarrow$ cover $\&_{I_j \in U \setminus I_i} \mathcal{D}(I_j)$ is the cover if one would include all unbound items except I. If $m \not\subseteq \mathcal{D}(I_i)$ then $m \cap \mathcal{D}(I_i)$ would be a smaller set than m and hence I_i is mandatory to obtain the smallest cover size *lb*. For the example of Fig. 1, let C be included then lb = 1, ub = 3. Let max(c) = 1 then A is mandatory: not including it results in $m = \{2,4\}, m \not\subseteq \{1,3,4\}$ because of transaction 2. This condition is equivalent to Rule 4 but slightly more efficient to compute as it does not require to consider every non-zero words by returning true as soon as one word of m is not included.

Algorithm 1. Class RSparseBitSet. t[0] denotes the first element of array t and 0^k denotes a sequence of k bits set to 0.

```
1  words: array of rlong                // reversible longs, array length = p
2  index: array of int                  // array length = p
3  limit: rint                          // a reversible integer
4  Method intersect (m: array of long)
       /* this ← this & m                                              */
5      foreach i from limit downto 0 do
6          o ← index[i]
7          w ← words[o] & m[o]                          // bitwise AND
8          words[o] ← w
9          if w = 0^64 then
10             swap(index[i], index[limit])
11             limit ← limit − 1

12 Method contains(m: array of long): bool
       /* m ⊆ this                                                     */
13     foreach i from 0 to limit do
14         o ← index[i]
15         if (words(o) & ~ m[o]) ≠ 0^64 then
16             return false

17     return true
```

---

[1] This is similar to the update of *currTable* in [14] for filtering table constraints.

Algorithm 2. Class CoverSize($[I_1, \ldots I_n], \mathcal{D}, c$)

```
1  cover: RSparseBitSet                                    // Current cover
2  N,U                          // New bound variables, Unbound variables
3  D                                         // D[Iᵢ] = bit-set for item Iᵢ
```

4 **Method** propagate()

    /* update current cover   */

5   **foreach** *variable* $I_i \in$ N **do**

6   **if** $I_i = 1$ **then**

7   cover $\leftarrow$ cover & $\mathcal{D}[I_i]$

    /* remove items that if included induce cover $< \min(c)$ */

8   **foreach** *variable* $I_i \in$ U **do**

9   **if** size(cover & $\mathcal{D}[I_i]$) $< \min(c)$ **then**

10            $I_i \leftarrow 0$

    /* cover bounds   */

11   ub $\leftarrow$ size(cover); max(c) $\leftarrow \min(\max(c), ub)$

12   lb $\leftarrow$ size(cover $\&_{I_i \in U} \mathcal{D}[I_i]$); min(c) $\leftarrow \max(\min(c), lb)$

    /* propagating maximum size   */

13   **if** lb $<$ ub $\wedge$ lb $= \max(c)$ **then**

14   **foreach** *variable* $I_i \in$ U **do**

            /* include items mandatory for a cover size $= $ lb   */

15   m $\leftarrow$ cover $\&_{I_j \in U \setminus I_i} \mathcal{D}[I_j]$

16   **if** m $\not\subseteq \mathcal{D}[I_i]$ **then**

17                $I_i \leftarrow 1$

The time complexity for executing propagate is $O(|\mathcal{I}| \times m/64)$ with $|\mathcal{I}|$ the number of items and $m/64$ the number of words necessary to represent the cover. In practice, the reversible sparse bitset will only iterate on the non-zero words in the cover bitvector. The space complexity is $O(|\mathcal{I}| \times m)$ similar to that of other approaches and due to the space needed to store the database.

Since domain consistency *CoverSize* is NP-hard we can unfortunately not clearly characterize[2] the filtering of Algorithm 2. Only in the case of an unconstrained max(c) (for instance for the frequent itemset problem), the filtering reaches domain consistency.

3.2 Closed Itemsets: The *CoverClosure* Constraint

The idea of mining for *closed* frequent itemsets is to reduce the set of extracted itemsets to a smaller, more interesting one. The intuitive idea is that if a frequent pattern has a cover that is exactly the same as a super pattern, then only the super pattern should be enumerated.

---

[2] As for many NP-hard global constraints like bin-packing, cumulative, circuit, etc.

An itemset is hence a *closed* itemset if there is no superset with the same cover: $\nexists I' \supset I : \{(t,T) \in \mathcal{H} \mid I' \subseteq T\} = \{(t,T) \in \mathcal{H} \mid I \subseteq T\}$. Hence, the *closure* of an itemset can be computed by verifying which items could be added to the itemset without changing the cover:

$$clo_{\mathcal{H}}(I) = I \cup \{j \notin I \mid \{(t,T) \in \mathcal{H} \mid I \cup \{j\} \subseteq T\} = \{(t,T) \in \mathcal{H} \mid I \subseteq T\}\} \quad (1)$$

As argued in [6] there are two ways of interpreting the *closed* property when combined with other constraints: (1) of all closed itemsets, keep only those that satisfy the constraints (2) take the closure such that the new itemset satisfies all constraints

This can have far-reaching consequences. Let us take the database in Fig. 1, where item B is in all transactions and so will be in all closed itemsets. If we now add the constraint $'B' \notin I$, then under interpretation 1 there would not be any valid itemset, while under interpretation 2 there is D, C, AD and ACD namely all closed itemsets when ignoring B. It should be clear that interpretation (1) should not be taken by default. On the other hand, enforcing interpretation (2) requires one to reason not in terms of local constraints but over valid solutions to the CSP, for example with dominance properties [28]. Nonetheless, interpretation (1) is valid as long as all constraints have the property that adding a cover-preserving item to the set can never violate another constraint; for example, one can freely add a minimum size or maximum frequency constraint [6]. In case of such constraints it must be expressed as a preference over solutions of the CSP, e.g. by adding constraints each time a solution is found [28]. We hence propose to offer the widely used unconstrained closure operator as a separate *CoverClosure* constraint.

Another argument for separating the closure constraint is that in case of discriminative itemset mining, we may want to enforce closedness on only one of the two databases or on the entire database [19]. A separate constraint allows this freedom.

CoverClosure Propagator. Two filtering rules are enforced similar to [24]:

1. (*Rule* 5) Closure inclusion. This rule checks for each unbound item I_i if including it would result in an unchanged cover. If yes, this item should be included in the final pattern. More formally

$$\forall I_i \in U : (\bigcap_{I_j \in P} \mathcal{D}(I_j) \cap \mathcal{D}(I_i)) = \bigcap_{I_j \in P} \mathcal{D}(I_j) \implies I_i = 1$$

2. (*Rule* 6) Closure exclusion. This rule detects if extending the pattern with an item would result in a cover for which there is an already excluded item that should be added by the closure operator. Hence, including the first item would lead to an inconsistency and so it should be excluded. More formally, assuming $I_k \in \mathcal{I}, I_k = 0$ represents the excluded items:

$$\forall I_i \in U, I_k \in \mathcal{I}, I_k = 0 : ((\bigcap_{I_j \in P} \mathcal{D}(I_j)) \cap \mathcal{D}(I_i)) \subseteq ((\bigcap_{I_j \in P} \mathcal{D}(I_j)) \cap \mathcal{D}(I_k)) \implies I_i = 0$$

Algorithm 3 implements the domain consistent filtering for the *CoverClosure* constraint. This constraint also uses the `RSparseBitSet` data structure to store the cover. It has a complexity of $O(|\mathcal{I}|^2 \times m/64)$.

A faster (but not domain consistent) filtering is obtained by replacing Rule 6 with a consistency check verifying that for each discarded item ($I_i = 0$), including it changes the cover: $\forall I_k \in \mathcal{I}, I_k = 0 : (\bigcap_{I_j \in P} \mathcal{D}(I_j) \cap \mathcal{D}(I_k)) = \bigcap_{I_j \in P} \mathcal{D}(I_j) \Longrightarrow$ *fail*. This version has a complexity of $O(|\mathcal{I}| \times m/64)$, similar to *CoverSize*.

4   Frequency-Based Itemset Mining with *CoverSize* and *CoverClosure*

4.1   Frequent Itemset Mining

Our model for frequent itemset mining contains just one *CoverSize* and a constraint that the size of the cover is above a fixed minimum frequency θ:

$$\text{enumerate}\quad CoverSize([I_1, \ldots, I_n], \mathcal{D}, c) \wedge c \geq \theta$$

Notice that as c is a variable, one can also add a maximum frequency constraint, or use it in branch-and-bound to search for the most frequent itemsets under constraints such as a minimum itemset cardinality:

$$\text{maximize}\quad c, \quad \text{s.t.}\quad CoverSize([I_1, \ldots, I_n], \mathcal{D}, c) \wedge \sum_i I_i \geq \beta$$

4.2   *Closed* Frequent Itemset Mining

Looking for the frequent closed itemset amounts to adding *CoverClosure*:

$$\text{enumerate}\quad CoverSize([I_1, \ldots, I_n], \mathcal{D}, c) \wedge c \geq \theta \wedge CoverClosure([I_1, \ldots, I_n], \mathcal{D})$$

As explained in Sect. 3.2, other constraints should only be added if they do not constrain the addition of (frequency-preserving) items.

4.3   Discriminative (Closed) Itemset Mining

Given a split database $\mathcal{D}^+$ and $\mathcal{D}^-$ containing positive (+) and negative (−) transactions defined on a same set of items, the objective is to find the highest scoring itemsets (discriminating one class over another) w.r.t. a correlation (discriminative) measure such as accuracy, information gain, χ^2 measure, Gini index, etc. Those itemsets are interesting as classification rules directly [9,11,16], or as features (if the itemset is present or not) for another classifier [10,15].

Using *accuracy* as a discriminative measure leads to the following problem:

$$\text{maximize}\quad p - n, \quad \text{s.t.}$$
$$CoverSize([I_1, \ldots, I_n], \mathcal{D}^+, p) \wedge CoverSize([I_1, \ldots, I_n], \mathcal{D}^-, n)$$

Algorithm 3. Class CoverClosure($[I_1, \ldots I_n], \mathcal{D}$)

```
1  cover: RSparseBitSet                                    // Current cover
2  N,U                        // New bound variables, Unbound variables
3  D                                       // D[Iᵢ] = bit-set for item Iᵢ
4  Method propagate()
        /* update current cover                                        */
5       foreach variable Iᵢ ∈ N do
6           if Iᵢ = 1 then
7               cover ← cover & D[Iᵢ]

        /* Rule 5                                                       */
8       foreach variable Iᵢ ∈ U do
9           if cover = (cover & D[Iᵢ]) then
10              Iᵢ ← 1

        /* Rule 6                                                       */
11      foreach variable Iᵢ ∈ U do
12          foreach variable Iₖ ∈ I with Iₖ = 0 do
13              if (cover & D[Iᵢ]) ⊆ (cover & D[Iₖ]) then
14                  Iᵢ ← 0; break
```

Another standard discriminative measure is the χ^2 one depicted on Fig. 1d. As explained in [31] the standard discriminative functions such as χ^2 have the property that they are zero on the diagonal (relative to the possible values of p, n) and convex (denoted by ZDC). A general ZDC-based model for discriminative itemset mining is composed of

- two constraints $CoverSize([I_1, \ldots, I_n], \mathcal{D}^+, p)$ and $CoverSize([I_1, \ldots, I_n], \mathcal{D}^-, n)$ to compute the cover size on the positive and negative transactions;
- a Zero Diagonal Convex constraint $ZDC(|\mathcal{D}^+|, |\mathcal{D}^-|, p, n, score)$ that links p, n and $score$ using a discriminative function (such as χ^2) to maximize;
- optionally $CoverClosure([I_1, \ldots, I_n], \mathcal{D}^- \cup \mathcal{D}^+)$ to obtain the closedness property. Note that posting $CoverClosure$ separately on the positive (negative) can decrease p (n) and is hence not allowed for symmetric ZDC measures.

This approach which employs a separate ZDC constraint that takes only the cardinalities p and n as input, is novel and favors reusability in a different context: it is itemset-agnostic, meaning that it could also be used for example to find discriminating sequences instead of itemsets. In [31] the authors also employ a global constraint for the discriminative itemset mining problem, but one that reasons at the transaction level with one variable per transaction. The filtering they achieve is stronger than our decomposition into three constraints. They perform what they call a redundant *look-ahead* filtering[3] on each item separately.

---

[3] A related generic technique in CP is *shaving* [26].

We briefly describe our filtering for *ZDC*, illustrated in Fig. 1e. Because of the ZDC property, the minimum and maximum is located at one of the four corners of the box $[\min(p), \max(p)] \times [\min(n), \max(n)]$, and hence only these extremes need to be computed for pruning $\min(score)$. Given a minimum value for $\min(score)$, for example as enforced during branch-and-bound maximization, the value of $\max(p)$ and $\max(n)$ can be reduced as illustrated on Fig. 1e. On this figure, the iso-curve corresponding to $\min(score)$ is visualized. The ZDC property implies that any larger score must lay outside of the region enclosed by the iso-curves. The gray zone on Fig. 1e corresponds to inconsistent combinations for p and n, hence discovering the new minimum for p requires to find v such that $ZDC(|\mathcal{D}^+|, |\mathcal{D}^-|, v, \max(n)) = \min(score)$. Any value larger than v for p would be inconsistent. The upper-cardinality of p is constrained and therefore the filtering of $CoverSize([I_1, \dots, I_n], \mathcal{D}^+, p)$ based on this upper cardinality is important to prune the search tree. A similar reasoning is used to prune $\min(n)$.

5   Experiments

In this section, we report the experimental results on frequent, closed as well as discriminative itemset mining. Each experiment is driven by a concrete question. All experiments were run in the JVM with maximum memory set to 8 GB on PCs with Intel Core i5 64 bits processor (2.7 GHz) and 8 GB of RAM running Linux Mint 17.3. Execution time is limited to 1000 s.

Datasets and Mining Algorithms. We use data from the FIMI[4] repository and from the CP4IM[5] website. The properties of the datasets are presented in Table 1 (first column) and Table 2a (these latter are labelled positive/negative datasets). We compare with the following methods:

– Frequent Itemset Mining: *FIMCP* [19] using the Gecode solver [17], *DMCP* [30] a custom CP bitvector solver, and four dedicated algorithms namely Borgelt's *Apriori* and *Eclat* implementations [7], *Nonordfp* [36] and LCMv3[6] [39].
– Closed Frequent Itemset Mining: *FIMCP*, *DMCP*, Borgelt's Apriori and LCMv3 again, as well as *ClosedPattern* [24] using the or-tools solver [18].
– Discriminative Itemset Mining: *CIMCP* [19] based on Gecode and the specialised algorithm *corrmine* [31].

We denote our approach by *CoverSize* and it uses the OscaR solver [33][7].

Q1: What is the impact of using a reversible *"sparse"-bitset* over a reversible non-sparse one? In Table 1 *CoverSize-bitset* is the same implementation as *CoverSize* but using a reversible bitset implementation that does not check for zero words. The results on *Frequent* in Table 1 convincingly show

---

[4] http://fimi.ua.ac.be/data/.
[5] https://dtai.cs.kuleuven.be/CP4IM/datasets/.
[6] http://research.nii.ac.jp/~uno/codes.htm (v3 is fastest of all versions in our experiments).
[7] https://sites.uclouvain.be/cp4dm/fim/.

Table 1. CPU runtime for several algorithms vs *CoverSize*. (TO ≡ TimeOut; * ≡ CoverSize + CoverClosure; $\rho \equiv density = \frac{1}{|\mathcal{T}| \times |\mathcal{I}|} \sum_{t \in \mathcal{T}, i \in \mathcal{I}} \mathcal{D}_{ti}$)

Name $\|\mathcal{T}\| \times \|\mathcal{I}\|$ ρ(%)	θ	Frequent								Closed						
		CP-based				Specialized				CP-based				Specialized		
		FIMCP	DMCP	CoverSize-bitset	CoverSize	Apriori	Eclat	Nonordfp	LCM3	FIMCP	ClosedPattern	CoverSize-DC*	CoverSize*	DMCP	Apriori-close	LCM3
retail 88162×16470 (ρ=0.06)	80	TO	6.91	25.76	5.33	0.60	5.81	0.98	**0.21**	TO	TO	394.48	45.09	16.95	0.82	**0.26**
	60	TO	10.45	33.87	7.37	0.71	8.26	1.31	**0.24**	TO	TO	952.83	67.74	25.10	1.03	**0.31**
	40	TO	15.96	65.13	11.19	0.77	11.42	1.83	**0.27**	TO	TO	TO	125.67	41.78	1.29	**0.49**
	20	TO	26.81	132.53	19.74	1.10	17.86	2.56	**0.43**	TO	TO	TO	226.32	94.24	1.61	**0.48**
	10	TO	40.03	191.05	37.08	1.73	24.63	3.68	**0.39**	TO	TO	TO	366.83	238.71	2.48	**0.66**
online-retails 541909×2603 (ρ=0.1)	70	TO	11.00	54.19	8.27	2.75	14.27	**0.31**	1.59	TO	TO	242.80	98.67	11.00	**1.28**	1.43
	40	TO	11.33	59.60	8.00	4.78	15.07	**0.40**	1.43	TO	TO	497.14	111.34	12.06	**1.28**	1.51
	10	TO	11.49	86.66	8.49	2.15	15.61	**0.43**	1.51	TO	TO	907.68	131.94	13.31	**1.36**	1.52
	5	TO	15.64	84.05	8.82	2.13	14.56	**0.31**	1.32	TO	TO	TO	148.29	13.72	**1.31**	1.65
	1	TO	TO	TO	TO	2.18	15.66	**0.42**	1.22	TO	TO	TO	TO	14.99	**1.44**	1.60
BMSWebView1 59602×497 (ρ=0.5)	48	TO	1.92	1.51	0.69	0.08	0.51	0.11	**0.03**	TO	TO	3.10	2.54	48.04	0.29	**0.11**
	36	TO	17.87	7.23	1.87	0.88	0.37	0.30	**0.11**	TO	TO	3.77	3.74	512.09	1.55	**0.26**
	34	TO	63.93	23.07	8.12	7.04	0.43	0.60	**0.10**	TO	TO	3.99	4.57	746.61	10.46	**0.37**
	32	TO	TO	TO	TO	TO	0.68	60.63	**0.28**	TO	TO	5.12	8.24	TO	TO	**0.47**
	30	TO	TO	TO	TO	TO	**0.53**	TO	1.22	TO	TO	6.61	13.50	TO	TO	**0.67**
T10I4D100K 100000×870 (ρ=1.0)	500	TO	1.07	3.58	0.79	0.48	3.32	0.80	**0.36**	TO	8.32	8.20	7.81	1.04	0.81	**0.34**
	400	TO	0.98	4.1	1.03	0.49	3.74	0.58	**0.29**	TO	13.06	9.27	8.88	1.12	0.67	**0.43**
	300	TO	1.39	4.96	1.27	0.69	4.14	0.94	**0.30**	TO	25.28	11.12	10.51	1.30	0.85	**0.41**
	200	TO	2.65	6.59	1.28	0.63	3.95	0.70	**0.39**	TO	77.20	12.95	10.98	1.61	1.07	**0.38**
	100	TO	2.35	7.27	1.78	1.02	5.08	1.01	**0.53**	TO	125.26	15.79	15.07	2.11	1.15	**0.58**
pumsb-star 49046×2088 (ρ=2.0)	18000	302.06	1.02	1.26	0.84	6.34	0.77	0.25	**0.21**	TO	56.55	0.99	0.89	2.11	5.11	**0.22**
	16000	375.07	2.57	2.55	1.52	18.81	1.04	0.27	**0.22**	TO	120.33	0.80	0.93	4.04	15.59	**0.27**
	14000	563.21	9.14	4.72	3.36	81.93	1.34	0.31	**0.26**	TO	275.23	1.65	2.25	6.22	57.99	**0.30**
	12000	TO	33.66	20.12	10.16	285.36	1.95	0.62	**0.38**	TO	601.72	4.51	5.55	14.04	284.39	**0.44**
	10000	TO	TO	TO	TO	TO	3.45	164.76	**0.45**	TO	TO	10.37	13.97	26.96	TO	**0.54**
pumsb 49046×2113 (ρ=3.0)	40000	237.09	2.82	2.02	1.51	1.45	0.87	**0.14**	0.15	TO	212.83	1.84	2.14	7.22	1.46	**0.15**
	35000	889.08	33.87	13.20	10.26	15.92	3.36	0.21	**0.20**	TO	TO	12.71	16.76	43.48	15.98	**0.27**
	30000	TO	220.03	124.65	63.91	356.90	10.55	0.54	**0.27**	TO	TO	62.81	82.47	121.24	370.78	**0.60**
	25000	TO	TO	TO	602.72	TO	66.46	3.56	**1.04**	TO	TO	468.99	611.62	TO	TO	**1.54**
accidents 340183×468 (ρ=7.0)	300000	50.02	0.78	0.16	**0.05**	1.23	1.02	0.23	0.21	221.80	0.13	0.08	**0.07**	0.30	1.48	0.19
	250000	55.69	1.08	0.18	**0.17**	1.75	1.26	0.33	0.40	253.90	1.71	0.17	**0.14**	1.16	2.00	0.21
	200000	112.55	0.99	0.34	**0.33**	2.46	1.76	0.35	0.62	302.97	8.57	**0.56**	0.68	1.41	2.55	0.56
	150000	386.51	2.87	1.52	1.32	17.83	3.12	**0.52**	1.06	575.32	61.60	5.00	5.09	3.71	18.91	**0.99**
	100000	TO	18.08	10.69	7.79	116.26	7.32	**0.74**	1.73	TO	570.38	47.55	43.75	23.63	140.26	1.47
mushroom 8124×119 (ρ=19.0)	600	33.06	0.85	2.14	1.92	14.61	0.30	0.05	**0.04**	8.38	0.62	0.73	0.84	0.16	10.09	**0.06**
	400	106.71	3.48	5.52	5.43	58.46	0.29	0.09	**0.04**	14.86	1.08	0.70	0.67	0.20	36.11	**0.11**
	200	449.85	16.77	23.37	20.10	133.54	0.37	0.27	**0.07**	20.12	2.35	0.75	1.56	0.35	112.82	**0.17**
	100	TO	67.09	96.13	68.46	264.69	0.65	0.95	**0.10**	27.10	4.30	1.11	2.91	0.63	284.15	**0.24**
soybeans 630×50 (ρ=32.0)	16	1.21	0.47	1.02	1.03	0.45	0.03	0.02	**0.00**	0.31	0.20	0.36	0.35	0.08	0.71	**0.02**
	13	1.57	0.70	1.40	1.44	0.44	0.03	0.02	**0.01**	0.32	0.25	0.32	0.28	0.12	0.95	**0.02**
	10	2.54	0.75	1.69	1.44	0.71	0.04	0.03	**0.02**	0.34	0.29	0.40	0.29	0.11	1.20	**0.02**
	7	3.54	1.46	2.35	2.07	0.84	0.05	0.03	**0.01**	0.47	0.33	0.23	0.22	0.14	1.61	**0.03**
	4	7.05	3.86	4.02	3.87	1.69	0.05	0.07	**0.02**	0.42	0.42	0.25	0.20	0.18	2.98	**0.03**
chess 3196×75 (ρ=49.0)	2000	3.71	0.30	1.59	1.39	3.14	0.11	0.01	**0.01**	1.85	1.71	1.11	1.02	0.42	3.48	**0.09**
	1500	46.05	3.05	4.81	3.88	77.85	0.39	0.11	**0.07**	14.06	15.53	4.73	3.42	1.88	69.56	**0.59**
	1000	577.29	35.16	52.60	44.94	849.49	2.15	0.68	**0.37**	101.68	96.65	28.11	22.76	14.96	885.14	**4.90**
	500	TO	959.16	TO	TO	TO	19.06	12.35	**2.96**	882.16	900.45	304.74	282.50	144.04	TO	**51.66**
	250	TO	TO	TO	TO	TO	72.69	129.05	**14.42**	TO	TO	TO	TO	580.64	TO	**211.99**

that using the sparse data structure is always better and sometimes an order of magnitude faster, especially on large and sparse datasets.

For closed, we can also compare *Coversize-DC** to *ClosedPattern* [24] which uses the same filtering rules but in a single global constraint and with a different solver and non-reversible non-sparse bitsets [24]. We only have the binaries and though different solvers will perform differently, or-tools has won MiniZinc challenge [38] gold medals and so the remarkable difference in runtime with our method provides strong evidence that the reversible sparse bitset is a well suited and very scalable data structure for itemset propagators.

Q2: Is domain consistency interesting for closed frequent itemset mining? In [24] the authors concluded that using the domain consistent version of Rule 6 dominates the simpler non-domain consistent one because of the resulting reduction in number of explored nodes. We reran the same experiment, *CoverSize-DC** and *CoverSize** in Table 1, and the conclusion changes when using reversible sparse bitsets: while on pumsb and pumsb-star the runtime increases when using the simpler non-DC version, on the other datasets it is similar or faster to use the simpler one. For the largest and sparsest datasets retail and online-retails, the difference is even up to an order of magnitude.

Q3: How does *CoverSize* compare with existing approaches? The *CoverSize* approach clearly outperforms the decomposition-based *FIMCP*. For frequent, *CoverSize* is on par (sometimes somewhat better or worse) with *DMCP*, the dedicated CP solver which uses bitvector variables. By profiling the execution we observed that for the instances where *DMCP* was faster (such as *mushroom*) only 1% of the time was spent in *CoverSize*. The remaining time is devoted to the solver (propagation management, search, trailing, ...), which a dedicated solver like *DMCP* has less overhead in. Hence, here we show that similar performance can be achieved with a generic solver, through the use of global constraints with carefully designed data structures.

For closed, our approach outperforms *FIMCP*, and also *ClosedPattern* as discussed in Q2. The differences between *CoverSize* and *DMCP* become more varied and pronounced, for example for the sparsest retail and online-retails dataset in favor of DMCP, and for BMSWebView1 in favor of *CoverSize*.

Specialised Algorithms. There remains a significant gap between CP-based methods and specialised methods though, and especially the highly praised LCMv3 algorithm lives up to its reputation. It should be pointed out that these algorithms do not allow any constraints, and for example a version of LCM (LCMv5) that allow some constraints is also remarkably slower. For denser datasets, our method does typically outperform Apriori.

Q4: What is the difference in performance for discriminative itemset mining with *CoverSize*? The state-of-the-art for this problem is the generic CP-based CIMCP method and the specialized *corrmine* method which implement the same bounds [31]. Table 2b shows a comparison using Information Gain as the ZDC measure, which is the one implemented in *corrmine*. Despite the stronger filtering of CIMCP, *CoverSize* outperforms *CIMCP* for the most difficult instances showing the importance of globals with a good data structure. *corrmine* is superior though specialized to this specific problem.

Table 2. Runtimes, in seconds, for discriminative itemset mining

a) Dataset features.				b) Discriminative			a)				b)		
Name	Dense	Trans	Item	CIMCP	CoverSize	cormine	Name	Dense	Trans	Item	CIMCP	CoverSize	cormine
anneal	0.45	812	93	**0.167**	0.24	0.014	letter	0.5	20000	224	54.255	**4.547**	0.367
australian-cr	0.41	653	125	**0.166**	0.195	0.012	mushroom	0.18	8124	119	15.979	**0.069**	0.025
breast-wisc	0.5	683	120	**0.193**	0.345	0.037	pendigits	0.5	7494	216	2.939	**1.196**	0.138
diabetes	0.5	768	112	**1.564**	1.769	0.28	primary-t	0.48	336	31	**0.02**	0.058	0.003
german-cr	0.34	1000	112	**1.521**	1.659	0.09	segment	0.5	2310	235	1.154	**0.15**	0.052
heart-clevel	0.47	296	95	**0.175**	0.221	0.055	soybean	0.32	630	50	**0.046**	0.048	0.003
hypothyroid	0.49	3247	88	0.592	**0.118**	0.016	splice-1	0.21	3190	287	22.341	**0.113**	0.025
ionosphere	0.5	351	445	1.047	**0.336**	0.23	vehicle	0.5	846	252	0.551	**0.55**	0.094
kr-vs-kp	0.49	3196	73	0.698	**0.145**	0.01	yeast	0.49	1484	89	3.386	**1.366**	0.818

Average. when found | 5.933 | 0.729 | 0.126

6   Conclusion and Perspectives

We showed that compared to the *ClosedPattern* approach [24] of using a global constraint for frequent closed itemset mining, both generality and efficiency can be significantly improved. Generality can be improved by a separation of concerns in terms of global constraints. We propose to use one global constraint that exposes the frequency through a decision variable which can then be used in other constraints. For example frequency constraints, objective functions or discrimination scores. Another global constraint can be used to enforce the closure property, though care has to be taken when combining it with other constraints.

Efficiency-wise we showed the connection with a well-known constraint that also has to handle a lot of data: the table constraint. Using the *Reversible Sparse bitset* data structure that was recently proposed [14] allows our global constraints to scale to even larger and sparser datasets while still in a generic CP solver. This is relevant not just for frequency-based itemset mining, but also for other existing as well as novel data mining problems in CP, and perhaps beyond.

Acknowledgments. The research is supported by the FRIA-FNRS (Fonds pour la Formation à la Recherche dans l'Industrie et dans l'Agriculture, Belgium) and FWO (Research Foundation – Flanders). We also thank Willard Zhan for his help with the reduction proof.

References

1. Aggarwal, C.C.: An introduction to frequent pattern mining. In: Aggarwal, C.C., Han, J. (eds.) Frequent Pattern Mining, pp. 1–17. Springer, Cham (2014). doi:10. 1007/978-3-319-07821-2_1

2. Agrawal, R., Imieliński, T., Swami, A.: Mining association rules between sets of items in large databases. Int. Conf. Manag. Data (SIGMOD) **22**(2), 207–216 (1993)

3. Aoga, J.O.R., Guns, T., Schaus, P.: An efficient algorithm for mining frequent sequence with constraint programming. In: Frasconi, P., Landwehr, N., Manco, G., Vreeken, J. (eds.) ECML PKDD 2016. LNCS, vol. 9852, pp. 315–330. Springer, Cham (2016). doi:10.1007/978-3-319-46227-1_20

4. Aoga, J.O., Guns, T., Schaus, P.: Mining time-constrained sequential patterns with constraint programming. Constraints **22**, 1–23 (2017)
5. Bessiere, C., Régin, J.C.: Arc consistency for general constraint networks: preliminary results. In: International Joint Conference on Artificial Intelligence (IJCAI) (1997)
6. Bonchi, F., Lucchese, C.: On closed constrained frequent pattern mining. In: Fourth IEEE International Conference on Data Mining, ICDM 2004, pp. 35–42, November 2004
7. Borgelt, C.: Efficient implementations of Apriori and Eclat. In: FIMI: Workshop on Frequent Itemset Mining Implementations (2003)
8. Borgelt, C.: Frequent item set mining. Wiley Interdisc. Rev.: Data Min. Knowl. Discov. **2**(6), 437–456 (2012)
9. Bringmann, B., Zimmermann, A.: TREE ^{2} – decision trees for tree structured data. In: Jorge, A.M., Torgo, L., Brazdil, P., Camacho, R., Gama, J. (eds.) PKDD 2005. LNCS, vol. 3721, pp. 46–58. Springer, Heidelberg (2005). doi:10.1007/11564126_10
10. Bringmann, B., Zimmermann, A., Raedt, L., Nijssen, S.: Don't be afraid of simpler patterns. In: Fürnkranz, J., Scheffer, T., Spiliopoulou, M. (eds.) PKDD 2006. LNCS, vol. 4213, pp. 55–66. Springer, Heidelberg (2006). doi:10.1007/11871637_10
11. Cheng, H., Yan, X., Han, J., Philip, S.Y.: Direct discriminative pattern mining for effective classification. In: IEEE 24th International Conference on Data Engineering, ICDE 2008, pp. 169–178. IEEE (2008)
12. Cheng, K.C., Yap, R.H.: An MDD-based generalized arc consistency algorithm for positive and negative table constraints and some global constraints. Constraints **15**(2), 265–304 (2010)
13. De Raedt, L., Guns, T., Nijssen, S.: Constraint programming for itemset mining. In: International Conference on Knowledge Discovery and Data Mining (SIGKDD), pp. 204–212 (2008)
14. Demeulenaere, J., Hartert, R., Lecoutre, C., Perez, G., Perron, L., Régin, J.-C., Schaus, P.: Compact-table: efficiently filtering table constraints with reversible sparse bit-sets. In: Rueher, M. (ed.) CP 2016. LNCS, vol. 9892, pp. 207–223. Springer, Cham (2016). doi:10.1007/978-3-319-44953-1_14
15. Deshpande, M., Kuramochi, M., Wale, N., Karypis, G.: Frequent substructure-based approaches for classifying chemical compounds. IEEE Trans. Knowl. Data Eng. **17**(8), 1036–1050 (2005)
16. Fan, W., Zhang, K., Cheng, H., Gao, J., Yan, X., Han, J., Yu, P., Verscheure, O.: Direct mining of discriminative and essential frequent patterns via model-based search tree. In: Proceedings of the 14th ACM SIGKDD International Conference on Knowledge Discovery and Data Mining, pp. 230–238. ACM (2008)
17. Gecode Team: Gecode: generic constraint development environment (2006). http://www.gecode.org
18. Google: Google optimization tools (2015). https://developers.google.com/optimization/
19. Guns, T., Nijssen, S., De Raedt, L.: Itemset mining: a constraint programming perspective. Artif. Intell. **175**(12–13), 1951–1983 (2011)
20. Jabbour, S., Sais, L., Salhi, Y.: The top-k frequent closed itemset mining using top-k SAT problem. In: Blockeel, H., Kersting, K., Nijssen, S., Železný, F. (eds.) ECML PKDD 2013. LNCS, vol. 8190, pp. 403–418. Springer, Heidelberg (2013). doi:10.1007/978-3-642-40994-3_26
21. Jabbour, S., Sais, L., Salhi, Y.: Mining top-k motifs with a sat-based framework. Artif. Intell. **244**, 30–47 (2017)

22. Kemmar, A., Loudni, S., Lebbah, Y., Boizumault, P., Charnois, T.: PREFIX-PROJECTION global constraint for sequential pattern mining. In: Pesant, G. (ed.) CP 2015. LNCS, vol. 9255, pp. 226–243. Springer, Cham (2015). doi:10.1007/978-3-319-23219-5_17

23. Knuth, D.: The Art of Computer Programming: Combinatorial Algorithms, vol. 4. Addison-Wesley, Upper Saddle River (2015)

24. Lazaar, N., Lebbah, Y., Loudni, S., Maamar, M., Lemière, V., Bessiere, C., Boizumault, P.: A global constraint for closed frequent pattern mining. In: Rueher, M. (ed.) CP 2016. LNCS, vol. 9892, pp. 333–349. Springer, Cham (2016). doi:10.1007/978-3-319-44953-1_22

25. Lecoutre, C.: STR2: optimized simple tabular reduction for table constraints. Constraints 16(4), 341–371 (2011)

26. Lhomme, O.: Quick shaving. In: Proceedings of the 20th National Conference on Artificial Intelligence, vol. 1. pp. 411–415. AAAI Press (2005)

27. Morishita, S., Sese, J.: Traversing itemset lattice with statistical metric pruning. In: Vianu, V., Gottlob, G. (eds.) Proceedings of the Nineteenth ACM SIGMOD-SIGACT-SIGART Symposium on Principles of Database Systems, 15–17 May 2000, Dallas, Texas, USA, pp. 226–236. ACM (2000)

28. Negrevergne, B., Dries, A., Guns, T., Nijssen, S.: Dominance programming for itemset mining. In: 2013 IEEE 13th International Conference on Data Mining Data Mining (ICDM), pp. 557–566. IEEE (2013)

29. Negrevergne, B., Guns, T.: Constraint-based sequence mining using constraint programming. In: Michel, L. (ed.) CPAIOR 2015. LNCS, vol. 9075, pp. 288–305. Springer, Cham (2015). doi:10.1007/978-3-319-18008-3_20

30. Nijssen, S., Guns, T.: Integrating constraint programming and itemset mining. In: Balcázar, J.L., Bonchi, F., Gionis, A., Sebag, M. (eds.) ECML PKDD 2010. LNCS, vol. 6322, pp. 467–482. Springer, Heidelberg (2010). doi:10.1007/978-3-642-15883-4_30

31. Nijssen, S., Guns, T., De Raedt, L.: Correlated itemset mining in ROC space: a constraint programming approach. In: International Conference on Knowledge Discovery and Data Mining (SIGKDD), pp. 647–656. ACM (2009)

32. Nijssen, S., Zimmermann, A.: Constraint-based pattern mining. In: Aggarwal, C.C., Han, J. (eds.) Frequent Pattern Mining, pp. 147–163. Springer, Cham (2014). doi:10.1007/978-3-319-07821-2_7

33. OscaR Team: OscaR: Scala in OR (2012). https://bitbucket.org/oscarlib/oscar

34. Pasquier, N., Bastide, Y., Taouil, R., Lakhal, L.: Discovering frequent closed itemsets for association rules. In: Beeri, C., Buneman, P. (eds.) ICDT 1999. LNCS, vol. 1540, pp. 398–416. Springer, Heidelberg (1999). doi:10.1007/3-540-49257-7_25

35. Perez, G., Régin, J.-C.: Improving GAC-4 for table and MDD constraints. In: O'Sullivan, B. (ed.) CP 2014. LNCS, vol. 8656, pp. 606–621. Springer, Cham (2014). doi:10.1007/978-3-319-10428-7_44

36. Rácz, B.: nonordfp: an FP-growth variation without rebuilding the FP-tree. In: FIMI: Workshop on Frequent Itemset Mining Implementations (2004)

37. de Saint-Marcq, V.l.C., Schaus, P., Solnon, C., Lecoutre, C.: Sparse-sets for domain implementation. In: CP workshop on - Techniques foR Implementing Constraint programming Systems (TRICS), pp. 1–10 (2013)

38. Stuckey, P.J., Becket, R., Fischer, J.: Philosophy of the minizinc challenge. Constraints 15(3), 307–316 (2010)

39. Uno, T., Kiyomi, M., Arimura, H.: LCM Ver. 3: collaboration of array, bitmap and prefix tree for frequent itemset mining. In: Proceedings of the 1st International Workshop on Open Source Data Mining: Frequent Pattern Mining Implementations (OSDM 2005), pp. 77–86. ACM (2005)

Operations Research and CP Track

A Column-Generation Algorithm for Evacuation Planning with Elementary Paths

Mohd. Hafiz Hasan$^{(\boxtimes)}$ and Pascal Van Hentenryck

University of Michigan, Ann Arbor, MI 48109, USA
{hasanm,pvanhent}@umich.edu

Abstract. Evacuation planning algorithms are critical tools for assisting authorities in orchestrating large-scale evacuations while ensuring optimal utilization of resources. To be deployed in practice, these algorithms must include a number of constraints that dramatically increase their complexity. This paper considers the zone-based non-preemptive evacuation planning problem in which each evacuation zone is assigned a unique evacuation path to safety and the flow of evacuees over time for a given zone follows one of a set of specified response curves. The starting point of the paper is the recognition that the first and only optimization algorithm previously proposed for zone-based non-preemptive evacuation planning may produce non-elementary paths, i.e., paths that visit the same node multiple times over the course of the evacuation. Since non-elementary paths are undesirable in practice, this paper proposes a column-generation algorithm where the pricing subproblem is a least-cost path under constraints. The paper investigates a variety of algorithms for solving the subproblem as well as their hybridization. Experimental results on a real-life case study show that the new algorithm produces evacuation plans with elementary paths of the same quality as the earlier algorithm in terms of the number of evacuees reaching safety and the completion time of the evacuation, at the expense of a modest increase in CPU time.

Keywords: Column generation · Evacuation planning · k-shortest paths · Mixed-integer programming · Constraint programming

1 Introduction

Large-scale evacuations are critical to the preservation of safety and lives of residents in regions threatened by man-made or natural disasters like floods, hurricanes, and wildfires. Once rare events, evacuations have been increasing in frequency, with emergency services around the world struggling to define practical evacuation plans, especially for urban environments with significant population growth.

This paper studies prescriptive evacuations where emergency services provide detailed instructions on when and how to evacuate. Prescriptive evacuations (e.g., [1–4,7,10,11,14–16]) are gaining traction around of the world thanks to

© Springer International Publishing AG 2017
J.C. Beck (Ed.): CP 2017, LNCS 10416, pp. 549–564, 2017.
DOI: 10.1007/978-3-319-66158-2_35

their ability to manage the flow of evacuees more effectively compared to self-evacuations, which are often sub-optimal and unpredictable. For prescriptive evacuations, emergency services are almost always interested in designing zone-based evacuation plans which assign a unique evacuation path to each residential zone. These plans are attractive for two main reasons: They make it possible to (1) communicate clear evacuation instructions to the community and (2) control the evacuation process accurately, while balancing the load on the transportation network over time. Zone-based evacuation planning however is computationally challenging and has been approached using optimization algorithms based on column generation and Benders decomposition.

Almost all zone-based evacuation planning algorithms are preemptive: They treat evacuees in a zone as flows and assume their evacuation rates can be changed and controlled easily. As a result, they typically produce preemptive evacuation schedules with widely fluctuating departure rates for a given zone, which makes them impractical to implement. To our knowledge, Pillac et al. [13] were the first to address this issue; they proposed an evacuation algorithm that utilizes response curves to control the evacuation flow of a zone over time. The resulting non-preemptive plan can then be enforced in practice by using a mobilization force to ensure the proper response for each evacuation zone. Pillac et al. [13] proposed a column-generation algorithm [5] for non-preemptive zone-based evacuations. They show that the resulting evacuation plans achieve roughly the same quality as preemptive plans in terms of the number of evacuees and the clearance time. *Unfortunately, we discovered that some of the paths generated by their algorithm are non-elementary*, i.e., they visit the same node multiple times. Figure 1 illustrates one such evacuation path generated for the

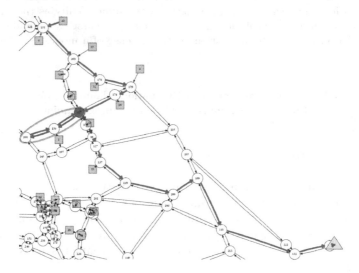

Fig. 1. The evacuation path (in red) for zone 43 (top-left) to a safe node (green triangle) containing a cycle (circled in green). (Color figure online)

Hawkesbury-Nepean region, west of Sydney. Such non-elementary paths are attractive from an optimization standpoint: They may be used to artificially delay evacuees so they arrive at a node (transit node 267 in the figure) at the right time to maximize the use of a bottleneck (the blue colored arc). Unfortunately, emergency services will never adopt such evacuation paths and evacuees will not comply with such plans in practice.

This paper remedies this shortcoming and proposes a pricing subproblem that generates elementary evacuation paths. The subproblem is a constrained least-cost path problem in the time-expanded graph that is significantly more challenging computationally. To solve the pricing subproblem, the paper explores three different approaches and hybridizes them, exploiting their respective strengths. Experimental results on a real-life case study show that the approach produces effective plans with an acceptable overhead in computation time.

The rest of this paper is organized as follows. Section 2 introduces the notation and concepts used throughout the paper. Section 3 reviews the column-generation algorithm proposed in [13]. Section 4 details our proposed solution for computing elementary evacuation paths. Section 5 presents the experimental results. Finally, Sect. 6 provides some concluding remarks.

2   Notation and Preliminaries

Figure 2 depicts a scenario where square 0 represents an evacuation node (e.g. a residential zone), triangles A and B represent safe nodes (final evacuation destinations), circles 1–3 represent transit nodes (road intersections), and arcs represent roads connecting the nodes. Times on each arc indicate when each road will become unavailable (e.g. due to being flooded) and the time on the evacuation node indicates its evacuation deadline.

The scenario can be abstracted into a static evacuation graph $\mathcal{G} = (\mathcal{N} = \mathcal{E} \cup \mathcal{T} \cup \mathcal{S}, \mathcal{A})$ where $\mathcal{E}$, $\mathcal{T}$, and $\mathcal{S}$ are the set of evacuation, transit, and safe nodes respectively, and $\mathcal{A}$ is the set of all arcs. Each evacuation node $k \in \mathcal{E}$

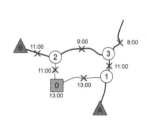

Fig. 2. A sample evacuation scenario.

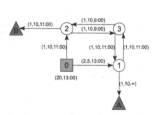

Fig. 3. The static evacuation graph of the scenario in Fig. 2.

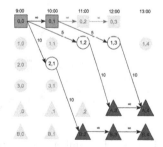

Fig. 4. The time-expanded graph of the static graph in Fig. 3.

has a demand d_k representing the number of vehicles to be evacuated and an evacuation deadline f_k, while each arc $e \in \mathcal{A}$ has a travel time s_e, a capacity u_e, and a deadline f_e after which the road becomes unavailable. Figure 3 shows the static evacuation graph for the scenario of Fig. 2. The evacuation node is labeled with its demand and evacuation deadline while the arcs are labeled with their travel times, capacities, and deadlines. Note that the evacuation node has no incoming arcs and the safe nodes have no outgoing arcs.

To reason about traffic flows over time, the static graph is converted into a time-expanded graph $\mathcal{G}^x = (\mathcal{N}^x = \mathcal{E}^x \cup \mathcal{T}^x \cup \mathcal{S}^x, \mathcal{A}^x)$. The conversion is performed by discretizing the time horizon $\mathcal{H}$ into time steps of identical length t, creating a copy of all nodes at each time step, and replacing each arc $e = (i, j)$ with arcs $e_t = (i_t, j_{t+s_e})$ for each time step where e is available. Figure 4 shows the time-expanded graph constructed from the static graph of Fig. 3, in which each arc is labeled with its capacity. Infinite capacity arcs are introduced connecting the evacuation and safe nodes at each time step to model evacuees staying at these locations. Nodes that cannot be reached from any evacuation or safe node within the time horizon are removed from the graph (they are grayed out in Fig. 4).

The algorithm uses response curves to model evacuation behaviors [12]. A response curve f is a function that realistically models the flow of evacuees over time. For a response curve f, the number of evacuees $D_k(t, t_0)$ departing evacuation node k at time t given that evacuation of k started at start time t_0 is defined as

$$D_k(t, t_0) = \begin{cases} 0 & \text{if } t < t_0 \\ f(t - t_0) & \text{if } t \geq t_0. \end{cases} \tag{1}$$

$D_k(t, t_0)$ precisely specifies a non-preemptive evacuation schedule for evacuation node k. Figure 5 shows $D_k(t, t_0)$ for four different types of response curves. The step response curve, in which evacuees depart at a constant rate after t_0 until the region is completely evacuated, is the type of response curve used in our experiments, since its simplicity also makes it attractive for emergency services. The results however apply to arbitrary curves.

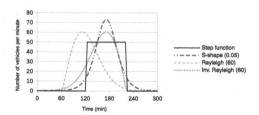

Fig. 5. The number of evacuees $D_k(t, t_0)$ departing evacuation node k as a function of time using four different types of response curves (from [13]).

An evacuation plan contains two components: (1) a set of evacuation paths, each of which is a sequence of connected nodes in the static graph from an evacuation node to a safe node specifying the route to be taken by residents of each evacuation node, and (2) a set of evacuation schedules indicating the number of vehicles that need to depart from each evacuation node at each time step. The Evacuation Planning Problem (EPP) amounts to designing an evacuation plan that maximizes the number of evacuees reaching safety.

3   The Column-Generation Algorithm

The column-generation algorithm consists of a master problem (MP) and pricing subproblem (PSP). The MP is a mixed-integer program (MIP) and the column-generation algorithm uses its linear relaxation. After convergence of the column-generation process, the MP is solved exactly as a MIP. The final MP solution is an upper bound of the optimum and the column generation provides a lower bound. Together, they can be used to compute the duality gap.

The Master Problem. The MP selects time-response evacuation plans from the set Ω' of available plans in order to maximize the number of evacuees reaching safety and minimize total evacuation time. A time-response evacuation plan $p = \langle k, f, P, t_0 \rangle$ consists of an evacuation node k, a response curve $f \in \mathcal{F}$ (where $\mathcal{F}$ is a set of predefined response curves), a path P in $\mathcal{G}$ from k to a safe node, and an evacuation start time t_0. The MP uses a binary variable x_p to indicate whether a plan $p \in \Omega'$ is selected. In the formulation, c_p denotes the cost of plan p, $\Omega_k \subset \Omega'$ is the subset of plans for evacuation node k, $\omega(e) \subset \Omega'$ is the subset of plans that utilize arc e, and a_{p,e_t} denotes the flow of evacuees along arc e at time t induced by plan p. The MP can then be formulated as

$$\min \sum_{p \in \Omega'} c_p \cdot x_p \tag{2}$$

subject to

$$\sum_{p \in \Omega_k} x_p = 1 \qquad \forall k \in \mathcal{E} \tag{3}$$

$$\sum_{p \in \omega(e)} a_{p,e_t} \cdot x_p \leq u_{e_t} \qquad \forall e \in \mathcal{A}, \forall t \in \mathcal{H} \tag{4}$$

$$x_p \geq 0 \qquad \forall p \in \Omega'. \tag{5}$$

The cost c_p linearly penalizes the late arrival times of evacuees at the safe nodes and heavily penalizes the number of evacuees that cannot reach safety (non-evacuees). More precisely, c_p is given by:

$$c_p = \sum_{e \in p} \sum_{t \in \mathcal{H}} c_{e_t} \cdot a_{p,e_t} + \bar{c} \cdot \bar{a}_p \tag{6}$$

where $\bar{a}_p$ denotes the number of non-evacuees in plan p, and c_{e_t} and $\bar{c}$ are defined as follows:

$$c_{e_t} = c_{(i,j)_t} = \begin{cases} \frac{t}{|\mathcal{H}|} & \text{if } j \in \mathcal{S} \\ 0 & \text{otherwise} \end{cases} \tag{7}$$

$$\bar{c} = 100 \max_{e \in \mathcal{A}, t \in \mathcal{H}} \{c_{e_t}\} \cdot \max_{k \in \mathcal{E}} \{d_k\} \tag{8}$$

The cost c_p represents a lexicographic objective function that first minimizes the number of non-evacuees and then late evacuations. Constraints (3) ensure only one plan is selected for each evacuation node and Constraints (4) enforce all arc capacities. The variables (5) are continuous during the column-generation procedure; they become integral in the final iteration of the MP.

The Original Pricing Subproblem. The PSP is responsible for identifying new time-response evacuation plans with negative reduced cost. The reduced cost r_p of plan p is given by:

$$r_p = -\pi_k + \bar{c} \cdot \bar{a}_p + \sum_{e \in p} \sum_{t \in \mathcal{H}} (c_{e_t} - \pi_{e_t}) \cdot a_{p,e_t} \tag{9}$$

where the π_k and π_{e_t} values denote the duals of Constraints (3) and (4) respectively. The PSP finds a plan p that minimizes the last two terms of Eq. (9) for each $k \in \mathcal{E}$ and $f \in \mathcal{F}$. We denote these two terms as $\text{cost}(p)$:

$$\text{cost}(p) = \bar{c} \cdot \bar{a}_p + \sum_{e \in p} \sum_{t \in \mathcal{H}} (c_{e_t} - \pi_{e_t}) \cdot a_{p,e_t} \tag{10}$$

Moreover, since these plans are independent of each other, the PSP for each $k \in \mathcal{E}$ and $f \in \mathcal{F}$ can be solved concurrently in parallel.

An evacuation plan that minimizes $\text{cost}(p)$ can be obtained from a dedicated time-expanded graph $\mathcal{G}^x$ with carefully chosen arc costs and a virtual super-sink v to which all safe nodes $s \in \mathcal{S}^x$ connect with the arcs in $\mathcal{A}_s^x = \{(s, v) \mid s \in \mathcal{S}^x\}$. Let $\mathcal{A}_w^x$ denote the set of all infinite capacity arcs used to model evacuees waiting at the evacuation nodes. Then, for a given $k \in \mathcal{E}$ and $f \in \mathcal{F}$, a path P^x in $\mathcal{G}^x$ from node k_0 (node k at time 0) to v precisely specifies a plan $p = \langle k, f, P, t_0 \rangle$, where P is given by the sequence of nodes visited by P^x excluding v, and t_0 corresponds to the time of the first non-waiting arc leaving evacuation node k. For instance, path P^x represented by the red colored arcs in Fig. 6 corresponds to $P = \langle 0, 1, A \rangle$ and $t_0 = 10{:}00$ (other arcs in $\mathcal{A}_s^x$ are omitted from the figure to reduce clutter). Arc costs $c_{e_t}^{\text{sp}}$ for each arc $e_t \in \mathcal{A}^x$ are then defined as follows:

$$c_{e_t}^{\text{sp}} = \sum_{t'=t}^{|\mathcal{H}|} (c_{e_{t'}} - \pi_{e_{t'}}) \cdot f(t' - t) \qquad \forall e_t \in \mathcal{A}^x \setminus (\mathcal{A}_w^x \cup \mathcal{A}_s^x) \tag{11}$$

$$c_{e_t}^{\text{sp}} = \bar{c} \cdot (d_k - F(|\mathcal{H}| - t)) \qquad \forall e_t \in \mathcal{A}_s^x \tag{12}$$

$$c_{e_t}^{\text{sp}} = 0 \qquad \forall e_t \in \mathcal{A}_w^x. \tag{13}$$

Equation (11) aggregates all the costs along arc e over time for a plan p that visits arc e first at time t. Equation (12) accounts for the cost of non-evacuees

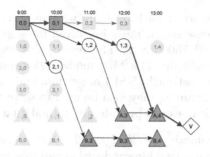

Fig. 6. Path P^x in the dedicated time-expanded graph. (Color figure online)

for plans that end with that arc (F is the cumulative density function associated with f).

With these arc costs, the total cost of any path P^x from k_0 to v using f is equivalent to $\mathrm{cost}(\langle k, f, P, t_0 \rangle)$:

$$\sum_{e_t \in P^x} c^{\mathrm{sp}}_{e_t} = \bar{c} \cdot (d_k - F(|\mathcal{H}| - t)) + \sum_{e_t \in P^x \setminus \mathcal{A}^x_s} \sum_{t'=t}^{|\mathcal{H}|} (c_{e_{t'}} - \pi_{e_{t'}}) \cdot f(t' - t) \quad (14)$$

$$= \bar{c} \cdot \bar{a}_p + \sum_{e \in p} \sum_{t \in \mathcal{H}} (c_{e_t} - \pi_{e_t}) \cdot a_{p,e_t} \quad (15)$$

Hence, the PSP can be solved by any shortest-path algorithm in order to find a path P^x of minimal cost for each $k \in \mathcal{E}$ and $f \in \mathcal{F}$. Observe that the algorithm only needs to find a path from k_0 (node k at time 0), since evacuation node k at time t is connected with an uncapacitated arc to node k at time $t+1$.

4   Revisiting the Pricing Subproblem

The original PSP does not preclude least-cost paths from visiting the same transit node in $\mathcal{G}^x$ at multiple time steps. While such paths are acyclic since $\mathcal{G}^x$ is acyclic by construction, their projection onto $\mathcal{G}$ results in non-elementary paths as the same transit node is visited more than once. Application of the original algorithm on a real-world evacuation scenario from the Hawkesbury-Nepean region revealed that approximately 44% of plans in the MP had non-elementary paths, with some being selected by the final MIP. This is problematic since these paths do not lead to realistic evacuation plans in practice.

This section outlines a new PSP that only generates time-response evacuation plans with elementary paths and proposes several algorithms for this task. Let $\Lambda(i)$ denote the set of time-expanded nodes in $\mathcal{G}^x$ for a node $i \in \mathcal{T}$, i.e., $\Lambda(i) = \{i_t \mid t \in \mathcal{H}\}$. A path P^x in $\mathcal{G}^x$ corresponds to an elementary path P in $\mathcal{G}$ if and only if P^x visits at most a single node in $\Lambda(i)$ for each node $i \in \mathcal{T}$. As a result, instead of finding a least-cost path in $\mathcal{G}^x$ with arc costs defined by (11)–(13), the revised PSP must find a least-cost path that corresponds to an elementary path in the static graph.

This problem is a shortest-path problem with resource constraints [8]. It is NP-hard and, for our case study, must be solved for a graph $\mathcal{G}^x$ with approximately 30000 nodes and 75000 arcs. The rest of this section explores several methods to solve this problem: (1) a MIP model, (2) a constraint programming (CP) model, (3) a k-shortest-path (KSP) algorithm, and (4) a hybrid algorithm. A general labeling algorithm [6] may also be used to solve the problem, however it was not explored as it does not provide any formal guarantees of optimality.

A Mixed-Integer Programming Method for the PSP. The PSP can be solved with a MIP model, which uses a binary decision variable x_{e_t} to indicate whether arc e_t is selected in the optimal path. Let $\delta^-(i)$ and $\delta^+(i)$ to denote the set of incoming and outgoing arcs of node i respectively. The MIP model is defined in Fig. 7. Objective function (16) specifies that the objective of the problem is to minimize total cost of the path. Constraint (17) specifies that exactly one path should emanate from source node k_0, while Constraint (19) ensures the path ends at super-sink node v. Constraints (18) enforce path continuity at every node other than the source and the sink. Finally, Constraints (20) are the elementary path constraints which guarantee that each transit node is visited by the path at most once over the time horizon.

$$\min \sum_{e_t \in \mathcal{A}^x} c_{e_t}^{\text{sp}} \cdot x_{e_t} \tag{16}$$

subject to

$$\sum_{e_t \in \delta^+(k_0)} x_{e_t} = 1 \tag{17}$$

$$\sum_{e_t \in \delta^-(i)} x_{e_t} - \sum_{e_t \in \delta^+(i)} x_{e_t} = 0 \quad \forall i \in \mathcal{N}^x \setminus \{k_0, v\} \tag{18}$$

$$\sum_{e_t \in \delta^-(v)} x_{e_t} = 1 \tag{19}$$

$$\sum_{i_t \in \Lambda(i)} \sum_{e_t \in \delta^+(i_t)} x_{e_t} \leq 1 \quad \forall i \in \mathcal{T} \tag{20}$$

$$x_{e_t} \in \{0, 1\} \quad \forall e_t \in \mathcal{A}^x \tag{21}$$

Fig. 7. The MIP model for the PSP.

A Constraint Programming Method for the PSP. Figure 8 presents a constraint program to find a constrained least-cost path. The model uses two decision variables: the float variable sp[j] that denotes the cost of a path from k_0 to node j and the Boolean variable absent[i] that indicates whether node i is absent or present in the optimal path. $\alpha^-(j)$ denotes the set of predecessors of node j. The objective function specifies that the goal is to minimize cost of the path from k_0 to v.[1]

---

[1] The model does not use the traditional predecessor and successor variables, since the solver runs out of memory due to the large number of element constraints.

```
dvar float sp[N^x];
dvar boolean absent[N^x];
minimize
    sp[v];
subject to {
    absent[v] = 0;
    absent[k_0] = 0;
    sp[k_0] = 0.0;
    forall(j in N^x \ {k_0})
        sp[j] = min(i in α^-(j)) sp[i] + c_(i,j) + M*absent[i];
    forall(j in N^x \ {k_0})
        sum(i in α^-(j)) absent[i] >= |α^-(j)| - 1;
    forall(i in T)
        sum(n in Λ(i)) absent[n] >= |Λ(i)| - 1;
};
```

Fig. 8. The CP model for the PSP.

The first three constraints of the model specify that v and k_0 must be present in the optimal path and that the least-cost path to the source k_0 is zero. The next set of constraints

```
forall(j in N^x \ {k_0})
    sp[j] = min(i in α^-(j)) sp[i] + c_(i,j) + M*absent[i];
```

specify the least-cost path from k_0 to a node j in terms of the least-cost paths to the predecessors of j. The constraints use a large number M for the predecessors that are not in the least-cost path and $c_{(i,j)}$ denotes the arc costs defined by Eqs. (11)–(13). The next set of constraints:

```
forall(j in N^x \ {k_0})
    sum(i in α^-(j)) absent[i] >= |α^-(j)| - 1;
```

specify that for each node j, at most one of its predecessors can be present in the optimal path. Finally, the last set of constraints

```
forall(i in T)
    sum(n in Λ(i)) absent[n] >= |Λ(i)| - 1;
```

specify that each transit node may only occur once in the optimal path over the time horizon.

A k-Shortest-Path Algorithm for the PSP. The third algorithm is indirect; it uses a k-shortest-path algorithm to generate paths in increasing order of costs. The algorithm completes as soon it generates an elementary path in the static graph. Our implementation uses the Recursive Enumeration Algorithm (REA) [9] to enumerate shortest paths from k_0 to v. The REA computes k shortest paths from a source s to a sink t in a graph $\mathcal{G} = (\mathcal{N}, \mathcal{A})$ in $\mathcal{O}(|\mathcal{A}| + k|\mathcal{N}|\log(|\mathcal{A}|/|\mathcal{N}|))$

time. It does so by recursively solving a set of equations which generalize the Bellman equations for the shortest-path problem.

Let $P^k(j)$ denote the kth-shortest path from source s to node j, $P^k(i) \cdot j$ denote a path from s to j formed by appending arc (i, j) to the end of path $P^k(i)$, $C^k(j)$ be the set of candidate paths from s to j from which $P^k(j)$ can be chosen, and $L(P)$ be the length of path P. The key intuition behind the REA is that if $\Pi^k(j)$ is the set of k shortest paths from s to j, then each path in $\Pi^k(j)$ must reach j through some node $i \in \alpha^-(j)$. Therefore, to find $P^k(j)$, one needs to consider all shortest paths from s to $i \in \alpha^-(j)$ and concatenate them with arc (i, j) while making sure that they are different from $P^{k-1}(j)$. To facilitate this, the algorithm constructs, for each node j, a set of candidates $C^k(j)$ from which $P^k(j)$ can be chosen. This set contains at most one path from s to $i \in \alpha^-(j)$ appended with arc (i, j). $P^k(j)$ can then be obtained by selecting the shortest path in $C^k(j)$.

The KSP algorithm first computes $P^1(j)$ for all $j \in \mathcal{N}$ using a one-to-all shortest-path algorithm (e.g., Bellman-Ford). If this already produces a path from s to t that satisfies the elementary path constraints, then the algorithm terminates. Otherwise, the procedure NEXTPATH(t, k) (Algorithm 1) is called repeatedly, with increasing values of k, to compute the kth-shortest path to the sink t. Procedure NEXTPATH(j, k) is recursive and constructs $C^k(j)$ from $C^{k-1}(j)$. Lines 2–3 handle the base case where j is the source in which no new path is produced. Line 5 initializes the candidate set for the second shortest path to j using the original shortest paths. Lines 6–10 represent the recursive case and compute a new candidate path by refining the $k - 1^{\text{th}}$-shortest path. They select the predecessor i of j that led to the $k - 1^{\text{th}}$-shortest path to j (i.e., $P^{k'}(i) \cdot j = P^{k-1}(j)$ for some $k' < k$) and generate a new path to i. This path, if it exists, is concatenated with j and inserted in $C^k(j)$ together with the paths in $C^{k-1}(j)$. Lines 12–16 select the shortest path $P^k(j)$ and remove the shortest path from $C^k(j)$, which will be used when computing $C^{k+1}(j)$. Note that, in the actual implementation, it is sufficient to keep a single set $C(j)$ for all the shortest-path candidates to j. Algorithm 1 can easily be adapted to do so.

A Hybrid Algorithm for the PSP. Empirical evaluations indicate that the KSP algorithm is very fast in finding an optimal elementary path in most cases. However, there are PSP instances for which the algorithm needs to enumerate an extremely large number of paths ($k > 10^6$). Moreover, in some instances, k is so large that the algorithm runs out of memory before finding an elementary path. These observations indicate that a hybridization of MIP or CP with the KSP algorithm should exploit the strengths of both approaches. The hybrid algorithm first enlists the KSP algorithm to enumerate all the k-shortest paths up to a fixed k-threshold. If no elementary path is found before the threshold is reached, the MIP or CP model is solved. As long as the threshold is set appropriately, the hybrid algorithm harnesses the speed benefits of the KSP algorithm while addressing its limitation.

Algorithm 1. Next Shortest Path Algorithm

Require: $P^1(j), P^2(j), \ldots, P^{k-1}(j), k > 1$
1: **procedure** NextPath(j, k)
2:     **if** $j = s$ **then**
3:         **return** $\perp$
4:     **if** $k = 2$ **then**
5:         $C^{k-1}(j) \leftarrow \{P^1(i) \cdot j \mid i \in \alpha^-(j)\} \setminus \{P^1(j)\}$
6:     Let i be a node and k' be an index such that $P^{k'}(i) \cdot j = P^{k-1}(j)$
7:     **if** $P^{k'+1}(i)$ has not been computed **then**
8:         $P^{k'+1}(i) \leftarrow$ NextPath$(i, k' + 1)$
9:     **if** $P^{k'+1}(i) \neq \perp$ **then**
10:         $C^k(j) \leftarrow C^{k-1}(j) \cup \{P^{k'+1}(i) \cdot j\}$
11:     **if** $C^k(j) \neq \emptyset$ **then**
12:         $P^k(j) \leftarrow \arg\min_{P \in C^k(j)} L(P)$
13:         $C^k(j) \leftarrow C^k(j) \setminus \{P^k(j)\}$
14:         **return** $P^k(j)$
15:     **else**
16:         **return** $\perp$

---

5 Experimental Results

The algorithms were evaluated on a real case study: the evacuation scenario of the Hawkesbury-Nepean (HN) region located north-west of Sydney, Australia, in case of a breach of the Warragamba dam. The evacuation graph consists of 80 evacuation nodes, 184 transit nodes, 5 safe nodes, and 580 arcs. While the region has a total of 38343 vehicles to be evacuated, the experimental results also consider scenarios with increased demand by linearly scaling the vehicle count by factor x $\in$ [1.0, 3.0]. We refer to a particular scenario as HN80-Ix with x representing the scaling factor. These scenarios are realistic given the significant population growth in the West Sydney region.

Each evacuation plan is given a time horizon of $\mathcal{H} = 10$ h discretized into 5 min time steps. $\mathcal{F}$ is populated with step response curves with flow rates of $\gamma \in \{2, 6, 10, 25, 50\}$ vehicles per time step. The original PSP is solved using the Bellman-Ford algorithm as in [13]. All algorithms were implemented in C++ with GUROBI 6.5.2 being invoked to solve all LPs and MIPs. In each iteration of the column-generation procedure, the PSPs for each $k \in \mathcal{E}$ and $f \in \mathcal{F}$ are solved in parallel using OpenMP, and all negative reduced cost columns found are added to the MP. IBM CP Optimizer 12.6.2 is used for the CP method. All experiments were conducted on a high-performance computing cluster with 8 cores of a 2.5 GHz Intel Xeon E5-2680v3 processor and 64 GB of RAM. CPU time limits of 96 h and 24 h are imposed on the column-generation phase and the final iteration of the master problem.

5.1 Comparing the Three Approaches

The results first compare the performance of the KSP, MIP, and CP methods for solving the PSPs for instance HN80-I1.0. The results report the time spent for solving each PSP, together with the k value of the KSP algorithm. The average computation times are summarized in Fig. 9. The KSP algorithm was terminated when k reached 10^6 due to its memory issues. The KSP algorithm is consistently the fastest across the various ranges of k values, although its average solution time experiences a sharp increase for $k > 10^5$. The MIP method comes in second while the CP method is always the slowest. These results led us to combine the KSP algorithm with the MIP method in the hybrid algorithm. Figure 10 shows the fraction of PSP instances solved for each range of k values. The figure indicates that the majority of PSP instances are solvable by the KSP algorithm with small k values and only approximately 10% of them require $k > 10^6$. This makes the prospect of utilizing the hybrid algorithm very promising. With a properly tuned k-threshold, the hybrid method will solve a large fraction of the PSP instances with the fast KSP algorithm.

5.2 Tuning k-Threshold for the Hybrid Algorithm

Figure 11 shows the average solution times of the KSP-MIP hybrid algorithm with k-threshold values of 10^5 and 10^6 for instance HN80-I1.0. The figure illustrates the trade-off involved when selecting the threshold. The algorithm with a 10^6-threshold solves the PSP faster when $k \in (10^5, 10^6]$ as its k-threshold is not exceeded, whereas the algorithm with the 10^5-threshold resorts to using the slower MIP method. However, when $k > 10^6$, the situation is reversed. While the thresholds of both algorithms are exceeded, the 10^5 algorithm spends less time trying to solve the PSP using the KSP algorithm. The 10^6 algorithm applies the KSP algorithm until $k = 10^6$ before switching to the MIP method and hence wastes more time. This suggests that an algorithm with a 10^5-threshold would

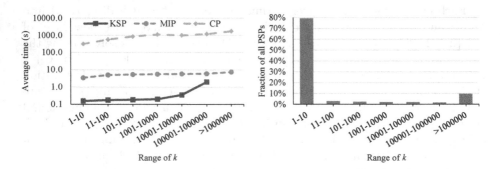

Fig. 9. Average computation times of the algorithms as a function of the range of k in the KSP algorithm.

Fig. 10. Distribution of the k value needed by the KSP algorithm for finding an elementary path.

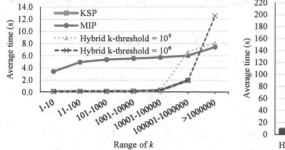

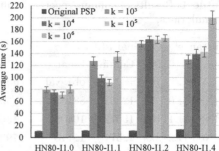

Fig. 11. Computation times of the KSP-MIP algorithm with thresholds $k = 10^5$ and $k = 10^6$.

Fig. 12. Computation times for the KSP-MIP algorithm during column generation.

be faster overall if the fraction of PSP instances with $k > 10^6$ is larger than those with $k \in (10^5, 10^6]$.

To get a better sense of threshold values that would result in the shortest PSP times, we executed the column-generation algorithm with the KSP-MIP algorithm with k-threshold $\in \{10^3, 10^4, 10^5, 10^6\}$ on HN80-Ix instances with x $\in \{1.0, 1.1, 1.2, 1.4\}$ and recorded the total time spent solving only the PSP in each iteration. The experiment was not extended to larger instances due to their prohibitively expensive computation times. Figure 12 reports the results with the error bars describing the standard error. The average times of the original PSP (i.e., without the elementary restrictions) are also included as a reference baseline. For instances HN80-I1.0 and HN80-I1.2, the differences in average computation times across various threshold values are not statistically significant. However, in the other instances, a k-threshold of 10^5 resulted in either the smallest or among the smallest average computation times. Therefore, a k-threshold of 10^5 is used for the hybrid KSP-MIP algorithm in all subsequent experiments.

5.3 Overall Performance of the Evacuation Planning Algorithm

Results of the column-generation phase of the algorithm with the original and elementary path formulations on all HN80-Ix instances are summarized in Table 1. The table shows the number of iterations, the number of columns generated, the final objective values, and the CPU time consumed during the column-generation phase of both algorithms. *The key result is that the new formulation successfully produces time-response evacuation plans with elementary paths while maintaining the same final objective values as the original in all instances.* The minor difference in objective values for some instances (e.g. HN80-I2.5) could be attributed to the column-generation phase being terminated before convergence. This means that the elementary path constraints do not induce any quality loss for the scenarios tested. Furthermore, for instances where the phase converged

Table 1. Results of column-generation phase using original and elementary path PSP formulations.

Instance	Original PSP				Elementary path PSP			
	Iter #	Column #	Final obj. val.	CPU time (mins)	Iter #	Column #	Final obj. val.	CPU time (mins)
HN80-I1.0	79	12251	8816	39	79	11678	8816	124
HN80-I1.1	104	15072	10405	136	95	13853	10405	218
HN80-I1.2	229	22571	12116	799	190	19543	12116	834
HN80-I1.4	152	20184	15935	690	108	17048	15935	404
HN80-I1.7	178	21871	22635	1312	120	19883	22635	2760
HN80-I2.0	197	31418	30490	5760	145	25051	30490	5760
HN80-I2.5	121	22806	46189	5760	129	23513	46188	5760
HN80-I3.0	132	31726	1.96×10^9	5760	87	21233	1.96×10^9	5760

before the CPU time limit (i.e., x ∈ [1.0, 1.7]), the new formulation generally converges in fewer iterations and produces fewer columns than the original formulation. This can be attributed to the new formulation considering a smaller subset of all possible paths (only elementary ones) while the original considers all possible paths. The CPU time of the new formulation is consistently higher in almost all instances. This is not surprising since the new PSP formulation involves solving an NP-hard problem whereas the original is solved using a low polynomial-time algorithm. *The challenge of solving a harder PSP is alleviated by the reduced number of iterations and the efficiency of the hybrid algorithm, which eventually results in only a modest increase in CPU time.*

The results of the final MIP are summarized in Table 2 for both the original and new formulations. It shows the duality gap, evacuation percentage, evacuation end time, and total CPU time (column generation plus final MIP) consumed by both formulations on all HN80-Ix instances. The duality gap is calculated using $\frac{z_{MP,MIP} - z_{MP,LP}}{z_{MP,MIP}}$ where $z_{MP,LP}$ and $z_{MP,MIP}$ are the final objective values of the MP LP and MP MIP respectively, the evacuation percentage is the fraction of evacuees reaching the safe nodes at the end of the time horizon, and the evacuation end time is simply the time at which the last evacuee reaches its safe destination. Aside from the final MIP of the new formulation successfully producing final evacuation plans with only elementary paths in all instances, the results across both formulations are approximately even. The 24 h time limit for the final MIP was reached in all instances for both formulations. Both also attained 100% evacuation in the first four instances, producing duality gaps of less than 15% in the first three. Both formulations have 100% duality gaps for x ∈ [1.7, 2.5], which can be attributed to the huge penalty incurred in the objective value of the final MIP due to not being able to evacuate everyone when its linear relaxation did. The new formulation also manages to complete evacuation earlier in two instances, HN80-I1.0 and HN80-I1.2.

Table 2. Results of entire column-generation algorithm using original and elementary path PSP formulations.

Instance	Original PSP				Elementary path PSP			
	Duality gap (%)	Evac. perc. (%)	Evac. end (mins)	CPU time (mins)	Duality gap (%)	Evac. perc. (%)	Evac. end (mins)	CPU time (mins)
HN80-I1.0	13.7	100.0	580	1479	12.3	100.0	550	1564
HN80-I1.1	13.6	100.0	510	1576	12.3	100.0	510	1658
HN80-I1.2	14.1	100.0	575	2239	14.7	100.0	565	2274
HN80-I1.4	15.1	100.0	600	2130	16.3	100.0	600	1844
HN80-I1.7	100.0	98.3	-	2752	100.0	96.2	-	4200
HN80-I2.0	100.0	89.9	-	7200	100.0	91.0	-	7200
HN80-I2.5	100.0	94.7	-	7200	100.0	92.5	-	7200
HN80-I3.0	69.1	77.3	-	7200	59.6	82.7	-	7200

6 Conclusion

This paper considered zone-based non-preemptive evacuation plans, where each zone is assigned a unique path to safety and the flow of evacuees over time in the zone follows one of a set of specified response curves. It proposes a generalization of the column-generation approach in [13] which was discovered to contain a critical shortcoming: The algorithm can produce non-elementary evacuation paths, i.e., evacuation paths that visit the same node multiple times over the course of an evacuation. Non-elementary paths are undesirable in practice and evacuees are highly unlikely to follow such plans.

This paper addressed this limitation and formulated a novel pricing subproblem, which can be modeled as a constrained least-cost path in the time-expanded graph. The paper also explored four approaches to solve the pricing subproblem: a MIP model, a CP model, and an algorithm based on the k-shortest-path problem, as well as their hybridization. Experimental results on a real case study showed that the hybridization of a k-shortest-path algorithm and a MIP model is most effective in practice. The paper also shows that the threshold to switch from the k-shortest-path algorithm to the MIP model can be tuned to provide an optimal switching point. Experimental results on the real case study of the Hawkesbury-Nepean region show that the new optimization algorithm generates evacuation plans with elementary paths that match the objective value and duality gap of the original formulation in [13]. Moreover, these practical evacuation plans can be obtained at the cost of a modest increase in CPU time.

This paper thus provided the first optimization approach to produce zone-based, non-preemptive evacuation plans with elementary paths. These evacuation plans are inherently practical. They provide paths that evacuees will feel comfortable complying to. Moreover, emergency services often have mobilization teams to ensure that the response curves are implemented properly as is the case in New South Wales. Future research will be devoted to studying how to integrate the proposed algorithm with convergent plans [7,15] and expansion planning, and speeding up the constraint program using a global shortest-path constraint.

References

1. Bish, D.R., Sherali, H.D., Hobeika, A.G.: Optimal evacuation planning using staging and routing. J. Oper. Res. Soc. **65**(1), 124–140 (2014)
2. Bish, D.R., Sherali, H.D.: Aggregate-level demand management in evacuation planning. Eur. J. Oper. Res. **224**(1), 79–92 (2013)
3. Bretschneider, S., Kimms, A.: Pattern-based evacuation planning for urban areas. Eur. J. Oper. Res. **216**(1), 57–69 (2012)
4. Cova, T.J., Johnson, J.P.: A network flow model for lane-based evacuation routing. Transp. Res. Part A: Policy Pract. **37**(7), 579–604 (2003)
5. Desaulniers, G., Desrosiers, J., Solomon, M.M.: Column Generation, vol. 5. Springer Science & Business Media, Berlin (2006)
6. Desrochers, M., Desrosiers, J., Solomon, M.: A new optimization algorithm for the vehicle routing problem with time windows. Oper. Res. **40**(2), 342–354 (1992)
7. Even, C., Pillac, V., Van Hentenryck, P.: Convergent plans for large-scale evacuations. In: Proceedings of the Twenty-Ninth AAAI Conference on Artificial Intelligence, pp. 1121–1127. AAAI Press (2015)
8. Irnich, S., Desaulniers, G.: Shortest path problems with resource constraints. In: Desaulniers, G., Desrosiers, J., Solomon, M.M. (eds.) Column Generation, pp. 33–65. Springer, Boston (2005). doi:10.1007/0-387-25486-2_2
9. Jiménez, V.M., Marzal, A.: Computing the K shortest paths: a new algorithm and an experimental comparison. In: Vitter, J.S., Zaroliagis, C.D. (eds.) WAE 1999. LNCS, vol. 1668, pp. 15–29. Springer, Heidelberg (1999). doi:10.1007/3-540-48318-7_4
10. Lim, G.J., Zangeneh, S., Baharnemati, M.R., Assavapokee, T.: A capacitated network flow optimization approach for short notice evacuation planning. Eur. J. Oper. Res. **223**(1), 234–245 (2012)
11. Miller-Hooks, E., Sorrel, G.: Maximal dynamic expected flows problem for emergency evacuation planning. Transp. Res. Record: J. Transp. Res. Board **2089**, 26–34 (2008)
12. Pel, A.J., Bliemer, M.C.J., Hoogendoorn, S.P.: A review on travel behaviour modelling in dynamic traffic simulation models for evacuations. Transportation **39**(1), 97–123 (2012)
13. Pillac, V., Cebrian, M., Van Hentenryck, P.: A column-generation approach for joint mobilization and evacuation planning. Constraints **20**(3), 285–303 (2015)
14. Pillac, V., Van Hentenryck, P., Even, C.: A conflict-based path-generation heuristic for evacuation planning. Transp. Res. Part B **83**, 136–150 (2016)
15. Romanski, J., Van Hentenryck, P.: Benders decomposition for large-scale prescriptive evacuations. In: Proceedings of the Thirtieth AAAI Conference on Artificial Intelligence, pp. 3894–3900. AAAI Press (2016)
16. Xie, C., Lin, D.Y., Waller, S.T.: A dynamic evacuation network optimization problem with lane reversal and crossing elimination strategies. Transp. Res. Part E: Logist. Transp. Rev. **46**(3), 295–316 (2010)

Job Sequencing Bounds from Decision Diagrams

J.N. Hooker[✉]

Carnegie Mellon University, Pittsburgh, USA
jh38@andrew.cmu.edu

Abstract. In recent research, decision diagrams have proved useful for the solution of discrete optimization problems. Their success relies on the use of relaxed decision diagrams to obtain bounds on the optimal value, either through a node merger or a node splitting mechanism. We investigate the potential of node merger to provide bounds for dynamic programming models that do not otherwise have a practical relaxation, in particular the job sequencing problem with time windows and state-dependent processing times. We prove general conditions under which a node merger operation yields a valid relaxation and apply them to job sequencing. Computational experiments show that, surprisingly, relaxed diagrams prove the optimal value when their size is only a small fraction of the size of an exact diagram. On the other hand, a relaxed diagram of fixed size ceases to provide a useful bound as the instances scale up.

1 Introduction

Decision diagrams have historically been used for circuit design and verification [1,12,22,23] and a variety of other purposes [24,27]. Recent research indicates that decision diagrams provide an alternative approach to discrete optimization and constraint solving [2,9,14]. Of special interest to optimization is the fact that decision diagrams are well suited to dynamic programming (DP) formulations, because the diagrams are essentially state transition graphs. Viewing DP from the perspective of decision diagrams opens the door to new techniques for solving DP problems, such as branch-and-bound methods [9].

A key element of this development is the use of *relaxed* decision diagrams to derive a bound on the optimal value. Optimization bounds are useful not only in branch-and-bound methods, but for assessing the quality of solutions obtained by heuristics, and perhaps for proving their optimality. Because the exact (non-relaxed) diagram tends to grow exponentially with the number of variables, it is vital to find relaxed diagrams of limited size.

Two methods for constructing relaxed decision diagrams of limited size were introduced in [2]: *node merger* and *node splitting*. Node merger reduces the size of the diagram by introducing some infeasible solutions. Node splitting works in the opposite direction: by beginning with a diagram that represents all possible solutions and creating new nodes to exclude some infeasible solutions.

We investigate here the potential for *node merger* to relax DP formulations, particularly those for which good relaxations are not currently available. While

© Springer International Publishing AG 2017
J.C. Beck (Ed.): CP 2017, LNCS 10416, pp. 565–578, 2017.
DOI: 10.1007/978-3-319-66158-2_36

a great advantage of DP is that it does not presuppose convexity, linearity, or even a closed-form description of the problem, this very generality often makes it difficult to find a good relaxation. The problem may also be difficult to solve due to the exponential growth of the state space, known as the "curse of dimensionality".

We focus on a job sequencing problem in which processing times are state-dependent. Due to this state dependency, the problem does not have a practical mixed-integer or other formulation for which a relaxation is readily available. We wish to determine whether relaxed decision diagrams of limited size can provide a useful bound and therefore assist in finding an optimal solution.

Node merger has not previously been applied to job sequencing problems, or apparently to any problems for which no effective relaxation is known. While various types of sequencing problems are solved by decision diagrams in [14], the relaxed diagrams are created by node splitting rather than merger. It is therefore important to investigate the potential of node merger as a relaxation technique for DP. Because the job sequencing problem studied here is a simple representative of a broad class of sequencing problems, the results could have wider implications.

After a brief review of previous work, we begin with a description of exact and relaxed decision diagrams and how they relate to DP, using the job sequencing problem as an example. We then prove general sufficient conditions under which node merger yields a valid relaxed diagram, since conditions of this kind have never appeared in the literature. They allow us to show that a proposed merger rule for the job sequencing problem produces a valid relaxation. We then describe alternative heuristics for selecting nodes to merge, because an effective merger heuristic is essential to obtaining tight bounds.

We report computational experiments that show the somewhat surprising result that bounds obtained from relaxed diagrams reach the optimal value rather quickly, when the diagrams are only a small fraction of the size of an exact diagram. This allows relaxed diagrams to prove the optimality of a heuristically obtained solution in much less time than would otherwise be necessary. On the other hand, relaxed diagrams of any given fixed size cease to prove a useful bound as the instances scale up. In the conclusion, we suggest a research direction that addresses this issue.

2   Previous Work

Decision diagrams were first proposed as an optimization method by [16,20]. Other early applications to optimization include cut generation for integer programming [4], post-optimality analysis [15,16], and vertex and facet enumeration [5]. The idea of a relaxed decision diagram was introduced by [2] for constraint programming. Subsequent applications are described by [6,17–19]. Relaxed diagrams were used to obtain optimization bounds in [7,11]. Branching within a relaxed diagram was introduced by [9]. Connections between decision diagrams and deterministic dynamic programming, including nonserial dynamic programming, are discussed in [21]. A comprehensive survey of decision diagrams as an optimization technique appears in [8].

Relaxation of a decision diagram is superficially related to state space relaxation in dynamic programming [3,13,25,26], but there are several fundamental differences. A relaxed decision diagram is created by splitting or merging nodes rather than mapping the state space into a smaller space. It can be tightened by filtering techniques. It is constructed dynamically as the diagram is built, rather than by defining a mapping a priori. It uses the same state variables as the exact formulation, which allows the relaxed diagram to be used as a branching framework for an exact branch-and-bound method. The relaxation can be calibrated to provide a bound of any desired quality, up to optimality, simply by adjusting the maximum width to be observed by the diagram while it is created.

3   Decision Diagrams

For our purposes, a decision diagram can be defined as a directed, acyclic multigraph in which the nodes are partitioned into *layers*. Each arc of the graph is directed from a node in layer j to a node in layer $j + 1$ for some $j \in \{1, \ldots, n\}$. Layers 1 and $n + 1$ contain a single node, namely the root r and the terminus t, respectively. Each layer j is associated with a finite-domain variable $x_j \in D_j$. The arcs leaving any node in layer j have distinct *labels* in D_j, representing possible values of x_j at that node. A path from r to t defines an assignment to the tuple $x = (x_1, \ldots, x_n)$ as indicated by the arc labels on the path. The decision diagram is *weighted* if there is a length (cost) associated with each arc.

Any discrete optimization problem with variables $x_1, \ldots x_n$ and a separable objective function $\sum_j f_j(x_j)$ can be represented by a weighted decision diagram.[1] The diagram is constructed so that its r–t paths correspond to the feasible solutions of the problem, and the cost of an arc with label v_j leaving layer j is $f_j(v_j)$. The length (cost) of any r–t path is the objective function value of the corresponding solution. If the objective is to minimize, the optimal value is the length of a shortest r–t path.

Many different diagrams can represent the same problem, but for a given variable ordering, there is a unique *reduced* diagram that represents it [12,21]. A diagram is reduced when for any pair of nodes u, u' in a given layer j, the set of u–t paths and their costs is different from the set of u'–t paths and their costs. That is, the two sets correspond to different sets of assignments to $x_j, \ldots, x_n$ or different costs.

As an example, consider a small instance of the job sequencing problem with time windows (Table 1). The jobs must be sequenced so that each job j begins processing no earlier than the release time r_j and requires processing time p_j. We assume that for a given sequencing, the start time of job j is $s_j = \max\{r_j, s_i + p_i\}$, where i is the immediately preceding job in the sequence, and the first job

---

[1] Problems with nonseparable objective functions can also be represented, as described in [21], but to simplify exposition we omit this possibility.

starts at its release time. The objective is to minimize total tardiness, where the tardiness of job j is $\max\{0, s_j + p_j - d_j\}$, and d_j is the job's due date.

Table 1. A small instance of a job scheduling problem.

j	r_j	p_j	d_j
1	0	3*	4
2	1	2	3
3	1	2	5

*2 when job 2 has previously been processed.

We can make the processing time state-dependent by supposing that it depends on which jobs have already been processed. This frequently occurs in practice, as processing one job may involve the fabrication of parts that can be used when processing another job. In the example, we suppose that the processing time p_1 for job 1 is 2 (rather than 3) when job 2 has been processed.

Figure 1 shows a reduced decision diagram that represents the problem. Variable x_j represents the jth job in the sequence. Each arc indicates its label and immediate cost (the latter in parentheses). Each r–t path encodes a feasible schedule, and any shortest (minimum-cost) r–t path indicates an optimal solution of the problem.

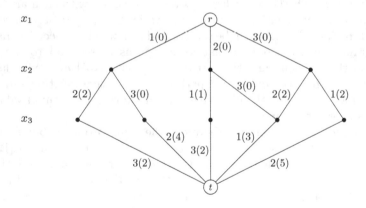

Fig. 1. Decision diagram for the job sequencing instance of Table 1.

When a problem is formulated recursively, a simple top-down compilation procedure yields a decision diagram that represents the problem. A general recursive formulation can be written

$$h_j(S_j) = \min_{x_j \in X_j(S_j)} \left\{ c_j(S_j, x_j) + h_{j+1}(\phi_j(S_j, x_j)) \right\} \tag{1}$$

Here, S_j is the *state* in stage j of the recursion, $X_j(S_j)$ is the set of possible *controls* (values of x_j) in state S_j, ϕ_j is the *transition function* in stage j, and $c_j(S_j, x_j)$ is the *immediate cost* of control x_j in state S_j. We assume there is single initial state S_1 and a single final state S_{n+1}, so that $h_{n+1}(S_{n+1}) = 0$ and $\phi_n(S_n, x_n) = S_{n+1}$ for all states S_n and controls $x_n \in X_n(S_n)$. The quantity $h_j(S_j)$ is the *cost-to-go* for state S_j in stage j, and an optimal solution has value $h_1(r)$.

In the job sequencing problem, the state S_j is the tuple (V_j, f_j), where V_j is the set of jobs scheduled so far, and f_j is the finish time of the last job scheduled. Thus the initial state is $r = (\emptyset, 0)$, and $X_j(S_j)$ is $\{1, \ldots, n\} \setminus V_j$. The transition function $\phi_j(S_j, x_j)$ is given by

$$\phi_j\big((V_j, f_j), x_j\big) = \big(V_j \cup \{x_j\}, \ \max\{r_{x_j}, f_j\} + p_{x_j}(V_j)\big)$$

Note that the processing time $p_{x_j}(V_j)$ depends on the current state V_j as well as the control x_j. The immediate cost $c_j((V_j, f_j), x_j)$ is the tardiness that results from scheduling job x_j in state (V_j, f_j), namely $(\max\{r_{x_j}, f_j\} + p_{x_j}(V_j) - d_{x_j})^+$, where $\alpha^+ = \max\{0, \alpha\}$.

We recursively construct a decision diagram D for the problem by associating a state with each node of D. The initial state S_1 is associated with the root node t and the final state S_{n+1} with the terminal node t. If state S_j is associated with node u in layer j, then for each $v_j \in X_j(S)$ we generate an arc with label v_j leaving u. The arc terminates at a node associated with state $\phi_j(S, v_i)$. Nodes on a given layer are identified when they are associated with the same state.

The process is illustrated for the job sequencing example in Fig. 2. Each node is labeled by its state (V_j, f_i), followed (in parentheses) by the minimum cost-to-go at the node. The cost-to-go at the terminus t is zero.

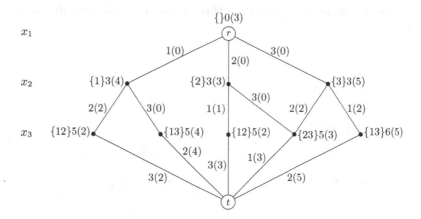

Fig. 2. Decision diagram, with states and minimum costs-to-go, for the job sequencing instance of Table 1.

4   Relaxed Decision Diagrams

A weighted decision diagram D' is a *relaxation* of diagram D when D' represents every solution in D with equal or smaller cost, and perhaps other solutions as well. To make this more precise, suppose layers $1, \ldots, n$ of both D and D' correspond to variables $x_1, \ldots, x_n$ with domains $X_1, \ldots, X_n$. Then D' is a relaxation of D if every assignment to x represented by an r–t path P in D is represented by an r–t path in D' with length no greater than that of P. The shortest path length in D' is a lower bound on the optimal value of the problem represented by D. We will refer to a diagram that has not been relaxed as *exact*.

We can construct a relaxed decision diagram by top-down compilation, again based on the recursive model (1). The procedure is as before, except that rather than simply identify nodes in each layer that are associated with the same states, we may also *merge* some nodes. That is, we may identify some nodes that are associated with different states. The object is to keep the width of the diagram (the maximum number of nodes in a layer) within a predetermined bound W. When we merge nodes with states S and T, we associate a state $S \oplus T$ with the resulting node. The operator $\oplus$ is chosen so as to yield a valid relaxation of the given recursion.

It is frequently necessary to introduce additional state variables to define a suitable merger operation [9], and this is the case in the job sequencing example. The state at a node will consist of (V, U, f), where V and f are as before, and U contains the jobs that occur along some path from the root. The processing time $p_{x_j}(U)$ of a job x_j depends on the jobs in U. The transition function is

$$\phi_j((V, U, f), x_j) = \left(V \cup \{x_j\}, U \cup \{x_j\}, \max\{r_{x_j}, f\} + p_{x_j}(U)\right)$$

and the immediate cost is $c_j((V, U, f), x_j) = (\max\{r_{x_j}, f\} + p_{x_j}(U) - d_{x_j})^+$. Merging states (V, U, f) and (V', U', f') results in state $(V \cap V', U \cup U', \min\{f, f'\})$. We will see in the next section that this merger operation results in a valid relaxation.

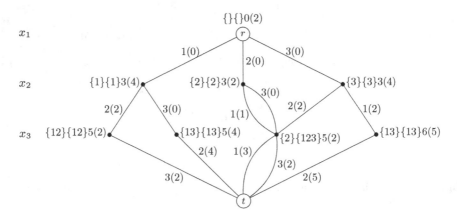

Fig. 3. A relaxation of the decision diagram in Fig. 2.

The merger operation is illustrated in Fig. 3, which is the result of merging states $(\{1,2\},5)$ and $(\{2,3\},5)$ in layer 3 of Fig. 2. The expanded states (V,U,f) are shown at each node, followed by the minimum cost-to-go in parentheses. The shortest path now has cost 2, which is a lower bound on the optimal cost of 3 in Fig. 2.

5   Conditions for Node Merger

We now develop general sufficient conditions under which node merger results in a relaxed decision diagram. Such conditions have apparently not be explicitly stated in the literature. It is shown in [19] that it suffices for the merged state to be a union of the states merged, but it is not useful to represent the merged state as a union of states. The merged state must be given in terms of the state variables in the states merged, so that the construction of the relaxed diagram can proceed with the same transition function and immediate cost function as at other nodes.

We will say that a state S' in layer j *relaxes* a state S in layer j when (a) all feasible controls in state S are feasible in state S', and (b) the immediate cost of any given feasible control in S is no less than its immediate cost in S'. That is, $X_j(S) \subseteq X_j(S')$, and $c_j(S, x_j) \geq c_j(S', x_j)$ for all $x_j \in X_j(S)$. Then node merger results in a valid relaxation when two conditions are satisfied. One is a condition on the transition function generally: when one state relaxes another, this must continue to hold when the same control is applied to both states. That is,

(C1) If state S' relaxes state S, then given any control v that is feasible in S, $\phi(S', v)$ relaxes $\phi(S, v)$.

The second condition places a requirement on the merger operation specifically. Namely, when two states are merged, the resulting state relaxes both of the states that are merged.

(C2) $S \oplus T$ relaxes both S and T.

We can now prove the relevant theorem. Let $c(P)$ be the cost (length) of path P in a decision diagram. It is also convenient to let D'_k be the first k layers of the relaxed diagram D' obtained during top-down compilation, but just before identifying and merging nodes in layer k. Thus $D'_{n+1} = D'$.

Theorem 1. *If conditions (C1) and (C2) are satisfied, the merger of nodes with states S and T within a diagram D results in a valid relaxation of D.*

Proof. Let D be the exact decision diagram that results from top-down compilation, and let D' be the diagram that results from top-down compilation with node merger. It suffices to show claim (H_k) inductively for $k = 1, \ldots, n+1$:

(H_k) Consider any path P from r to any u in layer k of D. Then D'_k contains a path P', from r to a node u' in layer k, that represents the same assignment to $(x_1, \ldots, x_{k-1})$. Furthermore $c(P) \geq c(P')$, and the state S' at u' relaxes the state S at u.

Claim (H_1) is trivially true. We therefore suppose (H_k) is true and show (H_{k+1}). Consider a path $\bar{P}$ from r to $\bar{u}$ in layer $k+1$ of D. We must show that D'_{k+1} contains a path $\bar{P}'$, from r to $\bar{u}'$, that represents the same assignment to $(x_1, \ldots, x_k)$. Furthermore, we must show that $c(\bar{P}) \geq c(\bar{P}')$, and that the state $\bar{S}'$ at $\bar{u}'$ relaxes the state $\bar{S}$ at $\bar{u}$.

Let P be the portion of path $\bar{P}$ that extends from r to a node u in layer k of D. By the induction hypothesis (H_k), D'_k contains a path P', terminating at some node u' in layer k, that represents the same assignment to $(x_1, \ldots, x_{k-1})$. Now D'_{k+1} is formed by identifying and merging nodes in layer k of D'_k and generating arcs from the resulting nodes. Thus if S' is the state at u' in D'_k, S' is merged with zero or more other states to form a state T. If v is a control that extends P to $\bar{P}$, then v is likewise a feasible control in state T. This is because T relaxes S' by condition (C2), and S' relaxes S by (H_k), which implies that T relaxes S. Now we can let $\bar{P}'$ be the path in D'_{k+1} that results from extending P' at state T with control v. To see that $c(\bar{P}) \geq c(\bar{P}')$, note that

$$c(\bar{P}) = c(P) + c_k(S, v) \geq c(P') + c_k(S', v) \geq c(P') + c_k(T, v) = c(\bar{P}')$$

where the first inequality is due to (H_k), and the second inequality is due to condition (C2). To show that $\bar{S}'$ relaxes $\bar{S}$, we note again that T relaxes S. This and condition (C1) imply that $\bar{S}'$ relaxes $\bar{S}$, as desired. $\qquad\square$

The merger operation defined earlier for the job sequencing problem satisfies conditions (C1) and (C2). To see that (C1) is satisfied, suppose (V', U', f') relaxes (V, U, F), which means that $V' \subseteq V$, $U' \supseteq U$, and $f' \leq f$. Then if control v is applied to either state, we have $V' \cup \{v\} \subseteq V \cup \{v\}$ and $U' \cup \{v\} \supseteq U \cup \{v\}$. Also

$$\min\{r_v, f'\} + p_v(U') \leq \min\{r_v, f\} + p_v(U)$$

because $f' \leq f$ and $p_v(U') \leq p_v(U)$, the latter due to the fact that $U' \supseteq U$. So $\phi_j((V', U', f'), v)$ relaxes $\phi_j((V, U, f), v)$, and (C1) follows. To show (C2), recall that (V, U, f) and (V', U', f') are merged to form $(V \cap V', U \cup U', \min\{f, f'\})$. The merger relaxes the two states that are merged because $V \cap V' \subseteq V, V'$, $U \cup U' \supseteq U, U'$, and $\min\{f, f'\} \leq f, f'$.

6   Merging Heuristics

We now address the question of which nodes to merge in a given layer so as to reduce the width to W. The merger strategy should be designed so that the diagram remains exact, to the extent possible, along paths that are likely to be optimal. Merging nodes in the remainder of the diagram will not affect the optimal solution and therefore the bound. The merger strategy is particularly important when processing times are state-dependent, because the additional state variable U results in smaller values for the state variable f and therefore shorter paths in the diagram, yielding a weaker bound.

We therefore need some indication of whether a given node is likely to lie on a shortest path. The most readily available indication is the state variable f, since it represents the finish time of the most recent job processed. If f is large, then the cost of a path through the node is more likely to be large. We can merge two nodes with the largest values of f and repeat the process until the width is reduced to W. We will refer to this as the *finish time heuristic*.

Another possibility is to compute the shortest path to a given node from the root. If the shortest path is already long, it is likely to be still longer by the time it reaches the terminus. We therefore merge nodes to which the shortest path from the root is longest. We will refer to this as the *shortest path heuristic*.

In the next section, we compare the effectiveness of these heuristic against a control heuristic that consists of randomly selecting nodes to merge.

7   Computational Experiments

The aim of the computational experiments is to determine how the quality of the bound depends on the width of the relaxed decision diagram. To our knowledge, there are no benchmark sets for sequence-dependent processing times, and so we ran tests on randomly generated problem instances. A meaningful assessment of the bound quality requires that we be able to solve the instances exactly, because the optimal values vary widely from zero to a rather large number. We must therefore generate instances that are small enough to be solved exactly in reasonable time.

Dynamic programming is the only viable method for exact solution of problems with general state dependence, including problems with sequence-dependent processing times. Since the state space becomes impracticably large for instances with more than 14 jobs, we generated and solved instances with 12 and 14 jobs. We found that the results are quite consistent over these instances, which suggests that the pattern may continue for larger instances.

Instances are generated as follows. The normal processing time p_j is drawn uniformly from the interval $[p_{\min}, p_{\max}]$. To make the processing time state dependent, we reduce it to $p_j/2$ when j is even and job $j-1$ has already been processed. The release time is drawn uniformly from $[0, np_{min}/2]$, where n is the number of jobs. The due date is $d_j = r_j + p_j + \text{slack}_j$, where slack_j is drawn uniformly from $[kp_{min}, kp_{\max}]$. We used $k = 4$ for 12 jobs and $k = 5$ for 14 jobs to obtain minimum tardiness values that are generally positive but not unrealistically large. We also set $[p_{\min}, p_{\max}] = [10, 16]$. The first 5 random instances are used for each problem size, after discarding those with zero minimum tardiness, because they provide no information about the quality of the bound.

Figures 4 and 5 show the results, using the finish time heuristic. The bound is plotted against the width of the relaxed diagram, where the latter is on a logarithmic scale. The bounds are sampled at 10 points per factor of 10 for 12

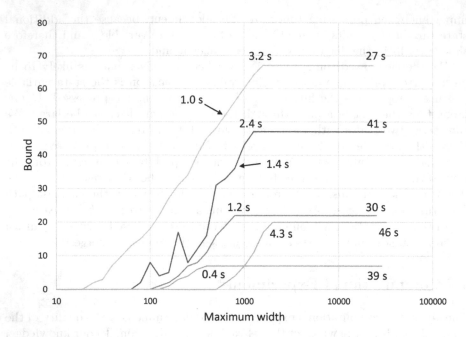

Fig. 4. 12-job instances: relaxation bound versus decision diagram width, up to the width of an exact diagram. Selected computation times are shown.

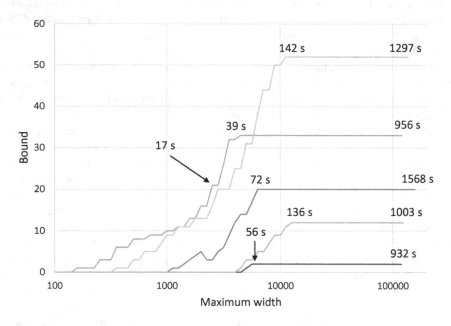

Fig. 5. 14-job instances: relaxation bound versus decision diagram width, up to the width of an exact diagram. Selected computation times are shown.

jobs, and 20 points per factor of 10 for 14 jobs. The far right end of each curve represents an exact decision diagram for the instance, which is equivalent to the state transition graph for the dynamic programming formulation of the problem. Selected computation times are shown on the curves. The computation time for a given width is very similar across all instances of a given size. The computation time for the exact diagram indicates the time necessary to solve the problem by dynamic programming.

The curves follow a different pattern than those reported for other types of problems, which tend to approach the optimal value asympotically [7]. Once the curves begin to rise above zero, they increase rapidly and level off at the optimal value when the width is less than one-tenth the width of an exact diagram. Specifically, when there are 12 jobs, the optimal value is achieved for diagrams that are between 1/32 and 1/15 the width of an exact diagram, and when there are 14 jobs, between 1/26 and 1/10 the width of an exact diagram. On the other hand, the bound does not rise above zero until the width of the relaxed diagram is roughly 1/1000 to 1/25 the width of an exact diagram. This indicates that diagrams of a fixed maximum width, such as 1000 or 10000, cease to provide useful bounds as the instances scale up. The curves are not always monotonic because the the bound depends on the merging heuristic, which may happen to perform better for a smaller width than a slightly larger one.

Figure 6 compares the performance of three merging heuristics for one of the 12-job instances, namely the finish time, shortest path, and random selection heuristics. The choice of heuristic is clearly key, as the finish time heuristic is vastly superior to the others. Examination of the shortest paths from the root at various nodes reveals why the shortest path heuristic fails. The shortest path length tends to remain at zero in the upper layers of the diagram, which means there is no guidance for node merger, resulting in bad merger decisions that

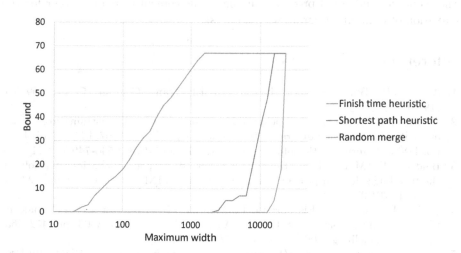

Fig. 6. Comparison of bound quality of three node merging heuristics on a 12-job instance.

propagate through the remainder of the diagram. Random merger is even worse, resulting in no useful bounds until the diagram is nearly exact.

8   Conclusions

We undertook a preliminary investigation of the potential of node merger as a relaxation mechanism for dynamic programming problems, particularly those for which no practical relaxation method exists. We focused on the job sequencing problem with time windows and state-dependent processing times, because it has no known mixed integer programming or other model that yields a useful relaxation.

We first proved two conditions that are jointly sufficient for a node merger operation to yield a valid relaxation, one a condition on the transition function of the dynamic programming model, and one a condition on the merger operation itself. We then formulated a merger rule for the job sequencing problem and used the conditions to show that it results in a relaxed diagram.

Computational testing revealed that relaxed diagrams for this problem have different characteristics than have been reported for other types of problems. The relaxed diagram yields the optimal value when its width is a small fraction of the width of an exact diagram. This allows a relaxed diagram to prove the optimality of a solution obtained heuristically, using much less computation than would otherwise be necessary.

On the other hand, diagrams of a fixed maximum width cease to provide useful bounds as the instances scale up. A intriguing line of research would be to use Lagrangian methods to strengthen the bounds provided by smaller diagrams. These methods adjust the arc costs in the relaxed diagram to exclude poor solutions while retaining a valid bound [10]. They have been used successfully in other contexts and could prove a valuable enhancement of node merger for the relaxation of dynamic programming models.

References

1. Akers, S.B.: Binary decision diagrams. IEEE Trans. Comput. **C–27**, 509–516 (1978)
2. Andersen, H.R., Hadzic, T., Hooker, J.N., Tiedemann, P.: A constraint store based on multivalued decision diagrams. In: Bessière, C. (ed.) CP 2007. LNCS, vol. 4741, pp. 118–132. Springer, Heidelberg (2007). doi:10.1007/978-3-540-74970-7_11
3. Baldacci, R., Mingozzi, A., Roberti, R.: New state-space relaxations for solving the traveling salesman problem with time windows. INFORMS J. Comput. **24**(3), 356–371 (2012)
4. Becker, B., Behle, M., Eisenbrand, F., Wimmer, R.: BDDs in a branch and cut framework. In: Nikoletseas, S.E. (ed.) WEA 2005. LNCS, vol. 3503, pp. 452–463. Springer, Heidelberg (2005). doi:10.1007/11427186_39
5. Behle, M., Eisenbrand, F.: 0/1 vertex and facet enumeration with BDDs. In: Proceedings of the Workshop on Algorithm Engineering and Experiments (ALENEX), pp. 158–165. SIAM (2007)

6. Bergman, D., Ciré, A.A., van Hoeve, W.-J., Hooker, J.N.: Variable ordering for the application of BDDs to the maximum independent set problem. In: Beldiceanu, N., Jussien, N., Pinson, É. (eds.) CPAIOR 2012. LNCS, vol. 7298, pp. 34–49. Springer, Heidelberg (2012). doi:10.1007/978-3-642-29828-8_3

7. Bergman, D., Ciré, A.A., van Hoeve, W.J., Hooker, J.N.: Optimization bounds from binary decision diagrams. INFORMS J. Comput. **26**, 253–268 (2013)

8. Bergman, D., Ciré, A.A., van Hoeve, W.J., Hooker, J.N.: Decision Diagrams for Optimization. Springer, Heidelberg (2016). doi:10.1007/978-3-319-42849-9

9. Bergman, D., Ciré, A.A., van Hoeve, W.J., Hooker, J.N.: Discrete optimization with binary decision diagrams. INFORMS J. Comput. **28**, 47–66 (2016)

10. Bergman, D., Ciré, A.A., van Hoeve, W.J.: Lagrangian bounds from decision diagrams. Constraints **20**, 346–361 (2015)

11. Bergman, D., van Hoeve, W.-J., Hooker, J.N.: Manipulating MDD relaxations for combinatorial optimization. In: Achterberg, T., Beck, J.C. (eds.) CPAIOR 2011. LNCS, vol. 6697, pp. 20–35. Springer, Heidelberg (2011). doi:10.1007/978-3-642-21311-3_5

12. Bryant, R.E.: Graph-based algorithms for boolean function manipulation. IEEE Trans. Comput. **C–35**, 677–691 (1986)

13. Christofides, N., Mingozzi, A., Toth, P.: State-space relaxation procedures for the computation of bounds to routing problems. Networks **11**(2), 145–164 (1981)

14. Ciré, A.A., van Hoeve, W.J.: Multivalued decision diagrams for sequencing problems. Oper. Res. **61**, 1411–1428 (2013)

15. Hadžić, T., Hooker, J.N.: Postoptimality analysis for integer programming using binary decision diagrams. Carnegie Mellon University, Technical report (2006)

16. Hadžić, T., Hooker, J.N.: Cost-bounded binary decision diagrams for 0-1 programming. In: Hentenryck, P., Wolsey, L. (eds.) CPAIOR 2007. LNCS, vol. 4510, pp. 84–98. Springer, Heidelberg (2007). doi:10.1007/978-3-540-72397-4_7

17. Hadžić, T., Hooker, J.N., O'Sullivan, B., Tiedemann, P.: Approximate compilation of constraints into multivalued decision diagrams. In: Stuckey, P.J. (ed.) CP 2008. LNCS, vol. 5202, pp. 448–462. Springer, Heidelberg (2008). doi:10.1007/978-3-540-85958-1_30

18. Hadžić, T., Hooker, J.N., Tiedemann, P.: Propagating separable equalities in an MDD store. In: Perron, L., Trick, M.A. (eds.) CPAIOR 2008. LNCS, vol. 5015, pp. 318–322. Springer, Heidelberg (2008). doi:10.1007/978-3-540-68155-7_30

19. Hoda, S., van Hoeve, W.-J., Hooker, J.N.: A systematic approach to MDD-based constraint programming. In: Cohen, D. (ed.) CP 2010. LNCS, vol. 6308, pp. 266–280. Springer, Heidelberg (2010). doi:10.1007/978-3-642-15396-9_23

20. Hooker, J.N.: Discrete global optimization with binary decision diagrams. In: GICOLAG 2006, Vienna, Austria, December 2006

21. Hooker, J.N.: Decision diagrams and dynamic programming. In: Gomes, C., Sellmann, M. (eds.) CPAIOR 2013. LNCS, vol. 7874, pp. 94–110. Springer, Heidelberg (2013). doi:10.1007/978-3-642-38171-3_7

22. Hu, A.J.: Techniques for efficient formal verification using binary decision diagrams. Thesis CS-TR-95-1561, Stanford University, Department of Computer Science, December 1995

23. Lee, C.Y.: Representation of switching circuits by binary-decision programs. Bell Syst. Tech. J. **38**, 985–999 (1959)

24. Loekito, E., Bailey, J., Pei, J.: A binary decision diagram based approach for mining frequent subsequences. Knowl. Inf. Syst. **24**(2), 235–268 (2010)

25. Mingozzi, A.: State space relaxation and search strategies in dynamic programming. In: Koenig, S., Holte, R.C. (eds.) SARA 2002. LNCS, vol. 2371, p. 51. Springer, Heidelberg (2002). doi:10.1007/3-540-45622-8_4
26. Righini, G., Salani, M.: New dynamic programming algorithms for the resource constrained shortest path problem. Networks 51, 155–170 (2008)
27. Wegener, I.: Branching Programs and Binary Decision Diagrams: Theory and Applications. SIAM Monographs on Discrete Mathematics and Applications. Society for Industrial and Applied Mathematics (2000)

Branch-and-Check with Explanations for the Vehicle Routing Problem with Time Windows

Edward Lam[1,2]([✉]) and Pascal Van Hentenryck[3]

[1] CSIRO Data61, Eveleigh, NSW 2015, Australia
[2] University of Melbourne, Parkville, VIC 3010, Australia
ed@ed-lam.com
[3] University of Michigan, Ann Arbor, MI 48109-2117, USA

Abstract. This paper proposes the framework of branch-and-check with explanations (BCE), a branch-and-check method where combinatorial cuts are found by general-purpose conflict analysis, rather than by specialized separation algorithms. Specifically, the method features a master problem that ignores combinatorial constraints, and a feasibility subproblem that uses propagation to check the feasibility of these constraints and performs conflict analysis to derive nogood cuts. The BCE method also leverages conflict-based branching rules and strengthens cuts in a post-processing step. Experimental results on the Vehicle Routing Problem with Time Windows show that BCE is a potential alternative to branch-and-cut. In particular, BCE dominates branch-and-cut, both in proving optimality and in finding high-quality solutions quickly.

1 Introduction

Vehicle Routing Problems (VRPs) generalize the Travelling Salesman Problem (TSP). The Capacitated Vehicle Routing Problem (CVRP) is a basic variant that aims to design routes of minimal travel distance that deliver all requests from a single depot while respecting vehicle capacity constraints. The Vehicle Routing Problem with Time Windows (VRPTW) additionally requires requests to be delivered within a given time window.

VRPs have been studied extensively over the past several decades, resulting in significant computational progress (e.g., [31]). Solution techniques include constraint programming (e.g., [7,8,28]), branch-and-bound, branch-and-cut (e.g., [4,18,21]), branch-and-price (e.g., [10]), and combinations thereof (e.g., [3,13,16,25,26]). Branch-and-cut (BC) methods are of particular interest to this paper. Their key idea is to omit difficult constraints from the original formulation and to remove solutions that violate these constraints using cuts generated by separation algorithms. Separation algorithms are typically problem-specific, which limits their applicability and reuse in other problems. Furthermore, developing and implementing separation algorithms often require significant expertise, hindering their use in many applications.

© Springer International Publishing AG 2017
J.C. Beck (Ed.): CP 2017, LNCS 10416, pp. 579–595, 2017.
DOI: 10.1007/978-3-319-66158-2_37

This paper addresses the following research question: *Is it possible to use a general-purpose mechanism to generate cuts, and hence, avoid the difficult aspects of BC*. This paper proposes *branch-and-check with explanations* (BCE) as one possible answer to this question. BCE divides an optimization problem into a master problem that ignores a number of difficult constraints, and a subproblem that checks the feasibility of these constraints and generates cuts using conflict analysis from constraint programming (CP) and Boolean satisfiability (SAT). More precisely, BCE uses CP for three purposes: (1) to fix variables in the master problem through propagation; (2) to generate cuts in the master problem using conflict analysis; and (3) to probe the feasibility of linear programming (LP) relaxation solutions and to derive additional cuts through conflict analysis if the probing process fails. Since the master problem does not operate on the same decision variables as the subproblem, the conflict analysis needs to continue until the variables involved in a nogood appear in the master problem.

BCE opens some interesting opportunities. First, it has the advantage of relying on a general-purpose CP engine for inference and cut separation. Second, it permits conflict-based branching rules. Finally, BCE can recognize special classes of cuts after conflict analysis and then strengthen them using well-known techniques. As a result, BCE offers a natural integration of LP, CP and SAT.

The BCE method is evaluated on the VRPTW. Experimental results indicate that BCE outperforms a BC algorithm: it proves optimality on more instances and finds significantly better solutions to instances for which BC cannot prove optimality. The results also show that a conflict-based branching rule is particularly effective in BCE and that cut strengthening produces interesting improvements to the lower bounds.

The rest of this paper is structured as follows. Section 2 reviews relevant methods for solving the VRPTW. Section 3 develops the BCE model. Section 4 discusses cut strengthening. Section 5 presents experimental results that compare the BCE model with the BC model. Section 6 discusses the limitations and potential improvements of the BCE approach for the VRPTW, as well as its relevance to branch-and-price. Section 7 concludes this paper.

2   Background

BC algorithms for VRPs are often based on a two-index flow model, which generalizes the standard formulation of the TSP. The two-index model omits the subtour elimination, vehicle capacity and time window constraints, which are added as required through cutting planes. At every node of the search tree, BC solves separation subproblems to determine if the LP relaxation solution is feasible with respect to the omitted constraints. If the solution is infeasible, the solution is discarded using a cut, forcing the solver to find another candidate solution. Branching and cutting are repeated until the search tree is explored, upon which the solver proves optimality or infeasibility.

Branch-and-Cut. BC models of the VRPTW rely on several types of cuts. The BC model in [4] inherits the capacity cuts from BC models of the CVRP. Capacity cuts generalize the subtour elimination cuts of the TSP to consider vehicle capacity. Hence, they serve the purpose of excluding both subtours and partial paths that exceed the vehicle capacity. This model also implements infeasible path cuts to exclude partial paths that violate the time windows. Infeasible path cuts require at least one arc in an infeasible partial path to be unused. The BC algorithm from [18] uses subtour elimination constraints from the TSP instead of the capacity cuts. Vehicle capacity constraints are enforced by the same infeasible path cuts that enforce the time windows. The authors also prove that both the subtour elimination cuts and the infeasible path cuts can be strengthened using ideas conceived in [22].

Branch-and-Check. Branch-and-check [5,29] is a form of logic-based Benders decomposition [15]. The method divides a problem into a master and checking problem. The master problem is first solved to find a candidate solution, which is checked using the checking subproblem. If the checking subproblem is infeasible for a candidate solution, a constraint prohibiting this solution, and hopefully many others, is added to the master problem. Branch-and-check iterates between the master problem and the checking subproblem until a globally optimal solution is found. It has been used successfully in various applications (e.g., [14,30]). The key difference between BC and branch-and-check is that checking subproblems encompass an entire optimization problem, whereas separation subproblems only check specific aspects of the problem (i.e., they find cuts from one family).

Constraint Programming. CP with large neighborhood search was instrumental in finding many best solutions to VRPs more than a decade ago [7,8,28]. The main difficulty with CP is proving optimality since VRP objective functions are usually linear, which are known to have weak propagators. This limitations can be alleviated using the WEIGHTEDCIRCUIT global constraint [6], for example.

Conflict Analysis. Conflict analysis has a long history in artificial intelligence and CP (e.g., [9,17]). Its popularity grew in the last two decades through the development of SAT solvers (e.g., [11,23]) and their integration in CP solvers (e.g., [12,24]). In CP solvers, propagators generate clauses that explain the inferences for an underlying SAT solver. When a propagator fails, the SAT solver performs conflict analysis, i.e., it walks the implication graph to derive a constraint, known as a nogood, that prevents the same failure from reoccurring in other parts of the search tree. Conflict analysis can also be implemented in mixed integer programming (MIP) solvers but its performance is still an open question [1].

3   The Branch-and-Check Model of the VRPTW

This section proposes the BCE model of the VRPTW. The model is organized around a MIP master problem and a CP checking subproblem.

Table 1. Data and decision variables of the two-index flow model of the VRPTW. Sets enclosed in braces (resp. square brackets) are integer-valued (resp. real-valued).

Name	Description
$T > 0$	Time horizon
$\mathcal{T} = [0, T]$	Time interval
$Q \geq 0$	Vehicle capacity
$\mathcal{Q} = [0, Q]$	Range of vehicle load
$R \in \{1, \ldots, \infty\}$	Number of requests
$\mathcal{R} = \{1, \ldots, R\}$	Set of requests
$s = 0$	Start node
$e = R + 1$	End node
$\mathcal{N} = \mathcal{R} \cup \{s, e\}$	Set of all nodes
$\mathcal{A}$	Arcs of the network. Defined in Eq. 4
$c_{i,j} \in \mathcal{T}$	Distance cost and travel time along arc $(i, j) \in \mathcal{A}$
$q_i \in \mathcal{Q}$	Vehicle load demand of $i \in \mathcal{N}$
$a_i \in \mathcal{T}$	Earliest service start time at $i \in \mathcal{N}$
$b_i \in \mathcal{T}$	Latest service start time at $i \in \mathcal{N}$
$x_{i,j} \in \{0, 1\}$	Decision variable indicating if a vehicle traverses $(i, j) \in \mathcal{A}$

The MIP Master Problem. The BCE model includes the traditional two-index model. Its data and decision variables are listed in Table 1. The arcs in the network are given by

$$\mathcal{A} = \{(s, i) | i \in \mathcal{R}\} \cup$$
$$\{(i, j) | i, j \in \mathcal{R}, i \neq j, a_i + c_{i,j} \leq b_j, q_i + q_j \leq Q\} \cup \qquad (4)$$
$$\{(i, e) | i \in \mathcal{R}\}.$$

The initial constraints of the model are shown in Fig. 1. The objective function, Eq. (1), minimizes the total distance cost. Constraints (2) and (3) require every request to be visited exactly once. Through its LP relaxation, the MIP master problem provides the lower bounds to the objective value and generates candidate solutions to be tested in the CP subproblem.

The CP Checking Subproblem. The BCE model uses a CP subproblem to check the solutions found by the master problem for feasibility of the subtour elimination, vehicle capacity and time window constraints. The decision variables are listed in Table 2. Using the y binary variables, instead of the conventional successor and predecessor variables, provides a one-to-one mapping between the y variables and the x variables of the master problem. The initial constraints (without the nogoods) are presented in Fig. 2. Constraints (5) to (8) ensure that every request is visited exactly once. Constraint (9) is a global constraint that prevents subtours. Its propagator is a simple checking algorithm that prevents

$$\min \sum_{(i,j)\in \mathcal{A}} c_{i,j} x_{i,j} \tag{1}$$

subject to

$$\sum_{h:(h,i)\in \mathcal{A}} x_{h,i} = 1 \qquad\qquad \forall i \in \mathcal{R}, \tag{2}$$

$$\sum_{j:(i,j)\in \mathcal{A}} x_{i,j} = 1 \qquad\qquad \forall i \in \mathcal{R}. \tag{3}$$

Fig. 1. Initial constraints of the two-index model of the VRPTW.

the head of a partial path from connecting to its tail. Constraints (10) and (11) enforce the vehicle capacity and travel time constraints.

Table 2. Decision variables of the CP subproblem.

Name	Description
$y_{i,j} \in \{0, 1\}$	Decision variable indicating if a vehicle traverses $(i, j) \in \mathcal{A}$
$l_i \in [q_i, Q] \subseteq \mathcal{Q}$	Vehicle load after service at request $i \in \mathcal{N}$
$t_i \in [a_i, b_i] \subseteq \mathcal{T}$	Time that a vehicle begins service at request $i \in \mathcal{N}$

$$\bigvee_{h:(h,i)\in \mathcal{A}} y_{h,i} \qquad\qquad \forall i \in \mathcal{R}, \tag{5}$$

$$\bigvee_{j:(i,j)\in \mathcal{A}} y_{i,j} \qquad\qquad \forall i \in \mathcal{R}, \tag{6}$$

$$\neg y_{h,i} \vee \neg y_{h,j} \qquad \forall h, i, j \in \mathcal{N} : (h, i) \in \mathcal{A}, (h, j) \in \mathcal{A}, i \neq j, \tag{7}$$

$$\neg y_{h,j} \vee \neg y_{i,j} \qquad \forall h, i, j \in \mathcal{N} : (h, j) \in \mathcal{A}, (i, j) \in \mathcal{A}, h \neq i, \tag{8}$$

$$\text{NoSubtour}(y), \tag{9}$$

$$y_{i,j} \rightarrow l_j \geq l_i + q_j \qquad\qquad \forall (i, j) \in \mathcal{A}, \tag{10}$$

$$y_{i,j} \rightarrow t_j \geq t_i + c_{i,j} \qquad\qquad \forall (i, j) \in \mathcal{A}. \tag{11}$$

Fig. 2. Initial constraints of the CP subproblem.

Communication Between the Two Models. The two models communicate in three ways: (1) variable assignments in the CP model are transmitted to the MIP model, (2) candidate solutions from the LP relaxation are probed using the CP model to determine if they are valid for the VRPTW and (3) nogoods found by conflict analysis in the CP model are translated into cuts in the MIP model.

Extended Conflict Analysis. When a failure occurs in the CP solver, conflict analysis derives a First Unique Implication Point (1UIP) nogood that is added to the CP subproblem. This constraint should also be added to the master problem but sometimes it cannot be translated into a cut for the master problem because it contains variables that do not appear in the master problem (i.e., the load and time variables). As a result, the BCE algorithm features an extended conflict analysis that continues explaining the failure until the the nogood only contains variables in master problem. This nogood has the form

$$\bigvee_{(i,j)\in\mathcal{C}_1} y_{i,j} \vee \bigvee_{(i,j)\in\mathcal{C}_2} \neg y_{i,j},$$

where $\mathcal{C}_1, \mathcal{C}_2 \subseteq \mathcal{A}$ are sets of arcs. This nogood can be rewritten as the cut

$$\sum_{(i,j)\in\mathcal{C}_1} x_{i,j} + \sum_{(i,j)\in\mathcal{C}_2} (1 - x_{i,j}) \geq 1.$$

It is always possible to obtain these cuts since the solver only branches on variables in the master problem. *Observe that the BCE algorithm provides a general-purpose mechanism to separate cuts in the master problem via the extended conflict analysis. These cuts, which we call MIP-1UIP nogoods, are automatically generated and do not rely on specialized separation algorithms.*

Probing the LP Relaxation. The BCE algorithm probes whether the current LP solution is feasible with respect to the subtour elimination, vehicle capacity and time constraints. It temporarily assigns every $y_{i,j}$ variable to the value of its corresponding $x_{i,j}$ variable in the LP relaxation, provided that this value is integral. The resulting tentative assignment can then be propagated by the CP solver. If a failure occurs, conflict analysis generates nogoods for both the CP and MIP models. The MIP cut will exclude the current LP solution, forcing it to find another candidate solution and improving the lower bound.

The Search Algorithm. The BCE algorithm, detailed in Fig. 3, includes the components described earlier. It blends depth-first and best-first search since best-first search is more effective for hard optimization problems, such as VRPs, but complicates the implementation of CP solvers with conflict analysis. The node selection strategy selects the node with the lowest lower bound from the set of open nodes and then explores the node subtree using depth-first search until it reaches a limit on the maximum number of open nodes per subtree. Once it reaches this limit, all unsolved siblings in the subtree are moved into the set of open nodes, and then the algorithm starts a new depth-first search from the node with the next lowest lower bound. Section 6 explains the rationale behind this search procedure.

Once a node is selected (step 1), the CP subproblem infers the implications of the decision (step 2). In the case of failure, the CP solver generates nogoods for both models and then backtracks (step 5b). If the test succeeds, the BCE algorithm checks for suboptimality using the LP relaxation (step 3). If the node

1. **Node Selection:** Select an open node. Terminate if no open nodes remain.
2. **Feasibility Check:** Solve the CP model to determine the implications of the branching decision of the node. If propagation fails, perform conflict analysis, add the 1UIP and the MIP-1UIP nogoods to both the CP and MIP models, and go to step 5b. Otherwise, fix $x_{i,j}$ in the MIP model to the values of the $y_{i,j}$ variables.
3. **Suboptimality Check:** Solve the LP relaxation. If the objective value is worse than the incumbent solution, go to step 5b.
4. **LP Probing:** For all $x_{i,j}$ variables with a value of 0 or 1 in the LP relaxation, temporarily fix the $y_{i,j}$ variables in the CP model to the same value. Propagate the CP model. If it fails, perform conflict analysis, generate the 1UIP and the MIP-1UIP nogoods and go back to step 3.
5. **Branching and Backtracking:** If all $x_{i,j}$ variables are integral, store the LP relaxation solution as the incumbent solution and go to step 5b. Otherwise, go to step 5a because the node is fractional.
   (a) **Branching:** Create two children nodes from a fractional $x_{i,j}$ variable. Fix the variable to 0 in one child node and to 1 in the other.
   (b) **Backtracking:** If the number of nodes in the current subtree exceeds the limit or if the subtree is entirely solved, move all unsolved siblings in the subtree to the set of open nodes and go back to step 1. Otherwise, backtrack to an ancestor with an unsolved child node, select the child node and go to step 2.

Fig. 3. The BCE search algorithm.

is suboptimal, it backtracks (step 5b). Otherwise, the BCE algorithm checks the LP relaxation solution against the omitted constraints and separates cuts using conflict analysis if necessary (step 4). The BCE algorithm iterates between the LP relaxation and the feasibility test until no cuts are generated. Then, if the node is fractional and not suboptimal, the BCE algorithm executes a branching step (step 5a). Two branching rules are implemented. The first selects the most fractional variable and the second selects the variable with the highest activity, which is defined as the number of nogoods in which the variable has previously appeared. This branching rule, known as activity-based search or variable state independent decaying sum (VSIDS) in the literature, guides the search tree towards subtrees that can be quickly pruned due to infeasibility.

Illustrating the Extended Conflict Analysis. The following discussion illustrates the extended conflict analysis procedure using the example in Figs. 4 and 5. Literals shown in a grey are fixed by the data at the root level, and hence, are always true. They are discarded in the explanations but are shown for clarity.

The BCE solver first branches on $\neg y_{4,6}$, making it true. The travel time constraint (Constraint (11)) propagates $[\![t_6 \geq 30]\!]$ with the reason

$$\neg y_{4,6} \wedge [\![t_3 \geq 25]\!] \wedge [\![c_{3,6} = 10]\!] \wedge [\![t_5 \geq 20]\!] \wedge [\![c_{5,6} = 10]\!] \rightarrow [\![t_6 \geq 30]\!]$$

because the predecessor of request 6 must be either 3 or 5, and the earliest time to reach 6 is at time $\min(\min(t_3) + c_{3,6}, \min(t_5) + c_{5,6}) = 30$. The BCE solver then

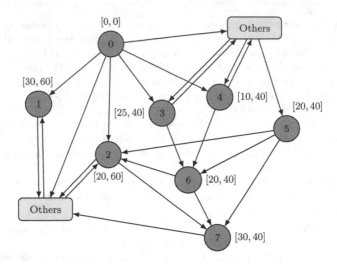

Fig. 4. Example of a network. Next to every request is its time window. The travel time across any arc is 10 units of time. The load demands are not shown as they are not relevant to the discussion.

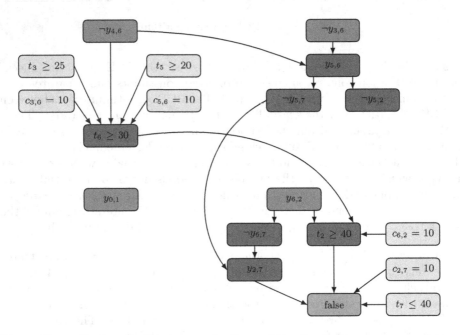

Fig. 5. Example of an implication graph after making the decisions $\neg y_{4,6}$, $\neg y_{3,6}$, $y_{0,1}$ and $y_{6,2}$ on the network in Fig. 4. Yellow literals are branching decisions. Blue literals are propagations. Grey literals are propagated at the root level, and hence, can be excluded from the nogoods since they are always true. (Color figure online)

branches on $\neg y_{3,6}$. Constraint (5) requires every request to have a predecessor, which leads to the assignment of $y_{5,6}$ with the reason

$$\neg y_{3,6} \wedge \neg y_{4,6} \rightarrow y_{5,6}.$$

Constraint (8) then propagates

$$y_{5,6} \rightarrow \neg y_{5,2}$$

and

$$y_{5,6} \rightarrow \neg y_{5,7}.$$

The BCE solver then branches on $y_{0,1}$, which does not produce any inference, and then branches on $y_{6,2}$, which produces the inferences:

$$y_{6,2} \rightarrow \neg y_{6,7},$$
$$\neg y_{6,7} \wedge \neg y_{5,7} \rightarrow y_{2,7},$$
$$y_{6,2} \wedge [\![t_6 \geq 30]\!] \wedge [\![c_{6,2} = 10]\!] \rightarrow [\![t_2 \geq 40]\!].$$

Then, the travel time propagator fails with

$$y_{2,7} \wedge [\![t_2 \geq 40]\!] \wedge [\![c_{2,7} = 10]\!] \wedge [\![t_7 \leq 40]\!] \rightarrow \text{false}.$$

Conflict analysis deduces the following:

$$y_{2,7} \wedge [\![t_2 \geq 40]\!] \wedge [\![c_{2,7} = 10]\!] \wedge [\![t_7 \leq 40]\!] \rightarrow \text{false}$$
$$y_{2,7} \wedge [\![t_2 \geq 40]\!] \wedge \text{true} \wedge \text{true} \rightarrow \text{false}$$
$$(\neg y_{6,7} \wedge \neg y_{5,7}) \wedge (y_{6,2} \wedge [\![t_6 \geq 30]\!] \wedge [\![c_{6,2} = 10]\!]) \rightarrow \text{false}$$
$$y_{6,2} \wedge \neg y_{5,7} \wedge [\![t_6 \geq 30]\!] \wedge \text{true} \rightarrow \text{false}$$
$$y_{6,2} \wedge \neg y_{5,7} \wedge [\![t_6 \geq 30]\!] \rightarrow \text{false}. \tag{12}$$

This explanation contains exactly one literal $(y_{6,2})$ at the current depth, and hence, is rewritten as the 1UIP clause

$$\neg y_{6,2} \vee y_{5,7} \vee [\![t_6 < 30]\!],$$

which is added to the CP model. Conflict analysis must continue because the nogood contains a time literal. It explains $[\![t_6 \geq 30]\!]$ in Eq. (12), which results in the MIP-1UIP explanation

$$y_{6,2} \wedge \neg y_{5,7} \wedge \neg y_{4,6} \rightarrow \text{false}.$$

This explanation is rewritten into the disjunction

$$\neg y_{6,2} \vee y_{5,7} \vee y_{4,6}, \tag{13}$$

and then into the cut

$$(1 - x_{6,2}) + x_{5,7} + x_{4,6} \geq 1.$$

Note that the literal $y_{5,7}$ was not assigned by the search.

4   Nogood Strengthening

The BCE algorithm presented so far uses a completely general-purpose mechanism for cut separation. Despite its generality, conflict analysis routinely discovers classical cuts. These cuts can be strengthened using proven techniques whenever they are recognized. This section presents a post-processing step that recognizes then strengthens several types of cuts.

Infeasible Path Cuts. Failure of the load or time constraints (Constraints (10) and (11)) frequently results in an infeasible partial path cut. Let $P = i_1, i_2, \ldots, i_k$, with all $i_1, \ldots, i_k \in \mathcal{N}$ distinct, be a partial path. The partial path P is infeasible with respect to the load constraint if $\sum_{u=1}^{k} q_{i_u} > Q$, and it is is infeasible with respect to the time constraint if $t_{i_k} > b_{i_k}$, where $t_{i_1} = a_{i_1}$ and $t_{i_u} = \max(a_{i_u}, t_{i_{u-1}} + c_{i_{u-1},i_u})$ for $u = 2, \ldots, k$. When a load or time window constraint fails, conflict analysis will usually produce the nogood

$$\bigvee_{(i,j) \in A(P)} \neg y_{i,j}, \tag{14}$$

where $A(P) = \{(i_1, i_2), \ldots, (i_{k-1}, i_k)\}$ is the arcs of P. This nogood requires one arc of P to be unused. It can be written equivalently as requiring at least one arc that exits P, i.e.,

$$\bigvee_{(i,j) \in \Delta^+(P)} y_{i,j},$$

where $\Delta^+(P) = \bigcup_{u=1}^{k-1}\{(i_u, j) \in \mathcal{A} \mid j \neq i_{u+1}\}$. This nogood can be translated into the cut

$$\sum_{(i,j) \in \Delta^+(P)} x_{i,j} \geq 1.$$

Using existing techniques [18], such a cut can be strengthened into

$$\sum_{(i,j) \in \tilde{\Delta}^+(P)} x_{i,j} \geq 1,$$

where

$$\tilde{\Delta}^+(P) = \bigcup_{u=1}^{k-1} \left(\left\{ (i_u, j) \in \mathcal{A} : i_u \in \mathcal{R}, j \in \mathcal{R}, j \neq i_1, \ldots, i_{u+1}, \right.\right.$$
$$\left.\left. \sum_{v=1}^{u} q_{i_v} + q_j \leq Q, t_{i_u} + c_{i_u,j} \leq b_j \right\} \cup \{(i_u, e) \in \mathcal{A}\} \right)$$

is the arcs that branch off P to a feasible request. In other words, the strengthening discards arcs that are not feasible when taking into account the load and time window constraints.

Subtour Elimination Cuts. The propagator of Constraint (9) will fail if the solution contains a subtour $S = i_1, i_2, \ldots, i_k$, where $i_1 = i_k$ and all $i_1, i_2, \ldots, i_{k-1} \in \mathcal{R}$ are distinct. Conflict analysis will usually find the nogood

$$\bigvee_{(i,j) \in A(S)} \neg y_{i,j}, \tag{15}$$

where $A(S) = \{(i_1, i_2), \ldots, (i_{k-1}, i_k)\}$ is the arcs of S. Using the same reasoning as for the infeasible path cuts, this nogood can be rewritten as the cut

$$\sum_{(i,j) \in \Delta^+(S)} x_{i,j} \geq 1. \tag{16}$$

If $a_j + c_{j,i} > b_i$, then no vehicle can depart j for i while respecting the time windows. Hence, i must precede j with respect to time, written as $i \prec j$. Let $\pi(j) = \{i \in \mathcal{N} | i \prec j\}$ be the set of requests that precedes j with respect to time. Proposition 1 strengthens Constraint (16) using these precedence relations [18]. Constraint (16) can also be similarly strengthened using the precedence relations in reverse, i.e., successor relations.

Proposition 1. *Let $\bar{S} = \mathcal{N} \setminus S$ be the nodes not in a subtour S, then for any $u \in S$, Constraint (16) can be strengthened to*

$$\sum_{\substack{(i,j) \in A: \\ i \in S \setminus \pi(u), \\ j \in \bar{S} \setminus \pi(u)}} x_{i,j} \geq 1.$$

Proof. Consider a subtour S and a feasible path F that visits the request u. Let $v \in \mathcal{R}$ be the last request of F visited by S. By definition, v is visited by S, i.e., $v \in S$. Furthermore, since F is a feasible path, v cannot precede u with respect to time, i.e., $v \notin \pi(u)$. Hence, $v \in S \setminus \pi(u)$. Now consider the successor of v, denoted by $succ(v) \in \mathcal{N}$. By the definition of v, $succ(v)$ cannot be visited by S, i.e., $succ(v) \notin S$. Again, $succ(v)$ cannot precede u with respect to time since F is a feasible path. Hence, $succ(v) \in \bar{S} \setminus \pi(u)$. Considering every request in S as v results in the proposition.

General Cuts. Conflict analysis can derive cuts that are do not have the form of Constraint (14) nor Constraint (15). These cuts contain both true literals and false literals, such as those of Constraint (13). They originate from fixing an arc to be unused (i.e., setting $x_{i,j} = 0$ for some $(i,j) \in \mathcal{A}$), which can result in tightening the bounds of a time or load variable. Consequently, an assigned arc can become infeasible. Hence, the originating nogood will contain both true and false literals. We are not aware of VRP cuts in the literature that mix true literals and false literals. This is possibly because tightening bounds is too costly for every call to a separation algorithm. CP maintains the bounds internally as part of propagation, and hence, the bounds are readily available. Because of this, these cuts seem to be fundamentally linked to CP. It is an open research issue to understand whether these cuts can be strengthened.

5   Experimental Results

The Solvers. The BCE solver includes a small CP solver and calls Gurobi 6.5.2 to solve the LP relaxations. The algorithm presented in Fig. 3 has a limit of 500 nodes for the depth-first search. This number was chosen experimentally as it was superior to limits of 100, 1,000, 5,000, and 10,000 nodes. The experiments consider four versions of the solver: with and without cut strengthening, and with the two branching rules. The four versions are compared against published results of a BC model [18], as well as a pure CP model and a pure MIP model. The CP model is the standard VRPTW model based on successor variables (e.g., [19,27]), and is solved using Chuffed. The MIP model is the three-index flow model (e.g., [31]), and is solved using Gurobi. The reported results for the BC model are given an hour of CPU time on a Pentium III CPU at 600 MHz. To be fair, our solvers are run for 10 min on a Xeon E5-2660 V3 at 2.6 GHz.

The Results. The solvers are tested on the Solomon benchmarks with 100 requests. The results are reported in Table 3. The pure CP model failed to find any feasible solution and is omitted from the table. The pure MIP model proves optimality on only one instance and finds poor solutions to three other instances. These results were expected and are given to confirm the need for the other approaches. The rest of this section compares the BC and BCE approaches.

Upper Bounds. The four BCE methods find the same or better solutions than the BC algorithm for all instances except C204. Of the best solutions found, all but two (R201, C204) can be found using the activity-based branching rule. For the C instances, BC and BCE with activity-based search and cut strengthening are comparable since they both dominate on seven of the eight instances. For the R and RC instances, BCE with activity-based search improves upon the BC method, which generally finds solutions with costs about five times higher.

Lower Bounds. First observe that BCE with activity-based search and cut strengthening proves optimality on one more instance (RC201) than BC, which is quite remarkable. The bounds found by the BC model are superior to those from all BCE methods except for instance RC201, on which BCE with activity-based search and cut strengthening finds a tighter bound. This is not surprising since the BC algorithm implements families of cuts not present in the BCE model. These families of cuts capture logic that the constraints in the checking subproblem do not. As will be mentioned in Sect. 6, stronger dual bounds should be available once the BCE model is expanded with optimization constraints.

The Impact of Branching Rules. Activity-based branching performs significantly better than most-fractional branching. Without cut strengthening, activity-based branching finds solutions better than most-fractional branching on all instances except C201, on which all four BCE methods prove optimality. With cut strengthening, activity-based search performs better on 19 of the 27 instances, and worse on only one instance. This is not surprising given that

Table 3. Solutions to the Solomon instances with 100 requests. The table reports the lower bound, upper bound and time to prove optimality for each of the solvers. The best upper bound for each instance is shown in bold. The CP model is omitted as it is unable to find feasible solutions to any instance.

| | Branch-and-check – most-fractional | | | | | | Branch-and-check – activity-based | | | | | | Branch-and-cut | | | MIP | | |
| | No strengthening | | | With strengthening | | | No strengthening | | | With strengthening | | | | | | | | |
Instance	LB	UB	Time	LB	UB	Time	LB	UB	Time	LB	UB	Time	LB	UB	Time	LB	UB	Time
R201	1055.8	1198.0	-	1117.7	**1143.3**	-	1054.7	1177.6	-	1114.3	1149.9	-	1132.7	1155.6	-	975.6	-	-
R202	762.8	1213.0	-	852.1	1219.6	-	763.2	1133.4	-	850.7	**1109.3**	-	888.6	4980.0	-	715.3	-	-
R203	660.1	1244.8	-	709.8	1253.6	-	659.9	**1025.2**	-	707.7	1052.2	-	748.1	4980.0	-	620.3	-	-
R204	625.3	1166.7	-	639.2	1193.3	-	625.8	**858.4**	-	638.3	887.4	-	661.9	4980.0	-	584.9	-	-
R205	796.3	1222.0	-	889.6	1069.9	-	794.3	1091.3	-	876.9	**1052.5**	-	900.0	4980.0	-	732.3	-	-
R206	686.3	1171.6	-	751.4	1157.4	-	686.0	1040.1	-	745.3	**1018.9**	-	783.6	4980.0	-	644.8	-	-
R207	648.1	1187.5	-	681.5	1168.5	-	647.5	**940.7**	-	685.8	941.4	-	714.8	4980.0	-	603.1	-	-
R208	623.2	1097.4	-	633.5	1187.7	-	623.4	855.0	-	635.3	**832.5**	-	651.8	4980.0	-	577.2	-	-
R209	687.5	1238.1	-	756.6	1172.5	-	686.5	**1046.6**	-	753.1	1073.8	-	785.8	4980.0	-	648.2	-	-
R210	679.7	1225.6	-	749.9	1240.3	-	679.8	1105.6	-	750.8	**1024.9**	-	798.3	4980.0	-	636.6	-	-
R211	621.2	1335.5	-	633.0	1355.9	-	621.2	**1004.2**	-	632.1	1065.1	-	645.1	4980.0	-	577.2	4224.9	-
C201	589.1	**589.1**	0.0	589.1	**589.1**	0.0	589.1	**589.1**	0.0	589.1	**589.1**	0.0	589.1	**589.1**	11.5	589.1	**589.1**	15.2
C202	548.7	679.8	-	589.1	**589.1**	131.2	548.2	629.9	-	589.1	**589.1**	12.6	589.1	**589.1**	202.9	524.3	-	-
C203	526.5	948.3	-	563.4	672.2	-	524.7	686.5	-	565.9	**601.2**	-	586.0	632.3	-	507.3	-	-
C204	516.3	946.7	-	552.9	1086.7	-	514.7	884.5	-	555.9	660.9	-	584.4	**597.1**	-	488.3	-	-
C205	546.9	685.8	-	586.4	**586.4**	0.2	546.5	613.1	-	586.4	**586.4**	16.3	586.4	**586.4**	334.4	511.4	-	-
C206	539.9	776.9	-	586.0	**586.0**	11.8	538.2	702.6	-	586.0	**586.0**	10.6	586.0	**586.0**	419.0	504.7	4997.5	-
C207	542.7	851.1	-	585.8	**585.8**	20.6	538.3	635.2	-	585.8	**585.8**	8.3	585.8	**585.8**	527.5	503.9	-	-
C208	534.5	857.2	-	585.8	**585.8**	60.0	533.1	652.4	-	585.8	**585.8**	11.2	585.8	**585.8**	569.7	500.3	-	-
RC201	1086.5	1403.6	-	1245.8	**1261.8**	-	1081.0	1338.3	-	1261.8	**1261.8**	44.5	1250.1	1288.2	-	938.3	-	-
RC202	704.5	1465.9	-	912.7	1418.6	-	699.2	1204.2	-	916.9	**1152.3**	-	940.1	6609.4	-	641.0	-	-
RC203	615.0	1402.7	-	750.2	1359.6	-	610.8	1149.0	-	748.8	**1117.6**	-	781.6	6609.4	-	563.1	-	-
RC204	583.9	1410.2	-	657.8	1352.4	-	581.0	1007.1	-	657.0	**923.5**	-	692.7	6609.4	-	532.4	-	-
RC205	822.5	1511.6	-	1075.5	1307.0	-	818.8	1249.8	-	1055.8	**1240.9**	-	1081.7	6609.4	-	746.3	-	-
RC206	785.4	1485.4	-	964.3	1273.9	-	784.9	1270.2	-	950.7	**1202.8**	-	974.8	6609.4	-	698.2	-	-
RC207	647.3	1486.3	-	794.6	1424.5	-	642.9	1193.5	-	800.9	**1172.1**	-	832.4	6609.4	-	594.9	-	-
RC208	572.7	1629.8	-	624.0	1776.1	-	573.9	**1039.5**	-	624.3	1078.4	-	647.7	6609.4	-	527.1	5299.5	-

branching on the most fractional variable is known to perform worse than random selection [2].

The Impact of Cut Strengthening. Cut strengthening improves the lower bounds for both branching rules. For the C instances, cut strengthening is critical for proving optimality. For the RC instances except RC208, BCE with activity-based branching and cut strengthening finds solutions better than the other methods. Cut strengthening interferes with the activity-based branching rule for about half of the R instances. The cause of this interference is not yet understood.

The results indicate that BCE is an interesting avenue for solving hard VRPs. The BCE model finds superior primal solutions despite its simplicity and the fact that it is missing many families of cuts and that the checking subproblem does not reason about optimality nor variables with fractional values in the LP relaxation solution. For practitioners without the expertise in BC, BCE provides an interesting and practically appealing alternative.

6   Future Research Directions

The BCE algorithm, presented in this paper as a proof-of-concept, can be improved in many ways. This section explores some potential improvements.

Branching. The branching rules simply assign a fractional variable to 0 in one child and 1 in the other. These branching rules make the search tree highly unbalanced, considerably degrading the performance of the solver. Future implementations should test branching on cutsets, which is the standard branching rule seen in BC models of VRPs. It would also be interesting to test branching on variables in the CP model (e.g., branching on time windows) by propagating these decisions and enforcing the implications in the MIP model.

Search Strategy. VRPs greatly benefit from best-first search. For simplicity, the BCE implementation uses depth-first search, which allows literals to be stored in a stack data structure. It is obviously possible to implement conflict analysis in best-first search but efficient implementations remain an open question today. As explained in Sect. 3, the BCE implementation blends depth-first search with periodic best-first selection to explore attractive parts of the search tree.

Subtour Elimination. The propagator of Constraint (9) is extremely simple and only eliminates assignments that would create a cycle. This contrasts with separation algorithms, which are able to separate cuts using fractional solutions. It would be highly desirable to study the impact of more advanced propagators and explanations for subtour elimination in CP.

Cut Strengthening. The CP model contains all the omitted constraints; namely, the subtour elimination, vehicle capacity and time window constraints. As a result, conflict analysis can deduce nogoods based on the combined infeasibility of multiple constraints. In contrast, separation algorithms only reason about one family of cuts. It is an open question whether conflict analysis can automatically strengthen the cuts by reasoning about a conjunction of constraints. This will reduce the need to develop dedicated cut strengthenings.

Optimization Constraints. The objective function has been omitted from the checking subproblem because propagators for linear function are known to be weak. Sophisticated propagators for the WEIGHTEDCIRCUIT constraint should be implemented, as they may produce considerably stronger nogoods.

Application to Branch-and-Price. Branch-and-cut-and-price, which includes column generation and cut generation, is the current state-of-the-art exact method for solving classical VRPs. Preliminary experiments with an existing branch-and-price solver show that BCE is not beneficial with column generation for the VRPTW as nogoods will not be generated in step 4 of Fig. 3 because the paths already respect the time and capacity constraints. However, step 2 can fail due to incompatibility between the branching decisions of a node. This infeasibility cannot be detected by the pricing problem because it has no knowledge of the global problem, nor detected by the master problem until all paths are generated because artificial variables satisfy the constraints in the interim. Incompatible branching decisions can induce nogoods but this seldom occurs in branch-and-price because its LP relaxation bound is asymptotically tight, allowing it to discard nodes due to suboptimality much earlier than infeasibility. Hence, branch-and-price-and-check is unlikely to prove useful in solving classical VRPs. It is, however, useful for rich VRPs with inter-route constraints (e.g., [20]) because the pricing subproblem, being a shortest path problem, has no knowledge of the interactions between routes in the parent problem.

7   Conclusion

This paper proposed the framework of branch-and-check with explanations (BCE) as a step towards the grand unification of linear programming, constraint programming and Boolean satisfiability. BCE finds cuts using general-purpose conflict analysis instead of specialized separation algorithms. The method features a master problem, which ignores a number of constraints, and a checking subproblem, which uses inference to check the feasibility of the omitted constraints and conflict analysis to derive nogood cuts. It also leverages conflict-based branching rules and can strengthen cuts using traditional insights from branch-and-cut in a post-processing step.

Experimental results on the Vehicle Routing Problem with Time Windows show that BCE is a viable alternative to branch-and-cut. In particular, BCE dominates branch-and-cut, both in proving optimality (with cut strengthening) and in finding high-quality solutions.

BCE offers an intalternative to existing branch-and-cut approaches. By using a general-purpose constraint programming solver to derive cuts, BCE can greatly simplify the modelling of problems that traditionally use branch-and-cut. This, in turn, avoids the need for dedicated separation algorithms. BCE is also capable of identifying well-known classes of cuts and strengthening them in a post-processing step. Finally, BCE significantly benefits from conflict-based branching rules, opening further opportunities typically not available in branch-and-cut.

References

1. Achterberg, T.: Conflict analysis in mixed integer programming. Discrete Optim. **4**(1), 4–20 (2007)
2. Achterberg, T., Koch, T., Martin, A.: Branching rules revisited. Oper. Res. Lett. **33**(1), 42–54 (2005)
3. Baldacci, R., Mingozzi, A., Roberti, R.: New route relaxation and pricing strategies for the vehicle routing problem. Oper. Res. **59**(5), 1269–1283 (2011)
4. Bard, J.F., Kontoravdis, G., Yu, G.: A branch-and-cut procedure for the vehicle routing problem with time windows. Transp. Sci. **36**(2), 250–269 (2002)
5. Beck, J.C.: Checking-up on branch-and-check. In: Cohen, D. (ed.) CP 2010. LNCS, vol. 6308, pp. 84–98. Springer, Heidelberg (2010). doi:10.1007/978-3-642-15396-9_10
6. Benchimol, P., Hoeve, W.J., Régin, J.C., Rousseau, L.M., Rueher, M.: Improved filtering for weighted circuit constraints. Constraints **17**(3), 205–233 (2012)
7. Bent, R., Van Hentenryck, P.: A two-stage hybrid local search for the vehicle routing problem with time windows. Transp. Sci. **38**(4), 515–530 (2004)
8. Bent, R., Van Hentenryck, P.: A two-stage hybrid algorithm for pickup and delivery vehicle routing problems with time windows. Comput. Oper. Res. **33**(4), 875–893 (2006)
9. Dechter, R.: Learning while searching in constraint-satisfaction-problems. In: Proceedings of the 5th National Conference on Artificial Intelligence, Philadelphia, PA, 11–15 August 1986. Science, vol. 1, pp. 178–185 (1986)
10. Desrochers, M., Desrosiers, J., Solomon, M.: A new optimization algorithm for the vehicle routing problem with time windows. Oper. Res. **40**(2), 342–354 (1992)
11. Eén, N., Sörensson, N.: An extensible SAT-solver. In: Giunchiglia, E., Tacchella, A. (eds.) SAT 2003. LNCS, vol. 2919, pp. 502–518. Springer, Heidelberg (2004). doi:10.1007/978-3-540-24605-3_37
12. Feydy, T., Stuckey, P.J.: Lazy clause generation reengineered. In: Gent, I.P. (ed.) CP 2009. LNCS, vol. 5732, pp. 352–366. Springer, Heidelberg (2009). doi:10.1007/978-3-642-04244-7_29
13. Fukasawa, R., Longo, H., Lysgaard, J., De Aragão, M.P., Reis, M., Uchoa, E., Werneck, R.F.: Robust branch-and-cut-and-price for the capacitated vehicle routing problem. Math. Program. **106**(3), 491–511 (2006)
14. Gendron, B., Scutellà, M.G., Garroppo, R.G., Nencioni, G., Tavanti, L.: A branch-and-benders-cut method for nonlinear power design in green wireless local area networks. Eur. J. Oper. Res. **255**(1), 151–162 (2016)
15. Hooker, J.N.: Logic-based methods for optimization. In: Borning, A. (ed.) PPCP 1994. LNCS, vol. 874, pp. 336–349. Springer, Heidelberg (1994). doi:10.1007/3-540-58601-6_111
16. Jepsen, M., Petersen, B., Spoorendonk, S., Pisinger, D.: Subset-row inequalities applied to the vehicle-routing problem with time windows. Oper. Res. **56**(2), 497–511 (2008)

17. Jussien, N., Barichard, V.: The palm system: explanation-based constraint programming. In: Proceedings of TRICS: Techniques foR Implementing Constraint Programming Systems, a Post-conference Workshop of CP 2000, pp. 118–133 (2000)
18. Kallehauge, B., Boland, N., Madsen, O.B.G.: Path inequalities for the vehicle routing problem with time windows. Networks 49(4), 273–293 (2007)
19. Kilby, P., Prosser, P., Shaw, P.: A comparison of traditional and constraint-based heuristic methods on vehicle routing problems with side constraints. Constraints 5(4), 389–414 (2000)
20. Lam, E., Van Hentenryck, P.: A branch-and-price-and-check model for the vehicle routing problem with location congestion. Constraints 21(3), 394–412 (2016)
21. Lysgaard, J., Letchford, A.N., Eglese, R.W.: A new branch-and-cut algorithm for the capacitated vehicle routing problem. Math. Program. 100(2), 423–445 - 0025–5610 (2004)
22. Mak, V.: On the asymmetric travelling salesman problem with replenishment arcs. Ph.D. thesis, University of Melbourne (2001)
23. Moskewicz, M.W., Madigan, C.F., Zhao, Y., Zhang, L., Malik, S.: Chaff: engineering an efficient SAT solver. In: Proceedings of the 38th Annual Design Automation Conference, pp. 530–535. ACM (2001)
24. Ohrimenko, O., Stuckey, P.J., Codish, M.: Propagation via lazy clause generation. Constraints 14(3), 357–391 (2009)
25. Pecin, D., Pessoa, A., Poggi, M., Uchoa, E.: Improved branch-cut-and-price for capacitated vehicle routing. In: Lee, J., Vygen, J. (eds.) IPCO 2014. LNCS, vol. 8494, pp. 393–403. Springer, Cham (2014). doi:10.1007/978-3-319-07557-0_33
26. Ropke, S., Cordeau, J.F.: Branch and cut and price for the pickup and delivery problem with time windows. Transp. Sci. 43(3), 267–286 (2009)
27. Rousseau, L.M., Gendreau, M., Pesant, G.: Using constraint-based operators to solve the vehicle routing problem with time windows. J. Heuristics 8(1), 43–58 (2002)
28. Shaw, P.: Using constraint programming and local search methods to solve vehicle routing problems. In: Maher, M., Puget, J.-F. (eds.) CP 1998. LNCS, vol. 1520, pp. 417–431. Springer, Heidelberg (1998). doi:10.1007/3-540-49481-2_30
29. Thorsteinsson, E.S.: Branch-and-check: a hybrid framework integrating mixed integer programming and constraint logic programming. In: Walsh, T. (ed.) CP 2001. LNCS, vol. 2239, pp. 16–30. Springer, Heidelberg (2001). doi:10.1007/3-540-45578-7_2
30. Tran, T.T., Araujo, A., Beck, J.C.: Decomposition methods for the parallel machine scheduling problem with setups. INFORMS J. Comput. 28(1), 83–95 (2016)
31. Vigo, D., Toth, P.: Vehicle Routing: Problems, Methods, and Applications, 2nd edn. Society for Industrial and Applied Mathematics, Philadelphia (2014)

Solving Multiobjective Discrete Optimization Problems with Propositional Minimal Model Generation

Takehide Soh[1($\boxtimes$)], Mutsunori Banbara[1], Naoyuki Tamura[1],
and Daniel Le Berre[2]

[1] Information Science and Technology Center, Kobe University, Kobe, Japan
soh@lion.kobe-u.ac.jp, {banbara,tamura}@kobe-u.ac.jp
[2] CRIL-CNRS, Université d'Artois, Lens, France
leberre@cril.fr

Abstract. We propose a propositional logic based approach to solve MultiObjective Discrete Optimization Problems (MODOPs). In our approach, there exists a one-to-one correspondence between a Pareto front point of MODOP and a *P*-minimal model of the CNF formula obtained from MODOP. This correspondence is achieved by adopting the order encoding as CNF encoding for multiobjective functions. Finding the Pareto front is done by enumerating all P-minimal models. The beauty of the approach is that each Pareto front point is blocked by a single clause that contains at most one literal for each objective function. We evaluate the effectiveness of our approach by empirically contrasting it to a state-of-the-art MODOP solving technique.

1 Introduction

Due to recent remarkable improvements in efficiency, propositional logic based approaches, e.g., Satisfiability Testing (SAT), Max-SAT, Answer Set Programming (ASP), Pseudo Boolean (PB), and Satisfiability Modulo Theories (SMT) solvers, have succeeded in solving many combinatorial (optimization) problems in diverse areas, such as scheduling, automated planning, constraint satisfaction, model checking, robotics, systems biology, etc. [7].

Especially for constraint satisfaction, encoding finite linear Constraint Satisfaction Problems (CSPs) into Conjunctive Normal Form (CNF) formulas and solving them by using CDCL solvers has proven to be highly effective method by the award-winning CP solver *Sugar*[1]. The encoding of *Sugar* relies on the order encoding [11,38,39], one of the most studied CNF encodings for CSP solving in recent years. Moreover, single-objective discrete optimization problems like Max-CSP can be efficiently solved by the order encoding [40].

P-minimal model [21,28] is a model which satisfies subset minimality with regard to a specific set of atoms (referred as to *P*). More precisely, let M and

---

[1] http://bach.istc.kobe-u.ac.jp/sugar/.

© Springer International Publishing AG 2017
J.C. Beck (Ed.): CP 2017, LNCS 10416, pp. 596–614, 2017.
DOI: 10.1007/978-3-319-66158-2_38

M' be models of a CNF formula over a set of propositional variables V. Let P be a set such that $P \subseteq V$. Suppose that there is a model M such that there is no model M' that satisfies $M' \cap P \subset M \cap P$. Then, $M \cap P$ is called a P-minimal model. P-minimal model generation has been used for solving single-objective job-shop scheduling problems [21].

MultiObjective Discrete Optimization Problem. (MODOP; [14]) is a problem involving multiple objective functions that should be considered separately and optimized simultaneously. MODOP is therefore well suited for modeling many real world applications involving multiple criteria, such as decision making [4,15], scheduling [10,18], automated planning [1,44], product design [42], etc. In this paper, we consider a problem of solving finite domain MODOP, and refer to it as MODOP.

The goal of MODOP is finding the Pareto front (viz. the set of Pareto front points) defined by Pareto optimality. Finding the Pareto front is known to be difficult. Several methods have been proposed for MODOP solving, such as approximation methods [12,13,45,46], quality-guaranteed approximation methods [25,33], exact methods [6,8,19,23,24,33,34,41], and many others. However, so far, little attention has been paid to using propositional logic based approaches to MODOP.

In this paper, we describe an approach to solve MODOPs based on propositional P-minimal model generation. From the viewpoint of propositional logic, we gain insights into the relation between Pareto optimality and subset minimality. In our approach, there exists a one-to-one correspondence between a Pareto front point of MODOP and a P-minimal model of the CNF formula obtained from MODOP. This correspondence is achieved by adopting the order encoding as CNF encoding for multiobjective functions[2]. More precisely, propositional variables obtained from multiobjective functions by the order encoding are used as P in P-minimal model generation. Finding the Pareto front is done by enumerating all P-minimal models. For this, due to the nice property of the order encoding, each Pareto front point can be elegantly blocked by a single clause contains at most one literal for each objective function.

Our propositional logic based approach opens up the possibilities of new application field of minimal model generation to MODOP. However, the question is whether it competes with state-of-the-art MODOP solving techniques in performance. We answer this question by empirically contrasting our approach to an efficient MODOP solving technique based on Binary Decision Diagram (BDD) proposed in [6].

In the sequel, we assume some familiarity with CNF encodings as well as minimal model generation. A comprehensive survey for CNF encodings can be found in [20]. Although we provide a brief introduction to MODOP in the next section, we refer the reader to the literature [14] for a broader perspective.

---

[2] MODOP other than multiobjective functions can be encoded by existing CNF encodings: direct, multivalue, support, log, and hybrid encodings.

2 Preliminaries

We begin with the definition of CSPs and then define MODOPs. We also give a brief introduction to P-minimal models.

Definition 1 (Finite Linear CSP). A tuple (X, D, C) is called a CSP when:

- X is a finite set of variables,
- D is a function specifying each variables' *domain* (a set of possible values) as a bounded Integer interval,
- C is a finite set of constraints over X.

We here consider CSPs that consist of linear expressions on finite domain Integers. We use $lb(x)$ and $ub(x)$ as the lower and upper bounds of $D(x)$ respectively, for each variable $x \in X$. An *assignment* of a CSP (X, D, C) is a function $\alpha : X \to \mathbb{Z}$ such that it satisfies $\alpha(x) \in D(x)$ for all variables $x \in X$. A *solution* of a CSP (X, D, C) is an assignment which satisfies all constraints $C \in C$. A CSP is *satisfiable* if there exists a solution, otherwise *unsatisfiable*.

For illustration, consider a CSP (X, D, C) with $X = \{x, y, z\}$ where $D(x) = D(y) = \{0, 1, 2, 3, 4\}$, and $D(z) = \{0, 1\}$, and $C = \{x + y \geq 3, (x \neq 1) \lor (y \neq 2) \lor (z = 1), (x \neq 2) \lor (y \neq 1) \lor (z = 1)\}$. We have 31 solutions, one of which is given by $\alpha(x) = 0$, $\alpha(y) = 3$, and $\alpha(z) = 0$. For convenience, we represent a solution α as a vector of Integer values, i.e., $(x, y, z) = (0, 3, 0)$.

Definition 2 (Finite domain MODOP). A tuple $(\overrightarrow{o}, (X, D, C))$ is called a finite domain MODOP when:

- *Multiobjective variables* $\overrightarrow{o} = (o_1, \ldots, o_m)$ is a finite vector of m objective variables to be minimized simultaneously.
- (X, D, C) is a CSP.

Without loss of generality, we use objective variables instead of objective functions to make our approach more understandable[3].

We simply refer to finite domain MODOPs as MODOPs. The goal of MODOP is to find the Pareto front defined by the concept of Pareto optimality. A *feasible solution* of a MODOP $(\overrightarrow{o}, \text{CSP})$ is a solution of CSP.

Definition 3 (Dominance). A feasible solution α *dominates* α' ($\alpha \prec \alpha'$) if α and α' satisfies the following conditions:

- $\alpha(o_i) \leq \alpha'(o_i)$ for all $i \in \{1, \ldots, m\}$,
- $\alpha(o_i) < \alpha'(o_i)$ for some $i \in \{1, \ldots, m\}$.

We extend the definition of assignment to multiobjective variables $\overrightarrow{o}$. For an assignment α, we use $\alpha(\overrightarrow{o})$ to express a *cost vector*, i.e., $\alpha(\overrightarrow{o}) = (\alpha(o_1), \ldots, \alpha(o_m))$. We also use $\alpha(\overrightarrow{o}) \prec \alpha'(\overrightarrow{o})$ to express that $\alpha(\overrightarrow{o})$ dominates $\alpha'(\overrightarrow{o})$.

---

[3] Note that there is no essential differences between them.

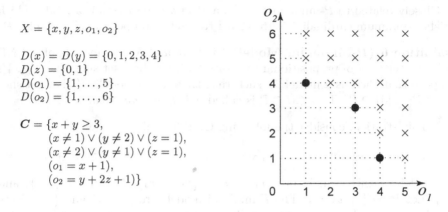

$X = \{x, y, z, o_1, o_2\}$

$D(x) = D(y) = \{0, 1, 2, 3, 4\}$
$D(z) = \{0, 1\}$
$D(o_1) = \{1, \ldots, 5\}$
$D(o_2) = \{1, \ldots, 6\}$

$C = \{x + y \geq 3,$
$\quad (x \neq 1) \vee (y \neq 2) \vee (z = 1),$
$\quad (x \neq 2) \vee (y \neq 1) \vee (z = 1),$
$\quad (o_1 = x + 1),$
$\quad (o_2 = y + 2z + 1)\}$

Fig. 1. An example of MODOP $((o_1, o_2), (X, D, C))$

Definition 4 (Pareto Optimality). A feasible solution α is *Pareto optimal* if there is no other feasible solution α' such that $\alpha'(\overrightarrow{o}) \prec \alpha(\overrightarrow{o})$. A cost vector $\alpha(\overrightarrow{o})$ is a *Pareto front point* (also called *nondominated point*) if α is Pareto optimal. The *Pareto front* is the set of all Pareto front points.

Figure 1 shows an example of MODOP having two objective variables. Again, we have 31 feasible solutions, since the main part is the same as the previous CSP example. Among them, feasible solutions $(x, y, z) = (0, 3, 0)$, $(2, 2, 0)$, and $(3, 0, 0)$ correspond to cost vectors $(o_1, o_2) = (1, 4)$, $(3, 3)$ and $(4, 1)$ respectively. These cost vectors marked by $\bullet$ in Fig. 1 are Pareto front points. Since there is no other Pareto front points, we have the nonconvex Pareto front $\{(1, 4), (3, 3), (4, 1)\}$.

*P***-Minimal Model.** A *propositional variable* is a variable whose value is either 0 (*false*) or 1 (*true*). A *literal* is a propositional variable v or its negation $\neg v$. A *clause* is a disjunction of literals and is also identified with a set of literals. A *Conjunctive Normal Form (CNF) formula* is a conjunction of clauses and is also identified with a set of clauses. An *assignment* of a CNF formula is a function $\beta : V \to \{0, 1\}$ where V is the set of propositional variables of the formula. A *model* of a CNF formula is an assignment which satisfies all clauses of the formula. A CNF formula is *satisfiable* if there exists a model, otherwise *unsatisfiable*. For convenience, we represent a model by a set of propositional variables which are assigned to 1s. For instance, consider a model β such that $\beta(v_1) = 0$, $\beta(v_2) = 1$, and $\beta(v_3) = 0$. This model is represented by $\{v_2\}$. A *P-minimal model* is a model which satisfies subset minimality with regard to a set of propositional variables P. The following definition is based on [21, 28].

Definition 5 (*P*-Minimal Model). Let M and M' be models of a CNF formula over a set of propositional variables V. Let P be a set such that $P \subseteq V$. Suppose that there is a model M such that there is no model M' that satisfies $M' \cap P \subset M \cap P$. Then, $M \cap P$ is called a *P-minimal model*.

Closely related to P-minimal models, a P-*minimum model* is a model which satisfies minimum cardinality with regard to a set of propositional variables P.

Definition 6 (P-Minimum Model). Let M and M' be models of a CNF formula over a set of propositional variables V. Let P be a set such that $P \subseteq V$. Suppose that there is a model M such that there is no model M' that satisfies $|M' \cap P| < |M \cap P|$. Then, $M \cap P$ is called a P-*minimum model*.

For illustration, consider the following CNF formula:

$$(\neg v_1 \vee \neg v_2 \vee v_3 \vee v_4) \wedge (v_1 \vee v_2 \vee \neg v_3) \wedge (v_2 \vee v_3) \wedge (\neg v_2 \vee \neg v_3)$$

All models of this formula are $\{v_1, v_2, v_4\}$, $\{v_1, v_3, v_4\}$, $\{v_2, v_4\}$, $\{v_1, v_3\}$, and $\{v_2\}$. Let P be $\{v_1, v_2, v_3\}$. The P-minimal models are $\{v_1, v_3\}$ and $\{v_2\}$. Note that $\{v_1, v_3\}$ and $\{v_1, v_3, v_4\}$ are not distinct when we consider P-minimal models. In contrast, the P-minimum model of the formula is $\{v_2\}$.

3   A Propositional Logic Based Approach to MODOP

In this section, we first give a brief introduction to the order encoding and show some properties. We then prove that there exists a one-to-one correspondence between a Pareto front point of MODOP and a P-minimal model of CNF formula obtained from MODOP. Finally, we present a method for finding the Pareto front by enumerating all P-minimal models.

3.1   Order Encoding and Some Properties

Order encoding is a method that encodes finite linear CSPs into CNF formulas [11,38,39]. The benefit of the order encoding is the natural representation of the order relation on Integers. Encoding CSP into CNF formulas by the order encoding and solving them by modern SAT solvers is an effective method for CSP solving in a sense that it keeps the bounds consistency of CSP by unit propagation. In recent years, the order encoding and related ones have succeeded in solving a wide range of CSPs [2,3,11,16,26,29].

In the order encoding, for a given CSP (X, D, C), a propositional variable $p_{x,d}$ is introduced for each Integer variable $x \in X$ and each domain value $d \in D(x)$ except $lb(x)$. The variable $p_{x,d}$ is intended to express $x \geq d$. The variable $p_{x,lb(x)}$ is unnecessary, since $x \geq lb(x)$ is always true. We use $\delta(x)$ to express the set of $p_{x,d}$'s, i.e., $\delta(x) = \{p_{x,d} \mid lb(x) + 1 \leq d \leq ub(x)\}$.

Then, the following *axiom clauses* are introduced to ensure that each Integer variable $x \in X$ takes an exactly one value from $D(x)$. The axiom clauses consists of two literals and represent the order relation on Integer variables.

$$p_{x,d} \vee \neg p_{x,d+1} \quad (d = lb(x) + 1 \ldots ub(x) - 1) \tag{1}$$

Intuitively, this clause means that x is greater than or equal to d if x is greater than or equal to $d + 1$.

For illustration, let us consider an Integer variable x with $D(x) = \{0, \ldots, 4\}$ in Fig. 1. For this, four variables $\delta(x) = \{p_{x,1}, p_{x,2}, p_{x,3}, p_{x,4}\}$ are introduced, and this example is encoded into the following axiom clauses:

$$p_{x,1} \vee \neg p_{x,2} \qquad p_{x,2} \vee \neg p_{x,3} \qquad p_{x,3} \vee \neg p_{x,4}$$

The possible models of the axiom clauses are shown in a table below.

Models	$p_{x,1}$	$p_{x,2}$	$p_{x,3}$	$p_{x,4}$	Assignment
$\{\}$	0	0	0	0	$x = 0$
$\{p_{x,1}\}$	1	0	0	0	$x = 1$
$\{p_{x,1}, p_{x,2}\}$	1	1	0	0	$x = 2$
$\{p_{x,1}, p_{x,2}, p_{x,3}\}$	1	1	1	0	$x = 3$
$\{p_{x,1}, p_{x,2}, p_{x,3}, p_{x,4}\}$	1	1	1	1	$x = 4$

Each model represents an assignment of Integer value x, e.g., a model $\{p_{x,1}, p_{x,2}\}$ representing an assignment $x = 2$. Moreover, we can observe in the table that the order relation between Integer values is represented by the subset relation between models, and vice versa. For instance, it is easy to see that $\{p_{x,1}\} \subseteq \{p_{x,1}, p_{x,2}\}$ means $1 \leq 2$. In general, the order encoding has the following property:

Property 1. Let x be a Integer variable, ψ be a CNF formula obtained from x by the order encoding. Let α and α' be assignments of x respectively obtained from models M and M' of ψ. Then, $\alpha(x) \leq \alpha'(x) \iff M \subseteq M'$ holds.

From this property, minimizing an Integer variable x can be done by finding the minimal model of ψ. In our example of x with $D(x) = \{0, \ldots, 4\}$, the minimal model $\{\}$ represents the minimum value $x = 0$.

Constraints are encoded into clauses representing conflict regions rather than conflict points. For illustration, consider a constraint $x + y \geq 3$ with $D(x) = D(y) = \{0, 1, 2, 3, 4\}$ in Fig. 1. Encoding Integer variables x and y is done in a similar way. The constraint is encoded into the following three clauses:

$$p_{x,1} \vee p_{y,3} \qquad p_{x,2} \vee p_{y,2} \qquad p_{x,3} \vee p_{y,1}$$

These clauses represents conflict regions $x \leq 0 \wedge y \leq 2$, $x \leq 1 \wedge y \leq 1$, and $x \leq 2 \wedge y \leq 0$ from left to right respectively. We obtain nineteen models from the clauses. Among them, for instance, a model $\{p_{x,1}, p_{y,1}, p_{y,2}, p_{y,3}\}$ represents a solution $(x, y) = (1, 3)$ of the original constraint $x + y \geq 3$.

Property 1 can be extended to a whole CSP as follows:

Property 2. Let (X, D, C) be a CSP, ψ be a CNF formula obtained from CSP, and P be the set of propositional variables obtained from $x \in X$ by the order encoding, i.e., $P = \delta(x)$. Let α and α' be solutions of CSP respectively obtained from models M and M' of ψ. Then, $\alpha(x) \leq \alpha'(x) \iff M \cap P \subseteq M' \cap P$ holds.

From this property, finding the optimal value of CSP (X, D, C) with a single objective variable $o \in X$ can be done by finding the P-minimal model of ψ with $P = \delta(o)$.

3.2 Correspondence Between a Pareto Front Point and a P-Minimal Model

We now present a one-to-one correspondence between a Pareto front point of MODOP and a P-minimal model of CNF-encoded formula from MODOP. This correspondence relies on the order encoding to encode multobjective variables into CNF.

First, we extend Property 2 to MODOPs as follows:

Lemma 1. Let $\Omega = (\overrightarrow{o}, CSP)$ be a MODOP, ψ be a CNF formula obtained from CSP, and P be the set of propositional variables from multiobjective variables $\overrightarrow{o} = (o_1, \ldots, o_m)$ by the order encoding, i.e., $P = \delta(o_1) \cup \cdots \cup \delta(o_m)$. Let α and α' be feasible solutions of Ω respectively obtained from models M and M' of ψ. Then, $\alpha(\overrightarrow{o}) \prec \alpha'(\overrightarrow{o}) \Longleftrightarrow M \cap P \subset M' \cap P$ holds.

Proof. ($\Longleftarrow$) From the hypotheses $M \cap P \subset M' \cap P$, the following holds.

- $M \cap \delta(o_i) \subseteq M' \cap \delta(o_i)$ for all i ($1 \leq i \leq m$),
- $M \cap \delta(o_i) \subset M' \cap \delta(o_i)$ for some i ($1 \leq i \leq m$).

By Property 2, the following holds from the above.

- $\alpha(o_i) \leq \alpha'(o_i)$ for all i ($1 \leq i \leq m$),
- $\alpha(o_i) < \alpha'(o_i)$ for some i ($1 \leq i \leq m$).

By Definition 3 (Dominance), $\alpha(\overrightarrow{o}) \prec \alpha'(\overrightarrow{o})$ holds.

($\Longrightarrow$) From the hypothesis $\alpha(\overrightarrow{o}) \prec \alpha'(\overrightarrow{o})$ and Definition 3 (Dominance), the following holds:

- $\alpha(o_i) \leq \alpha'(o_i)$ for all i ($1 \leq i \leq m$),
- $\alpha(o_i) < \alpha'(o_i)$ for some i ($1 \leq i \leq m$).

By Property 2, the following holds from the above.

- $M \cap \delta(o_i) \subseteq M' \cap \delta(o_i)$ for all i ($1 \leq i \leq m$),
- $M \cap \delta(o_i) \subset M' \cap \delta(o_i)$ for some i ($1 \leq i \leq m$).

Then, $M \cap P \subset M' \cap P$ holds.   □

From this lemma, it can be seen that the $\prec$-relation between feasible solutions is represented by the $\subset$-relation between sets of the intersection of models and $P = \delta(o_1) \cup \cdots \cup \delta(o_m)$, and vice versa. As mentioned, any existing CNF encodings can be used to encode CSP of Ω by adding channeling constraints between P and propositional variables of a target encoding if necessary. By using Lemma 1, the following proposition holds.

Proposition 1. Let $\Omega = (\overrightarrow{o}, CSP)$ be a MODOP, ψ be a CNF formula obtained from CSP, and P be a set of propositional variables from multiobjective variables $\overrightarrow{o} = (o_1, \ldots, o_m)$ by the order encoding, i.e., $P = \delta(o_1) \cup \cdots \cup \delta(o_m)$. Then, there is one-to-one correspondence between a Pareto front point of Ω and a P-minimal model of ψ.

Proof. ($\Longleftarrow$)
Let $M' \cap P$ be a P-minimal model of ψ, and α' be a feasible solution of Ω obtained from M' of ψ. By Definition 5 (*P*-Minimal Model), there is no model M of ψ that satisfies $M \cap P \subset M' \cap P$. Then, by Lemma 1, there is no feasible solution α that satisfies $\alpha(\overrightarrow{o}) \prec \alpha'(\overrightarrow{o})$. Thus, by Definition 4 (Pareto Optimality), the feasible solution α' is Pareto optimal, and a cost vector $\alpha'(\overrightarrow{o})$ is a Pareto front point.

($\Longrightarrow$) Let $\alpha'(\overrightarrow{o})$ be a Pareto front point of Ω, and M' be any model of ψ representing the feasible solution α' of Ω. By Definition 4 (Pareto Optimality), there is no feasible solution α that satisfies $\alpha(\overrightarrow{o}) \prec \alpha'(\overrightarrow{o})$. Then, by Lemma 1, there is no model M of ψ that satisfies $M \cap P \subset M' \cap P$. Therefore, by Definition 5 (*P*-Minimal Model), $M' \cap P$ is a P-minimal model.   □

For illustration, let us go back to an example in Fig. 1, a MODOP $((o_1, o_2), CSP)$ with $o_1 \in \{1, \ldots, 5\}$ and $o_2 \in \{1, \ldots, 6\}$. Table 1 shows all possible P-minimal models of a CNF formula obtained from CSP and corresponding Pareto front points, where

$$P = \{p_{o_1,2}, p_{o_1,3}, p_{o_1,4}, p_{o_1,5}, p_{o_2,2}, p_{o_2,3}, p_{o_2,4}, p_{o_2,5}, p_{o_2,6}\}.$$

It can be seen that there is one-to-one correspondence between a Pareto front point and a P-minimal model: $\{p_{o_2,2}, p_{o_2,3}, p_{o_2,4}\}$ and $(1,4)$, $\{p_{o_1,2}, p_{o_1,3}, p_{o_2,2}, p_{o_2,3}\}$ and $(3,3)$, and $\{p_{o_1,2}, p_{o_1,3}, p_{o_1,4}\}$ and $(4,1)$.

3.3   Finding the Pareto Front

Due to one-to-one correspondence from Proposition 1, finding the Pareto front is done by enumerating all P-minimal models. Here, the important thing is how we block[4] each P-minimal model.

Let us consider a MODOP $((o_1, \ldots, o_m), CSP)$. Let ψ be a CNF formula obtained from CSP. Suppose that we obtain the following P-minimal model which represent a Pareto front point $(d_1, \ldots, d_m)$, where $lb(o_i) \leq d_i \leq ub(o_i)$ for each i.

$$\{p_{o_1,lb(o_1)+1}, \ldots, p_{o_1,d_1}\} \cup \cdots \cup \{p_{o_m,lb(o_m)+1}, \ldots, p_{o_m,d_m}\}$$

---

[4] The terminology "block" is often used in SAT/MaxSAT communities to denote adding a constraint which prevents a solution to be found again (to "block" it) in an iterative process.

Table 1. Possible P-minimal models for a MODOP in Fig. 1

P-minimal models	$p_{o_1,2}$	$p_{o_1,3}$	$p_{o_1,4}$	$p_{o_1,5}$	$p_{o_2,2}$	$p_{o_2,3}$	$p_{o_2,4}$	$p_{o_2,5}$	$p_{o_2,6}$	Pareto front points
$\{p_{o_2,2}, p_{o_2,3}, p_{o_2,4}\}$	0	0	0	0	1	1	1	0	0	(1,4)
$\{p_{o_1,2}, p_{o_1,3}, p_{o_2,2}, p_{o_2,3}\}$	1	1	0	0	1	1	0	0	0	(3,3)
$\{p_{o_1,2}, p_{o_1,3}, p_{o_1,4}\}$	1	1	1	0	0	0	0	0	0	(4,1)

Region Ⓐ blocked by $\neg p_{o_2,4}$

Region Ⓑ blocked by $\neg p_{o_1,3} \vee \neg p_{o_2,3}$

Region Ⓒ blocked by $\neg p_{o_1,4}$

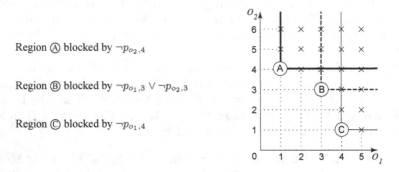

Fig. 2. Regions pruned by blocking clauses

Naively, this P-minimal model is blocked by the following clause.

$$\bigvee_{i=1}^{m} \bigvee_{j=lb(o_i)+1}^{d_i} \neg p_{o_i,j}$$

By taking advantage of axiom clauses (1) of the order encoding, we can reduce it to the following clause which contains at most one literal for each objective variable. Note that $\neg p_{o_i,lb(o_i)}$ is treated as false.

$$\bigvee_{i=1}^{m} \neg p_{o_i,d_i}$$

The beauty of the approach is that each P-minimal model (i.e., Pareto front point) is blocked by the above reduced clause.

For instance, the first P-minimal model $\{p_{o_2,2}, p_{o_2,3}, p_{o_2,4}\}$ representing the Pareto front point (1, 4) in Table 1 is naively blocked by $\neg p_{o_2,2} \vee \neg p_{o_2,3} \vee \neg p_{o_2,4}$. This blocking clause is reduced to $\neg p_{o_2,4}$ by resolution with axiom clauses $p_{o_2,2} \vee \neg p_{o_2,3}$ and $p_{o_2,3} \vee \neg p_{o_2,4}$. In a similar way, $\{p_{o_1,2}, p_{o_1,3}, p_{o_2,2}, p_{o_2,3}\}$ representing (3, 3) and $\{p_{o_1,2}, p_{o_1,3}, p_{o_1,4}\}$ representing (4, 1) are respectively blocked by the reduced clauses $\neg p_{o_1,3} \vee \neg p_{o_2,3}$ and $\neg p_{o_1,4}$. Figure 2 illustrates the regions of cost vectors pruned by blocking clauses. The upper right region of Ⓐ dominated by a Pareto front point (1, 4) is pruned by a blocking clause $\neg p_{o_2,4}$. In a similar way, the upper right regions of Ⓑ dominated by $(3,3)$ and Ⓒ dominated by $(4,1)$ are pruned by blocking clauses $\neg p_{o_1,3} \vee \neg p_{o_2,3}$ and $\neg p_{o_1,4}$ respectively.

Input: a MODOP $\Omega = ((o_1, \ldots, o_m), CSP)$
Output: the Pareto front of Ω
1: $\psi :=$ a CNF formula obtained from CSP
2: $P := \delta(o_1) \cup \cdots \cup \delta(o_m)$
3: $\mathcal{Y}_{\mathcal{N}} := \emptyset$
4: **while (findP-MinimalModel(ψ, P))**
5: $\quad \lfloor \quad \mathcal{Y}_{\mathcal{N}} := \mathcal{Y}_{\mathcal{N}} \cup \{M_P\}$ # add an obtained P-minimal model to $\mathcal{Y}_{\mathcal{N}}$
6: $\quad \lfloor \quad \psi := \psi \wedge \mathbf{block}(M_P)$
7: **return decode($\mathcal{Y}_{\mathcal{N}}$)**

Fig. 3. Basic algorithm for finding the Pareto front

A basic algorithm for finding the Pareto front is shown in Fig. 3. For a given MODOP $\Omega = ((o_1, \ldots, o_m), CSP)$, a CNF formula obtained from CSP is set to ψ in Line 1. Propositional variables obtained from multiobjective variables $(o_1, \ldots, o_m)$ by the order encoding is set to P in Line 2. Line 3 initializes $\mathcal{Y}_{\mathcal{N}}$ that stores P-minimal models. The method **findP-MinimalModel(ψ, P)** in Line 4 returns true if a P-minimal model of ψ is found, otherwise false. The loop in Line 4-6, every time a P-minimal model M_P is found, puts it to $\mathcal{Y}_{\mathcal{N}}$ and then adds the blocking clause **block(M_P)** to ψ. Finally, Line 7 returns the Pareto front of Ω when no more P-minimal model is found.

3.4 P-Minimal Model Versus P-Minimum Model

Obviously, *Single-Objective Discrete Optimization Problem* (SODOP) is a special case of MODOP where the number of objective variables is restricted to only one. Several SODOPs have been successfully solved so far by using the order encoding and incremental solving techniques [5,17,27,37,39].

In the order encoding, finding the optimal value of SODOP, i.e., CSP (X, D, C) with a single objective variable $o \in X$, can be done by finding the P-minimum model as well as the P-minimal model of CNF formula obtained from CSP with $P = \delta(o)$.

However, this is not the case for MODOPs having multiple objective variables. Indeed P-minimum model generation can provide some Pareto front points of MODOP, but it cannot always provide the Pareto front. For instance, a Pareto front point $(3,3)$ of Fig. 1 cannot be obtained by P-minimum model generation, since its cardinality (6) is greater than the ones of the other two Pareto front points $(1,4)$ and $(4,1)$.

4   Experiments

We implemented the proposed approach presented in the previous section. The resulting solver sucre reads a MODOP instance and encodes it into PB constraints, which are subsequently solved by a PB-based P-minimal solver that returns P-minimal models representing the Pareto front as well as its subset of the original MODOP instance. Note that PB constraints can be considered as a generalization of CNF formula.

For PB encoding, we choose a hybrid encoding integrating the order and log encodings [36]. This is because the recent results in [36] showed that the hybrid encoding scales to larger CSP instances than the order encoding. We implemented the hybrid encoding in Scala, where multiobjective variables are encoded by the order encoding. For finding P-minimal models, we implemented a P-minimal solver based on PB optimization. We use the clasp solver[5] as a back-end PB solver.

To evaluate the effectiveness of our approach, we carry out experiments on the *Multicriteria Set Covering Problem* (MSCP). MSCP has four parameters, the number of items, sets, objective functions, and the range of coefficients of objective functions.

#Items	#Sets	#Objectives	Range of Coefficients
$\{100, 150\}$	$\{20, 30, 50, 100, 200\}$	$\{3, 4, 5\}$	$\{[1, 10], [1, 100]\}$

We use the combination of parameter values shown above for our comprehensive evaluation[6]. This combination is an extension of one used in a related work [6] with #Sets = $\{50, 100, 200\}$ and Range of Coefficients = $[1, 10]$. For the six combinations from $(100, 20, \{3, 4, 5\}, [1, 100])$ and $(150, 30, \{3, 4, 5\}, [1, 100])$, we use 60 instances (10 instances per combination) which are available from the web[7]. For other combinations, we uniformly generate 10 instances at random for each combination in the similar way with [6].

We contrast our approach with a state-of-the-art MODOP solving technique based on Binary Decision Diagram (BDD) proposed by Bergman and Cire [6]. It first constructs a BDD representing all feasible solutions of MODOP and then computes possible candidates of Pareto front points on BDD, which are subsequently guaranteed to be Pareto optimal. The results in [6] shows that the BDD-based solver can outperform existing exact methods [19,31].

We ran the instances on a Mac OS equipped with Intel Core i7 3 GHz CPU and 16 GB RAM. We impose a time-limit of 3,600 s for each run. For the BDD-based solver, we compiled its C++ source code with clang-800.0.42.1 and linked it to ILOG CPLEX 12.7.1[8].

---

[5] https://potassco.org/.

[6] #Sets = 20 is used only for #Items = 100. #Sets = 30 is used only for #Items = 150.

[7] http://www.andrew.cmu.edu/user/vanhoeve/mdd/.

[8] We confirmed in our computational environment that the BDD solver was able to solve 55 instances out of the 60 MSCP instances, compared with 53 in [6].

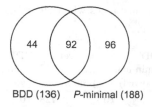

BDD (136)   *P*-minimal (188)

Fig. 4. Venn diagram for solved instances

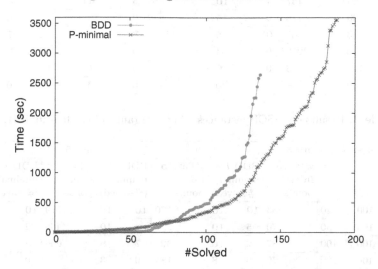

Fig. 5. Cactus plot of benchmark results on MSCP

First, we analyze the difference between the proposed approach and the BDD-based one. Overall, out of all 480 instances, our *P*-minimal-based solver `sucre` found the Pareto fronts of 188 instances, compared with 136 of the BDD-based solver. The inclusion relation of the solved instances obtained by the two solvers is shown in Figure 4. Out of 232 solved instances, 96 of them are obtained only by `sucre`, 44 obtained only by the BDD-based solver, and 92 commonly obtained by both solvers. For CPU time, a cactus plot of the results is shown in Fig. 5. In the following, we discuss more details of our experimental results.

Tables 2 and 3 contrast the results obtained from the `sucre` solver (denoted by *P*-minimal) and the BDD-based solver (denoted by BDD). In both tables, the number of items, sets, and objective functions are given at the top. For each combination, the average size of the Pareto front (viz. the number of Pareto front points) and the number of solved instances are shown. For example, `sucre` found the Pareto fronts of all 10 instances for a combination $(100, 50, 3, [1, 100])$, and the BDD-based solver found the ones of 4 instances. The symbol "—" means that both solvers failed to find the Pareto fronts of all 10 instances. We highlight the best result for each combination. The difference of the two tables is the range

Table 2. Results on MSCP instances when the range of coefficients is $[1, 100]$

#Items	#Sets	m = 3			m = 4			m = 5		
		#Pareto front points	BDD	P-minimal (#solved)	#Pareto front points	BDD	P-minimal (#solved)	#Pareto front points	BDD	P-minimal (#solved)
100	20	117	**10**	10	428	**10**	7	1171	**10**	4
100	50	186	4	**10**	515	2	**4**	2336	**3**	0
100	100	157	4	**10**	466	3	**4**	1783	**3**	1
100	200	178	0	**5**	—	0	0	—	0	0
150	30	305	**10**	5	1178	**8**	0	5182	**7**	0
150	50	252	0	**3**	—	0	0	7124	**2**	0
150	100	213	0	**6**	344	0	**1**	—	0	0
150	200	—	0	0	—	0	0	—	0	0

Table 3. Results on MSCP instances when the range of coefficients is $[1, 10]$

#Items	#Sets	m = 3			m = 4			m = 5		
		#Pareto front points	BDD P-minimal (#solved)		#Pareto front points	BDD P-minimal (#solved)		#Pareto front points	BDD P-minimal (#solved)	
100	20	53	**10**	10	170	**10**	10	519	**10**	8
100	50	70	5	**10**	227	3	**10**	466	2	**4**
100	100	87	2	**10**	323	7	**9**	696	2	**4**
100	200	79	0	**8**	184	0	**3**	—	0	0
150	30	114	4	**8**	334	**4**	2	369	1	1
150	50	121	0	**7**	230	0	**2**	—	0	0
150	100	101	0	**7**	342	0	**1**	—	0	0
150	200	119	0	**4**	—	0	0	—	0	0

of coefficients of objective functions which affects the size of the Pareto front: Table 2 for $[1, 100]$ and Table 3 for $[1, 10]$.

In Table 2, out of 24 combinations, both solvers gave the best results of 9 combinations and failed to find the Pareto fronts of all instances for 7 combinations marked by "—". The sucre solver performs better for combinations with #Set = $\{50, 100, 200\}$ and $m = \{3, 4\}$ than the BDD-based solver. We can observe that sucre scales to #Set = 200 and solved 5 instances for a combination (100, 200, 3, $[1, 100]$), while the BDD-based solver gave no solution. We note that the BDD-based solver failed to construct BDDs for 9 instances and did not complete the computation for 1 instance in the time-limit.

On the other hand, sucre does not match the BDD-based solver in performance for combinations with #Set = $\{20, 30\}$ or $m = 5$. We can observe that the BDD-based solver solved 7 instances for a combination (150, 30, 5, $[1, 100]$), while sucre gave no solution. One of the reasons for this limitation is that sucre cannot complete the enumeration of Pareto optimal points in the time-limit, since the number of objective functions is higher, the size of Pareto front becomes large.

Table 3 shows the results on MSCP instances when the range of coefficients of objective functions is $[1, 10]$. The sucre solver outperforms the BDD-based solver. Compared with $[1, 100]$, sucre solved more instances, but the BDD-based solver solved less instances. Both solvers failed to find the Pareto fronts of all instances for 5 combinations marked by "—".

Table 4. Size of partial Pareto fronts of hard MSCP instances

#Items	#Sets	#Objectives	Range of coefficients	#Pareto front points		
				Average	Minimum	Maximum
100	200	4	$[1, 100]$	216	119	446
100	200	5	$[1, 100]$	204	91	431
150	50	4	$[1, 100]$	178	76	359
150	100	5	$[1, 100]$	157	46	375
150	200	3	$[1, 100]$	169	43	335
150	200	4	$[1, 100]$	97	8	209
150	200	5	$[1, 100]$	77	16	215
100	200	5	$[1, 10]$	115	35	210
150	50	5	$[1, 10]$	221	45	469
150	100	5	$[1, 10]$	157	17	321
150	200	4	$[1, 10]$	102	13	261
150	200	5	$[1, 10]$	112	20	241

It is often impractical to find the Pareto front for hard MODOP instances because of a large number of Pareto front points. Actually, as can be seen in Tables 2 and 3, both solvers failed to find the Pareto fronts of all instances for 12 combinations (120 instances in total). From a practical point of view, finding a subset of Pareto front is extremely useful for such instances. For this, the proposed approach has an obvious advantage of a one-to-one correspondence between a Pareto front point and a P-minimal model.

Table 4 shows the size of partial Pareto fronts (a set of Pareto front points) obtained from those hard instances by the sucre solver. In Table 4, the number of items, sets, objective functions, and the range of coefficients for objective functions are given at first. Then for each combination, the average, minimum, and maximum number of Pareto front points for 10 instances are shown. The sucre solver succeeded in computing around 150 Pareto front points in average for the hard instances, which were unsolvable by the BDD-based solver.

5   Related Work

At first, to the best of our knowledge, so far, no approach has been proposed to solving MODOPs based on a one-to-one correspondence between a Pareto front point of MODOP and a P-minimal model of CNF formula obtained from MODOP via the order encoding.

A study by Koshimura et al. [21] is the most relevant related work. They solved single-objective job-shop scheduling problems by using P-minimal model generation. As we discussed in Sect. 3.4, finding optimal values of single-objective optimization problems can be done by finding the P-minimum model as well as the P-minimal model. The study is therefore a special case of the proposed approach presented in this paper.

There have been several proposals for solving MODOPs such as approximation methods and exact methods. The majority of the approximation methods is based on metaheuristics-based algorithms [12,13,45,46]. Those methods can deal with large instances but cannot guarantee the Pareto optimality of obtained solutions. To obtain qualified solutions, some quality-guaranteed approximation methods have been proposed such as [25,33]. Among them, the best-first AND/OR search algorithm [25] computes a relaxed Pareto front using ϵ-dominance [32].

On the other hands, several exact methods have been proposed [6,8,19,24, 31,33,34]. Among them, the methods recently proposed in [19,31] are extensions of ϵ-constraint method [22], which iteratively searches on sub-problems having subsets of objective functions. The results in [6] showed that a BDD-based solver, empirically contrasted to the proposed approach in this paper, can outperform those two exact methods [19,31]. Very recently, Boland et al. have proposed an approach to MODOP solving based on Integer Programming (IP) [8]. In our preliminary experiments on MSCP, the IP-based solver was able to find Pareto fronts for more instances compared with sucre and a BDD-based solver [6]. In future work, we will investigate the pros and cons of these solvers on a wide range of MODOP instances.

From the viewpoint of propositional logic based methods, a SAT-based one [23] solves multiobjective PB problems. This method is based on a standard iterative procedure for finding a feasible solution and for pruning its dominated region while maintaining candidates of Pareto front points. Moreover, as with the BDD-based method above, it cannot guarantee Pareto optimality of candidates until a whole procedure terminates.

From the viewpoint of declarative programming paradigm, there have been several studies on *MultiObjective Constraint Optimization Problems* (MO-COP) [24,25,30,41,43] in Constraint Programming. For Answer Set Programming, asprin [9] has been implemented for *preference* handling, which is closely related to multiobjective optimization.

6   Conclusion

We presented an approach to solve MODOPs based on propositional P-minimal model generation. We proved that there exists a one-to-one correspondence between a Pareto front point of MODOP and a P-minimal model of CNF formula obtained from MODOP. This correspondence is achieved by adopting the order encoding as CNF encoding for multiobjective functions. We also showed that finding the Pareto front is done by enumerating all P-minimal models. In particular, each Pareto front point can be elegantly blocked by a single clause contains at most one literal for each objective function due to the nice property of the order encoding.

The resulting solver sucre relies on the correspondence we proved and delegates the solving task to a general-purpose P-minimal solver. Our empirical analysis used Multicriteria Set Covering Problems (MSCPs) and confirmed that sucre performs well on MSCPs which involves the large number of sets as well as the small range of coefficients of objective functions. We contrasted the performance of sucre with the results obtained by an efficient exact MODOP solver based on BDD. The sucre solver demonstrated that the proposed approach allows to compete with state-of-the-art MODOP solving techniques. The source code of sucre and the benchmark instances will be available from the web[9].

Modern real-world applications in Artificial Intelligence involve multiple criteria that should be considered separately and optimized simultaneously. It is therefore often impractical to find the whole Pareto front. From this point of view, the proposed approach has a great advantage of finding a subset of Pareto front. Our empirical analysis confirmed that sucre is extremely useful to find such partial Pareto front on hard MSCP instances.

Our future work includes experiments with real-world applications as well as classical problems in diverse areas. Our P-minimal-based approach can be applied to a wide range of propositional logic based approaches, such as SAT, Max-SAT, ASP, PB, SMT, etc. Moreover, it can be further extended in selecting a representative subset of Pareto front (e.g., [35]) by introducing variable selection heuristics for blocking clauses in back-end P-minimal solvers. In future, we will investigate such possibilities, and the results will be applied to solving more practical MODOPs.

References

1. Alarcon-Rodriguez, A., Ault, G., Galloway, S.: Multi-objective planning of distributed energy resources: a review of the state-of-the-art. Renew. Sustain. Energy Rev. **14**(5), 1353–1366 (2010)
2. Ansótegui, C., Manyà, F.: Mapping problems with finite-domain variables to problems with boolean variables. In: Hoos, H.H., Mitchell, D.G. (eds.) SAT 2004. LNCS, vol. 3542, pp. 1–15. Springer, Heidelberg (2005). doi:10.1007/11527695_1

---

[9] http://kix.istc.kobe-u.ac.jp/~soh/sucre/.

3. Bailleux, O., Boufkhad, Y.: Efficient CNF encoding of boolean cardinality constraints. In: Rossi, F. (ed.) CP 2003. LNCS, vol. 2833, pp. 108–122. Springer, Heidelberg (2003). doi:10.1007/978-3-540-45193-8_8

4. Ballestero, E., Bravo, M., Pérez-Gladish, B., Parra, M.A., Plà-Santamaria, D.: Socially responsible investment: a multicriteria approach to portfolio selection combining ethical and financial objectives. Eur. J. Oper. Res. **216**(2), 487–494 (2012)

5. Banbara, M., Matsunaka, H., Tamura, N., Inoue, K.: Generating combinatorial test cases by efficient SAT encodings suitable for CDCL SAT solvers. In: Fermüller, C.G., Voronkov, A. (eds.) LPAR 2010. LNCS, vol. 6397, pp. 112–126. Springer, Heidelberg (2010). doi:10.1007/978-3-642-16242-8_9

6. Bergman, D., Cire, A.A.: Multiobjective optimization by decision diagrams. In: Rueher, M. (ed.) CP 2016. LNCS, vol. 9892, pp. 86–95. Springer, Cham (2016). doi:10.1007/978-3-319-44953-1_6

7. Biere, A., Heule, M., van Maaren, H., Walsh, T. (eds.): Handbook of Satisfiability, Frontiers in Artificial Intelligence and Applications (FAIA), vol. 185. IOS Press, Amsterdam (2009)

8. Boland, N., Charkhgard, H., Savelsbergh, M.W.P.: A new method for optimizing a linear function over the efficient set of a multiobjective integer program. Eur. J. Oper. Res. **260**(3), 904–919 (2017)

9. Brewka, G., Delgrande, J.P., Romero, J., Schaub, T.: asprin: customizing answer set preferences without a headache. In: Proceedings of the 29th National Conference on Artificial Intelligence (AAAI 2015), pp. 1467–1474 (2015)

10. Burke, E.K., Li, J., Qu, R.: A pareto-based search methodology for multi-objective nurse scheduling. Ann. Oper. Res. **196**(1), 91–109 (2012)

11. Crawford, J.M., Baker, A.B.: Experimental results on the application of satisfiability algorithms to scheduling problems. In: Proceedings of the 12th National Conference on Artificial Intelligence (AAAI 1994), pp. 1092–1097 (1994)

12. Deb, K., Agrawal, S., Pratap, A., Meyarivan, T.: A fast elitist non-dominated sorting genetic algorithm for multi-objective optimization: NSGA-II. In: Schoenauer, M., Deb, K., Rudolph, G., Yao, X., Lutton, E., Merelo, J.J., Schwefel, H.-P. (eds.) PPSN 2000. LNCS, vol. 1917, pp. 849–858. Springer, Heidelberg (2000). doi:10.1007/3-540-45356-3_83

13. Deb, K., Agrawal, S., Pratap, A., Meyarivan, T.: A fast and elitist multiobjective genetic algorithm: NSGA-II. IEEE Trans. Evol. Comput. **6**(2), 182–197 (2002)

14. Ehrgott, M.: Multicriteria Optimization. Springer, Heidelberg (2005). doi:10.1007/3-540-27659-9

15. Figueira, J., Greco, S., Ehrgott, M.: Multiple Criteria Decision Analysis: State of the Art Surveys. International Series in Operations Research & Management Science. Springer, Heidelberg (2005). doi:10.1007/b100605

16. Gent, I.P., Nightingale, P.: A new encoding of alldifferent into SAT. In: Proceedings of the 3rd International Workshop on Modelling and Reformulating Constraint Satisfaction Problems (2004)

17. Inoue, K., Soh, T., Ueda, S., Sasaura, Y., Banbara, M., Tamura, N.: A competitive and cooperative approach to propositional satisfiability. Discrete Appl. Math. **154**(16), 2291–2306 (2006)

18. Iturriaga, S., Dorronsoro, B., Nesmachnow, S.: Multiobjective evolutionary algorithms for energy and service level scheduling in a federation of distributed datacenters. Int. Trans. Oper. Res. **24**(1–2), 199–228 (2017)

19. Kirlik, G., Sayin, S.: A new algorithm for generating all nondominated solutions of multiobjective discrete optimization problems. Eur. J. Oper. Res. **232**(3), 479–488 (2014)

20. Knuth, D.E.: The Art of Computer Programming, Volume 4, Fascicle 6: Satisfiability. Addison-Wesley Professional, Boston (2015)
21. Koshimura, M., Nabeshima, H., Fujita, H., Hasegawa, R.: Minimal model generation with respect to an atom set. In: Proceedings of the the 7th International Workshop on First-Order Theorem Proving (FTP 2009), pp. 49–59 (2009)
22. Laumanns, M., Thiele, L., Zitzler, E.: An efficient, adaptive parameter variation scheme for metaheuristics based on the epsilon-constraint method. Eur. J. Oper. Res. **169**(3), 932–942 (2006)
23. Lukasiewycz, M., Glaß, M., Haubelt, C., Teich, J.: Solving multi-objective pseudo-boolean problems. In: Marques-Silva, J., Sakallah, K.A. (eds.) SAT 2007. LNCS, vol. 4501, pp. 56–69. Springer, Heidelberg (2007). doi:10.1007/978-3-540-72788-0_9
24. Marinescu, R.: Exploiting problem decomposition in multi-objective constraint optimization. In: Gent, I.P. (ed.) CP 2009. LNCS, vol. 5732, pp. 592–607. Springer, Heidelberg (2009). doi:10.1007/978-3-642-04244-7_47
25. Marinescu, R.: Best-first vs. depth-first AND/OR search for multi-objective constraint optimization. In: Proceedings of the 22nd IEEE International Conference on Tools with Artificial Intelligence (ICTAI 2010), pp. 439–446 (2010)
26. Metodi, A., Codish, M., Lagoon, V., Stuckey, P.J.: Boolean equi-propagation for optimized SAT encoding. In: Lee, J. (ed.) CP 2011. LNCS, vol. 6876, pp. 621–636. Springer, Heidelberg (2011). doi:10.1007/978-3-642-23786-7_47
27. Nabeshima, H., Soh, T., Inoue, K., Iwanuma, K.: Lemma reusing for SAT based planning and scheduling. In: Proceedings of the International Conference on Automated Planning and Scheduling 2006 (ICAPS 2006), pp. 103–112 (2006)
28. Niemelä, I.: A tableau calculus for minimal model reasoning. In: Miglioli, P., Moscato, U., Mundici, D., Ornaghi, M. (eds.) TABLEAUX 1996. LNCS, vol. 1071, pp. 278–294. Springer, Heidelberg (1996). doi:10.1007/3-540-61208-4_18
29. Ohrimenko, O., Stuckey, P.J., Codish, M.: Propagation via lazy clause generation. Constraints **14**(3), 357–391 (2009)
30. Okimoto, T., Joe, Y., Iwasaki, A., Matsui, T., Hirayama, K., Yokoo, M.: Interactive algorithm for multi-objective constraint optimization. In: Milano, M. (ed.) CP 2012. LNCS, pp. 561–576. Springer, Heidelberg (2012). doi:10.1007/978-3-642-33558-7_41
31. Ozlen, M., Burton, B.A., MacRae, C.A.G.: Multi-objective integer programming: an improved recursive algorithm. J. Optim. Theory Appl. **160**(2), 470–482 (2014)
32. Papadimitriou, C.H., Yannakakis, M.: On the approximability of trade-offs and optimal access of web sources. In: Proceedings of the 41st Annual Symposium on Foundations of Computer Science (FOCS 2000), pp. 86–92 (2000)
33. Rollon, E., Larrosa, J.: Bucket elimination for multiobjective optimization problems. J. Heuristics **12**(4–5), 307–328 (2006)
34. Rollon, E., Larrosa, J.: Multi-objective russian doll search. In: Proceedings of the 22nd National Conference on Artificial Intelligence (AAAI 2007), pp. 249–254 (2007)
35. Schwind, N., Okimoto, T., Konieczny, S., Wack, M., Inoue, K.: Utilitarian and egalitarian solutions for multi-objective constraint optimization. In: Proceedings of the 26th IEEE International Conference on Tools with Artificial Intelligence (ICTAI 2014), pp. 170–177. IEEE Computer Society (2014)
36. Soh, T., Banbara, M., Tamura, N.: Proposal and evaluation of hybrid encoding of CSP to SAT integrating order and log encodings. Int. J. Artif. Intell. Tools **26**(1), 1–29 (2017)

37. Soh, T., Inoue, K., Tamura, N., Banbara, M., Nabeshima, H.: A SAT-based method for solving the two-dimensional strip packing problem. Fundam. Inf. **102**(3–4), 467–487 (2010)
38. Tamura, N., Banbara, M., Soh, T.: PBSugar: Compiling pseudo-boolean constraints to SAT with order encoding. In: Proceedings of the 25th IEEE International Conference on Tools with Artificial Intelligence (ICTAI 2013), pp. 1020–1027. IEEE, November 2013
39. Tamura, N., Taga, A., Kitagawa, S., Banbara, M.: Compiling finite linear CSP into SAT. Constraints **14**(2), 254–272 (2009)
40. Tanjo, T., Tamura, N., Banbara, M.: Sugar++: a SAT-based Max-CSP/COP solver. In: Proceedings of the 3rd International CSP Solver Competition, pp. 77–82 (2008)
41. Ugarte, W., Boizumault, P., Crémilleux, B., Lepailleur, A., Loudni, S., Plantevit, M., Raïssi, C., Soulet, A.: Skypattern mining: from pattern condensed representations to dynamic constraint satisfaction problems. Artif. Intell. **244**, 48–69 (2017). https://doi.org/10.1016/j.artint.2015.04.003
42. Wang, L., Ng, A.H.C., Deb, K. (eds.): Multi-objective Evolutionary Optimisation for Product Design and Manufacturing. Springer, Heidelberg (2011). doi:10.1007/978-0-85729-652-8
43. Wilson, N., Razak, A., Marinescu, R.: Computing possibly optimal solutions for multi-objective constraint optimisation with tradeoffs. In: Proceedings of the 24th International Joint Conference on Artificial Intelligence (IJCAI 2015), pp. 815–822 (2015)
44. Yi, D., Goodrich, M.A., Seppi, K.D.: MORRF*: sampling-based multi-objective motion planning. In: Proceedings of the 24th International Joint Conference on Artificial Intelligence (IJCAI 2015), pp. 1733–1741 (2015)
45. Zitzler, E., Laumanns, M., Thiele, L.: SPEA2: improving the strength pareto evolutionary algorithm for multiobjective optimization. In: Proceedings of the 4th International Conference of Evolutionary Methods for Design, Optimisation and Control with Application to Industrial Problems (EUROGEN 2001), pp. 95–100 (2002)
46. Zitzler, E., Künzli, S.: Indicator-based selection in multiobjective search. In: Proceedings of the 8th International Conference on Parallel Problem Solving from Nature (PPSN VIII), pp. 832–842 (2004)

Analyzing Lattice Point Feasibility in UTVPI Constraints

K. Subramani and Piotr Wojciechowski$^{(\boxtimes)}$

LDCSEE, West Virginia University, Morgantown, WV, USA
k.subramani@mail.wvu.edu, pwojciec@mix.wvu.edu

Abstract. This paper is concerned with the design and analysis of a time-optimal and space-optimal, *certifying* algorithm for checking the lattice point feasibility of a class of constraints called Unit Two Variable Per Inequality (UTVPI) constraints. A UTVPI constraint has at most two non-zero variables and the coefficients of the non-zero variables belong to the set $\{+1, -1\}$. These constraints occur in a number of application domains, including but not limited to program verification, abstract interpretation, and operations research. As per the literature, the fastest known model-generating algorithm for checking lattice point feasibility in UTVPI constraint systems runs in $O(m \cdot n + n^2 \cdot \log n)$ time and $O(n^2)$ space, where m represents the number of constraints and n represents the number of variables in the constraint system. In this paper, we design and analyze a new algorithm for checking the lattice point feasibility of UTVPI constraints. The presented algorithm runs in $O(m \cdot n)$ time and $O(m + n)$ space. Additionally, it is certifying in that it produces a satisfying assignment in the event that it is presented with feasible instances and a refutation in the event that it is presented with infeasible instance. Our approach for the lattice point feasibility problem in UTVPI constraint systems is fundamentally different from existing approaches for this problem.

1 Introduction

In this paper, we propose a new certifying algorithm for checking the lattice point (integer) feasibility of a conjunction of Unit Two Variable Per Inequality (UTVPI) constraints. A UTVPI constraint is a linear constraint of the form: $a_i \cdot x_i + a_j \cdot x_j \leq c_{ij}$, where $a_i, a_j \in \{-1, 0, 1\}$ and c_{ij} is an integer constant. A conjunction of such constraints is called a UTVPI constraint system. Observe

---

K. Subramani—This work was supported by the Air Force Research Laboratory under US Air Force contract FA8750-16-3-6003. The views expressed are those of the authors and do not reflect the official policy or position of the Department of Defense or the U.S. Government.

P. Wojciechowski—This research was supported in part by the National Science Foundation through Award CCF-1305054.

© Springer International Publishing AG 2017
J.C. Beck (Ed.): CP 2017, LNCS 10416, pp. 615–629, 2017.
DOI: 10.1007/978-3-319-66158-2_39

that UTVPI systems subsume difference constraint systems [NO05], since in the latter, a_i and a_j must have opposite signs.

This paper deals with the integer feasibility problem in UTVPI systems. UTVPI constraint systems find applications in a number of problem domains, including but not limited to real-time scheduling, program verification and operations research (see Sect. 3).

The algorithm that we present runs in $O(m \cdot n)$ time and uses $O(m+n)$ space on a UTVPI constraint system with n variables over m constraints. This is a marked improvement over the current state-of-the-art model-generating algorithm which runs in $O(m \cdot n + n^2 \cdot \log n)$ time and $O(n^2)$ space [LM05]. We note that the fastest known strongly polynomial time algorithm for checking linear feasibility in difference constraints is the Bellman-Ford procedure (or one of its variants), which runs in $O(m \cdot n)$ time and $O(m + n)$ space. It is important to note that unlike difference constraints linear feasibility does not imply lattice point feasibility in UTVPI constraints (see Sect. 2).

We reiterate the fact that our algorithm is certifying, i.e., in the event that the given UTVPI system is feasible, we provide a satisfying assignment and in the event that it is infeasible, we provide a refutation, which explains the infeasibility.

The important contributions of this paper are:

1. a certifying algorithm (**IA**) for checking integer feasibility in UTVPI constraint systems, and
2. establishing that determining integer feasibility of an UTVPI constraint system can be determined by examining a $\frac{1}{2}$-neighborhood of a linear solution.

The rest of this paper is organized as follows: Sect. 2 formally specifies the problem under consideration in this paper. Section 3 enumerates the domains in which UTVPI constraints occur and also motivates the need for certifying algorithms. Section 4 describes the related work in the literature. Section 5 details the new algorithm for the lattice point feasibility problem in UTVPI constraint systems. A detailed proof of correctness of this algorithm is provided in Sect. 6.

2   Statement of Problem

In this section, we formally define the integer feasibility problem in UTVPI constraints and also define the various terms that will be used in the rest of the paper.

Definition 1. *A constraint of the form $a_i \cdot x_i + a_j \cdot x_j \le c_{ij}$ is said to be a Unit Two Variable Per Inequality (UTVPI) constraint if $a_i, a_j \in \{-1, 0, +1\}$ and $c_{ij} \in \mathbb{Z}$.*

Definition 2. *A constraint of the form $x_i \le c_i$ or $-x_i \le c_i$, where $c_i \in \mathbb{Z}$, is called an absolute constraint.*

Observe that an absolute constraint is a UTVPI constraint, in which one of the coefficients (a_i or a_j) is 0. Such a constraint can be converted into constraints of the form: $a_i \cdot x_i + a_j \cdot x_j \leq c_{ij}$, where both a_i and a_j are non-zero [SW17].

Definition 3. *A conjunction of UTVPI constraints is called a UTVPI constraint system and can be represented in matrix form as* $\mathbf{A} \cdot \mathbf{x} \leq \mathbf{c}$*. If the constraint system has m constraints over n variables, then $\mathbf{A}$ has dimensions $m \times n$.*

A UTVPI system defines a polyhedron in n-dimensional space. Given such a system, we are interested in the following question: Does the defined polyhedron enclose a lattice point? This problem is called the *Integer Feasibility problem* (IF).

Several authors ([LM05, Rev09]) have used the following two inference rules in analyzing UTVPI constraints:

1.

$$\frac{a_i \cdot x_i + a_j \cdot x_j \leq c_{ij} \qquad\qquad -a_j \cdot x_j + a_k \cdot x_k \leq c_{jk}}{a_i \cdot x_i + a_k \cdot x_k \leq c_{ij} + c_{jk}}$$

This rule is called the transitive inference rule and it is solution preserving. The constraints generated by the transitive inference rule correspond to the constraints generated by Fourier-Motzkin elimination.

2.

$$\frac{a_i \cdot x_i + a_j \cdot x_j \leq c_{ij} \qquad\qquad a_i \cdot x_i - a_j \cdot x_j \leq c'_{ij}}{a_i \cdot x_i \leq \lfloor \frac{c_{ij} + c'_{ij}}{2} \rfloor}$$

This rule is called the tightening rule and it is lattice-point preserving. The constraints generated by the transitive and tightening inference rules correspond to the constraints generated by Fourier-Motzkin with rounding.

Our goal is to design a certifying algorithm for the IF problem. In other words, our algorithm should produce models (satisfying solutions) for feasible instances and refutations for infeasible instances. Our algorithm incorporates the following properties of UTVPI constraints:

(i) Fourier-Motzkin with rounding (FMR) is a sound and complete procedure for detecting integer feasibility in UTVPI constraints (see [Sub04]).

(ii) For a UTVPI constraint system which is integer infeasible, there exists a certificate of integer infeasibility consisting of at most $2 \cdot n$ constraints.

(iii) Given a solution to the LF problem in a UTVPI system, we can obtain a lattice point solution (or establish that none exists) in $O(m \cdot n)$ time, by a specialized rounding procedure (see Sects. 5 and 6).

3 Motivation

In this section, we briefly motivate the study of UTVPI constraints and discuss the importance of certifying algorithms.

UTVPI constraints occur in a number of problem domains including but not limited to program verification [LM05], abstract interpretation [Min06, CC77], real-time scheduling [GPS95] and operations research. Indeed many software and hardware verification queries are naturally expressed using this fragment of integer linear arithmetic. For instance, when the goal is to model indices of an array or queues in hardware or software, rational solutions are meaningless [LM05]. Other application areas include spatial databases [SS00] and theorem proving. UTVPI constraints are also known as Generalized 2SAT constraints [Sub04] and are the invariants of the octagon abstract domain in [Min06].

The field of certifying algorithms is concerned with validating the results of implementations of algorithms. A method for generating both feasibility and infeasibility certificates for general linear programs is covered in [FMP13].

4 Related Work

In this section, we briefly review some of the important milestones in the design of algorithms for checking integer feasibility in UTVPI constraint systems.

The first known decision procedure for checking the integer feasibility of a system of UTVPI constraints is detailed in [JMSY94]. This algorithm runs in $O(m \cdot n^2)$ time and uses $O(n^2)$ space. Furthermore, it is not certifying. [HS97] improves on the approach in [JMSY94] from an ease-of-implementation standpoint, by combining the transitive and tightening closures into a single step. However, the additional wrinkle does not improve the asymptotic complexity of the algorithm in [JMSY94]; nor does it provide certificates.

A rather different approach was used in [Sub04] to decide integer feasibility in UTVPI systems, while also producing a model. This algorithm uses Fourier-Motzkin elimination [DE73] to project the polyhedral representation of a system of UTVPI constraints to a single variable in a solution-preserving manner, thereby determining bounds for that variable. The algorithm then works in reverse order to assign values to the rest of the variables. The algorithm takes $O(n^3)$ time and $O(n^2)$ space.

Recently, there has been some work on *incremental* satisfiability of UTVPI constraints. For instance, [SS10] describes an algorithm for incremental (integer) satisfiability checking in UTVPI constraints. Their algorithm adds a single constraint to a set of UTVPI constraints in $O(m + n \cdot \log n)$ time. Incremental algorithms are extremely important from the perspective of SAT Modulo Theories [NOT04].

5   Checking for Integer Feasibility

In this section, we introduce our algorithm for integer feasibility for systems of UTVPI constraints. This algorithm extends the linear feasibility algorithm in [SW17].

We use PRODUCE-INTEGER-SOLUTION() to determine if a system of UTVPI constraints, $\mathbf{U}$, encloses a lattice point. The principal idea underlying our approach is the following: We start with a half-integral solution $\mathbf{a}$, then perform a rounding procedure which finds a lattice point within a $\frac{1}{2}$-neighborhood of $\mathbf{a}$. (A $\frac{1}{2}$-neighborhood of a point $\mathbf{a}$, is the set of all points $\mathbf{b}$, such that $a_i - \frac{1}{2} \leq b_i \leq a_i + \frac{1}{2}, \forall i = 1, 2, \ldots, n$.) Furthermore, if no such lattice point exists, then $\mathbf{U}$ does not enclose a lattice point.

Let $\mathbf{x} = \mathbf{a}$ denote a half-integral solution to the UTVPI system $\mathbf{U} : \mathbf{A} \cdot \mathbf{x} \leq \mathbf{c}$. We can obtain $\mathbf{a}$ in $O(m \cdot n)$ time [SW17]. There are three types of roundings used by PRODUCE-INTEGER-SOLUTION(), viz.,

1. *Forced* roundings : These are roundings in which one of the possible roundings of a variable, x_i, causes an immediate contradiction. These are the roundings performed by FORCED-ROUNDING().
2. *Optional* roundings : These are roundings in which a variable x_i can be set to either $\lceil a_i \rceil$ or $\lfloor a_i \rfloor$, without causing an immediate contradiction. These are the roundings performed by OPTIONAL-ROUNDINGS().
3. *Resultant* roundings : These are roundings which are necessitated by a forced rounding or an optional rounding. Note that a resultant rounding could cause subsequent roundings; these roundings are also called resultant roundings. These are the roundings performed by CHECK-DEPENDENCIES().

A rounding causes an immediate contradiction, if rounded value violates a constraint derivable from $\mathbf{U}$ by a single application of the tightening rule.

Example 1. Let $\mathbf{U}$ denote the following system of UTVPI constraints: $l_1 : x_1 + x_2 \leq 0$ and $l_2 : x_1 - x_2 \leq 1$. Assume that $x_1 = a_1 = \frac{1}{2}$. Rounding x_1 up sets $x_1 = 1$. Note that the constraint $l_3 : x_1 \leq 0$ is obtained from l_1 and l_2 by the tightening rule. Clearly, the new value for x_1 violates l_3. This means that x_1 is *forced* to be rounded down.

After rounding a variable x_i, we check to see if any of the variables sharing a constraint with x_i needs to be rounded, in order to satisfy all the constraints involved.

If rounding x_i in one direction eventually causes a contradiction, then x_i is rounded in the other direction. If that rounding also results in a contradiction, then the system is declared infeasible.

After a variable has been successfully rounded, and all the resultant roundings have been performed, no future roundings will violate any constraint containing any of these variables (see Lemma 4). Thus, x_i will not be rounded again. This is true on account of the structure of UTVPI constraint systems. Observe that a general integer program does not have such a structure.

5.1   Algorithms

The task of finding an integer solution is handled by the following sub procedures.

1. PRODUCE-INTEGER-SOLUTION() - This algorithm either returns a feasible integer solution or a certificate of integer infeasibility.
2. FORCED-ROUNDING() - This algorithm rounds a variable x_i, only if the rounding is a direct consequence of the tightening inference rule. The roundings performed by this algorithm are forced roundings.
3. OPTIONAL-ROUNDINGS() - This algorithm attempts to round each variable that has been left unrounded by FORCED-ROUNDING() and CHECK-DEPENDENCIES(). The roundings performed by this algorithm are optional roundings.
4. CHECK-DEPENDENCIES() - This algorithm rounds all variables that need to be rounded as a consequence of performing a forced or optional rounding on a variable. The roundings performed by this algorithm are resultant roundings. As indicated previously, a resultant rounding could result in additional (resultant) roundings.

The algorithm PRODUCE-INTEGER-SOLUTION() finds an integral solution to the input system of UTVPI constraints $\mathbf{U}$, or demonstrates that none exists. It starts with a half-integral solution $\mathbf{a}$, and proceeds to round the variables until a solution is found, or a contradiction is established.

The algorithm creates an array Z to store the integer solution being constructed. In the algorithm, the variable M simply represents an arbitrary value that is much larger than $||\mathbf{a}||_\infty$. Note that it is necessary to check whether or not a variable is rounded and Z aids in this determination. The queue Q is used to store the variables that need to be rounded due to resultant roundings. The algorithm also creates a tree structure T, which will be used to return the constraints that demonstrate the integer infeasibility of the system. The root of T contains the new variable x_0. Each subsequent node of T consists of a variable x_i of the original system that has been rounded, and the set constraints that were used to round x_i. This set contains either one constraint or three constraints. It contains three constraints, if x_i was rounded because of a forced rounding, and one constraint, if it was rounded because of an optional rounding or a resultant rounding.

The parent of a node x_i, represents the rounding that necessitated the rounding of x_i. The children of the node represent all of the resultant roundings which stem from rounding x_i. Since each variable is rounded at most once, each node will occur at most once in the tree.

PRODUCE-INTEGER-SOLUTION() does not alter the integer values of the linear solution $\mathbf{a}$, since they will also be part of the rounded solution. On the fractional values of $\mathbf{a}$, it calls FORCED-ROUNDING(), to perform forced roundings, as needed.

The algorithm FORCED-ROUNDING() checks to see if a variable takes part in a forced rounding. If the variable is forced to be rounded, then that rounding

---

Function PRODUCE-INTEGER-SOLUTION (system of UTVPI constraints **U**, and linear solution **a** of **U**)

1: {This is the main function that calls all the other functions. It returns either a feasible integral solution or a proof that none exist.}
2: **for** (each variable x_i) **do**
3:    $Z_i = M$.
4: **end for**
5: Create tree T of constraints with node x_0 at the root.
6: Create empty queue Q of variables.
7: **for** (each variable x_i) **do**
8:    **if** (a_i is an integer) **then**
9:       $Z_i = a_i$. {x_i already has an integer value and so does not need to be rounded.}
10:    **else**
11:       FORCED-ROUNDING(x_i, Z, **a**, T, **U**, Q).
12:    **end if**
13: **end for**
14: CHECK-DEPENDENCIES (Z, **a**, T, **U**, Q). {Perform all roundings that are a consequence of previous roundings.}
15: **for** (every constraint in **U**) **do**
16:    **if** (constraint is violated by current assignments to some Z_i and Z_j) **then**
17:       **return** (violated constraint and constraints obtained by backtracking in T from x_i to x_0 and x_j to x_0).
18:    **end if**
19: **end for**
20: $S \leftarrow$ Subset of **U** restricted to constraints consisting of only variables with $Z_i = M$.
21: $O \leftarrow$ OPTIONAL-ROUNDINGS(S, Z, T, **a**, Q).
22: {Try to round all remaining unrounded variables. This function either returns a satisfying lattice point or a proof that no satisfying lattice point exists.}
23: **return** O.

Algorithm 1: PRODUCE-INTEGER-SOLUTION

is performed and the appropriate constraints are added to the tree T. CHECK-DEPENDENCIES() then checks to see if other variables need to be rounded as a consequence of variables being rounded by FORCED-ROUNDING().

Once the forced and resultant roundings have been performed, PRODUCE-INTEGER-SOLUTION() checks to see if any constraint is violated. If a constraint involving the variables x_i and x_j is violated, then that constraint and all the constraints that caused x_i and x_j to be rounded to the current values, are returned as proof of integer infeasibility. To determine which constraints caused variable x_i to be rounded, the algorithm starts with node x_i in the tree T and proceeds to traverse up the tree until the root node is reached returning all of the constraints stored in the nodes traversed. This is then repeated for variable x_j.

If no constraint is violated, then OPTIONAL-ROUNDINGS() is called to perform optional roundings.

The algorithm OPTIONAL-ROUNDINGS() handles the rounding of variables that were left unaffected by the forced roundings and the subsequent resultant roundings. It first rounds a variable (say x_i) down and then calls

---

Function FORCED-ROUNDING (variable x_i, variable values Z, linear solution $\mathbf{a}$, constraint tree T, system $\mathbf{U}$, queue Q)

1: **for** (each x_j that is involved in a constraint with x_i) **do**
2:     Define R as the set of constraints in $\mathbf{U}$ involving both x_i and x_j.
3:     **if** ($\{x_i + x_j \leq a_i + a_j, \ x_i - x_j \leq a_i - a_j\} \subseteq R$) **then**
4:     {We know that $2 \cdot x_i \leq 2 \cdot a_i$.}
5:         $Z_i = \lfloor a_i \rfloor$. {The tightening rule forces x_i to be rounded down.}
6:         Create branch x_i from x_0 in T.
7:         Add $\{x_i + x_j \leq a_i + a_j, \ x_i - x_j \leq a_i - a_j, \text{ and } x_i \leq \lfloor a_i \rfloor$ to T under $x_i\}$.
8:         Add x_i to Q.
9:     **end if**
10:    **if** ($\{-x_i - x_j \leq -a_i - a_j, \ x_j - x_i \leq a_j - a_i\} \subseteq R$) **then**
11:    {We know that $-2 \cdot x_i \leq -2 \cdot a_i$.}
12:        $Z_i = \lceil a_i \rceil$. {The tightening rule forces x_i to be rounded up.}
13:        Create branch x_i from x_0 in T.
14:        Add $\{-x_i - x_j \leq -a_i - a_j, \ x_j - x_i \leq a_j - a_i, \text{ and } -x_i \leq -\lceil a_i \rceil$ to T under
                $x_i\}$.
15:        Add x_i to Q.
16:    **end if**
17: **end for**

Algorithm 2: FORCED-ROUNDING

CHECK-DEPENDENCIES() to evaluate all of the resultant roundings. Subsequently, it stores all of the new values in a temporary version of Z called Z^T. For each rounding performed (the initial rounding of x_i and all subsequent resultant roudings), the constraints responsible for that rounding are stored in T as discussed previously.

If this rounding of x_i succeeds, then the temporary values are made permanent and the algorithm proceeds onto the next unrounded variable. If rounding x_i down fails, then the algorithm stores the constraints that cause a contradiction when $x_i \leq \lfloor a_i \rfloor$ to the structure L and clears the temporary assignments.

OPTIONAL-ROUNDINGS() then attempts to round x_i up, again evaluating all of the resultant roundings. This time, T^T, a temporary version of the tree T, is used to store the constraints responsible for each rounding. If this rounding of x_i succeeds, then all temporary assignments are made permanent and the algorithm proceeds onto the next unrounded variable. If this rounding also fails, then the constraints that cause a contradiction when $-x_i \leq -\lceil a_i \rceil$ are added to L, and L is returned as a certificate of integer infeasibility.

The list of constraints L can be divided into two parts. The constraints in the first part of L can be added together to form the constraint $-2 \cdot x_i \leq -2 \cdot a_i$, thereby showing that the system $\mathbf{U} \cup \{x_i \leq \lfloor a_i \rfloor\}$ is inconsistent. Likewise, the constraints in the second part show that the system $\mathbf{U} \cup \{-x_i \leq -\lceil a_i \rceil\}$ is inconsistent.

The algorithm CHECK-DEPENDENCIES() checks to see if rounding any of the variables in Q results in a constraint being violated. If a violation occurs, other variables are rounded and added to Q.

Function OPTIONAL-ROUNDINGS (set S of constraints, variable values Z, tree T, linear solution **a**, queue Q)

1: {Recall that S is the subset of **U** restricted to constraints consisting of only variables with $Z_i = M$.}
2: Create tree T^T of constraints with node x_0 at the root.
3: Create array Z^T of temporary variable assignments.
4: Create list L of constraints to be returned in case of infeasibility.
5: **for** (each variable x_i) **do**
6:   **if** $(Z_i = M)$ **then**
7:     **for** (each variable x_j) **do**
8:       $Z_j^T = M$.
9:     **end for**
10:    $Z_i^T = \lfloor a_i \rfloor$. {Try to round x_i down.}
11:    Create branch x_i from x_0 in T.
12:    Add $x_i \le \lfloor a_i \rfloor$ to T under x_i.
13:    Add x_i to Q.
14:    CHECK-DEPENDENCIES $(Z^T, \mathbf{a}, T, S)$.
15:    {Perform all roundings that are a consequence of rounding x_i down.}
16:    **for** (each constraint l in S) **do**
17:      **if** (l is violated by current assignments to some Z_j^T and Z_k^T) **then**
18:        Add the violated constraint l and the constraints in T along paths from x_j to x_i and x_k to x_i to L.
19:        **for** ($j = 1$ **to** n) **do**
20:          $Z_j^T = M$. {Reset temporary variable assignments.}
21:        **end for**
22:        $Z_i^T = \lceil a_i \rceil$. {Try to round x_i up, since rounding down failed.}
23:        Create branch x_i from x_0 in T^T.
24:        Add $-x_i \le -\lceil a_i \rceil$ to T^T under x_i.
25:        Add x_i to Q.
26:        CHECK-DEPENDENCIES $(Z^T, \mathbf{a}, T^T, S)$.
27:        {Perform all roundings that are a consequence of rounding x_i up.}
28:        **for** (each constraint l in S) **do**
29:          **if** (l is violated by current assignments to some Z_j^T and Z_k^T) **then**
30:            Add the violated constraint l and the constraints in T^T along paths from x_j to x_i and x_k to x_i to L.
31:            **return** (set L of constraints.) {Both attempts to round x_i failed.}
32:          **end if**
33:        **end for**
34:        **for** ($j = 1$ **to** n) **do** {Rounding x_i up succeeded.}
35:          **if** $(Z_j^T \ne M)$ **then**
36:            $Z_j \leftarrow Z_j^T$. {Make temporary assignments permanent.}
37:          **end if**
38:        **end for**
39:        **break**{Exits **for** loop commencing on Line 16.}
40:      **end if**
41:    **end for**

```
42:        if (no constraints violated as a result of rounding xᵢ down) then
43:            for (j = 1 to n) do {Rounding xᵢ down succeeded.}
44:                if (Zⱼᵀ ≠ M) then
45:                    Zⱼ ← Zⱼᵀ {Make temporary assignments permanent.}
46:                end if
47:            end for
48:        end if
49:        Empty L. Empty Tᵀ and T and set them to single-node trees with root x₀.
50:    end if
51: end for
52: return (Z as a valid integer solution.)
```

Algorithm 3: OPTIONAL-ROUNDINGS

5.2 Resource Analysis

In this subsection, we analyze the time and space complexity of our lattice point feasibility algorithm. We first establish the time complexities of the three constituent algorithms.

(i) FORCED-ROUNDING() - Observe that a single call to FORCED-ROUNDING() (say on variable x_i) takes $\mathcal{O}(n)$ time, since the call merely examines all the constraints which involve x_i.

(i) CHECK-DEPENDENCIES() - As each variable is rounded, the resultant roundings of the new assignment are deduced using CHECK-DEPENDENCIES(). During these deductions each variable is assigned a value at most once. Furthermore, each constraint is examined at most twice. It follows that CHECK-DEPENDENCIES)() takes $\mathcal{O}(m)$ time.

(i) OPTIONAL-ROUNDINGS() - For each variable, OPTIONAL-ROUNDINGS() calls CHECK-DEPENDENCIES() at most twice (once when rounding up and once when rounding down).

If both roundings for a single variable fail, then OPTIONAL-ROUNDINGS() backtracks along the tree T to obtain a negative cycle. This takes at most $O(n)$ time and only happens once. It follows that OPTIONAL-ROUNDINGS() runs in $\mathcal{O}(m \cdot n)$ time.

We are now ready to analyze the resources taken by PRODUCE-INTEGER-SOLUTION(). Observe that the initialization steps can be accomplished in $\mathcal{O}(n)$ time. Although a single call to FORCED-ROUNDING() takes $\mathcal{O}(n)$ time, a sequence of n successive calls to FORCED-ROUNDING(), one for each variable, takes $\mathcal{O}(m)$ time. This is because each constraint is examined at most twice during the successive calls, once for each variable involved in defining that constraint.

The call to CHECK-DEPENDENCIES() takes $\mathcal{O}(m)$ time. If these roundings variable fail, then PRODUCE-INTEGER-SOLUTION() backtracks along the tree T to obtain a negative cycle. This takes at most $O(n)$ time and only happens once. We finally note that the call to OPTIONAL-ROUNDINGS() takes $\mathcal{O}(m \cdot n)$ time. It thus follows that PRODUCE-INTEGER-SOLUTION() takes $\mathcal{O}(m \cdot n)$ time.

Function CHECK-DEPENDENCIES (vector Z^T of assignments, **a** linear solution, tree T of UTVPI constraints, system **U** of constraints, queue Q)

  1: **while** (Q is not empty) **do**
  2:     Let x_i be the first element of Q.
  3:     Remove x_i from Q.
  4:     **for** (each variable x_j involved in a constraint with x_i) **do**
  5:       Set R to be the set of constraints involving both x_i and x_j.
  6:       **if** ($Z_i^T = \lfloor a_i \rfloor$) **then** $\{x_i$ was rounded down.$\}$
  7:         **if** ($-x_i + x_j \le -a_i + a_j \in R$ and $Z_j^T = M$) **then**
  8:           $Z_j \leftarrow \lfloor a_j \rfloor$. $\{x_j$ needs to be rounded down.$\}$
  9:           Create branch x_j from x_i in T.
10:           Add $-x_i + x_j \le -a_i + a_j$ to T under x_j.
11:           Add x_j to Q.
12:         **end if**
13:         **if** ($-x_i - x_j \le -a_i - a_j \in R$ and $Z_j^T = M$) **then**
14:           $Z_j \leftarrow \lceil a_j \rceil$. $\{x_j$ needs to be rounded up.$\}$
15:           Create branch x_j from x_i in T.
16:           Add $-x_i - x_j \le -a_i - a_j$ to T under x_j.
17:           Add x_j to Q.
18:         **end if**
19:       **end if**
20:       **if** ($Z_i^T = \lceil a_i \rceil$) **then** $\{x_i$ was rounded up.$\}$
21:         **if** ($x_i + x_j \le a_i + a_j \in R$ and $Z_j^T = M$) **then**
22:           $Z_j \leftarrow \lfloor a_j \rfloor$. $\{x_j$ needs to be rounded down.$\}$
23:           Create branch x_j from x_i in T.
24:           Add $x_i + x_j \le a_i + a_j$ to T under x_j.
25:           Add x_j to Q.
26:         **end if**
27:         **if** ($x_i - x_j \le a_i - a_j \in R$ and $Z_j^T = M$) **then**
28:           $Z_j \leftarrow \lceil a_j \rceil$. $\{x_j$ needs to be rounded up.$\}$
29:           Create branch x_j from x_i in T.
30:           Add $x_i - x_j \le a_i - a_j$ to T under x_j.
31:           Add x_j to Q.
32:         **end if**
33:       **end if**
34:     **end for**
35: **end while**

Algorithm 4: CHECK-DEPENDENCIES

The only auxiliary data structures used by PRODUCE-INTEGER-SOLUTION() are the variable structures Z and Z^T, the trees T and T^T, the structure storing the proof of infeasibility L, and the queue of variables Q. All of these structure use $\mathcal{O}(m + n)$ space. Additionally, $\mathcal{O}(n)$ space is used for auxiliary storage. It follows that PRODUCE-INTEGER-SOLUTION() can be implemented in $\mathcal{O}(m + n)$ space.

6   Correctness of Integer Feasibility Algorithm

In this section, we establish the correctness of PRODUCE-INTEGER-SOLUTION().
Note that PRODUCE-INTEGER-SOLUTION() starts with a feasible, half-integral
solution and performs a sequence of roundings as needed, on variables which are
not integral. In the proof, we shall demonstrate that every rounding results from
a deducible constraint. Furthermore, we shall show that if a variable cannot be
rounded without creating an inconsistency, then the input UTVPI system does
not enclose a lattice point.

Lemma 1. *Each forced rounding results from a constraint that can be deduced
from* **U** *by the tightening inference rule.*

Proof. Let x_i be the variable that undergoes a forced rounding. Assume that the
initial value of x_i is a_i.

If x_i is rounded down by FORCED-ROUNDING(), then there must exist a
variable x_j, with initial value a_j, such that there exist constraints $x_i - x_j \le a_i - a_j$
and $x_i + x_j \le a_i + a_j$. Using the tightening inference rule we can deduce the
constraint $x_i \le \lfloor \frac{a_i - a_j + a_i + a_j}{2} \rfloor = \lfloor a_i \rfloor$. This is the deduced constraint that causes
x_i to be rounded down.

Likewise, if x_i is rounded up by FORCED-ROUNDING(), then there must exist
a variable x_j, with initial value a_j, such that there exist constraints $-x_i - x_j \le
-a_i - a_j$ and $-x_i + x_j \le -a_i + a_j$. Using the tightening inference rule we
can deduce the constraint $-x_i \le \lfloor \frac{-a_i - a_j - a_i + a_j}{2} \rfloor = -\lceil a_i \rceil$. This is the deduced
constraint that causes x_i to be rounded up.   □

Lemma 2. *Every resultant rounding caused by rounding x_k down results from
a constraint that can be deduced from* $\mathbf{U} \cup \{x_k \le \lfloor a_k \rfloor\}$ *by the transitive inference
rule.*

Proof. All the resultant roundings caused by rounding x_k correspond to nodes in
the tree T that are descendants of the node x_k (see CHECK-DEPENDENCIES()).
Let T_k denote the sub-tree of T rooted at x_k. We will show that the rounding of
each variable in T_k corresponds to a constraint deducible from $\mathbf{U} \cup \{x_k \le \lfloor a_k \rfloor\}$.

Rounding x_k down results from the constraint $x_k \le \lfloor a_k \rfloor$ which is trivially
deducible from $\mathbf{U} \cup \{x_k \le \lfloor a_k \rfloor\}$.

We now assume that all the roundings of all variables at depth d in T_k result
from constraints deducible from $\mathbf{U} \cup \{x_k \le \lfloor a_k \rfloor\}$. Let x_j be a variable of depth
$(d+1)$ in T_k, and let x_i be the parent of x_j. Since x_i is at a depth of d in T_k, we
know that its rounding results from a constraint deducible from $\mathbf{U} \cup \{x_k \le \lfloor a_k \rfloor\}$.
Let a_i and a_j denote the initial values of x_i and x_j respectively. We can assume
without loss of generality that both a_i and a_j are odd multiples of $\frac{1}{2}$.

There are four cases that need to be considered. We establish the proof
in one case. The proofs for the remaining cases are analogous. Assume that
x_j is rounded down as a result of x_i being rounded down. As per CHECK-
DEPENDENCIES(), there must be a constraint of type $-x_i + x_j \le -a_i + a_j$
in **U**. Since x_i was rounded down, the constraint $x_i \le \lfloor a_i \rfloor = a_i - \frac{1}{2}$ is deducible

from $\mathbf{U} \cup \{x_k \leq \lfloor a_k \rfloor\}$. Using the transitive inference rule, we get the constraint $x_j \leq -a_i + a_j + a_i - \frac{1}{2} = a_j - \frac{1}{2} = \lfloor a_j \rfloor$. This is the deduced constraint that caused x_j to be rounded down.

Thus, all the resultant roundings caused by rounding x_k down, can be deduced from $\mathbf{U} \cup \{x_k \leq \lfloor a_k \rfloor\}$.   □

Lemma 3. *Each resultant rounding caused by rounding x_k up results from a constraint that can be deduced from $\mathbf{U} \cup \{-x_k \leq -\lceil a_k \rceil\}$ by the transitive inference rule.*

The proof of this lemma is analogous to the proof of Lemma 2. Lemmas 1, 2, and 3 lead to the following theorem.

Theorem 1. *Each resultant rounding caused by a forced rounding results from a constraint that can be deduced from $\mathbf{U}$ by the transitive and tightening inference rules.*

Let x_j be a variable that was rounded, as a result of a forced rounding or an optional rounding. Let $\mathcal{V}$ be the set containing x_j and all of the variables rounded as a result of x_j being rounded.

Let x_i be any variable that remains unrounded even after x_j was rounded. It is clear that a_i is an odd multiple of $\frac{1}{2}$. We will show that x_i can be rounded up or rounded down, without violating a constraint involving variables in $\mathcal{V}$. Thus, if no constraint is violated as a result of rounding x_j, and performing all subsequent resultant roundings, the values of the variables in $\mathcal{V}$ can be considered permanent. This follows, since no subsequent roundings will violate a constraint involving any of these variables.

Lemma 4. *Any unrounded variable not in $\mathcal{V}$ can be rounded up or down, without violating any constraints shared with a variable in $\mathcal{V}$.*

Proof. Let x_i be an unrounded variable, such that $x_i \notin \mathcal{V}$. Let $x_k \in \mathcal{V}$. Let us assume the contrary, i.e., we assume that a constraint involving x_i and x_k is violated, when x_i is rounded in a certain direction. Clearly, we are concerned *only* with constraints of the form: $\pm x_i \pm x_k \leq c_{ik}$.

There are four cases that need to be considered. We establish the proof in one case. The proofs for the remaining cases are analogous.

Assume that x_k was rounded down, and that rounding x_i down results in a violation. In this case, there is a constraint that was satisfied when $x_i = a_i$ and $x_k = a_k$, but violated when $x_i = a_i - \frac{1}{2}$ and $x_k = a_k - \frac{1}{2}$. Thus, the violated constraint is of the form $-x_i - x_k \leq c_{ik}$ (since rounding down results in constraint violation). We know that $-a_i - a_k \leq c_{ik}$ and that $-a_i - a_k + 1 > c_{ik}$. This means that $c_{ik} = -a_i - a_k$. Thus, the violated constraint *must* be $-x_i - x_k \leq -a_i - a_k$. However, this constraint would cause CHECK-DEPENDENCIES() to round x_i up, as a result of rounding x_k. But this contradicts our assumption that x_i was unrounded to begin with.

Since all four cases result in a contradiction, it follows that no constraint involving x_k can be violated when x_i is rounded and the lemma follows.   □

Theorem 2. *If* PRODUCE-INTEGER-SOLUTION *() declares the input UTVPI system* **U** *to be feasible, then the system has integral solutions.*

Proof. PRODUCE-INTEGER-SOLUTION() declares **U** to be feasible, only if an integer value is assigned to each variable. In PRODUCE-INTEGER-SOLUTION() and OPTIONAL-ROUNDINGS(), we check every constraint in **U** to see if it is violated by the current assignment. We only return an integer solution, if every constraint in **U** is satisfied. Thus, the integer solution returned by PRODUCE-INTEGER-SOLUTION() is a valid integral solution.   □

Theorem 3. *If* PRODUCE-INTEGER-SOLUTION *() declares the input UTVPI system* **U** *to be infeasible, then* **U** *does not enclose a lattice point, and L is a proof of infeasibility for* **U**.

Proof. The algorithms can declare the system infeasible as a result of a forced rounding, and the subsequent resultant roundings, or as a result of an optional rounding and the subsequent resultant roundings.

  If the system is declared infeasible as a result of a forced rounding (and the subsequent resultant roundings), then there is a constraint between some x_i and x_j that is violated, when all the resultant roundings have been computed. Since we started with a valid linear solution, if both x_i and x_j are unrounded, then the constraint is still satisfied. Similarly, by the proof of Lemma 4, if only one of x_i or x_j is unrounded, the constraint is still satisfied. This means that both x_i and x_j must already have been rounded. There are four cases which need to be considered, depending on the type of constraint violated. We establish this for one case. The proofs for the remaining cases are analogous.

  Assume that the violated constraint is of the form $l_1 : x_i + x_j \leq c_{ij}$. Since the initial (linear) solution, **a**, was valid, we know that $a_i + a_j \leq c_{ij}$. Thus, for l_1 to be violated, x_i and x_j must both have been rounded up. It follows that $c_{ij} < a_i + a_j + 1$.

  Since x_i and x_j were rounded up, from Theorem 1, we know that the constraints $l_2 : -x_i \leq -\lceil a_i \rceil = -a_i - \frac{1}{2}$ and $l_3 : -x_j \leq -\lceil a_j \rceil = -a_j - \frac{1}{2}$, are deducible from **U**. When the constraints l_1, l_2, and l_3 are added together, we get $0 \leq c_{ij} - a_i - a_j - 1 < 0$, which is a contradiction. This contradiction establishes the integer infeasibility of **U**.

  By construction, the constraint $x_i + x_j \leq c_{ij}$, the constraints used to derive $-x_i \leq -\lceil a_i \rceil$, and the constraints used to derive $-x_j \leq -\lceil a_j \rceil$ are all in L. Thus L is a proof of integer infeasibility for **U**.

  If the system is declared infeasible as a result of an optional rounding (and the subsequent resultant roundings), then for some variable x_k, rounding x_k and performing all subsequent resultant roundings result in or more violated constraints. Note that this holds when x_k is rounded up and when x_k is rounded down. From Lemmas 2 and 3, and the arguments made above, we know that both $\mathbf{U} \cup \{x_k \leq \lfloor a_k \rfloor\}$ and $\mathbf{U} \cup \{-x_k \leq -\lceil a_k \rceil\}$ are infeasible. We can also conclude that the constraints used to establish these infeasibilities are in L. Since all possible integer values of x_k are covered by these the two systems, we can conclude that **U** has no integer solutions.   □

As discussed above, PRODUCE-INTEGER-SOLUTION() starts with an arbitrary half integral solution and always maintains $\lfloor a_i \rfloor \leq Z_i \leq \lceil a_i \rceil$, for each $Z_i \neq M$. Thus, we have the following corollary.

Corollary 1. *If a system* **U** *of UTVPI constraints is integer feasible, and* **a** *is a valid half-integral solution to* **U**, *then there exists an integral solution* **Z** *such that for each* $i = 1 \ldots n$, $\lfloor a_i \rfloor \leq Z_i \leq \lceil a_i \rceil$.

References

[CC77] Cousot, P., Cousot, R.: Abstract interpretation: a unified lattice model for static analysis of programs by construction or approximation of fixpoints. In: POPL, pp. 238–252 (1977)

[DE73] Dantzig, G.B., Eaves, B.C.: Fourier-Motzkin elimination and its dual. J. Comb. Theory (A) **14**, 288–297 (1973)

[FMP13] Fouilhe, A., Monniaux, D., Périn, M.: Efficient generation of correctness certificates for the abstract domain of polyhedra. In: Logozzo, F., Fähndrich, M. (eds.) SAS 2013. LNCS, vol. 7935, pp. 345–365. Springer, Heidelberg (2013). doi:10.1007/978-3-642-38856-9_19

[GPS95] Gerber, R., Pugh, W., Saksena, M.: Parametric dispatching of hard real-time tasks. IEEE Trans. Comput. **44**(3), 471–479 (1995)

[HS97] Harvey, W., Stuckey, P.J.: A unit two variable per inequality integer constraint solver for constraint logic programming. In: Proceedings of the 20th Australasian Computer Science Conference, pp. 102–111 (1997)

[JMSY94] Jaffar, J., Maher, M.J., Stuckey, P.J., Yap, R.H.C.: Beyond finite domains. In: Borning, A. (ed.) PPCP 1994. LNCS, vol. 874, pp. 86–94. Springer, Heidelberg (1994). doi:10.1007/3-540-58601-6_92

[LM05] Lahiri, S.K., Musuvathi, M.: An efficient decision procedure for UTVPI constraints. In: Gramlich, B. (ed.) FroCoS 2005. LNCS, vol. 3717, pp. 168–183. Springer, Heidelberg (2005). doi:10.1007/11559306_9

[Min06] Miné, A.: The octagon abstract domain. Higher-Order Symb. Comput. **19**(1), 31–100 (2006)

[NO05] Nieuwenhuis, R., Oliveras, A.: DPLL(T) with exhaustive theory propagation and its application to difference logic. In: Etessami, K., Rajamani, S.K. (eds.) CAV 2005. LNCS, vol. 3576, pp. 321–334. Springer, Heidelberg (2005). doi:10.1007/11513988_33

[NOT04] Nieuwenhuis, R., Oliveras, A., Tinelli, C.: Abstract DPLL and abstract DPLL modulo theories. In: Baader, F., Voronkov, A. (eds.) LPAR 2005. LNCS, vol. 3452, pp. 36–50. Springer, Heidelberg (2005). doi:10.1007/978-3-540-32275-7_3

[Rev09] Revesz, P.Z.: Tightened transitive closure of integer addition constraints. In: SARA (2009)

[SS00] Sitzmann, I., Stuckey, P.J.: O-trees: a constraint-based index structure. In: Australasian Database Conference, pp. 127–134 (2000)

[SS10] Schutt, A., Stuckey, P.J.: Incremental satisfiability and implication for UTVPI constraints. INFORMS J. Comput. **22**(4), 514–527 (2010)

[Sub04] Subramani, K.: On deciding the non-emptiness of 2SAT polytopes with respect to first order queries. Math. Log. Q. **50**(3), 281–292 (2004)

[SW17] Subramani, K., Wojciechowski, P.J.: A combinatorial certifying algorithm for linear feasibility in UTVPI constraints. Algorithmica **78**(1), 166–208 (2017)

A Constraint Composite Graph-Based ILP Encoding of the Boolean Weighted CSP

Hong Xu[✉][ID], Sven Koenig, and T.K. Satish Kumar

University of Southern California, Los Angeles, CA 90089, USA
{hongx,skoenig}@usc.edu, tkskwork@gmail.com

Abstract. The weighted constraint satisfaction problem (WCSP) occurs in the crux of many real-world applications of operations research, artificial intelligence, bioinformatics, etc. Despite its importance as a combinatorial substrate, many attempts for building an efficient WCSP solver have been largely unsatisfactory. In this paper, we introduce a new method for encoding a (Boolean) WCSP instance as an integer linear program (ILP). This encoding is based on the idea of the constraint composite graph (CCG) associated with a WCSP instance. We show that our CCG-based ILP encoding of the Boolean WCSP is significantly more efficient than previously known ILP encodings. Theoretically, we show that the CCG-based ILP encoding has a number of interesting properties. Empirically, we show that it allows us to solve many hard Boolean WCSP instances that cannot be solved by ILP solvers with previously known ILP encodings.

1 Introduction

The weighted constraint satisfaction problem (WCSP) is a combinatorial optimization problem. It is a generalization of the constraint satisfaction problem (CSP) in which the constraints are no longer "hard." Instead, each tuple in a constraint—i.e., an assignment of values to all variables in that constraint—is associated with a non-negative weight (sometimes referred to as "cost"). The goal is to find a complete assignment of values to all variables from their respective domains such that the total weight is minimized [2], called an optimal solution.

More formally, the WCSP is defined by a triplet $\langle \mathcal{X}, \mathcal{D}, \mathcal{C} \rangle$, where $\mathcal{X} = \{X_1, X_2, \ldots, X_N\}$ is a set of N variables, $\mathcal{D} = \{D(X_1), D(X_2), \ldots, D(X_N)\}$ is a set of N domains with discrete values, and $\mathcal{C} = \{C_1, C_2, \ldots, C_M\}$ is a set of M weighted constraints. Each variable $X_i \in \mathcal{X}$ can be assigned a value in its associated domain $D(X_i) \in \mathcal{D}$. Each constraint $C_i \in \mathcal{C}$ is defined over a certain subset of the variables $S(C_i) \subseteq \mathcal{X}$, called the scope of C_i. C_i associates a non-negative weight with each possible assignment of values to the variables in $S(C_i)$. The goal is to find a complete assignment of values to all variables in $\mathcal{X}$ from their respective domains that minimizes the sum of the weights specified by each constraint in $\mathcal{C}$ [2]. This combinatorial task can equivalently be characterized by having to compute

$$\arg\min_{a \in A(\mathcal{X})} \sum_{C_i \in \mathcal{C}} E_{C_i}(a|S(C_i)), \tag{1}$$

© Springer International Publishing AG 2017
J.C. Beck (Ed.): CP 2017, LNCS 10416, pp. 630–638, 2017.
DOI: 10.1007/978-3-319-66158-2_40

where $A(\mathcal{X})$ represents the set of all $|D(X_1)| \times |D(X_2)| \times \ldots \times |D(X_N)|$ complete assignments to all variables in $\mathcal{X}$. $a|S(C_i)$ represents the projection of a complete assignment a onto the subset of variables in $S(C_i)$. E_{C_i} is a function that maps each $a|S(C_i)$ to its associated weight in C_i. The Boolean WCSP is the WCSP with only variables of domain size 2, i.e., $\forall X \in \mathcal{X} : |D(X)| = 2$. It is representationally as powerful as the WCSP.

There are many ways to solve a given WCSP instance. The state-of-the-art methods include best-first AND/OR search [10] and branch-and-bound algorithms that exploit soft arc consistencies [4]. Unfortunately, none of these WCSP solvers make use of the power of integer linear programming (integer LP, ILP) solvers, such as the Gurobi Optimizer [3] and lp_solve [1]. ILP solvers are highly optimized and are extensively used for solving problems in operations research. An efficient ILP encoding of the WCSP would therefore create an important nexus between constraint programming and operations research.

An ILP encoding of the WCSP can be borrowed from the probabilistic reasoning community. Here, the WCSP arises as the max-a-posteriori (MAP) problem.[1] Although this ILP encoding is popularly used in probabilistic reasoning [5, Sect. 13.5], it does not scale to large instances since it creates an unwieldy number of variables and constraints. In the rest of the paper, we refer to this ILP encoding as the "direct" ILP encoding.

In this paper, we introduce a new ILP encoding of the WCSP that is based on the idea of the constraint composite graph (CCG) [7–9]. We refer to this encoding as the "CCG-based" ILP encoding. We compare it with the direct ILP encoding in [5, Sect. 13.5] for the Boolean WCSP. We first derive and compare the theoretical bounds on the number of variables, the number of constraints and the number of variables in each constraint in the ILPs generated by these two ILP encodings. We then experimentally compare the efficiency of solving the ILPs generated by the two ILP encodings. Finally, we establish an important theoretical property of the CCG-based ILP encoding.

2   ILP Encodings of the WCSP

In this section, we describe two methods to encode a given WCSP instance as an ILP: One is the direct ILP encoding adapted from [5, Sect. 13.5] and the other one is our proposed CCG-based ILP encoding. For notational convenience, throughout this section, we consider the WCSP instance $\mathcal{B} = \langle \mathcal{X} = \{X_1, X_2, \ldots, X_N\}, \mathcal{D} = \{D(X_1), D(X_2), \ldots, D(X_N)\}, \mathcal{C} = \{C_1, C_2, \ldots, C_M\}\rangle$.

2.1   Direct ILP Encoding

For each $C \in \mathcal{C}$ and $a \in A(S(C))$, we introduce an ILP variable q_a^C. Here, $A(S(C))$ is the set of all assignments of values to variables in constraint C

---

[1] A MAP problem instance on a probabilistic graphic model, such as a Belief Network, can be formulated as a WCSP instance by taking the negative logarithm on the individual probabilities.

(therefore $|A(S(C))| = \prod_{X \in S(C)} |D(X)|$). q_a^C is either 0 or 1: If $q_a^C = 1$, then the assignment a to the variables in C is part of the to-be-determined optimal solution a^*, i.e., $a^*|S(C) = a$; otherwise it is not. The direct ILP encoding of $\mathcal{B}$ is

$$\underset{q_a^C : q_a^C \in q}{\text{minimize}} \sum_{C \in \mathcal{C}} \sum_{a \in A(S(C))} w_a^C q_a^C \tag{2}$$

$$\text{s.t.} \quad q_a^C \in \{0, 1\} \qquad\qquad \forall q_a^C \in q \tag{3}$$

$$\sum_{a \in A(S(C))} q_a^C = 1 \qquad\qquad \forall C \in \mathcal{C} \tag{4}$$

$$\sum_{a \in A(S(C)):a|S(C) \cap S(C')=s} q_a^C = \sum_{a' \in A(S(C')):a'|S(C) \cap S(C')=s} q_{a'}^{C'} \qquad \forall C, C' \in \mathcal{C} \text{ and} \tag{5}$$

$$s \in A(S(C) \cap S(C')),$$

where $q = \{q_a^C \mid C \in \mathcal{C} \wedge a \in A(S(C))\}$, w_a^C denotes the weight of assignment a specified by constraint C, and $a|S(C) \cap S(C')$ is the projection of the complete assignment a onto the set of common variables in C and C'. The cardinality of q is $\sum_{C \in \mathcal{C}} \prod_{X \in S(C)} |D(X)|$. Here,

- Equation (3) represents the ILP constraints that enforce the Boolean property for all q_a^C's. It consists of $\sum_{C \in \mathcal{C}} \prod_{X \in S(C)} |D(X)| = \mathcal{O}\left(|\mathcal{C}|\hat{D}^{\hat{C}}\right)$ ILP constraints, where $\hat{C} = \max_{C \in \mathcal{C}} |S(C)|$ and $\hat{D} = \max_{X \in \mathcal{X}} |D(X)|$.
- Equation (4) represents the ILP constraints that enforce a unique assignment of values to variables in each WCSP constraint. It consists of $|\mathcal{C}|$ ILP constraints, each of which has $|A(S(C))| = \prod_{X \in S(C)} |D(X)| = \mathcal{O}\left(\hat{D}^{\hat{C}}\right)$ variables.
- Equation (5) represents the ILP constraints which enforce that every two assignments in two WCSP constraints must be consistent on their shared variables. It consists of $\mathcal{O}\left(|\mathcal{C}|^2\hat{D}^{\hat{C}}\right)$ ILP constraints. Each of these ILP constraints has $\mathcal{O}\left(\hat{D}^{\hat{C}}\right)$ variables.

Therefore, if $\mathcal{B}$ is a Boolean WCSP instance, the direct ILP encoding has $|q| = \mathcal{O}\left(|\mathcal{C}|\hat{D}^{\hat{C}}\right) = \mathcal{O}\left(|\mathcal{C}|2^{\hat{C}}\right)$ variables and $\mathcal{O}\left(|\mathcal{C}|^2\hat{D}^{\hat{C}}\right) = \mathcal{O}\left(|\mathcal{C}|^2 2^{\hat{C}}\right)$ ILP constraints. Each of these ILP constraints has $\mathcal{O}\left(\hat{D}^{\hat{C}}\right) = \mathcal{O}\left(2^{\hat{C}}\right)$ variables.

2.2   CCG-Based ILP Encoding

The CCG [7–9] is a combinatorial structure associated with the WCSP. It provides a unifying framework for simultaneously exploiting the graphical structure of the variable interactions in the WCSP as well as the numerical structure of the constraints in it. The task of solving the WCSP can be reformulated as the task of finding a minimum weighted vertex cover (MWVC) (namely the MWVC problem) on its associated CCG [7–9]. CCGs can be constructed in polynomial time and are always tripartite [7–9]. A subclass of the WCSP has instances with bipartite CCGs. This subclass is tractable since an MWVC can be found in polynomial time on bipartite graphs using a staged maxflow algorithm [6]. The CCG also enables the use of kernelization methods for the MWVC problem, such as the Nemhauser-Trotter reduction, for solving the WCSP [14]. Empirically, the min-sum message passing algorithm often produces better solutions for the MWVC problem on the CCG than directly on the WCSP [14].

We can encode a WCSP instance as an ILP after transforming it to an equivalent MWVC problem instance on its CCG $G = \langle V, E, w \rangle$. The resulting CCG-based ILP encoding is

$$\text{minimize}_{x_i : v_i \in V} \sum_{i=1}^{|V|} w_i x_i \tag{6}$$

$$\text{s.t.} \quad x_i \quad \in \{0, 1\} \quad \forall\, v_i \in V \tag{7}$$

$$x_i + x_j \quad \geq 1 \quad \forall\, (v_i, v_j) \in E, \tag{8}$$

where variable x_i represents the presence of v_i in the MWVC, i.e., $x_i = 1$ and $x_i = 0$ indicate that v_i is and is not in the MWVC, respectively [13]. The numbers of ILP variables and constraints are determined by the CCG. We now assume that $\mathcal{B}$ is a Boolean WCSP instance. We can compute the number of vertices and edges in the CCG by following the CCG construction procedure in [7].[2] A constraint C can be represented by the multivariate polynomial

$$\sum_{T \in \mathcal{P}(S(C))} \left[c_T \prod_{X \in T} X \right], \tag{9}$$

where $\mathcal{P}(S(C))$ is the power set of $S(C)$ and the c_T's are constants. The CCG gadget corresponding to term $c_T \prod_{X \in T} X$ has $\mathcal{O}(|T|)$ vertices and edges. The CCG gadget corresponding to constraint C therefore has an upper bound of

$$\mathcal{O}\left(\sum_{T \in \mathcal{P}(S(C))} |T| \right) = \mathcal{O}\left(\sum_{|T|=0}^{|S(C)|} \binom{|S(C)|}{|T|} |T| \right) = \mathcal{O}\left(2^{|S(C)|-1} |S(C)| \right) \tag{10}$$

vertices and edges. Therefore, if $\mathcal{B}$ is a Boolean WCSP instance, the CCG has $\mathcal{O}\left(|\mathcal{C}| 2^{\hat{C}} \hat{C} \right)$ vertices and edges constituting the ILP variables (Eq. (7)) and constraints (Eq. (8)), respectively, with each of these ILP constraints having at most 2 variables.

2.3   Comparison

We compare various parameters of the two ILP encodings for the Boolean WCSP in Table 1. For any non-trivial Boolean WCSP instances, the CCG-based ILP

---

[2] As shown in [8], our techniques can also be generalized to the WCSP with variables of domain sizes larger than 2. However, for a proof of concept, this paper focuses on the Boolean WCSP.

encoding has a huge advantage over the direct ILP encoding with respect to the number of variables per constraint. This is true even if $\hat{C}$ is bounded because, in the direct ILP encoding, the number of variables in an ILP constraint corresponding to a WCSP constraint C in Eq. (4) is $2^{|S(C)|} \geq 2$. For the number of constraints, while different values of the parameters lead to different trade-offs, the most interesting real-world applications of the WCSP have a large number $|\mathcal{C}|$ of constraints and a bounded arity $\hat{C}$ of the individual constraints. Under such assumptions, the CCG-based ILP encoding is more advantageous than the direct ILP encoding with respect to the number of constraints as well. For the number of variables, the CCG-based ILP encoding loses by a factor of $\hat{C}$. However, as argued before, in many real-world applications, $\hat{C}$ is bounded, and therefore the number of variables for both ILP encodings are of the same order. In general, when $\hat{C}$ is bounded, the CCG-based ILP encoding retains the same order of the number of variables as the direct ILP encoding and significantly wins on the number of constraints and the number of variables per constraint.

Table 1. Shows the numbers of variables, constraints, and variables per constraint in the two ILP encodings of Boolean WCSP instance $\mathcal{B} = \langle \mathcal{X}, \mathcal{D}, \mathcal{C} \rangle$.

Encoding	Direct	CCG-based
Number of variables	$\mathcal{O}\left(|\mathcal{C}|2^{\hat{C}}\right)$	$\mathcal{O}\left(|\mathcal{C}|2^{\hat{C}}\hat{C}\right)$
Number of constraints	$\mathcal{O}\left(|\mathcal{C}|^2 2^{\hat{C}}\right)$	$\mathcal{O}\left(|\mathcal{C}|2^{\hat{C}}\hat{C}\right)$
Number of variables per constraint	$\mathcal{O}\left(2^{\hat{C}}\right)$	≤ 2

3   Experimental Evaluation

In this section, we experimentally evaluate the efficiencies of solving the Boolean WCSP using the two ILP encodings. We refer to the two algorithms that use these ILP encodings as the direct algorithm and the CCG-based algorithm.

In our experiments, the benchmark instances were generated from the UAI 2014 Inference Competition[3]. Here, MAP inference queries with no evidence on the PR and MMAP benchmark instances can be formulated as Boolean WCSP instances by first taking the negative logarithms of the probabilities in each factor and then normalizing them. The experiments were performed on those 160 benchmark instances with only Boolean variables. We set a running time limit of 120 s for each algorithm on each benchmark instance.

We used the Gurobi Optimizer version 7.0.2 [3] as the ILP solver. All default settings of the Gurobi Optimizer were kept except that it was configured to use only one CPU thread. The ILP encoding procedures and the CCG construction algorithm were implemented in C++ and were compiled by the GNU Compiler Collection (GCC) 6.3.0 with the "-O3" option. We used the Boost

---

[3] http://www.hlt.utdallas.edu/~vgogate/uai14-competition/index.html.

graph library [11] to implement the graph representations and operations. We performed our experiments on a GNU/Linux workstation with an Intel Xeon processor E3-1240 v3 (8 MB Cache, 3.4 GHz) and 16 GB RAM.

Table 2 shows the number of benchmark instances on which both algorithms terminated within the running time limit. The number of benchmark instances on which only the CCG-based algorithm terminated is much larger than the number of benchmark instances on which only the direct algorithm terminated. On 118 out of 160 benchmark instances, both terminated within the running time limit. However, even in this category, the CCG-based algorithm was much more efficient.

Table 2. Shows the number of benchmark instances on which the direct and CCG-based algorithms terminated within a running time limit of 120 s.

Termination status	Total	CCG-based only	Direct only	Neither	Both
Number of benchmark instances	160	23	5	14	118

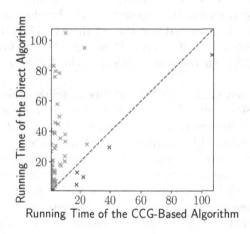

Fig. 1. Compares the efficiencies of the direct and CCG-based algorithms on the benchmark instances on which both algorithms terminated within a running time limit of 120 s. Each point represents a benchmark instance. The x and y coordinates of each point show the running times of the CCG-based and direct algorithm on its corresponding benchmark instance, respectively. The dashed diagonal line represents equal running times. Points above and below this line are colored red and blue, respectively. Red and blue points represent benchmark instances on which the CCG-based and direct algorithm terminated more quickly, respectively. There are 110 red points and 8 blue points. (Color figure online)

Figure 1 compares the efficiencies of the two algorithms on the 118 benchmark instances on which both of them terminated within the running time limit.

The CCG-based algorithm was more efficient on 110 benchmark instances (red points), and the direct algorithm was more efficient on 8 benchmark instances (blue points). Most red points are far from the dashed diagonal line, meaning that the gap between the running times of the two algorithms was very large for those benchmark instances on which the CCG-based algorithm was more efficient. On the other hand, all blue points are close to the dashed diagonal line, meaning that the direct algorithm only marginally outperformed the CCG-based algorithm on these benchmark instances in terms of running time.

4   A Theoretical Property of CCG-Based ILP Encoding

Since an ILP itself can be interpreted as a WCSP instance with an infinite weight marking the violation of an ILP constraint and unary constraints representing the ILP objective function, the concept of the CCG is well defined for ILPs. It can be constructed in polynomial time for an ILP and can be used to generate the CCG-based ILP encoding of the given ILP. A desirable property of the CCG-based ILP encoding is therefore its ability to preserve the integrality of the vertices of the feasible region of its LP relaxation.

ILPs can be relaxed to LPs by removing all integrality constraints on their variables. LPs have convex feasible regions and can therefore be solved efficiently (in polynomial time). If the feasible region of the LP relaxation of an ILP has only integer vertices (equivalent to an ILP having a totally unimodular (TUM) constraint matrix [12]), an optimal solution of the LP also yields an optimal solution of the ILP.

An ILP can be viewed as a WCSP instance as follows. Each ILP constraint translates to a WCSP constraint with weights of values zero or infinity. The ILP objective function translates to a set of unary WCSP constraints. The CCG-based ILP encoding of an ILP produces a new ILP. If the original ILP has only integer vertices in the feasible region of its LP relaxation, it is desirable for the new ILP to also have the same property. This would mean that, if the original ILP is solvable through LP relaxation, the new ILP is also solvable through LP relaxation. In this section, we show that this property of the CCG-based ILP encoding in fact holds for an important subclass of such ILPs, namely, ILPs that model MWVC problem instances on bipartite graphs.

The MWVC problem on a given vertex-weighted graph $G = \langle V, E, w \rangle$ is formulated as an ILP of the same form of Eqs. (6) to (8), where we simply associate a 0/1 variable x_i with each vertex $v_i \in V$ of non-negative weight w_i indicating the presence of v_i in the MWVC. If G is bipartite, its constraint matrix is TUM. Therefore, the LP relaxation of this ILP has only integer vertices in its feasible region [12]. We can formulate this ILP as a WCSP instance with the two types of constraints shown in Table 3.

Now we show that the CCG created for the MWVC problem on any given bipartite graph is also bipartite, which establishes that the LP relaxation of the CCG-based ILP encoding has only integer vertices in its feasible region. Consider an edge $(v_i, v_j) \in E$. The CCG gadget that represents the constraint of covering

Table 3. Shows the two types of WCSP constraints for the MWVC problem ((a)The binary constraint that represents the requirement of covering each edge $(v_i, v_j) \in E$,(b)The unary constraint for each vertex v_i that represents a term in the objective function of minimizing the total weight of the vertex cover)

x_i \ x_j	0	1
0	$+\infty$	0
1	0	0

(a) The binary constraint that represents the requirement of covering each edge $(v_i, v_j) \in E$

x_i	0	1
Value	0	w_i

(b) The unary constraint for each vertex v_i that represents a term in the objective function of minimizing the total weight of the vertex cover

this edge involves auxiliary vertices A and A' [7]. The CCG gadget itself has the edges (v_i, A), (A, A') and (A', v_j). If the original graph is bipartite, then its vertices can be colored using either of two colors, red and blue, such that every edge connects a red vertex and a blue vertex. Without loss of generality, we assume that v_i is colored red and v_j is colored blue. We then color A blue and A' red. Such a coloring of the vertices ensures that the edges of the gadgets also always connect a red vertex and a blue vertex. This means that the CCG is also bipartite. Hence, we establish the desired property of the CCG-based ILP encoding for the MWVC problem on any given bipartite graph.

5   Conclusions and Future Work

In this paper, we introduced the CCG-based ILP encoding of the WCSP. We compared it to the direct ILP encoding adapted from the probabilistic reasoning community. Theoretically, we showed that the CCG-based ILP encoding has several advantages over the direct ILP encoding with respect to the number of variables per constraint and the number of constraints. Empirically, we showed that the CCG-based algorithm significantly outperforms the direct algorithm with respect to the running time on benchmark instances. Finally, we showed that MWVC problem instances on bipartite graphs, whose corresponding ILPs have only integer vertices in the feasible regions of their LP relaxations, preserve this property in their CCG-based ILP encodings as well.

It is future research to prove properties of the CCG-based ILP encoding for ILPs with TUM constraint matrices, to use our techniques to make ILP-based approaches competitive with other approaches for solving the WCSP, and to extend our results to the WCSP with variables of domain sizes larger than 2.

Acknowledgment. The research at the University of Southern California was supported by the National Science Foundation (NSF) under grant numbers 1409987 and 1319966. The views and conclusions contained in this document are those of the authors and should not be interpreted as representing the official policies, either expressed or implied, of the sponsoring organizations, agencies or the U.S. government.

References

1. Berkelaar, M., Eikland, K., Notebaert, P.: lp_solve 5.5 open source (mixed integer) linear programming software (2004). http://lpsolve.sourceforge.net/5.5/
2. Bistarelli, S., Montanari, U., Rossi, F., Schiex, T., Verfaillie, G., Fargier, H.: Semiring-based CSPs and valued CSPs: frameworks, properties, and comparison. Constraints **4**(3), 199–240 (1999)
3. Gurobi Optimization Inc.: Gurobi optimizer reference manual (2017) http://www.gurobi.com
4. Hurley, B., O'Sullivan, B., Allouche, D., Katsirelos, G., Schiex, T., Zytnicki, M., de Givry, S.: Multi-language evaluation of exact solvers in graphical model discrete optimization. Constraints **21**(3), 413–434 (2016)
5. Koller, D., Friedman, N.: Probabilistic Graphical Models: Principles and Techniques. MIT Press, Cambridge (2009)
6. Kumar, T.K.S.: Incremental computation of resource-envelopes in producer-consumer models. In: Rossi, F. (ed.) CP 2003. LNCS, vol. 2833, pp. 664–678. Springer, Heidelberg (2003). doi:10.1007/978-3-540-45193-8_45
7. Kumar, T.K.S.: A framework for hybrid tractability results in Boolean weighted constraint satisfaction problems. In: Stuckey, P.J. (ed.) CP 2008. LNCS, vol. 5202, pp. 282–297. Springer, Heidelberg (2008). doi:10.1007/978-3-540-85958-1_19
8. Kumar, T.K.S.: Lifting techniques for weighted constraint satisfaction problems. In: The International Symposium on Artificial Intelligence and Mathematics (2008)
9. Kumar, T.K.S.: Kernelization, generation of bounds, and the scope of incremental computation for weighted constraint satisfaction problems. In: The International Symposium on Artificial Intelligence and Mathematics (2016)
10. Marinescu, R., Dechter, R.: Best-first AND/OR search for graphical models. In: The AAAI Conference on Artificial Intelligence, pp. 1171–1176 (2007)
11. Siek, J., Lee, L.Q., Lumsdain, A.: The Boost Graph Library: User Guide and Reference Manual. Addison-Wesley, Boston (2002)
12. Sierksma, G.: Linear and Integer Programming: Theory and Practice, 2nd edn. CRC Press, Boca Raton (2001)
13. Xu, H., Kumar, T.K.S., Koenig, S.: A new solver for the minimum weighted vertex cover problem. In: Quimper, C.-G. (ed.) CPAIOR 2016. LNCS, vol. 9676, pp. 392–405. Springer, Cham (2016). doi:10.1007/978-3-319-33954-2_28
14. Xu, H., Kumar, T.K.S., Koenig, S.: The Nemhauser-Trotter reduction and lifted message passing for the weighted CSP. In: Salvagnin, D., Lombardi, M. (eds.) CPAIOR 2017. LNCS, vol. 10335, pp. 387–402. Springer, Cham (2017). doi:10.1007/978-3-319-59776-8_31

Satisfiability and CP Track

Reduced Cost Fixing in MaxSAT

Fahiem Bacchus[1(✉)], Antti Hyttinen[2], Matti Järvisalo[2], and Paul Saikko[2]

[1] Department of Computer Science, University of Toronto, Toronto, Canada
fbacchus@cs.toronto.edu
[2] HIIT, Department of Computer Science, University of Helsinki, Helsinki, Finland
matti.jarvisalo@helsinki.fi

Abstract. We investigate utilizing the integer programming (IP) technique of reduced cost fixing to improve maximum satisfiability (MaxSAT) solving. In particular, we show how reduced cost fixing can be used within the implicit hitting set approach (IHS) for solving MaxSAT. Solvers based on IHS have proved to be quite effective for MaxSAT, especially on problems with a variety of clause weights. The unique feature of IHS solvers is that they utilize both SAT and IP techniques. We show how reduced cost fixing can be used in this framework to conclude that some soft clauses can be left falsified or forced to be satisfied without influencing the optimal cost. Applying these forcings simplifies the remaining problem. We provide an extensive empirical study showing that reduced cost fixing employed in this manner can be useful in improving the state-of-the-art in MaxSAT solving especially on hard instances arising from real-world application domains.

1 Introduction

Maximum satisfiability (MaxSAT) [17] is a thriving constraint optimization paradigm, successfully applied in a growing number of NP-hard real-world problem domains. The currently most successful MaxSAT solvers are SAT-based, i.e., rely on Boolean satisfiability solver technology [4]. In particular, they use SAT solvers to iteratively extract unsatisfiable cores (unsatisfiable sets of soft clauses) and block these cores from the search in the later iterations, until a solution is found. One of the currently most successful algorithmic approaches—as witnessed by the most recent MaxSAT Evaluations [2]—are solvers implementing the so-called implicit hitting set (IHS) approach for MaxSAT. IHS MaxSAT solvers [8,9,11,23,24] employ a hybrid approach that exploits both a SAT solver for core extraction and an integer programming (IP) solver for obtaining minimum cost hitting sets of the accumulated cores.

Work supported in part by Academy of Finland (grants 251170 COIN, 276412, 284591 and 295673), the Research Funds and DoCS Doctoral School in Computer Science of the University of Helsinki, and the Natural Sciences and Engineering Research Council of Canada.

© Springer International Publishing AG 2017
J.C. Beck (Ed.): CP 2017, LNCS 10416, pp. 641–651, 2017.
DOI: 10.1007/978-3-319-66158-2_41

Despite the success and recent algorithmic advances in MaxSAT solvers, the SAT-based MaxSAT solvers do not—as witnessed by the empirical results presented in this paper—currently harness the full potential of bounds-based problem simplification during search. Focusing on IHS as the approach which solved the most instances in the general weighted partial category of the 2016 MaxSAT Evaluation, we propose to take advantage of classical ideas from the realm of integer programming to further improve state-of-the-art MaxSAT solvers. In more detail, we show how to integrate *reduced cost fixing* [6,7,22], a standard technique in IP solving that uses bounds on the optimal cost derived during search for inferring variables whose values can be fixed while preserving at least one optimal solution. As we will explain in detail, in terms of MaxSAT search, reduced cost fixing amounts to using upper bounds obtained during search to harden or falsify specific soft clauses, i.e., to force them to be satisfied or falsified. The IHS approach to MaxSAT is a prime candidate for integrating reduced cost fixing since the reduced costs of soft clauses can be readily obtained by solving a linear (LP) relaxation of the hitting set problem maintained during IHS search. Putting this idea into practice, we extend the IHS solver MaxHS with reduced cost fixing, and provide an extensive empirical evaluation showing that reduced cost fixing considerably speeds up MaxHS.

In terms of related work, different techniques of using lower and lower bounds for speeding up MaxSAT solver have studied in varying contexts, including branch-and-bound for MaxSAT [15,16,18,19], use of bounds for MaxSAT solvers in general [13], and hardening based on SAT inferred costs of residual formulas in pure SAT-based core-guided MaxSAT solving [1,21]. However, to the best of our knowledge, linear programming relaxation based reduced cost fixing has not been previously proposed in the context of MaxSAT. There has, however, been a number of related works exploiting the technique of reduced cost fixing in constraint programming, IP/constraint logic programming, and IP/constraint programming, e.g., [12,26,28].

After background on MaxSAT (Sect. 2), we give a bounds-based view of the IHS approach to MaxSAT (Sect. 3), explain how to integrate reduced cost fixing into it (Sect. 4), and present empirical results on the effectiveness of reduced cost fixing in speeding up the IHS solver MaxHS (Sect. 5).

2   Maximum Satisfiability

We work with propositional formulas expressed in conjunctive normal form (CNF). Satisfaction of CNF formulas is defined as usual. Whenever convenient we treat a clause as a set of literals and a CNF formula as a set of clauses. An instance of (weighted partial) maximum satisfiability (MaxSAT) $F = (F_h, F_s, wt)$ consists of two CNF formulas: the hard clauses F_h, the soft clauses F_s and a weight function $wt \colon F_s \to \mathbb{Q}$ associating a positive rational weight to each soft clause. Given such an instance, any truth assignment τ that satisfies the hard clauses is a **solution** to F. The **cost** of a solution τ, $cost(F, \tau)$, is the sum of the weights of the soft clauses it falsifies: $cost(F, \tau) = \sum_{\{\tau \not\models C \mid C \in F_s\}} wt(C)$.

A solution τ is optimal if $cost(F, \tau) \leq cost(F, \tau')$ for all solutions τ'. Given an instance F the MaxSAT problem is to find an optimal solution to F. We denote the cost of optimal solutions to F by $opt\_cost(F)$. We also use $cost(S)$, for any set of soft clauses S, to denote the sum of weights of the soft clauses in S: $cost(S) = \sum_{c \in S} wt(c)$. For a MaxSAT instance $F = (F_h, F_s, wt)$, an unsatisfiable **core** of F is any subset $S \subseteq F_s$ of soft clauses such that $F_h \cup S$ is unsatisfiable.

3   The SAT-IP Implicit Hitting Set Approach to MaxSAT

IHS MaxSAT solvers [8–11,23,24] utilize the so-called implicit hitting set approach [14,20,25] to solve weighted partial MaxSAT. These solvers use a SAT solver to accumulate cores and an IP solver to compute a minimum-cost hitting set of the accumulated cores. Since each core is unsatisfiable, any solution must falsify at least one soft clause in every core, i.e., the set of soft clauses falsified by any solution must form a hitting set of the set of cores. Therefore, the cost of any solution is lower-bounded by the cost of the minimum-cost hitting. Further, if these costs are equal, then the solution must be optimal. As first described in [8] an iteration can be set up that ensures that the IHS solver finds an optimal solution after producing a finite number of cores.

Algorithm 1. The IHS approach to MaxSAT (generalized from [8])

---

1 **IHS-MaxSAT** $(F = (F_h, F_s, wt))$
2 $(sat, \kappa, \tau) \leftarrow$ **SolveSAT**(F_h) /* If unsat return a core κ, else a solution τ */
3 **if** *not* sat **then return** *"No solutions since F_h is UNSAT"*
4 $UB \leftarrow cost(F, \tau)$; $best\_\tau \leftarrow \tau$; $LB = 0$ /* Initial bounds */
5 **Optimizer**.$initialze(wt)$; $new\_cores \leftarrow \emptyset$; $hs\_is\_sat \leftarrow hs\_is\_opt \leftarrow false$
6 **while** $UB > LB$ **do**
7 $\quad$ $(hs\_is\_opt, HS) \leftarrow$ **Optimizer**(new_cores, UB)
8 $\quad$ **if** hs_is_opt **then** $LB = cost(HS)$
9 $\quad$ $(hs\_is\_sat, \kappa, \tau) \leftarrow$ **SolveSAT**$(F_h \cup (F_s \setminus HS))$
10 $\quad$ **if** *not* hs_is_sat **then**
11 $\quad\quad$ **repeat**
12 $\quad\quad\quad$ $new\_cores \leftarrow new\_cores \cup \kappa$
13 $\quad\quad\quad$ $(sat, \kappa, \tau) \leftarrow$ **SolveSAT**$(F_n \cup (F_s \setminus (HS \cup \bigcup_{\kappa \in new\_cores} \kappa)))$
14 $\quad\quad$ **until** sat
15 $\quad$ **if** $cost(\tau) < UB$ **then** $UB \leftarrow cost(\tau)$; $best\_\tau \leftarrow \tau$
16 **return** $best\_\tau$

This original algorithm does not, however, provide the upper bounds needed for reduced cost fixing. Upper bounds can be obtained by using non-optimal hitting sets as described in [11]. We give, in Algorithm 1, a new more general formalization of the algorithm described in [11] and a more general correctness condition.

The algorithm first computes an initial model by solving F_h. The returned τ is also a solution to F, and provides an initial upper bound once we check which clauses of F_s are satisfied by τ.

The **Optimizer** maintains the set of cores passed to it, adding the cores in *new_cores* to this set (line 7). It always returns a hitting set HS of its current set of cores, and a flag (hs_is_opt) indicating whether it has verified HS to be of minimum cost. (HS might be of minimum cost even if the **Optimizer** has not verified this.) If HS is of minimum cost, its cost is a valid lower bound on $opt\_cost(F)$ and we can update LB. Note that **Optimizer**'s set of cores can only grow so the cost of a minimum-cost hitting set cannot decrease and the updates never decrease LB.

The SAT solver tests if removing HS from F_s results in satisfiability; if not we obtain a new core, κ, and add it to the set of *new_cores*. We then enter a loop where we accumulate more cores, repeatedly removing all of the soft clauses in HS and all newly discovered cores (cf. the "disjoint, g" strategy in [24]). At each step a new core is found and the set of soft clauses passed to the SAT solver is further reduced. Since F_h is satisfiable, the loop must terminate as we will eventually remove enough soft clauses to obtain satisfiability. We then update the upper bound if the found solution has lower cost.

The algorithm terminates when it finds a solution whose cost achieves the lower bound. Such a solution must be optimal; hence Algorithm 1 always returns an optimal solution. We also have that Algorithm 1 must terminate as long as **Optimizer** satisfies the following general condition. During Algorithm 1 a sequence of calls are made to **Optimizer** (once every iteration of the while loop). In each call **Optimizer** computes a hitting set of the accumulated set of cores passed to it in the current and all previous calls, and during that call UB is the best known upper bound.

Definition 1 (Correctness Condition). *Optimizer* *always returns a hitting set of its accumulated set of cores. And, for every i there exists an $k > i$ such that the k'th call to* **Optimizer** *returns a hitting set, HS, such that either (a) $cost(HS) < UB$ or (b) HS is a minimum-cost hitting set.*

Theorem 1. *If* **Optimizer** *satisfies the correctness condition, then Algorithm 1 must eventually terminate returning an optimal solution.*

Proof. We show that the sequence of calls to **Optimizer** is finite, and thus the while loop must terminate. In fact, we need only consider the sub-sequence calls consisting of those calls where **Optimizer** returns a minimum-cost hitting set or a hitting set with cost less than the current upper bound. By the correctness condition this sub-sequence is infinite iff Algorithm 1 fails to terminate. We say that a hitting set HS returned by **Optimizer** is infeasible if $F_h \cup (F_s \setminus HS)$ is unsatisfiable, otherwise it is feasible. Note that when HS is feasible, we will have $cost(\tau) \leq cost(HS)$ for the returned model τ, and $UB \leq cost(HS)$ after line 15.

Optimizer cannot return an infeasible hitting set HS more than once: HS will cause a core to be added to **Optimizer** that HS does not hit, so HS will not be a hitting set for any subsequent calls. In the sub-sequence **Optimizer** can never return a feasible hitting set HS more than twice. After HS is returned we have that $UB \leq cost(HS)$. If **Optimizer** also returned $hs\_is\_opt = true$, then LB will become equal to UB and the algorithm will terminate. Otherwise, if $hs\_is\_opt = false$, the lowered UB implies that if HS is returned once more in the subsequence it must be with $hs\_is\_opt = true$, which will cause termination. There are only a finite number of hitting sets, so the sub-sequence must be finite, and Algorithm 1 must terminate. □

In the version of MaxHS reported on in our empirical evaluation **Optimizer** utilizes both a heuristic greedy solver and an exact IP solver (IBM CPLEX). It always uses the greedy solver unless it is passed an empty set new_cores (which happens when the previous call to **Optimizer** returned a feasible hitting set). For an empty new_cores it uses the IP solver to compute a hitting set. However, it does not ask the IP solver to compute a minimum-cost hitting set. Rather it stops the IP solver as soon as a hitting set with cost less than UB has been found. When UB is already equals the optimal cost, the IP solver will run to completion as a lower cost hitting set will never be found. In this case the IP solver will find a hitting set that it can verify to be of minimum cost, and this hitting set and $hs\_is\_opt = true$ is returned. This scheme is used to reduce the number of times the hitting set problem needs to be solved to optimality [11].

4 Reduced Cost Fixing

Reduced cost fixing is a standard technique in OR [6,7,22,27]. It uses an upper bound and reduced costs obtained from an LP relaxation to fix variables in an IP. Given a minimization IP P containing Boolean (0/1) variables, we can solve P as an LP by allowing the Boolean variables to take on intermediate values between 0 and 1. The cost of the LP solution will be a lower bound on the optimal cost of P. The LP solver also provides a reduced cost for the non-basic[1] variables set at 0 or 1 in the LP solution. These reduced costs specify the influence of changing a non-basic variable at 0 (1) to 1 (0) on the cost of the LP. Suppose we know a feasible IP solution to P with cost z. If changing a non-basic variable causes the LP solution to increase in cost beyond z, then we can fix that variable to the value it has in the LP solution. Since the LP solution is a lower bound, putting such variables at their opposite values would cause the cost of the IP to increase beyond the cost of an already known feasible solution.

Here we explain how this technique can be used within IHS MaxSAT solvers. In contrast with standard uses of reduced cost fixing we do not want to fix variables of the IP (our IP is the IP of the hitting set problem). Rather we want to fix variables of the MaxSAT problem from which the IP has been derived. This can be done as follows.

---

[1] The variables in the LP solution are either basic or non-basic. All of the non-basic variables will be at their upper or lower bounds in the LP solution [5].

Theorem 2. *For a MaxSAT problem $F = (F_h, F_s, wt)$, suppose we have (a) $B = \{b_1, \ldots, b_n\}$ a set of Boolean variables where each $b_i = 0$ ($b_i = 1$) represents the satisfaction (falsification) of soft clause $c_i \in F_s$, (b) IP_{HS} an IP over the b_i representing the minimum-cost hitting set problem over the current set of cores, (c) LP_{HS} the LP relaxation of IP_{HS}, (d) $best\_\tau$ a feasible solution to F, (e) an optimal solution to LP_{HS} with cost $z_{opt}^{LP_{HS}}$, and (f) LP reduced costs $rc(b_i)$ at the optimal basis.*

Then the following simplifications can be performed without changing $opt\_cost(F)$. (1) For every non-basic variable b_i set to 0 in the optimal LP_{HS} solution we can make soft clause c_i hard in F if $z_{opt}^{LP_{HS}} + rc(b_i) > cost(best\_\tau)$ or if $z_{opt}^{LP_{HS}} + rc(b_i) = cost(best\_\tau)$ and c_i is satisfied in $best\_\tau$. (2) For every non-basic variable b_i set to 1 in the optimal LP_{HS} we can make soft clause c_i false in F if $z_{opt}^{LP_{HS}} - rc(b_i) > cost(best\_\tau)$ or if $z_{opt}^{LP_{HS}} - rc(b_i) = cost(best\_\tau)$ and c_i is falsified in $best\_\tau$.

Proof. Let b_i be a non-basic variable at its lower bound in the optimal solution to LP_{HS}. Then either $b_i = 1$ is feasible in LP_{HS} or it is not.[2] If it is not, then, since LP_{HS} is a relaxation of IP_{HS}, $b_i = 1$ is also infeasible in IP_{HS}. Furthermore, since every core is a logical consequence of F, IP_{HS} is a relaxation of F and thus $c_i = false$ must be infeasible in F, and we can harden c_i. On the other hand, if $b_i = 1$ is feasible in LP_{HS}, then by the properties of reduced costs, forcing $b_i = 1$ will increase the optimal cost of LP_{HS} by at least $rc(b_i)$ [3]. Stated a different way, if LP^+ is LP_{HS} with the added constraint $b_i = 1$, then its optimal cost will be at least $z^{LP_{HS}} + rc(b_i)$. LP^+ is the linear relaxation of IP^+, which is IP_{HS} with the added constraint $b_i = 1$; and IP^+ is a relaxation of F^+ which is $F \cup \neg c_i$. Hence, $cost(F^+) >= z^{LP_{HS}} + rc(b_i)$ and if $z^{LP_{HS}} + rc(b_i) > cost(best\_\tau)$, or if $z^{LP_{HS}} + rc(b_i) = cost(best\_\tau)$ and c_i is satisfied in $best\_\tau$, then we can force c_i to be satisfied in F while still preserving at least one of the optimal solutions of F. The argument for b_i at its upper bound is analogous.   □

In Algorithm 1 reduced cost fixing can be utilized whenever $UB - LB$ decreases and is small enough to allow the forcing of some unforced soft clause. In particular, $rc(b_i)$ is upper-bounded by $wt(c_i)$ and hence c_i cannot be forced if $(UB - LB) > wt(c_i)$. We use CPLEX to solve the LP relaxation of the hitting set problem to obtain the reduced costs; we do this just before invoking CPLEX in **Optimizer**.

5   Experiments

We implemented reduced cost fixing in MaxHS v2.9.8 which entered the 2016 MaxSAT Evaluation. This version of MaxHS included a number of other features shown to improve the solver, described in [9,11,23]. We compare the performance of MaxHS with and without reduced cost fixing, with all other features

---

[2] In a hitting set problem $b_i = 1$ is always feasible. However, MaxHS can also add other constraints to the hitting set problem via a process of constraint seeding [9]. It is not difficult to show that all of our results continue to hold with seeding.

unchanged. We utilized IBM CPLEX v12.7 as the IP/LP solver, and ran our experiments on computing nodes with Xeon 2.8-GHz cores and 256-GB RAM. We limited MaxHS to 1800 s and 3.5 GB on each instance. We also report on longer 5-h (18,000 s), 5-GB runs on Xeon 2.0-GHz cores and 256-GB RAM.

We experimented with all non-random instances that have been collected by and made available by the MaxSAT Evaluation during the years 2008 to 2016. These include extra submitted benchmarks never used in the evaluation. After pruning duplicate instances this yielded 6290 MaxSAT instances (4361 unweighted, 1929 weighted). For the 5-h runs, however, we omitted 507 unweighted instances with no hard clauses (MS instances) most of which encode MaxCut on random graphs. Core-based solvers, including IHS solvers, perform poorly on such instances, and we did not expect any of these instances to complete in 5 h with or without reduced cost fixing. This left 5783 instances to run in these longer experiments (4361 unweighted, 1422 unweighted).

First we examine how frequently reduced cost fixing occurs in our benchmark suite. Figure 1 left shows a histogram of the instances grouped by the number of soft clauses that become fixed during solving. In 5024 of the 6290 instances no reduced cost fixing ever occurs (3953 unweighted, 1071 weighted), but in the remaining 1266 instances fixing can be quite common—in 791 of these instances 100 or more fixings occurred. In extreme cases over a million soft clauses were fixed by the technique (this makes average number of soft clauses fixed misleadingly large). There was little difference in the histograms between weighted and unweighted instances once the zero fixing instances were removed; in fact, the instance with the most fixings was unweighted.

The second question is how much overhead does reduced cost fixing incur, particularly since the LP is solved even when no fixing occurs. There were 26 instances where fixing took more than 100 s. However, 25 of these were not solvable with or without fixing (22 were MaxCut on random graphs). On one solved instance fixing required 214 s out of a total solve time of 835 s (this instance was solved in 416 s without fixing). Of the remaining 6264 instances, on 1782 instances fixing took zero seconds (LP solving was never invoked since the gap between UB and LB was never small enough), on 3746 instances fixing took less than 1 s, on 298 instances fixing took between 1 and 10 s, and on 438 instances fixing took more than between 10 and 100 s. Figure 1 right shows, however, that on these 438 instances fixing is well worth the time it takes. The scatter plot shows that fixing provides a significant speedup for most of these instances, especially on the harder instances.

In the rest of our plots we omit data from the 5024 instances on which no reduced cost fixing occurred. We omitted these instances because their run times will only vary by the overhead of fixing (and experimental variances induced by varying cluster loads), and we have already provided data in the previous paragraph showing that this overhead is not significant.

Figure 2 shows scatter plots for all instances, all unweighted instances and all weighted instances. The plots show that fixing generally provides a speedup, and that speedups occur on both weighted and unweighted instances.

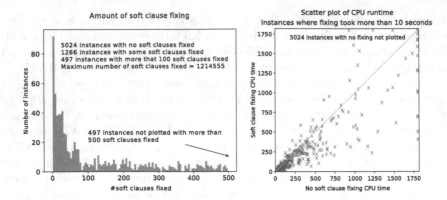

Fig. 1. Left: distribution of the frequency reduced cost fixing forces a soft clause to be relaxed or hardened. Right: scatter plot showing that fixing on instances where fixing takes significant time also pays off.

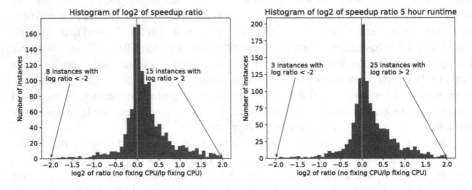

Fig. 2. Speedup histograms over instances on which reduced cost fixing would force some variables in terms of $\log_2$ of CPU time with fixing and without fixing. Left: under 30-min per-instance time limit, Right: under 5-h per-instance time limit.

In Fig. 3 we show in more detail the performance improvement obtained from reduced cost fixing. Here we computed the speedup ratio for each instance, i.e., the CPU time taken without reduced cost fixing divided by the CPU time taken when reduced cost fixing is used. As this ratio will be between 0 and 1, for instances that are slowed down by fixing we took $\log_2$ of this ratio which produces a symmetry between speedups and slowdowns. The plots are in the form of histograms showing for how many instances experience various ranges of the log speedup. Figure 3 left shows the log speedup ratio for all instances, while on the right we examine the 4361 instances that were run under a per-instance time limit of 5 h.

These histograms verify the value of our technique for exploiting reduced cost fixing in IHS based MaxSAT solvers. When we look at the data from the 5-h runs we see an even more pronounced effect with fewer instances being slowed down, and a smoother distribution for the instances being speeded up.

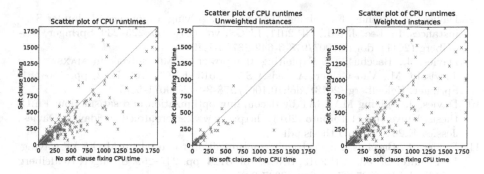

Fig. 3. Scatter plots of CPU times with and without reduced cost fixing, omitting instances 5024 where no fixing occurred. Left: all instances; Middle: unweighted instances; Right: weighted instances.

6   Conclusions

We proposed the use of reduced cost fixing—a standard approach in IP—in MaxSAT solving as a means of utilizing bounds information during search to infer knowledge of soft clauses which are satisfied or left falsified by some optimal solutions. We explained how reduced cost fixing can be integrated into the implicit hitting set approach to MaxSAT by performing reduced cost analysis directly on the LP relaxation of the hitting-set IP already utilized in the IHS search routine. We showed through an extensive empirical evaluation that reduced cost fixing can provide considerable speedups improving on the overall performance of MaxHS.

References

1. Ansótegui, C., Bonet, M.L., Gabàs, J., Levy, J.: Improving WPM2 for (weighted) partial MaxSAT. In: Schulte, C. (ed.) CP 2013. LNCS, vol. 8124, pp. 117–132. Springer, Heidelberg (2013). doi:10.1007/978-3-642-40627-0_12
2. Argelich, J., Li, C.M., Manyà, F., Planes, J.: MaxSAT evaluation (2016). http://maxsat.ia.udl.cat/introduction/. Accessed 27 Apr 2017
3. Bajgiran, O.S., Cire, A.A., Rousseau, L.-M.: A first look at picking dual variables for maximizing reduced cost fixing. In: Salvagnin, D., Lombardi, M. (eds.) CPAIOR 2017. LNCS, vol. 10335, pp. 221–228. Springer, Cham (2017). doi:10.1007/978-3-319-59776-8_18
4. Biere, A., Heule, M., van Maaren, H., Walsh, T. (eds.): Handbook of Satisfiability. Frontiers in Artificial Intelligence and Applications, vol. 185. IOS Press, Amsterdam (2009)
5. Chvátal, V.: Linear Programming. Freeman, New York (1983)
6. Crowder, H., Johnson, E.L., Padberg, M.: Solving large-scale zero-one linear programming problems. Oper. Res. **31**(5), 803–834 (1983)
7. Danzig, G., Fulkerson, D., Johnson, S.: Solution of a large-scale traveling-salesman problem. Oper. Res. **2**, 393–410 (1954)

8. Davies, J., Bacchus, F.: Solving MAXSAT by solving a sequence of simpler SAT instances. In: Lee, J. (ed.) CP 2011. LNCS, vol. 6876, pp. 225–239. Springer, Heidelberg (2011). doi:10.1007/978-3-642-23786-7_19

9. Davies, J., Bacchus, F.: Exploiting the power of MIP solvers in MAXSAT. In: Järvisalo, M., Van Gelder, A. (eds.) SAT 2013. LNCS, vol. 7962, pp. 166–181. Springer, Heidelberg (2013). doi:10.1007/978-3-642-39071-5_13

10. Davies, J.: Solving MAXSAT by decoupling optimization and satisfaction. Ph.D. thesis, University of Toronto (2013). http://www.cs.toronto.edu/~jdavies/Davies_Jessica_E_201311_PhD_thesis.pdf

11. Davies, J., Bacchus, F.: Postponing optimization to speed up MAXSAT solving. In: Schulte, C. (ed.) CP 2013. LNCS, vol. 8124, pp. 247–262. Springer, Heidelberg (2013). doi:10.1007/978-3-642-40627-0_21

12. Focacci, F., Lodi, A., Milano, M.: Cost-based domain filtering. In: Jaffar, J. (ed.) CP 1999. LNCS, vol. 1713, pp. 189–203. Springer, Heidelberg (1999). doi:10.1007/978-3-540-48085-3_14

13. Heras, F., Morgado, A., Marques-Silva, J.: Lower bounds and upper bounds for MaxSAT. In: Hamadi, Y., Schoenauer, M. (eds.) LION 2012. LNCS, pp. 402–407. Springer, Heidelberg (2012). doi:10.1007/978-3-642-34413-8_35

14. Karp, R.M.: Implicit hitting set problems and multi-genome alignment. In: Amir, A., Parida, L. (eds.) CPM 2010. LNCS, vol. 6129, p. 151. Springer, Heidelberg (2010). doi:10.1007/978-3-642-13509-5_14

15. Li, C.M., Manyà, F., Mohamedou, N.O., Planes, J.: Transforming inconsistent subformulas in MaxSAT lower bound computation. In: Stuckey, P.J. (ed.) CP 2008. LNCS, vol. 5202, pp. 582–587. Springer, Heidelberg (2008). doi:10.1007/978-3-540-85958-1_46

16. Li, C.M., Manyà, F., Planes, J.: Detecting disjoint inconsistent subformulas for computing lower bounds for Max-SAT. In: Proceedings of AAAI, pp. 86–91. AAAI Press (2006)

17. Li, C., Manyà, F.: MaxSAT, hard and soft constraints. In: Handbook of Satisfiability, pp. 613–631. IOS Press, Amsterdam (2009)

18. Lin, H., Su, K.: Exploiting inference rules to compute lower bounds for MAX-SAT solving. In: Proceedings of IJCAI, pp. 2334–2339 (2007)

19. Lin, H., Su, K., Li, C.M.: Within-problem learning for efficient lower bound computation in Max-SAT solving. In: Proceedings of AAAI, pp. 351–356. AAAI Press (2008)

20. Moreno-Centeno, E., Karp, R.M.: The implicit hitting set approach to solve combinatorial optimization problems with an application to multigenome alignment. Oper. Res. **61**(2), 453–468 (2013)

21. Morgado, A., Heras, F., Marques-Silva, J.: Improvements to core-guided binary search for MaxSAT. In: Cimatti, A., Sebastiani, R. (eds.) SAT 2012. LNCS, vol. 7317, pp. 284–297. Springer, Heidelberg (2012). doi:10.1007/978-3-642-31612-8_22

22. Nemhauser, G.L., Wolsey, L.A.: Integer and Combinatorial Optimization. Wiley-Interscience, Hoboken (1999)

23. Saikko, P., Berg, J., Järvisalo, M.: LMHS: a SAT-IP hybrid MaxSAT solver. In: Creignou, N., Le Berre, D. (eds.) SAT 2016. LNCS, vol. 9710, pp. 539–546. Springer, Cham (2016). doi:10.1007/978-3-319-40970-2_34

24. Saikko, P.: Re-implementing and extending a hybrid SAT-IP approach to maximum satisfiability. Master's thesis, University of Helsinki (2015). http://hdl.handle.net/10138/159186

25. Saikko, P., Wallner, J.P., Järvisalo, M.: Implicit hitting set algorithms for reasoning beyond NP. In: Proceedings of KR, pp. 104–113. AAAI Press (2016)

26. Thorsteinsson, E.S., Ottosson, G.: Linear relaxations and reduced-cost based propagation of continuous variable subscripts. Ann. Oper. Res. **115**(1–4), 15–29 (2002)
27. Wolsey, L.A.: Integer Programming. Wiley, Hoboken (1998)
28. Yunes, T.H., Aron, I.D., Hooker, J.N.: An integrated solver for optimization problems. Oper. Res. **58**(2), 342–356 (2010)

Weight-Aware Core Extraction in SAT-Based MaxSAT Solving

Jeremias Berg$^{(\boxtimes)}$ and Matti Järvisalo

HIIT, Department of Computer Science, University of Helsinki, Helsinki, Finland
{jeremias.berg,matti.jarvisalo}@helsinki.fi

Abstract. Maximum satisfiability (MaxSAT) is today a competitive approach to tackling NP-hard optimization problems in a variety of AI and industrial domains. A great majority of the modern state-of-the-art MaxSAT solvers are core-guided, relying on a SAT solver to iteratively extract unsatisfiable cores of the soft clauses in the working formula and ruling out the found cores via adding cardinality constraints into the working formula until a solution is found. In this work we propose weight-aware core extraction (WCE) as a refinement to the current common approach of core-guided solvers. WCE integrates knowledge of soft clause weights into the core extraction process, and allows for delaying the addition of cardinality constraints into the working formula. We show that WCE noticeably improves in practice the performance of PMRES, one of the recent core-guided MaxSAT algorithms using soft cardinality constraints, and explain how the approach can be integrated into other core-guided algorithms.

1   Introduction

Several recent breakthroughs in algorithmic techniques for the constraint optimization paradigm of maximum satisfiability (MaxSAT) are making MaxSAT today a competitive approach to tackling NP-hard optimization problems in a variety of AI and industrial domains, from planning, debugging, and diagnosis to machine learning and systems biology, see e.g. [8,11,13,14,17,25,34].

A great majority of the most successful MaxSAT solvers today are based on the so-called core-guided MaxSAT solving paradigm, see e.g. [4,12,16,29–31]. Such solvers iteratively use Boolean satisfiability (SAT) solvers for finding unsatisfiable cores, i.e., sets of soft clauses that together with the hard clauses are unsatisfiable, of the input MaxSAT instance. After finding a new core, the core is essentially compiled into the MaxSAT instance via adding a cardinality constraint enforcing that one of the soft clauses in the core cannot be satisfied. An in-built property of core-guided solvers is hence that the MaxSAT instance grows

Work supported by Academy of Finland (grants 251170 COIN, 276412, 284591); and DoCS Doctoral School in Computer Science and Research Funds of the University of Helsinki.

J.C. Beck (Ed.): CP 2017, LNCS 10416, pp. 652–670, 2017.
DOI: 10.1007/978-3-319-66158-2_42

at each iteration due to compiling a new core into the instance. This can lead to the instance becoming bloated, as many, possibly large, cores are compiled, intuitively making the job of the SAT solver increasingly difficult. One way of improving core-guided solvers is to develop more efficient ways of compiling the cores, decreasing the blow-up of the instance. Most recently, progress in core-guided solvers has been made by developing new ways of compiling the cores via *soft cardinality constraints* [12,30,31].

In this work we propose *weight-aware core extraction* (WCE) as a technique that refines the process of how cores are extracted and when they are compiled into the working formula during core-guided MaxSAT search. WCE integrates knowledge of soft clause weights into the core extraction process, and allows for delaying the addition of cardinality constraints into the working formula by enabling the extraction of more cores between compilation steps, thereby also intuitively making the job of the core extractor (SAT solver) easier. In this paper we explain in detail how a specific implementation of clause cloning allows integrating WCE into PMRES, the first algorithm making use of soft cardinality constraints [31]. We also show empirically that WCE noticeably improves the performance of PMRES in practice on standard weighted partial MaxSAT benchmarks from the most recent MaxSAT solver evaluation. Going beyond PMRES, we also explain how ideas behind WCE can be integrated into other core-guided algorithms employing soft cardinality constraints, and to what extent the presented ideas can be used in some of the other MaxSAT approaches utilizing SAT solvers for core extraction.

In terms of related work, ideas underlying WCE have been previously applied for computing lower bounds for MaxSAT instances [18,20–23]. Specifically, the lower bounds are applied in the context of core-guided MaxSAT solving before the actual search in [18]. Furthermore, WCE also bears some resemblance with the (weaker) approaches to obtaining bounds during branch-and-bound search for MaxSAT based on detecting unsatisfiable cores by e.g. unit propagation [20–23].

The rest of the paper is organized as follows. After necessary background on MaxSAT (Sect. 2) and a detailed description of the PMRES algorithm (Sect. 3), we present our main contributions, weight-aware core extraction in the context of PMRES (Sect. 4). We then present empirical results on the speed-ups obtained via WCE on PMRES (Sect. 5), and further, explain how and to what extent the presented technique can be integrated into other SAT-based MaxSAT algorithms (Sect. 6).

2   Maximum Satisfiability

For background on weighted partial maximum satisfiability (MaxSAT in short), recall that for a Boolean variable x, there are two literals, the positive x and the negative $\neg x$. A clause is a disjunction ($\vee$) of literals, and a CNF formula a conjunction ($\wedge$) of clauses. When convenient, we treat a clause as a set of literals and a CNF formula as a set of clauses. We assume familiarity with other logical connectives and denote by $\text{CNF}(\phi)$ a set of clauses logically equivalent to the formula ϕ; we can assume without loss of generality that the size of $\text{CNF}(\phi)$ is linear in the size of ϕ [33].

A MaxSAT instance consists of a set of hard clauses F_h, a set of soft clauses F_s, and a function $w\colon F_s \to \mathbb{N}$ that associates a positive integral cost to each of the soft clauses. We extend w to a set $S \subseteq F_s$ of soft clauses by $w(S) = \sum_{C \in S} w(C)$. Further, let $w_S^{\min} = \min_{C \in S}\{w(C)\}$, i.e., the smallest weight among the clauses in S. If $w(C) = 1$ for all $C \in F_s$, the instance is unweighted.

A truth assignment τ is a function from Boolean variables to true (1) and false (0). A clause C is satisfied by τ if $\tau(l) = 1$ for a positive or $\tau(l) = 0$ for a negative literal $l \in C$. A CNF formula is satisfied by τ if τ satisfies all clauses in the formula. If some τ satisfies a CNF formula, the formula is satisfiable, and otherwise unsatisfiable. An assignment τ is a solution to a MaxSAT instance $F = (F_h, F_s, w)$ if τ satisfies F_h. We denote the set of soft clauses not satisfied by τ by $F_{\bar\tau}$, i.e., $F_{\bar\tau} = \{C \in F_s \mid \tau(C) = 0\}$. The cost of τ is $w(F_{\bar\tau})$. A solution τ is optimal (for F) if $w(F_{\bar\tau}) \leq w(F_{\bar\tau'})$ for every solution τ' to F. We denote the cost of optimal solutions to F by $\text{COST}(F)$. Without loss of generality, we will assume that a MaxSAT instance always has a solution, i.e., that F_h is satisfiable.

A central concept in modern SAT-based MaxSAT algorithms is that of (unsatisfiable) cores. For a MaxSAT instance $F = (F_h, F_s, w)$, a subset $S \subseteq F_s$ of soft clauses is an unsatisfiable core of F iff $F_h \cup S$ is unsatisfiable. An unsatisfiable core S is minimal (an MUS) of F iff $F_h \cup S'$ is satisfiable for all $S' \subset S$.

3   The PMRES Algorithm

In order to explain weight-aware core extraction, we will use the PMRES algorithm [31]. Figure 1 gives PMRES in pseudo-code. When invoked on a MaxSAT

```
1  PMRES(F_h, F_s, w):                          1  RELAX(w_κ^min, R):
2  (F_h^w, F_s^w) ← (F_h, F_s)                   2  n ← |R|
3  while true do                                 3  F_h^w.add((r_1 ∨ ... ∨ r_n))
4     (result, κ, τ) ← SATSOLVE(F_h^w ∪ F_s^w)   4  for i=1...n-1 do
5     if result="satisfiable" then return τ;     5     F_h^w.add(CNF(d_i ↔ (r_{i+1} ∨ d_{i+1})))
6     else                                       6     F_s^w.add((¬r_i ∨ ¬d_i))
7        R ← ∅                                    7     w((¬r_i ∨ ¬d_i)) ← w_κ^min
8        w_κ^min ← min{w(C) | C ∈ κ}
9        for C_i ∈ κ do
10          F_s^w.remove(C_i)
11          if w(C_i) > w_κ^min then
12             F_s^w.add(CL(C_i))
13             w(CL(C_i)) ← w(C_i)−w_κ^min
14          F_h^w ← F_h^w.add((C_i ∨ r_i))
15          R.add(r_i)
16       RELAX(w_κ^min, R)
```

Fig. 1. The PMRES algorithm.

instance (F_h, F_s, w) PMRES works by iteratively calling a SAT solver (line 4) on a working formula, initialized to $F_h \cup F_s$, i.e., considering all hard and soft clauses of the input formula as a SAT instance (line 2). If the working formula is satisfiable (line 5), PMRES returns the satisfying assignment reported by the SAT solver, which is guaranteed to be an optimal solution to the MaxSAT instance [31]. Otherwise the SAT solver returns an unsatisfiable core κ of the working formula. PMRES then proceeds by removing all of the soft clauses in the core from the working formula and *cloning* a subset of them; clause cloning is a common way of extending MaxSAT algorithms from unweighted to weighted MaxSAT [3,6,12,30,31], and works as follows. First the minimum-weight $w_\kappa^{\min}$ of clauses in the core κ is determined (line 8). Then each clause in *core* is removed (line 10), and a soft clone $\mathrm{CL}(C)$ of each clause $C \in \kappa$ with $w(C) > w_\kappa^{\min}$ is introduced to the working formula and given the weight $w(C) - w_\kappa^{\min}$ (lines 11–13).

After clause cloning, PMRES extends each $C \in \kappa$ by a fresh relaxation variable r and adds the extended clause $C \vee r$ as hard to the working formula (line 14). The intuition here is that setting $r = 1$ allows for the corresponding soft clause to be left unsatisfied, while setting $r = 0$ forces the corresponding clause to be satisfied. Finally, PMRES *relaxes* the found core by adding a soft cardinality constraint over the introduced r variables via the function $\mathrm{RELAX}(w_\kappa^{\min}, \mathcal{R})$ (line 16). The added cardinality constraint is encoded as hard and soft clauses using additional new variables $d_1, \ldots, d_{|\kappa|-1}$, and essentially enforces that either exactly one of the introduced relaxation variables is set to true, or some soft clause corresponding to $(r_i \rightarrow \neg d_i)$ is falsified (lines 2–7 of RELAX). In order to see this, notice first that the hard clause $(r_1 \vee \ldots \vee r_{|\kappa|})$ forces at least one relaxation variable to be set to true, Assume then that two variables r_k and r_t for some $k < t$ are both set to true. Then the hard clauses of form $d_i \leftrightarrow (r_{i+1} \vee d_{i+1})$ imply that d_j is set to true for all $j < t$. Specifically the variable d_k is set to true, and the soft clause encoding $(r_k \rightarrow \neg d_k)$ will become falsified.

We end this section by discussing two improvements that have been proposed for PMRES and other similar core-guided MaxSAT algorithms; the so-called stratification and hardening rules [4,5,26]. Assume that PMRES in invoked on a MaxSAT instance $F = (F_h, F_s, w)$. The stratification rule aims at prioritizing the extraction of cores κ for which $w_\kappa^{\min}$ is large. Since the sum of the minimum weights of the extracted cores is a lower bound on the optimal cost of the MaxSAT instance, the goal in extracting cores with large minimum weights is to decrease the total number of iterations required for termination. More precisely, PMRES extended with stratification maintains a bound $w_{\max}$, initialized by a heuristic. During solving, PMRES does not invoke the SAT solver on $F_h^w \cup F_s^w$, i.e., all of the clauses of the working formula, but rather, only on a subset of them consisting of all hard clauses and the soft clauses with weight greater than $w_{\max}$. Whenever this subset of the working formula is satisfiable, the algorithm checks if the SAT solver was invoked on the whole working formula, i.e., whether $w_{\max} = 1$. If that is the case, the algorithm terminates. Otherwise the value of $w_{\max}$ is decreased heuristically, and the search continues.

Several different strategies for updating w_{max} have been proposed [4,5]. A fairly simple one is to initialize w_{max} to the maximum weight of the soft clauses, i.e., $w_{max} = \max\{w(C) \mid C \in F_s\}$ and update it by decreasing it to the highest weight of soft clauses that is lower than the current value of w_{max}.[1]

The *hardening rule* attempts to further exploit information that can be obtained from the satisfying assignments obtained during solving in conjunction with stratification. For some intuition, notice that all subsets of the working formula that PMRES with stratification invokes the SAT solver on, always include F_h. Hence whenever the SAT solver returns satisfiable, the returned assignment τ is a solution to the MaxSAT instance, and as such an upper bound on the optimal cost of the instance. The hardening rule exploits this fact by noting that any solution τ_2 that does not satisfy a clause $C \in F_s^w$ with $w(C) > w(F_{\bar{\tau}})$ will have $w(F_{\bar{\tau}_2}) > w(F_{\bar{\tau}})$ and as such can not be an optimal solution to F. Hence all such soft clauses have to be satisfied by any optimal solution to F and can therefore be hardened, i.e., turned into hard clauses.

Even though the presentation here is specific to PMRES, the stratification and hardening rules can be used in conjunction with several different core-guided MaxSAT algorithms. This is also the case for weight-aware core extraction presented next.

4   Weight-Aware Core Extraction for PMRES

We now describe weight-aware core extraction (WCE), a generic technique designed to improve performance of PMRES and other similar MaxSAT algorithms. WCE delays the addition of cardinality constraints to the working formula with the aim of extracting more valid cores (or "core mining") from the working instance before adding more constraints to the formula.

Clause Cloning Through Assumptions (Without Cloning). WCE requires clause cloning to be implemented in a specific way through *assumptions* which essentially avoid actual clause cloning (copying) altogether. A similar approach to clause cloning is taken in [1]. For more details, we first need to overview how core extraction is usually implemented in SAT-based MaxSAT solving. Several modern SAT solvers allow querying for the satisfiability of a CNF formula under a set of assumptions, represented as a partial assignment of the variables in the formula. Whenever the formula is unsatisfiable under those assumptions, the SAT solver returns some subset of the assumptions that are required in the proof of unsatisfiability. Notice that not only are unsatisfiable formulas unsatisfiable under all assumptions, but a satisfiable formula may also be unsatisfiable under some assumptions. For example, consider the CNF formula $F = \{(x \vee y), (\neg x)\}$. Although the formula is satisfiable, it is unsatisfiable when assuming $x = 1$ or $y = 0$.

---

[1] In our implementation used in the experiments of this work, we use the slightly more sophisticated *diversity heuristic* [5] which attempts to balance the number of new soft clauses introduced and the amount that w_{max} is decreased.

Core-guided MaxSAT solvers make use of the assumptions interface in SAT solvers by extending each soft clause $C \in F_s$ with a fresh assumption variable $\textsc{a}(C)$ and sending the extended clause $C \vee \textsc{a}(C)$ to the SAT solver. During each SAT solver call, all assumption variables are assumed false, thus reducing all extended clauses $C \vee \textsc{a}(C)$ to C. Whenever the working formula is unsatisfiable, the SAT solver will return the extracted core κ in terms of the subset of assumption variables corresponding to the clauses in κ. Importantly for the PMRES algorithm, this means that each clause $C_i \in \kappa$ is already extended with the variable $\textsc{a}(C_i)$ and that variable can be reused as the relaxation variable r_i that would otherwise be introduced on lines 15–16 in the algorithm described in Fig. 1. Notice that whenever the SAT solver is invoked without assuming the value of $\textsc{a}(C_i)$ and the variable only appears in the extended clause $C_i \vee \textsc{a}(C_i)$, it can be set to true by the SAT solver, thereby satisfying the extended clause and effectively removing the clause from the formula. The same argument does not hold as soon as other constraints involving $\textsc{a}(C_i)$ are added to the formula.

With this, clause cloning through assumption variables is implemented as follows. Assume that a clause $C \in \kappa$ extended with the assumption variable $\textsc{a}(C)$ needs to be cloned, i.e., it is a member of some extracted core κ and $w(C) > w_\kappa^{\min}$. A simple way of improving on the naive description of clause cloning in Sect. 3 is to introduce a new soft clause $C' = (\neg \textsc{a}(C))$ with weight $w(C') = w(C_i) - w_\kappa^{\min}$. The correctness of this follows by noting that the extended clause $(C \vee \textsc{a}(C))$ will be hard in all subsequent SAT solver calls. Thus satisfying C' forces the clause C to be satisfied as well, achieving the same effect as cloning the whole C. To further improve on this, notice that as C' would be added as a soft clause, it would also be extended with an assumption variable and the extended clause $C' \vee \textsc{a}(C')$ would be sent to the SAT solver. This creates the logical chain $\neg \textsc{a}(C') \rightarrow \neg \textsc{a}(C) \rightarrow C$. The basic form of clause cloning would then assume the variable $\textsc{a}(C')$ to false in subsequent SAT solver calls, thus forcing C as well. To refine this, note that the same affect is achieved by simply assuming the value of $\textsc{a}(C)$ to false instead, thus removing the need of introducing the clause C' at all. In more detail, when a core κ is extracted, the minimum weight $w_\kappa^{\min}$ is computed. Then the weight of each clause $C \in \kappa$ is decreased by $w_\kappa^{\min}$. In subsequent SAT calls, the assumption variable of each clause C with weight $w(C) > 0$ is assumed false, essentially treating that clause as soft. All other clauses are treated as hard. Refining clause cloning in this way blurs the line between hard and soft clauses. When discussing PMRES with clause cloning implemented through assumptions we say that a clause C is soft as long as the internal SAT solver is invoked assuming $\textsc{a}(C) = 0$, i.e., as long as $w(C) > 0$. When $w(C)$ drops to 0, the extended clause $(C \vee \textsc{a}(C))$ becomes hard. Notice that in order for $w(C)$ to become 0, the clause C has appeared in at least one core. Hence we have added a cardinality constraint over $\textsc{a}(C)$ so not assuming the value of it does not remove the clause C from the formula.

Except for removing the need of introducing clones to the formula, implementing clause cloning through assumptions also results in tighter cardinality constraints, as illustrated by the following example.

Example 1. Let $F = (F_h, F_s, w)$ be a MaxSAT instance $F_h = \{(x \vee y), (y \vee z)\}$, $F_s = \{C_1 = (\neg x), C_2 = (\neg y), C_3 = (\neg z)\}$, and $w(C_1) = 1$ and $w(C_2) = w(C_3) = 2$. Assume that we invoke the basic version of PMRES, i.e., the algorithm in Fig. 1, on F and that it first extracts the core $\{C_1, C_2\}$. After relaxing the core the working instance (F_h^w, F_s^w, w) consists of $F_h^w = F_h \wedge (C_1 \vee r_1) \wedge (C_2 \vee r_2) \wedge \mathrm{CNF}(r_1 + r_2 = 1)_h$, $F_s^w = \mathrm{CL}(C_2) \wedge C_3 \wedge \mathrm{CNF}(r_1 + r_2 = 1)_s$ with $w(\mathrm{CL}(C_2)) = 1, w(C_3) = 2$. Here we use $\mathrm{CNF}(r_1 + r_2 = 1)_h$ and $\mathrm{CNF}(r_1 + r_2 = 1)_s$ to denote the hard and soft clauses, respectively, introduced in the RELAX subroutine. If PMRES next extracts and relaxes the core $\{\mathrm{CL}(C_2), C_3\}$, the final working formula will have the hard clauses $F_h^w = F_h \wedge (C_1 \vee r_1) \wedge (C_2 \vee r_2) \wedge (\mathrm{CL}(C_2) \vee r_3) \wedge (C_3 \vee r_4) \wedge \mathrm{CNF}(r_1 + r_2 = 1)_h \wedge \mathrm{CNF}(r_3 + r_4 = 1)_h$ and the soft clauses $F_s^w = \mathrm{CL}(C_3) \wedge \mathrm{CNF}(r_1 + r_2 = 1)_s \wedge \mathrm{CNF}(r_3 + r_4 = 1)_s$. This instance is satisfiable by setting $r_2 = r_3 = 1$ and $r_1 = r_4 = 0$. In total there are 4 different ways of satisfying the added cardinality constraints. A similar argument holds even if we use the assumption variables of soft clauses in cores when encoding the cardinality constraints and introduce the negations of those variables as soft clauses when performing clause cloning.

If we instead use assumptions to implement clause cloning, the final working formula will have the hard clauses $F_h^w = F_h \wedge (C_1 \vee \mathrm{A}(C_1)) \wedge (C_2 \vee \mathrm{A}(C_2)) \wedge \mathrm{CNF}(\mathrm{A}(C_1) + \mathrm{A}(C_2) = 1)_h \wedge \mathrm{CNF}(\mathrm{A}(C_2) + \mathrm{A}(C_3) = 1)_h$ and the soft clauses $F_s^w = C_3 \wedge \mathrm{CNF}(\mathrm{A}(C_1) + \mathrm{A}(C_2) = 1)_s \wedge \mathrm{CNF}(\mathrm{A}(C_2) + \mathrm{A}(C_3) = 1)_s$ with $w(C_1) = w(C_2) = 0$ and $w(C_3) = 1$. As $w(C_3) > 0$, the final SAT call will be made assuming $\mathrm{A}(C_3) = 0$. Under the assumption, there is only a single way of satisfying the added cardinality constraints. Even disregarding the assumptions, there are only 2 ways of satisfying the added cardinality constraints with one of them resulting in the rest of the instance becoming satisfiable. ∎

WCE. Having discussed clause cloning in conjunction with WCE, we now turn to describing WCE in detail. For some intuition, consider the following example.

Example 2. Consider again the MaxSAT instance F from Example 1 and assume that PMRES with clause cloning implemented through assumptions first extracts $\{C_1, C_2\}$. The working formula (F_h^w, F_s^w, w) will then become $F_h^w = F_h \wedge (C_1 \vee \mathrm{A}(C_1)) \wedge \mathrm{CNF}(\mathrm{A}(C_1) + \mathrm{A}(C_2) = 1)_h$ and $F_s^w = C_2 \wedge C_3 \wedge \mathrm{CNF}(\mathrm{A}(C_1) + \mathrm{A}(C_2) = 1)_s$ with $w(C_2) = 1, w(C_3) = 2$. The only core of the instance is $\{C_2, C_3\}$. Notice, however, that ignoring the added cardinality constraints at this stage and invoking the SAT solver on the (simpler) subset of the working formula consisting of $F_h^w = F_h \wedge (C_1 \vee \mathrm{A}(C_1))$ and $F_s^w = C_2 \wedge C_3$ with $w(C_2) = 1, w(C_3) = 2$ would result in the exact same core being extracted. ∎

The pseudocode of PMRES extended with WCE is shown in Fig. 2. Before invoking its SAT solver, PMRES with WCE first adds an assumption $\mathrm{A}(C) = 0$ for all soft clauses C with $w(C) > 0$ (line 5). Then it invokes the SAT solver on the working formula with these assumptions. If a core κ is extracted, $w_\kappa^{\min}$ is computed and the weight of all clauses in the core decreased by $w_\kappa^{\min}$ (lines 14 and 16). However, instead of immediately calling $\mathrm{RELAX}(w_\kappa^{\min}, \mathcal{R})$, the tuple

```
1  PMRES+WCE(F_h, F_s, w):
2  (F_h^w, F_s^w) ← (F_h, F_s)
3  ℝ ← ∅
4  while true do
5  │   𝒜 ← {A(C) = 0 | C_i ∈ F_s^w, w(C) > 0}
6  │   (result, κ, τ) ← SATSOLVE(F_h^w ∪ F_s^w, 𝒜)
7  │   if result="satisfiable" AND |ℝ| = 0 then return τ;
8  │   else if result="satisfiable" then
9  │   │   for (ℛ, w_κ^min) ∈ ℝ do
10 │   │   │   RELAX(w_κ^min, ℛ)
11 │   │   ℝ ← ∅
12 │   else
13 │   │   ℛ ← ∅
14 │   │   w_κ^min ← min{w(C) | C ∈ κ}
15 │   │   for C ∈ κ do
16 │   │   │   w(C) ← w(C) − w_κ^min
17 │   │   │   ℛ ← ℛ ∪ {A(C)}
18 │   │   ℝ.add((ℛ, w_κ^min))
```

Fig. 2. PMRES+WCE, the PMRES algorithm with WCE. In the pseudocode, the assumption variable of a soft clause C used in core extraction is given by A(C).

$(\mathcal{R}, w_\kappa^{min})$ is added to the set $\mathbb{R}$ (line 18). Then the SAT solver in invoked again with a new set of assumptions. Notice that at each iteration, the weight of at least one soft clause C is dropped to 0. In subsequent SAT solver calls the value of A(C) is not assumed anymore, effectively removing that clause from the formula until the cardinality constraints are added, which is why the working formula will eventually become satisfiable. The algorithm then checks if new cores have been extracted since the last time cardinality constraints were added. If so, the corresponding cardinality constraints are added to the formula and the loop iterates (lines 8–11). If there are no new cores, the algorithm terminates and returns the satisfying truth assignment as an optimal MaxSAT solution (line 7).

Similarly to the stratification rule, all working formulas of PMRES extended with WCE contain the original hard clauses F_h. As such, whenever the working formula is satisfiable, the algorithm obtains an upper bound on the cost of the optimal solutions. The bound might in some cases allow PMRES with WCE to terminate even before all cardinality constraints have been added to the working formula.

Example 3. Consider the MaxSAT instance F from Examples 1 and 2. Invoke PMRES with WCE on F and assume that the first core it extracts is again $\kappa^1 = \{C_1, C_2\}$. Now the addition of cardinality constraints is delayed and the SAT solver is invoked on $F_h^w = F_h \wedge (C_1 \vee A(C_1))$ and $F_s^w = C_2 \wedge C_3$ with $w(C_1) = 0, w(C_2) = 1$ and $w(C_3) = 2$. As $w(C_1) = 0$, the variable A(C_1) is not assumed to any value and the clause $(C_1 \vee A(C_1))$ can be satisfied by setting A(C_1) = 1. Nevertheless, PMRES+WCE still extracts the core $\kappa^2 = \{C_2, C_3\}$. On the third iteration the SAT solver is invoked on $F_h^w = F_h \wedge (C_1 \vee \text{A}(C_1)) \wedge (C_2 \vee \text{A}(C_2))$ and $F_s^w = C_3$ with $w(C_1) = w(C_2) = 0$ and $w(C_3) = 1$ assuming $\text{A}(C_3) = 0$. This instance is satisfiable. Next the algorithm adds cardinality constraints to form the same (satisfiable) final working instance as shown in Example 1. Then it would invoke the SAT solver on that instance, find it satisfiable, and terminate.

However, by investigating the second to last SAT solver call we see that PMRES+WCE might be able to terminate without adding any cardinality constraints at all. First note that after extracting the cores κ^1 and κ^2 we know that $\text{COST}(F) \geq w_{\kappa^1}^{\min} + w_{\kappa^2}^{\min} = 1 + 1 = 2$. Now, the second to last SAT solver call is performed on the clauses $(x \vee y), (y \vee z), (\neg x \vee \text{A}(C_1)), (\neg y \vee \text{A}(C_2)), (\neg z \vee \text{A}(C_3))$ assuming $\text{A}(C_3) = 0$. The assumption propagates $z = 0$ which in turn propagates $y = 1$ and $\text{A}(C_2) = 1$. At this point, all clauses except for $(\neg x \vee \text{A}(C_1))$ are already satisfied. If the internal SAT solver now satisfies the clause by setting $x = 0$, the cost of the assignment it returns will be 2, thus proving that $\text{COST}(F) \leq 2$ and allowing the algorithm to terminate early. Although we in general can not guarantee early termination, empirically we found that it does happen.   ∎

The correctness of WCE is based on the correctness of PMRES. For more intuition, note that all cores that are extractable by PMRES with WCE are a subset of the cores that could be extracted by PMRES with clause cloning implemented through assumptions, and that the final working instance of both algorithms is the same.

Related Work. The method presented in [18] for computing MaxSAT lower bounds is equivalent to running Algorithm 2 until the working instance becomes satisfiable for the first time, returning the sum $\sum w_\kappa^{\min}$ over all of the cores extracted; already this lower bounding step is shown in [18] to improve the performance of specific MaxSAT algorithms compared to starting search with the trivial bound of 0. Alternatively, WCE can be seen as a more thorough integration of the bound computation and the MaxSAT algorithm itself by performing the lower bound computation *in-between* each core compilation step. Computation of lower bounds has also received significant interest in the context of branch-and-bound MaxSAT solvers [20–23], which rely heavily on good lower bounds in order to prune the search tree. For example, in [21] the authors propose a technique in which unit propagation is used to extract several cores of the working instance in order to compute a lower bound. The main difference to WCE is that WCE is not limited to cores detectable by unit propagation.

Integrating Stratification and Hardening. We end this section by discussing how the commonly used stratification and hardening rules can be integrated with WCE. There are two obvious ways of integrating stratification in conjunction with WCE. The first one is to prefer WCE to stratification: initialize the bound $w_{\max}$ heuristically [5] and assume the assumption variable $\text{A}(C)$ to false only for each of the clauses C with $w(C) \geq w_{\max}$. Then iteratively extract cores over the subset of soft clauses under consideration, delaying the addition of cardinality constraints until the instance becomes satisfiable. At that point, add the cardinality constraints and continue. Whenever the instance remains satisfiable after the addition of cardinality constraints, harden any possible clauses and decrease the bound w_{max}. The algorithm can terminate when no new cores can be extracted and the SAT solver has been invoked on all soft clauses.

Another, dual way of integrating the two is to prefer stratification to WCE, i.e., to delay the addition of cardinality constraints until w_{max} has been decreased to 1, at which point all cardinality constraints are added and the bound w_{max} is reinitialized. However, the choice between preferring stratification or WCE to the other influences the applicability of the hardening rule and the quality of the satisfiable assignments produced by WCE and stratification. Preferring WCE to stratification we know that, whenever the bound w_{max} needs to be lowered, all clauses C with $w(C) \geq w_{max}$ are satisfied, thus allowing for the hardening of several soft clauses. In contrast, when preferring stratification to WCE, sound use of the hardening rule requires considering the delayed soft cardinality constraints, of which the SAT solver has had no information during search. The empirical results presented next support this intuition, as preferring WCE to stratification leads to performance boosts within PMRES, while preferring stratification to WCE actually degrades performance compared to PMRES without using WCE.

5   Experiments

We investigate how WCE affects the performance of the PMRES algorithm in practice. Since the implementation of PMRES by the original authors—coined Eva500a as it participated in MaxSAT Evaluation 2014 [9]—is not available in open source, we re-implemented PMRES on top of the open-source core-guided MaxSAT solver Open-WBO [29], following the description in the paper introducing the algorithm [31] using Glucose [10] as the underlying incremental SAT solver.

In the experiments we compare the following MaxSAT algorithms.

- PMRES: our re-implementation of the PMRES algorithm using stratification, implemented using assumption variables on soft clauses, hardening, and clause cloning implemented through assumptions.
- PMRES+WCE: PMRES extended with WCE, preferring WCE to stratification.
- PMRES+WCE (S/to/WCE): PMRES extended with WCE, preferring stratification to WCE.
- Eva500a [31]: the closed-source implementation of PMRES that participated, and won the industrial category of the 2014 MaxSAT Evaluation.

For reference, we also provide a comparison with MSCG15b [30], a closed-source as the best-performing core-guided MaxSAT solver using soft cardinality constraints in 2016 MaxSAT evaluation. As we will explain later in Sect. 6, WCE could also be integrated into MSCG.[2] As benchmarks we used the weighted partial industrial (630) and crafted (331) instances from the 2016 MaxSAT Evaluation [9]. The experiments were run on 2.83-GHz Intel Xeon E5440 quad-core machines with 32-GB RAM and Debian GNU/Linux 8 using a per-instance timeout of 3600 s.

An overview of the results, comparing the performance of Eva500a and the variants PMRES, PMRES+WCE, and PMRES+WCE (S/to/WCE) of our implementation, is provided through Figs. 3, 4, and Table 1. The "cactus" plot of Fig. 3 gives the number of instances solved (x-axis) by the individual solvers under different per-instance time limits (y-axis) over all benchmarks. More detailed results are provided in Table 1, with the industrial and crafted benchmarks separated by domain, showing the number of instances from each domain, and the number of solved instances and the cumulative running time used for solving the solved instances for each solver. First, note that our PMRES reimplementation is competitive in terms of overall performance with Eva500a; on the industrial instances PMRES solves three more instances overall and uses cumulatively only 55% of the running time that Eva500a uses on the respectively solved instances. On the crafted instances, PMRES solves two instances less than Eva500a but still uses noticeably less time over all solved instances.

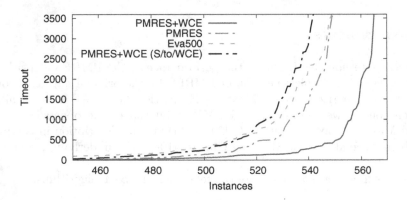

Fig. 3. Solver comparisons: number of instances solved (x-axis) by the individual solvers under different per-instance time limits (y-axis).

Turning to the influence of WCE on the performance of PMRES, we observe that PMRES+WCE (preferring WCE to stratification) has noticeably improved performance wrt PMRES (and thus also Eva500a), solving 11 more industrial and 3 more crafted instances (14 and 1 more than Eva500a). Notice that of all three solvers, PMRES+WCE is the best performing on both industrial and crafted benchmarks. Most interestingly, PMRES+WCE uses at the same time much less time on the solved instances; on the industrial instances

---

[2] Unfortunately, we do not have access to the source code of MSCG.

Table 1. Comparison of Eva500a, PMRES, and PMRES+WCE: number of solved instances (#) and the cumulative running time used for solving the instances (Σ) for the individual solvers, divided into the industrial (top) and crafted (bottom) benchmarks according to the individual domains with the number of instances from each domain given in parentheses.

	Eva500a		PMRES		PMRES+WCE	
	Solved	Time (s)	Solved	Time (s)	Solved	Time (s)
	#	Σ	#	Σ	#	Σ
Industrial domain (#instances)						
abstraction refinement (11)	6	3670	9	2842	**10**	**2147**
BTBNSL (60)	9	403	16	6142	**19**	679
correlation clustering (129)	18	17630	11	4559	**19**	**4499**
haplotyping pedigrees (100)	100	7321	100	3409	100	**1374**
hs-timetabling (14)	1	**477**	1	1858	1	2596
packup-wpms (99)	**99**	2981	95	969	94	191
preference planning (29)	29	1416	29	311	29	**264**
railway transport (11)	2	126	3	603	3	**340**
relational inference (9)	5	8391	8	1431	8	**1360**
timetabling (26)	12	1818	12	1846	12	**878**
upgradeability (100)	100	2996	100	**54**	100	59
wcsp_spot5_dir (21)	14	30	14	**13**	14	24
wcsp_spot5_log (21)	14	61	14	1975	14	**17**
Total industrial (630)	409	47319	412	26012	**423**	**14426**
Crafted domain (#instances)						
auctions (40)	**40**	6111	39	2691	38	1759
causal discovery (35)	**10**	9325	6	1267	6	1357
CSG (10)	7	610	8	**1056**	8	1370
frb (34)	**20**	**3310**	12	4461	17	4041
min-enc (48)	32	198	36	758	36	**455**
miplib (12)	5	1558	5	**247**	5	437
ramsey (15)	1	1	3	2282	3	**2073**
random-net (32)	13	4001	13	1196	13	**5**
set-covering (45)	9	2069	**10**	1520	9	155
staff-scheduling (12)	1	0	2	**1069**	2	1406
wmaxcut (48)	1	44	3	3208	3	**2960**
Total crafted (331)	139	27226	137	19756	**140**	**16018**

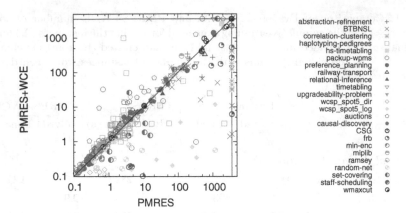

Fig. 4. Solver comparisons: per-instance running time comparison of PMRES (x-axis) and PMRES+WCE (y-axis), with the ticks below the $y = x$ line representing instances on which PMRES+WCE is faster than PMRES.

PMRES+WCE uses in total 55% of the time PMRES uses and 30% of the time Eva500a uses, even though PMRES+WCE solves more instances than PMRES and Eva500a individually. A similar observation can be made of the crafted instances; PMRES+WCE uses 81% of the time used by PMRES and 59% of the time used by Eva500a, again solving more instances than either one. The scatter plot of Fig. 4 gives a per-instance running time comparison on a log-log scale of PMRES+WCE and PMRES, with the ticks below the $y = x$ line representing instances on which PMRES+WCE is faster that PMRES. The colors of the ticks distinguish between the benchmark domains listed in Table 1.

Next, we consider the question of the relative influence of preferring stratification or WCE within PMRES. Here we observe that PMRES+WCE (S/to/WCE)—preferring stratification over WCE—actually harms the overall performance of PMRES noticeably, making it perform worse than Eva500a overall (see Fig. 3). This supports the earlier discussed intuition that preferring WCE to stratification assures that whenever the bound w_{max} needs to be lowered, all clauses C with $w(C) \geq w_{max}$ are satisfied, thus allowing for hardening several soft clauses. In contrast, when preferring stratification to WCE, sound use of the hardening rule requires considering the delayed soft cardinality constraints, of which the SAT solver has had no information during search.

Finally, we consider the relative performance of PMRES+WCE and MSCG15b. Here we note that this is not a direct comparison of the influence of WCE in the sense that MSCG15b does not implement the PMRES algorithm of Eva500a, but rather a different core-guided algorithm using soft cardinality constraints, OLL [2,30]. As we will explain later in Sect. 6, WCE can also be integrated into the OLL algorithm. However, we could not implement WCE directly to MSCG as MSCG is not available in open source. Nevertheless, a comparison of the performance of PMRES+WCE and MSCG15b is provided in Table 2. Overall MSCG15b solves 12 more industrial instances than PMRES+WCE. How-

Table 2. Comparison of MSCG15b and PMRES+WCE: percentage (%) and number (#) of solved instances and the cumulative running time used for solving the instances (Σ) for the individual solvers, divided into industrial (top) and crafted (bottom) benchmarks according to the individual domains with the number of instances from each domain given in parentheses.

	MSCG15b			PMRES+WCE		
	Solved		Time (s)	Solved		Time (s)
	%	#	Σ	%	#	Σ
Industrial domain (#instances)						
abstraction refinement (11)	90.9	10	17096	90.9	10	**2147**
BTBNSL (60)	21.7	13	885	**31.7**	**19**	**679**
correlation clustering (129)	**25.6**	**33**	19780	14.7	19	4499
haplotyping pedigrees (100)	100.0	100	**1343**	100.0	100	1374
hs-timetabling (14)	7.1	1	**167**	7.1	1	2596
packup-wpms (99)	**100**	**99**	410	95.0	94	191
preference planning (29)	100	29	2021	100	29	**264**
railway transport (11)	27.3	3	**283**	27.3	3	340
relational inference (9)	44.4	4	3167	**88.9**	**8**	**1360**
timetabling (26)	46.2	12	**764**	46.2	12	878
upgradeability (100)	100.0	100	118	100.0	100	**59**
wcsp_spot5_dir (21)	**81.0**	**17**	2776	66.7	14	24
wcsp_spot5_log (21)	66.7	14	19	66.7	14	**17**
Total industrial (630)		435	50333		423	14426
Crafted domain (#instances)						
auctions (40)	60.0%	24	313	**95.0%**	**38**	1759
causal discovery (35)	**82.9%**	**29**	6851	17.1%	6	1357
CSG (10)	**100.0%**	**10**	825	80.0%	8	1370
frb (34)	**73.5%**	**25**	4298	50.0%	17	4041
min-enc (48)	66.7%	32	27	**75.0%**	**36**	455
miplib (12)	41.7%	5	**56**	41.7%	5	437
ramsey (15)	13.3%	2	1335	**20.0%**	**3**	2073
random-net (32)	**100.0%**	**32**	288	40.6%	13	5
set-covering (45)	**46.7%**	**21**	1530	20.0%	9	155
staff-scheduling (12)	16.7%	2	**1244**	16.7%	2	1406
wmaxcut (48)	**10.4%**	**5**	3651	6.3%	3	2960
Total crafted (331)		187	20417		140	16018

ever, at the same time MSCG used considerably more time per solved instance; this can be observed by inspecting the total cumulative running times: while PMRES+WCE uses 14416 s to solve 423 instances, MSCG15b uses a noticeable 35917 s more to solve additional 12 instances. Looking more closely at the results on a benchmark domain basis, we notice that the main advantage of MSCG15b is within the correlation clustering domain, where the solver also uses a noticeably amount of time to solve an additional 14 instances; furthermore, notice that the correlation clustering domain is over-represented among the full benchmark set with 129 instances. On the other hand, PMRES solves twice as many instances as MSCG15b within the relational inference domain, using at the same time only 43% of the cumulative running time of MSCG15b. The abstraction refinement domain provides another example where PMRES+WCE solves instances cumulatively noticeably faster than MSCG15b: here the solvers solve the same number of instances, but the cumulative running time of PMRES+WCE is less than 13% of that of MSCG15b (i.e., an 8x speed-up relative to MSCG15b). Turning to the crafted domains, we observe that MSCG15b clearly dominates on several of them. This is an interesting observation, also in that Eva500a never participated in the crafted MaxSAT evaluation categories.

6   WCE and Other SAT-Based MaxSAT Algorithms

In this section we discuss WCE in a more general setting and the question of to what extent it could be applied to other recently proposed MaxSAT algorithms.

The key to integrating WCE with a core-guided MaxSAT algorithm is whether clause cloning can be implemented through assumptions in the algorithm. As far as we understand, the reason clause cloning through assumptions can be added to PMRES is the fact that no clause ever appears in a core more than once. In more detail, let κ be a core extracted by PMRES during solving and assume $C \in \kappa$ needs to be cloned, i.e., that $w(C) > w_\kappa^{\min}$. Since the extended clause $C \vee \mathrm{A}(C)$ is added to the working formula as hard, that clause is not going to be extracted in any subsequent cores, but rather, only its clone $\mathrm{CL}(C)$ or some of the soft clauses added in $\mathrm{RELAX}(w_\kappa^{\min}, \mathcal{R})$. This simple observation is a key to implementing clause cloning through assumptions.

WPM1. As an example of an algorithm in which clause cloning seems to be difficult to implement through assumptions, consider the WPM1 algorithm [3,24]. WPM1 works similarly to PMRES in the sense that it uses a SAT solver to extract and relax unsatisfiable cores of the input instance F. Given a core κ, WPM1 clones its clauses similarly to PMRES and extends each clause $C_i \in \kappa$ (now of weight $w_\kappa^{\min}$) with a fresh relaxation variable r_i. For WCE, the key difference between PMRES and WPM1 is that WPM1 leaves all extended clauses $C_i \vee r_i$ as *soft* in the working formula and adds a cardinality constraint $\mathrm{CNF}(\sum r_i = 1)$ as *hard* clauses. Hence the extended clause might appear in subsequent cores, making it difficult if not impossible to also reuse it as its own clone.

Finally, we point out MaxSAT algorithms to which WCE can be integrated.

OLL and K. Two algorithms that closely resemble PMRES[3] are OLL [2,30] and K [1]. Both extract cores iteratively, harden and clone the clauses in cores, and compile them into the formula using soft cardinality constraints. In contrast to PMRES, OLL makes use of cardinality networks in order to dynamically modify the previously added cardinality constraints while K uses parametrized constraints for bounding their size. In both cases the clauses in the extracted cores are hardened and do not appear in subsequent cores, and as such WCE could be incorporated into both algorithms. Indeed, at least the K algorithm does implement clause cloning through assumptions [1]. The MSCG15b MaxSAT solver considered in Sect. 5 implements OLL.

WPM3 [7] maintains a set of at-most constraints, initialized to not allow any soft clauses to be falsified. During solving all clauses are treated as hard and the at-most constraints as soft, and hence all cores are subsets of these constraints. After finding a new core, WPM3 performs clause cloning and then merges the constraints to form new ones that make effective use of the global core structure. In contrast to OLL and PMRES, the cardinality constraints in the extracted cores are not hardened, but instead removed from the instance. To the best of our understanding, the version of WPM3 presented in [7] does not use the SAT solver iteratively, but instead rebuilds it on each iteration. Hence the idea of implementing clause cloning through assumptions is not applicable to this version of WPM3, even though it might be if the algorithm is extended with incremental cardinality constraints in the spirit of [28]. However, the discussed requirement of a clause in a core not appearing in any subsequent cores is satisfied by the algorithm. Thereby delaying the modification of cardinality constraints could be incorporated to the presented version of WPM3.

WMSU3 [27] maintains a single cardinality constraint $\sum_{r \in \mathcal{R}} r = \lambda$ over the set $\mathcal{R}$ of relaxation variables of clauses appearing in cores extracted so far. When a new core κ is extracted, all the assumption variables $\mathrm{A}(C)$ of clauses $C \in \kappa$ are reused as relaxation variables, i.e., added into the set $\mathcal{R}$, after which a new bound λ is computed and the solver invoked again. Here λ is a lower bound on the optimal cost of the instance. As noted in Sect. 4, WCE can be viewed as an extension of the lower bounding technique from [18], the difference being that WMSU3 extended with WCE would perform such a core mining step in between each modification of the cardinality constraint, not only before the cardinality constraint is added.

SAT-IP Hybrids. Finally, also the SAT-IP hybrid solvers MaxHS [15] and LMHS [32], based on the implicit hitting set approach to MaxSAT, could potentially make use of specific ideas related to WCE. Specifically, WCE could be

---

[3] Originally in [31] PMRES was formalized as a special case of the so-called MAXRES [19] rule; the specific special case is equivalent to the formalization of PMRES used here.

incorporated into the *disjoint core extraction phase*—that is very important in terms of performance in practice [15]—in this context in a straight-forward way: instead of ruling out each clause C in a core κ from the working instance during the disjoint phase, lower the weight of all clauses in the core by w_κ^{min}, and rule out only those clauses whose weight is lowered to 0.

7   Conclusions

We proposed weight-aware core extraction (WCE) as a refinement to the approach taken by various core-guided MaxSAT solvers for compiling the cores extracted at each iteration of search. WCE allows for extracting multiple cores of the same working formula by taking into account the residual weights of the current soft clauses, thereby postponing the compilation step and allowing the SAT solver to work on a less bloated working formula. We detailed WCE in the context of PMRES, a representative of the most recent line of core-guided MaxSAT solvers that use soft cardinality constraints in the compilation step, and showed empirically that WCE noticeably improves the performance of PMRES on standard weighted partial MaxSAT benchmarks. We also outlined how to integrate ideas behind WCE into other core-guided MaxSAT algorithms. The empirical results obtained for PMRES suggests that integrating WCE into other recent MaxSAT solvers may provide further improvements to the state of the art.

References

1. Alviano, M., Dodaro, C., Ricca, F.: A MaxSAT algorithm using cardinality constraints of bounded size. In: Proceedings of IJCAI, pp. 2677–2683. AAAI Press (2015)
2. Andres, B., Kaufmann, B., Matheis, O., Schaub, T.: Unsatisfiability-based optimization in clasp. In: Proceedings of ICLP Technical Communications, LIPIcs, vol. 17, pp. 211–221. Schloss Dagstuhl - Leibniz-Zentrum fuer Informatik (2012)
3. Ansótegui, C., Bonet, M.L., Levy, J.: Solving (weighted) partial MaxSAT through satisfiability testing. In: Kullmann, O. (ed.) SAT 2009. LNCS, vol. 5584, pp. 427–440. Springer, Heidelberg (2009). doi:10.1007/978-3-642-02777-2_39
4. Ansótegui, C., Bonet, M., Levy, J.: SAT-based MaxSAT algorithms. Artif. Intell. **196**, 77–105 (2013)
5. Ansótegui, C., Bonet, M.L., Gabàs, J., Levy, J.: Improving SAT-based weighted MaxSat solvers. In: Milano, M. (ed.) CP 2012. LNCS, pp. 86–101. Springer, Heidelberg (2012). doi:10.1007/978-3-642-33558-7_9
6. Ansótegui, C., Didier, F., Gabàs, J.: Exploiting the structure of unsatisfiable cores in MaxSAT. In: Proceedings of IJCAI, pp. 283–289. AAAI Press (2015)
7. Ansótegui, C., Gabàs, J., Levy, J.: Exploiting subproblem optimization in SAT-based MaxSAT algorithms. J. Heuristics **22**(1), 1–53 (2016)
8. Argelich, J., Le Berre, D., Lynce, I., Marques-Silva, J., Rapicault, P.: Solving Linux upgradeability problems using Boolean optimization. In: Proceedings of LoCoCo. Electronic Proceedings in Theoretical Computer Science, vol. 29, pp. 11–22 (2010)

9. Argelich, J., Li, C.M., Manyà, F., Planes, J.: MaxSAT Evaluations. http://maxsat. ia.udl.cat/

10. Audemard, G., Lagniez, J.-M., Simon, L.: Improving glucose for incremental SAT solving with assumptions: application to MUS extraction. In: Järvisalo, M., Van Gelder, A. (eds.) SAT 2013. LNCS, vol. 7962, pp. 309–317. Springer, Heidelberg (2013). doi:10.1007/978-3-642-39071-5_23

11. Berg, J., Järvisalo, M., Malone, B.: Learning optimal bounded treewidth Bayesian networks via maximum satisfiability. In: Proceedings of AISTATS, JMLR Workshop and Conference Proceedings, vol. 33, pp. 86–95 (2014). JMLR.org

12. Bjørner, N., Narodytska, N.: Maximum satisfiability using cores and correction sets. In: Proceedings of IJCAI, pp. 246–252. AAAI Press (2015)

13. Bunte, K., Järvisalo, M., Berg, J., Myllymäki, P., Peltonen, J., Kaski, S.: Optimal neighborhood preserving visualization by maximum satisfiability. In: Proceedings of AAAI, pp. 1694–1700. AAAI Press (2014)

14. Chen, Y., Safarpour, S., Marques-Silva, J., Veneris, A.: Automated design debugging with maximum satisfiability. IEEE Trans. Comput. Aided Des. Integr. Circuits Syst. 29(11), 1804–1817 (2010)

15. Davies, J., Bacchus, F.: Exploiting the power of MIP solvers in MaxSaT. In: Järvisalo, M., Van Gelder, A. (eds.) SAT 2013. LNCS, vol. 7962, pp. 166–181. Springer, Heidelberg (2013). doi:10.1007/978-3-642-39071-5_13

16. Fu, Z., Malik, S.: On solving the partial MAX-SAT problem. In: Biere, A., Gomes, C.P. (eds.) SAT 2006. LNCS, vol. 4121, pp. 252–265. Springer, Heidelberg (2006). doi:10.1007/11814948_25

17. Guerra, J., Lynce, I.: Reasoning over biological networks using maximum satisfiability. In: Milano, M. (ed.) CP 2012. LNCS, pp. 941–956. Springer, Heidelberg (2012). doi:10.1007/978-3-642-33558-7_67

18. Heras, F., Morgado, A., Marques-Silva, J.: Lower bounds and upper bounds for MaxSAT. In: Hamadi, Y., Schoenauer, M. (eds.) LION 2012. LNCS, pp. 402–407. Springer, Heidelberg (2012). doi:10.1007/978-3-642-34413-8_35

19. Larrosa, J., Heras, F.: Resolution in Max-SAT and its relation to local consistency in weighted CSPs. In: Proceedings of IJCAI, pp. 193–198. Professional Book Center (2005)

20. Li, C.M., Manyà, F., Mohamedou, N.O., Planes, J.: Resolution-based lower bounds in MaxSAT. Constraints 15(4), 456–484 (2010)

21. Li, C.M., Manyà, F., Planes, J.: Exploiting unit propagation to compute lower bounds in branch and bound Max-SAT solvers. In: van Beek, P. (ed.) CP 2005. LNCS, vol. 3709, pp. 403–414. Springer, Heidelberg (2005). doi:10.1007/11564751_31

22. Li, C.M., Manyà, F., Planes, J.: Detecting disjoint inconsistent subformulas for computing lower bounds for Max-SAT. In: Proceedings of AAAI, pp. 86–91. AAAI Press (2006)

23. Lin, H., Su, K., Li, C.M.: Within-problem learning for efficient lower bound computation in Max-SAT solving. In: Proceedings of AAAI, pp. 351–356. AAAI Press (2008)

24. Manquinho, V., Marques-Silva, J., Planes, J.: Algorithms for weighted Boolean optimization. In: Kullmann, O. (ed.) SAT 2009. LNCS, vol. 5584, pp. 495–508. Springer, Heidelberg (2009). doi:10.1007/978-3-642-02777-2_45

25. Marques-Silva, J., Janota, M., Ignatiev, A., Morgado, A.: Efficient model based diagnosis with maximum satisfiability. In: Proceedings of IJCAI, pp. 1966–1972. AAAI Press (2015)

26. Marques-Silva, J., Argelich, J., Graça, A., Lynce, I.: Boolean lexicographic optimization: algorithms & applications. Ann. Math. Artif. Intell. **62**(3–4), 317–343 (2011)
27. Marques-Silva, J., Planes, J.: On using unsatisfiability for solving maximum satisfiability. CoRR abs/0712.1097 (2007)
28. Martins, R., Joshi, S., Manquinho, V., Lynce, I.: Incremental cardinality constraints for MaxSAT. In: O'Sullivan, B. (ed.) CP 2014. LNCS, vol. 8656, pp. 531–548. Springer, Cham (2014). doi:10.1007/978-3-319-10428-7_39
29. Martins, R., Manquinho, V., Lynce, I.: Open-WBO: a modular MaxSAT solver'. In: Sinz, C., Egly, U. (eds.) SAT 2014. LNCS, vol. 8561, pp. 438–445. Springer, Cham (2014). doi:10.1007/978-3-319-09284-3_33
30. Morgado, A., Dodaro, C., Marques-Silva, J.: Core-guided MaxSAT with soft cardinality constraints. In: O'Sullivan, B. (ed.) CP 2014. LNCS, vol. 8656, pp. 564–573. Springer, Cham (2014). doi:10.1007/978-3-319-10428-7_41
31. Narodytska, N., Bacchus, F.: Maximum satisfiability using core-guided MaxSAT resolution. In: Proceedings of AAAI, pp. 2717–2723. AAAI Press (2014)
32. Saikko, P., Berg, J., Järvisalo, M.: LMHS: a SAT-IP hybrid MaxSAT solver. In: Creignou, N., Le Berre, D. (eds.) SAT 2016. LNCS, vol. 9710, pp. 539–546. Springer, Cham (2016). doi:10.1007/978-3-319-40970-2_34
33. Tseitin, G.S.: On the complexity of derivation in propositional calculus. In: Siekmann, J.H., Wrightson, G. (eds.) Automation of Reasoning. 2: Classical Papers on Computational Logic 1967–1970. Symbolic Computation, pp. 466–483. Springer, Heidelberg (1983)
34. Zhu, C., Weissenbacher, G., Malik, S.: Post-silicon fault localisation using maximum satisfiability and backbones. In: Proceedings of FMCAD, pp. 63–66. FMCAD Inc. (2011)

Optimizing SAT Encodings for Arithmetic Constraints

Neng-Fa Zhou[1](✉) and Håkan Kjellerstrand[2]

[1] CUNY Brooklyn College & Graduate Center, New York, USA
zhou@sci.brooklyn.cuny.edu
[2] hakank.org, Malmö, Sweden

Abstract. The log encoding has been perceived to be unsuited to arithmetic constraints due to its hindrance to propagation. The surprising performance of PicatSAT, which is a pure eager SAT compiler based on the log encoding, in the MiniZinc Challenge 2016 has revived interest in the log encoding. This paper details the optimizations used in PicatSAT for encoding arithmetic constraints. PicatSAT adopts some well-known optimizations from CP systems, language compilers, and hardware design systems for encoding constraints into compact and efficient SAT code. PicatSAT is also empowered by a novel optimization, called equivalence reasoning, for arithmetic constraints, which leads to reduction of code size and execution time. In a nutshell, this paper demonstrates that the optimized log encoding is competitive for encoding arithmetic constraints.

1 Introduction

The drastic enhancement of SAT solvers' performance has made SAT a viable backbone for general CSP (Constraint Satisfaction Problem) solvers [8,21,30,33, 34]. Many real-world combinatorial problems involve arithmetic constraints, and it remains a challenge to efficiently encode arithmetic constraints into SAT. The *sparse* encoding [20,36] and *order* encoding [13,25,34] can easily blow up the code size, and the *log* encoding [18,22] is perceived to be a poor choice, despite its compactness, due to its failure to maintain arc consistency, even for binary constraints. This dilemma of the *eager* approach has led to the emergence of the *lazy* approach, as represented by SMT solvers that use integer arithmetic as a theory [6,14,26] and the *lazy clause generation* (LCG) solver that combines SAT and constraint propagation [17,29]. Both the eager and lazy approaches have strengths and weaknesses [27]. For problems that require frequent checking of arithmetic constraints the lazy approach may not be competitive due to the overhead, even when checking is done incrementally and in a priori manner. From an engineering perspective, the eager approach also has its merit, just like the separation of computer hardware and language compilers is beneficial.

© Springer International Publishing AG 2017
J.C. Beck (Ed.): CP 2017, LNCS 10416, pp. 671–686, 2017.
DOI: 10.1007/978-3-319-66158-2_43

The surprising performance of the PicatSAT compiler in the MiniZinc Challenge 2016 is thought-provoking.[1] PicatSAT is a pure SAT compiler that translates CSPs into log-encoded SAT code. PicatSAT adopts the sign-and-magnitude log encoding for domain variables. For a domain with the maximum absolute value n, it uses $log_2(n)$ Boolean variables to encode the domain. If the domain contains both negative and positive values, then another Boolean variable is employed to encode the sign. Each combination of values of the Boolean variables represents a valuation for the domain variable. The addition constraint is encoded as logic adders, and the multiplication constraint is encoded as logic adders using the *shift-and-add* algorithm.

PicatSAT adopts some well-known optimizations from CP systems, language compilers, and hardware design systems for encoding constraints into compact and efficient SAT code: it preprocesses constraints before compilation in order to remove no-good values from the domains of variables whenever possible; it eliminates common subexpressions so that no primitive constraint is duplicated; it uses a logic optimizer to generate optimized code for adders. These optimizations significantly improve the quality of the generated code.

This paper proposes a new optimization, named *equivalence reasoning*, for log-encoded arithmetic constraints. Equivalence reasoning identifies information about if a Boolean variable is 0 or 1, if a Boolean variable is equivalent to another Boolean variable, or if a Boolean variable is the negation of another Boolean variable. This optimization can reduce both the number of Boolean variables and the number of clauses in the CNF code.

The experimental results show that equivalence reasoning reduces code sizes, and for some benchmarks, significantly reduces the solving time. The MiniZinc Challenge 2016 results show that PicatSAT outperformed some of the fastest CP solvers in the competition. Our new comparisons of PicatSAT with fzn2smt, an SMT-based CSP solver, and Chuffed, a cutting-edge LCG solver, also reveal the competitiveness of PicatSAT.

2   The PicatSAT Compiler

PicatSAT is offered in Picat as a module named `sat`. In addition to the `sat` module, Picat offers two other solver modules, named `cp` and `mip`, respectively. All these three modules implement the same set of basic linear constraints over integer domains and Boolean constraints. The `cp` and `sat` modules also implement non-linear and global constraints, and the `mip` module also supports real-domain variables. The common interface that Picat provides for the solver modules allows seamless switching from one solver to another.

In order to give the reader a complete picture of the PicatSAT compiler, we give in this section an overview of the compiler, including the adopted optimizations. A description of a preliminary version of PicatSAT with no optimizations is given in [38].

---

[1] PicatSAT with the Lingeling SAT solver won two silver medals and one bronze medal in the competition (http://www.minizinc.org/challenge2016/results2016.html).

2.1 Preprocessing and Decomposition

In general, a constraint model consists of a set of *decision variables*, each of which has a specified domain, and a set of *constraints*, each of which restricts the possible combinations of values of the involved decision variables. A constraint program normally poses a problem in three steps: (1) generate variables; (2) generate constraints over the variables; and (3) call `solve` to invoke the solver in order to find a valuation for the variables that satisfies the constraints and possibly optimizes an objective function.

PicatSAT preprocesses the accumulated constraints when the `solve` predicate is called. For binary equality constraints, PicatSAT excludes no-good values from the domains to achieve arc consistency, unless the domains are too big.[2] For instance, for the constraint $X = 9 * B + 1$, where B is a Boolean variable that has the domain 0..1, PicatSAT narrows X's domain to $\{1, 10\}$. For other types of constraints, including binary equality constraints that involve large-domain variables, PicatSAT narrows the bounds of domains to archieve interval consistency.

PicatSAT decomposes arithmetic constraints into basic constraints, including *primitive*, *reified*, and *implication* constraints. A primitive constraint is one of the following: $\Sigma_i^n B_i \; r \; c$ (r is =, $\geq$, or $\leq$, and c is 1 or 2),[3] $X \; r \; Y$ (r is =, $\neq$, >, $\geq$, <, or $\leq$), $X + Y = Z$, and $X \times Y = Z$, where B_i is a Boolean variable, and X, Y, and Z are integers or integer domain variables. A reified constraint, after decomposition, takes the form $B \Leftrightarrow C$, and an implication constraint has the form $B \Rightarrow C$, where B is a Boolean variable, and C is a primitive constraint.

For a linear constraint, PicatSAT sorts the terms by the variables. This ordering facilitates merging terms of the same variables, but is hardly optimal for generating efficient SAT code. PicatSAT breaks down Pseudo-Boolean (PB) constraints, including cardinality constraints, in the same way as other linear constraints, unless they are cardinality constraints of the form $\Sigma_i^n B_i \; r \; c$ (c =1 or 2). For example, the cardinality constraint $U + V + W + X \leq 3$, where all the variables are Boolean, is split into the following primitive constraints:

$$U + V = T_1$$
$$W + X = T_2$$
$$T_1 + T_2 = T_3$$
$$T_3 \leq 3$$

This method of decomposing PB constraints is simple, and generates compact code. For a cardinality constraint that has n variables, this method introduces $O(n)$ auxiliary integer-domain variables, which require a total number of $O(n \times log_2(n))$ Boolean variables to encode.[4]

---

[2] The default bounds of small domains are −3200 and 3200, which can be reset by using the built-in `fd_vector_min_max(LB,UB)`.

[3] The cardinality constraint is treated as a normal linear constraint when $c > 2$.

[4] This information is disclosed here in order to give the reader a complete picture of PicatSAT. A study is underway to investigate how this adder-based encoding compares with other encodings, such as sorting networks [16], totalizers [5], BDDs [7], and the decomposition method for adders by [37].

During decomposition, PicatSAT introduces auxiliary variables to combine terms in the same way as language compilers break expressions into triplets. PicatSAT makes efforts not to create variables with both positive and negative values in their domains, if possible, because such a domain requires a sign bit and cannot be encoded compactly. For example, for the constraint $U - V + W - X \leq 3$, PicatSAT merges U with W, and V with X so that no auxiliary variables have negative values in their domains if the domains of the original variables do not contain negative values.

PicatSAT eliminates common subexpressions in constraints. Whenever Picat-SAT introduces an auxiliary variable for a primitive constraint, it tables the constraint. When the same primitive constraint is encountered again, Picat-SAT reuses the auxiliary variable, rather than introducing a new variable for the constraint. For example, when a reification constraint $B \Leftrightarrow C$ is generated, PicatSAT tables it and reuses the variable B, rather than introducing a new variable for the primitive constraint C, when C is encountered again in another constraint. The algorithm used in PicatSAT for identifying common subexpressions is not as sophisticated as those used in [4,28]. It incurs little overhead on compilation, but fails to eliminate common subexpressions in many cases.

2.2   The Sign-and-Magnitude Log Encoding

PicatSAT employs the *log-encoding* for domain variables. For a domain variable, $\lceil log_2(n) \rceil$ Boolean variables are used, where n is the maximum absolute value in the domain. If the domain contains both negative and positive values, then another Boolean variable is employed to represent the sign. In this paper, for a log-encoded domain variable X, $X.\mathsf{s}$ denotes the sign, $X.\mathsf{m}$ denotes the magnitude, which is a vector of Boolean variables $< X_{n-1} X_{n-2} \ldots X_1 X_0 >$.

This *sign-and-magnitude* encoding requires a clause to disallow negative zero if the domain contains values of both signs. Each combination of values of the Boolean variables represents a valuation for the domain variable: $X_{n-1} \times 2^{n-1} + X_{n-2} \times 2^{n-2} + \ldots + X_1 \times 2 + X_0$. If there are holes in the domain, then not-equal constraints are generated to disallow assigning those hole values to the variable. Also, inequality constraints ($\geq$ and $\leq$) are generated to prohibit assigning out-of-bounds values to the variable if either bound is not $2^k - 1$ for some k.

For small-domain variables, PicatSAT calls the logic optimizer, Espresso [9], to generate an optimal or near-optimal CNF formula. For example, for the domain constraint $X :: [-2, -1, 2, 1]$, one Boolean variable, S, is utilized to encode the sign, and two variables, X_1 and X_0, are employed to encode the magnitude. A naive encoding with *conflict clauses* [18] for the domain requires four clauses:

$$\neg S \vee \neg X_1 \vee \neg X_0$$
$$\neg S \vee X_1 \vee X_0$$
$$S \vee X_1 \vee X_0$$
$$S \vee \neg X_1 \vee \neg X_0$$

These clauses correspond to four no-good values: -3, -0, 0, and 3, where -0 denotes the negative 0. Espresso only returns two clauses for the domain:

$$X_0 \lor X_1$$
$$\neg X_0 \lor \neg X_1$$

Note that the sign variable is optimized away.

2.3   Encoding Basic Constraints

The encodings for the addition and multiplication constraints will be described in later sections. This subsection briefly describes the Booleanization of other basic constraints.

The *at-least-one* constraint $\Sigma_i^n B_i \geq 1$ is encoded into one CNF clause:

$$B_1 \lor B_2 \lor \ldots \lor B_n$$

The *at-least-two* constraint $\Sigma_i^n B_i \geq 2$ is converted into n at-least-one constraints: for each $n - 1$ variables, the sum of the variables is at least one. The *at-most-one* constraint $\Sigma_i B_i \leq 1$ is encoded into CNF by using the two-product algorithm [12]. The *at-most-two* constraint is converted into n at-most-one constraints. The *exactly-one* constraint $\Sigma_i^n B_i = 1$ is converted into a conjunction of an at-least-one constraint and an at-most-one constraint. The *exactly-two* constraint is compiled similarly.

A recursive algorithm is utilized to compile binary primitive constraints. For example, consider $X \geq Y$. This constraint is translated to the following:

$$X.s = 0 \land Y.s = 1 \lor$$
$$X.s = 1 \land Y.s = 1 \Rightarrow X.m \leq Y.m \lor$$
$$X.s = 0 \land Y.s = 0 \Rightarrow X.m \geq Y.m$$

PicatSAT simplifies the formula if the variables' signs are known at compile time. Let $X.m = <X_{n-1}X_{n-2}\ldots X_1X_0>$, $Y.m = <Y_{n-1}Y_{n-2}\ldots Y_1Y_0>$.[5] PicatSAT introduces auxiliary variables $T_0, T_1, \ldots, T_{n-1}$ for comparing the bits:

$$T_0 \Leftrightarrow (X_0 \geq Y_0)$$
$$T_1 \Leftrightarrow (X_1 > Y_1) \lor (X_1 = Y_1 \land T_0)$$
$$\vdots$$
$$T_{n-1} \Leftrightarrow (X_{n-1} > Y_{n-1}) \lor (X_{n-1} = Y_{n-1} \land T_{n-2})$$

PicatSAT then encodes the constraint $X.m \geq Y.m$ as T_{n-1}. When either X or Y is a constant, PicatSAT compiles the constraint without introducing any auxiliary variables.[6]

---

[5] The two bit strings are made to have the same length after *padding with zeros*.

[6] This is done by using a recursive algorithm. When X or Y is a constant, the number of clauses in the generated code is still $O(n)$ even though no auxiliary variables are used.

The reified constraint $B \Leftrightarrow C$ is equivalent to $B \Rightarrow C$ and $\neg B \Rightarrow \neg C$, where $\neg C$ is the negation of C. Let $C_1 \wedge \ldots \wedge C_n$ be the CNF formula of C after Booleanization. Then $B \Rightarrow C$ is encoded into $C'_1 \wedge \ldots \wedge C'_n$, where $C'_i = (C_i \vee \neg B)$ for $i = 1, \ldots, n$.

3   Equivalence Reasoning

Equivalence reasoning is an optimization that reasons about a possible value for a Boolean variable or the relationship between two Boolean variables. This reasoning exploits the properties of constraints. For example, consider the domain constraint $X :: [2, 6]$. The magnitude of X is encoded with three Boolean variables $X.\mathrm{m} = <X_2, X_1, X_0>$. PicatSAT infers $X_1 = 1$ and $X_0 = 0$ from the fact that the binary representations of both 2 and 6 end with 10. With this reasoning, PicatSAT does not generate a single clause for this domain.

The following gives several constraints on which PicatSAT performs equivalence reasoning:

$$X = \mathrm{abs}(Y) \quad \Rightarrow X.\mathrm{m} = Y.\mathrm{m},\ X.\mathrm{s} = 0$$
$$X = -Y \quad\quad\ \Rightarrow X.\mathrm{m} = Y.\mathrm{m}$$
$$X = Y \bmod 2^K \Rightarrow X_0 = Y_0,\ X_1 = Y_1,\ \ldots,\ X_{k-1} = Y_{k-1}$$
$$X = Y \operatorname{div} 2^K \Rightarrow X_0 = Y_K,\ X_1 = Y_{K+1},\ \ldots$$

Note that the constraint $X.\mathrm{m} = Y.\mathrm{m}$ is enforced by unifying the corresponding Boolean variables of X and Y at compile time. Equivalence reasoning considerably eases encoding for some of the constraints. For example, the following clause encodes the constraint $X = -Y$, regardless of the sizes of the domains:[7]

$$\neg X.\mathrm{s} \vee \neg Y.\mathrm{s}$$

and no clause is needed to encode the constraint $X = \mathrm{abs}(Y)$.

Equivalence reasoning can be applied to those addition and multiplication constraints that involve constants. We call this kind of equivalence reasoning *constant propagation*. The remaining of this section gives the propagation rules. In order to make the description self-contained, we include the encoding algorithms for addition and multiplication constraints described in [38].

3.1   Constant Propagation for $X + Y = Z$

For the constraint $X + Y = Z$, if either X or Y has values of mixed signs in its domain, then PicatSAT generates implication constraints to handle different sign combinations [38]. In the following we assume that both operands X and Y are non-negative (i.e., $X.\mathrm{s} = 0$ and $Y.\mathrm{s} = 0$), so the constraint $X + Y = Z$ can be rewritten into the unsigned addition $X.\mathrm{m} + Y.\mathrm{m} = Z.\mathrm{m}$.

Let $X.\mathrm{m} = X_{n-1} \ldots X_1 X_0$, $Y.\mathrm{m} = Y_{n-1} \ldots Y_1 Y_0$, and $Z.\mathrm{m} = Z_n \ldots Z_1 Z_0$. The unsigned addition can be Booleanized by using logic adders as follows:

---

[7] Recall that since no negative zeros are allowed, the domain constraints already guarantee that $X.\mathrm{m} = 0 \Rightarrow \neg X.\mathrm{s}$ and $X.\mathrm{m} = 0 \Rightarrow \neg Y.\mathrm{s}$.

$$X_{n-1} \ldots X_1 \ X_0$$
$$+ \ \ Y_{n-1} \ \ldots Y_1 \ Y_0$$
$$\overline{Z_n \ Z_{n-1} \ \ldots Z_1 \ Z_0}$$

A half-adder is employed for $X_0 + Y_0 = C_1 Z_0$, where C_1 is the carry-out. For each other position i ($0 < i \le n-1$), a full adder is employed for $X_i + Y_i + C_i = C_{i+1} Z_i$. The top-most bit of Z, Z_n, is equal to C_n. This encoding corresponds to the *ripple-carry adder* used in computer architectures.

The full adder $X_i + Y_i + C_{in} = C_{out} Z_i$ is encoded with the following 10 CNF clauses when all the operands are variables:

$$X_i \vee \neg Y_i \vee C_{in} \vee Z_i$$
$$X_i \vee Y_i \vee \neg C_{in} \vee Z_i$$
$$\neg X_i \vee \neg Y_i \vee C_{in} \vee \neg Z_i$$
$$\neg X_i \vee Y_i \vee \neg C_{in} \vee \neg Z_i$$
$$\neg X_i \vee C_{out} \vee Z_i$$
$$X_i \vee \neg C_{out} \vee \neg Z_i$$
$$\neg Y_i \vee \neg C_{in} \vee C_{out}$$
$$Y_i \vee C_{in} \vee \neg C_{out}$$
$$\neg X_i \vee \neg Y_i \vee \neg C_{in} \vee Z_i$$
$$X_i \vee Y_i \vee C_{in} \vee \neg Z_i$$

If any of the operands is a constant, then the code can be simplified. For example, if C_{in} is 0, then the full adder becomes a half adder, which can be encoded with 7 CNF clauses.

For the half adder $X_i + Y_i = C_{out} Z_i$, if any of the operands is a constant, PicatSAT infers that the other two variables are equal or one variable is the negation of the other. PicatSAT performs the following inferences when X or Z is a constant:

Rule-1: $X_i = 0 \Rightarrow C_{out} = 0 \wedge Z_i = Y_i$.
Rule-2: $X_i = 1 \Rightarrow C_{out} = Y_i \wedge Z_i = \neg Y_i$.
Rule-3: $Z_i = 0 \Rightarrow C_{out} = X_i \wedge X_i = Y_i$
Rule-4: $Z_i = 1 \Rightarrow C_{out} = 0 \wedge X_i = \neg Y_i$.

Similar inference rules apply when Y_i is a constant.

For example, consider the addition:

$$X_2 \ X_1 \ X_0$$
$$+ \ \ 1 \ \ \ 0 \ \ \ 0$$
$$\overline{Z_3 \ Z_2 \ Z_1 \ Z_0}$$

PicatSAT infers the following equivalences:

$$X_0 = Z_0$$
$$X_1 = Z_1$$
$$\neg X_2 = Z_2$$
$$X_2 = Z_3$$

Consider, as another example, the addition:

$$\begin{array}{cccc} & X_2 & X_1 & X_0 \\ + & Y_2 & Y_1 & Y_0 \\ \hline 1 & 0 & 1 & 1 \end{array}$$

PicatSAT infers the following equivalences:

$$\neg X_0 = Y_0$$
$$\neg X_1 = Y_1$$
$$X_2 = Y_2$$
$$X_2 = 1$$
$$Y_2 = 1$$

The last two equivalences, $X_2 = 1$ and $Y_2 = 1$, are obtained by Rule-3 and the fact that $Z_3 = 1$.

When PicatSAT infers an equivalence between two variables or between one variable and another variable's negation, PicatSAT only uses one variable in the CNF code for the two variables, and eliminates the two CNF clauses for the equivalence.

4   Constant Propagation for $X \times Y = Z$

PicatSAT adopts the *shift-and-add* algorithm for multiplication. Let X.m be $X_{n-1} \ldots X_1 X_0$. The *shift-and-add* algorithm generates the following conditional constraints for $X \times Y = Z$.

$$X_0 = 0 \Rightarrow S_0 = 0$$
$$X_0 = 1 \Rightarrow S_0 = Y$$
$$X_1 = 0 \Rightarrow S_1 = S_0$$
$$X_1 = 1 \Rightarrow S_1 = (Y << 1) + S_0$$
$$\vdots$$
$$X_i = 0 \Rightarrow S_i = S_{i-1}$$
$$X_i = 1 \Rightarrow S_i = (Y << i) + S_{i-1}$$
$$\vdots$$
$$X_{n-1} = 0 \Rightarrow S_{n-1} = S_{n-2}$$
$$X_{n-1} = 1 \Rightarrow S_{n-1} = (Y << (n-1)) + S_{n-2}$$
$$Z = S_{n-1}$$

The operation $(Y << i)$ shifts the binary string of Y to left by i positions. Let the length of the binary string of Y be u. The length of S_0 is u, that of S_1 is $u + 1$, and so on. So the total number of auxiliary Boolean variables that are created to hold the sums is $\Sigma_{i=u}^{(n+u-2)} i$ plus the number of auxiliary variables used for carries in the additions. Note that because S_{n-1} is the same as Z, S_{n-1} is never created.

If either X or Y is a constant, the basic algorithm can be improved to reduce the number of auxiliary variables. In the following of this subsection, we assume that X is a constant.

In the conditional constraints:

$$X_i = 0 \Rightarrow S_i = S_{i-1}$$
$$X_i = 1 \Rightarrow S_i = (Y << i) + S_{i-1}$$

If X_i is 0, then S_i is the same as S_{i-1}, and the variables in S_{i-1} can be reused for S_i. If X_i is 1, then the lowest i bits of S_i are the same as the lowest i bits of S_{i-1}, and these i variables can be copied from S_{i-1} to S_i. The following two rules perform these propagation:

Rule 5: $X_i = 0 \Rightarrow$ copy all of the bits of S_{i-1} into S_i.
Rule 6: $X_i = 1 \Rightarrow$ copy the lowest i bits of S_{i-1} into S_i.

Let X_i be the lowest bit of X that is 1, meaning that $X_j = 0$ for $j \in 0..i-1$. Then Z_i is the same as Y_0, the lowest bit of Y, and $Z_k = 0$ for $k \in 0..i-1$. Here is the rule that performs this propagation:

Rule 7: $X.m = <X_{n-1} \ldots X_i 0 \ldots 0> \wedge X_i = 1 \Rightarrow$
$Z_i = Y_0 \wedge Z_k = 0$ for $k \in 0..(i-1)$.

In particular, if X is a power of 2, then Z is the result of shifting Y to left by $n-1$ positions, and no auxiliary variables are needed.

The number of additions performed by the shift-and-add algorithm is the number of 1s in the binary string of X. If X is a constant that is not a power of 2 but is close to a power of 2, then PicatSAT converts the multiplication into an addition. The following rules perform this optimization:

Rule-8: $X = 2^K - 1 \Rightarrow$
rewrite $X \times Y = Z$ into $2^K \times Y = Z + Y$.
Rule-9: $X = 2^K - 2 \Rightarrow$
rewrite $X \times Y = Z$ into $2^K \times Y = Z + 2 \times Y$.

For example, PicatSAT converts the constraint $7 \times X = Z$ to $8 \times X = Z + X$, which requires one addition while the original constraint requires three additions.

5   Experimental Results

PicatSAT, which is implemented in Picat, has about 8,000 lines of code, excluding comments. PicatSAT connects to Lingeling SAT solver (version 587f) through a C interface.

We have done two experiments in order to evaluate the compiler using Picat version 2.1.[8] In the first experiment, we ran PicatSAT on several benchmark problems that involve arithmetic constraints, and measured the code size and

---

[8] http://picat-lang.org.

the execution time of each of the benchmarks. These results show how effective equivalence reasoning is on reducing the code size and the execution time.

In the second experiment, which was conducted on Linux Ubuntu 14.04LTS with an Intel i7-5820K 3.30 GHz CPU and 32 GHz RAM, we compared Picat-SAT with Chuffed[9] and fzn2smt version 2.0.02.[10] Chuffed is a cutting-edge LCG solver. The **fzn2smt** solver, which took the second place in MiniZinc Challenge 2012, translates FlatZinc[11] into SMT that is solved by Yices[12].

This section does not include comparisons of PicatSAT with many other state-of-the-art CSP solvers. In the free search category of the MiniZinc Challenge 2016, an old version of PicatSAT outperformed some of the fastest CSP solvers, and was only second to HaifaCSP[13] by a small margin.

Table 1 evaluates how effective equivalence reasoning is on reducing the code size (#vars, the number of variables, and #cls, the number of clauses). The first three benchmarks are well known puzzles in the CP community. The next three are instances of the magic square problem for the grid sizes of 7×7, 8×8, and 9×9. The remaining 5 benchmarks are integer programming benchmarks

Table 1. Evaluation of effectiveness of equivalence reasoning (code size)

Benchmark	PicatSAT^{nor}		PicatSAT	
	#vars	#cls	#vars(%)	#cls(%)
crypta	3445	15374	1893 (**54**)	11537 (**75**)
eq10	10212	46087	6043 (59)	36696 (79)
eq20	19292	86469	11397 (59)	68869 (79)
magic_square_7	3543	81588	3463 (97)	81428 (99)
magic_square_8	6882	56324	6864 (99)	56288 (99)
magic_square_9	10306	86268	10226 (99)	85908 (99)
maxclosed_10_100_10	3778	23219	3091 (83)	21150 (91)
maxclosed_10_100_100	25420	162110	21128 (83)	148580 (91)
maxclosed_10_200_10	3066	18864	2559 (83)	17418 (92)
maxclosed_20_100_1000	221454	1577283	198137 (89)	1464492 (92)
maxclosed_30_200_1000	417816	3078098	379477 (90)	2881243 (93)

---

[9] https://github.com/chuffed/chuffed, released in December 2016.

[10] The fzn2smt solver (http://ima.udg.edu/recerca/lap/fzn2smt) has not been updated since 2012. There have been no significant speedups in the past five years in SAT solvers, on which both PicatSAT and SMT are based. PicatSAT uses Lingeling version 587f, which was released in February 2011 but is still faster than recent versions on most MiniZinc Challenge benchmarks. Therefore, this comparison is still relevant, if not completely up to date.

[11] http://www.minizinc.org.

[12] http://yices.csl.sri.com.

[13] http://strichman.net.technion.ac.il/haifacsp/.

taken from [30].[14] The column PicatSAT[nor] shows the results obtained with the equivalence reasoning optimization disabled, and the column PicatSAT shows the results obtained when the optimization was enabled.

As can be seen, the optimization led to certain amounts of reduction in the code size for each of the benchmarks. For *crypta*, the reduction in code size is most significant; the number of variables is reduced to 54%, and the number of clauses is reduced to 75%. For *magic_square*, whose code size is dominated by an *all-different* constraint, the reduction in code size is only 1%.

Table 2 evaluates how effective equivalence reasoning is on reducing the compile time (comp) and the solving time (solve). The solving time of the CNF code was measured using Lingeling version 587f on a Cygwin notebook computer with 2.60 GHz Intel i7 and 64 GB RAM. Interestingly, while the code reduction for *magic_square* is the least significant, the speedups are the most significant. Overall, the optimization reduces both the compile time and the solving time, and the results show that the equivalence reasoning optimization is worthwhile to incorporate.

Table 2. Evaluation of effectiveness of equivalence reasoning (time, seconds)

Benchmark	PicatSAT[nor]		PicatSAT	
	comp	solve	comp(%)	solve(%)
crypta	0.18	0.337	0.17 (94)	0.310 (91)
eq10	0.32	1.177	0.26 (81)	1.000 (84)
eq20	0.61	1.239	0.48 (80)	1.134 (91)
magic_square_7	0.50	2.414	0.47 (94)	1.020 (**42**)
magic_square_8	0.57	14.908	0.56 (98)	6.835 (45)
magic_square_9	0.66	37.501	0.63 (95)	25.326 (67)
maxclosed_10_100_10	0.43	0.420	0.40 (93)	0.393 (93)
maxclosed_10_100_100	2.50	2.041	2.36 (94)	1.877 (91)
maxclosed_10_200_10	0.23	0.353	0.25 (108)	0.306 (86)
maxclosed_20_100_1000	29.00	14.809	28.47 (98)	14.352 (96)
maxclosed_30_200_1000	42.713	31.030	39.58 (92)	28.584 (92)

Table 3 compares PicatSAT with fzn2smt and Chuffed on the instances used in the MiniZinc Challenge 2012.[15] All of the instances were translated from MiniZinc into FlatZinc using each individual solver's global constraints.[16] The time limit was set to 15 min per instance, which limits the total of the conversion

---

[14] Instances that can be solved by the preprocessor are not included.

[15] The benchmarks of MiniZinc 2012 were used because fzn2smt took the second place in that competition, and didn't compete ever since.

[16] fzn2smt, which does not have any solver-specific globals, uses MiniZinc's default decomposer.

time and the solving time. For each benchmark, the number in the parentheses is the total number of instances, and the number in each column indicates the number of completely solved instances by the solver. For optimization problems, an instance is considered solved if an optimal solution was given and its optimality was proven. PicatSAT solved 77 instances, while fzn2smt solved 52, and Chuffed solved 67 instances. This experiment demonstrates the competitiveness of PicatSAT in comparison with two lazy-approach-based solvers. When equivalence reasoning was disabled, PicatSAT solved 75 instances. This result also shows the worthiness of the optimization.

Table 3. A comparison of three CSP solvers (solved instances)

Benchmark	PicatSAT	PicatSAT^{nor}	fzn2smt	Chuffed
amaze (6)	6	6	5	5
amaze2 (6)	6	6	6	6
carpet-cutting (5)	1	0	0	2
fast-food (5)	5	5	5	4
filters (5)	4	3	2	3
league (6)	3	3	2	2
mspsp (6)	6	6	6	6
nonogram (5)	5	5	5	5
parity-learning (7)	7	7	0	2
pattern-set-mining-k1 (2)	1	1	0	0
pattern-set-mining-k2 (3)	2	2	1	1
project-planning (6)	5	5	0	6
radiation (5)	3	3	0	2
ship-schedule (5)	5	5	5	5
solbat (5)	5	5	5	5
still-life-wastage (5)	5	5	3	5
tpp (7)	7	7	7	7
train (6)	1	1	0	1
vrp (5)	0	0	0	0
total (100)	77	75	52	67

6   Related Work

A wide variety of problems have been encoded into SAT and solved by SAT solvers. SAT is also the backbone of many logic language systems, such as formal methods [23], answer set programming [10,19], and NP-SPEC [11]. PicatSAT is a compiler that translates high-level constraints into SAT. Other SAT compilers include BEE [25], FznTini [21], meSAT [33], and Sugar [34]. PicatSAT, like FznTini, adopts the log encoding, while the other compilers are based on the order encoding.

Despite the compactness of the log encoding, it has received little attention for CSP solving, probably because of its weak propagation strength and its requirement of engineering efforts. FznTini was the only known log-encoding based CSP solver before PicatSAT. FznTini employs the 2's complement encoding for domain variables, while PicatSAT uses the sign-and-magnitude encoding. FznTini demonstrated the promise of the log encoding, but it lacks optimizations, and is not considered competitive with recent CP solvers [30].

The perception that eager SAT-encoding approaches are not suited to arithmetic constraints has motivated the development of lazy approaches such as SMT [6,27] and lazy clause generation [17,29] that integrates CP and SAT solving techniques. Recent MiniZinc competitions have been dominated by LCG solvers; the HaifaCSP solver [35] also performs learning during search.

Encodings, such as BDDs [1,7] and sorting networks [16], have been proposed for special form of arithmetic constraints, such as Boolean cardinality constraints and Pseudo-Boolean (PB) constraints. Linear arithmetic constraints can be compiled into SAT through PB constraints [2]. This encoding through PB constraints is less compact than the adder-based log encoding, but has stronger propagation power. The hybrid encoding that integrates order and log encodings [32] compiles linear constraints that involve large-domain variables into SAT through PB constraints.

The idea of identifying equivalences and exploiting them to reduce code sizes has been explored in both SAT compilation and SAT solving. The BEE compiler [25] performs *equi-propagation*, which takes advantage of the properties of the order-encoding to infer equivalences. Although equivalences could be detected to some extent by SAT solvers, it is always beneficial to do the reasoning at compile time, because SAT solvers are unaware of the meaning of the original constraints, and it is expensive to detect equivalences at preprocessing time [15] or reason about them at solving time [24].

PicatSAT embodies optimizations used in CP systems for processing constraints [31], in language compilers for eliminating sub-expressions [3], and in hardware design systems for optimizing logic circuits [9]. The equivalence reasoning optimization, which is specific to the log encoding, has not been implemented in any other SAT compilers.

7 Conclusion

In this paper we have presented the PicatSAT compiler and its optimizations. PicatSAT employs the log encoding, which has received little attention by SAT-based CSP solvers for its lack of propagation strength. PicatSAT adopts optimizations from CP systems (preprocessing constraints to narrow the domains of variables), language compilers (decomposing constraints into basic ones and eliminating common subexpressions), and hardware design systems (using a logic optimizer to optimize codes for adders and other basic constraints). Furthermore, PicatSAT reasons about equivalences in arithmetic constraints, and exploit them to eliminate variables and clauses. With these optimizations, PicatSAT is able to generate more compact and faster code for arithmetic constraints.

This work has shed a light on the debate between the eager and lazy approaches to constraint solving with SAT. The failed attempts to find efficient encodings for arithmetic constraints have motivated the development of the lazy approaches. This paper has shown that the eager approach based on the optimized log encoding is not as bad for arithmetic constraints as it was perceived before. A comparison with an SMT-based CSP solver and a LCG solver shows the competitiveness of PicatSAT.

In addition to the optimizations, including the novel equivalence reasoning optimization, and the engineering effort, the success of PicatSAT is also attributed to Picat, the implementation language. The log encoding is arguably more difficult to implement than the sparse and order encodings. Picat's features, such as attributed variables, unification, pattern-matching rules, and loops, are all put into good use in the implementation. There are hundreds of optimization rules, and they can be described easily as pattern-matching rules in Picat. Logic programming has been proven to be suitable for language processing in general, and for compiler writing in particular; PicatSAT has provided another testament.

The success of PicatSAT does not in any way undermine the lazy approaches: there are certainly many problems for which the lazy approaches prevail, and many theories incorporated in SMT solvers do not yet have efficient SAT encodings. A comprehensive comparison of eager encoding and lazy approaches is on the stack for future work.

PicatSAT still has plenty of room for improvement, especially concerning global constraints and special constraints. One direction for future work is to carry out these improvements.

The optimizations reported in this paper could be only the tip of the iceberg. Numerous algorithms and optimizations have been proposed for hardware design systems, such as multi-bit and multi-operand adders and multipliers. When multiple bits are considered at once, there will be more reasoning opportunities opening up. Another direction for future work is to investigate these algorithms for SAT encodings.

Acknowledgement. The authors would like to thank the anonymous reviewers for helpful comments. This project was supported in part by the NSF under grant number CCF1618046.

References

1. Abío, I., Nieuwenhuis, R., Oliveras, A., Rodríguez-Carbonell, E., Mayer-Eichberger, V.: A new look at BDDs for Pseudo-Boolean constraints. J. Artif. Intell. Res. (JAIR) **45**, 443–480 (2012)

2. Abío, I., Stuckey, P.J.: Encoding linear constraints into SAT. In: O'Sullivan, B. (ed.) CP 2014. LNCS, vol. 8656, pp. 75–91. Springer, Cham (2014). doi:10.1007/978-3-319-10428-7_9

3. Aho, A.V., Lam, M.S., Sethi, R., Ullman, J.D.: Compilers: Principles. Techniques. Addison-Wesley, Tools Boston (2007)

4. Araya, I., Neveu, B., Trombettoni, G.: Exploiting common subexpressions in numerical CSPs. In: Stuckey, P.J. (ed.) CP 2008. LNCS, vol. 5202, pp. 342–357. Springer, Heidelberg (2008). doi:10.1007/978-3-540-85958-1_23

5. Bailleux, O., Boufkhad, Y., Roussel, O.: New Encodings of Pseudo-Boolean Constraints into CNF. In: Kullmann, O. (ed.) SAT 2009. LNCS, vol. 5584, pp. 181–194. Springer, Heidelberg (2009). doi:10.1007/978-3-642-02777-2_19

6. Barrett, C.W., Sebastiani, R., Seshia, S.A., Tinelli, C.: Satisfiability modulo theories. In: Handbook of Satisfiability, pp. 825–885 (2009)

7. Bartzis, C., Bultan, T.: Efficient BDDs for bounded arithmetic constraints. Int. J. Softw. Tools Technol. Transf. (STTT) 8(1), 26–36 (2006)

8. Bordeaux, L., Hamadi, Y., Zhang, L.: Propositional satisfiability and constraint programming: a comparative survey. ACM Comput. Surv. 38(4), 1–54 (2006)

9. Brayton, R.K., Hachtel, G.D., McMullen, C.T., Sangiovanni-Vincentelli, A.L.: Logic Minimization Algorithms for VLSI Synthesis. Kluwer Academic Publishers, Dordrecht (1984)

10. Brewka, G., Eiter, T., Truszczyński, M.: Answer set programming at a glance. Commun. ACM 54(12), 92–103 (2011)

11. Cadoli, M., Schaerf, A.: Compiling problem specifications into SAT. Artif. Intell. 162(1–2), 89–120 (2005)

12. Chen, J.: A new SAT encoding of the at-most-one constraint. In: Proceedings of the 9th International Workshop of Constraint Modeling and Reformulation (2010)

13. Crawford, J.M., Baker, A.B.: Experimental results on the application of satisfiability algorithms to scheduling problems. In: AAAI, pp. 1092–1097 (1994)

14. Dutertre, B., Moura, L.: A fast linear-arithmetic solver for DPLL(T). In: Ball, T., Jones, R.B. (eds.) CAV 2006. LNCS, vol. 4144, pp. 81–94. Springer, Heidelberg (2006). doi:10.1007/11817963_11

15. Eén, N., Biere, A.: Effective preprocessing in SAT through variable and clause elimination. In: Bacchus, F., Walsh, T. (eds.) SAT 2005. LNCS, vol. 3569, pp. 61–75. Springer, Heidelberg (2005). doi:10.1007/11499107_5

16. Eén, N., Sörensson, N.: Translating Pseudo-Boolean constraints into SAT. JSAT 2(1–4), 1–26 (2006)

17. Feydy, T., Stuckey, P.J.: Lazy clause generation reengineered. In: Gent, I.P. (ed.) CP 2009. LNCS, vol. 5732, pp. 352–366. Springer, Heidelberg (2009). doi:10.1007/978-3-642-04244-7_29

18. Gavanelli, M.: The log-support encoding of CSP into SAT. In: Bessière, C. (ed.) CP 2007. LNCS, vol. 4741, pp. 815–822. Springer, Heidelberg (2007). doi:10.1007/978-3-540-74970-7_59

19. Gebser, M., Kaufmann, B., Neumann, A., Schaub, T.: Conflict-driven answer set solving. In: IJCAI, p. 386 (2007)

20. Gent, I.P.: Arc consistency in SAT. In: ECAI, pp. 121–125 (2002)

21. Huang, J.: Universal booleanization of constraint models. In: Stuckey, P.J. (ed.) CP 2008. LNCS, vol. 5202, pp. 144–158. Springer, Heidelberg (2008). doi:10.1007/978-3-540-85958-1_10

22. Iwama, K., Miyazaki, S.: SAT-varible complexity of hard combinatorial problems. In: IFIP Congress, vol. 1, pp. 253–258 (1994)

23. Jackson, D., Abstractions, S.: Logic, Language, and Analysis. MIT Press, Cambridge (2012)
24. Li, C.M.: Integrating equivalency reasoning into davis-putnam procedure. In: AAAI, pp. 291–296 (2000)
25. Metodi, A., Codish, M.: Compiling finite domain constraints to SAT with BEE. Theor. Pract. Log. Program. **12**(4–5), 465–483 (2012)
26. Nieuwenhuis, R.: The intsat method for integer linear programming. In: O'Sullivan, B. (ed.) CP 2014. LNCS, vol. 8656, pp. 574–589. Springer, Cham (2014). doi:10.1007/978-3-319-10428-7_42
27. Nieuwenhuis, R., Oliveras, A., Tinelli, C.: Solving SAT and SAT modulo theories: from an abstract davis-putnam-logemann-loveland procedure to DPLL(T). J. ACM **53**(6), 937–977 (2006)
28. Nightingale, P., Spracklen, P., Miguel, I.: Automatically improving SAT encoding of constraint problems through common subexpression elimination in savile row. In: Pesant, G. (ed.) CP 2015. LNCS, vol. 9255, pp. 330–340. Springer, Cham (2015). doi:10.1007/978-3-319-23219-5_23
29. Ohrimenko, O., Stuckey, P.J., Codish, M.: Propagation via lazy clause generation. Constraints **14**(3), 357–391 (2009)
30. Petke, J.: Bridging Constraint Satisfaction and Boolean Satisfiability. Artificial Intelligence: Foundations, Theory, and Algorithms. Springer, Heidelberg (2015)
31. Rossi, F., van Beek, P., Walsh, T.: Handbook of Constraint Programming. Elsevier, Amsterdam (2006)
32. Soh, T., Banbara, M., Tamura, N.: Proposal and evaluation of hybrid encoding of CSP to SAT integratin order and log encodings. Int. J. Artif. Intell. Tools **26**(1), 1–29 (2017)
33. Stojadinovic, M., Maric, F.: meSAT: multiple encodings of CSP to SAT. Constraints **19**(4), 380–403 (2014)
34. Tamura, N., Taga, A., Kitagawa, S., Banbara, M.: Compiling finite linear CSP into SAT. Constraints **14**(2), 254–272 (2009)
35. Veksler, M., Strichman, O.: Learning general constraints in CSP. Artif. Intell. **238**, 135–153 (2016)
36. Walsh, T.: SAT v CSP. In: Dechter, R. (ed.) CP 2000. LNCS, vol. 1894, pp. 441–456. Springer, Heidelberg (2000). doi:10.1007/3-540-45349-0_32
37. Warners, J.P.: A linear-time transformation of linear inequalities into conjunctive normal form. Inf. Process. Lett. **68**(2), 63–69 (1998)
38. Zhou, N.-F., Kjellerstrand, H.: The picat-SAT compiler. In: Gavanelli, M., Reppy, J. (eds.) PADL 2016. LNCS, vol. 9585, pp. 48–62. Springer, Cham (2016). doi:10.1007/978-3-319-28228-2_4

Test and Verification and CP Track

Constraint-Based Synthesis of Datalog Programs

Aws Albarghouthi[1]([✉]), Paraschos Koutris[1], Mayur Naik[2], and Calvin Smith[1]

[1] University of Wisconsin–Madison, Madison, USA
aws@cs.wisc.edu
[2] Unviersity of Pennsylvania, Philadelphia, USA

Abstract. We study the problem of synthesizing recursive Datalog programs from examples. We propose a constraint-based synthesis approach that uses an SMT solver to efficiently navigate the space of Datalog programs and their corresponding *derivation trees*. We demonstrate our technique's ability to synthesize a range of graph-manipulating recursive programs from a small number of examples. In addition, we demonstrate our technique's potential for use in *automatic construction of program analyses* from example programs and desired analysis output.

1 Introduction

The program synthesis problem—as studied in verification and AI—involves constructing an executable program that satisfies a specification. Recently, there has been a surge of interest in *programming by example* (PBE), where the specification is a set of input–output examples that the program should satisfy [2,7,11,14,22]. The primary motivations behind PBE have been to (*i*) allow end users with no programming knowledge to automatically construct desired computations by supplying examples, and (*ii*) enable automatic construction and repair of programs from tests, e.g., in test-driven development [23].

In this paper, we present a constraint-based approach to synthesizing Datalog programs from examples. A Datalog program is comprised of a set of *Horn clauses* encoding *monotone, recursive constraints* between relations. Our primary motivation in targeting Datalog is to expand the range of synthesizable programs to the new domains addressed by Datalog. Datalog has been used in information extraction [28], graph analytics [4,26,32], and in specifying static program analyses [29,33], amongst others. We believe a PBE approach to Datalog has the potential to simplify programming in an exciting range of applications. We demonstrate how our approach can automatically synthesize a popular static analysis from examples. We envision a future in which developers will be able to automatically synthesize static analyses by specifying examples of information they would like to compute from their code. For instance, the synthesizer can live in the background of an IDE and learn what kind of information a developer likes to extract. Our approach is a concrete step towards realizing these goals.

To synthesize Datalog programs, we exploit a key technical insight: We are searching for a Datalog program whose least fixpoint—maximal *derivation tree*—includes all the positive examples and none of the negative ones. Encoding the

© Springer International Publishing AG 2017
J.C. Beck (Ed.): CP 2017, LNCS 10416, pp. 689–706, 2017.
DOI: 10.1007/978-3-319-66158-2_44

search space of all Datalog programs and their fixpoints in some first-order theory results in a complex set of constraints. Instead, we construct a set of quantifier-free constraints that encode (i) all sets of clauses up to a given size and (ii) all *derivations*—proof trees—of a fixed size for all those clauses. In other words, we encode *underapproximations of the least fixpoints*. We then employ an *inductive synthesis loop* (as shown in Fig. 1) to ensure the program is correct and restart otherwise.

Our choice of a constraint-based synthesis technique is advantageous in (i) simulating execution of Datalog programs and (ii) steering synthesis towards desirable programs. First, we exploit the axioms of the McCarthy's first-order theory of arrays [18] to encode Datalog proof trees. Second, we define the notion of *clause templates*: additional constraints that impose a certain structure on synthesized clauses. Clause templates (i) constrain the search space and (ii) steer the synthesizer towards programs satisfying certain properties: for example, if we want programs in the complexity class NC—i.e., efficiently parallelizable—we can apply a template that ensures that all clauses are *linear*.

The field of inductive logic programming (ILP) has extensively studied the problem of inducing logic programs from examples [8, 20]. Generally, the emphasis there has been on synthesizing classifiers, and therefore more examples are used and not all examples need be classified correctly. Our emphasis here is on programming-by-example, where the user provides a small number of examples and we want to match all of them. Our technical contribution can be viewed as a novel ILP technique that completely delegates the combinatorial search to an off-the-shelf SMT solver. To the best of our knowledge, this is the first such use of SMT solvers in synthesizing logic programs. We refer to Sect. 7 for a detailed comparison with related works.

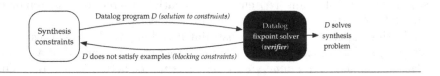

Fig. 1. High-level view of inductive synthesis loop.

Contributions. First, we demonstrate that constraint solving can be applied to the problem of synthesizing recursive Datalog programs. Second, we demonstrate how to constrain the search using logical encodings of clause templates. Third, we implement our approach and use it to synthesize a collection of recursive Datalog programs. In addition to efficiently synthesizing a range of standard Datalog programs, we demonstrate our approach's novel ability to synthesize program analyses from examples.

2   Overview and Examples

2.1   Datalog Overview

Datalog is a logic programming language where programs are composed of a set
of Horn clauses over relations. For illustration, let us fix the domain (*universe*)
to be $\mathcal{U} = \{a, b, c, d, e, \ldots\}$. Suppose that we are given the binary relation $E =
\{(a, b), (b, c), (c, d), (d, c)\}$. We can think of relations as (hyper-)graphs, where
nodes are elements of the universe $\mathcal{U}$ and (hyper-)edges denote that a tuple is in
the relation. Pictorially, we can view E as representing a graph, where there is
an edge from node x to node y iff $(x, y) \in E$, i.e., $E(x, y)$ is a *fact*.

To compute the transitive closure of the *input relation*
E, we can write the Datalog program in Fig. 3(a), where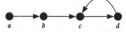
T is an *output relation* that will contain the transitive
closure after executing the program. X, Y, and Z are interpreted as universally
quantified variables. For instance, the second clause says: *for all values of X, Y
and Z picked from $\mathcal{U}$, if $(X, Z) \in E$ and $(Z, Y) \in T$, then (X, Y) must also be
in T*.

One can view the execution of a Datalog program as a sequence of *derivations*, where in each step we add a new tuple to the output relation, until we
reach a fixpoint. Figure 2 pictorially illustrates the process of deriving the transitive closure for our example. T starts out as the empty set, denoted T_0. By
instantiating variables in the first Horn clause with constants, we can derive the
edge (a, b) and add it to T, resulting in T_1. After 9 derivations, we arrive at the
fixpoint, T_9, which is the full transitive closure.

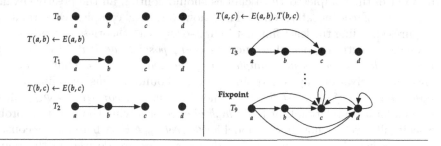

Fig. 2. Derivation sequence for transitive closure example.

2.2   Illustrative Examples

Transitive Closure. Assume we have the same input relation E as above—the
graph for which we want to compute the transitive closure. We can now supply
positive and *negative* examples of what edges should appear in T. For instance,

$$Ex^+ = \{(a, b), (b, c), (a, c), (a, d)\}, \qquad Ex^- = \{(a, a)\}$$

The synthesis problem is: Find a set of Horn clauses C, defining the relation T, such that: (i) $Ex^+ \subseteq T$, and (ii) $Ex^- \cap T = \emptyset$. In other words, we want all positive examples to appear in T, but none of the negative ones.

Our synthesis technique employs an inductive synthesis loop, where in each iteration (i) a set of clauses C are *synthesized*, and (ii) C are *verified* to ensure that they derive all positive examples and none of the negative ones. We illustrate two iterations below.

Iteration 1: Synthesis Phase. To synthesize a set of clauses C, we fix the maximum number of clauses in C and the maximum number of atoms in the body of a clause. Assume we fix the number of clauses to 2 and the number of atoms to 2. Then we are looking a set of two clauses, where each clause is of the form $\blacklozenge(\bullet, \bullet) \leftarrow \blacklozenge(\bullet, \bullet), \blacklozenge(\bullet, \bullet)$. Intuitively, we would like to replace the $\blacklozenge$'s with relation symbols and the $\bullet$'s

(a) $T(X,Y) \leftarrow E(X,Y).$
        $T(X,Z) \leftarrow E(X,Y), T(Y,Z).$

(b) $T(X,Y) \leftarrow E(X,Y).$
        $T(X,Z) \leftarrow E(X,Y), E(Y,Z).$

(c) $T(X,Y) \leftarrow E(X,Y).$
        $T(X,Z) \leftarrow T(X,Y), T(Y,Z).$

Fig. 3. Transitive closure example.

with variables. To do so, the synthesis phase constructs a set of constraints Φ_{cl}, where every model of Φ_{cl} is one possible *completion* of the above. A naïve way to proceed here is to simply sample models of Φ_{cl} and verify whether the completion derives all positive examples and none of the negative ones. This *guess-and-check* process would take a very long time due to the large number of possible Datalog programs comprised of two clauses.

Therefore, we need to be able to add a new constraint specifying that the least fixpoint of the completed Horn clauses should contain all the positive examples and none of the negative ones. This, however, is a complex constraint, as it requires encoding the least fixpoint in first-order SMT theories. Instead, we create a weaker constraint, one that encodes *every possible derivation of some finite length d, for every possible completion of the above clauses.* That is, instead of encoding derivations up to fixpoint, we fix a bound d, thus encoding an *under-approximation of the least fixpoint.* These *simulation constraints* Φ_{sim} allow us to look for a completion that has a *high chance* of solving the synthesis problem. Specifically, we can now find a model for $\Phi_{cl} \wedge \Phi_{sim} \wedge Ex$, where Ex is a constraint that specifies that none of the negative examples are derived in the bounded derivation, and *most* of the positive examples are derived—in other words, we want to maximize the number of derived positive examples. This is because not all positive examples may be derivable in the bounded derivation. A possible solution for the above constraints is the set of clauses shown in Fig. 3(b).

Iteration 1: Verification Phase. The verification phase will compute the fixpoint of these clauses and determine that they do not derive all the positive examples. As a result, the verification phase produces a *blocking constraint* Φ_{neg} that avoids all similar sets of clauses.

Iteration 2: In the second iteration, the synthesis phase computes a new set of clauses that satisfy the following constraints: $\Phi_{cl} \wedge \Phi_{sim} \wedge Ex \wedge \Phi_{neg}$. As a result, it might synthesize the correct set of clauses in Fig. 3(c).

Notice that the second clause is *non-linear*, meaning that an output relation, T, appears more than once in its body. Due to the symbolic encoding, it is simple to impose additional constraints that steer synthesis towards programs of a specific form: we call these constraints *clause templates*. For instance, if we impose a template specifying that all clauses are linear, then we synthesize the equivalent transitive closure program in Fig. 3(a).

Andersen's Pointer Analysis. In static program analysis, many analyses are routinely written as Datalog programs. A given program to be analyzed is represented as a set of input relations. The Horn clauses then *compute* the results of the static analysis from these input relations. Pointer analysis is a popular target for Datalog, where the output relation is an over-approximation of which variables point to which other variables.

We can specify a slice of the desired output of a static analysis, and have our synthesizer *automatically* detect and produce the desired analysis in the form of a Datalog pro-

$$pt(X,Y) \leftarrow addressOf(X,Y).$$
$$pt(X,Z) \leftarrow assign(X,Y), pt(Y,Z).$$
$$pt(W,Z) \leftarrow store(X,Y), pt(Y,Z), pt(X,W).$$
$$pt(X,W) \leftarrow load(X,Y), pt(Y,Z), pt(Z,W).$$

gram. Indeed, we show that our technique is able to synthesize *Andersen's pointer analysis* [3] from examples (shown above). Specifically, here we specify examples of tuples that should or should not appear in the relation pt, where $pt(a,b)$ specifies that variable a points to (the location of) variable b in the program. The rest of the relations are input relations specifying the program to be analyzed. For example, $addressOf(a,b)$ indicates that there is a statement in the program of the form a = &b.

3   Preliminaries

Horn Clauses. A *term* t is either a variable $X, Y, Z, \ldots$, or a constant $a, b, c, \ldots$. A *predicate symbol* P is associated with an arity $arity(P)$. An *atom* is an application of a predicate symbol to a vector of variables and constants, e.g., $P(X,Y,a)$ for a predicate P with arity 3. A *ground atom* is an application of a predicate symbol to constants, e.g., $P(a_1, \ldots, a_n)$, where $\{a_i\}_i$ are constants. A *substitution* θ is a mapping from variables to constants. Applying θ to an atom yields a ground atom. For example, if $\theta = \{X \mapsto a, Y \mapsto b\}$, then $H(X,Y)\theta$ is the ground atom $H(a,b)$. When clear from context, we simplify notation to $H\theta$.

A *Horn clause* c is of the form: $H(\boldsymbol{X}) \leftarrow B_1(\boldsymbol{X_1}) \wedge \ldots \wedge B_n(\boldsymbol{X_n})$, where $H(\boldsymbol{X}), B_1(\boldsymbol{X_1}), \ldots, B_n(\boldsymbol{X_n})$ are atoms. The atom $H(\boldsymbol{X})$ is called the *head* of the clause, denoted $head(c)$; the set of atoms $\{B_i(\boldsymbol{X_i})\}$ are the *body* of the clause, denoted $body(c)$. As is standard, we replace conjunctions ($\wedge$) in the body of with commas (,). We use C to denote a finite set of Horn clauses $\{c_1, \ldots, c_n\}$.

Herbrand Interpretations. We define the semantics of Horn clauses using *Herbrand interpretations*. First, assume we have a fixed *Herbrand Universe* $\mathcal{U}$, which is a set of *constants* that can appear in atoms, e.g., $\mathcal{U} = \{a, b, c, \ldots\}$. A *Herbrand interpretation* I of a set of Horn clauses C is a set of ground atoms with constants drawn from $\mathcal{U}$. For example, an interpretation I of our transitive closure example (Sect. 2) could be: $\{T(a, b), E(b, c), T(c, d)\}$.

Definition 1 (Herbrand models and minimality). *A Herbrand interpretation M for a set of clauses C is a* Herbrand model *for C iff for every clause $H \leftarrow B_1, \ldots, B_n \in C$, for all substitutions θ, if $\{B_1\theta, \ldots, B_n\theta\} \subseteq I$, then $H\theta \in I$. A Herbrand model M is a* minimal model *for C iff for all $M' \subset M$, M' is not a Herbrand model of C.*

Datalog Programs. A *Datalog program* C is a finite set of Horn clauses. The predicates of the program can be partitioned into two disjoint sets, $\mathcal{R}_{in}(C)$ and $\mathcal{R}_{out}(C)$: $\mathcal{R}_{in}(C)$ are the predicates that appear only in the bodies of clauses in C, and are called the *input relations*. $\mathcal{R}_{out}(C)$ are the predicates that appear at the heads of clauses in C, and are called the *output relations*.

Semantics and Derivations. The input of a Datalog program C is a finite set of *facts* F, which are ground atoms over the input relations $\mathcal{R}_{in}(C)$. A Herbrand model M of C with input F is a model of C such that $F \subseteq M$. The interpretation of a Datalog program C with input F is its minimal model M. Computing the minimal model M is done using the clauses in C to *derive* all possible facts until the least fixpoint is reached. We denote the minimal model M as $C(F)$.

There always exists a unique minimal model, thus, semantics are well-defined. Figure 4 encodes the least fixpoint computation of $C(F)$ as two rules that monotonically populate the model M with more facts. The rule INIT initializes M to the set of facts F. The rule DERIVE uses clauses in C to derive a new fact to be added to M. Observe that (i) the set M is monotonically increasing and (ii) the fixpoint computation eventually terminates, as the derived facts can only contain constants from the set of facts F, which is finite.

Definition 2 (Derivation sequence). *Given a Datalog program C with input F, a* derivation sequence *is a sequence of sets of ground atoms: $M_0 \xrightarrow{c_{i_1}, \theta_1} M_1 \xrightarrow{c_{i_2}, \theta_2} M_2 \xrightarrow{c_{i_3}, \theta_3} \cdots \xrightarrow{c_{i_n}, \theta_n} M_n$, where $M_0 = F$, and M_j is the set of ground facts resulting from applying DERIVE to M_{j-1} with the clause $c_{i_j} \in C$ and substitution θ_j. A* maximal derivation sequence *is one where DERIVE cannot be applied to M_n, i.e., $M_n = C(F)$.*

Datalog Synthesis Problem. A *Datalog synthesis problem* S, or synthesis problem for short, is a tuple $(\mathcal{R}, F, E)$, where: (i) $\mathcal{R} = (\mathcal{R}_{in}, \mathcal{R}_{out})$ is a pair of input and output predicate sets that are disjoint. (ii) F is a finite set of facts— ground atoms over predicates in $\mathcal{R}_{in}$; (iii) $E = (E^+, E^-)$: E^+ is a finite set of *positive examples*, which are ground atoms over predicates in $\mathcal{R}_{out}$. E^- is a finite set of *negative examples*, which are also ground atoms over predicates in $\mathcal{R}_{out}$. We assume that $E^+ \cap E^- = \emptyset$.

$$\frac{}{M \leftarrow F} \text{ INIT} \qquad \frac{H \leftarrow B_1, \ldots, B_n \in C \qquad \{B_1\theta, \ldots, B_n\theta\} \subseteq M \qquad H\theta \notin M}{M \leftarrow M \cup \{H\theta\}} \text{ DERIVE} \qquad \theta \text{ is an arbitrary substitution}$$

Fig. 4. Rules for deriving minimal Herbrand model.

Definition 3 (Solution to Datalog synthesis problem). *A solution to a synthesis problem* $S = (\mathcal{R}, F, E)$ *is a Datalog program* C *with* $\mathcal{R}_{in}(C) = \mathcal{R}_{in}$ *and* $\mathcal{R}_{out}(C) = \mathcal{R}_{out}$ *such that the following two conditions hold: (i)* $E^+ \subseteq C(F)$, *i.e., all positive examples are in the minimal model of* C; *and (ii)* $E^- \cap C(F) = \emptyset$, *i.e.,* C *does not derive any of the negative examples.*

4 Constraint-Based Synthesis Algorithm

We now formally define our synthesis algorithm. Recall that, given a synthesis problem $S = (\mathcal{R}, F, E)$, our goal is to discover a set of clauses C where the least fixpoint $C(F)$ contains all positive examples and none of the negative ones.

To avoid encoding least fixpoints, we will encode derivations of a fixed size—i.e., we encode under-approximations of the least fixpoint—and search for a set of clauses with a bounded derivation that derives *most* positive examples and none of the negative ones. In Sects. 4.1 and 4.2, we show how to encode the space of clauses and bounded derivations. In Sect. 4.3, we present an inductive synthesis loop that alternates between synthesizing clauses and verifying them until arriving at a solution.

4.1 Clause Constraints

Preliminaries. We describe here the *clause constraints*, a set of first-order constraints that define the space of all possible Datalog programs of a given size.

Throughout this section we shall assume a fixed Datalog synthesis problem $S = (\mathcal{R}, F, E)$, where $\mathcal{R}_{in} = \{R_1, \ldots, R_n\}$ and $\mathcal{R}_{out} = \{R_{n+1}, \ldots, R_m\}$. Without loss of generality, we shall assume that all predicates are of arity 2. In Sect. 6, we describe how we implement the algorithm for arbitrary arities. Our goal is to synthesize a solution C. We shall fix the *maximum number of clauses*, $n_c > 0$, to appear in C and a *maximum number of body atoms per clause*, $n_b > 0$. We will construct the clause constraints such that they capture every set of n_c clauses $C = \{c_1, \ldots, c_{n_c}\}$, where for each $c_i \in C$, $|body(c_i)| = n_b$.

Variables and Constraints. For each clause c_i, for $i \in [1, n_c]$, we will introduce the following integer variables:

$$h_i, b_{i,1}, \ldots, b_{i,n_b} \tag{$\mathcal{V}_{\text{PREDS}}$}$$

$$vh_{i,1}, vh_{i,2}, vb_{i,1}, vb_{i,2}, \ldots, vb_{i,2n_b-1}, vb_{i,2n_b} \tag{$\mathcal{V}_{\text{ARGS}}$}$$

The variables h_i denote the predicate symbol in the head of the clause c_i; similarly, $b_{i,j}$ denote the j'th predicate symbol in the body of c_i. Specifically, the value of the variable will be the index of the predicate symbol to appear at that location. For instance, if $h_2 = 5$, then the head of clause c_2 will be $R_5 \in \mathcal{R}_{out}$. The variables $\mathcal{V}_{\mathrm{ARGS}}$ denote the arguments (variables) in the atoms of the clauses. For instance, $vh_{i,1}$ and $vh_{i,2}$ denote the arguments to the head of clause c_i.

Since heads of clauses can only be output predicates, and body predicates can be any predicate in $\mathcal{R}_{in} \cup \mathcal{R}_{out}$, we formulate the following constraints:

$$\varphi_{cl}^h \triangleq \bigwedge_{i \in [1, n_c]} n+1 \le h_i \le m \qquad \varphi_{cl}^b \triangleq \bigwedge_{i \in [1, n_c]} \bigwedge_{j \in [1, n_b]} 1 \le b_{i,j} \le m$$

We do not impose any constraints on $\mathcal{V}_{\mathrm{ARGS}}$; we will simply use their values to partition arguments into equivalence classes. For instance, if $vh_{i,1} = vh_{i,2}$, the head of c_i will be an atom of the form $R(X, X)$, for some predicate $R \in \mathcal{R}_{out}$; otherwise, if $vh_{i,1} \ne vh_{i,2}$, it would be of the form $R(X, Y)$. Finally, the clause constraints are defined as follows: $\boxed{\Phi_{cl} \triangleq \varphi_{cl}^h \wedge \varphi_{cl}^b}$

Denotation and Properties. A model m of Φ_{cl}, denoted $m \models \Phi_{cl}$, maps every variable in $\mathcal{V}_{\mathrm{PREDS}} \cup \mathcal{V}_{\mathrm{ARGS}}$ to an integer. We now show how to transform a model m into a set of clauses C. We start by defining the function $[\![.]\!]_m$ as follows:

$$[\![h_i]\!]_m = R_{m(h_i)} \quad [\![b_i]\!]_m = R_{m(b_i)} \quad [\![vh_i]\!]_m = vmap(vh_i) \quad [\![vb_i]\!]_m = vmap(vb_i)$$

where $m(x)$ is the value of variable x in model m. Let us partition $\mathcal{V}_{\mathrm{ARGS}}$ into equivalence classes, defined by $m(.)$—i.e., $x, y \in \mathcal{V}_{\mathrm{ARGS}}$ are equivalent *iff* $m(x) = m(y)$. We shall now assign to each equivalence class a unique argument from the set $\{X, Y, Z, \ldots\}$. The function $vmap$ maps each variable in $\mathcal{V}_{\mathrm{ARGS}}$ to the argument assigned to its equivalence class. Using $[\![.]\!]_m$, the head of clause c_i is $[\![h_i]\!]([\![vh_{i,1}]\!], [\![vh_{i,2}]\!])$, and the j'th body atom of c_i is $[\![b_{i,j}]\!]([\![vb_{i,2j-1}]\!], [\![vb_{i,2j}]\!])$. We abuse notation and use $[\![m]\!]$ to represent the set of clauses C denoted by m.

Example 1. The above constraints and their solution are best demonstrated through a simple example. Suppose that $\mathcal{R}_{in} = \{R_1\}$ and $\mathcal{R}_{out} = \{R_2\}$, and suppose that $n_c = 1$ and $n_b = 2$. Φ_{cl} will then be $\varphi_{cl}^h \wedge \varphi_{cl}^b$, where $\varphi_{cl}^h \triangleq 2 \le h_1 \le 2$ and $\varphi_{cl}^b \triangleq 1 \le b_{1,1} \le 2 \wedge 1 \le b_{2,2} \le 2$. Suppose we solve Φ_{cl} and get the model $m \models \Phi_{cl}$:

$$m = \begin{bmatrix} h_1 \mapsto 2 \;\; b_{1,1} \mapsto 1 \;\; b_{1,2} \mapsto 2 \;\; vh_{1,1} \mapsto 1 \\ vh_{1,2} \mapsto 3 \;\; vb_{1,1} \mapsto 1 \;\; vb_{1,2} \mapsto 2 \;\; vb_{1,3} \mapsto 2 \;\; vb_{1,4} \mapsto 3 \end{bmatrix}$$

The denotation $[\![m]\!]$ is the clause $R_2(X, Z) \leftarrow R_1(X, Y), R_2(Y, Z)$. Observe that the first argument of the head and the first argument of the first body atom are the same; this is because $m(vh_{1,1}) = m(vb_{1,1})$. Observe also that the predicate symbol in the head is R_2 and the first symbol in the body is R_1; this is because $m(h_1) = 2$ and $m(b_{1,1}) = 1$.

Theorem 1. *Let C be the set of all Datalog programs with n_c clauses, n_b atoms per clause, and no constants in atoms. Let $\mathcal{L}$ be the set of models of Φ_{cl}. Then, for each $C \in \mathcal{C}$, there exists a model $m \in \mathcal{L}$ such that $[\![m]\!]$ is equivalent to C.*

4.2 Simulation Constraints

Arrays and Monotonic Derivations. The goal of the simulation constraints is to encode all derivation sequences of the set of clauses represented by the clause constraints, Φ_{cl}. Due to the complexity of encoding all maximal derivation sequences (least fixpoints), we place a bound d on the number of derivations.[1] That is, the simulation constraints will encode all derivations with exactly d steps. It is critical to recall that a derivation, as we define it in this paper, always produces a new fact. Contrast this with the standard database-theoretic definition, where we can derive the same fact multiple times.

Recall that, given a Datalog program C with input F, a derivation sequence of length d is $M_0 \xrightarrow{c_{i_1},\theta_1} M_1 \xrightarrow{c_{i_2},\theta_2} M_2 \xrightarrow{c_{i_3},\theta_3} \cdots \xrightarrow{c_{i_d},\theta_d} M_d$. Thus, we will create a set of constraints Φ_{sim} that encode *all possible* derivations of length d from *all possible* sets of clauses C defined by Φ_{cl}.

A key observation in our technique is that M_i grows monotonically, that is, $\forall i \in [0, d-1].M_i \subsetneq M_{i+1}$. We exploit this property of Datalog to encode the set of true facts after every derivation using *McCarthy's theory of arrays* [18]. An array $arr : X \to Y$ is a map from some domain X to another domain Y. The i'th element of an array is denoted $arr[i]$. We shall therefore use arrays to represent input and output relations. Specifically, the arrays will be of the type $\mathcal{U}^2 \to \mathbb{B}$, that is, from pairs of elements of the universe to a Boolean value indicating whether the pair is in the relation. The axiom of the theory of arrays that allows us to model derivations is *read-over-write*. Specifically, read-over-write allows us to model adding one element to the relation, without explicitly having to state the *frame condition*—that all other array elements remain unchanged.

We decompose the definition of simulation constraints into (i) constraints encoding the initial *state* of all relations, (ii) constraints encoding a single application of DERIVE, and (iii) constraints encoding derivation sequences.

Encoding the Initial State. For each $R_i \in \mathcal{R}_{in}$, we create an array variable

$$in_i : \mathcal{U}^2 \to \mathbb{B} \qquad\qquad (\mathcal{V}_{\text{INRELS}})$$

The universe $\mathcal{U}$ is set to be all constants appearing in facts F and examples E.[2] For each output relation $R_i \in \mathcal{R}_{out}$, we create a set of arrays:

$$out_{i,0}, \dots, out_{i,d} \qquad\qquad (\mathcal{V}_{\text{OUTRELS}})$$

---

[1] Encoding maximal derivations requires unrollings up to the size of the Herbrand base, along with universal quantification.

[2] For Datalog without constants, we can assume w.l.o.g. that the constants in the examples E are a subset of the constants in F.

where $out_{i,j}$ will represent what facts have been derived over R_i after the first j applications of DERIVE. The input and output arrays are constrained as follows:

$$\varphi_{init}^{\mathcal{R}_{in}} \triangleq \bigwedge_{R_i \in \mathcal{R}_{in}} \bigwedge_{d \in \mathcal{U}^2} in_i[d] \iff R_i(d) \in F \qquad \varphi_{init}^{\mathcal{R}_{out}} \triangleq \bigwedge_{R_i \in \mathcal{R}_{out}} \bigwedge_{d \in \mathcal{U}^2} \neg out_{i,0}[d]$$

Encoding a Single Derivation. We now show how to encode a single step of the derivation sequence (the i'th derivation): $M_{i-1} \longrightarrow M_i$. We define the formula $derive_{i,j}$ to encode the effect of applying clause j in the i'th derivation.

To formally define $derive_{i,j}$, we need to first introduce a set of variables representing the substitution θ_i that is used in the i'th derivation. Specifically, we introduce the following variables of type $\mathcal{U}$, for $i \in [1, d]$:

$$sh_{i,1}, sh_{i,2}, sb_{i,1}, sb_{i,2}, \ldots, sb_{i,2n_b - 1}, sb_{i,2n_b} \qquad (\mathcal{V}_{\text{SUBS}})$$

where $(sh_{i,1}, sh_{i,2})$ denote the substitutions to the arguments in the head of the clause used in the i'th derivation, and $(sb_{i,2j-1}, sb_{i,2j})$ denote the substitutions to the arguments in the j'th body atom of the clause used in the i'th derivation. We constrain these variables such that they adhere to the arguments of the clause used in the i'th derivation. For example, if a body atom is $R(X, X)$, then we want to ensure that any substitution is of the form $R(a, a)$, for $a \in \mathcal{U}$. Therefore, we introduce the constraint $latches_{i,j}$, which indicates that, if any two arguments in atoms of clause j are the same variable, then they should always get the same substitution at position i in the derivation:

$$latches_{i,j} \triangleq \bigwedge_{vx_{j,k}, vx_{j,l} \in \mathcal{V}_{\text{ARGS}}} vx_{j,k} = vx_{j,l} \Rightarrow \sigma(vx_{j,k}) = \sigma(vx_{j,l})$$

where the notation $vx_{j,k}$ denotes any variable $vh_{j,k}$ or $vb_{j,k}$ in $\mathcal{V}_{\text{ARGS}}$, and the function σ is defined such that $\sigma(vh_{j,k}) = sh_{i,k}$ and $\sigma(vb_{j,k}) = sb_{i,k}$; that is, σ encodes the correspondence between the argument variables of the j'th clause and the substitution variables in the i'th derivation.

Now, $derive_{i,j}$ is a conjunction of two constraints: (i) $derive_{i,j}^b$, which specifies that all ground atoms in the body of clause j should be *true* in M_{i-1}, and (ii) $derive_{i,j}^h$, which specifies the new fact derived by applying the clause j at point i of the derivation. (We ensure that no fact is derived more than once.)

$$derive_{i,j}^b \triangleq \bigwedge_{k \in [1,n]} \bigwedge_{l \in [1,n_b]} b_{j,l} = k \Rightarrow in_k[(sb_{i,2l-1}, sb_{i,2l})]$$

$$\wedge \bigwedge_{k \in [n+1,m]} \bigwedge_{l \in [1,n_b]} b_{j,l} = k \Rightarrow out_{k,i-1}[(sb_{i,2l-1}, sb_{i,2l})]$$

$$derive_{i,j}^h \triangleq \bigwedge_{k \in [n+1,m]} h_j = k \Rightarrow (\neg out_{k,i-1}[(sh_{i,1}, sh_{i,2})] \wedge out_{k,i}[(sh_{i,1}, sh_{i,2})] \mapsto true])$$

Encoding All Derivation Sequences. Now that we have defined how to encode a single step of the derivation, we can present the encoding of a derivation sequence of a fixed length d. First, we introduce the following integer variables:

$$s_1, \ldots, s_d \qquad (\mathcal{V}_{\text{DERIVCLS}})$$

where s_i encodes which clause is applied in the i'th point in the derivation sequence. Since Φ_{cl} fixes the number of clauses to n_c, we require the condition $\varphi^c_{sim} \triangleq \bigwedge_{i \in [1,d]} 1 \leq s_i \leq n_c$.

We now encode the effect of an application of DERIVE. The following constraint specifies, for every value s_i could take (from 1 to n_c), the effect on the output arrays in the i'th step of the derivation sequence.

$$\varphi^{der}_{sim} \triangleq \bigwedge_{i \in [1,d]} \bigwedge_{j \in [1,n_c]} s_i = j \Rightarrow derive_{i,j} \wedge latches_{i,j}$$

Finally, the simulation constraints are defined as follows:

$$\boxed{\Phi_{sim} \triangleq \varphi^{\mathcal{R}_{in}}_{init} \wedge \varphi^{\mathcal{R}_{out}}_{init} \wedge \varphi^c_{sim} \wedge \varphi^{der}_{sim}}$$

Correctness. The following theorem states correctness of simulation constraints by showing that, for a fixed set of clauses C, the models of Φ_{sim} have a one-to-one correspondence with the derivations of C of length d. Intuitively, the facts true in the output relation after d steps of a derivation are encoded in the input arrays in_i and final output arrays $out_{i,d}$. Given a model $m \models \Phi_{sim}$, we define $final(m)$ to denote the set of all facts at the end of the derivation defined by m: $final(m) = \{R_k(\boldsymbol{a}) \mid \boldsymbol{a} \in \mathcal{U}^2, m(out_{k,d}(\boldsymbol{a})) = true \vee m(in_k(\boldsymbol{a})) = true\}$.

Theorem 2. *Let $m \models \Phi_{cl}$. Let T be the set of all unique derivation sequences of $[\![m]\!]$ of length d. Let $\mathcal{L}$ be the set of all models $m' \models \Phi_{sim}$, where m' agrees with m on valuations of $\mathcal{V}_{\text{PREDS}}, \mathcal{V}_{\text{ARGS}}$. There is a bijection $f : \mathcal{L} \to T$ s.t., for all $m \in \mathcal{L}$ and $t \in T$, if $f(m) = t$ then $final(m) = M_d$ (final set of facts in t).*

4.3   Inductive Synthesis Loop

We now present our inductive synthesis loop (Fig. 5), given a synthesis problem $S = (\mathcal{R}, F, E)$. We fix the maximum number of clauses n_c, the maximum number of body atoms n_b, and we assume that the simulation is of length $d \leq |E^+|$ (otherwise, the simulation constraint may be UNSAT). SYNTH begins by constructing the clause and simulation constraints, Φ_{cl} and Φ_{sim}. It then employs a synthesize–verify loop.

Synthesis Phase. In line 6, SYNTH finds a model m for the constraints, which denotes a set of clauses $C = [\![m]\!]$. This can be performed using an off-the-shelf SMT solver. We impose two additional constraints. First, ψ^- ensures that no negative examples in E^- are derived in the d steps of the derivation sequence. Second, ψ^+_{soft} is a soft constraint that attempts to *maximize* the number of positive examples derived in the d steps of the derivation sequence. This is because *not all* positive examples may be derivable in d derivations.

Verification Phase. In line 8, SYNTH verifies whether C results in a solution to the synthesis problem. Specifically, it computes the fixpoint $C(F)$ and checks whether all positive examples are in the fixpoint and none of the negative ones. If so, a solution is found and SYNTH terminates. The verification step can be performed using an off-the-shelf Datalog solver.

Blocking Constraints. If verification fails, we create a set of constraints, $block(m)$, that removes sets of Horn clauses equivalent to $[\![m]\!]$. Specifically, we first characterize a set of models whose denotation is equivalent to m:

$$\bigwedge_{i\in[1,n_c]} \left(\bigwedge_{v\in\mathcal{V}^i_{\mathrm{PREDS}}} v = m(v) \wedge \bigwedge_{v,v'\in\mathcal{V}^i_{\mathrm{ARGS}},m(v)=m(v')} v = v' \right)$$

where $\mathcal{V}^i_{\mathrm{ARGS}}$ and $\mathcal{V}^i_{\mathrm{PREDS}}$ denote the respective subsets of $\mathcal{V}_{\mathrm{ARGS}}$ and $\mathcal{V}_{\mathrm{PREDS}}$ of the i'th clause. Therefore, the above constraint specifies all models whose denotation is syntactically equivalent to $[\![m]\!]$, modulo variable renaming. $block(m)$ is the negation of the above constraint. Note that characterizing all models whose denotation is equivalent to $[\![m]\!]$ is an undecidable problem [1].

The following theorem states soundness and completeness of SYNTH, relative to a fixed n_c and n_b. Note that, in point 2, if SYNTH terminates with no solution, then this means that we have proven non-existence of a solution with $\leq n_c$ clauses and $\leq n_b$ atoms. Point 2 is true because all programs that are smaller than n_c and n_b can be written as a program with exactly n_c clauses and n_b body atoms— simply by duplicating clauses and body atoms.

Theorem 3 (Soundness and completeness). *(1) If* SYNTH(S) *returns a Datalog program* D, *then* D *is a solution to* S. *(2) If* SYNTH(S) *terminates with no solution, then no solution exists with* $\leq n_c$ *clauses and* $\leq n_b$ *atoms per body of each clause. (3)* SYNTH(S) *terminates in finitely many steps.*

1: **function** SYNTH(Synthesis problem S)
2:      Construct Φ_{cl} and Φ_{sim} for S
3:      $\Phi_{neg} \leftarrow true$
4:      $\psi^- \leftarrow \bigwedge_{R_i(d)\in E^-} \neg out_{i,d}[d]$
5:      $\psi^+_{soft} \leftarrow$ maximize $|\{R_i(d) \in E^+ \mid out_{i,d}[d] = true\}|$
6:      **while** $\exists m \models \Phi_{cl} \wedge \Phi_{sim} \wedge \Phi_{neg} \wedge \psi^- \wedge \psi^+_{soft}$ **do**
7:          $C \leftarrow [\![m]\!]$
8:          **if** $E^+ \subseteq C(F)$ and $E^- \cap C(F) = \emptyset$ **then**
9:              **return** C
10:         $\Phi_{neg} \leftarrow \Phi_{neg} \wedge block(m)$
11:     **return** no solution exists for S

Fig. 5. Inductive synthesis loop.

5   Encoding Templates

We now present *clause templates:* additional constraints that exploit the use of the symbolic encoding to impose a certain structure on the synthesized clauses.

Non-recursive Clauses. The most natural clause template is the one that ensures that at least one of the clauses is a *base case*—with no output relation in the body. To define this template, we designate one of the clauses (say the first) to be the base case. Recall that the predicate symbols appearing in the body of the first clause are $b_{1,1}, \ldots, b_{1,n_b}$, where each b variable holds a value from 1 to m indicating the index of the predicate symbol. Since all indices of input relations are in $[1, n]$, all we need to impose is the following constraint: BASECASE $\triangleq \bigwedge_{i\in[1,n_b]} b_{1,i} \leq n$. If we specify that every clause is non-recursive, then we syntactically restrict the solution to be in the class of *Unions of Conjunctive Queries* (UCQs), a fundamental query class [1], since it captures the class of positive SQL queries.

Linear Clauses. A clause is *linear* when there is at most one occurrence of an output predicate in its body. Linear Datalog programs—a strict subset of Datalog—are in the complexity class NC (Nick's class): the set of problems solvable in polylogarithmic time with a polynomial number of processors. Informally, a problem in NC is inherently parallel. In addition to their theoretical niceties, linear Datalog programs have also proven useful in distributed processing [27]. In order to synthesize linear programs, we impose the following constraint:

$$\text{LINEAR} \triangleq \bigwedge_{i\in[1,n_c]} \neg \bigvee_{j,k\in[1,n_b],j\neq k} b_{i,j} \geq n+1 \wedge b_{i,k} \geq n+1$$

The above constraint states that for every clause i, no two predicate symbols in the body, $b_{i,j}$ and $b_{i,k}$, refer to output relations.

Connected Clauses. It is most often the case that arguments in the head of a clause also appear in its body. For instance, the clause $H(X,Y) \leftarrow B(Z,W)$ will end up deriving every possible tuple in $\mathcal{U}^2$ (assuming B is not empty), which is unlikely a program of interest. To avoid programs that are able to derive all possible tuples, we can impose the following constraint:

$$\text{CONN} \triangleq \bigwedge_{i\in[1,n_c]} \left(\bigvee_{j\in[1,2n_b]} vh_{i,1} = vb_{i,j} \wedge \bigvee_{k\in[1,2n_b]} vh_{i,2} = vb_{i,k} \right)$$

6   Implementation and Evaluation

We have implemented the presented synthesis technique in a new tool called *Zaatar*. Zaatar utilizes the open-source Z3 SMT solver [19] for satisfiability checking and evaluating Datalog programs (using Z3's fixpoint engine [13]).

Our implementation takes as input a synthesis problem where relations can be of arbitrary arities. The encoding in Sect. 4 assumes that all relations are binary. We extend the encoding to assume that all relations have the same arity, the maximum arity amongst all relations in $\mathcal{R}$. For example, if the maximum arity is 4, we describe a binary relation $R(X,Y)$ by disregarding the variables in $\mathcal{V}_{\text{ARGS}}$ that represent the third and fourth arguments.

Benchmarks. We collected a set of Datalog programs comprised of recursive and non-recursive programs. The benchmarks are fully listed in Table 1. The non-recursive benchmark programs include (*i*) path extraction programs (`path3` and `path4`, which extract all paths of length 3 or 4 from a graph); (*ii*) cycle and triangle extraction from a graph (`cycle` and `triangle`); and (*iii*) path extraction with alternating edge colors (`redblue` and `redblueUnd`). Our recursive benchmarks include standard Datalog programs, like transitive closure of a graph (`TC` and `TCUnd`) and *same generation* (`samegen`), which extracts all individuals of the same generation from a family tree. In addition to the standard graph-manipulating Datalog programs, we also synthesized (*i*) *least inductive invariant generation* (`leastInvariant`), which, given a finite-state program and its least inductive invariant, returns the *initiation* and *consecution* rules defining an inductive invariant [5]; and (*ii*) pointer analysis programs (`andersenFull` and `andersenSimple`).

Experimental Results. The experimental results are shown in Table 1. For each synthesis problem, we instantiated it with a small number of facts F in the input relations (3–8 facts per benchmark). Then, we supplied a small and sufficient number of positive and negative examples that describe the problem.

Table 1. Experimental results. Mac OS X 10.11; 4 GHz Intel Core i7; 16 GB RAM.

Benchmark	Time (s)	#Iters.	$|F|$	$|\mathcal{R}_{in}|$	$|\mathcal{R}_{out}|$	$|E^+|$	$|E^-|$	n_c	n_b	Description
Non-recursive benchmarks										
path3	0.13	1	4	1	1	1	0	1	3	Extract all pairs of vertices (x, y) where x reaches y in 3 steps
path4	0.23	1	5	1	1	2	0	1	4	Same as path3, but 4 steps
redblue	0.24	1	5	2	1	1	0	1	3	Extract all pairs of vertices (x, y) where x reaches y using one red edge followed by a blue edge
redblueUnd	1.19	1	5	2	1	2	0	2	3	Extract all pairs of vertices (x, y) where x (y) reaches y (x) using one red edge followed by a blue edge
triangle	1.32	6	5	1	1	2	8	1	3	Extract all triples (x, y, z) that form a triangle
cycles	0.01	1	6	1	1	3	0	1	2	Extract all vertices x where x is in a cycle of length 2
Recursive benchmarks										
TC	0.57	1	3	1	1	3	0	2	2	Compute transitive closure of a directed graph
TCUnd	4.25	1	3	1	1	6	0	3	2	Compute the undirected transitive closure of a directed graph
pathsEven	0.61	1	4	1	1	3	0	2	3	Extract all pairs of vertices (x, y) where x reaches y through a path of even length
pathsOdd	14.37	1	5	1	1	5	5	2	3	Same as pathsEven, but for odd length paths
pathsMod3	0.66	2	6	1	1	2	6	2	4	Extract all pairs of vertices (x, y) where x reaches y through a path of length divisible by 3
pathsMod4	3.43	5	8	1	1	2	5	2	5	Same as pathsMod3, but for paths divisible by 4
redblueRec	0.29	1	6	2	1	3	1	2	2	Extract all pairs of vertices (x, y) where x reaches y through a path with alternating red and blue edges
redblueRecSep	5.18	1	4	2	2	6	0	4	2	Extract two relations: one all pairs of vertices (x, y) where x reaches y through red edges only, and another with blue edges only
samegen	3.53	1	6	1	1	4	1	2	3	Extract all pairs of individuals (x, y) who are from the same generation in a family tree
leastInvariant	1.07	5	5	2	1	4	2	2	2	Compute the least inductive invariant of a finite-state program
andersenSimple	33.02	1	6	3	1	6	0	3	3	Andersen's pointer analysis (without load instructions)
andersenFull	115.87	1	7	4	1	7	1	4	3	Andersen's pointer analysis

The number of positive/negative examples required per benchmark are shown in Table 1 (see columns $|E^+|$ and $|E^-|$). For all benchmarks, we fixed the derivation bound d to be the number of positive examples $|E^+|$. The only templates we used were BASECASE (to force a base case in recursive programs) and CONN (see Sect. 5). (Without the CONN template imposed, most recursive benchmarks do not terminate within a reasonable amount of time: they keep synthesizing trivial programs with head arguments disjoint from body arguments, and therefore keep iterating through the synthesis loop.)

Our results indicate that our approach can synthesize non-trivial programs within a small amount of time. For most benchmarks, Zaatar synthesizes the correct solution within 0–5 s. The longest running benchmark—Andersen's analysis, andersenFull—requires around 2 min. Furthermore, the correct solution is usually discovered within the first iteration of SYNTH (see column #Iters. in Table 1). This result indicates that, using our approach, a small number of examples is sufficient to describe a non-trivial graph-based computation.

For most benchmarks, a very small number of negative examples is required—often none. We notice that the numbers of required negative examples increases with the size of the desired program (as defined by n_c and n_b) and the arities of relations. For example, in the `triangle` benchmark, there are many non-recursive programs correlating triples of vertices; we thus needed to supply 8 negative examples to ensure that the program is indeed only extracting triples that form triangles. Other benchmarks that require multiple negative examples are `pathsOdd` and `pathsMod3`, which have 3 and 4 atoms in the bodies of their clauses.

Discussion. Our results demonstrate the merit of our approach at synthesizing a range of different Datalog programs in a small amount of time and with a small number of examples. It is important, however, to state the limitations. First, our search process imposes an upper bound on the number of clauses and atoms that can appear in programs. Second, as the size of the desired program increases, the number of examples required increases (and, therefore, the size of the derivation bound d), thus stressing the SMT solver.

7   Related Work and Discussion

Synthesis of Recursive Programs. We are not the first to synthesize recursive functions. Recent synthesis works from the community have focused on functional programs—for example, [2,10,15,16,22,24,31]. Our work synthesizes recursive graph-/relation-manipulating programs. While we can encode relations in a functional programming language, the synthesis task becomes tedious. Datalog is a more natural fit for relation manipulation.

Inductive Logic Programming. Our synthesis target is closest in nature to the rich field of inductive logic programming (ILP) [8,20]. Generally speaking, a primary focus of ILP research has been on inducing *theories*, e.g., a program explaining a biological process. The formulation, however, is very similar to our setting, with the addition of *background knowledge* encoded as clauses. The search technique in ILP is often a bottom-up or top-down search for a theory. In the bottom-up setting, the synthesizer begins with no clauses, and incrementally grows the set of clauses (by climbing up a subsumption lattice of clauses) using the provided positive and negative examples; the top-down setting proceeds in the opposite direction. Our search strategy is rather different: we consider all examples at once and present a novel encoding that delegates the process to an SMT solver. Naturally, our approach directly benefits from future advances in SMT techniques. Additionally, ILP techniques often require a large number of examples. In our approach, our goal is to utilize only a small number of examples.

Synthesis of recursive clauses has not received as much attention in ILP. Flener and Yilmaz [9] provide a survey of ILP in the recursive setting. A recent line of work by Muggleton et al. presents a *meta-interpretive learning* (MIL) technique for synthesizing recursive clauses for various domains [6,17,21]. Compared to our technique, MIL requires *meta-rules* (similar to our clause templates) that can constrain the search towards recursive clauses. Meta-rules, however, are very specific: they (*i*) fix all the variables in the rule, (*ii*) exactly specify which relations in the rule are recursive, and (*iii*) fix the size of a rule. Thus, the only parameters of the search are the relations to appear in the body and the head of a clause. Our templates can encode meta-rules and are strictly more general than meta-rules. In practice, we do not restrict the variables at all (we only use CONN to eliminate ill-formed rules). Further, MIL tools like Metagol require a total ordering on the Herbrand base, which might not exist for certain examples, e.g., transitive closure and Andersen, where graphs are cyclic. Nonetheless, MIL also addresses *predicate invention*, the problem of introducing new predicates. This is an interesting and difficult problem that relates to synthesizing auxiliary procedures in the more general program synthesis setting.

Constraint-Based Synthesis. Our technique is inspired by symbolic synthesis techniques [12,14,25,30].Techniques like [12] encode all loop-free programs of up to a certain size, along with the examples, as a formula to be solved by an SMT solver. We encode the search for Horn clauses along with bounded derivations of the clauses as a first-order formula. Since efficient symbolic encodings can only express loop/recursion-free executions, we use bounded derivations to induce recursive programs symbolically, without having to encode the least fixpoint.

Acknowledgements. This work is supported by NSF awards 1566015, 1652140, and a Google Faculty Research Award.

References

1. Abiteboul, S., Hull, R., Vianu, V.: Foundations of Databases: The Logical Level. Addison-Wesley Longman Publishing Co., Inc., Boston (1995)
2. Albarghouthi, A., Gulwani, S., Kincaid, Z.: Recursive program synthesis. In: Sharygina, N., Veith, H. (eds.) CAV 2013. LNCS, vol. 8044, pp. 934–950. Springer, Heidelberg (2013). doi:10.1007/978-3-642-39799-8_67

3. Andersen, L.O.: Program analysis and specialization for the C programming language. Ph.D. thesis, University of Cophenhagen (1994)
4. Aref, M., ten Cate, B., Green, T.J., Kimelfeld, B., Olteanu, D., Pasalic, E., Veldhuizen, T.L., Washburn, G.: Design and implementation of the logicblox system. In: Proceedings of 2015 ACM SIGMOD International Conference on Management of Data, pp. 1371–1382. ACM (2015)
5. Bradley, A.R., Manna, Z.: The Calculus of Computation: Decision Procedures with Applications to Verification. Springer Science and Business Media, Heidelberg (2007). doi:10.1007/978-3-540-74113-8
6. Cropper, A., Muggleton, S.H.: Learning efficient logical robot strategies involving composable objects. In: Proceedings of 24th International Joint Conference Artificial Intelligence (IJCAI 2015), pp. 3423–3429 (2015)
7. Cropper, A., Tamaddoni-Nezhad, A., Muggleton, S.H.: Meta-interpretive learning of data transformation programs. In: Proceedings of 24th International Conference on Inductive Logic Programming (2015)
8. De Raedt, L.: Logical and Relational Learning. Springer Science and Business Media, Heidelberg (2008)
9. Flener, P., Yilmaz, S.: Inductive synthesis of recursive logic programs: achievements and prospects. JLP 41, 141–195 (1999)
10. Frankle, J., Osera, P.M., Walker, D., Zdancewic, S.: Example-directed synthesis: a type-theoretic interpretation. In: POPL. ACM (2016)
11. Gulwani, S., Harris, W.R., Singh, R.: Spreadsheet data manipulation using examples. CACM 55, 97–105 (2012)
12. Gulwani, S., Jha, S., Tiwari, A., Venkatesan, R.: Synthesis of loop-free programs. In: PLDI (2011)
13. Hoder, K., Bjørner, N., De Moura, L.: μZ–an efficient engine for fixed points with constraints. In: Gopalakrishnan, G., Qadeer, S. (eds.) CAV 2011. LNCS, vol. 6806, pp. 457–462. Springer, Heidelberg (2011). doi:10.1007/978-3-642-22110-1_36
14. Jha, S., Gulwani, S., Seshia, S.A., Tiwari, A.: Oracle-guided component-based program synthesis. In: ICSE (2010)
15. Kitzelmann, E., Schmid, U.: Inductive synthesis of functional programs: an explanation based generalization approach. JMLR 7, 429–454 (2006)
16. Kneuss, E., Kuraj, I., Kuncak, V., Suter, P.: Synthesis modulo recursive functions. In: OOPSLA (2013)
17. Lin, D., Dechter, E., Ellis, K., Tenenbaum, J.B., Muggleton, S.: Bias reformulation for one-shot function induction. In: ECAI, pp. 525–530 (2014)
18. McCarthy, J.: Towards a mathematical science of computation. In: Colburn, T.R., Fetzer, J.H., Rankin, T.L. (eds.) Program Verification. SCS, vol. 14, pp. 35–56. Springer, Dordrecht (1993)
19. De Moura, L., Bjørner, N.: Z3: an efficient SMT solver. In: Ramakrishnan, C.R., Rehof, J. (eds.) TACAS 2008. LNCS, vol. 4963, pp. 337–340. Springer, Heidelberg (2008). doi:10.1007/978-3-540-78800-3_24
20. Muggleton, S.: Inductive logic programming. N. Gener. Comput. 8, 295–318 (1991)
21. Muggleton, S.H., Lin, D., Pahlavi, N., Tamaddoni-Nezhad, A.: Meta-interpretive learning: application to grammatical inference. Mach. Learn. 94, 25–49 (2014)
22. Osera, P., Zdancewic, S.: Type-and-example-directed program synthesis. In: PLDI (2015)
23. Perelman, D., Gulwani, S., Grossman, D., Provost, P.: Test-driven synthesis. In: PLDI (2014)

24. Polikarpova, N., Kuraj, I., Solar-Lezama, A.: Program synthesis from polymorphic refinement types. In: Proceedings of 37th ACM SIGPLAN Conference on Programming Language Design and Implementation, pp. 522–538. ACM (2016)
25. Reynolds, A., Deters, M., Kuncak, V., Tinelli, C., Barrett, C.: Counterexample-guided quantifier instantiation for synthesis in SMT. In: Kroening, D., Păsăreanu, C.S. (eds.) CAV 2015. LNCS, vol. 9207, pp. 198–216. Springer, Cham (2015). doi:10.1007/978-3-319-21668-3_12
26. Seo, J., Guo, S., Lam, M.S.: Socialite: datalog extensions for efficient social network analysis. In: 2013 IEEE 29th International Conference on Data Engineering (ICDE), pp. 278–289. IEEE (2013)
27. Shaw, M., Koutris, P., Howe, B., Suciu, D.: Optimizing large-scale semi-naïve datalog evaluation in hadoop. In: Barceló, P., Pichler, R. (eds.) Datalog 2.0 2012. LNCS, vol. 7494, pp. 165–176. Springer, Heidelberg (2012). doi:10.1007/978-3-642-32925-8_17
28. Shen, W., Doan, A., Naughton, J.F., Ramakrishnan, R.: Declarative information extraction using datalog with embedded extraction predicates. In: Proceedings of 33rd international conference on Very large data bases, pp. 1033–1044. VLDB Endowment (2007)
29. Smaragdakis, Y., Balatsouras, G., et al.: Pointer analysis. Found. Trends Program. Lang. **2**, 1–69 (2015)
30. Solar-Lezama, A., Tancau, L., Bodík, R., Seshia, S.A., Saraswat, V.A.: Combinatorial sketching for finite programs. In: ASPLOS (2006)
31. Suter, P., Köksal, A.S., Kuncak, V.: Satisfiability modulo recursive programs. In: Yahav, E. (ed.) SAS 2011. LNCS, vol. 6887, pp. 298–315. Springer, Heidelberg (2011). doi:10.1007/978-3-642-23702-7_23
32. Wang, J., Balazinska, M., Halperin, D.: Asynchronous and fault-tolerant recursive datalog evaluation in shared-nothing engines. Proc. VLDB Endow. **8**, 1542–1553 (2015)
33. Whaley, J., Lam, M.S.: Cloning-based context-sensitive pointer alias analysis using binary decision diagrams. In: PLDI, pp. 131–144. ACM (2004)

Search Strategies for Floating Point Constraint Systems

Heytem Zitoun[1](✉), Claude Michel[1], Michel Rueher[1], and Laurent Michel[2]

[1] CNRS, I3S, Université Côte d'Azur, Nice, France
{heytem.zitoun,claude.michel,michel.rueher}@i3s.unice.fr
[2] University of Connecticut, Storrs, CT 06269-2155, USA
ldm@engr.uconn.edu

Abstract. The ability to verify critical software is a key issue in embedded and cyber physical systems typical of automotive, aeronautics or aerospace industries. Bounded model checking and constraint programming approaches search for counter-examples that exemplify a property violation. The search of such counter-examples is a long, tedious and costly task especially for programs performing floating point computations. Indeed, available search strategies are dedicated to finite domains and, to a lesser extent, to continuous domains. In this paper, we introduce new strategies dedicated to floating point constraints. They take advantage of the properties of floating point domains (e.g., domain density) and of floating point constraints (e.g., floating point arithmetic) to improve the search for floating point constraint problems. First experiments on a set of realistic benchmarks show that such dedicated strategies outperform standard search and splitting strategies.

1 Introduction

A key issue while verifying programs with floating point computations is the search of floating point arithmetic errors that produce results quite different from the expected result over the reals. Consider foo, a program doing floating point computations:

```
void foo(){
    float a = 1e8f;
    float b = 1.0f;
    float c = -1e8f;
    float r = a + b + c;
    if(r >= 1.0f)
        doThenPart();
    else doElsePart();
}
```

This work was partially supported by ANR COVERIF (ANR-15-CE25-0002).

J.C. Beck (Ed.): CP 2017, LNCS 10416, pp. 707–722, 2017.
DOI: 10.1007/978-3-319-66158-2_45

Over the reals, r is equal to 1.0 and the `doThenPart` function is called. However, over the floats with a "round to the nearest" rounding mode, an absorption phenomenon occurs: $a + b$ is equal to a and, thus, r is assigned to 0. As a result, the `doThenElse` function is called instead of the `doElsePart` function. This simple example illustrates how the flow of a very simple program over the floats ($\mathbb{F}$) can differs from the expected flow over the reals ($\mathbb{R}$). Such a flow discrepancy might have critical consequences if, for instance, the condition is related to decide whether to brake or not in an ABS system.

Constraint programming has been used to verify such properties [5, 16] in a bounded model checking framework [6, 7]. However, the search of such counterexamples is a long, tedious and costly task especially for programs performing floating point computations. The use of standard search technique to solve constraints over $\mathbb{F}$ lacks efficiency. Numerous search strategies over finite domains have been proposed [4, 8, 14, 15, 17] and, to a lesser extent, over continuous domains [11, 12]. But, such strategies do not adapt well to floating point numbers. A subset of integers bounded by two integers is a finite and uniformly distributed set which can be enumerated. A subset of reals bounded by two floating point numbers is an infinite set of reals that cannot be enumerated and thus, search strategies over continuous domains rely on interval arithmetic, bisection and mathematical properties to prove the existence of solutions in some small interval [1]. A contrario, the set of floating point numbers is a finite set with a huge cardinality and a non-uniform distribution (half of the floating point numbers belongs to the interval $[-1, 1]$). The aforementioned technique like enumeration are not well suited to floating point number density and distribution. Though floating point number approximate real numbers, they do not benefit from the same properties such as continuity. It is thus difficult to reuse search strategies designed for the reals with floating point variables.

The purpose of this paper is to introduce new search strategies dedicated to floating point numbers to ease and, perhaps more importantly, speed-up the solving of verification problems. Preliminary experiments performed on a limited but realistic set of benchmarks show that such dedicated strategies outperform standard search and splitting strategies.

2 Notations and Definitions

2.1 Floating Point Numbers

Floating point numbers were introduced to approximate real numbers. The IEEE754-2008 standard for floating point numbers [10] sets floating point formats, as well as, some floating point arithmetic properties. The two most common formats defined in the IEEE754 standard are *simple* and *double* floating point number precision which, respectively, use 32 bits and 64 bits. A floating point number is a triple (s, m, e) where $s \in \{0, 1\}$ represents the sign, the p bits m, the significant or mantissa and, e the exponent [9]. A *normalized* floating point number is defined by:

$$(-1)^s 1.m \times 2^e$$

To allow gradual underflow, IEEE754 introduces de-normalized numbers whose value is given by:

$$(-1)^s 0.m \times 2^0$$

Note that simple precision are represented with 32 bits and a 23 bits mantissa ($p = 23$) while doubles use 64 bits and a 52 bits mantissa ($p = 52$).

2.2 Absorption

Absorption occurs when adding two floating point numbers with different order of magnitude. The result of such an addition is the furthest from zero. For instance, in C, using simple floating point numbers with a rounding mode set to "round to nearest", $10^8 + 1.0$ evaluates to 10^8. Thus, 1.0 is absorbed by 10^8.

2.3 Cancellation

Cancellation occurs when most of the most significant bits are lost. For instance, it appears when subtracting the close results of two operations. Consequences of cancellation increase with the accumulation of rounding errors. Such a phenomenon is highlighted by subtracting two close operands [18].

For instance, evaluating[1] `((1.0f - 1.0e-7f) - 1.0f) * 1.0e+7f` in C using simple floating point numbers and a rounding mode sets to "round to nearest" yields 1.1920928955078125 instead of -1.0. Indeed, over $\mathbb{F}$ subtracting 1.0 to the result of $1.0 - 10^{-7}$ leads to loose the most significant bits. The subtraction result is then used in a product that amplifies this loss in the mantissa.

2.4 Notations

In the sequel, x, y and z denote variables and $\mathbf{x}$, $\mathbf{y}$ and $\mathbf{z}$, their respective domains. When required, $x_{\mathbb{F}}$, $y_{\mathbb{F}}$ and $z_{\mathbb{F}}$ denote variables over $\mathbb{F}$ and $\mathbf{x}_{\mathbb{F}}$, $\mathbf{y}_{\mathbb{F}}$ and $\mathbf{z}_{\mathbb{F}}$, their respective domains while $x_{\mathbb{R}}$, $y_{\mathbb{R}}$ and $z_{\mathbb{R}}$ denote variables over $\mathbb{R}$ and $\mathbf{x}_{\mathbb{R}}$, $\mathbf{y}_{\mathbb{R}}$ and $\mathbf{z}_{\mathbb{R}}$, their respective domains. Note that $\mathbf{x}_{\mathbb{F}} = [\underline{x}_{\mathbb{F}}, \overline{x}_{\mathbb{F}}] = \{x_{\mathbb{F}} \in \mathbb{F}, \underline{x}_{\mathbb{F}} \leq x_{\mathbb{F}} \leq \overline{x}_{\mathbb{F}}\}$ with $\underline{x}_{\mathbb{F}} \in \mathbb{F}$ and $\overline{x}_{\mathbb{F}} \in \mathbb{F}$. Likewise, $\mathbf{x}_{\mathbb{R}} = [\underline{x}_{\mathbb{R}}, \overline{x}_{\mathbb{R}}] = \{x_{\mathbb{R}} \in \mathbb{R}, \underline{x}_{\mathbb{R}} \leq x_{\mathbb{R}} \leq \overline{x}_{\mathbb{R}}\}$ with $\underline{x}_{\mathbb{R}} \in \mathbb{F}$ and $\overline{x}_{\mathbb{R}} \in \mathbb{F}$. Let $x_{\mathbb{F}} \in \mathbb{F}$, then $x_{\mathbb{F}}^+$ is the smallest floating point number strictly superior to $x_{\mathbb{F}}$ and $x_{\mathbb{F}}^-$ is the biggest floating point number strictly inferior to $x_{\mathbb{F}}$. In addition, given a constraint c, $vars(c)$ denotes the set of floating point variables appearing in c.

3 Properties of Floating Point Domains, Variables and Constraints

This section defines properties on floating point domains and constraints that are useful to build dedicated search strategies. Domain properties like cardinality or

---

[1] One must take care to annotate all literals with 'f' to force floating point constants and to decompose the expression into elementary arithmetic operations to prevent the compiler from evaluating at compile time.

density capture the structure of the domains of the floating point variables. Constraint properties take into account floating point arithmetic properties like absorption or cancellation. They also capture structural properties by, for instance, taking advantage of the derivative.

3.1 Properties of Floating Point Domains and Variables

Definition 1 (Width). *Let $w(\mathbf{x}_\mathbb{F})$ the width of domain $\mathbf{x}_\mathbb{F}$ be defined as*

$$w(\mathbf{x}_\mathbb{F}) = \overline{x}_\mathbb{F} - \underline{x}_\mathbb{F}$$

The domain width is defined by the distance between its two bounds. It is a rather historical criteria. On finite domains, many strategies rely on this criteria, especially one of the most widespread, namely $minDom$ [14]. Selecting variables with the smallest domain aims at focusing on the most constrained variables. However, over the floats, this criteria is questionable because of the non uniformly distributed floating point values. Here, a smaller width does not necessarily mean a smaller number of values.

Example 1 (Width versus size). Let $x_\mathbb{F}$ and $y_\mathbb{F}$ be two simple floating point variables and $\mathbf{x}_\mathbb{F} = [1, 2]$, $\mathbf{y}_\mathbb{F} = [10, 12]$ be their respective domains. While $w(\mathbf{x}_\mathbb{F}) = 1$ and $w(\mathbf{y}_\mathbb{F}) = 2$, $\mathbf{x}_\mathbb{F}$ contains 8388608 values and $\mathbf{y}_\mathbb{F}$ contains 2097152 values. Thus, the most constrained variable is $y_\mathbb{F}$ rather than $x_\mathbb{F}$.

Definition 2 (Cardinality). *Let $|\mathbf{x}_\mathbb{F}|$ denotes the cardinality of domain $\mathbf{x}_\mathbb{F}$. Given $\mathbf{x}_\mathbb{F} = [\underline{x}_\mathbb{F}, \overline{x}_\mathbb{F}]$ with $\underline{x}_\mathbb{F} \geq 0$ one can define $|\mathbf{x}_\mathbb{F}|$ with*

$$|\mathbf{x}_\mathbb{F}| = 2^p * (e_{\overline{x}_\mathbb{F}} - e_{\underline{x}_\mathbb{F}}) + m_{\overline{x}_\mathbb{F}} - m_{\underline{x}_\mathbb{F}} + 1$$

where $e_{\overline{x}_\mathbb{F}}$ and $e_{\underline{x}_\mathbb{F}}$ are the exponents of, respectively, $\overline{x}_\mathbb{F}$ and $\underline{x}_\mathbb{F}$, and $m_{\overline{x}_\mathbb{F}}$ and $m_{\underline{x}_\mathbb{F}}$ are the mantissa of, respectively, $\overline{x}_\mathbb{F}$ and $\underline{x}_\mathbb{F}$ while p is the length of the mantissa.

This formula can be extended to other cases by exploiting symmetries. Figure 1 illustrates how the cardinality of a floating point interval is computed. The bold double ended arrow represents the interval $\mathbf{x}_\mathbb{F}$. The main idea is to compute the number of floating point values contained in the interval $[2^{e_{\underline{x}_\mathbb{F}}}, 2^{e_{\overline{x}_\mathbb{F}}}]$ (computed by $2^p * (e_{\overline{x}_\mathbb{F}} - e_{\underline{x}_\mathbb{F}}) + 1$) represented by the simple double ended arrow. Then, it withdraws the number of floats in $[2^{e_{\underline{x}_\mathbb{F}}}, \underline{x}_\mathbb{F})$ (i.e., $m_{\underline{x}_\mathbb{F}}$ floats) and adds the number of floats in $(2^{e_{\overline{x}_\mathbb{F}}}, \overline{x}_\mathbb{F}]$ (i.e., $m_{\overline{x}_\mathbb{F}}$ floats).

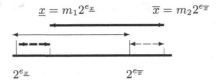

$$\underline{x} = m_1 2^{e_{\underline{x}}} \qquad \overline{x} = m_2 2^{e_{\overline{x}}}$$

$$2^{e_{\underline{x}}} \qquad 2^{e_{\overline{x}}}$$

Fig. 1. Computing cardinality of floating point intervals

Notice that, over finite domains, width and cardinality return nearly the same values (especially when there are no 'holes' in the finite domains). However, over the floats, these two properties are not correlated. Width and cardinality play different roles over the floats. Cardinality could be used to identify either the domain with the smallest number of floats, i.e., the variable that constraint the most the problem, or the domain with the biggest number of floats, i.e., the variable with a high potential of solutions.

Definition 3 (Density). *Let $\rho(\mathbf{x}_{\mathbb{F}})$ the density of $\mathbf{x}_{\mathbb{F}}$ be defined as*

$$\rho(\mathbf{x}_{\mathbb{F}}) = \frac{|\mathbf{x}_{\mathbb{F}}|}{w(\mathbf{x}_{\mathbb{F}})}$$

Intuitively, density captures the proximity of floating point values within a given domain. It helps identifying domains that have a small number of values on a big domain or a big number of values on a small domain (with respect to the width). The former allows to reach easily values that should correspond to various behaviors while the latter potentially contains many values corresponding to the same behavior. Remember that, over the floats, density increases near zero.

Definition 4 (Magnitude). *Let $mag(\mathbf{x}_{\mathbb{F}})$ be the magnitude of $\mathbf{x}_{\mathbb{F}}$ and defined as*

$$mag(\mathbf{x}_{\mathbb{F}}) = \frac{e_{\underline{x}_{\mathbb{F}}} + e_{\overline{x}_{\mathbb{F}}}}{2 \cdot e_{max}}$$

where e_{max} is the biggest exponent in $\mathbb{F}$.

In practice, the magnitude of $[0, 1]$ should be near zero while magnitude of $[10^{36}, 10^{37}]$ should be near 1. In essence, the property helps identifying domains that mainly hold big values or small values. More precisely, magnitude has a dual purpose. First, the property helps selecting variables involved in an absorption, for instance when a big magnitude domain and a small magnitude domain are both involved in an addition. This is easier to implement but less precise than the dedicated property defined in the upcoming Definition 8. Second, this property might help selecting domains with extreme values. Extreme values are those that are often associated to undesirable behaviors.

Definition 5 (Degree). *Let $degree(x_{\mathbb{F}})$ denote the degree of a variable $x_{\mathbb{F}}$ and be defined as the number of constraints in which $x_{\mathbb{F}}$ appears. It is defined as*

$$degree(x_{\mathbb{F}}) = \sum_{c \in C} (x_{\mathbb{F}} \in vars(c))$$

where C is the set of constraints.

Naturally, the degree definition mirrors its counterpart in finite-domain solvers. It is a static property. The higher the degree of $x_{\mathbb{F}}$, the more $x_{\mathbb{F}}$ plays an important role in the solving process. Many strategies over finite domains take advantage of this property like the weighted degree strategy [4].

Definition 6 (Occurrences). *Let $occur(x_\mathbb{F})$ denote the maximum number of occurrences of $x_\mathbb{F}$ among all constraints in a set C be defined as*

$$occur(x_\mathbb{F}) = \max_{c \in C} count(x_\mathbb{F}, c)$$

where $count(x_\mathbb{F}, c)$ is the number of $x_\mathbb{F}$ occurrences in constraint c.

Multiple occurrences is a recurring problem in handling floating point variables. While solutions have been proposed to handle this problem [2,13], identifying variables with multiple occurrences, might help by, for instance, choosing a more adapted filtering process and fixing these variables as soon as possible.

3.2   Properties of Floating Point Constraints

This section introduces properties that take advantage of floating point arithmetic operators used within constraints. The properties will be helpful to define constraint-driven branching strategies.

To appreciate the first property, consider a floating point addition constraint $z_\mathbb{F} = x_\mathbb{F} \oplus y_\mathbb{F}$ in which the rounding more is set to "round to nearest even". If the domain $x_\mathbb{F}$ has a significantly larger magnitude than $y_\mathbb{F}$, some values in $y_\mathbb{F}$ may simply be absorbed when carrying out the addition. Measuring which *fraction* of $y_\mathbb{F}$ is obliterated in this way is the purpose of the absorption property.

Definition 7 (Absorption). *Let $absorb(\mathbf{y}_\mathbb{F}, \mathbf{x}_\mathbb{F})$ denote the absorption of $\mathbf{y}_\mathbb{F}$ by $\mathbf{x}_\mathbb{F}$ and be defined as:*

$$absorb(\mathbf{y}_\mathbb{F}, \mathbf{x}_\mathbb{F}) = \frac{|[-2^{e_{max}-p-1}, 2^{e_{max}-p-1}] \cap \mathbf{y}_\mathbb{F}|}{|\mathbf{y}_\mathbb{F}|}$$

Namely, it is the number of $\mathbf{y}_\mathbb{F}$ values that are absorbed by at least a value of $\mathbf{x}_\mathbb{F}$. In the above, e_{max} is the exponent of $max\{abs(\underline{x}_\mathbb{F}), abs(\overline{x}_\mathbb{F})\}$.

Note how $\mathbf{u}_\mathbb{F} = [-2^{e_{max}-p-1}, 2^{e_{max}-p-1}] \cap \mathbf{y}_\mathbb{F}$ captures the part of $\mathbf{y}_\mathbb{F}$ that is absorbed by the biggest value in magnitude in $\mathbf{x}_\mathbb{F}$. Thus, if none of the values of $\mathbf{y}_\mathbb{F}$ are absorbed by $\mathbf{x}_\mathbb{F}$, $\mathbf{u}_\mathbb{F}$ will be empty and $absorb(\mathbf{y}_\mathbb{F}, \mathbf{x}_\mathbb{F})$ will be equal to 0. On the contrary, when all values of $\mathbf{y}_\mathbb{F}$ are absorbed by $\mathbf{x}_\mathbb{F}$, $\mathbf{u}_\mathbb{F}$ will be equal to $\mathbf{y}$ and $absorb(\mathbf{y}_\mathbb{F}, \mathbf{x}_\mathbb{F})$ will be equal to 1. Selecting variables that are involved in an absorption could help improving the quality of the software and providing counter-examples that instantiate an absorption (see Fig. 2).

The next property only applies to subtraction constraints.

Definition 8 (Cancellation). *Given a floating point subtraction constraint $z_\mathbb{F} = x_\mathbb{F} \ominus y_\mathbb{F}$ where $\ominus$ is the floating point subtraction with the rounding more set to "round to nearest even", let cancellation denote the number of bits canceled by the subtraction and be defined as*

$$cancellation = max\{e_{\underline{x}_\mathbb{F}}, e_{\overline{x}_\mathbb{F}}, e_{\underline{y}_\mathbb{F}}, e_{\overline{y}_\mathbb{F}}\} - min\{e_{z_\mathbb{F}}, z_\mathbb{F} \in \mathbf{z}_\mathbb{F}\}$$

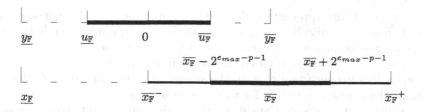

Fig. 2. Illustration of absorption phenomena

The *cancellation* definition was extracted from [3]. It increases with the number of canceled bits and whenever it becomes strictly positive, some bits are potentially lost.

Definition 9 (Derivative). *Given a constraint* $c : e_1 \Diamond e_2$ *in which* $\Diamond \in \{=, \leq , \geq, <, >\}$, *c can be rewritten as* $f : e_1 - e_2 \Diamond 0$. *If f is a monovariate function, its derivate can be evaluated using interval arithmetic and gives rise to the definition of c's derivative as*

$$derive(c) = \mathbf{f}'(\mathbf{x}) \approx \frac{\mathbf{f}(\mathbf{x} + h) - \mathbf{f}(\mathbf{x})}{h}$$

The approach generalizes to the case where f is a multivariate function. Its jacobian J gives the variation of each of the variables of f according to other variables of f. Using this matrix, either component-wise or by computing an aggregation of the variation of each variable according to the others, the involvement of a variable of f in the variation of f can be estimated.

Over the floats, a big variation of f might introduce some holes in the representation of the function while a small variation is often represented by the same floating point value. It thus provides useful information to drive the search.

4   Search Strategies for the Floats

As usual, search strategies over floats are based on a combination of variable selection heuristics and splitting techniques. The next subsection introduces different variable selection heuristics based on the above-mentioned properties. The subsection wraps up with four splitting techniques used in the experiments.

4.1   The Choice of a Variable

Single Property Strategies
Single property based strategies select the variable that either maximizes or minimizes the chosen properties. For instance, one can choose the variable that maximizes the domain density or the one that minimizes this density. That's to say, $maxDens = \max_{x_\mathbb{F} \in X} \rho(x_\mathbb{F})$ (ditto for $minDens$).

Other constraint properties deserve a more specialized approach. For instance, *absorb* or *cancellation* can be maximized or minimized while the minimization or maximization of *derive* should be done according to its absolute value. The *absorb* property is based on constraints of the form $z = x + y$. So, to implement this property, we pick up the subset of constraints from C that are additions (form $z = x + y$) and for which $absorb(y, x) > 0$.

Finally, *degree* and *occur* are static properties whose value stays the same along the search tree.

Multi Property Strategies

In the following, we define two strategies that are based on two properties: `absWDens` and `densWAbs`.

`absWDens` selects the variable that maximizes density from a subset of variables that are involved in an absorption ($absorb > 0$).

`densWAbs` selects the variable that maximizes absorption among the subset of variables that satisfies $density \geq \frac{maxDens + minDens}{2}$.

4.2 Domain Splitting Strategies

Problems over the floating point numbers are characterized by huge domains and non uniformly distributed values. As a result, an enumeration strategy like the one often used in finite domains would fail to quickly find a solution, spending most of the time to exhaustively enumerate all possible combinations of values. It would also fail by missing the opportunity to reduce the size of the domains which is offered when a classical domain splitting strategy like a simple bisection is followed by a filtering process. However, in the presence of a lot of solutions, a simple bisection (Fig. 3a) quickly reaches its limits, the filtering applied after each bisection being unable to reduce domain sizes. On the other hand, problems with no solution should benefit from a simple bisection.

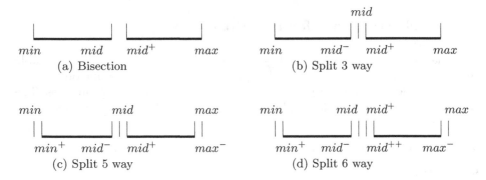

Fig. 3. Different splitting strategies

To overcome these difficulties, we use 3 splitting strategies that mix bisection and enumeration ad that are derived from the strategies introduced in [5]. Instead of just splitting the domain in two parts, some of the floating point values at the boundaries of the split are isolated and used as enumerated values. Figures 3b–d illustrate the new splitting strategies that combine bisection and enumeration. These combinations begin always with the enumeration of the selected values before handling the two remaining sub-domains. These splitting strategies are called *partial enumeration* splittings.

4.3 Semi-dynamic and Dynamic Strategies

Two alternatives are possible when it comes to composing variable selection and domain splitting. The *semi-dynamic* strategy can first choose a variable and then recursively split that variable until it becomes bound. This approach does not reconsider other variables until the chosen one is grounded. Note that it is possible to leverage any splitting strategy, including the partial enumeration. The *dynamic* strategy adopts a more permissive view. At each node of the search tree, it selects a variable, splits its domain according to some strategy and moves on to possibly select a different variable at the next node. It does not insist on fully instantiating the chosen variable.

5 Experiments

We combined the different variable selection heuristics and splitting techniques on a set of 8 realistic benchmarks. A standard strategy based on a lexicographic order variable selection and *dynamic 2 way* split (i.e., a classical bisection) serves as reference value.

All the experiments were carried out on a MacBook Pro i7 2.3 GHz with 8 GB of memory. All strategies have been implemented in the Objective-CP solver enhanced with floating point constraints. All floating point computations are done with simple precision floats and a rounding mode set to "nearest even".

5.1 Benchmarks

The benchmarks used in these experiments come from test and verification of floating point software.

Heron. The heron function compute the area of a triangle from the lengths of its sides a, b, and c with Heron's formula: $\sqrt{s * (s - a) * (s - b) * (s - c)}$ where $s = (a + b + c)/2$. The next C program implements this formula, where a is the longest side of the triangle.

```
// Precondition: a > 0 and b > 0 and c > 0 and  a > b and b > c
float heron(float a, float b, float c) {
    float s, squared_area;

    squared_area = 0.0f;
    if ((a + b >= c) && (b + c >= a) && (a + c >= b)) {
```

```
    s = (a + b + c) / 2.0f;
    squared_area = s*(s-a)*(s-b)*(s-c);
  }
  return sqrt(squared_area);
}
```

The first benchmark verifies that if $a \in (5.0, 10.0]$, $b \in (0.0, 5.0]$ and $c \in (0.0, 5.0]$, then $squared\_area < 10^5$. The second verifies that with the same input domains, $squared\_area > 156.25 + 10^{-5}$) [5].

Optimized Heron. Optimized_heron is a variation of heron which uses a more reliable floating point expression to compute squared_area.

```
// Precondition: a > 0 and b > 0 and c > 0 and a > b and b > c
float optimized_heron(float a, float b, float c) {
  float s, squared_area;

  squared_area = 0.0f;

  if ((a + b >= c) && (b + c >= a) && (a + c >= b)) {
    squared_area = (((a+(b+c))*(c-(a-b))*
                    (c+(a-b))*(a+(b-c)))/16.0f);
  }

  return sqrt(squared_area);
}
```

Here, one test verifies that if $a \in (5.0, 10.0]$, $b \in (0.0, 5.0]$ and $c \in (0.0, 5.0]$, then $squared\_area < 10^5$ while the second verifies that with the same input domains, $squared\_area > 156.25 + 10^{-5}$). Note that the latter benchmark has no solution.

Cubic. The solve_cubic benchmark was extracted from the Gnu Scientific Library. It seeks a set of input values that reach the first condition of the program.

```
int solve_cubic (double a, double b, double c,
                 double *x0, double *x1, double *x2) {
  double q = (a * a - 3 * b);
  double r = (2 * a * a * a - 9 * a * b + 27 * c);
  double Q = q / 9;
  double R = r / 54;
  double Q3 = Q * Q * Q;
  double R2 = R * R;
  double CR2 = 729 * r * r;
  double CQ3 = 2916 * q * q * q;
  if (R == 0 && Q == 0) {
    ...
```

Square 2. The next benchmark checks that the square product of a float cannot be equal to 2.

```
// inv_square_int_true-unreach-call.c
int f(int x) {
    float y, z;
    // assume(x >= -10 && x <= 10);
    y = x*x - 2.f;
    // assert(y != 0.f);
    z = 1.f / y;
    return 0;
}
```

As a matter of fact, there is no simple floating point value whose square equal to 2 with the standard rounding mode. Thus, this bench has no solution.

Square 4. A variation checks that the square of a float cannot be equal to 4.

```
// float_int_inv_square_false-unreach-call.c
int g(int x) {
    float y, z;
    // assume(x >= -10 && x <= 10);
    y = x*x - 4.f;
    // assert(y != 0.f);
    z = 1.f / y;
    return 0;
}
```

A solution for this problem is well known and this benchmark has solutions.

Slope. The slope function computes an approximation of the derivative of the square function.

```
float slope(float x0, float h) {
    float x1 = x0 + h;    float x2 = x0 - h;
    float fx1 = x1*x1;    float fx2 = x2*x2;
    float res = (fx1 - fx2) / (2.0*h);
    return res;
}
```

The benchmark checks that for $x0 = 13$, the result is always inferior or equal to 25 for all value of $h \in [10^{-9}, 10^{-6}]$.

5.2 Results

In the tables, the variable choice column contains two columns, the strategy column (short name "strat.") and the dynamic column (short name "dyn."). The strategy column specifies the kind of strategy used to choose the variable whose domain will be split. It is a minimization or a maximization of the defined properties (noted "min" or "max" followed by the first letters of the property name) or one of the combinations of density and absorption that we have defined. Note that we have not implemented the derivate property yet. The dynamic column takes the value "full" when a different variable is chosen at each node of the search tree or "semi" when the variable choice is postponed until the curent variable is fully instantiated.

Column "split" gives the number of generated values and subdomains. Thus, 2 stands for Fig. 3a, that is to say, a classical bisection, 3 stands for Fig. 3b, 5 for Fig. 3c and 6 for Fig. 3d. Column $\sum t$ gives the total amount of milliseconds required to solve all the benchmarks, or all the benchmarks with or without solutions, according to the selected strategies. When available, the "#OUT" column gives the number of timeout and memory out. Note that the timeout is 180 s and that each memory out is accounted as a time out.

Table 1 gathers three subtables that give the total amount of time required to solve all the benchmarks (Table 1a), all the benchmarks with solutions (Table 1b) and all the benchmarks without solution (Table 1c) according to a given combination of variable choice and splitting strategy. Note that these tables reports only the ten best cases and the ten worst cases among the 144 combinations of variable choice and splitting strategies tested, as well as the time required to solve the related set of benchmarks using the reference strategy.

Tables 2, 3 and 4 give the total amount of time to solve all benchmarks, all benchmarks with solutions, and all benchmarks without solution according to one of the criteria introduced in our search strategies, i.e., respectively, the variable choice strategy, the nature of the variable choice (semi- or fully-dynamic)

Table 1. Total time to solve benchmarks according to variable choice and splitting

variable choice		split.	$\sum t$	variable choice		split.	$\sum t$	variable choice		split.	$\sum t$
strat.	dyn.		(ms)	strat.	dyn.		(ms)	strat.	dyn.		(ms)
maxAbs	semi	6	4883	maxAbs	semi	6	187	maxAbs	semi	2	2376
maxAbs	full	6	4930	maxCard	semi	6	189	maxAbs	full	2	2379
maxDens	semi	6	5059	densWAbs	semi	6	191	maxAbs	full	3	2410
densWAbs	full	6	7517	densWAbs	full	6	196	maxDens	semi	2	2439
maxCard	semi	6	180191	maxAbs	full	6	202	maxCard	full	3	4405
densWAbs	semi	6	180194	maxDens	semi	6	217	maxAbs	semi	5	4451
maxDegree	full	6	180307	maxDegree	full	6	305	maxAbs	full	5	4467
maxDegree	semi	6	180310	maxDegree	semi	6	307	maxCard	full	2	4594
maxAbs	full	5	184613	maxWidth	full	6	31244	maxDens	semi	5	4626
maxDens	semi	5	184796	minDens	full	6	38332	maxAbs	semi	6	4696
		...				...				...	
ref			550988	ref			540011	ref			10977
		...				...				...	
minDegree	semi	3	906285	minDens	semi	3	720005	maxMagn	semi	3	360000
minOcc	semi	3	906285	minDegree	semi	2	720005	minMagn	semi	3	360000
maxWidth	semi	3	906607	minDegree	full	2	720005	maxDegree	semi	3	360000
minCard	semi	3	911526	minAbs	full	3	720005	minDegree	semi	3	360000
maxMagn	semi	3	1077852	maxDens	full	3	720006	minOcc	semi	3	360000
absWDens	full	3	1080002	minOcc	semi	2	720006	absWDens	semi	2	360000
maxWidth	semi	2	1080004	minAbs	full	2	720006	absWDens	semi	3	360000
minDens	semi	3	1080005	absWDens	full	5	720147	absWDens	semi	5	360000
absWDens	full	5	1080147	maxWidth	semi	2	900002	absWDens	semi	6	360000
absWDens	full	5	1440000	absWDens	full	5	1080000	densWAbs	semi	3	360000
(a) all				(b) with solutions				(c) without solution			

Table 2. Total time to solve benchmarks according to variable choice strategy

Variable choice	All		With solution		Without solution	
Strat.	$\sum$ t (ms)	#OUT	$\sum$ t (ms)	#OUT	$\sum$ t (ms)	#OUT
maxWidth	4330019	21	2962680	14	1367339	7
minWidth	3762938	19	3470297	18	292641	1
maxCard	3231581	16	1962573	9	1269008	7
minCard	4315427	25	4023103	24	292324	1
maxDens	2573614	13	2323093	12	250521	1
minDens	4936905	27	3316894	18	1620011	9
maxMagn	4881081	24	3261049	15	1620011	9
minMagn	3722681	19	2761916	14	960765	5
maxDegree	3413676	17	1793656	8	1620020	9
minDegree	5259904	27	3639886	18	1620018	9
maxOcc	3360986	17	3071415	16	289571	1
minOcc	5259433	27	3639415	18	1620018	9
maxAbs	2728996	15	2521099	14	207897	1
minAbs	3784212	20	3492984	19	291228	1
maxCan	3360698	17	3071986	16	288712	1
minCan	3356934	17	3068356	16	288578	1
absWDens	5065493	27	3391300	18	1674193	9
densWAbs	2344948	11	1418791	6	926157	5

Table 3. Total time to solve benchmarks according to semi or full dynamic search

Variable choice	All		With solution		Without solution	
Dyn.	$\sum$ t (ms)	#OUT	$\sum$ t (ms)	#OUT	$\sum$ t (ms)	#OUT
Semi	37278081	192	27144829	138	10133252	54
Full	33851636	173	27485876	141	6365760	32

Table 4. Total time to solve benchmarks according to splitting strategy

Split.	All		With solution		Without solution	
	$\sum$ t(ms)	#OUT	$\sum$ t(ms)	#OUT	$\sum$ t (ms)	#OUT
2	23444527	124	20325639	108	3118888	16
3	23539280	123	17177874	89	6361406	34
5	13654772	72	10155196	54	3499576	18
6	10491138	46	6971996	28	3519142	18

and the number of fragments created by splits. Thus, each line of Table 2 sum 64 cases, each line of Table 3, 576 cases and each line of Table 4 288 cases.

5.3 Analysis

As shown in Table 1a, the best strategy outperforms the standard strategy by a factor of more than 110. These performances are even better for problems with solution (see Table 1b) where the gain factor is of more than 2800. On the other hand, the improvement for benchmarks without solution is only 4 times. Thus, the best tested strategies can significantly improve the search of a first solution whenever such a solution exist.

Combining wisely two properties can also be helpful to select useful solutions: the densWabs combination improves the *density* property while selecting solution that provide an absorption phenomena.

Thanks to Table 2, we can compare the different variable choice strategies. Here, the tested combination of strategies have the overall best behavior, especially, on benchmarks with solutions.

Table 3 shows that the fully dynamic strategy brings the best results on average, though the semi dynamic strategy is slightly better on benchmarks with solutions. However, these results are somewhat unbalanced by two of the variable strategies, namely the *occurence* and *degree* strategies. Such properties are static properties whose values stay the same along the search tree. As a consequence, once a variable is chosen according to this property, it will be chosen in the next node of the search tree until it cannot be chosen anymore, i.e., when fully instantiated. Thus, these two properties, whether maximized or minimized, behave alike the semi-dynamic strategy and penalize the fully dynamic results.

Table 4 confirms that the 2 splits or bisection is a better choice for problem without solution while the 6 splits have better performances on problems with solutions.

On the whole the most successful strategies are based on the absorption property, a purely floating point property. The goal when maximizing the absorption is to generate floating point errors, that's to say values for which the control flow over the floats differs from the expected flow over the reals. This is precisely the case of the benchmarks derived from Heron's formula.

6 Conclusion

This paper introduced a set of properties to choose a variable in a search for solving constraints over the floating point numbers. These maximized or minimized properties have been used to choose a variable during the search and combined with a semi dynamic and fully dynamic choice of variable, as well as, 4 splitting strategies. Preliminary experiments have shown that some of these combinations outperforms the standard strategy by two order of magnitude for all kind of benchmarks and three order of magnitude for benchmarks with solutions. Further works include experimenting on a broader set of benchmarks, exploring other properties and evaluating which combination of properties could benefit to the search.

References

1. Alefeld, G.E., Potra, F.A., Shen, Z.: On the existence theorems of Kantorovich, Moore and Miranda. In: Alefeld, G., Chen, X. (eds.) Topics in Numerical Analysis: With Special Emphasis on Nonlinear Problems, vol. 15, pp. 21–28. Springer, Vienna (2001). doi:10.1007/978-3-7091-6217-0_3
2. Belaid, M.S., Michel, C., Rueher, M.: Boosting local consistency algorithms over floating-point numbers. In: Milano, M. (ed.) CP 2012. LNCS, pp. 127–140. Springer, Heidelberg (2012). doi:10.1007/978-3-642-33558-7_12
3. Benz, F., Hildebrandt, A., Hack, S.: A dynamic program analysis to find floating-point accuracy problems. In: ACM SIGPLAN Conference on Programming Language Design and Implementation, PLDI 2012, Beijing, China, 11–16 June 2012, pp. 453–462 (2012)
4. Boussemart, F., Hemery, F., Lecoutre, C., Sais, L.: Boosting systematic search by weighting constraints. In: ECAI 2004, pp. 146–150 (2004)
5. Collavizza, H., Michel, C., Rueher, M.: Searching critical values for floating-point programs. In: Wotawa, F., Nica, M., Kushik, N. (eds.) ICTSS 2016. LNCS, vol. 9976, pp. 209–217. Springer, Cham (2016). doi:10.1007/978-3-319-47443-4_13
6. Collavizza, H., Rueher, M., Van Hentenryck, P.: CPBPV: A constraint-programming framework for bounded program verification. Constraints 15(2), 238–264 (2010)
7. Collavizza, H., Le Vinh, N., Rueher, M., Devulder, S., Gueguen, T.: A dynamic constraint-based BMC strategy for generating counterexamples. In: 26th ACM Symposium On Applied Computing (2011)
8. Gay, S., Hartert, R., Lecoutre, C., Schaus, P.: Conflict ordering search for scheduling problems. In: Pesant, G. (ed.) CP 2015. LNCS, vol. 9255, pp. 140–148. Springer, Cham (2015). doi:10.1007/978-3-319-23219-5_10
9. Goldberg, D.: What every computer scientist should know about floating-point arithmetic. ACM Comput. Surv. 23(1), 5–48 (1991)
10. IEEE: IEEE standard for binary floating-point arithmetic. ANSI/IEEE Standard, 754 (2008)
11. Jussien, N., Lhomme, O.: Dynamic domain splitting for numeric CSPs. In: ECAI, pp. 224–228 (1998)
12. Kearfott, R.B.: Some tests of generalized bisection. ACM Trans. Math. Softw. 13(3), 197–220 (1987)
13. Lhomme, O.: Consistency techniques for numeric CSPs. In: Proceedings of 13th International Joint Conference on Artifical Intelligence, IJCAI 1993, vol. 1, pp. 232–238. Morgan Kaufmann Publishers Inc., San Francisco (1993)
14. Linderoth, J.T., Savelsbergh, M.W.P.: A computational study of search strategies for mixed integer programming. INFORMS J. Comput. 11(2), 173–187 (1999)
15. Michel, L., Van Hentenryck, P.: Activity-based search for black-box constraint programming solvers. In: Beldiceanu, N., Jussien, N., Pinson, É. (eds.) CPAIOR 2012. LNCS, vol. 7298, pp. 228–243. Springer, Heidelberg (2012). doi:10.1007/978-3-642-29828-8_15

16. Ponsini, O., Michel, C., Rueher, M.: Verifying floating-point programs with constraint programming and abstract interpretation techniques. Autom. Softw. Eng. **23**(2), 191–217 (2016)
17. Refalo, P.: Impact-based search strategies for constraint programming. In: Wallace, M. (ed.) CP 2004. LNCS, vol. 3258, pp. 557–571. Springer, Heidelberg (2004). doi:10.1007/978-3-540-30201-8_41
18. Sterbenz, P.H.: Floating-Point Computation. Prentice-Hall Series in Automatic Computation. Prentice-Hall, Upper Saddle River (1973)

Author Index

Albarghouthi, Aws 689
Amadini, Roberto 3
Aoga, John O.R. 529
Arafailova, Ekaterina 21, 38

Babaki, Behrouz 495
Bacchus, Fahiem 641
Bailey, James 477
Banbara, Mutsunori 596
Beldiceanu, Nicolas 21, 38
Belov, Gleb 321
Berg, Jeremias 443, 652
Bilgory, Erez 55
Bin, Eyal 55
Bofill, Miquel 71
Bouchard, Mathieu 512
Boussemart, Frederic 129
Briant, Olivier 114

Cambazard, Hadrien 114
Carlsson, Mats 387
Chabert, Maxime 460
Chiarandini, Marco 354
Coll, Jordi 71
Cruz, Waldemar 189
Czauderna, Tobias 321

Dalmau, Víctor 80
de Givry, Simon 97
Deville, Yves 297
Di Cosmo, Roberto 370
Dries, Anton 495
Dzaferovic, Amel 321

Feydy, Thibaut 308
Fioretto, Ferdinando 278

Gabbrielli, Maurizio 370
Gange, Graeme 3
Ganji, Mohadeseh 477
Gao, Xin 405
Garcia de la Banda, Maria 321
German, Grigori 114

Glorian, Gael 129
Goldwaser, Adrian 338
Gotlieb, Arnaud 387
Guns, Tias 529

Ham, Lucy 139
Hasan, Mohd. Hafiz 549
Hooker, J.N. 565
Hyttinen, Antti 641

Jackson, Marcel 139
Järvisalo, Matti 443, 641, 652
Jin, Jiwei 405
Johnson, Greg 189
Jost, Vincent 114

Katsirelos, George 97
Kilby, Philip 414
Kimmig, Angelika 495
Kjellerstrand, Håkan 671
Knudsen, Anders Nicolai 354
Koenig, Sven 630
Koutris, Paraschos 689
Kumar, T.K. Satish 630

Lagerkvist, Victor 157
Lagniez, Jean-Marie 129, 172
Lam, Edward 579
Larsen, Kim S. 354
Latour, Anna L.D. 495
Le Berre, Daniel 596
Le, Tiep 278
Lecoutre, Christophe 129, 297
Liu, Fanghui 189
Liu, Tong 370

Ma, Chujiao 189
Ma, Feifei 405
Marquis, Pierre 172
Mauro, Jacopo 370
Mazure, Bertrand 129
McCreesh, Ciaran 206
Meling, Hein 387
Michel, Claude 707

Michel, Laurent   189, 707
Mossige, Morten   387

Naik, Mayur   689
Nijssen, Siegfried   495

O'Sullivan, Barry   262
Oikarinen, Emilia   443

Pan, Linjie   405
Paparrizou, Anastasia   172
Perez, Guillaume   226
Picard-Cantin, Émilie   512
Pralet, Cédric   243
Prosser, Patrick   206
Puolamäki, Kai   443

Quimper, Claude-Guy   512

Régin, Jean-Charles   226
Rueher, Michel   707

Saikko, Paul   641
Schaus, Pierre   297, 529
Schutt, Andreas   308, 338
Siala, Mohamed   262
Simonis, Helmut   21, 38
Simpson, Kyle   206
Smith, Calvin   689
Soh, Takehide   596
Solnon, Christine   460
Spieker, Helge   387
Stuckey, Peter J.   3, 477

Subramani, K.   615
Sun, Wei   405
Suy, Josep   71
Sweeney, Jason   512

Tabakhi, Atena M.   278
Tack, Guido   3
Tamura, Naoyuki   596
Tayur, Sridhar   431
Trimble, James   206

Urli, Tommaso   414

Van den Broeck, Guy   495
Van Hentenryck, Pascal   549, 579
van Hoeve, Willem-Jan   431
Verhaeghe, Hélène   297
Villaret, Mateu   71

Wahlström, Magnus   157
Wallace, Mark   321
Wojciechowski, Piotr   615
Wybrow, Michael   321

Xu, Hong   630

Yeoh, William   278
Yin, Minghao   405
Young, Kenneth D.   308

Zhang, Jian   405
Zhou, Neng-Fa   671
Zitoun, Heytem   707
Ziv, Avi   55

Printed in the United States
By Bookmasters